Urban Runoff Pollution

NATO ASI Series

Advanced Science Institutes Series

A series presenting the results of activities sponsored by the NATO Science Committee, which aims at the dissemination of advanced scientific and technological knowledge, with a view to strengthening links between scientific communities.

The Series is published by an international board of publishers in conjunction with the NATO Scientific Affairs Division

A Life Sciences **B Physics**	Plenum Publishing Corporation London and New York
C Mathematical and Physical Sciences	D. Reidel Publishing Company Dordrecht, Boston and Lancaster
D Behavioural and Social Sciences **E Applied Sciences**	Martinus Nijhoff Publishers Boston, The Hague, Dordrecht and Lancaster
F Computer and Systems Sciences **G Ecological Sciences**	Springer-Verlag Berlin Heidelberg New York Tokyo

Series G: Ecological Sciences Vol. 10

Urban Runoff Pollution

Edited by

Harry C. Torno

Senior Staff Engineer, U.S. Environmental Protection Agency
Science Advisory Board (A-101-F), Washington, D.C. 20460, USA

Jiri Marsalek

Research Scientist, National Water Research Institute
867 Lakeshore Road, Burlington, Ontario L7R4A6, Canada

Michel Desbordes

Maître de Conférences, Laboratoire d'Hydrologie Mathématique
Université des Sciences et Techniques du Languedoc
F-34060 Montpellier Cedex, France

Springer-Verlag Berlin Heidelberg New York Tokyo
Published in cooperation with NATO Scientific Affairs Division

Proceedings of the NATO Advanced Research Workshop on "Urban Runoff Pollution" held at the Hotel Frantel, Montpellier, France, August 26–30, 1985

ISBN 3-540-16090-6 Springer-Verlag Berlin Heidelberg New York Tokyo
ISBN 0-387-16090-6 Springer-Verlag New York Heidelberg Berlin Tokyo

Library of Congress Cataloging in Publication Data. Urban runoff pollution. (NATO ASI series. Series G, Ecological sciences ; vol. 10) Proceedings of the NATO Advanced Research Workshop on "Urban Runoff Pollution," held Aug. 26–30, 1985, Montpellier, France, sponsored by the Scientific Affairs Division of the North Atlantic Treaty Organization. Includes indexes. 1. Urban runoff—Congresses. 2. Water quality management—Congresses. I. Torno, Harry C. II. Marsalek, Jiri, 1940-. III. Desbordes, Michel, 1943-. IV. NATO Advanced Research Workshop on "Urban Runoff Pollution" (1985 : Montpellier, France) V. North Atlantic Treaty Organization. Scientific Affairs Division. VI. Series: NATO ASI series. Series G, Ecological sciences ; no. 10. TD657.U75 1986 628.1'68 86-3930
ISBN 0-387-16090-6 (U.S.)

Printed in Germany

Printing: Beltz Offsetdruck, Hemsbach; Bookbinding: J. Schäffer OHG, Grünstadt
2131/3140-543210

TABLE OF CONTENTS

ACKNOWLEDGEMENTS

This Workshop was one of a series of Advanced Research Workshops sponsored by the Scientific Affairs Division of the North Atlantic Treaty Organization (NATO), and their assistance is gratefully acknowledged.

The Workshop organizers would like to express their sincere appreciation to the other organizations who provided financial and administrative support, without which the Workshop would not have been a success. The following organizations in France provided substantial funding support, in addition to contributing materially to the Workshop in the form of papers and discussion from members of their staff:

Plan Urbain
64, rue de la Federation
75015 Paris

Universite des Sciences et Techniques du Languedoc
place Eugene Bataillon
34060 Montpellier

VERSEAU
Laboratoire d'Hydrologie Mathematique
Universite des Sciences et Techniques du Languedoc
34060 Montpellier

Nancy Centre International de l'Eau
149, rue Gabriel
54500 Vandoeuvre

Centre National de la Recherche Scientifique
Programme PIREN
15, quai Anatole France
75700 Paris

Additional funding support for the travel of U. S. participants was received from the U. S. Environmental Protection Agency and the U. S. National Science Foundation.

Recognizing the need for a Workshop dealing exclusively with urban runoff pollution, the Council organized a NATO Advanced Research Workshop on Urban Runoff Pollution, held in Montpellier, France, August 26-30, 1985. This book is the Proceedings of that Workshop.

The organization of the Proceedings follows the agenda for the Workshop. For each major topic there is a review paper, accompanied by one or more papers on some specific aspect of the subject, and by a summary of the discussions which followed the presentation of the papers. There is also a summary of research needs expressed by the participants. Each of the papers in the Workshop was reviewed and accepted for publication by the Workshop co-Chairmen, and represents original work by the author(s).

We extend our sincere appreciation to all Workshop participants for their active and enthusiastic involvement and lively discussion. Special thanks go to the session leaders and to authors, who produced an outstanding set of papers on relatively short notice.

FOREWORD

In most of the developed countries of the World, significant efforts to control the pollution of surface waters have been underway for decades, and particularly the last 10-15 years. These efforts have focused mainly on eliminating or mitigating the effects of point sources of pollution. In many cases, however, it is clear that we have achieved only limited improvement in water quality, and that non-point sources of pollution are going to control any further improvement. It has long been known that urban runoff is a major non-point source, and much research has been done in an attempt to understand the mechanisms and processes which govern this source and to reduce or eliminate its impacts. Many urban jurisdictions have adopted urban runoff pollution control measures, in spite of the fact that there is a great deal that we still do not know, and without really being able to quantify the benefits achieved. A major problem is that while a great deal of work is being done, both in Europe and North America, it is very difficult to keep abreast of new developments.

The Urban Water Resources Research Council of the American Society of Civil Engineers has for many years had as one of its major objectives the transfer of urban runoff technology among researchers and practitioners in the field, as well as to those engineers who are not in the forefront but who nonetheless need the information on the latest developments. This has been principally accomplished by a series of International Symposia, Engineering Foundation Conferences and technical sessions at various professional meetings. In each case a proceedings or series of papers has been published.

POLLUTIONAL ASPECTS OF URBAN RUNOFF

J. Bryan Ellis
Urban Pollution Research Centre
Middlesex Polytechnic
Queensway, Enfield, EN3 4SF, U.K.

Introduction

It is widely recognised that surface waters in urbanised areas are water quality limited and that secondary treatment of dry weather flows is insufficient to improve receiving water quality or to achieve and maintain long term quality objectives. The impact of urban runoff is perceived by many at the local municipal or borough level as limiting and impairing the full and beneficial uses of receiving waters in urban environments. Existing sewer design, operation and management singularly fails to embrace or adequately incorporate the quality dimension and the water industry in both Europe and America has been slow to undertake post-project appraisal in terms of evaluations of storm sewer performance. This is despite the accepted fact that, for example, some 35% of the total annual pollutant levels discharged to receiving waters in the U.K. comes from combined or storm sewer overflows which only operate some 2-3% of the time. It is also a fact that total effluent discharges represent a very significant proportion of most urban river flows (1). A recent US Environmental Protection Agency report (2) stated that urban nonpoint source water quality problems affect some 20% of the nation's river mileage whilst Heaney and Huber (3) have indicated that 36% of 248 urbanised areas in the US suffer priority runoff problems.

NATO ASI Series, Vol. G10
Urban Runoff Pollution.
Edited by H. C. Torno, J. Marsalek, and M. Desbordes

The US EPA Nationwide Urban Runoff Program (4) as well as recent studies in the UK and Sweden (5) have demonstrated frequent and prolonged periods when urban runoff discharges exceed toxic water standards. In the latter studies, for example, 53%, 63%, and 95% of all events monitored were found to possess average total Zn, Pb and Cu concentrations respectively above EEC threshold limits for freshwater biota. Indeed 11%, 16% and 80% were defined as being above "significant mortality" levels, and it is not surprising that most urban rivers are so grossly polluted that they support very few fish as well as a reduced aquatic fauna. Wet weather runoff flows, especially when overflows are operating, can also lead to substantial violations of bacterial standards. In one UK study (6) nearly 50% of the recorded annual storm sewer events to an urban river resulted in discharges that violated instream EEC mandatory recreational standards of 10,000 MPN/100 ml and 2000 MPN/100 ml for total and faecal coliforms respectively.

The objective evaluation of urban runoff quality and its receiving stream impacts is nevertheless an elusive problem because it requires description of a phenomenon that is a function of several extremely variable and complex inputs and regulatory control parameters which are in practise difficult to analyse independently. The urban drainage system can be conceived as comprising various interacting hydraulic and pollutant reservoirs or temporary storage phases which are linked by complex transport and transformational pathways. The indices of pollutant removal efficiency in this structured network are normally represented by delivery ratio factors that relate the available mass of a particular pollutant to that of the total transported suspended solids mass, which can be in turn functionally related to specific hydrometeorological citeria. This "black box" approach encourages the use of lumped parameter modelling at the expense of real time dynamic simulation. It also concentrates attention on above-ground and end-of-pipe conditions, and overlooks the potential below ground, in-pipe processes controlling pollutant storage, mobilisation and resuspension. Similarly, the obvious and overt slug effects of transient urban runoff and other effluent discharges mask their covert but undoubtedly significant delayed, residual impacts upon receiving waters; apart from dissolved oxygen regimes, these latter quality effects have received relatively little attention.

Despite the many compelling and expressed local, as well as national, institutional and legislative, commitments to improve urban runoff quality, very few control structures installed or designed over the past decade have had their quality performance rigorously field tested. The pollutant removal capability of urban detention basins both in the US (7,8) and UK (9) would appear to be highly variable and negative efficiencies for solids, heavy metals and hydrocarbons are not uncommonly recorded. The effectiveness of alternative source control methods, including street sweeping, are either speculative or questionable (10). The adoption of infiltration schemes, including porous paving, whilst appearing at first consideration to be attractive propositions (11), have yet to be evaluated in terms of effectiveness, durability and cost.

Therefore, the characterisation and quantification of the quality impact of urban runoff discharges is both tenuous and controversial and the identification, calibration and choice of optimum quality control and management procedures is also fraught with technical and legal difficulties. A successful planning strategy for urban runoff quality control will require an adquate definition of source inputs as well as investigations and evaluations of best management practises. Given current economic, political and professional constraints it is likely to be some considerable time before these objectives can be fully realised, integrated and implemented into sewer design protocol.

Quantity-Quality Relations: Some Basic Concepts and Considerations

Runoff Quantity: Design considerations and Quality Control

Traditional hydraulic design of storm sewers is firmly based on the concepts of rapid and efficient removal and conveyance of effluent from the urban surface or industrial/domestic source with the design procedure based on long return period event criteria. This approach ensures an inbuilt spare capacity in the operational design and reflects a professional antipathy towards the idea of surface flooding which might involve public and political problems of both quantity and quality. In the UK most designs utilise standard diameter increments between pipe sizes which means that for small urban developments, very large increases in flow capacity occur between adjacent pipe sizes e.g. an increment from 150 mm to 225 mm pipes will represent a trebling of discharge capacity which further increases by seven fold for 375 mm diameter pipe sizes. The existing overdesign of storm sewers produces an average capacity of at least 20% above design conditions with a return period of pipe full flow which "may be several times the intended return period" (12). It is also becoming increasingly evident that part of the overdesign, and a major source of design calibration error, results from the overestimation of the runoff coefficient for the sewered or effective impervious area. A lumped parameter approach in impermeability assessment is inevitably subject to considerable error; even relatively simple surfaces may have substantial time- and storm-varying responses.

Typical drainage design values that are used to assess catchment impermeability are given in Table 1. Although observations over a two year period for a residential development in NW London of 20 dwellings/hectare gave a mean value of 40%, the range of results from differing storm events (with rainfall intensities ranging from 0.45 to 2.01 mm hr^{-1} and peak flows of 8.35 to 37.45 ls^{-1}) varied from 10 to 70%. A design flow based on the recommended 40% tabled value

Development Density	Percentage Impermeability (RC x paved area)
Industrial	50-1000) according
Spatial areas eg retail precincts, goods yards etc.	40-60) to type
70 dwellings/hectare	80
60 dwellings/hectare	70
50 dwellings/hectare	60
40 dwellings/hectare	50
15-40 dwellings hectare	40
5-10 dwellings/hectare	10
Open spaces	10

TABLE 1. Impermeability Assessment

would be double or two thirds that of the real flow. A large part of this discrepancy is due to the high percentage water losses that can occur on urban surfaces from detention storage evaporation and infiltration such that rainfall-runoff volumes can vary from 15 to 90% (13,14,15). Effective rainfall depths of 0.5 to 1.5 mm are required to initiate surface runoff and on rough textured, rutted or cracked asphalt or block surfaces, particularly if open jointed gutters and kerbs are present, these initial losses can be very high. If such "storage" losses are not included in the surface routing then overprediction of runoff volumes by some 50-70% can be forced into modelling routines.

The implications of these considerations in terms of runoff quality are firstly to enable large quantities of solids to accumulate on impermeable surfaces as well as effect fine particle aggregation prior to the achievement of the critical flow depths required for surface runoff. Much of the observed differences in solids and associated pollutant loadings delivered to roadside inlets during storm events can be explained by reference to these differences in threshold flow depths (16). Thus the nature, conditions and gradient of the contributing surface - as demonstrated by Sartor and Boyd (17) in their classic 1972 studies - is of major significance in controlling pollutant delivery to the sewer system. This surface loss factor also helps to explain the consistently weak correlations that many workers have noted between antecedant conditions and surface pollutant washoff rates (15,18,19) as well as explaining a large proportion of the variability in the quality control effectiveness of street sweeping operations. The second major implication of inherent design oversizing and overcapacity is to increase the capacity for in-pipe deposition particularly during the recessional phase of storm flow. Ackers (20) has shown that self cleansing velocities must increase by 50% for a five fold increase in sediment concentrations or by 25% for each doubling of pipe diameter. The effect of transported sediment will increase friction losses in the sewer pipe by as much as 400 to 500% as a varying effect during storms. Progressive deposition will therefore occur once the transporting capacity of the clean invert pipe is exceeded, and this is achieved much earlier in oversized sewers. Thus extensive and long term dry weather pipe deposits will be characteristic of all urban sewer systems, and can take up 20 to 45% of the inline storage capacity (21). Apart from the obvious hydraulic implications of increased boundary roughness and surcharging the slimed lag deposits act as important secondary pollutant sources as well as stimulating sulphate reducing bacterial (SRB) corrosion of sewer linings and jointing compounds.

There can be no doubt that the flow factor plays a dominant role in urban runoff pollution dynamics as it is the driving force in the mobilisation, transport and deposition of pollutants. This necessarily implies that there is a fundamental requirement to undertake comprehensive surface water management as a first step in urban pollution control.

Hydrograph - Pollutant Relationships

Ellis (22) has recognised two major pollutograph patterns in urban storm sewers which can be described as flushing and dilution - exhaustion regimes respectively (Fig.1a). The latter represents an inconsistency introduced as a result of either variations in material supply, transporting capacity or flow volume. Given the transport-limited surface controls previously discussed under surface storage effects and the successive reductions in downstream solids concentrations that occur as a result of exceedance of pipe transporting capacity, it can be reasonably assumed that some of the observed pollutant decline is brought about by sedimentation rather than through dilution or exhaustion.

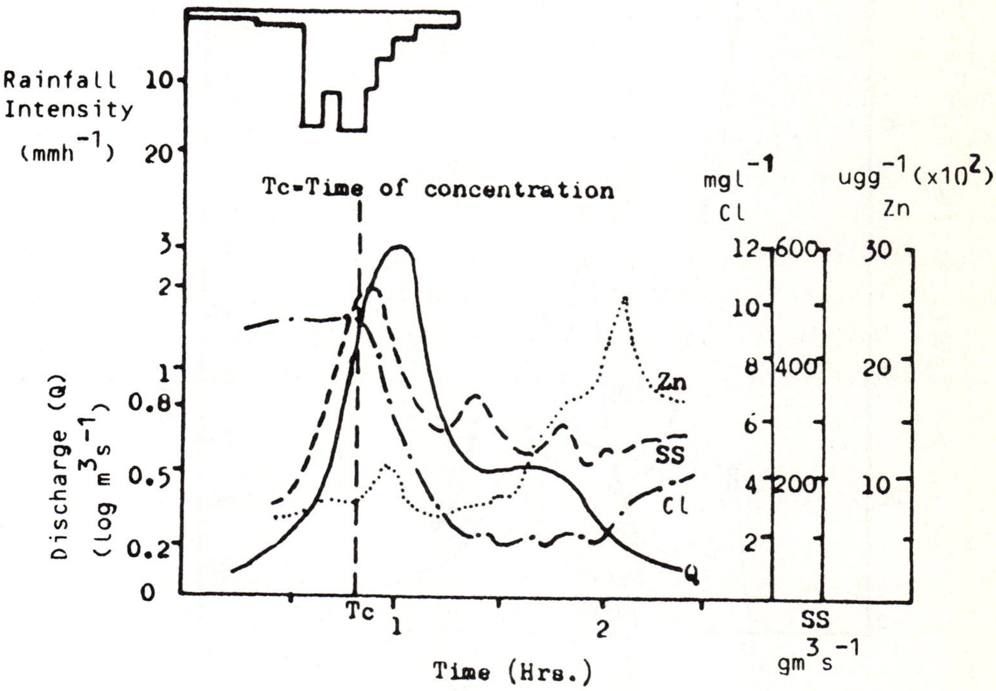

Figure 1a. Pollution Regimes in Storm Sewers

The occurrence and timing of flushing events is a highly variable phenomenon. The most common explanation of the prime flushing event is ascribed to the early removal of materials accumulated in the system during intervening dry periods (24).

Subsidiary or delayed peaks on the recessional storm limb have been attributed to fresh material entering the drainage system from roads and paved surfaces. Combined systems show similar patterns (Fig.1b) to those of separate storm sewers. The initial flush can be equivalent to or in excess of dry weather flow but rapidly declines until the delayed surface

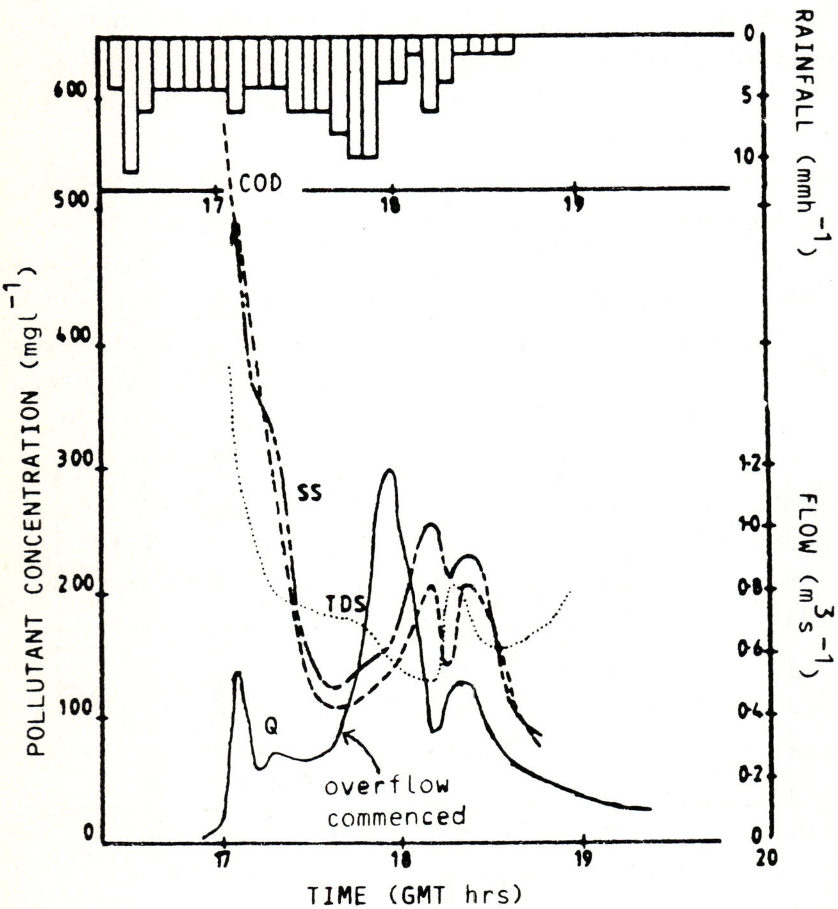

[From Saul et al., (23)]

Figure 1b. CSO Pollution Regimes

sediment-pollutant wave is received. The excess COD, BOD and solids load represents the scouring effects of organically enriched, faecal materials that have been deposited in the sewer during preceding dry weather periods and from previous storm events. However, it must be stressed that decreases in pollutant concentrations from peak levels are due to sedimentation as much as to dilution effects.

The occurrence of a first flush is not a consistent feature of either separate or combined storm sewer systems. In general, some 30 to 60% of the total pollutant load might be expected to be discharged by 30% of the water volume up to the time of concentration of the flow event for the system (25). Combined sewers have been observed to discharge between 40 to 85% of their storm load within the time of concentration (23,26). These figures however, must be viewed with considerable caution as wide variations have been reported. Peak pollutant concentrations can have a wide spread around the peak discharge (Fig.2) and thus mass pollutant flows can be significant after the time of concentration has elapsed. This is clearly seen by inspection of Fig.3 which plots some cumulative percentage mass flows observed at the stormwater outfall of a 247 ha residential catchment in NW London.

Similar results have been demonstrated by Geiger (28) for a combined system in Munich where positive TSS first flush only occurred some 25% of the time. Soluble pollutants show the least tendency towards a first flush and delayed flushes of solids and heavy metals (particularly Zn and Cd) are quite common as can also be seen from Fig.1a. Where such lag hysteresis is a dominant feature of the sewer regime the use of lumped linear approaches will inevitably result in considerable modelling inefficiency. It has also been suggested (29) that the first flush will occupy a volume of runoff equivalent to 0.5 - 0.9 mm of effective rainfall on the impervious catchment with excesses above this threshold being normally associated with some three quarters of the mass of pollutants available for export.

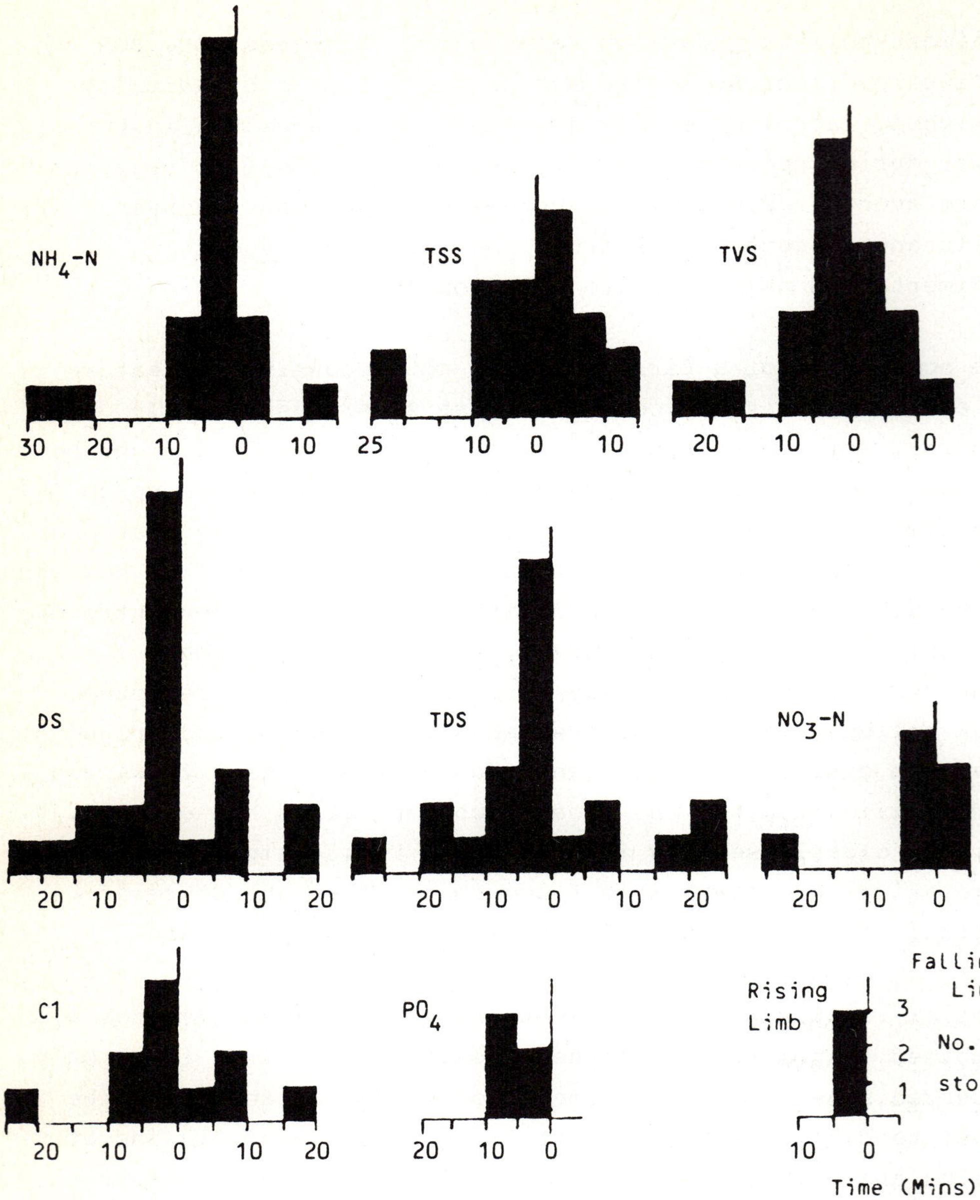

[From Harrop, (27)]

Figure 2. Relation of Peak Concentrations and Flow

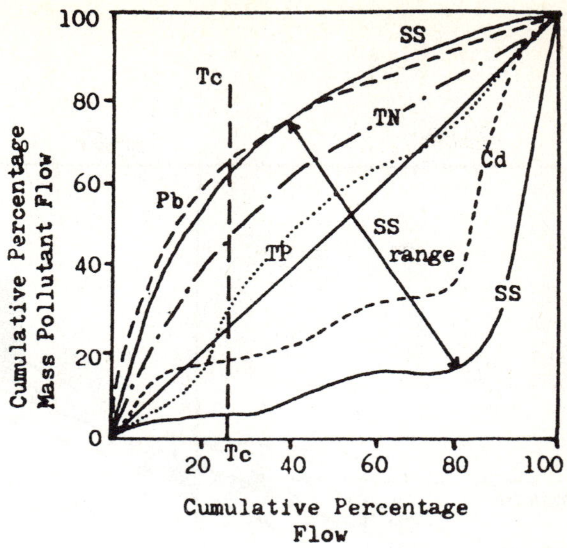

Figure 3. Pollutant Mass Flow Analysis

Whilst pollutant concentrations are normally independent of flow rate, they exhibit mass flow relationships and it is now becoming increasingly conventional to express storm pollutant yields in terms of flow weighted event mean concentrations (EMC's). Mass loadings, even though of considerable importance for receiving water quality are strongly affected by the magnitude of storm events. Both the EMC's as well as maximum concentrations, which are generally a factor of two greater, have been shown to approximate log normal distributions (4,30,31). The statistical consistency of this log normal distribution provides a basis for a probabilistic modelling of urban runoff quality (32,33) which allows specification of concentration return periods (defined as the reciprocal of the probability) as well as exceedance values for various water quality or ecological criterial as illustrated in Fig.4.

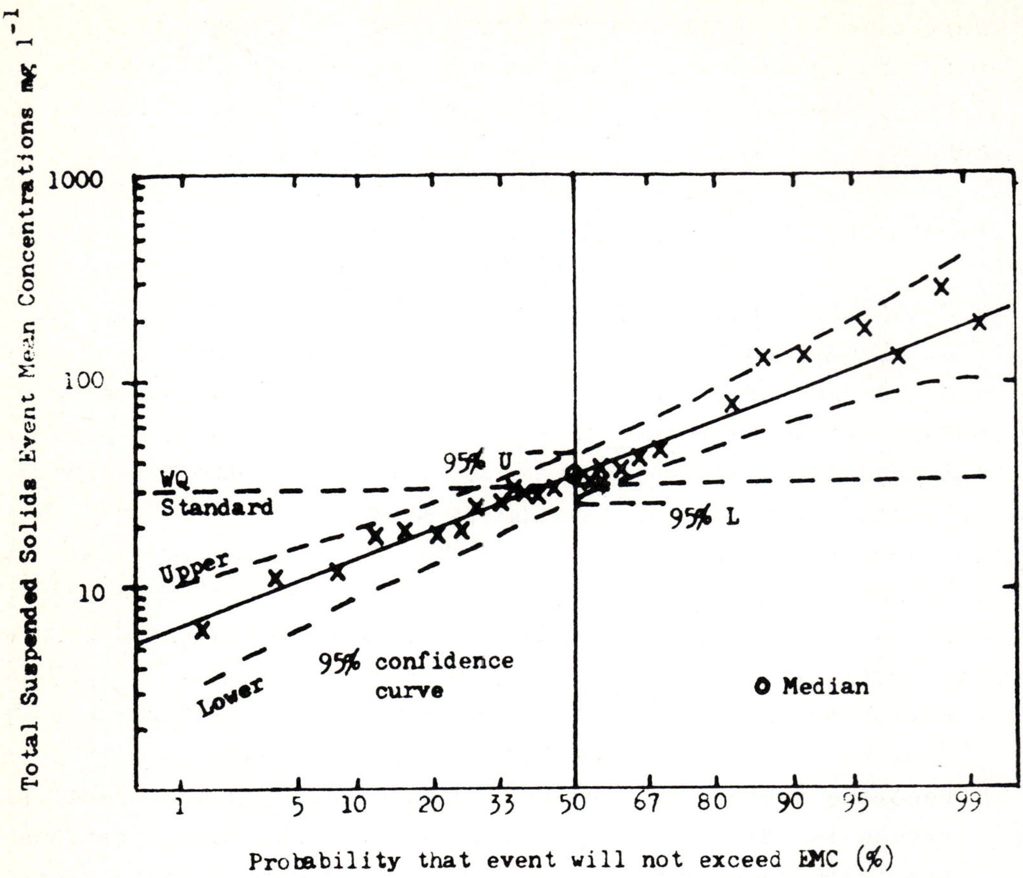

Figure. 4. Probability Distribution of Pollutant EMC's. [from Ellis, 1984 (33)]

The use of quality event statistics for CSO design has been demostrated by Hvitved-Jacobsen and Harremoes (34) in their studies of critical overflow oxygen concentrations and receiving stream capacity. Other workers (35) have advocated the use of a derived beta probability distribution to approximate the rescaled pollutant concentrations given the rather cumbersome convolution of log normal quality with exponential flow and hydrologic related density functions. Both of these distributional approaches provide relatively simple simulation modelling methodologies of use in preliminary planning and design of urban runoff quality. In addition the approaches offer a convenient framework for transferring data in an easily understandable and interpretable manner.

Runoff Quality: Priority Needs and Problems

The 1969 ASCE review of information needs in urban hydrology, (36) as well as the various national urban hydrological modeling and catchment research reviews published under the UNESCO state-of-the-art reports on urban hydrology as part of the International Hydrological Programme (1975-1977), provided only very limited considerations of urban runoff quality. The more recent update of the latter national reviews (37) paid more attention to the quality dimension reflecting in part the increased information available from the US EPA NURP (4). This latter programme has been summarised by Torno (38) and the more significant quality problems identified:

- solids: both inorganic and organic solids are carried in storm runoff in either particulate or colloidal form with the suspended solids fraction being normally considered to be the most significant in terms of induced turbidity, pollutant adsorption, substrate smothering and benthal accumulation.
- oxygen demand: the discharge of organic and other oxidizable materials exerts substantial oxygen demands on receiving waters which are conventionally measured in terms of BOD or COD. The latter parameter, or some such alternative as TOC/TOD, is a preferable determin and of the oxygen status as BOD is inhibited by the inherent toxic properties of storm runoff. Additionally, the traditional BOD test fails to consider the very substantial delayed oxygen demands resulting from benthal deposition in receiving waters (34). The potential and long term degradation that can result from such delayed oxygen depletion is overlooked in the NURP reviews.

- toxicity: stormwater runoff from urban surfaces is highly charged with priority pollutants such as heavy metals, hydrocarbons, pesticides, PCB's etc. However, of the 24 priority pollutants detected in at least 10% of the urban runoff samples collected by the NURP, only two PAH's and five metals exceeded EPA health criteria at the 10^{-5} risk level although many more were found at levels exceeding freshwater acute aquatic life criteria (38). The relative toxicity of these pollutants, however, is very dependent on their speciation, the ambient water pH, ionic strength and organic content, as well as on the biological species being affected. In addition, the accumulative and synergistic toxic effects of these micropollutants in benthal sediments make them available over extended periods of time. Neither the chronic persistency or the available toxicity effects were considered by NURP and their conclusions that priority pollutants associated with urban runoff pose little risk at the detected levels is both questionable and short-sighted.
- bacteria: urban runoff contains large concentrations of bacteria, viruses and pathogens, and with sediment-water bacterial ratios of 45:1, these micro-organisms accumulate in both dry weather pipe deposits as well as in benthal receiving waters and has implications in terms of public health and recreational use (6,40,41).
- nutrients: phosphorous and nitrogen loadings accelerate eutrophication problems in quiescent reaches of urban water bodies and can exert subtoxic effects on aquatic organisms as well as producing highly organic sediments. The accumulation of macrophytic growths during prolonged dry summer periods can also cause severe depletion of and variations in dissolved oxygen concentrations. Mean annual nutrient loads however, are an order of magnitude less than those from secondary treatment plants, and benthal sediments act as very efficient sinks for both phosphorous and nitrogen (42,43). Elevated "slug" chloride levels can also cause short term toxicity problems in receiving streams following winter applications of road salt (43,44).

Once contaminants enter the below ground drainage system they are rapidly degraded and altered by biochemical and geochemical processes. Roadside gully chambers (or catchpits) have been identified as a major source of stormwater pollution with the anaerobic digestion of the settled organically enriched solids contributing COD, NH_4-N and BOD, mainly during summer and autumn (29,45). The septic supernatant in the gully pot is washed away early in the storm event and is superimposed on the inpipe first flush. Solids cannot be regarded as permanent pollution sinks, for although the majority of contaminants may be particulate-associated, many nutrients, metals, hydrocarbons and other organic micropollutants can be easily transferred to the water phase. Grizzard and Randall (46) have observed as much as 50-80% of total nitrogen and phosophorous contained in storm runoff to be in a dissolved form. Morrison et al (47) have shown that 63% of total zinc, 77% of cadmium and 66% of lead found in urban runoff exists in an available and potentially toxic form. Furthermore, Ellis et al (48) have demonstrated a progressive downstream increase in PAH throughout an urban catchment with benthal sludges providing an easily mobilised reservoir of micropollutants. Similar observations have been made by Ellis (6) for bacterial sediment reservoirs that provide a continuous innoculation to the receiving stream. In consequence, treatment of initial runoff volumes alone is unlikely to provide a panacea for urban runoff quality control. The problem is undoubtedly exacerbated by the fine modal 15 to 20 micron sizes that are characteristic of runoff suspended solids and with which much of the pollution potential is associated (49). These contaminated fines are readily mobilised and thus transformation and translocation of pollutants through, and between, the sediment and water phases are not inhibited by conventional engineering structures - which considerably reduces their removal efficiency in quality terms.

There can be no doubt that a hard core of professional scepticism exists about the design utility, transferability and theoretical completeness of the field data so far reported on urban runoff quality. Is there a sufficient available data base to provide appropriate and reliable generalisations of engineering applicability or is the existing data bank sufficient as well as being confusing, contradictory and site specific? The UK review of urban runoff quality undertaken by Mance (29) tends to adopt the viewpoint that urban runoff does not constitute a pollution problem. This conclusion, which is shared by many, would therefore imply that stormwater and public health engineers are wasting time and money to develop designs and installations to alter or control runoff quality. The dilution capacity and sediment sinks provided by and within urban receiving water bodies can, on the above reasoning, be considered to be effective quality control buffers. However, even if it could be accepted that there is enough corroborative evidence to substantiate this philosphy - which many would fundamentally disagree with - the question of impact still remains. The quality performance, and impact upon receiving waters, of urban discharges is still speculative and only vaguely quantified. There remains a need to identify and plug the data gaps in our understanding of engineering systems that will enable an effective control of urban runoff pollutants. Some of these research needs and priorities have been outlined in the preceding sections. The argument presented here would assert that if we possess only a vague knowledge of the physico-biochemical descriptors and mechanisms of runoff impacts, it is not possible to proceed to develop adequate or "correct" dimensioning of engineering controls. The latter require an input from the bioenvironmental scientist as much as from the hydraulic engineer.

Pollutant Sources and Pathways

The various inputs, outputs and pathways followed by water and contaminants from both natural and anthropogenic sources encountered in urban catchments are summarised in Fig.5 which depicts the drainage area as an "open" system. The type of pollutants and the magnitude of the outfall loadings are a

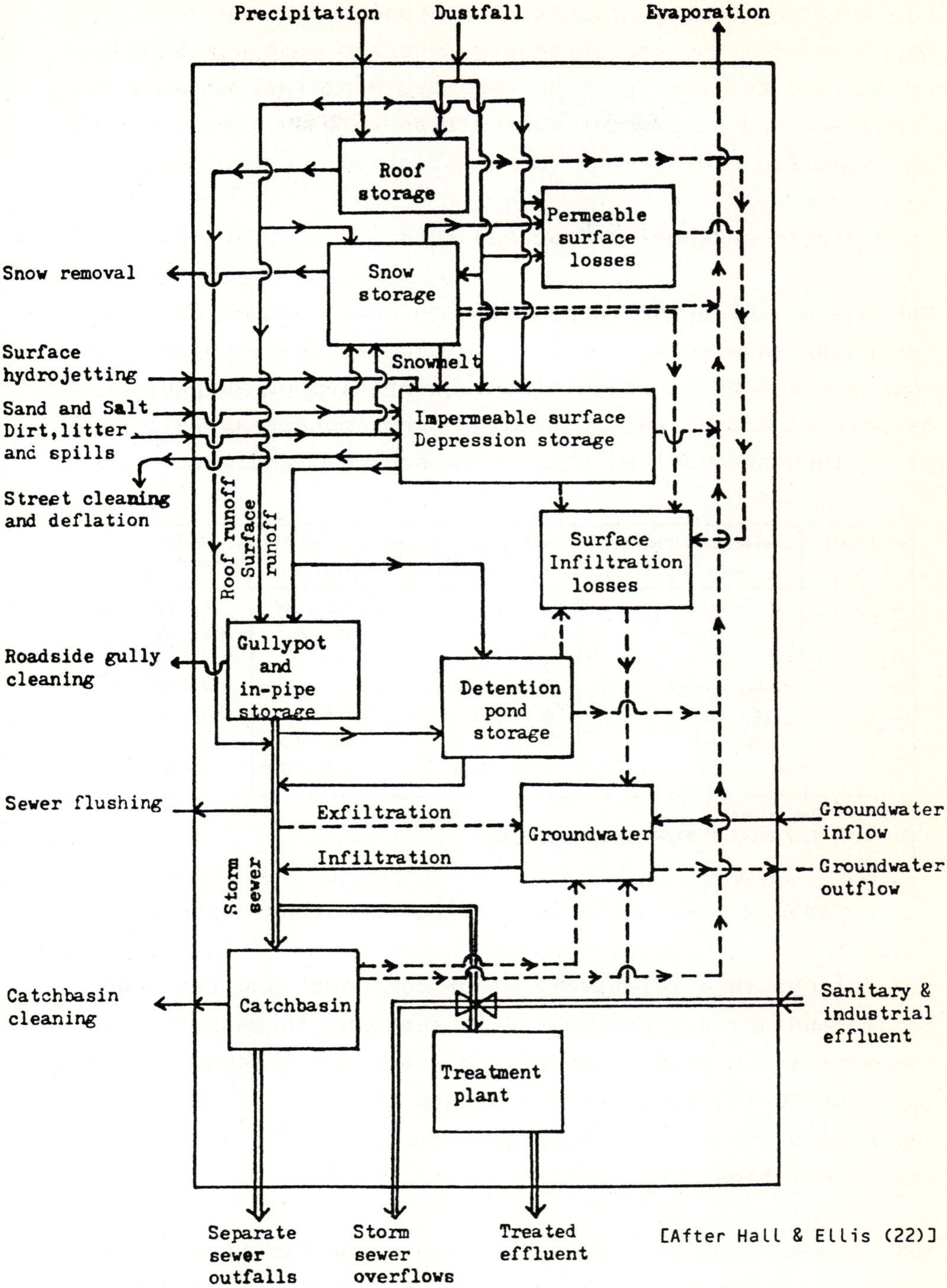

Figure 5. Water and Pollutant Sources and Pathways in Urban Catchments

combined and complex function of atmospheric quality, urban land use activity and intensity, surface composition and condition, traffic type and density, municipal cleaning practises and stormwater controls as well as a function of the specific storm event characteristics.

Atmospheric and Roof Surface Sources

The sources of aerial deposition include industrial and vehicular emissions as well as windblown dust, pollen and vegetative debris. Significant quantities of sulphur, ammoniacal and nitrate nitrogen, heavy metals as well as pesticides are derived from these sources (Table 2).

Pollutant	Total Deposition Rate ($g\ m^{-2} yr^{-1}$)	Wet Deposition Rate (mgl^{-1})	Snowmelt (mgl^{-1})	% Contribution to Runoff
SS	8.4 - 36.2	5 - 70	263 - 690	10 - 25
COD	0.44 - 31.6	8 - 27	15 - 25	15 - 30
SO_4	6 - 15	4.8 - 46.1		31 - 100
Total P	0.021 - 0.204	0.02 - 0.37		17 - 140
NO_3-N	1.8 - 8.2	0.5 - 4.4	4.1 - 5.7	30 - 94
Pb	0.04 - 4.0	0.03 - 0.12	0.3 - 0.4	15 - 54
Zn	0.1 - 1.3	0.05 - 0.38	0.35 - 0.41	20 - 62

(Reported mean values)
(Sources: 13,27,28,42,49,50,51,52,53,54,55,57,58)

Table 2. Atmospheric Loadings in Urban Catchments

Dry fallout is a relatively minor component and reflects local land use activities and is seasonal in character whilst wet deposition predominates with a wet:total deposition ratio of about 60-70% for most pollutants (13,51). In addition, wet deposition tends to be more uniform in quality and loading characteristics.

Whilst some workers (51,52,53) have suggested that atmospheric inputs may account for as much as 50-70% of the mass runoff loads, especially of metals and nutrients, others (54,55) attribute generally less than 15% of any pollutant to precipitation. The latter studies have also drawn attention to regional climatic differences with higher loadings determined for the humid midwest and southeastern areas of the U.S. Some of the variation noted in Table 2 may also be ascribed to turbulent resuspension and deflation caused by traffic densities and movements in urban areas (56,57). Snowfall has an increased pollutant scavenging efficiency because of the large snowflake surface area, and the increased ionic strength - particularly associated with salting activities - of the consequent meltwater provides enhanced metal exchange capabilities (5).

Atmospheric contaminants are washed out during the very early stages of a rainfall event, and therefore the resultant impermeable surface loading tends to be independent of both precipitation volume and intensity (53), with the pollutant load tending to increase with the length of the antecedent dry period and with local traffic density (43,52). Assuming that washout is function of the particulate concentration, the specific weight and particle diameter then:

$$T = 1 - [1 - e_d\ (D^n)]$$

where T = amount of particles washed from atmosphere

e_d = effectiveness of collision of particles with diameter d with raindrops of diameter D

n = number of raindrops per unit area.

As shown in Table 3, roof runoff can also contribute significant loadings of solids and nitrogen to the storm drainage. The E. Coli/streptococci ratio of less than unity (6) implies that the bacterial content of roof runoff results from bird droppings, confirmation of which is provided by the high nutrient levels. Roof surfaces can add substantial quantities of inert solids to stormwater runoff; some 15-30% of total mass flows coming from this source (25,56). Malmquist (43)

Table 3. Pollutant Discharges in Urban Sewers and Catchments

Sewer Type	Mean Pollutant Concentrations ($mg\text{-}l^{-1}$)								Mean Pollutant Loadings ($kg\ ha^{-1}yr^{-1}$)					
	TS	TVS	BOD	COD	NH_4	Pb	PAH[a] B(a)P	E.Coli[b]	TS	TVS	BOD	COD	NH 4	Pb
Separate Storm Sewer Runoff	21 - 582	26 - 149	7 - 22	33 - 265	0.2 - 4.6	0.03 - 3.1	29 - 200	10^2 - 10^4	347 - 2340	90 - 127	35 - 172	22 - 703	1.2 - 25.1	0.09 - 1.91
Combined Storm Sewer Overflows	237 - 635	N/A	43 - 95	120 - 560	2.9 - 4.8	0.15 - 2.9	12 - 215	10^4 - 10^6	1230 - 4917	N/A	505 - 1345	1760 - 3256	52 - 85	N/A
Highway Runoff	28 - 1178	18 - 86	12 - 32	128 - 171	0.02 - 2.1	0.15 - 2.9	365 - 60000	10 - 10^3	121 - 6289	45 - 851	90 - 172	181 - 3865	0.8 - 6.1	0.65 - 13.0
Roof Runoff	12.3 - 216	40 - 88	2.8 - 8.1	57.9 - 80.6	0.4 - 3.8	0.001 - 0.030	N/A	10^2						
Gulley Pot Liquors	15 - 840	185	6.8 - 241	25 - 109	0.7 - 1.39	0.06 - 0.85	N/A	10 - 10^2						
Residential Areas	112 - 1104	28 - 124	7 - 56	37 - 120	0.3 - 3.3	0.09 - 0.44	N/A	10 - 10^3	620 - 2300	N/A	5 - 76.8	22 - 761	N/A	0.06 - 1.91
Commercial Areas	230 - 1894	75 - 85	5 - 17	74 - 160	0.03 - 5.1	0.1 - 0.4	N/A	10^2 - 10^4	50 - 840	N/A	43 - 87	1000 - 1029	N/A	0.17 - 6.84
Light Industrial Areas	45 - 375	35 - 72	8 - 12	40 - 70	0.2 - 1.1	0.6 - 1.2	N/A	10	400 - 1700	N/A	N/A	N/A	N/A	2.2 - 7.0

Notes

a. ng l^{-1}

b. MPN/100 ml

N/A No data available

suggests that as much as 70-90% and 50-70% of the total mass discharges of Cu and Zn respectively can be derived from corrosion of roofing materials, and galvanised guttering can provide high Fe contributions to the sewer system (59). Some UK studies have suggested that roof surfaces, with their fast response times can dominate initial stormwater flows and together with the inpipe, dry weather deposits are responsible for a major proportion of any first flush (14,27). If this conclusion can be more widely supported, then infiltration or downpipe detention, such as suggested by Schilling (60), of roof runoff could both help alleviate the quality problem as well as attenuate and reduce the peak flow volumes by some 20 to 30%.

Impermeable Surface Accumulation and Washoff

According to Sartor and Boyd (17), end-of-pipe pollutant concentrations may be regarded as a function of contaminant accummulation on street surfaces, rainfall-runoff washoff and removal by road sweeping. Loading curves were found to be exponential in form as shown in Figure 6, achieving a steady state after several days unless disrupted by either a rainfall event or a cleaning operation. Highway studies in NW London have suggested that the equilibrium level is reached within 4-5 days when vehicle induced resuspension is dominant (56), but that where resuspension is a negligible factor, such as in residential side streets, the mass accumulates linearly with time. Indeed when total pollutant loads rather than time are plotted against daily accumulation rates, there is little evidence to support diminishing accumulation rates with increasing street loads (15). However, the conceptual exponential relationship has been adopted for several of the available mathematical models of the quantity and quality of urban runoff, such as the Storage, Treatment and Overflow Runoff Model, STORM (61), the Stormwater Management Model, SWMM (62) and the Management of Urban Non-point Pollution

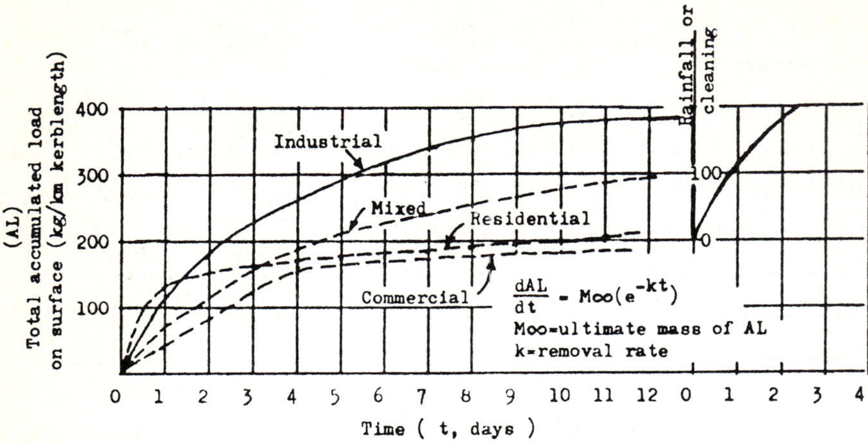

Figure 6. Solids Accumulation for Various Urban Land Uses [after Sartor and Boyd (17)]

model, MUNP (63). However, this approach does not allow for regional differences in aerial deposition or, more importantly, for traffic intensities. In addition, statistical and field verification of the exponential relationship as an adequate description of the washoff process is sadly lacking. Indeed, Jewell and Adrian (18) have asserted that generalised pollutant washoff functions cannot be derived for either storm event total loadings or instantaneous fluxes, and that in order to predict stormwater pollution washoff rates, formulations must be obtained on a catchment-by-catchment basis.

Table 4 presents a summary of mean values for dirt and dust accumulation and pollutant concentrations that have been reported for different urban land uses. Some 20-40 per cent of the total solids is organic. Much of this material is not easily biodegradable, being derived from leaf and wood litter, and rubber and bituminous road surface materials (24), thereby accounting for the high ratio of COD to Biochemical Oxygen Demand (BOD). The high bacterial values recorded in residential and commercial districts can be largely ascribed to animal faeces, whereas the 5 to 6-fold increase in metals for highway

TABLE 4. Solids Accumulation and Associated Pollutant Concentrations in Urban Areas.

		RESIDENTIAL Low Density Single Housing	RESIDENTIAL High Density Multi-Housing	URBAN COMMERCIAL	LIGHT INDUSTRIAL	HIGHWAYS (inc. motorways)
SOLIDS ACCUMULATION kg/km (kerb length)		10-182	30-210	13-180	80-288	13-1100
POLLUTANT CONCENTRATIONS (ppm)	BOD_5	5260	3370	7190	2920	2300-10000
	COD	39300-40000	40000-42000	39000-61730	25100	53650-80000
	N_{tot}	460-480	550-610	410-420	430	225-1600
	Pb	1570	1980	2330	1590	450-2346
	Cd	3.2	2.7	2.9	3.6	2.1-10.2
	Faecal Coliforms MPN/g	60570-82500	25621-31800	36900	30700	18768-38000

surfaces (64) reflects vehicle emissions and corrosion. The strong and distinctive metal content of street surface solids gives them characteristic magnetic properties which enables them to be readily "fingerprinted" and discriminated from atmospheric and roof surface solids (59). Ellis (24) has also shown that the surface solids are log-normally distributed with 70-85 per cent falling within the 2-2000 micron size range. However, although only 4-8 per cent are less than 50 microns in size, these particulates account for 25 per cent of the oxygen demand, 50-60 per cent of grease and rubber, 30 per cent of heavy metals, 30-50 per cent of nitrogen and phosphorous and 10-15 per cent of total coliforms (Fig.7). This confirms the classic baseline studies of Sartor and Boyd (17) and if one takes a higher cut-off of 246 microns, then some 35-43 per cent of the total solids accounts for some three quarters of most pollutants.

The removal of pollutants from impermeable surfaces is determined by a number of factors of which the following are the most important:

- type and condition of the street surface; as shown by Fig.7b more dirt is washed from relatively smooth concrete surfaces. The importance of surface condition on pollutant

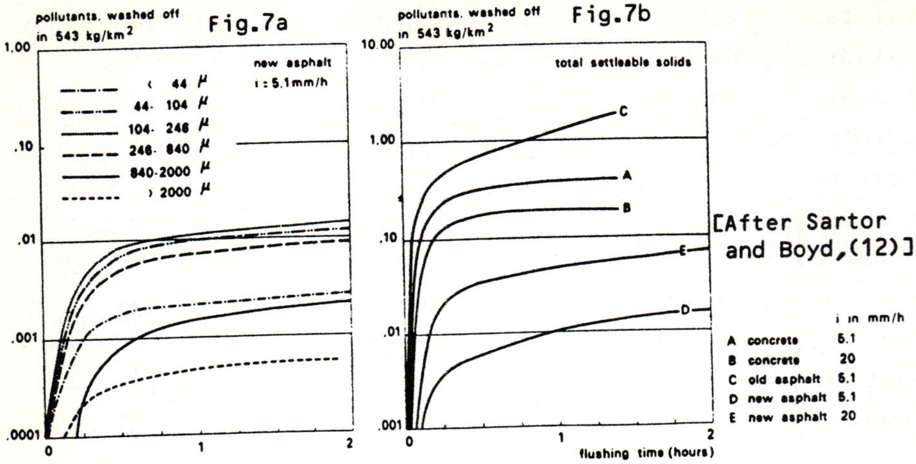

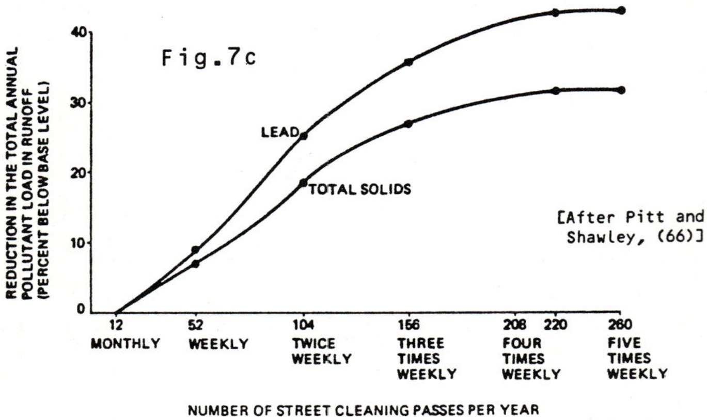

Figures 7a, b, c. Pollutant Washoff and Particle Size, Surface Type and Street Cleaning Controls

yield is also obvious as can be seen from inspection of the removal curve for worn asphalt. Poor surface condition therefore not only accounts for a reduced runoff coefficient but also increases the solids storage capacity and means that pollutant washoff yield is rarely deposition or availability limited.

- rainfall intensity and volume; particle entrainment through raindrop impact and overland flow provides the main surface removal mechanisms with Price and Mance (65) favouring the predominance of the former process. Hydraulic flow depths must exceed the drag force exerted by surface boundary roughness before any major quantities of solids can be transported. These threshold forces will be higher on poor surfaces where the amplitude of the microtopographic variation will be greatest imposing maximum shear friction. Low rainfall intensities and volumes never exceed the boundary shear stress and encourage particle aggregation and uptake of soluble materials.
- particle size; if one assumes similar specific weights, it might be expected that the contaminated fines would be entrained more easily than larger size fractions. However, inspection of Fig.7a suggests that there are only relatively minor differences in removal rates for particles of different sizes. This may reflect the "light" organic nature of many large particles in comparison to the "heavy" inert character of small clay size particles (49). There is also some evidence that organic particulates, as well as clay-silt particles, occur most frequently during the early stages of stormflow (24).
- runoff acidity; low precipitation pH is rapidly buffered to neutral values during overland flow across the impermeable surface, but initial inputs may well be sufficiently acid to solubilise weakly bound pollutants such as heavy metals and nutrients.
- street cleaning; apart from the cosmetic removal of coarse litter, street cleaning has been shown to be relatively ineffective as a means of pollution control with overall removals of 30-50% being typical and accounting for only 10-20% of the contaminated fines (17,24,67 and Fig. 8). Cleaning interval is the dominant influence on effectiveness (Fig.7c) and achievement of optimum performance requires a sweeping frequency of at least the average time between storms.

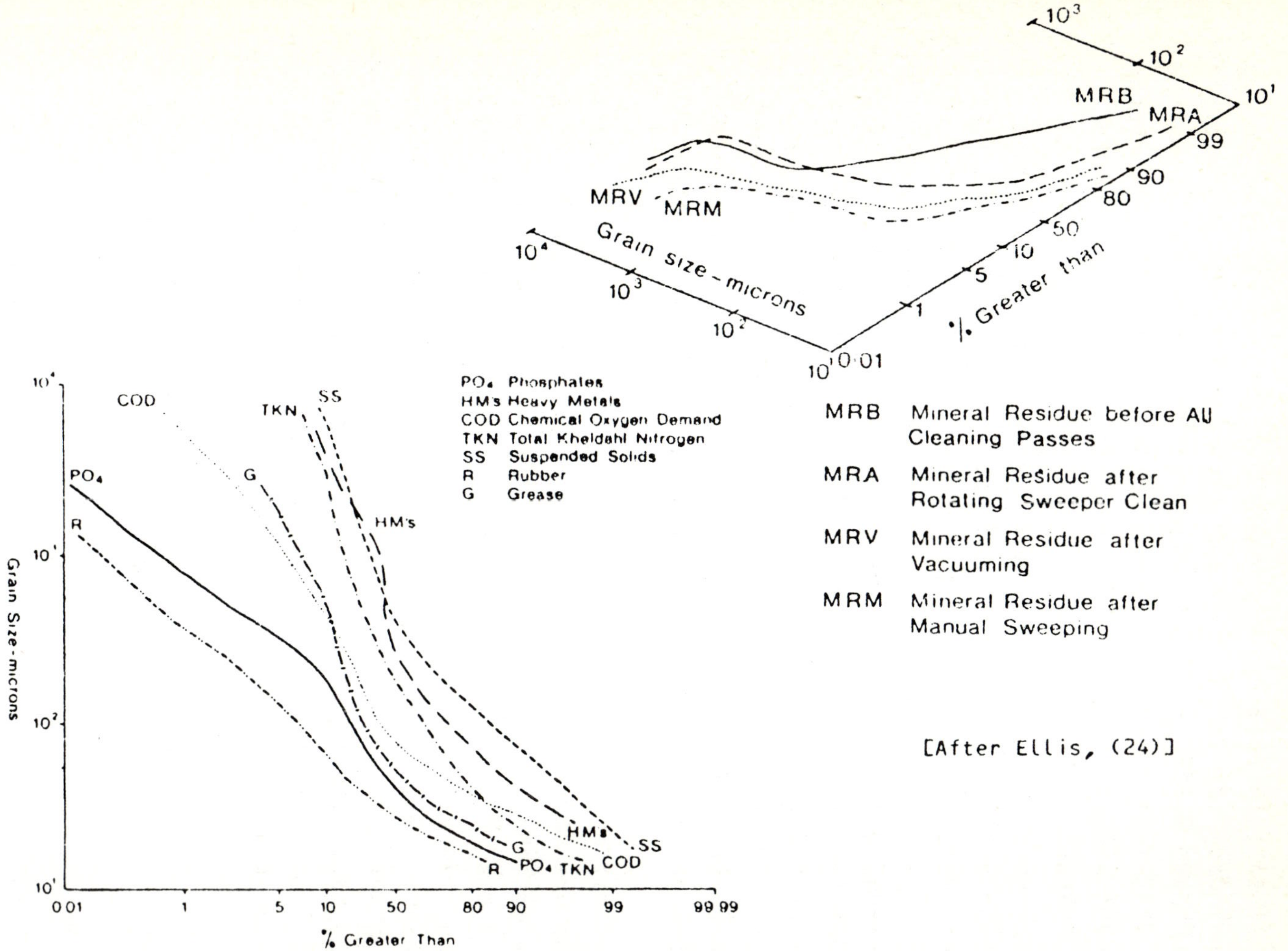

Figure 8. Particle Size Distribution of Street Surface Pollutants and Cleaning Efficiency

- traffic densities; resuspension, deflation and spray account for substantial removals of solids from the impermeable surface. Spray losses of 50% during rainfall events of less than 1mm have been recorded on motorways when the wetted surface depth has been insufficient to support overland flow (68).

Surface washoff to roadside gully pots or catchpit chambers represent a further pollution source at the above-below ground interface (19,27,45,58,69). Anaerobic digestion of organically enriched solids, especially during summer and autumn, contributes high BOD, COD, NH_4-N, hydrocarbons and dissolved metals. The septic supernatant in the chamber is washed out early in the storm event, and contributes to the first flush phenomenon. Pratt and Adams (69) have suggested that gully storage volume needs to be doubled to provide adequate oxygen recharge.

Runoff Loadings

In a recent review, Heaney and Huber (3) have concluded that in the US the quantity of urban runoff is approximately equal to the quantities of sewage and van de Ven (70) has quoted a similar ratio for the Netherlands. Table 3 provides a summary of event mean pollutant discharge data which has been compiled from US, European and Australian sources. Clearly, stormwater runoff solids, metals and hydrocarbons can be regarded as being of sewage quality although BOD and COD are relatively low. The bacterial quality of runoff, especially from residential areas, can be poor as a result of animal/bird faeces and wrong connections and as indicated by Ellis (6) pathogenic strains are similar to those found in secondary treatment effluents. Highway runoff can be of particularly poor quality, discharging high metal and hydrocarbon loadings to receiving waters. Indeed, some 50% of solids, 40-75% of metals and 70% of the total PAH receiving stream input budgets can be derived from highway runoff sources (71).

The flow weighted event mean concentrations of Table 3 probably provide a better picture of pollutional characteristics than the loadings which are based on short (monthly/annual) duration studies and which contain both quantitative and qualitative data that has been "assembled" into one figure. Such loading data cannot adequately allow for the marked variations in meteorological conditions that can exist between individual years and load is strongly influenced by the magnitude of the storm event. Land use types show some common features such as high solids and metal concentrations and low BOD, COD although commercial districts and highways yield above average loadings than residential areas.

Comparison of Tables 3 and 4 shows that in-pipe particulate (especially the fine suspended solids less that 250 microns) has metal concentrations which are nearly two orders of magnitude over levels normally found in raw street sediments. These latter have a strong exchangeable metal affinity which is independent of particle size and in order Cd>Zn>Cu>Pb; outfall discharges show Pb to be dominantly exchangeable with Cd, and Zn present mainly in the dissolved form (72).

The NURP final report (4) quantifies inter-storm variability by the site median and coefficient of variation but as site variability is log normally distributed, probability analysis as previously described can be applied to characterise a specific land use type. However, it must be realised that overlapping confidence limits for median values implies that no statistically significant differences can be placed against a particular land use type or activity.

Ellis and Harrop (16) have demonstrated that 70-95% of the variance in total runoff loadings can be explained by total runoff volume, storm duration and 5 minute rainfall intensities. Although both Wilkinson (73) and Hedley and King (26) found that the strength of pollutants increased with the length of the antecedent dry period, Waller and Counter (44), Jewell and Adrian (18), Ellis, Harrop and Revitt (15) and many others have been unable to support this conclusion. Simple bivariate analysis consistently produces poor correlations irrespective of the length of record available but improved relationships result when multiple regression is applied. The STORM model utilises both land use and storm event characteristics to determine total loadings in the general form:

$$Ms = \alpha_{(ij)} \cdot P \cdot Pd \cdot Ssw$$

where Ms = pollutant mass (j) generated per unit area of land use (i) (kg ha $^{-1}$ yr $^{-1}$)

α = land use pollutant factor (based on normalisation of data such as included in Tables 3 and 4.

P = annual precipitation (cm yr^{-1})

Pd = population density function (based on value 1.0 for commercial area and 0.1 for open spaces).

Ssw = street sweeping effectiveness factor (with value of 1.0 for interval greater than 20 days).

The SWMM model assumes that the amount of pollutant washed off at any time interval is a function of an availability factor and is proportional to the amount remaining on the surface. This results in a first order differential equation which integrates to the form:

$$Po - P = AoPo\,(1-e^{kt})$$ (See Insert Fig.6)

where Po = initial amount of pollutant per unit area

P = pollutant remaining per unit area at time t

k = removal rate

A_o = availability factor defined as (Po) pollutant % available for removal by the specific storm runoff event.

More simplistic approaches have been adopted by many workers, such as Colston (74) in his Tulsa studies and Desbordes and Servat (75) in their French investigations; all advocate consideration of only storm event characteristics for the estimation of loading rates e.g.:

$$TS = 3.35Q^{1.14}\ TSS^{-0.18} \quad \text{(Colston: 74)}$$

where TS = total solids loading (lb/min)

Q = runoff (inches)

TSS = time since commencement of storm (hrs)

and $$SS = 125.52 + 58.7\ DTS + 13.49\ IMAX5$$

(Desbordes and Servat: 75)

where SS = mean suspended solids concentration (mg l^{-1})

DTS = preceding dry weather duration (days)

IMAX5 = mean maximum 5 minute rainfall intensity (mm h^{-1})

A stepwise multiple regression model developed by Ellis, Harrop and Revitt (15) for a residential road in Oxhey, Herts., produced a work equation of the following form:

$$Ts = 114.86 + 3.589Q + 0.346(ADP) + 4.257\ (RV)$$

where Ts = total solids loading (g ha^{-1})

Q = total surface runoff (litres)

ADP = antecedent dry period length (days), expressed as duration less than effective rainfall depth (h)

RV = total rainfall volume (mm)

The extrapolation of such formulations to other unmonitored urban catchments must be done with extreme care and, as previously stated, there is a need to obtain specific data from each site/basin if satisfactory washoff predictions are to be obtained. Apart from their site specific calibrative nature, the formulations cannot be statistically verified as adequate models of the pollutant washoff process. However, if one accepts the validity of invidiual approaches for the sites in question, then it is feasible to derive first order approximations for loadings of sediment associated pollutants such as heavy metals, nutrients or hydrocarbons.

In the case of the road study at Oxhey, Herts. quoted above, total metal removal rates of 67.4 ug m^{-2} day^{-1}, 99.2 ug m^{-2} day^{-1}, 34.1 ug m^{-2} day $^{-1}$ and 0.76 ug m^{-2} day^{-1} were predicted for Pb, Zn, Cu and Cd respectively (15). Using Fig. 8 it is also feasible to estimate the pollutant loading rate associated with varying particle size fractions and to evaluate the reduction in loadings to be expected from various street cleaning methods. Thus the particle size fraction less than 60 microns is responsible for Pb loadings of some 22.5 ug m day $^{-1}$, which might be reduced to 19.7 ug m^{-2} day-1 following a single cleaning run by a rotating brush sweeper.

It is obvious that characterisation of urban runoff quality with its inherent variability and transient nature presents a difficult task although it is equally obvious that short term management efforts need to focus on high runoff, high pollutant concentration areas such as car parks, roof and road surfaces, roadside gully chambers and inpipe dry weather deposits. The three former pollution sources might be considerably reduced through the introduction of local porous paving and infiltration schemes with removal (or reduction) of the conventional kerb-grating-gully systems, especially in residential areas (1,11). The latter source requires a more careful consideration of hydraulic design criteria to ensure better in line quality control and sediment transport. Generalised management programmes should be avoided in favour of specific site schemes with expensive structural urban runoff controls reserved for locations of particular environmental priority.

References

1. Ellis, J.B. Urban Runoff Quality and Control. 234-240 in Tebbutt, T.H.Y. (Ed), Advances in Water Engineering, Elsevier Applied Science Publ. Ltd., London. 1985.

2. US EPA. Non Point Pollution in the US: Report to Congress. US EPA Summary Report, Water Planning Div., Washington, D.C. 1984.

3. Heaney, J.P., Huber, W.C. Nationwide Assessment of Urban Runoff Impact on Receiving Water Quality. Water Res. Bull. 20 (1): 35-42; 1984.

4. US EPA. Final Report of the Nationwide Urban Runoff Program. Vol. I. US EPA, Washington, D.C. 1983.

5. Morrison, G.M.P., Revitt, D.M., Ellis, J.B., Balmer, P., Svensson, G. The Transport Mechanisms and Phase Interactions of Bioavailable Heavy Metals in Runoff. 226-229 in Sly, P. (Ed): Interactions Between Sediments and Water. CEP Ltd, Edinburgh. 1984.

6. Ellis, J.B. Water and Sediment Microbiology of Urban Rivers and their Public Health Implications. Publ. Hlth. Eng. 13 (2): 95-107; 1985.

7. Grizzard, T.J., Weand, B.L., Randall, C.W. An Evaluation of Stormwater Management Ponds for the Control of Urban Runoff Pollution. 135-148 in Featherstone, R.E., James, A. (Eds): Urban Drainage Systems., Pitmans, London. 1982.

8. Ferrara, R.A., Witkowski, P. Stormwater Quality Characteristics in Detention Basins. Journ. Env. Eng. Div., ASCE, 109 (2) 428-444; 1983.

9. Ellis, J.B. Structural Control for Urban Stormwater Quality. In: Hydraulics of Floods and Flood Control, British Hydrodynamics Res. Assoc., Cranfield, Bedford. 1985.

10. Sartor, J.D., Gaboury, D.R. Street sweeping as a Water Pollution Control Measure: Lessons Learned over the past ten years. Science Tot. Env., 33: 171-183; 1984.

11. Pratt, C.J. Design Limits on Pollution. Science Tot. Env., 33; 161-170; 1984.

12. Hydraulics Research Station. The Design and Analysis of Urban Storm Drainage. Report DE 50, Vol. I. Wallingford, Oxon. 1980.

13. Oberst, G.L. Water Resources Management: Non-Point Source Pollution. Tech. Rept. 10-82-016. Metropolitan Council Publ., St. Paul, Minnesota. 1982.

14. Pratt, C.J., Harrison, J.J., Adams, J.R.W. Storm Runoff Simulation in Runoff Quality Investigations. 285-294 in Balmer, P., Malmquist, P.A., Sjoberg, A. (Eds): Analysis & Design of Stormwater Systems. Chalmers Univ. of Tech., Gothenburg. 1984.

15. Ellis, J.B. Harrop, D.O., Revitt, D.M. Hydrological Controls of Pollutant Removal from Highway Surfaces. Water Res., 19: 1985. (In Press).

16. Ellis, J.B., Harrop, D.O. Variations in Solids Loadings to Roadside Gully Pots. Science Tot. Env., 33: 203-211; 1984.

17. Sartor, J.D., Boyd, G. Water Pollution Aspects of Street Surface Contaminants. Rept. EPA-R2-72-081. US EPA., Washington, D.C. 1972.

18. Jewell, T.K., Adrian, P.A. Improved Techniques for Modelling Stormwater Quality. 156-165 in Yen, B.C. (Ed): Urban Stormwater Quality, Management & Planning. Water Res. Publ. Ltd., Littleton, Colorado. 1982.

19. Waller, D.H. Factors that Influence Variations in the Composition of Urban Surface Runoff. Water Poll. Res. in Canada, 7: 68-95; 1972.

20. Ackers, P. Urban Drainage: The Effects of Sediment on Performance and Design Criteria. 535-545 in Helliwell, P.R. (Ed): Urban Storm Drainage. Pentech Press, London, 1978.

21. Lindholm, O., Balmer, P. Pollution in Storm Runoff and Combined Sewer Overflows. 575-585 in Helliwell, P.R. (Ed): Urban Storm Drainage. Pentech Press, London. 1978.

22. Ellis, J.B. Blights and Benefits of Urban Stormwater Quality Control. 39-51 in Featherstone, R.E., James, A. (Ed): Urban Drainage Systems. Pitmans, London. 1982.

23. Saul, A.J., Thornton, R.C., Henderson, R. The Flood Control Characteristics of a Storage Volume in a Sewer System based on Field Measurements and Computer Simulation. In: Hydraulics of Floods and Flood Control, British Hydrodynamics Res. Assoc., Cranfield, Bedford. 1985.

24. Ellis, J.B. The Nature and Sources of Urban Sediments and their relation to Water Quality. 199-216 in Hollis, G.E. (Ed): Man's Impact on the Hydrological Cycle in the U.K. Geobooks, Norwich. 1979.

25. Ellis, J.B. The Characterisation of Particulate Solids and Quality of Water discharged from an Urban Catchment. 283-291 in Effects of Urbanisation and Industrialisation on Hydrological Regime and on Water Quality. IAHS/AISH Publ. No. 123, Unesco, Amsterdam.

26. Hedley, G., King, M.V. Suggested Correlations between Storm Sewage Characteristics and Storm Overflow Performance. Proc. Inst. Civ. Eng. 48: 399-411; 1971.

27. Harrop, D.O. Stormwater Runoff and Water Quality in Urban Catchments. Unpublished Ph.D. Thesis, Urban Poll. Res. Centre, Middlesex Polytechnic, Enfield, UK. 1984.

28. Geiger, W.F. Characteristics of Combined Sewer Runoff. 851-860 in Balmer P., Malmquist, P.A., Sjoberg, A. (Eds): Planning & Control of Urban Storm Drainage, Chalmers Univ. of Tech., Gothenburg. 1984.

29. Mance, G. The Quality of Urban Storm Discharges: A Review. Tech. Rept. 192-M, Water Research Center, Stevenage. 1981.

30. Ellis, J.B. Sediments and Water Quality of Urban Stormwater. Water Services, 80: 730-734; 1976.

31. Marsalek, J. Characteristics of Runoff from an Urban Commercial Area. Science et Tech. de L'eau, 17 (2): 163-167; 1984.

32. Di Toro, D.M. Probability Model of Stream Quality due to Runoff. Journ. Env. Eng., ASCE, 110 (3): 607-628; 1984.

33. Ellis, J.B. Pollutant Loads from Separate Sewer Systems. 1613-1623 in Balmer, P., Malmquist, P.A., Sjoberg, A. (Edits): Urban Storm Drainage. Vol. 4. (Supplements), Chalmers Univ. of Tech., Gothenburg,. 1984.

34. Hvited-Jacobsen, T., Harremoes, P. Impact of CSO on DO in Receiving Streams. 226-235 in Yen, B.C. (Ed): Urban Stormwater Quality, Management & Planning, Water Res. Publ. Ltd., Littleton, Colorado. 1982.

35. Loganathan, G.V., Delleur, J.W. Effects of Urbanisation on Frequencies of Overflow and Pollutant Loadings from Storm Sewer Overflow. Water Res. Res. 20 (7): 857-865; 1984.

36. ASCE Urban Hydrology Research Council. Basic Information Needs in Urban Hydrology. ASCE, New York. 1969.

37. Delleur, J.W., Torno, H.C. Urban Hydrology. ASCE, New York. 1983.

38. Torno, H.C. The Nationwide Urban Runoff Program. 1475-1474 in Balmer, P., Malmquist, P.A., Sjoberg, A. (Eds): Urban Storm Drainage. Vol IV (Supplements). Chalmers University of Tech., Gothenburg. 1984.

39. Cole, R.H., Frederick, R.E., Healy, R.P., Rolan, R.G. Preliminary Findings of the Priority Pollutant Monitoring Project of the NURP. Journ. Wat. Poll. Cont. Fed., 56(7): 898-908; 1984.

40. Geldreich, E.E., Best, L.C., Kenner, B.A., Van Donsel, D.J. The Bacteriological Aspects of Stormwater Pollution. Journ. Wat. Poll. Cont. Fed., 40: 1861-1870; 1968.

41. Olivieri, V.P., Kruse, C.W., Kawata, K. Micro-organisms in Urban Stormwater. Rept. 600/2-77-087, US EPA., Cincinnati, Ohio. 1977.

42. Hvitved-Jacobsen, T., Yousef, Y.A., Wanielista, M.P., Pearce, D.B. Fate of Phosphorous and Nitrogen in Ponds Receiving Highway Runoff. Science Tot. Env., 33: 259-270; 1984.

43. Malmquist, P.A. Urban Stormwater Pollutant Sources. Chalmers Univ. of Tech., Gothenburg. 1983.

44. Waller, D.J., Coulter, W.A. Winter Runoff from an Urban Catchment. Res. Rpt. 41. Env. Canada, Ottowa. 1974.

45. Fletcher, I.J., Pratt, C.J., Elliott, G.E.P. An Assessment of the Importance of Roadside Gully Pots in Determining the Quality of Stormwater Runoff. 586-602 in Helliwell, P.R. (Ed): Urban Storm Drainage, Pentech Press, London. 1978.

46. Grizzard, T.J., Randall, C.W. Occoquan-Four Mile Run Runoff Pollution Field Study. Rpt., Dept of Civ. Eng., Virginia State Inst., Blocksburg, Virginia. 1978.

47. Morrison, G.M.P., Revitt, D.M., Ellis, J.B., Svensson, G., Balmer, P. The Physico-Chemical Speciation of Zn, Pb, Cd and Cu in Urban Stormwater. 989-1000 in Balmer, P., Malmquist, P.A., Sjoberg, A. (Eds): Planning & Control of Urban Storm Drainage. Chalmers Univ. of Tech., Gothenburg, 1984.

48. Ellis, J.B., Revitt, D.M., Gavens, A. Polyaromatic Hydrocarbons in Sediments of an Urban Catchment. Int. Journ. Env. Anal. Chem., 13: 1985 (In Press).

49. Ellis, J.B., Hamilton, R.S., Roberts, A.H. Sedimentary Characteristics of Suspensions in London Stormwater. Sediment Geol., 33: 147-154; 1982.

50. Hall, M.J., Ellis, J.B. Water Quality Problems of Urban Areas. Geo Journal. 1985 (In Press).

51. MeLanen, M. Quality of Runoff Water in Urban Areas. 123-190, Publ. 42., Water Research Inst., National Board of Waters, Helsinki. 1981.

52. Goettle, A. Atmospheric Contaminants, Fallout and their Effects on Stormwater Quality. Prog. Water Tech., 10: 455467; 1978

53. Randall, C.W., Helsel, D.R., Grizzard T.J., Hoehn, R.C. The Impact of Atmospheric Contaminants on Stormwater Quality in an Urban Area. Prog. Water Tech., 10: 417-431; 1978.

54. Novotny, V., Chesters, V. Handbook of Non-Point Pollution: Sources and Management. Van Nostrand Reinhold Co., New York. 1981.

55. Pitt, R.E., Bazeman, M. Sources of Urban Runoff Pollution and its Effects on an Urban Creek. Rpt. 6002/2-82-090., US EPA., Washington, D.C. 1982.

56. Hamilton, R.S., Revitt, D.M., Warren, R.S., Duggan, M.J. Resuspension of Heavy Metals from Road Surface Dust. In: Heavy Metals in the Environment, CEP Ltd., Edinburg. 1985 (In Press).

57. Klein, H. Die Beschaffenheit der Niederschlagsabflusse von Autobahnen. Wasserwirtschaft. 72 (2): 37-43; 1982.

58. Mance, G., Harman, I.I.M. The Quality of Urban Stormwater Runoff. 603-618 in Helliwell, P.R. (Ed): Urban Storm Drainage, Pentech Press, London. 1978.

59. Beckwith, P.R., Ellis, J.B., Revitt, D.M., Oldfield, F. Heavy Metal and Magnetic Relationships for Urban Source Sediments. Phys. of the Earth & Planetary Int., 1985. (In Press).

60. Schilling, W. Cisterns against Storms. 49-60 in Featherstone, R.E., James, A. (Eds): Urban Drainage Systems. Pentech Press, London. 1982.

61. US Army Corps of Engineers. Storage, Treatment, Overflow, Runoff Model STORM. USACE Hydrologic Eng. Centre, Davis, California. 1974.

62. Lager, J.A., Shubinski, R.P., Russell, L.W. Development of a Simulation Model for Stormwater Management. Journ. Wat. Poll. Cont. Fed., 43: 2424-2435; 1971.

63. Sutherland, R.C., McCuen, R.J. Simulation of Urban Non-Point Source Pollution. Wat. Res. Bull., 14: 409-428; 1978.

64. Ellis, J.B., Revitt. D.M. Incidence of Heavy Metals in Street Surface Sediments. Water, Air & Soil Poll., 17; 87-100; 1982.

65. Price, R.K.. Mance, G. A Suspended Solids Model for Stormwater Runoff. 546-555 in Helliwell, P.R. (Ed): Urban Storm Drainage, Pentech Press, London. 1978.

66. Pitt, R.E., Shawley, G. San Francisco Bay Area NURP. Rpt., US EPA., Washington. D.C. 1981.

67. Bender, G.M., Terstriep, M.L. Effectiveness of Street Sweeping in Urban Runoff Pollution Control. Science Total Env., 33: 185;192; 1984.

68. Jones, A.D. Tinker, J.A. Quantified Appraisal of Pollutants Dispersed from Road Surfaces by Airborne Mechanisms. Science Total Env., 33: 193-201; 1984.

69. Pratt C.J., Adams., J.R.W. Sediment Supply and Transmission via Roadside Gully Pots. Science Total Env., 33; 213-224; 1984.

70. van de Ven, F.H.M. Amount and Quality of Stormwater Runoff in Lelystad and its Influence on Receiving Waters. H_2O, 16: 481-487; 1983.

71. Perry, R., McIntyre, A.E. Impact of Motorway Runoff upon Surface Water Ouality. In Solbe, J. (Ed): Effects of Land Use on Fresh Water, Water Research Centre, Medmenham. 1985. (In press).

72. Ellis, J.B., Revitt, D.M., Shutes, R.B.E., Hamilton, R.S. The Effects of Urbanisation on Receiving Water Quality: Heavy Metal Toxicity. In Solbe, J. (Ed): Effects of Land Use on Fresh Water, Water Research Centre, Medmenham. 1985. (In Press).

73. Wilkinson, R. The Quality of Urban Rainfall-Runoff Water from a Housing Estate. Journ. Inst. Pub. Hlth. Eng., 55: 70-84; 1956.

74. Colston, N.V. Characteristics and Treatment of Urban Land Runoff. Rept. 670/2-74-096, US EPA., Washington, D.C. 1974.

75. Desbordes, M., Servat, E. Solids in Urban Runoff: Statistical Analysis of French Experimental Data. 947-956 in Balmer, P., Malmquist, P.A., Sjoberg, A. (Eds): Planning & Control of Urban Storm Drainage. Chalmers Univ. of Tech.. Gothenburg. 1984.

TOXIC CONTAMINANTS IN URBAN RUNOFF: A CASE STUDY

J. Marsalek, National Water Research Institute, Canada Centre for Inland Waters, 867 Lakeshore Road, Burlington, Ontario, Canada, L7R 4A6

Introduction

Concerns about adverse effects of toxic substances on the environment and human health have increased in recent years as a result of mounting experimental evidence indicating carcinogenic, mutagenic and teratogenic impacts of these substances throughout the biosphere. Such concerns have been further heightened by the widespread distribution of toxics in the environment and by indications of bioaccumulation and biomagnification of many toxic substances. Although in many cases the impacts of relatively low-level concentrations of toxics are not yet fully understood, a conservative approach to this problem and the need to improve the understanding of underlying processes are generally advocated. Consequently, the Canadian Department of Environment has initiated a research program on toxic substances. The main goals of this program include the compilation of inventories of toxics in the environment; identification of their sources, pathways and impacts; and, the development of control and remedial measures.

The problem of toxic substances in the Great Lakes Basin has been addressed earlier by the International Joint Commission (IJC) which noted recurring pollutional problems in some parts of the basin. In these parts, which were designated as "areas of concern", high levels of such toxic substances as heavy metals, PCB's, organochlorine pesticides, and industrial organics have been found and identified as primary causes of environmental degradation. These substances originated from the following six sources identified by the Commission (7):

NATO ASI Series, Vol. G10
Urban Runoff Pollution.
Edited by H. C. Torno, J. Marsalek, and M. Desbordes

Municipal and industrial discharges
Waste disposal sites
Combined sewer overflows
Urban land runoff
Agricultural land runoff, and
In-place pollutants

To evaluate the above sources, a number of studies has been conducted in recent years. Preliminary results of one of these studies, which dealt with toxics in urban runoff, are presented in the paper that follows.

Experimental Program

The main study effort focussed on extensive field investigations of selected toxic substances in urban runoff in the study area. The planning and conduct of these investigations differed from earlier studies of conventional pollutants in urban runoff and it is, therefore, desirable to explain such differences before proceeding with the description of the field program.

Conventional pollutants found in urban runoff, for example suspended solids or nutrients, originate mostly from local sources and occur in relatively high concentrations (10^{-5} - 10^{-3}) which exhibit large variations during runoff events. It is feasible to design a program for the monitoring of pollutants in runoff by considering local sources of pollutants and variations in pollutant concentrations during runoff events. There are no severe restrictions on the number of samples analyzed, because conventional analyses are relatively inexpensive. The data collected are fairly site specific, because they reflect the local composition of pollutant sources.

By contrast, toxic trace organics found in urban runoff typically originate from remote sources and occur in relatively low concentrations (10^{-12} - 10^{-9}). Sources of trace organics are rather dispersed and, for many substances, fairly rare. To some extent, urban runoff provides a transport mechanism (or a pathway) for toxics imported into the catchment mostly with wet precipitation. Consequently, observed concentrations of trace organics exhibit low variations when comparing data from various urban centres, unless strong point sources are present. Such sources can be readily identified from pollution source inventories (mostly industrial inventories) which are compiled by pollution control enforcement agencies. It follows from the above discussion that the trace organics data may be transferable to other areas without strong point sources of organics.

Although trace organics are found in urban runoff in low concentrations, these concentrations may be important because of high toxicity of many of these substances. Low concentrations and the nature of these substances lead to certain problems. In particular, special sampling procedures have to be used to avoid sample cross contamination and the analytical costs are rather high. The costs of sample collection and analysis are further increased by the fact that trace organics are transported by both water and sediment which may have to be investigated separately. High analytical costs then place a constraint on the number of samples collected in a particular study.

With the present state of knowledge, the interest in toxics in urban runoff is generally limited to their loadings for relatively long time intervals. For comparisons of various sources of toxics, monthly or even yearly loadings in urban runoff are adequate. Such loadings can be determined from mean event concentrations which are determined for composite event samples. Thus, there is no obvious need for costly sequential sampling of toxics in urban runoff.

The occurrence of toxic metals in urban runoff is similar to that of conventional pollutants. Metals originate mostly from local sources (e.g., traffic) and occur in relatively high concentrations (up to 10^{-6}). Their levels may be related to local sources (e.g., land use) and in this sense, the data on toxic metals may be less readily transferable to other areas without observations. On the other hand, the data on metals in urban runoff are available from numerous sources (2, 4, 5, 8, 11 and 12) and that should help in transposition of data to other areas.

Recognizing the above discussed characteristics of toxic substances in urban runoff, a special field progrm for their monitoring was designed and conducted. A brief description of this program follows.

Toxic Substances Studied

The selection of toxic substances to be studied was initially based on the list of priority pollutants which was developed by the U.S. Environmental Protection Agency (17). This list was reduced to 50 persistent substances for which the samples could be analyzed in the supporting laboratories. These substances, which are listed in Table 1, can be summarized into the following five groups: Polychlorinated biphenyls (PCB's), organochlorine pesticides, polyaromatic hydrocarbons (PAH's), chlorinated benzenes, and trace elements.

Study area

The study area encompassed 12 large urban centres which, with two exceptions, are located in the Canadian part of the Lower Great Lakes Basin. Field investigations were conducted in 12 urban centres which are listed in Table 2 and their locations are shown in Fig. 1. In each centre, several sampling sites were established in areas with various urban land use. The types of land use sampled included residential, commercial, institutional and industrial land. In total, runoff and/or sediment samples were collected at more than 80 sites. At each site, from three to fifteen runoff events were sampled.

Sampling Procedures

Flow measurement and sampling methods have been well established for studies of conventional pollutants in urban runoff.

Table 1. Toxic Substances Studied, Their Detection Limits and Frequencies of Detection

Substance	Detection Limit Water (ppb)	Detection Limit Sediment (ppm)	Frequency of Detection (%) Water	Frequency of Detection (%) Sediment
PCB's				
Total PCB's	.009	.09	46	85
Organochlorine Pesticides				
Hexachlor benzene	.0004	.004	45	64
α-BHC	.0004	.004	98	28
Lindane	.0004	.004	86	18
Heptachlor	.0004	.004	6	16
Aldrin	.0004	.004	1	10
Heptachlor epoxide	.0004	.004	27	35
γ-Chlordane	.0004	.004	20	35
α-Chlordane	.0004	.004	11	18
α-Endosulfan	.0004	.004	26	51
p,p' - DDE	.0004	.004	19	21
Dieldrin	.0004	.004	26	16
Endrin	.0004	.004	23	26
o,p' - DDT	.0004	.004	18	18
p,p' - TDE	.0004	.004	28	40
p,p' - DDT	.0004	.004	22	3
β- Endosulfan	.0004	.004	26	10
Mirex	.0004	.004	3	10
Methoxychlor	.0004	.004	21	12
Polyaromatic Hydrocarbons				
Indene	.05	.05	2	13
1,2,3,4, Tetrahydro-naphthalene	.05	.05	2	8
2, Methylnaphthalene	.05	.05	19	0
Quinoline	.05	.05	4	17
1, Methylnaphthalene	.05	.05	9	1
β- Chloronaphthalene	.05	.05	15	3
Acenaphthylene	.05	.05	17	10
Acenaphthene	.05	.05	13	1
Fluorene	.05	.05	2	8
Phenanthrene	.05	.05	6	30
Fluoranthene	.05	.05	13	37
Pyrene	.05	.05	17	28
Chlorinated Benzenes				
1,3 Dichlorobenzene	.005	.05	18	11
1,4 Dichlorobenzene	.005	.05	25	13
1,2 Dichlorobenzene	.005	.05	48	26
1,3,5 Trichlorobenzene	.001	.005	15	30
1,2,4 Trichlorobenzene	.001	.005	21	31

Table 1. (continued)

Substance	Detection Limit Water (ppb)	Detection Limit Sediment (ppm)	Frequency of Detection (%) Water	Frequency of Detection (%) Sediment
1,2,3 Trichlorobenzene	.001	.005	5	21
1,2,4,5, + 1,2,3,5, Tetrachlorobenzene	.001	.005	9	19
1,2,3,4 Tetrachlorobenzene	.001	.005	4	17
Pentachlorobenzene	.001	.005	14	53
Trace Elements				
Arsenic	.1	.05	87	100
Cadmium	1.0	10	12	70
Copper	1.0	10	93	94
Cobalt	1.0	10	27	58
Chromium	1.0	.5	44	91
Lead	1.0	.5	78	94
Mercury	.05	.1	66	80
Nickel	1.0	10	87	88
Selenium	.1	.05	86	84
Zinc	1.0	.1	98	100

Table 2. Study Sites

Urban Centre	Approximate Population (inh.)	Annual Precipitation (mm)	Number of Sampling Sites
Burlington	115,000	800	2
Cornwall	48,000	680	8
Fort Erie	24,000	1000	2
Guelph	71,000	830	4
Hamilton	306,000	830	11
Kingston	53,000	900	5
Niagara Falls	71,000	940	4
Ottawa	310,000	660	6
Sarnia	51,000	890	14
Stratford	26,000	1030	4
Toronto (City)	600,00	800	12
Welland	45,000	940	9

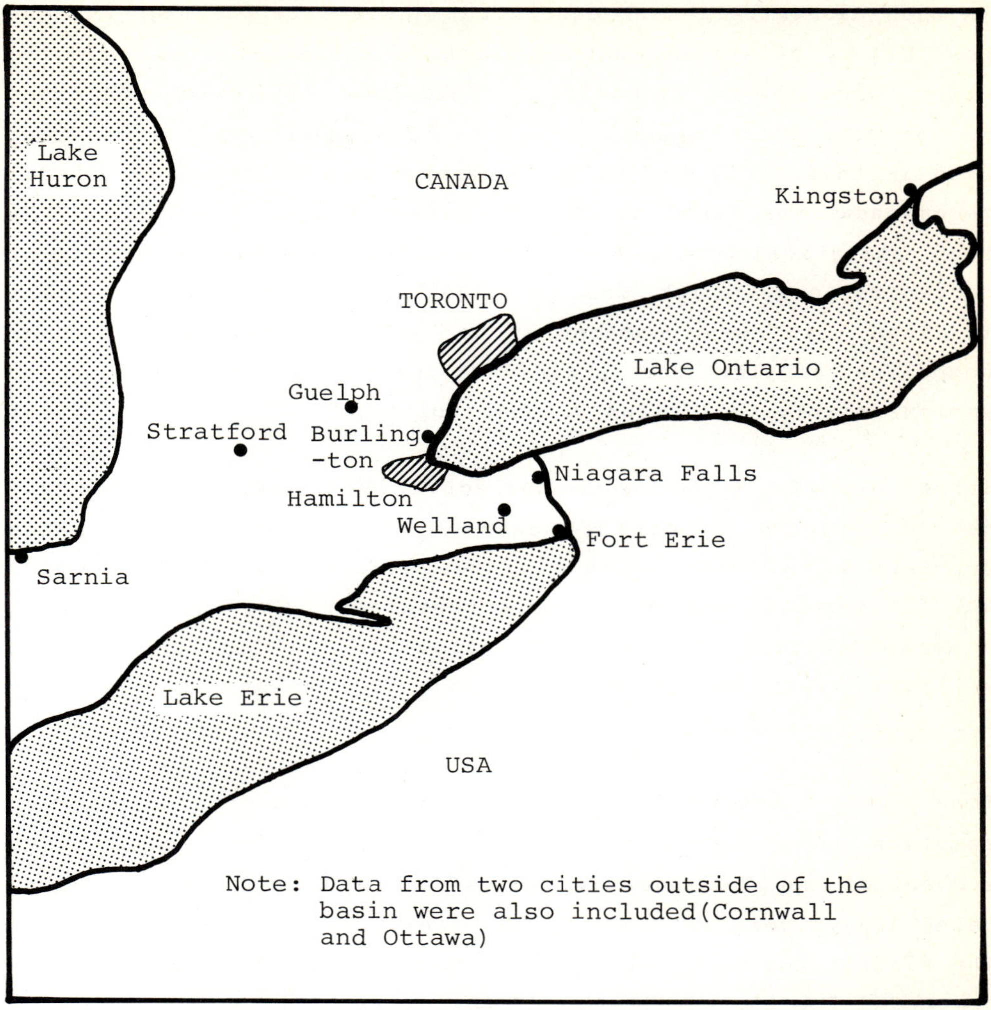

Figure 1. Sampling Sites: Location Map

In such cases, pollutant concentrations are fairly high and the possibility of sample contamination is rather limited. On the other hand, the corresponding methodology for toxic substances is not well established and some procedures as well as devices used in this study had to be developed on the basis of general recommendations found in the literature (16). Considering the high analytical costs and the interest in mean contaminant concentrations, which can be readily used to produce event loadings, it was desirable to collect flow-proportional composite samples for individual events. Descriptions of various sample collection methods follow.

Street runoff samples were collected by a simple custom made device referred to as the sewer inlet sampler. The sampler consists of a large stainless steel funnel, which fits under the inlet grate, and of a sample container. The funnel diverts a small fraction of the total inflow via a teflon tube into a 22-litre glass bottle which serves as the sample container (12).

Some runoff samples were collected by automatic wastewater samplers operated in the sequential mode. After each event, sequential samples were composed proportionately to the flow rates and volumes recorded by an electromagnetic flow recorder. The sampler employed was the ISCO model 2100 which has been recommended for the monitoring of toxics (9). It is a sequential sampler with internal plumbing made of stainless steel or teflon, except for a short piece of tubing (0.6 m) leading through the peristaltic pump. This tubing has to be very flexible for good pump operation and, consequently, a medical-grade silicone rubber tubing is used for this purpose. This brand of tubing does not contribute any organics to the samples (9).

Sediment samples were collected by two methods. In the first one, suspended solids samples were obtained by filtration of runoff samples. The filtration process was aided by vacuum or, for large volumes, by compressed gas. To obtain a sufficient quantity of solids, from 5 to 15 litres of stormwater had to be filtered. Filtered solids were placed in glass jars and submitted for analysis. Samples of street surface sediments were collected by dry vacuuming. Some fractions of such samples were further analyzed to obtain a particle size distribution.

Preservation of Samples

Samples were removed from sampling devices as soon as possible after the sampled events and placed in glass bottles or jars with tops covered by aluminum foil. In the laboratory, the samples were transferred into laboratory bottles, preserved as required by the analytical laboratory (18) and submitted for analysis. Sediment samples were delivered in glass jars and subsequently frozen. Empty sample containers were washed with soap and hot water, rinsed with hot and distilled water and, after draining, further rinsed with analytical grade acetone and petroleum ether. The final rinse was made with pesticide residue grade ethyl acetate and hexane.

Analytical Methods

Details of analytical methods used can be found elsewhere (18). A brief summary of these methods follows.

Trace Organics - Water samples were extracted by the methylene chloride water extraction method. Sediment samples were extracted by ultrasonic extraction. Because of high amounts of coextracted interferring compounds present, an additonal cleanup step involving gel permeation was added. The analysis of extracts was done by gas chromatography. For PCB's and pesticide analysis, high-resolution chromatography with two columns was used. For the remaining substances, single-column chromatography was used. Two detectors were used - a flame ionization detector for PAH's and an electron capture detector for the remaining substances.

Trace Elements - Runoff samples were analyzed by atomic absorption spectroscopy. Sediment analyses were done by cold vapour atomic absorption for mercury, flameless atomic absorption for arsenic and selenium, and the bomb digestion method for the remaining elements.

Results and Discussion

The presentation of results starts with frequencies of detection, followed by mean concentrations and partitioning of loadings in water and sediment phases of runoff.

Frequency of Detection

The frequency of detection is a useful statistic which indicates prevalence of a particular substance in the medium studied. This statistic, however, is fully meaningful only in combination with the detection limit concentration which applies to the analytical method used. When less expensive and less sensitive methods are used, lower detection frequencies are obtained and vice versa.

The detection frequencies discussed below correspond to the detection limits shown in Table 1. It should be also noted that detection frequencies are reported here separately for runoff and sediment samples.

As a group, the trace elements were found to be the most prevalent toxic contaminants in urban runoff and sediment. For water samples, the detection limit was typically 1 ppb with the exception of arsenic, mercury and selenium. At this level, the most frequently detected elements were zinc (98% of all water samples), copper (93%), nickel (87%), and lead (78%). Such results agree quite well with those reported in the NURP program (2). For arsenic, mercury and selenium, the detection limit was substantially lower (0.1 ppb). At this level, the frequencies of detection varied from 66% to 87%.

The frequencies of detection of trace elements in sediment samples were higher than those found for water samples and typically ranged from 58% to 100%.

Among trace organics, the detection frequencies varied widely. Pesticides were the most prevalent group. Two substances, α-BHC and γ-BHC (lindane) were found in 98% and 86% of all water samples, respectively. This finding is in agreement with the earlier reported prevalence of these substances in rainwater (14). For the remaining substances, the detection frequencies in water samples varied from 1% to 45%. The two lowest frequencies were found for aldrin (1%) and mirex (3%). The highest frequency was found for hexachlor benzene (45%) and the remaining frequencies varied from 6% to 28%. The detection frequencies reported here were generally higher than those found in the NURP program, but the general trends were quite similar (2). The above differences were possibly caused by different detection limits used in both programs.

The detection frequencies in sediment samples were more uniformly distributed. At the detection level of 0.04 ppm, the frequencies varied from 3% to 64%. The highest frequencies were found for hexachlor benzene (64%) and α-endosulfan (51%), the lowest for p,p'-DDT (3%). The frequency of detection of total PCB's was 46% and 85% for water and sediment samples, respectively.

The detection frequencies of chlorinated benzenes varied from 4% to 48% in water samples and from 1% to 53% in sediment samples. The least prevalent group were PAH's, particularly in water samples. The frequencies of PAH's varied from 2% to 19% in water samples and from 1% to 37% in sediment samples. These results were affected by somewhat higher detection limits used in PAH's analyses.

Mean Concentrations

The mean concentrations reported here were defined as arithmetic means of mean concentrations determined for composite event samples. The decision to use the arithmetic mean as opposed to the sometimes used geometric mean was based on limited variations in mean event concentrations.

For data below the detection limit, the concentration was set equal to 0.5 of the detection limit. This expedient barely affected the final results for substances with high detection frequencies and low detection limits. The only group where the above assumption may have caused significant uncertainties in the final results were PAH's. In their case, the detection limit was somewhat higher and the detection frequencies were therefore somewhat lower. For reasons of brevity, only the salient data are presented here. A complete listing of mean concentrations can be found elsewhere (12).

The highest concentrations were found for trace elements as a group. Among the individual elements, the highest mean concentrations were found for zinc (440 ppb in water, 400 ppm in sediment), lead (90 ppb in water, 470 ppm in sediment), copper (19 ppb in water, 67 ppm in sediment), and nickel (16 ppb in water, 50 ppm in sediment). Such values are generally consistent with those reported earlier by others (4, 5, 8, and 11).

The remaining substances were found in significantly lower (by several orders of magnitude) concentrations. The highest values were found for PAH's, but these may have contained relatively significant uncertainties caused by their low frequencies of detection and many readings below the level of detection. The remaining trace organics were found in water samples in mean concentrations ranging from .001 ppb to 0.04 ppb. The highest concentrations were found for 1,2 dichlorobenzene (.04 ppb), α-BHC (.02 ppb) and the total PCB's (.014 ppb). In sediment samples, the highest concentration was found for PCB's (1 ppm). Other concentrations varied from .001 ppm to .12 ppm. High values were found for several chlorinated benzenes, chlordane, and some derivatives of DDT.

Water/Sediment Partitioning of Loadings

The division or partitioning of contaminant loadings into the adsorbed and dissolved components is of interest for the planning of future studies and control measures. In general, such a division depends on numerous physico-chemical factors and may vary with time. Two approaches are commonly used to obtain water/sediment partitioning - a theoretical approach based on equations of the sorption process and an empirical approach based on separate analyses of water and sediment components of collected samples. Both approaches have some disadvantages. In the former case, several simplifying assumptions have to be made. In the latter case, an equilibrium distribution may not have been reached. Considering these factors, the empirical approach was considered more appropriate and therefore adopted in this study.

The division of water/sediment loadings was calculated for a reference concentration of suspended solids of 200 mg/L which corresponds to the mean concentration established in an earlier study (15). The calculated divisions of loadings needed to be further evaluated. In this connection, the substances whose loadings were strongly associated with either water or sediment were of particular interest. Considering the uncertainties involved in this evaluation, the strong association was defined somewhat conservatively as the case where 80% or more of the total loading was associated with either water or sediment. Using this definition, loading partitionings were evaluated for all 50 substances studied. It was noted that the results obtained for PAH's were not meaningful, because they were strongly influenced by data below the detection limits. Consequently, PAH's were eliminated from this analysis. Among the remaining 38 substances, 12 exhibited loadings strongly associated with either water or sediment. These substances are listed in Table 3. For the remaining 26 substances, loading divisions were more uniform.

Although the concentrations of toxics in sediment were greater than those in water by a factor ranging from 10^2 to 10^5, the

Table. 3. Partitioning of Selected Toxic Loadings Between the Liquid and Solid Phase of Runoff

Substance Loadings Associated with the Liquid (% of total)		Substance Loadings Associated With the Sediment (% of total)	
α-BHC	95%	Hexachlor benzene	95%
γ-BHC	90%	α-chlordane	84%
Mercury	99%	α-endosulfan	92
Selenium	96%	p,p'-DDE	83%
Zinc	85%	o,p'-DDT	88%
		p,p'-DDT	90%
		PCB's	93%

loadings calculated for five substances (see Table 3) were carried mostly by water. Because two of these substances, mercury and zinc, are known for their affinity for suspended solids (17), the calculations for those two substances may not reflect equilibrium conditions. Under such conditions, both substances would be associated mostly with solids. The remaining three substances, BHC isomers and selenium, are known for their low affinity for suspended solids (17) and the calculated results are therefore plausible. All seven substances whose loadings were carried mostly by sediment are known for their high affinity for suspended solids (17) and this is correctly reflected by the calculated results in Table 3.

Sources of Toxic Contaminants in Urban Runoff

The most prevalent group, trace elements, have been studied most extensively among all the contaminants investigated in this study. This is particularly true for the most commonly occurring metals - zinc, lead and copper. An extensive discussion of sources of these metals in runoff can be found elsewhere (11). Only a brief listing of common sources follows.

The most significant sources of zinc in urban runoff include atmospheric fallout, corrosion processes (particularly in galvanized metal sewers), tires, pavement wear, automobile exhausts, exterior paint, road salt and possibly some terrestrial sources (11). The sources of lead include gasoline products and by-products of their combustion, and exterior paints and stains. The sources of copper include corrosion of copper plumbing, electroplating wastes, some algicides, brake linings and asphalt pavement wear (11).

PCB's were used extensively in many industrial applications, particularly in fire-resistant transformers and insulating condensers. Although their use has been recently restricted, they are commonly found in the environment.

Many organochlorine pesticides were found in runoff samples. The two most prevalent pesticides, α-BHC and γ-BHC, are used commonly in soil treatment to eliminate nematodes and other pests. The remaining pesticides are much less prevalent and were found in low concentrations.Some of these substances have been banned for use in Ontario and their residues came either from atmospheric sources or from soil residues. The composition of rainwater in the Great Lakes region (14) indicated that PCB's, α-BHC and γ-BHC were most prevalent and occurred in concentrations which exceeded those of other pesticides by an order of magnitude. Similar trends were reflected in the composition of water in the Lower Great Lakes (13).

PAH's are widely distributed in the environment. Besides their natural sources (some forms of biosynthesis in phytoplankton, algae and plants; natural forest fires (1)), there are anthropogenic sources associated with combustion of fossil fuels. PAH's were reported in urban runoff studies undertaken in Great Britain (1, 8), Norway (3, 10), Sweden (11) and USA (2). It appears that PAH's are commonly found in highway runoff and their origin can be traced to combustion byproducts. PAH's tend to accumulate in soils (8, 10) and snowbanks (10) along highways. Typical dustfall deposition rates of PAH's along highways vary from 0.3 ng/day/m^2 to 12 ng/day/m^2. PAH's accumulated in snowbanks are suddenly released to the receiving waters during the snowmelt period and create a shock loading (10).

Chlorinated benzenes originate primarily from industrial sources. No other information on their occurrence in urban runoff has been found in the literature.

Environmental Implications

The significance of the reported levels of toxic substances needs to be evaluated in terms of environmental criteria. Such criteria may be based on a whole variety of considerations, including the human health, drinking water standards and toxicity to aquatic life. From these considerations, general criteria for water quality in the receiving water bodies are developed. In the study described here, the water quality criteria used were established by the International Joint Commission and referred to as water quality objectives for the Great Lakes (6). The discussion that follows is based on those objectives.

IJC established water quality objectives for the seven trace elements listed in Table 4. It is apparent from this table that such objectives were exceeded by the mean concentrations of four elements - cadmium, lead, mercury and zinc. The reported exceedances are highly significant, because the mean observed concentrations varied from 3.6 to 17 times the objective values. Considering the fact that the reported levels represent means of event means, temporary exceedances for limited durations will be even higher than indicated in Table 4.

A PCB's concentration of 0.01 ppb has been suggested as an objective on the basis of their bioaccumulation in fish (6). Such an objective was slightly exceeded (by 40%) by the mean concentration reported here. This concentration reported for urban runoff (0.014 ppb) compared quite well with the values reported for the rainwater in the study area - 0.11-0.031 ppb

TABLE 4. Comparison of Mean Trace Element Concentrations With the IJC Water Quality Objectives

Element	IJC Objective (ppb)	Mean Observed Concentrations (ppb)	Objective Exceeded
Arsenic	50	1.7	No
Cadmium	.2	1.3	Yes
Chromium	50	6.4	No
Lead	25	90	Yes
Mercury	.2	3.4	Yes
Selenium	10	1.6	No
Zinc	30	440	Yes

(14). It would appear that the above objective for PCB's would be exceeded even by the levels reported for the rainwater.

Water quality objectives for eight organochlorine pesticides were also developed by IJC and listed in Table 5. Such objectives were exceeded only in the case of heptachlor epoxide. Temporal exceedances, in some locations, can be expected for aldrin/dieldrin, $\sum$DDT, endosulfan and lindane. No objectives were available for the most prevalent pesticide - α-BHC. This pesticide was found in the NURP study (2) to exceed the EPA criteria for the protection of human life from carcinogenesis at the 10^{-5} risk level.

Conclusions

Among the toxic contaminants studied, trace elements and particularly heavy metals were the most prevalent. The highest levels were found for lead and zinc. Four elements were found at levels which exceeded the ICJ water quality objectives. The next group in terms of levels were PAH's. Although they were infrequently detected in water samples (partly because of their detection limit of .001 ppb), they seem to be quite widespread

Table 5. Comparison of Observed Pesticide Concentrations With the IJC Water Quality Objectives

Element	IJC Objective (ppb)	Mean Observed Concentration (ppb)	Objective Exceeded
Aldrin/Dieldrin	.001	.0006	No
Chlordane	.060	.0012	No
$\sum$DDT	.003	.0012	No
Endosulfan	.003	.0011	No
Endrin	.002	.0008	No
Heptachlor Epoxide	.001	.0011	Yes
Lindane	.010	.0065	No
Methoxychlor	.040	.0015	No

in sediments in urban areas. Among the pesticides studied, α-BHC and γ-BHC occurred most frequently and at the highest levels. Another pesticide, heptachlor expoxide, was found at the level exceeding the IJC water quality objective. PCB's were detected rather frequently and their mean levels also exceeded the IJC objective. Finally, chlorinated benzenes were found at the levels comparable to those reported for pesticides. Their occurrence was related to industrial sources.

References

1. Butler, J.D., V. Butterworth, S.C. Kellow, H.G. Robinson. Some Observations on the Polycyclic Aromatic Hydrocarbon (PAH) Content of Surface Soils in Urban Areas. Science Tot. Env., 33:75-84, 1984.
2. Cole, R.H., R.E. Frederick, R.P. Healy, R.G. Rolan. Preliminary Findings of the Priority Pollutant Monitoring Project of the NURP. Journ. Wat. Poll. Cont. Fed., 56(7):898-908, 1984.
3. Gjessing, E., E. Lygren, T. Berglind, T. Bulbrandsen, R. Skaane. Effect of Highway Runoff on Lake Water Quality. Science Tot. Env., 33:245-257, 1984.

4. Hamilton, R.S., D.M. Revitt, R.S. Warren. Levels and Physico-Chemical Associations of Cd, Cu, Pb and Zn in Road Sediment.Science Tot. Env., 33:59-74, 1984.
5. Harrison, R.M., S.J. Wilson. The Chemical Composition of Highway Drainage Waters I. Major Ions and Selected Trace Metals. Science Tot. Env., 43:63-77, 1985.
6. International Joint Commission. 1982 Report on Great Lakes Water Quality. ICJ Report, Windsor, Ontario, 1982.
8. Johnston, W.R., R.M. Harrison. Deposition of Metalic and Organic Pollutants Alongside the M6 Motorway. Science Tot. Env., 33:119-127, 1984.
9. ISCO Ltd. Instruction Manual Model 2100 Wastewater Sampler. ISCO, Lincoln, Nebraska, 1980.
10. Lygren, E., E. Gjessing, L. Berglind. Pollution Transport from a Highway. Science Tot. Env., 3:147-159, 1984.
11. Malmqvist, P.A. Urban Stormwater Pollutant Sources. Chalmers University of Technology, Goteborg, 1983.
12. Marsalek, J., H.O. Schroeter. Loadings of Selected Toxic Substances in the Canadian Great Lakes Basin. Nat. Water Res. Inst., Burlington, Ontario, 1984.
13. McCrea, R.C., J.D. Fischer, K.W. Kuntz. Partitioning of Organochlorine Pesticides and PCB's in the Lower Great Lakes Region. Draft report, IWD-Ontario Region, Burlington, Ont., 1983.
14. Strachan, W.M.J., H. Huneault, W.M. Schertzer, F.C. Elder. Organochlorines in Precipitation in the Great Lakes Region. 387-396 in Hydrocarbons and Halogenated Hydrocarbons in the Aquatic Environment, Afghan, B.K. and D. MacKay, Editors, Plenum Publishing Corp., New York, N.Y., 1978.
15. Sullivan, R.H., W.D. Hurst, T.M. Kipp. Evaluation of the Magnitude and Significance of Pollution from Urban Storm Water Runoff in Ontario. Res. Report 81, Canada-Ontario Agreement on Great Lakes Water Quality, Ottawa, 1978.
16. Versar, Inc. Monitoring of Toxic Pollutants in Urban Runoff. Draft report for U.S. EPA, Washington, D.C., 1980.
17. Versar, Inc. Water-Related Environmental Fate of 129Priority Pollutants. Report EPA-440/4-79-029. U.S. EPA, Washington, D.C., 1979.
18. Water Quality Branch, IWD. Analytical Methods Manual. Water Quality Branch, Ottawa, Ontario, 1979.

SOLIDS, NUTRIENTS, AND CHLORIDES IN URBAN RUNOFF

Donald H. Waller, and William C. Hart
Centre for Water Resources Studies
Technical University of Nova Scotia
P. O. Box 1000
Halifax, Nova Scotia, Canada
B3J 2X4

Introduction

This presentation draws on Canadian experience to characterize and discuss solids, nutrients, and road salts in urban storm water.

Surface runoff and combined sewage are significant pollution sources in Canada. New development each year changes large areas of forest and farm land to urban land uses, with significant hydrologic and water quality impacts. With a few notable exceptions, the sewerage systems of every major Canadian community include combined sewers; more than 30 percent of the urban population of Canada is served by combined sewerage systems (1).

Significance of Solids, Nutrients, and Chlorides

In Canada the impact of combined sewer overflows is primarily on coastal estuaries, the Great Lakes, and on large rivers such as the St. Lawrence. Urban surface runoff also contributes pollutants to these receiving waters, but primary impacts of surface runoff are on smaller streams and on urban lakes. The effects of solids, nutrients, and chlorides in surface runoff are most likely to be experienced in small lakes, or bays and harbours, where rates of water movement are relatively small and opportunities for dilution and dispersion are limited.

The water quality effects of these pollutants have been well documented e.g. (2) (3) (4); only a brief review

NATO ASI Series, Vol. G10
Urban Runoff Pollution.
Edited by H.C. Torno, J. Marsalek, and M. Desbordes

follows.

Suspended solids affect the appearance of water, and reduce visibility and light penetration. The value of a receiving water as an aesthetic amenity is reduced, its recreational potential is affected, and algal productivity that determines fisheries potential is also reduced. Low concentrations (greater than 25 mg/l) of suspended solids can result in fish mortality. Water treatment costs are increased. Deposited solids may smother aquatic bottom life, including fish eggs and larvae. In large quantities sediments may fill reservoir storage. Resuspension of sediment during storm periods may create serious short-term effects for fish or for downstream dissolved oxygen levels. Chemicals associated with sediment, and the role of sediments in inhibiting effective water disinfection, are also causes of concern.

Phosphorus is the limiting algal nutrient in most fresh water systems, while nitrogen is the usual cause of concern in marine waters. Effects of nutrient enrichment include: increases in particulate organic matter, such as phytoplankton, zooplankton, bacteria, fungi, and detritus; shifts of algal populations towards less desirable types such as blue-greens; increases in dissolved organic substances with a variety of undesirable characteristics; increased pH and pH variations; and chronic oxygen depletion at the sediment-water interface, resulting in release of dissolved and gaseous products of anaerobic decomposition of algal material, and solublization of metals. Results of these water quality changes include all of the effects of increased turbidity, and serious effects on water supply operations and treatment costs. Aesthetic and recreational impacts are aggravated by the unsightly appearance of algal accumulations and of decomposition products, and shift of fish populations towards less desirable species.

Salt applied for highway ice control contributes both sodium and chloride ions to water. The most important effects on a lake system may be establishment of density gradients that prevent lake mixing, lead to stagnation of bottom layers, and aggravate effects of decomposition of organic matter in the stagnant layer. Other effects may include pH increases and shifts in ecological communities, including encouragement of blue-green algal growth. Effects of increased salinity on raw water quality include reduced suitability for many industrial purposes. Chloride concentrations in excess of 250 mg/l may impart undesirable taste to water. Sodium concentrations greater than 20 mg/l may represent an excessive contribution to the intake of persons on sodium-restricted diets.

Canadian Examples

The effects of these pollutants have been of concern in Canada. Nutrient enrichment is one of the problems that exist in a number of harbours on the Canadian side of the Great Lakes (5). Phosphorus loads in urban runoff and combined sewage discharged by Ontario communities are 1.9 times loads included in wet weather discharges from treatment plants. Suspended solids load and storm water discharges, which on an annual basis are 74 percent of total municipal loads, in wet weather represent 19 times the load discharged from treatment plants. On the other hand, nutrient loads from storm water are not significant on a lake-wide basis, because phosphorus in storm water represents only 4 percent of total phosphorus loads to the Great Lakes.

Cultural euthrophication has for many years been a potential problem in eastern Canada, where lakes in previously undeveloped areas are exposed to development effects (6).

Effects of urban surface runoff and high salt loadings on a small lake in Halifax, Nova Scotia, that is surrounded by urban residential development, and improved lake water quality associated with altered flow regimes that reduce lake flushing times, have been described (7).

An example of concern about potential effects of storm water on a marine system is Halifax Harbour, for which a detailed hydrodynamic/water quality model is being prepared. Effects of nutrient enrichment in one portion of this estuary have already been identified. An important aspect of this study will be preparation of a regional pollution loadings model, which will identify inputs of pollutants, including suspended solids and nutrients, to shoreline segments, and will assist in identifying the effects of pollution control alternatives that may affect discharges from tributary combined and separate sewerage systems.

Characterization

Loadings and Concentrations

Table 1 presents typical loads and concentrations in storm water, compared with raw and treated sewage, based on data from the Province of Ontario (5). Figure 1 illustrates the relationships on which Table 1 is based, and serves to compare relative wet and dry weather phosphorus loads from the various sources. Relative loads of the other parameters are presented in Table 2. The differences in per hectare sewage loads between combined and separate systems in Figure 1 reflect the fact that combined sewers generally serve older areas of higher population density than those served by separate systems.

It is clear that for the four parameters considered in Table 2 each system discharges significant loads in wet weather. For both combined and separate areas wet weather loads of solids and chlorides exceed dry weather values; dry weather nutrient loads exceed wet weather discharges.

Table 1 Annual Mean Urban Pollutant Concentrations and Loads for Ontario

	Suspended Solids[1]		Total Nitrogen[1]		Total Phosphorus[1]		Chloride[2]	
	mg/l	kg/ha-yr	mg/l	kg/ha-yr	mg/l	kg/ha-yr	mg/l	kg/ha-yr
Raw sewage	225	1971	30	275	6.5	76	47	439
Treated sewage[3]	25.8	226	17	156	1.0	11.7	71.4	695
Surface runoff	170	515	3.5	10.1	0.35	1.1	976	3,000
Combined sewer overflow	190	568	8.3	24.7	1.4	4.5	664	1,963

1. Suspended solids, total nitrogen, and total phosphorus for Ontario from (5).
2. Chloride concentration in sewage from (3). Chloride load in surface runoff assumed. combined loads and overflows after (5).
3. Treated sewage values are means for combined and separate systems.

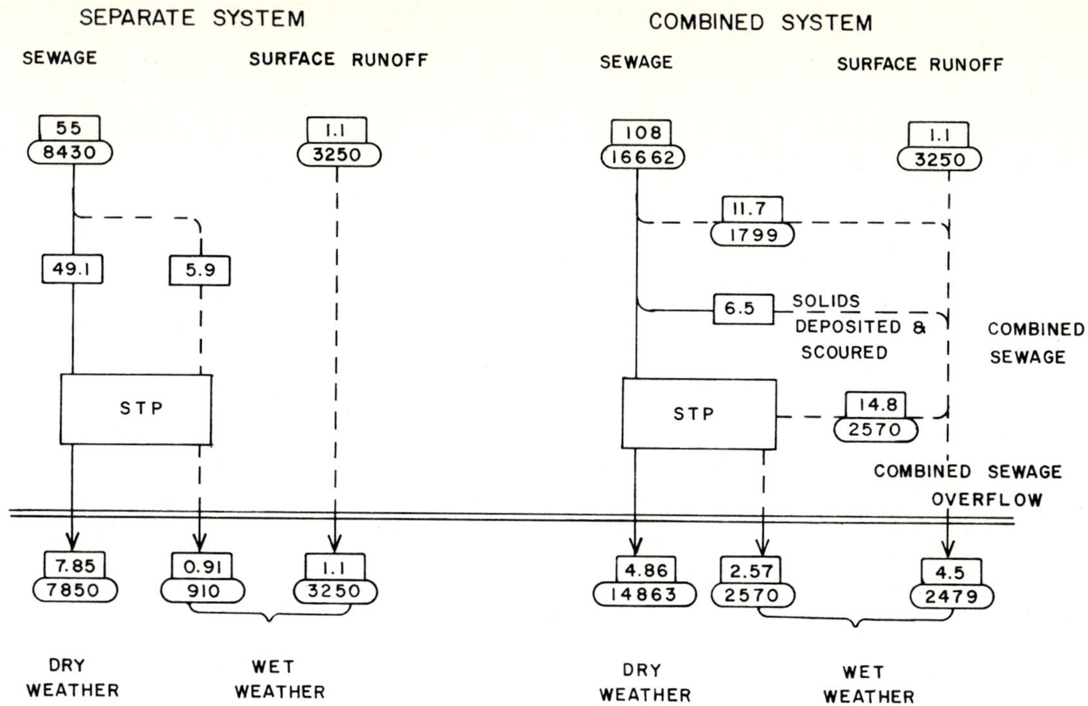

1. Dry weather flow — — —, wet weather flow ———, kg/ha-yr total phosphorus ☐, m^3/ha-yr water ⬭.
2. Sewage loads and flows do not correspond with Table 1, which is based on average values from separate and combined systems in Ontario.
3. Sewage loads for separate and combined systems differ because of typically higher population densities in combined areas.

FIGURE 1 Comparison of Hypothetical Discharges of Phosphorus From Separate and Combined Sewerage Systems

Table 2 Relative Pollutant Loading (kg/ha-yr) Discharged From the Sources Shown in Figure 1

	Separate Systems			Combined Systems		
	Dry Weather STP Effluent	Wet Weather STP Effluent	Surface Runoff	Dry Weather STP Effluent	Wet Weather STP Effluent	Combined Sewer Overflow
SS	194	24	553	383	66	490
TN	133	16	11	253	44	25
TP	8	0.9	1.1	15	2.5	4.5
CL	353	43	3000	698	1117	1967

Components

Table 3 summarizes information about components of solids, nitrogen, and phosphorus. Chloride is not included because in aqueous solutions it is normally present as Cl^- ions.

The chemical form of nitrogen or phosphorus may determine its availability as an algal nutrient. Cowen and Lee (8) assumed that all filterable phosphorus will ultimately be converted to the soluble inorganic form, and determined that a maximum of 30 percent of non-filterable phosphorus would become available. Cowen _et al_. (9) assumed that all inorganic nitrogen (NH_3, NO_2, NO_3) is available, and concluded that at least 30 percent of total nitrogen in urban runoff would be resistant to bacterial mineralization. Most nutrient loading models are based on total nitrogen or total phosphorus. Lee _et al_. (10) noted that phosphorus loading models implicitly consider the available phosphorus load, because many of the receiving waters to which they are applied have relatively constant ratios of available phosphorus to total phosphorus.

A particular reason for interest in nitrogen forms is their differing impacts on the dissolved oxygen resources of receiving streams: organic nitrogen and ammonia represent nitrogenous oxygen demand, whereas nitrate represents an oxygen reserve.

One application of information about forms of solids is the use of volatile/total suspended solid values in the interpretation of combined sewage composition. Variations in this ratio have provided useful indicators of relative proportions of sewage (VSS/SS = 0.8) and surface runoff (VSS/SS = 0.3).in combined sewage flows (11).

The values in Table 3 for sewage, and for solids and nitrogen components in runoff, are consistent with values determined in earlier work in Halifax (11), and by others (12)(13). Phosphorus composition in runoff is more

Table 3 Components of Solids, Nitrogen, and Phosphorus in Sewage and Surface Runoff (Percent)

A. Solids

	Dissolved		Suspended		Total Solids
	Volatile	Non-volatile	Volatile	Non-volatile	
Sewage	32	32	26	10	100
Surface Runoff	5.3	8.9	22.5	63.3	100

B. Nitrogen

	Kejldahl		Nitrite & Nitrate	Total Nitrogen
	Organic	Ammonia		
Sewage	28	72	--	100
Surface Runoff	59.6	0.6	39.8	100

C. Phosphorus

	Filterable		Non-Filterable		Total Phosphorus
	Organic	Inorganic	Organic	Inorganic	
Sewage	14	69	17		100
Surface Runoff	0.3	3.9	11.6	84.2	100

Sewage solids from (12). Sewage nitrogen and phosphorus, and surface runoff values, from analyses at Halifax, Nova Scotia.

variable. The Table 3 values may reflect seasonal influences: they are based on samples collected in late August through September; samples collected in June include significantly smaller proportions of non-filterable phosphorus (38%), and of inorganic phosphorus (66%)(14). Cowen and Lee found that non-filterable phosphorus forms accounted for 13 to 97 percent of total phosphorus in their runoff samples (8). Nearly one-half of runoff events monitored at a site in Ontario produced mean non-filterable phosphorus concentrations that were 90 percent or more of total phosphorus (15).

Sources

Understanding of pollution sources is important both for prediction and control of loads from these sources. Research at Halifax has emphasized sources of phosphorus in urban surface runoff. Waller (16) demonstrated that, for a small residential area in Halifax, vegetation was the primary source.

Subsequent work (14) has helped to substantiate the importance of vegetation as a nutrient source. Samples of the first flush of runoff in storms at six sites at Halifax, shown in Table 4, indicated that concentrations of total phosphorus were consistently higher for sites affected by local vegetation inputs.

Table 4 Effect of Vegetation on Phosphorus in Surface Runoff

	First Flush Concentrations of Total Phosphorus (mg/l)	
Land Use	Vegetation	No Vegetation
Residential	0.88	0.68
Commercial	2.87	1.29
Institutional	0.97	0.30

Source (14)

Information from which to determine sources of phosphorus from all of these sites is incomplete, but preliminary consideration of available information from the "vegetated" and "non-vegetated" sites suggests the relative inputs indicated in Table 5 (14).

The first column of Table 5 is based on the results of water and chemical mass balances described by Waller (16), modified to take into account results of recent and more complete measurement of total (wet + dry) phosphorus deposition (17) (18) and traffic counts. Phosphorus balances for the other sites are based on less complete information but they are believed to accurately illustrate the relative contributions from the sources shown.

Fertilizer is not a significant input for these watersheds, where local measurements, supported by evidence from other areas, indicate that pervious surfaces do not contribute significantly to annual runoff composition. Sediment from construction, or erosion of adjacent areas, is not a factor in the watersheds considered here.

The potential significance of vegetation is supported by information about leaching of phosphorus from vegetation. Reported total phosphorus content of tree leaves and seeds ranges from 1.6 to 11 mg/g (19) (20). Significant quantities of this material may be leached into runoff. Hart et al. (21) reported leaching of a mean concentration of 0.08 mg/g from fresh leaves. Leaching of phosphorus from tree canopies at rates of 0.1 to 0.3 kg/ha-yr has been reported (22).

Decomposition of vegetation releases larger amounts of phosphorus. Hart et al. (21) leached 0.2 to 0.3 mg/g from leaves collected from gutters, and measured concentrations of up to 5.3 mg/l in puddles containing leaves.

Construction sediment, which is not a problem in the mature

Table 5 Hypothesized sources of Total Phosphorus in Surface Runoff (kg/ha-yr based on impervious areas)[1]

Source	Residential Vegetation Low Traffic	Commercial No Vegetation High Traffic	Commercial Vegetation Moderately High Traffic	Institutional No Vegetation Low Traffic
Vehicles	0.003	2.00	0.74	0.10
Animals	0.30	0.20	0.90	0.20
Vegetation	5.94	--	4.00	--
Atmosphere	0.12	0.12	0.12	0.12
Less Sweeping	0.90	0.30	0.10	--
Less Leaf Removal	3.60	--	1.70	--
Net total	1.86	2.02	3.98	0.42
Measured Mean First Flush Concentration	0.88	1.29	2.87	0.30

[1]Source - (14)

urban developments in Table 5, is a potentially serious problem in other areas of the city. Measurement of suspended solids concentrations and flows upstream and downstream of a previously undeveloped 14 hectare site that was stripped and exposed to runoff for several months provided estimates of increased suspended solids, from 25 kg/ha-yr to more than 2100 kg/ha-yr. These values may be compared with an average of 441 kg/ha-yr of suspended solids recorded at developed urban residential sites in Halifax (23).

Sediment includes substantial quantities of nutrients, but the proportions of available nutrients are small. Values cited by McElroy (24) are in the range of 0.02 mg of phosphorus per gram of sediment, which may be compared with a mean value 1.0 mg of total phosphorus per gram of suspended sediment in urban surface runoff . Applied to the sediment yields above, these concentrations correspond to 0.04 kg/ha-yr and 0.43 kg/ha-yr of phosphorus from construction sites and urban development respectively.

A number of authors (e.g. (26)) have expressed concern that the contribution of street solids has been emphasized to the exclusion of pollutant loads from off-street surfaces. Pitt (25) concluded that off-street surfaces contributed to runoff, based on significantly higher ratios of phosphorus/total solids in runoff compared with street deposits. For the 6 sites in Table 4, total phosphorus in runoff was higher (mean 1.0 mg/g) than in runoff from the same sites (mean 0.68 mg/g). Pitt observed larger differences, primarily due to more phosphorus-rich runoff solids. The sources in Table 5 reflect the importance of contributions from off-street surfaces.

The composition of runoff at Halifax indicates that loads in runoff from a 25 mm storm could account for most or all of solids and phosphorus measured in street solids accumulations at individual sites (14). Most of the phosphorus is removed in particulate form: leaching tests removed only 0.02 to 3.0 percent of total phosphorus in street solids, and most of the phosphorus in the runoff samples during the same period is non-filterable (See Table 2).

In those communities that experience winter icing conditions most of the chlorides in urban surface runoff originate in road salt. Based on estimates of city-wide salt application rates in Halifax, annual average chloride loadings in runoff are in the order of 3,000 kg/ha-yr. The median chloride concentration in winter grab samples was 786 mg/l (27). The mean summer concentration in runoff was 14 mg/l, which is higher than the concentration of 4.6 mg/l (70 kg/ha-yr) recorded in total atmospheric deposition at Halifax (17). The probable explanation is salt deposits resulting from dehydration on road surfaces, in crevices, and under road surfaces, which continue to leach into runoff throughout the summer (3).

The surface runoff loads reported here are from small areas tributary to street inlets, and do not recognize the effect of catch basins. To the extent that catch basins accumulate solids that are removed in significant quantities they may reduce runoff loads. In the Halifax area frequency of catch basin cleaning is not considered likely to affect quantities of pollutants discharged to sewerage systems (11) (28).

The City of Halifax adds sand to the salt it applies to bus routes and traffic arteries only when icing conditions are anticipated. An indication of the relative contribution from this source to winter solids loadings may be gained from an estimate, based on limited information from City records, that approximately 55 kg/curb km were applied to arterial streets during one winter season. This may be compared with total solids accumulations of 162 to 781 kg/curb km measured on an arterial street in winter in

Halifax (26).

Combined sewage loads are derived from surface runoff and sanitary sewage . The proportions from these sources that are discharged in loads in combined sewer overflows, or contained in wet or dry weather flows reaching sewage treatment plants, is affected (as illustrated by Figure 1) by duration of wet weather periods, the extent to which solids are deposited in dry weather and scoured in wet weather periods, and the capacity of interceptors and treatment plants to handle wet weather flows. Waller and Novak (5) determined that, for a hypothetical city having average characteristics of Ontario communities, combined sewer overflow would include 0.0226 of the annual wastewater and 0.647 of the annual surface runoff.

Sewage solids deposited in dry weather and scoured during storm periods can make an important contribution to wet weather loads. Based on examination of all of the sewers in a relatively steep 68 hectare combined sewer area,in Halifax, it was concluded that 3 percent of dry weather solids might be deposited and eventually scoured (11). Pisano indicated that average dry weather deposition rates are 10 to 15 percent of daily sewage loads (29). Figure 1 is based on the assumption that 6 percent of solids are deposited and scoured.

Another source of wet weather loads, which is not recognized in Figure 1 because it is difficult to quantify, is by-pass and overflow from overloaded sanitary sewerage systems and treatment plants. A study of three plants servicing separate sewer systems in Ontario indicated that, depending on the plant and the pollutant being measured, 1 to 12 percent of annual sewage loads are by-passed (5).

Variations

Seasonal

Reference has already been made to sand and salt, applied

for ice control, as seasonal contributions to urban streets. The Halifax experience is that most of the sand accumulates in gutters until the early spring, when it is removed by concerted sweeping programs. Early spring suspended solids concentrations from the Halifax sites have been higher than those from other seasons.

As indicated above, winter versus summer chloride concentrations of 786 and 14 mg/l respectively have been measured. Most of the salt passes through the system during the winter salting season. In those communities where large quantitites of snow from downtown streets are removed for disposal elsewhere, salt may be removed with it, but information from Ottawa suggests that the quantity of salt removed by this route is small (29).

Seasonal variations in phosphorus loadings will be apparent where the main source is vegetation. Most of the other potential sources contribute at relatively uniform rates during the year. Mean total phosphorus concentrations measured in individual storms in the summer at Halifax are higher (0.5 mg/l) than winter values measured at the same site (0.1 mg/l) (14)(27). If the annual phosphorus load from vegetation for Site 1 in Table 5 is included in summer-fall runoff only, and the remaining phosphorus is distributed throughout the year, corresponding mean summer and winter concentrations are 0.32 and 0.08 mg/l (14).

An obvious implication of the seasonal variations cited above is that extrapolation of loading data from a single season to the complete year may provide misleading estimates of annual loads.

Short Term

Short term variations in urban runoff composition -- the result of processes of pollution accumulation and wash off -- have been a subject of concern for those attempting to model urban storm water quality. Questions that have been addressed in algorithm development have been: rates of accumulation; effects on removal rates of time (presumably as a surrogate variable for runoff volume, cumulative removal, or pollutant remaining); and rate of flow; and differences in accumulation and removal rates for different parameters.

All of the work at Halifax that has considered variations in surface runoff quality, including street runoff under constant intensities of simulated rainfall, has indicated that concentrations of solids and nutrients decline with time during storms, and increase with rates of flow. Chloride concentrations did not vary significantly with time. Total phosphorus values were closely correlated with suspended solids, which is consistent with high proportions of non-filterable phosphorus reported in Table 3. Correlation of nitrogen with solids was weaker.

As indicated earlier, catch basins in Halifax are not cleaned with sufficient frequency to affect storm water loads, but they may affect the timing and form of discharges. Materials that are deposited in periods of low flow, e.g. sediment and decomposing leaves, are washed out in periods of increased runoff. Several sets of simultaneous measurements showed increasing concentrations of suspended solids when samples from a gutter, the catch basin, and the catch basin outlet to the storm sewer were compared (10). Uncertainty remains in North America about the effectiveness of catch basins. In 1972 a US EPA Report suggested that they could be eliminated (30); in 1983 another EPA study indicated significant pollutant removals, and suggested further research (31).

Variations in composition of combined sewage can be explained by consideration of the relative contributions from sanitary and storm sewage during storm periods. The amount of deposited solids available to be swept out during storm periods, and the time of concentration of the system (which determines the time over which the discharge of these materials will be distributed) will determine whether a "first flush" phenomenon is noticed. Relative contributions of sewage and storm water will be evident by consideration of parameters such as the ratio of volatile suspended solids/suspended solids, which decreases as sewage is diluted by storm water (11).

Control of Urban Storm Water Pollution

Measures for control of pollution from urban storm water will be discussed by others at this Workshop, but it may be useful to highlight some of the control-related points that arise from consideration of the sources and characteristics of solids, nutrients, and chlorides in runoff.

Chloride is the best example of arguments for control of pollution at the source. Once chloride becomes dissolved it remains in solution, and no practical means are available to remove it from runoff. Quantities entering storm water can, however, be reduced by control of salt application rates, and by attention to storage and handling of salt prior to application (3). Measurement and observation to determine sources of salt entering a small water supply reservoir in Halifax have made it clear that significant unnecessary quantities of salt can be deposited on roads in the immediate vicinity of a salt storage facility by departing salt trucks.

Erosion control, to prevent the entry of sediment from construction sites into storm water, is an important source control measure. Control on sizes of exposed areas, duration of exposure, and protection from upstream runoff are in many cases far more effective than attempts at sediment removal from storm water. An extreme example of this point is sediment derived from the construction site in Halifax referred to earlier: the site is located on drumlin soils with a high proportion of clay so fine that it passes the 1.2 micron filter used for suspended solids determination. Sedimentation ponds located on-site cannot remove this material, but the quantities included in storm water could be significantly reduced by on-site erosion control measures.

Programs of street sweeping (25), catch basin cleaning (31), and leaf removal (16), can play important roles in removing solids and nutrients before they enter urban surface runoff.

Analyses at Halifax indicate that a large proportion of street solids and nutrients may be removable by sedimentation. Initial tests measured settleable solids in runoff of 60 percent, and in combined sewage of 96 percent, of suspended solids (11). In later tests settleable/suspended solids ratios of 76 percent were observed in runoff samples. Settleable/total phosphorus ratios averaging 76 percent were measured, but seasonal variations may produce lower ratios (14). Observations of the performance of a combined sewage detention tank in Halifax during several storms indicated that suspended solid removals, depending on detention time, were 30 to 70 percent (23).

Information and Research Needs

One of the outputs from a 1978 conference on urban water problems (32) was the report of a discussion session, which recognized information needs on two levels:

1. Information to enable improved management of water quality in urbanizing areas, and
2. Information to enable a better understanding of fundamental processes.

An important comment was included: these two levels are related, the distinction being that better management does not necessarily lead to better understanding of processes. Clear understanding of processes, however, contributes to improved management, because it improves predictive ability by first recognizing relationships and then describing them in quantitative terms.

An example of an attempt to address the first of these information needs is a program currently underway at Halifax (33). Two lakes that were until recently protected as water supply sources are now threatened by urban development within their watersheds. One lake continues as a reserve water supply; the other is a major recreational resource. The plan of the research program is to study these lakes and their watersheds with the objective of contributing to more effective management of lakes, and their watersheds, that are exposed to urban development. Developments that have been started or proposed since early 1984 include two industrial parks, a major exhibition facility, a highway salt storage facility, residential development, and a regional park. All of the pollutants that are the subject of this paper -- construction sediment, chlorides from the storage facility and from local highways, and nutrients -- are of concern in these lakes, in addition to toxics, THM precursors, and bacteria.

One of the most striking aspects of this program, which is being carried on in close consultation with all interested and responsible government agencies, is that research is but one aspect of a spectrum of management concerns that will determine the ultimate condition of these watershed/lake systems. A completed study (34) has identified the jurisdictional complexities that hamper effective water quality management for these lakes: they and their watersheds are affected by, or regulated by, two municipalities, an industrial commission, a water utility, a provincial park department, and a provincial Department of the Environment. The extent to which institutional barriers -- related to regulation, enforcement, and cooperation among these agencies -- are overcome will have a profound effect on the water quality of these lakes.

Clear understanding of the consequences of management decisions will be a vitally important input to the making of effective decisions about these systems. The test of available water quality management tools, and the starting point for relevant research, will be the ability to reliabily predict the water quality impacts of land use decisions, and to predict the effects of water quality management measures designed to control these impacts.

Initial (and continuing) effort was directed at establishment of base line watershed water and chemical balances (35), and of the capability to detect and measure changes in quality and quantity. Current emphasis is on evaluating the effectiveness of erosion control measures in comparison with their design objectives. With completion of development and installation of permanent pollution control measures, attention will shift to measurement of development loadings, effects of pollution control measures, and extrapolation of this information to other planning and design applications.

Applied research of this type will not answer all of the needs related to urban storm water quality. Further research on sources and processes is needed. A sampling of questions to which answers are incomplete in terms of understanding or predictive ability are:

- prediction of sediment loads from construction sites;
- effects on these loads of specific control measures that can be included in construction specifications;
- solids forms and chemical species in street deposits and their interaction with rainfall to produce runoff composition;
- the role of hydrologically active off-street surfaces in contributing to runoff composition;
- seasonal variations in inputs to urban runoff, and their relationship to annual loads;

- roles of winter rainfall and snowmelt in rates of loadings of winter runoff;
- more complete development and/or testing of relationships to describe: effectiveness of sweeping, catch basin cleaning, and leaf removal; rates of deposition of solids in combined sewers, and relationship of the composition of scoured material to the composition of sewage; and traffic related pollution loads;
- results of systematic studies of existing pollution control methodologies, aimed at production of predictive relationships that can be associated with identifiable characteristics, e.g. settleable solids, in runoff.

Summary and Conclusions

This presentation deals with solids, nutrients, and chlorides in combined sewage and urban surface runoff. Its emphasis on experience in Canada, and in Halifax, Nova Scotia, in particular, and on phosphorus in surface runoff, reflects the authors' background and experience. Most of the following points, however, are generally applicable:

- solids, nutrients, and chlorides in surface runoff are in some situations significant pollution sources.
- solids and chloride loads are much higher than those in treated sewage; nutrient loads are lower (Table 1)
- loads from a combined sewerage system are higher than those from a separate system, reflecting in part the generally higher population densities in areas served by combined systems (Figure 1; Table 2)
- knowledge about the forms of solids and nutrients in storm water may be important; relative proportions in sewage and surface runoff differ greatly (Table 3)

- the proportion of non-filterable phosphorus in surface runoff may vary widely; seasonal variations are likely
- vegetation was established as the primary source of phosphorus in runoff from a small residential area in Halifax (Table 5)
- it has been possible to construct a convincing description of annual loads from other small areas, based on relative inputs from vegetation, bulk precipitation, animal wastes, and vehicles, and outputs via sweeping and leaf removal
- the significance of vegetation is supported by results of leaching tests, and measurements of phosphorus concentrations in puddles
- construction sediment from a Halifax site produced suspended solids values two orders of magnitude higher than pre-development values, and five times comparable post-development runoff concentrations
- construction sediment is not expected to be a significant nutrient source
- offstreet surfaces can be a significant source of surface runoff pollution
- solids and phosphorus loads measured in street deposits at Halifax are comparable with the amount removed in runoff from a single 25 mm storm
- it is likely that much of the phosphorus in the Halifax runoff is washed off in particulate form
- chloride concentrations in winter surface runoff, which originate with road salt, exceed summer values; summer values are increased somewhat by rehydration of salt deposited by evaporation
- sand applied for ice control may contribute significantly to solids in winter street deposits
- concentrations and loadings of pollutants in combined sewage can be estimated from information about sanitary sewage, surface runoff, and dry weather sewer deposits (Figure 1)
- extrapolation of surface runoff loadings data from a single season to a total year may be misleading:

chloride loadings vary due to salt application, and where vegetation is a significant phosphorus source seasonal variations in phosphorus loads can be expected

- concentrations of solids and nutrients decrease with time during storms; solids, nutrients, and chloride concentrations increase with rate of flow
- catch basins have been observed to contribute pollutants to runoff, but their role in pollution control is not fully understood
- source controls are essential for reduction of chlorides and construction sediment, and can play an important role in removal of solids and nutrients from surface runoff
- significant proportions of settleable solids have been observed in both surface runoff and combined sewage
- more effective decisions about urban land use and land use practices that affect water quality will be made when reliable quantitative predictions can be made about the water quality impacts of land use decisions, and about the effectiveness of measures designed to control these impacts
- in addition to better management tools, continuing research is needed on a variety of questions related to sources, processes, and loadings of urban runoff pollutants.

References

1. Waller, D. H., "Combined Sewers in Canada," Engineering Journal, 52, 6, 22-29; 1969 June.

2. Walker, W. W., "Significance of Eutrophication in Water Supply Reservoirs," J. Am. Wat. Wks. Assn., 38-41: 1983 January.

3. Edison Water Quality Laboratory, "Environmental Impact of Highway Deicing," U. S. Env. Prot. Agency, 120 pp; 1971 June.

4. Waller D. H., "Solids in Waste Waters," Water and Pollution Control, Ref. Ed., 1966 March.

5. Waller, D. H., and Z. Novak, "Pollution Loadings to the Great Lakes from Municipal Sources in Ontario," Jl. Wat. Poll. Cont. Fed., 387-395; 1980 May.

6. Waller, D. H., "Cultural Eutrophication in the Maritime Provinces," Water and Pollution Control, 1971 September.

7. Ramalingaiah, D. Thirumurthi, and W. C. Hart, "Chemistry of Bulk Deposition and its Effects on Chain and Chocolate Lakes," Water Poll. Res. J. Canada, 20:89-105; 1985.

8. Cowen, W. F., and G. R. Lee, "Phosphorus Availability in Particulate Matter Transported by Leaves," J. Wat. Poll. Cont. Fed., 48, 3, 580-590; 1976 March.

9. Cowen, W. F., K. Sirisinha, G. F. Lee, "Nitrogen Availability in Urban Runoff", 48,2, J. Wat. Poll. Cont. Fed., 1976 February.

10. Lee, G. F., R. A. Jones, and W. Best, "Availability of Phosphorus to Phytoplankton and its Implication for Phosphorus Management Strategies," in Loehr, R. C. et al. eds. "Phosphorus Management Strategies for Lakes," Ann Arbor, 259-308; 1980.

11. Waller, D. H., "Pollution Attributable to Surface Runoff and Overflows from Combined Sewerage Systems," Central Mortgage & Housing Corporation, 190 pp; 1971 March.

12. Novak, Z., pers. comm.

13. Fair, G. M., J. C. Geyer, and D. A. Okun, Water and Wastewater Engineering, Vol. 2, Wiley, 1968.

14. Thorburn, R., D. H. Waller, W. C. Hart, and R. F. McCurdy, "Phosphorus Fractions in Urban Road Deposits and Surface Runoff," in preparation.

15. J. F. McLaren Limited, "Bruce Wood Urban Test Catchment," Project No. 77-8-43, Ontario Ministry of the Environment, Toronto, Ontario, 58 pp.

16. Waller, D. H., "Effects of Urbanization on Phosphorus Flows in a Residential System," Proc. of UNESCO Symp. on "Effects of Urbanization and Industrialization on the Hydrological Regime and on Water Quality," Amsterdam, IAHS-AISH Publ. No. 123, 52-58; 1977.

17. Underwood, J. K., and D. H. Waller, "Major Ion Chemistry of Bulk Precipitation in Nova Scotia", in preparation.

18. Ramalingaiah, and D. Thirumurthi, "Baseline Water Quality Monitoring of Lakes with Potential for Partial Watershed Development", presented at the 18th Canadian Symposium on Water Pollution Research. Toronto, Ontario, 1983 February.

19. Likens, G. E., and F. H. Bormann, "Chemical Analysis of Plant tissue from the Hubbard Brook Ecosystem in New Hampshire", Yale University, School of Forestry Bulletin, 79, 25 pp; 1970.

20. Gunn, J. D., and R. E. Bailey, "Disposal of Swine Manure from Lagoon Storage to Forest Land", presented at Canadian Society of Agricultural Engineering, Brandon, 1975.

21. Hart, W. C., D. H. Waller, R. Thorburn, I. Holden, "Contributions of Phosphorous Fractions from Leaves and Standing Water to Urban Runoff", in preparation.

22. Rodin, L. L. and N. I. Bazilevich, "Production and Mineral Cycling in Terrestrial Vegetation", Oliver and Boyd, London, 1965.

23. Waller, D. H., "Cost and Performance of Retention Basins in the Treatment of Wet Weather Sewage Flows", in Advances in Water Pollution Research, (Proc. 6th International Conference, Jerusalem), Ed. S. H. Jenkins, Pergamon, 327-336; 1973.

24. McElroy, A. D., S. Y. Chin, J. W. Nebgen, A. Aleti, F. W. Bennett, "Loading Functions for Assessment of Water Pollution from Non-Point Sources", U.S. EPA-600/2-76-151, 1976 May.

25. Pitt, R., "Demonstration of Non-point Pollution Abatement Through Improved Street Cleaning Practices", U.S. EPA-600/2-79-161, 220 pp; 1979 March.

26. Sonnen, M. B., "Information Needs for the Quality of Urban Runoff", American Society of Civil Engineers, preprint 3789, Atlanta, 33-50; 1979 October.

27. Waller, D. H., and W. A. Coulter, "Winter Runoff from an Urban Catchment", Ontario Ministry of the Environment Pollution Control Branch, Toronto, Ontario. Project No. 74-8-3 1974.

28. Pisano, W. C., "Case Study: Best Management Practice (BMP) Solution for a Combined Sewer Problem," in Field, R., Ed., Urban Stormwater Management Workshop Proceedings, U.S. EPA-600/9-78-017, 1978 August.

29. Oliver, B. G., J. B. Milne and N. LaBarre, "Chloride and Lead in Urban Snow", J. WPCF, 766-767; 1974 April.

30. Sartor, J. D., and G. B. Boyd, "Water Pollution Aspects of Street Surface Contaminants", U.S. EPA-R2-72-081, 1972, November.

31. Aronson, G. L., D. S. Watson, and W. C. Pisano, "Evaluation of Catch Basin Performance for Urban Storm Water Pollution Control," U.S. EPA-600/S2-83-043, 1983 August.

32. Whipple, W., Jr., ed., Water Problems of Urbanizing Areas, American Society of Civil Engineers, New York, 348 pp., 1978.

33. Waller, D. H., W. C. Hart and D. Thirumurthi, "Halifax Urban Watersheds Program -- A Case Study in Urban Stormwater Quality Management", Proc. CSCE 6th Can. Hydrotech. Conf., Ottawa. 1001-1017; 1983 June 2-3.

34. Farah, G. , unpublished, 1985.

35. Waller, D. H., W. C. Hart, and R. S. Scott, "Preliminary Water and Chemical Mass Balance for Lakes in the Halifax Urban Watersheds Program," Proceedings of the CSCE Annual Conference, 905-915; 1984 May 23-25.

DISCUSSION - POLLUTIONAL ASPECTS OF URBAN RUNOFF

In the discussion which followed the presentation of the papers, the following points were raised:

1. Several participants commented that much more research needed to be done on particles, including particle structure (scanning electron microscope studies indicate particle structures much different than those generally assumed), processes of pollutant sorption and desorption, and deposition and resuspension both in-pipe and within the receiving stream environment.

2. Mr. Field pointed out the need to correlate pollutant loadings with human exposure and epidemiology information, as well as with ecological impacts. He also noted the fact that pollutant loadings tend to be very site-specific, and are difficult to generalize.

3. Mr. Ellis reinforced Mr. Field's comments on ecological impacts, emphasizing that much work remains to be done on the identification of bio-indicators of runoff pollutant impacts, particularly for toxins.

FIELD STUDIES OF URBAN RUNOFF QUALITY

Per-Arne Malmqvist,
VIAK AB,
Mölndalsvägen 85,
S-412 85 Göteborg,
Sweden

Urban Runoff Field Studies - Objectives and Planning

During the last two decades many studies of urban runoff quality have been undertaken throughout the world. The possible environmental effects of urban runoff were first emphasized in the USA in the 1960's, when the question was put: "When a city takes a bath, what do you do with the dirty water?" Field studies of urban runoff quality were initiated by the U.S. Environmental Protection Agency and by some universities. Pioneers were, among others, W. H. Espey, C. W. Morgan and F. D. Masch (16); S. R. Weibel, R. J. Andersson and R. L. Woodward (46, 47); R. J. Burm, D. F. Krawszyk and G. L. Harlow (8); E. H. Bryan (7); J. A. DeFilippi and C. S. Shih (13); J. D. Sartor and G. B. Boyd (36). However, most of the studies at that time dealt with combined sewer overflows.

In Europe studies were initiated some years later, especially in Germany (P. G. Brunner (6)), Sweden (G. Söderlund and H. Lehtinen (41); P. Lisper (26)), and Norway (O. Lindholm (25)). Towards the end of the 1970's the problem was also recognized in countries like Great Britain (G. Mance and M. Harman (29); B. Ellis (15)) and Denmark (P. Heise and J. Simonsen (20)). From 1980 and until now quality studies have been and are being initiated in several other countries in Europe and in other parts of the world. An overview of the status in different countries may be found in Urban Hydrology, edited by J. W. Delleur and H. C. Torno (14) and in the proceedings of the three conferences on Urban Storm Drainage in Southampton 1978, in Urbana 1981 and in Göteborg 1984.

NATO ASI Series, Vol. G10
Urban Runoff Pollution.
Edited by H. C. Torno, J. Marsalek, and M. Desbordes

The objectives of most of these early studies were explorative and aimed only at defining the problem and getting an idea of the magnitudes of some concentrations and mass flows. These limited objectives naturally meant that only rather simple measurements were made. For instance, grab sampling during office hours was the most common sampling method. But soon the need for more detailed data became evident. First the model builders asked for field data to validate their theoretical models which, although they seemed very realistic and logical at first sight, failed to simulate what really happened in an urban area and its storm sewers. Then some of the model builders asked for knowledge of the actual processes - that is, the source-inflow relationships and the accumulation - washoff processes. Other model builders, who were in favour of the regression techniques, asked for data from many different sites, preferably sampled and evaluated in some standardized manner. This led to other kinds of field studies - either studies in which the sources or the buildup and washoff processes were studied, or studies which comprised several urban areas with different land use. Examples of studies of the pollutant sources are those by A. Göttle (19), M. Melanen (32) and P-A. Malmqvist (28). Examples of studies of the buildup and washoff processes are the ones by R. C. Sutherland (40), Morrison et al (33) and B. Shivalingaiah and W. James (39).

In several studies attempts have been made to correlate pollutant concentrations or pollutant loads to hydrological parameters and land use parameters. Hydrological parameters have often been used for the description of runoff quality in single runoff events. Land use parameters have mostly been used to describe average concentrations over longer periods, seasons or years, or pollutant loads from urban areas. Examples of regression analyses are those made in Tulsa, Oklahoma (AVCO, (3)); Houston, Texas (Bedient et al 4); Durham, North Carolina (Colston (10)); Rochester, New York (Lager et al 23); München, West Germany (Göttle 19); Finland (Melanen 32); Madison, Wisconsin (Freund et al 17) and elsewhere.

Attempts to summarize and evaluate the stormwater pollution models, derived from different areas and different studies, have been made by Bradford (5), Jewell (22) and Ammon and Field (2).

The results of all known attempts to develop general predictive models for stormwater loads and concentrations, based on statistical analyses of hydrological or land use data, are on the whole discouraging. It is obvious that hydrological or land use data, as used in the regression equations, cannot sufficiently predict stormwater quality. Calibrated data from single catchment areas may give high R^2-values, but this method is of little interest for planning purposes. Two methods, in combination, seem to be able to lead further:

- The development of deterministic models, based on fully understood relationships for the accumulation and washoff of pollutants.

- The exploration and quantification of the pollutant sources.

Meanwhile, other people were also asking for field data. Between 1970 and 1980 it was often said that the models were much better than the input data. Be that as it may, more field studies were initiated at the request of the model builders and the model users, and by governmental agencies, environmental authorities, and city planners.

Towards the end of the 1970's a need seemed to develop in many countries to assess the stormwater problem and to come to grip with necessary control strategies. This need could be met in two ways - by initiating national studies like the national French study (Hemain and Deutsch, (21)) or the NURP-program in the USA (Torno (43)), or by compiling existing data into some kind of manual, like in Sweden (Malmqvist (27)) or in Norway (Reinertsen (35)). This task, however, has not been easy to fulfill, and the assessments and manuals made to date will probably be revised many times in the years to come.

In summary, today there is a need for three kinds of field data:

a) data for regional and local planning, and for the assessment of environmental impact and the need for control measures (planning objectives)

b) data for the verification and calibration of mathematical models of various kinds (analysis objectives)

c) data for the understanding and modelling of the sources and pathways of pollutants, and the real buildup and washoff processes (research objectives).

When the need for field data is recognized, it is of the greatest importance first to make clear the purpose of the data acquisition. Most requirements and specifications for the measurements, like the choice of pollutants to be studied, the accuracy and frequency of sampling, etc. follow naturally from knowledge of the purpose. These requirements affect not only the usefulness of the data, but also, to a great extent, the cost of the study. It is therefore of importance to adopt a suitable measurement strategy.

Field Studies for Planning Purposes

On a national level there is a need to know to what extent urban stormwater runoff affects the receiving waters, as compared to other discharges and to natural conditions and, if there are problems, to determine what control measures are appropriate in a stormwater management plan. This may also include the creation and analysis of a data base for such a management plan, and the comparison of the results of this analysis to the water quality criteria. There may also be a desire to instruct regional and local authorities along these lines.

The data needed for such national surveys should represent the most common land use categories, and investigation areas should thus be selected from among typical residential areas, industrial areas and so on, in typical climate regions. Of course a very large country like the USA needs many more survey sites than a small country like Sweden.

When environmental impacts are an issue, it is usually necessary to simplify the study. If there are no local conditions that require the study of certain pollutants (like discharges from certain industrial areas), a typical minimum list of substances to study might include solids (TSS), oxygen demand (COD), nutrients (total P, total N), heavy metals (such as Pb).

If this list is extended, discussions as to which form of a nutrient, which heavy metal, or which priority pollutant is to be included soon arise. For a national, general survey the list above should be sufficient. When the funds available for the measurements allow the inclusion of other substances this should, of course, be done.

On a <u>local</u> level the problem may be environmental but may also be one of public health. A minimum list of substances might therefore, in addition to the list above, include bacteria, such as fecal coliforms.

On a local level it is also necessary to take into consideration exactly which kind of industry discharges the stormwater. Ironworks, galvanic industries and chemical industries naturally contribute very different types of stormwater, which has to affect the measuring program. The question of determining whether indicator analyses (like COD, redox etc.) are to be used or if the suspected substances are to be analysed may then arise.

In all these kinds of studies the intention is to <u>demonstrate</u> or prove that the stormwater runoff contains certain amounts of certain pollutants or that it does not, that is, to indicate the possible harmfulness of the runoff. It is therefore important that the results of a study be credible. The study has to conform to the general rules for scientific research. One of the most important rules is: "The same results should be obtained if the study was repeated".

For studies of urban stormwater runoff, this is a rule that is difficult to follow when time and money set the limits. Stormwater quality is as random a phenomenon as is precipitation. You never know what kind of rain you have sampled until afterwards. This dilemma can only be dealt with by sampling in excess and by making statistical analyses of the results after the whole measuring program is finished.

One way of doing this is to compare the sampled events with some kind of normal series for the investigation area, such as the simplified method used by Lindholm (24). Normally, quite a number of events are needed to make the choice of events representative, covering the four seasons of the year, small and large rainstorms, etc.

When environmental impact is to be assessed, a probability approach may be better than a regression approach. The probability method also requires less data, since most of the variables that govern stormwater quality can be neglected. Mean concentrations and coefficients of variation are all that is needed. The method was developed by Dominic M. Di Toro (12) as a part of the NURP project and has proven successful, for instance in Vermont, USA (48). However, the data base that would be needed in order to develop proper values of arithmetic means and the standard deviations would have to be of the same magnitude as in the regression studies.

A method for determining the minimum measurement interval is described by Frans van de Ven in this book.

Field Studies for the Verification and Calibration of Models

When a stormwater model is developed or updated it has to be validated, that is, tested against measured data. Another need for data arises when a stormwater model is to be implemented for use in a certain urban area. It then has to be calibrated, which means that certain constants or input variables have to be given values that depend on the specific area. After this, verifications of the simulated results have to be made. These constants and variables have to be developed through actual mesurements in the area.

The number of stormwater models has grown rapidly during the past decade, and new and better versions of early models are continually being presented. Since they differ quite greatly from each other, it is not possible to give a general recommendation on what sort of data is necessary for validation, calibration and verification. Often some sort of accumulation rates are needed as input, and there are seldom field data to develop such rates. Because of this lack of data, and because of the generally poor performance of the deterministic parts of the quality models, the best results are most often obtained by calibrating the model, that is by adjusting the input variables so that the output concentrations come close to measured concentrations.

Field Studies for Research Purposes

Today most research on urban runoff quality is concentrated on discovering the physical and chemical processes that rule the pollutant sources-inflow relationship and the accumulation and washoff of pollutants. Often these kinds of studies require very small and homogenous areas, in which the researcher has full control of the processes. Many of the studies also have to be carried out in the laboratory. It is not possible to state any general recommendations for measurements aimed at acquiring data for research purposes, since the studies are all different. However, it should be borne in mind, that data acquired for research purposes are never meant to fill, and are very seldomly fit to fill, the needs of data for planning purposes.

Another research area of today is the monitoring of catchment areas for control purposes. Although in most cases this is a question of hydraulic control, it is possible to adopt a control strategy based on measurements of runoff quality. Such measurements do create requirements of a different kind - automatic sampling and analysis, data transmission, protection and maintenance of the instruments etc., which are not further dealt with here.

A more detailed description of how different field studies should be planned and performed has been made by Wolfgang Geiger (18).

Instrumentation and Data Evaluation

An old truism is that "to measure is to know". In field studies of urban runoff quality, as well as in all other studies, it is essential to establish, very early in the planning of the study, what we need to know (measure), and what we may estimate or guess. The purpose of the study, and the time and cost limits, will certainly help to define the measurement program. But there are still many questions of how, where and when to measure. These, and related questions have been answered during the 1970's in manuals and textbooks in various countries, for instance by Shelley and Kirkpatrick (37, 38); Marsalek (30, 31), Alley (1) and DIF Spildevandskomitee (The Danish Sewage Committee, (11)).

Data quality is of course of vital importance to the results and conclusions of the study. This matter has been discussed in some of the manuals, for instance by Alley (1): "The accuracy and precision of the data collected in a sampling program are difficult to assess. For model calibration and verification, it is often desirable that flow measurements be within ±5%, but lesser accuracies may be acceptable depending on program objectives. Precipitation measurements will be of lesser accuracy. Water-quality samples may only give constituent values within about ±25% of the actual in-stream values, at best."

Experience from research carried out at Chalmers University of Technology, Göteborg, tells us otherwise. The accuracies of both pipe flow and water quality are poorer, sometimes even far poorer, than Alley has proposed. In the hydraulics laboratory a concrete pipe with a diameter of 225 mm, a length of 24 m, and variable scope was constructed (H. Bäckman, G. Svensson (9)). Different flow meters were tested under controlled conditions. Some of the conclusions were as follows:

- The total error in flow was found to be up to ±50 per cent when the water level was measured using an ultrasonic meter in an open pipe. No calibration was carried out, but the pipe was perfectly constructed.

- The error in flow was found to be up to ±25 per cent for flows above 20% of the maximum flow, when the water level was measured using an ultrasonic meter in a Palmer-Bowlus flume. For smaller flows the error was larger.

- The error in flow was found to be less than ±25%, when the flow was measured using an air-bubbler and an electromagnetic velocity meter.

The accuracy of the water quality data depends on the representativeness of the storms sampled, the ability of the sampler to collect representative samples, the accuracy of the laboratory analyses, and a few other factors. The accuracy of the laboratory analyses is well defined for most substances, and maintained under laboratory conditions, for concentrations not too close to the detection limits. This source of error is probably the smallest. If all the errors mentioned are summed up, the total error will be larger than 25 per cent in most cases.

A discussion of what the effects on the concentration mean values of a measurement series would be if the largest storm had not been sampled is given by Malmqvist (28). On an average for the four investigated areas in Göteborg, Sweden, the decreases would have been 13 per cent for COD, 20 per cent for phosphorus, and 12 per cent for lead (for 18-27 values per area). This points clearly to the importance of the choice of events in a data series.

Fairly large errors have thus been shown to be inherent in field studies. This may, however, not be incompatible with the objectives of the study, if the errors are acknowledged in the data analyses.

Requirements for a proper data collection system are listed and discussed by Michael L. Terstriep in this book (42).

A Continuous On-line Monitoring System

A continuous on-line monitoring system for the measurement of urban stormwater pollutants has been running for some years at the storm sewer outlet from the catchment area Bergsjön in Göteborg, Sweden. The purpose of the field study is to examine some aspects of the stormwater quality, like the buffering of the runoff, and to develop long data series on stormwater quality.

This system has proved reliable and has eliminated many of the problems associated with discrete sample collection and off-line analysis.

The system is situated inside a shelter above the sewer outlet.

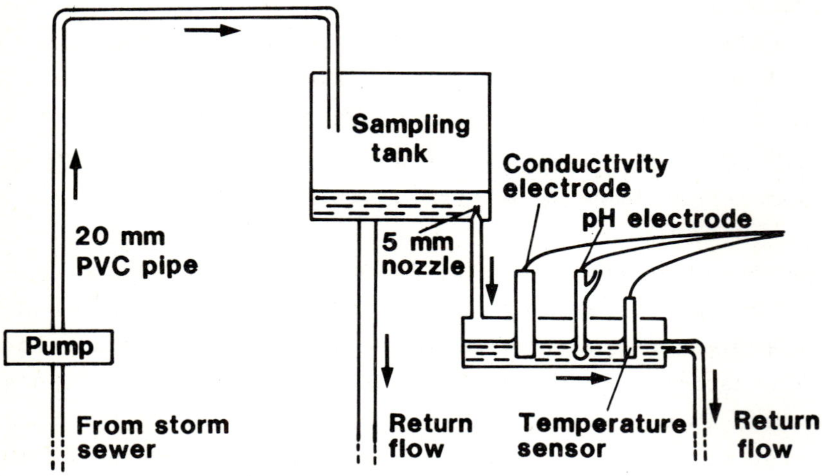

Source: Ref.4.

Before sampling, the turbidity in the sewer is measured to give some idea of the concentration of suspended solids. Water is continuously pumped up from the sewer at a rate of 34 l/min (about 2 m/s).

After passing a 5 mm nozzle the water flows through a PVC pipe of 70 mm diameter in which sensors are mounted. A conductivity electrode, a temperature sensor, and a pH electrode are inserted. The electrode signals are recorded by a 12-channel chart recorder, and also fed into a microcomputer (ABC 80). A program, which has been specially written for the system, includes options for both the analysis and output of data, and for computer controlled discrete sampling. The electrodes are calibrated once a week, but otherwise the system requires very little maintenance.

A similar system is used for the monitoring of the water quality in a gullypot. In this system a redox electrode is also used. The research project is carried out at Chalmers, and has been described in greater detail by Morrison, Sjölander, and Svensson (34).

References

1. Alley, W. M. (editor). Guide for collection, analysis, and use of urban stormwater data. ASCE Conference in Easton, Maryland, 1976.

2. Ammon, D. C., Field, R. Characterization of urban wet-weather discharges. The Second International Conference on Urban Storm Drainage. University of Illinois, Urbana, USA. 1981.

3. AVCO (American Economic Systems Corporation). Storm water pollution from urban land activity. FWQA, Water Pollution Control Research Series. PB 145281. Washington D. C. 1970.

4. Bedient, P. B. et al. Stormwater pollutant load-runoff relationships. Journal WPCF. Vol. 52, No 9, pp 2396-2404. Washington D. C. 1980.

5. Bradford, W. L. Urban stormwater pollutant loadings: a statistical summary through 1972. Journal WPCF, Vol. 49, No 4, pp 613-622, Washington D. C. 1977.

6. Brunner, P. G. Die Verschmutzung des Regenwasserabflusses im Trennverfahren. Untersuchungen unter besonderer Berücksichtigung der Niederschlagsverhältnisse im voralpinen Raum. Berichte aus Wassergütewirtschaft und Gesundheitsingenieurwesen. Technische Universität, München. No 9. 1975.

7. Bryan, E. H. Quality of stormwater drainage from urban land. 7th American Water Resources Association Meeting. Washington D. C. 1971.

8. Burm, R. J., Krawszyk, D. F., Harlow, G. L. Chemical and physical comparison of combined and separate sewer discharges. Journal WPCF, 40, No 1, pp 112-126. 1968.

9. Bäckman, H., Svensson, G. Flödesmätning i avloppsnät med portabla utrustningar. Urban Geohydrology Research Group, Chalmers University of Technology. Report No 69. Göteborg, 1983.

10. Colston, N. Jr. Characterization and treatment of urban land runoff. USEPA-670/2-74-096. Cincinnati, USA. 1974.

11. DIF Spildevannskomitee. Vandföringsmåling og prövetagning. Copenhagen 1976.

12. DiToro, D. M. Probability model of stream response to runoff. EPA, NURP. 1982.

13. DeFilippi, J. A., Shih, C. S. Characteristics of separated storm and combined sewer flows. Journal WPCF, 43, No 10, pp 2033-2058. 1971.

14. Delleur, J. W., Torno, H. C. Urban Hydrology. Proceedings of the International Symposium on Urban Hydrology. ASCE. 1983.

15. Ellis, B. J. The Characterization of particulate solids and quality of water discharge from an urban catchment. IAHS-AISH Publication No 123. London. 1977.

16. Espey, W. H., Morgan, C. W., Masch, F. D. A study of some effects of urbanization on storm runoff from a small watershed. Technical Report HYO 07-6501. Texas Water Comission. 1965.

17. Freund, A. P., Johnson, C. D. Comparison and relationships of stormwater quality and basin characteristics. International symposium of urban storm runoff. University of Kentucky, Lexington, USA. 1980.

18. Geiger, F. W. Use of field data in urban drainage planning. This book.

19. Göttle, A. Ursachen und Mechanismen der Regenwasserverschmutzung - Ein Beitrag zur Modellierung der Abflussbeschaffenheit in Städtischen Gebieten. Berichte aus Wassergütewirtschaft und Gesundheitsingenieurwesen. Technische Universität, München. No 23. 1978.

20. Heise, P. Simonsen, J. Gudenåundersögelsen 1973-1975. Regnvandundersögelser. Vandkvalitetsinstituttet. Rapport no 17. Hörsholm, Denmark. 1977.

21. Hemain, J. C. Deutsch, J. C. Main results of the French national program of urban runoff quality measurement. Proceedings of the Third International Conference on Urban Storm Drainage, Vol. 3, pp 939-946. Göteborg. 1984.

22. Jewell, J. A., Adrian, D. D. Statistical analysis to derive improved stormwater quality models. Journal WPCF, Vol. 54, No 5, pp 489-499. Washington D.C. 1982.

23. Lager, J. A. , et al. Development and application of a simplified stormwater management model. USEPA-600/2-76-218. Cincinnati, USA. 1976.

24. Lindholm, O. Valg av modellregn. Norsk institutt for vannforskning. PRA 6. Oslo 1975.

25. Lindholm, O. et al. Forurensning i overvann. NIVA 0-57/74. PRA 4.7. Oslo. 1976.

26. Lisper, P. On the composition of urban runoff and its variations (In Swedish). Ph. D. Thesis at Chalmers University of Technology. 1974.

27. Malmqvist, P-A. Estimations of urban stormwater quality. (In Swedish.) Urban Geohydrology Research Group. Chalmers University of Technology. Report No 66. Göteborg. 1982.

28. Malmqvist, P-A. Urban stormwater pollutant sources. An analysis of inflows and outflows of nitrogen, phosphorus, lead, zinc and copper in urban areas. Chalmers University of Technology. Göteborg. 1983.

29. Mance, G., Harman, M. The quality of urban stormwater runoff. In Helliwell PR (ed) 1978: Urban Storm Drainage, pp 603-617. 1978.

30. Marsalek, J. Instrumentation for field studies of urban runoff. Agreement on Great Lakes Water Quality, Environment Canada. Research Report No 42. 1978.

31. Marsalek, J. Sampling techniques in urban runoff quality studies. Special Technical Publication 573. ASTM. pp 526-542. 1975.

32. Melanen, M. Quality of runoff water in urban areas. Water Research Institute. No 42, pp 123-190. Helsinki. 1981.

33. Morrison, G. M. et al. The physico-chemical speciation of zinc, cadmium, lead and copper in urban stormwater. Proceedings of the Third International Conference on Urban Storm Drainage. Vol. 3, pp 1001-1008. Göteborg. 1984.

34. Morrison, G., Sjölander, B., Svensson, G. The continuous on-line monitoring of environmental pollutants in stormwater. VATTEN, vol. 2, pp 91-96. Lund 1985.

35. Reinertsen, T. Forurensninger i overvann. Vann og Avlöp. NTNF-PTV 15. Trondheim 1981.

36. Sartor, J. D., Boyd, G. B. Water pollution aspects of street surface contaminants. EPA-R2-72-081. 1972.

37. Shelley, P. E., Kirkpatrick, G. A. An assessment of automatic sewer flow samplers. USEPA-600/2-75-065. Cincinnati 1975.

38. Shelley, P. E., Kirkpatrick, G. A. Sewer flow measurement - a state of the art assessment. USEPA-600/2-75-027. Cincinnati 1975.

39. Shivalingaiah, B., James, W. Algorithms for buildup, washoff and routing pollutants in urban runoff. Proceedings of the Third International Conference on Urban Storm Drainage. Vol. 4, pp 1445-1456. Göteborg. 1984.

40. Sutherland, R. C. A mathematical model for estimating pollution loadings and removals from urban streets. University of Maryland, USA. 1975.

41. Söderlund, G., Lehtinen, H., Friberg, S. Physicochemical and microbiological properties of urban stormwater runoff. 5th International Water Pollution Research Conference. 1970.

42. Terstriep, M. L. Design of data collection systems. This book.

43. Torno, H. The EPA Nationwide urban runoff program (NURP). Proceedings of the Third International Conference on Urban Storm Drainage, Vol. 4, pp 1465-1474. Göteborg 1984.

44. USEPA. Final report of the Nationwide urban runoff program. Water Planning Division. Washington D. C. 1983.

45. Van de Ven, F. H. M. Data evaluation in field studies of urban runoff quality; Aspects of assessing the measurement interval. This book.

46. Weibel, S. R., Andersson, R. J., Wordward, R. L. Urban land runoff as a factor in stream pollution. Journal WPCF, 36:914. 1964.

47. Weibel, S. R. et al. Pesticides and other contaminants in rainfall and runoff. Journal AWWA, 58, No 8, pp 1075-1084. 1966.

48. Vermont Agency of Environmental Conservation of Stormwater Management. USA. 1983.

USE OF FIELD DATA IN URBAN DRAINAGE PLANNING

Wolfgang F. Geiger, Lehrstuhl fur Wassergutewirtschaft und Gesundheitsingenieurwesen
Technische Universitat Munchen
Am Coulombwall
8046 Garching bei Munchen
Federal Republic of Germany

Introduction

The trend towards continued urbanization results in more intensive industrial, commercial and residential land use. This in turn increases water consumption and storm runoff. The concept of direct and rapid discharge of urban waste waters to the nearest water course leads to downstream flooding, lowering of groundwater tables, changes in the microclimate of urban areas, and pollution of receiving waters. On the other hand the increasing population, along with increases in leisure time, increases the demand for water and water-based recreation.

It is well recognized that water pollution has increased enormously during the last 25 years although, in most countries, technical and financial efforts to control water pollution have been intensified steadily. Legal measures have also been introduced to dam the flood of pollution. International organizations such as WHO, OECD, UNESCO and others have tried to assist the fight against water pollution by international cooperation. At present, all these efforts have not yet been fully successful, although some progress has been made. One reason is that urban drainage planning frequently does not consider the pollutional aspects adequately. Modern drainage should provide for a balanced combination of natural and manmade drainage elements which are designed not only to provide adequate flood protection, but also to minimize the drainage

NATO ASI Series, Vol. G10
Urban Runoff Pollution.
Edited by H.C. Torno, J. Marsalek, and M. Desbordes

impact on receiving waters. Such increasing sophistication of drainage design requires application of sophisticated de-design tools and the availability of a variety of supporting data. Planning for preventative and remedial measures also requires an understanding of the causes of runoff pollution, the magnitude of pollution and its effect upon receiving waters.

Urban storm drainage is usually considered as a nonpoint source of receiving water pollution. Receiving water pollution, however, also results from municipal and industrial point sources and other nonpoint sources. Many human activities cause nonpoint pollution - farming, lumbering, pesticides and fertilizers applied to crops and lawns, disposal of waters on or under the ground, construction, and mining. This paper only deals with municipal nonpoint sources, i.e. storm runoff in separate and combined sewer systems.

The Rainfall-Runoff Process and Data Collection Constraints

Urban drainage design depends, to a large extent, on measured data. In order to make measurements, one must have an idea of the rainfall-runoff process. A natural concept is to follow the path of rain (and associated pollutants) as it becomes runoff. Figure 1 illustrates the major components of runoff development and flow transport in urban drainage systems. For runoff quantity, enough knowledge of the processes involved is available that planning methods of different detail and accuracy could be developed. Knowledge of runoff quantity is prerequisite for runoff quality considerations. The quantity and mass flux of pollutants in urban runoff depend on complex interactions between land use, pollutant accumulation, drainage system configuration, characteristics of precipitation and time factors such as season or length of time since the previous storm. Figure 2 shows pollutant pathways in urban runoff, including the pollutant origin, chemical changes and accumulation. Many of the individual components of the pollution process, however, have not been identified to a degree that they could be included in planning procedures similar to those for runoff

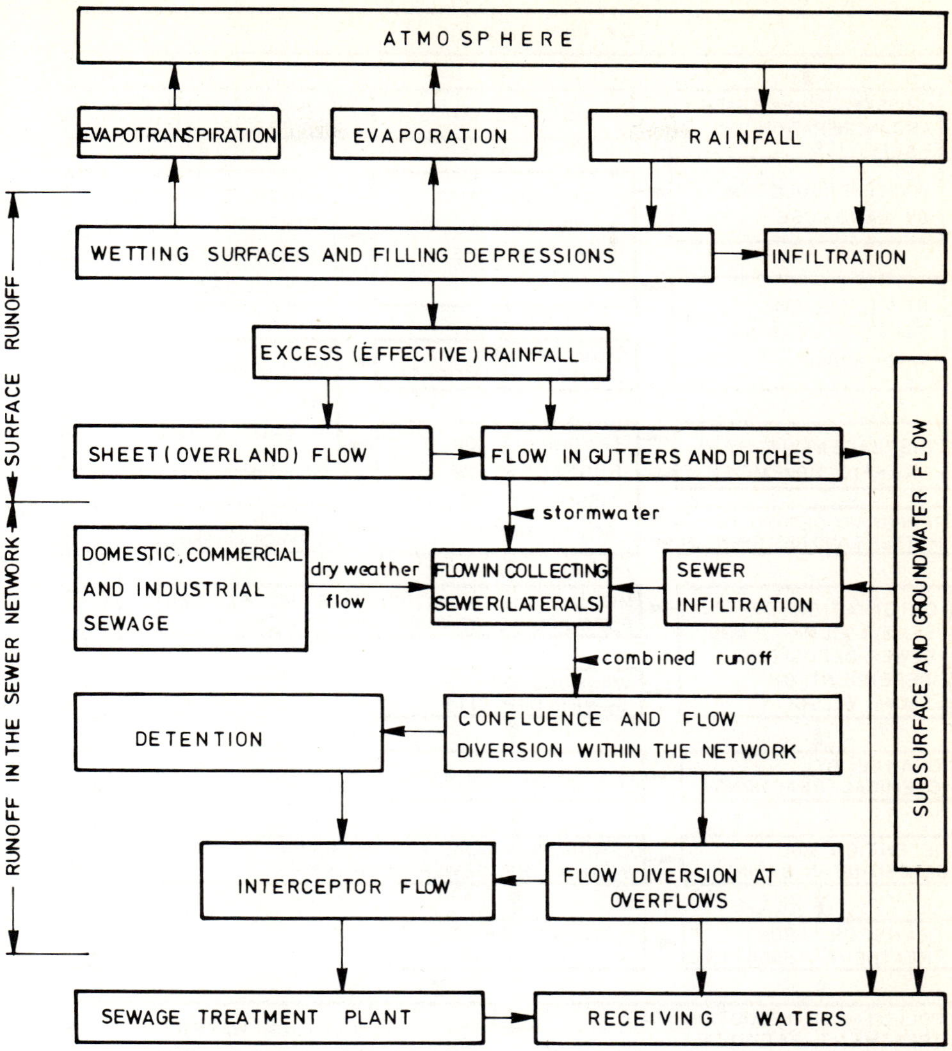

Figure 1: Components of runoff development and flow transport in urban drainage systems

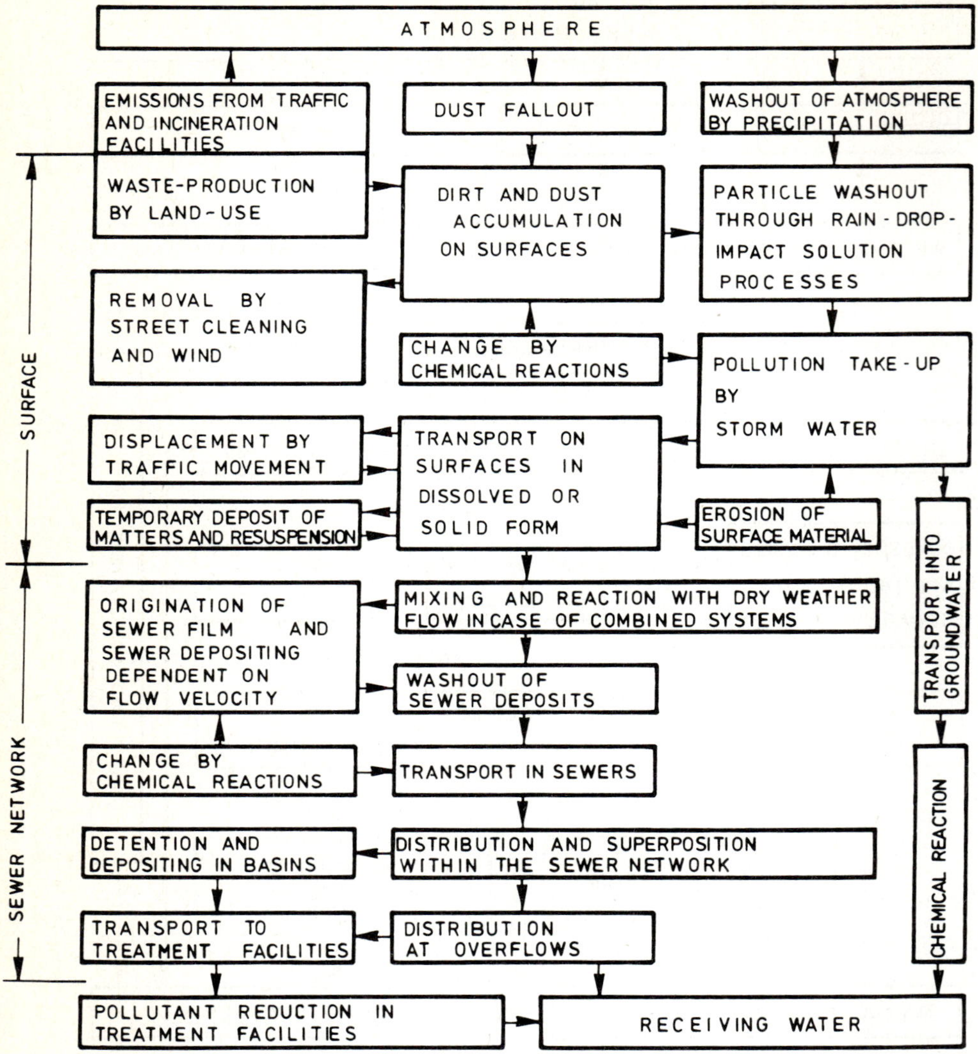

Figure 2: Development of runoff pollution in urban drainage systems

quantity. Most of the quality measurements have been taken at the downstream end of drainage systems. They only report overall effects, and reflect the complex interactions within the the system.

The effect of pollutant emissions on receiving waters depends on the severity of a runoff event, its antecedent conditions, and the characteristics and uses of receiving waters. In order to evaluate this effect, it is necessary to identify both runoff pollution and receiving water conditions, and to establish a cause and effect relationship between the two. The pollutants observed in municipal discharges include fecal coliform bacteria, oxygen-demanding materials, phosphorus, nitrogen, oil and grease. Municipal discharges can also contain an excess of suspended sediments, dissolved solids, heavy metals and toxic organic compounds. They may be contained in storm runoff and combined sewer overflows and can severely degrade receiving water quality. When storm runoff is added to normal sewage flows in systems with combined sewers, sewage treatment plants cannot handle all flow, and much of it bypasses treatment.

Urban runoff is highly intermittent, and its flow rate changes rapidly. Therefore, it cannot be measured by the periodic sampling normally used to monitor quality of rivers. Figure 3 shows the variation in runoff flow and pollutant concentration during the course of a storm event. This example also demonstrates that variations in flow and quality may not occur simultaneously. Among all waste waters, storm and combined sewer flows are obviously the most difficult to characterize accurately. This is due to:

- The wide range of flow rates, which change rapidly from virtually zero to the peak rate.
- The change from free to surcharged flow conditions.
- The possibility of backwater, or even flow reversal, in certain situations.
- The extremely wide range of pollutants that can be found.

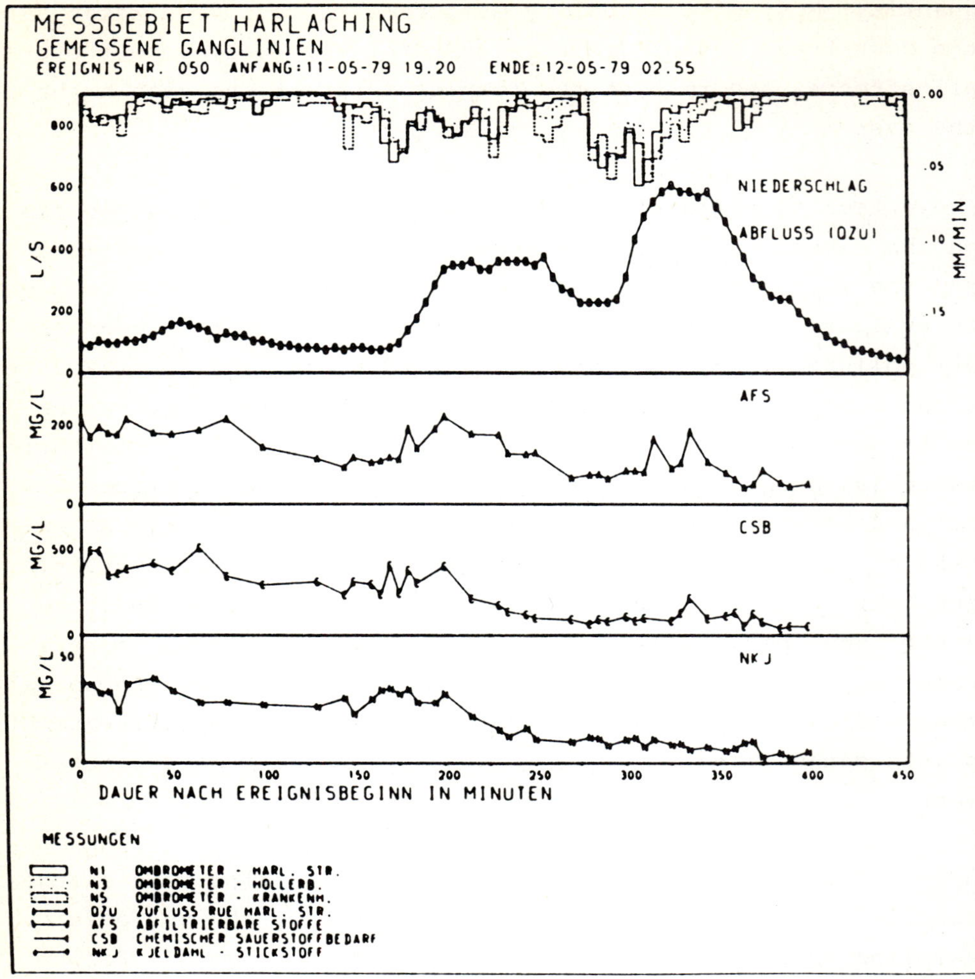

Figure 3: Variation of flow and pollutants during the course of a storm event (2)

- The possible spatial variation of these pollutants within a given cross-section.
- The frequent presence of significant bed loads which may be highly polluted.

For characterization of storm runoff, densely spaced data are necessary, which calls for automation of the monitoring process This automation, however, is limited by the following factors:

- The necessity to determine the majority of pollutional constituents by laboratory analyses.
- The limited laboratory capacities conflicting with the need to handle an unpredictabe number of samples.
- The damp and corrosive sewer environment, which necessitates frequent expert maintenance of all installations.
- The need for frequent and knowledgable maintenance and calibration of instruments.
- The contamination of the metered media by suspended solids, fibers and floating debris, which poses a physical threat to any sensors or sampler intakes.

Experience with a continuous quality monitoring program is reported in (1).

Dependence of Data Requirements on Study Objectives

Flow measurements and quality sampling are essential for pollution control. Research, planning, design, operation and maintenance of urban drainage systems, and enforcement of pertinent laws, are activities which rely on measurements for their effective conduct. The data required may be classified into two groups:

- Data leading to a better understanding of the rainfall-runoff process and its pollutional aspects, which consequently leads to the improvement of existing design procedures, the development of new planning approaches and the enhancement of regulations, and

- Data necessary to carry out the planning for a specific area, which includes the identification of existing conditions as well as data forecasts for planning horizons.

While local data are vital for planning and design of a specific urban drainage system, the understanding of the processes involved may be gained from data of any representative catchment. For some activities, very precise time synchronized, or continuous, records are needed. For others, infrequent rough estimates may suffice. Data requirements cannot be generalized, because some of them result from the study objectives.

Results of Research Studies as Basis for Urban Drainage Planning

Many recent research studies tried to collect data in order to:

- Identify the quality parameters most characteristic for urban runoff pollutants and their sources.
- Explain the mechanisms governing accumulation, wash-off, and transport of pollutants on surfaces and in sewers.
- Define the time-dependency of the rainfall-runoff-process and its pollution.
- Indicate the influence of areal distribution of rainfall on runoff and on runoff pollution.
- Describe the range of constituent concentrations and the factors influencing their variations.
- Determine means, medians, variances, and other statistics of quality parameters, and link these data to catchment characteristics and land-use.
- Compute mass balances and total mass emissions of quality constituentss yielding pollutant loadings such as kg/ha or kg/a.
- Define antecedent moisture conditions and pollution potentials.
- Find the most important characteristics of rainfall events related to the pollution impact on receiving waters.
- Assess the impact of man's activities on water quality in view of future needs.

- Initiate, construct and maintain an extensive data base.
- Check the transferability of individual data sets.

These objectives require:

- Location of the data collection sites in such a manner that the desired phase of the process can be identified.
- Data acquisition in time series, so that hydrographs and pollutographs may be derived.
- Exact time synchronization of precipitation, runoff and quality measurements,
- Consistent accuracy of measurements with data evaluation.
- Investigation of a broad variety of quality parameters.
- The analysis of samples from many points on the hydrograph.

Considering the rapid variation of pollution levels during storm runoff events, the use of single samples to characterize urban runoff is entirely unacceptable. The above research objectives also yield a number of criteria for the selection of a test catchment:

- The drainage area must be large enough to be representative for the process studied.
- The drainage area must have an interior position within an urbanized area to avoid boundary effects.
- Erosion of topsoil must be avoided.
- The effect of construction sites must be known.
- The point sources of pollution within the area must be known.
- The sewer network and the drainage area must be well defined.
- For valid correlation between rainfall and runoff, multiple diversions or a large number of interception points are not desirable.
- The catchment should have representative slopes.
- The monitoring sites must be accessible, so that equipment can be installed, operated and maintained.
- Traffic and public disruption should be minimized.

Further details on the collection and analysis of urban stormwater data are given in (3).

Better understanding of the sources of pollution and of the mechanisms of runoff pollution development and transport leads to:

- An assessment of pollutional impacts caused by separate and combined systems as well as by different land uses.
- Revision of abatement measures, which aim for first flush retainment and storm water treatment.
- Improvement of existing procedures for drainage system design.
- Development of mathemdtical algorithms of a deterministic, parametric and stochastic nature, which may also be applied to planning.
- Advancement of the planning approach which would replace the design storm concept.
- Enhancement of receiving water standards accounting for the hydrological process and water use.

Data are needed, especially for mathematical models, to define their parameters and validate model structures. The characterization of stormwater and combined runoff with respect to the geographic location, population density, land use and other parameters, together with mathematical modeling, makes it possible to produce reasonably accurate estimations without measurements.

Direct Use of Data in Urban Drainage Planning

The main objectives of urban drainage planning are to avoid flooding within the urbanized area and to provide the comfort of a waste water disposal system. At the same time, the linkage to the hydrological cycle prohibits downstream flooding and the overloading of receiving waters beyond their capacity. This requires a well balanced layout and design of drainage systems as a whole, as well as of individual control structures.

Planning should begin with an assessment of existing conditions which, for runoff quality, requires some field data. Most quality data consist of concentrations of various parameters. Some parameters may be measured in situ, such as pH, conductivity, temperature and dissolved oxygen, but the majority must be obtained by laboratory analysis of samples. At least summary parameters such as total suspended solids, BOD or COD should be analyzed, which give a good indication of the status of pollution. In individual studies, it may be necessary to include nitrogen, phosphorus, heavy metals, such as lead, cadmium, copper and zinc, chlorinated organic compounds and bacteriological indicators. Because the water quality measurements are costly and time consuming, it may be preferable to use a mathematical model which is verified by limited local measurements and then applied to the system under consideration to obtain an appreciation of the pollutional situation.

Time and budget constraints usually limit collection of any local data needed throughout the planning process. Obviously, data are indispensable for planning, such as:

- Size and topographic situation of the drainage area.
- Soil conditions, consolidated and hard surfaces, surface slopes and vegetative cover.
- Receiving water system layout and condition including floodplain and groundwater data.
- Sewer system configuration including storm runoff control and treatment facilities.
- Land-use, zoning , population density.
- Traffic and other supply facilites.

Many of these data must be forecasted, and will depend on other city planning efforts. Economic development, for instance, may well overrule the assumptions made for future drainage needs and, consequently, drainage master plans will become obsolete. This fact must be kept in mind when specifying the accuracy requirements for hydrological data needed in the planning process.

The major hydrological data input is rainfall. Because precipitation records collected within the drainage area usually do not exist, the rainfall input is derived from data of the closest weather bureau station. In rare cases intensity-duration curves of entirely entirely different areas have been used incorrectly to derive design storms. At this point the validity of the design storm concept is not discussed. Even if local rainfall records are available, the current design rainfall analyses require knowledge, time and money and often are not done. This results in rainfall input of low accuracy especially if the areal variation of rainfall and its consequences for runoff are considered.

Local rainfall and runoff data were collected in larger urban drainage studies which employed mathematical models for planing. These models usually represent relationships between rainfall, runoff, basin characteristics, and concentrations of pollutants in the runoff. The principal limiting factor in the development and testing of such models has been the scarcity of satisfactory data on the quantity and quality of runoff in urban areas. Nevertheless, such models are increasingly applied and the need for relevant data has intensified. Since most recent urban hydrologic models define the complete hydrograph or pollutograph during a storm event, measurement of only, say, peak flows or average concentrations is inadequate for calibration of these models. It is particularly difficult to obtain sequential and closely spaced water quality data so that mathematical models can be calibrated and verified. Another principal consideration in the use of quality data for urban runoff modeling is the method of samples collection. However, once a model adequately represents the rainfall runoff processes within a specific drainage area, the effectiveness of many systems and operational alternatives may be evaluated quickly. With long-term simulations, even total pollutant mass emissions from drainage systems may be obtained. The necessary monitoring programs are quite time consuming and costly. Therefore, there is a need for generally accessible data banks. To date, the lack of standardized data and formats has prevented the establishment of complete and generally usable data sets.

Data Required Beyond the Planning Stage

Modern drainage planning may address, in advance, on-line control of storm runoff, sewer system operation and maintenance. To develop such concepts in the planning stage, the data used for testing must be compatible, in their quantity and accuracy, to those collected during real operation. This applies particularly to the concept of on-line control involving rainfall data forecasts. However, data collection programs initiated for such purposes should be coordinated with the requirements for planning data, so that they can be used for later planning updates. To comply with receiving water quality or effluent regulations, sporadic measurements often are taken. For water quality aspects, it is sufficient to analyze flow-proportional composite samples, which can collected either by well designed automatic stations or by hand methods. In either case, the result is, for instance, a more or less accurate determination of the total pollutant loading for a given storm. These data could also be made available in a form to be integrated into later planning updates.

Urban drainage planning obviously relies on data. There is a need for centralized data banks. Such data banks, however, will only be useful if generalized format and standards for data presentation and storage are available and complied with. Data are useful only if the catchment characteristics and other relevant information, such as construction activities within the drainage area during test phases, are reported. Furthermore, the multi-purpose use of data should be considered. For past measurements it is quite difficult to provide all necessary information to make the data generally usable. For current and future monitoring efforts, the standardization of data reporting is of utmost importance.

Significance of Data for Urban Drainage Planning

Rainfall-runoff data are useful, for example, to calibrate and verify planning models, to check the validity of the design storm concept, to support continuous simulation as an alternative to the design storm concept, to provide the basis for progressive master drainage plans, to develop operational control strategies and to support sewer system operation and maintenance.

Application and Verification of Models used for Urban Drainage Planning

To verify different concepts of runoff modelling a number of models were applied to the combined system of Munich-Harlaching, Federal Republic of Germany, which drains an area of 540 ha. The results of three different, but typical modelling approaches represented by the ILLUDAS (Illinois Urban Drainage Area Simulator), HYBAG (Hydrologische Berechnung von Abflussganglinien) and SWMM-level III (Storm Water Management Model) models are presented. ILLUDAS (4) calculates runoff from pervious and impervious areas using time-area diagrams and routes flow through channels using an implicit method approximating the continuity equation. HYBAG (2), to the extend used in this comparison, is a hydrologic model which uses the unit hydrograph principle for both pervious and impervious areas as well as for other quantity and quality considerations. SWMM (5) calculates surface runoff using a nonlinear reservoir approach, combined with the Manning formula, and the flow routing in sewer networks is done by using either the kinematic wave equation or the full St. Venant Equations depending on whether or not backwater is considered. All models can handle dry weather flow.

The geometrical data on catchment surface elements and the sewer network were derived from maps and public works records. Obviously, the preparation of detailedl data for the SWMM model was very time consuming, while the preparation of the limited data needed for the lumped approach of the HYBAG model was relatively easy. Hydrologic and hydraulic model parameters were easily established. The depression depth, initial losses, and infiltration capacities were known from measured data. Hydraulic parameters, such as roughness coefficients, were selected within the established ranges. For calibration of model parameters used in both quantity and quality, considerations, up to 30 measured rainfall-runoff events were utilized.

13 measured rainfall-runoff events, different from the ones used for calibration, were chosen for verification. For each event, runoff was calculated with all three models using the measured rainfall as input. The observed and simulated hydrographs were compared by calculating the ratio, , and the relative error, , for runoff volume, peak and the time from start to peak of the runoff hydrograph. λ and ϵ are defined as follows:

$$\lambda = \frac{C}{M}, \text{ and } \quad \epsilon = \frac{C - M}{M}$$

where C and M are calculated and measured values, respectively.

For ϵ, in addition to the mean error, the positive and negative errors were averaged separately. The verification results for for runoff volume, peak runoff and time to peak are presented in Table 1, and are further subdivided into events with flow rates less than 1 m^3/s, greater than 1 m^3/s, and all events combined. While the runoff volume was overestimated by SWMM and ILLUDAS, the hydrologic approach tended to underestimate volume. For runoff peaks the opposite was true.

Furthermore, the accuracy of the simulation of the hydrograph shape was described by the root mean square error ϵ_{RMS}:

$$\epsilon_{RMS} = \sqrt{(t_A / V^2_M) \sum_i (Q_{C_i} - Q_{M_i})^2 \Delta t}, \text{ where}$$

t_A = runoff duration

V_M = measured runoff volume

Table 1. Comparison of Measured and Calculated Runoff Volumes, Peak Runoff, and Times to Peak.

model	runoff property	mean of λ			Std.Dev.of λ			ε in %		
		$\leq 1\frac{m^3}{s}$	$>1\frac{m^3}{s}$	all	$\leq 1\frac{m^3}{s}$	$>1\frac{m^3}{s}$	all	pos.	neg.	all
ILLUDAS	volume	1.01	1.28	1.13	0.04	0.22	0.20	18.30	2.67	+13.5
	peak	1.00	0.94	0.97	0.06	0.08	0.07	3.33	7.71	- 2.6
	time to peak	1.22	1.18	1.20	0.09	0.17	0.13	21.67	0.00	+20.0
HYBAG	volume	0.88	0.93	0.91	0.08	0.24	0.17	8.00	16.22	- 9.4
	peak	1.05	0.87	0.97	0.05	0.26	0.20	6.29	14.67	- 3.4
	time to peak	1.00	1.04	1.02	0.13	0.10	0.12	11.67	9.00	+ 1.9
SWMM	volume	1.00	1.16	1.07	0.05	0.13	0.12	14.25	3.80	+ 7.3
	peak	1.00	0.94	0.97	0.09	0.08	0.09	6.75	8.13	- 2.2
	time to peak	1.00	0.96	0.98	0.04	0.08	0.06	2.50	12.00	- 2.0

Q_{C_i} = calculated runoff rate for interval i

Q_{M_i} = measured runoff rate for interval i

Δt = time step

A root mean square error of zero indicates a complete match of the calculated and measured hydrographs. For all models the root mean square errors varied between .3 and .4. It was found that the calculated hydrographs often were often almost identical to the measured ones, but shifted in time by one or two time steps ahead of or after the measured ones. This resulted in high root mean square errors, although visual comparison was good. For practical model application, reliance on visual comparison of hydrograph shapes is suggested, since small time shifts are irrelevant for model application. In summary, all three models were equally verified for all flow ranges. Under heavy storm conditions backwater occurred. However, under under such condition convergency problems were encountered with SWMM, losing its benefits. All verification runs are reported in detail by Geiger (2). For quality simulations performed with SWMM and HYBAG, no consistent set of input parameters was found that yielded differences in measured and calculated concentrations and loads of less than 50%. For the individual storm events, adjustments of input parameters yielded almost exact matches of simulated and measured results.

Validity of the Design Storm Concept

The validity of the design storm concept has been argued for a long time (6). A 5-year continuous data base of rainfall and runoff (2) is used here to demonstrate that one can not derive runoff frequencies from rainfall analysis. Using an extreme value analysis of a partial series and assuming exponential distribution, the recurrence frequencies of rainfall and runoff properties differed randomly for individual events (Figure 4). The same was found when comparing the frequencies of runoff volumes and loads. In addition, the assumed exponential distribution was replaced by an extreme value type 3 distribution which gave nearly identical results. This indicates that the standard procedures used for sewer design should be reconsidered.

To show the influence of the design storm duration and frequency on design safety, the above mentioned model HYBAG and the extreme value analysis of the 5-year continuous data base were used. For design storms the Sifalda storm (7) and the Chicago storm (8) were chosen. The required storm model parameters were derived from local rainfall data of 24 years duration. The calculated runoff peaks for storms of different frequencies and durations are compared in Figure 5 with the extreme values of the partial series. For comparison the corresponding values calculated from the rational method for the intensity duration curves of the same rainfall record are also included in Figure 5. The results obtained for the Chicago and Sifalda storms are different, because the peak intensity value remains the same for all storm model durations for the Chicago storm, while it varies for the Sifalda storm. However, in practice the design storm duration would be selected to yield the highest peak

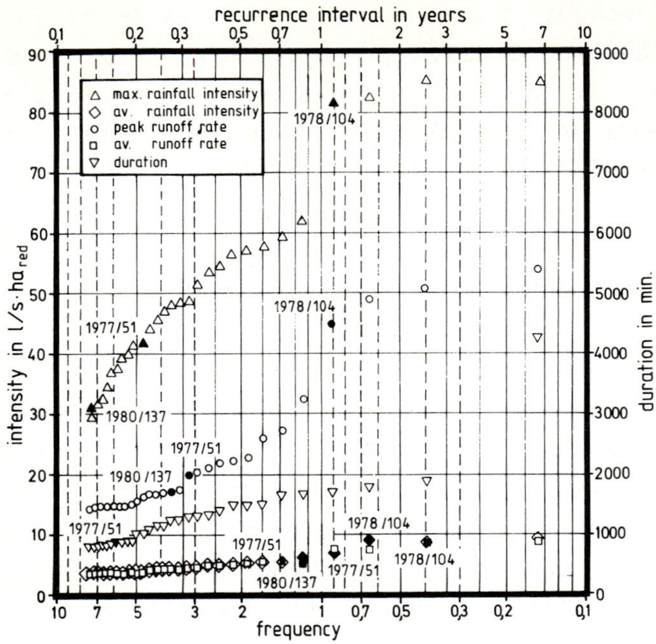

Figure 4. Identification of individual events within the partial series of peak and mean intensities of rainfall and runoff respectively and of runoff duration(2)

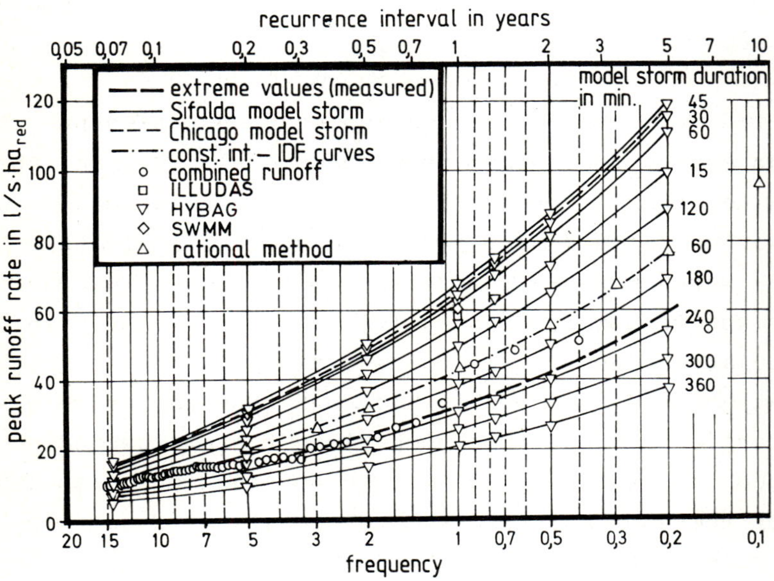

Figure 5. Comparison of measured (partial series of extreme values) and calculated (design storm based) peak runoff of different frequency (2)

which would result in design safety factors up to 2.1, depending on the design frequency. It was surprising that the rational method produced results which were safe, but the safety factors were much lower than the model approach.

A similar investigation was performed for the runoff volume (2). All volumes obtained were underestimated. This is not surprising, since the original storm models were meant for peak flow only. However, their application is frequently extended to volume considerations. This is incorrect, because they neglect the multi-peak hyetograph shape which is typically encountered in long duration storms, and which results in large volumes. Consequently, for dimensioning of storage facilities, entirely different design storms must be used. Unfortunately, this is not a general practice, although done occasionally.

Reliability of continuous Simulations and their Use in Master Planning

Continuous simulation is defined here as a continual calculation of storm or combined runoff for rainfall records which are several years long. It aims for statistical analysis of runoff properties and represents an alternative to the design storm concept. Furthermore, overflow frequencies and durations provide a better measure of the effectiveness of design alternatives than peak magnitudes derived with design storms. For cost reasons, continuous simulations are frequently performed with simpler models in which simplifying assumptions are made for the consideration of dry spells. Therefore, the reliability of continuous simulations has been questioned.

For the 1977-80 rainfall record measured in Munich-Harlaching, the runoff was simulated with HYBAG. By comparing the simulations with the measured runoff, it was proven that an assumption of average antecedent moisture conditions at the start of each storm is fully acceptable, and may be used instead of complicated water balance calculations between storms. On the average, the error ϵ's for runoff volume, peak runoff and time to peak were 2.4%, 11.2% and 1.3%, respectively. The root mean square error, however, was 0.66, due to frequent parallel shifts of the measured and calculated hydrographs (2).

It is obvious that since even the design storm frequency differs from that of runoff, the conventional design storms cannot yield any information for runoff quality considerations. Despite the difficulties in general, verifications of quality calculations using continuous simulations for total suspended solids and COD were performed with HYBAG (Table 2), using estimates for average pollutant potentials at the beginning of storms. The results demonstrate that, on the average, calculated concentrations approximated those measured in runoff by an order of magnitude. Satisfactory calibration of quality simulations for a set of individual storms proved to be difficult. Continuous quality simulations seemed to statistically equalize this deficiency.

Table 2. Comparison of Continuous Quality Simulations (HYBAG) with Measured Concentrations for 1977-1981 (2).

parameter	dim.	average of interval values		average of all events of load/volume	
		measured	calculated	measured	calculated
TSS	mg/l	153	145	163	137
COD	mg/l	275	305	274	262

The general drainage plan of the city of Munich, presently in effect, may serve as an example of progressive master drainage planning using continuous simulations. In this case, a measured, long-term precipitation series was used for computations with a rather sophisticated continuous simulation model, the QQS (QuantityQuality-Simulation) model (9). Significant estimates of combined sewer overflows, treatment plant inflows, and the pollutional impact on the receiving water system were obtained. The plan was prepared from 1976 to 1980 and refers to conditions expected around the year 2016, when there will be

3.2 million inhabitants (or equivalents) occupying the 20,500 hectares of tributary area (10). The plan upgrading program, whose structures are now being built step by step, has also shown that a re-analysis of the existing system and optimization of the actual combined overflow conditions are necessary at regular intervals. The most important practical purpose of such reanalysis is in periodic adjustments of the major structures, which were built in the meantime, but which were designed for future flows. Such adjustments will guarantee optimal functioning under changing interim conditions (Figure 6). In the case of Munich, this means that the "Rest-Isar", a highly sensitive section of the receiving water in terms of quality, must be kept largely free from combined sewer overflows. This will be achieved by discharging excess combined flows into more capacious receiving water body sections according to these reanalyses (11).

Data as the Basis for Real-Time Control and Sewer System Operation

Urban drainage systems are usually designed to hold runoff from storms of low frequency, and therefore provide tremendous storage volumes. These storage capacities are not utilized during most storm events. In order to minimize overflow loads, to direct overflows to receiving water portions with higher dissolved oxygen capacities, and to improve the local effectiveness of detention basins, operational control is increasingly being introduced into urban drainage planning. For individual storms the effect of operational control may be enormous, as the simulation results in Figure 7 demonstrate. For the catchment of Munich-Harlaching under investigation, the control strategies employed were ineffective for some storms. These examples are taken from a research study funded by the German Research Society (DFG). The layout and operation of such controls require local data. Past control strategies have been developed on a trial and error basis, or were predetermined by mathematical simulations. Because improper operation of the system may cause damage, models and data used must be precise and the effectiveness of operational controls must be proven

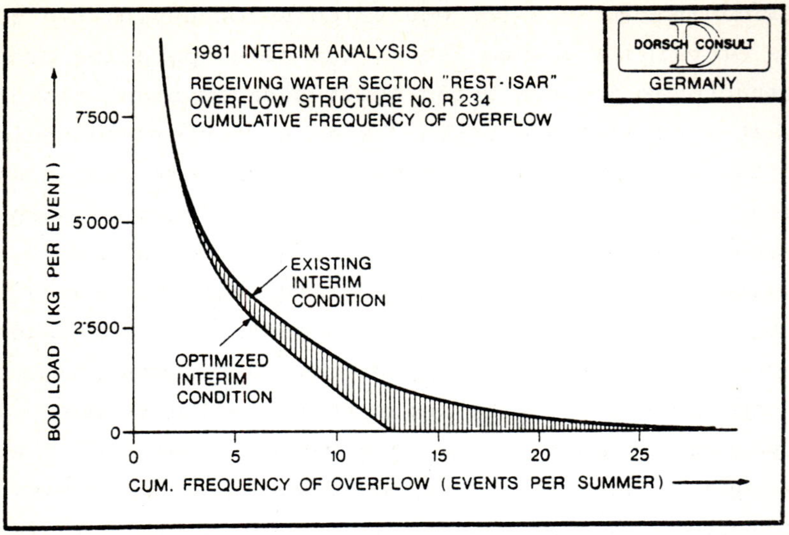

Figure 6: Predicted reduction of overflow optimizing interim conditions, City of Munich (11)

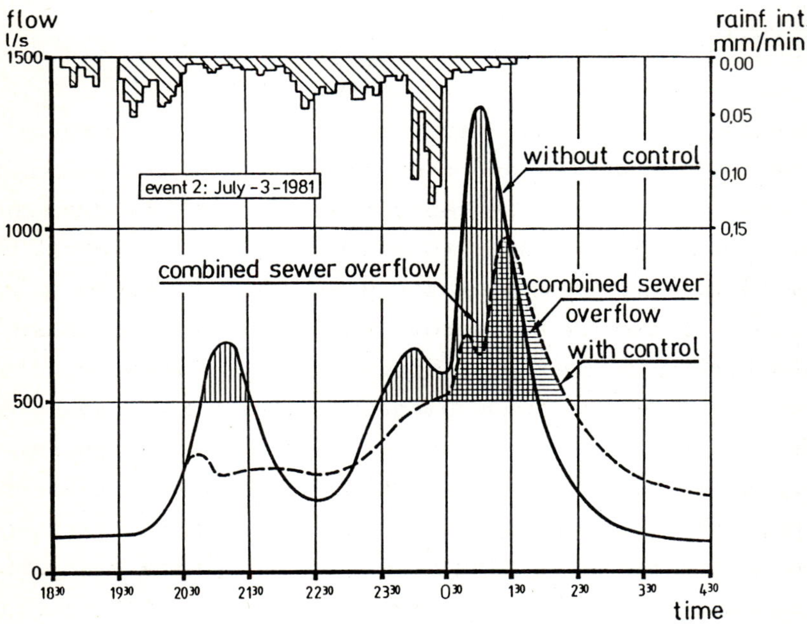

Figure 7: Retardation of flow by controlled utilization of sewer volume

on a longterm basis. Successful long-term operation of controlled sewer systems has been demonstrated in a few cities.

For the city of Hamburg, Federal Republic of Germany, the impact of depositions on sewer operation has been demonstrated (12). It is shown that the amount of depositions occurring under certain local conditions cannot be removed by commonly used cleansing devices. The measured and calculated velocities presented lead to the conclusion that mathematical models may be used not only to specify critical velocities in sewer reaches, but also to find operational control measures in order to flush sections susceptible to deposition of solids. This, however, also requires a continuous survey of actual velocities or depositions in the sewer network.

Conclusions

Data are necessary for all steps in planning, maintenance and operation of urban drainage system. Because special features of urban drainage systems, there will always be a need for local data. This applies particularly to data describing the system layout and the rainfall input. Accuracy and detail of data depends on the study objectives. However, considering the multi-purpose use of field data for planning of urban drainage systems, the data requirements should be standardized and the data be catalogued including the local climatological, land use, system and operational conditions. Unless this data is implemented, the use of data use must be supervised by someone who is familiar with the catchment and data collection conditions. Use of data without this knowledge may lead to misinterpretation and poor designs. A summary of the major characteristics of urban test catchments and monitoring programs in different countries and climatological zones is presently being prepared under the UNESCO-IHP Project A.2.9, and will be included in a manual on urban drainage.

References

1. Geiger, W.F. Continuous quality monitoring of storm runoff. Wat.Sci.Tech.Vol. 13, Munich, pp.117-123. IAWPR/Pergamon Press Ltd. 1981

2. Geiger, W.F. Mischwasserabfluss und dessen Beschaffenheit - ein Beitrag zur Kanalnetzplanung. Berichte aus Wassergutewirtschaft und Gesundheitsingenieurwesen der Technischen Universitat Munchen. H. 50, pp.399. 1984

3. Alley, W.M. Guide for collection analysis, and use of urban stormwater data. A conference report. Published by ASCE, New York. pp.115. 1977

4. Terstriep, M.L., and Stall, J.B. The Illinois Urban Drainage Simulator, ILLUDAS. Illinois State Water Survey, Urbana. pp.90. 1974

5. Huber, W.C., Heaney, J.P., Nise, S.J., Dickinson, R.E., and Polmann, D.J. Storm Water Management Model - Users Manual, version III. U.S. Environmental Protection Agency, Cincinnatti, Ohio. 1981

6. Patry, G., and Mc Pherson, M.B. The design storm concept. Ecole Polytechnique de Montreal. 1979

7. Sifalda, V. Entwicklung eines Berechnungsregens fur die Bemessung von Kanalnetzen. gwf-Wasser/Abwasser, Vol. 114, No. 9, pp.435-440. 1973

8. Kiefer, C.J., and Chu, H.H. Synthetic Storm Pattern for Drainage Design. Journal of the Hydraulics Division, Proc. of the ASCE, Vol. 83, No. HY4, pp.1332, 1-25. 1957

9. Geiger, W.F., and Dorsch, H.R. Quantity-Quality-Simulation (QQS): A Detailed Continuous Planning Model for Urban Runoff Control. Vol. I. Model description, testing, and applications. pp.87. Vol. II. Users Manual. pp.346. U.S. Environmental Protection Agency, Cincinnati, Ohio. 1980

10. Mevius, F., and Wirth, H. Sanierung der Regenentlastungen Munchens. Korrespondenz Abwasser, Vol. 28, No. 2, pp.79-85. 1981

11. Geiger, W.F. Progress in urban hydrology in the Federal Republic of Germany Since 1979. Proceedings of the International symposium on urban hydrology. Published by ASCE, New York. pp.54-78. 1983

12. Broeker, H.W. Impact of depositions on sewer operation. Proceedings of the Third International Conference on Urban Storm Drainage. Edited by Per-Arne Malmqvist, Peter Balmer, and Anders Sjoberg. Goteborg. pp.579-887. 1984

DESIGN OF DATA COLLECTION SYSTEMS

Michael L. Terstriep
Illinois State Water Survey
2204 Griffith Drive
Champaign, IL 61820 U.S.A.

Background

Until the 1960's most field hydrologists practiced their trade in agricultural areas. When it was necessary to measure flow, one could usually find a suitable bridge opening or uniform open channel at which to make a measurement or install a stilling well and stage recorder. Things didn't happen too quickly, and with a little planning one could reach the site after a thunderstorm and make a flow measurement while the flow was still relatively high or maybe still increasing. The nearest recording raingage may have been several miles away and operating on its own clock, but this was no real problem since a few minutes difference between rainfall and runoff records could be attributed to storm travel or aerial distribution of the rainfall.

With a few exceptions, there was not much interaction between these hydrologists and the biologists and chemists who had interest in the water quality of those same streams. Water quality concerns were usually related to sewage treatment plants and other point sources of pollution. Water quality samples were often collected without regard to flow since the concentration of various pollutants was all that was really needed to compare with water quality standards or fish tolerance limits.

When these same scientists and engineers began to work in the urban environment and to deal with non-point sources of pollution, it was truly a whole new world. Conventional equipment and techniques were inadequate for measuring and sampling flow in a pipe system. Even if the flow happened to be in an open channel, time of concentrations were measured in minutes rather than hours. The pollutants of interest often required special han-

NATO ASI Series, Vol. G10
Urban Runoff Pollution.
Edited by H. C. Torno, J. Marsalek, and M. Desbordes

dling if lab results were to be meaningful. Concentrations were no longer enough since one had to know the quantity or load of the pollutant and how it varied with flow. The question of "first flush" and whether it existed or how significant it was required sampling intervals to drop to one minute or less.

To their credit, the engineering and scientific community responded quickly with new tools and techniques designed to cope with the demanding conditions and difficult questions surrounding non-point pollution from urban areas. Today the question is often not how to collect enough data to answer the questions, but how to limit the data collected to fit within a realistic budget.

Introduction

The design of a data collection system is dependent on many factors including the objectives of the particular study, the physiography of the study area, the duration of the data collection phase, the expertise of the staff, and the funding level of the project. With so many variables, it is difficult to address system design in general terms. The data collection system used in an urban runoff study in 1980-1982 in Champaign, Illinois is therefore used as an example. The various aspects of the data collection design including the problems encountered and possible alternatives are discussed.

In March 1982, the Illinois State Water Survey (ISWS) completed a three year study for the Illinois Environmental Protection Agency (IEPA) to determine the effects of municipal street sweeping on urban storm runoff quality (1). The project was part of the Nationwide Urban Runoff Program sponsored by the U.S. Environmental Protection Agency (2).

Four urban drainage basins featuring separate sewers and ranging in size from 6.5 to 22 ha were instrumented for runoff event monitoring and sampling. In addition, a single inlet within one of the basins was instrumented. Data collection included recording of precipitation and storm runoff, and sampling of atmospheric fallout. A telemetry system was developed and installed to link the remote sampling sites to a small computer which automatically controlled the collection and storage of data and issued the instructions for sampling. During 1980 and 1981, samples were collected and analyzed for total concentrations of solids, nutrients, and metals for 90 events. A street dirt sampling program designed to measure the total loads of material on the streets was conducted over the same time span. Determinations were made of the particle size distributions and constituent concentrations in the street loads. The street dirt sampling portion of this study will not be considered in this paper since it is somewhat specialized.

Figure 1 shows the instrumented sites in relation to each other and the city of Champaign, Illinois. Average annual rainfall in this portion of Illinois is about 914 mm and about 25 mm of rainfall would be expected during a 1 hour 5 year frequency event. The topography is very gently sloping with only 13 meters difference in elevation on the steepest of the four basins shown.

Site Criteria and Selection

The site selection process is the key to a successful project. Too often site selection must be made before the project is completely designed or methodology determined. Site selection can not be done independently of instrument selection, since instrument requirements such as power, space, and environmental requirements must be considered during the site selection process. Normally, however, study objectives will dictate site selection or at least have more weight than instrument requirements. Site requirements relative to study objectives should be listed completely and given highest priority when the time comes to compromise on actual site selection.

In the Champaign study, at least two and possibly three basins were to be selected. Each was to have the following criteria: drainage area of 15 to 60 ha - separate sewers, with no cross connections or illegal discharges - one distinct land use - streets of uniform types of surface with curbs and gutters, all in good condition - and sufficient surface grade to prevent deposition of suspended materials in gutters or sewers. Since this particular study was to be operated with basin pairs with one

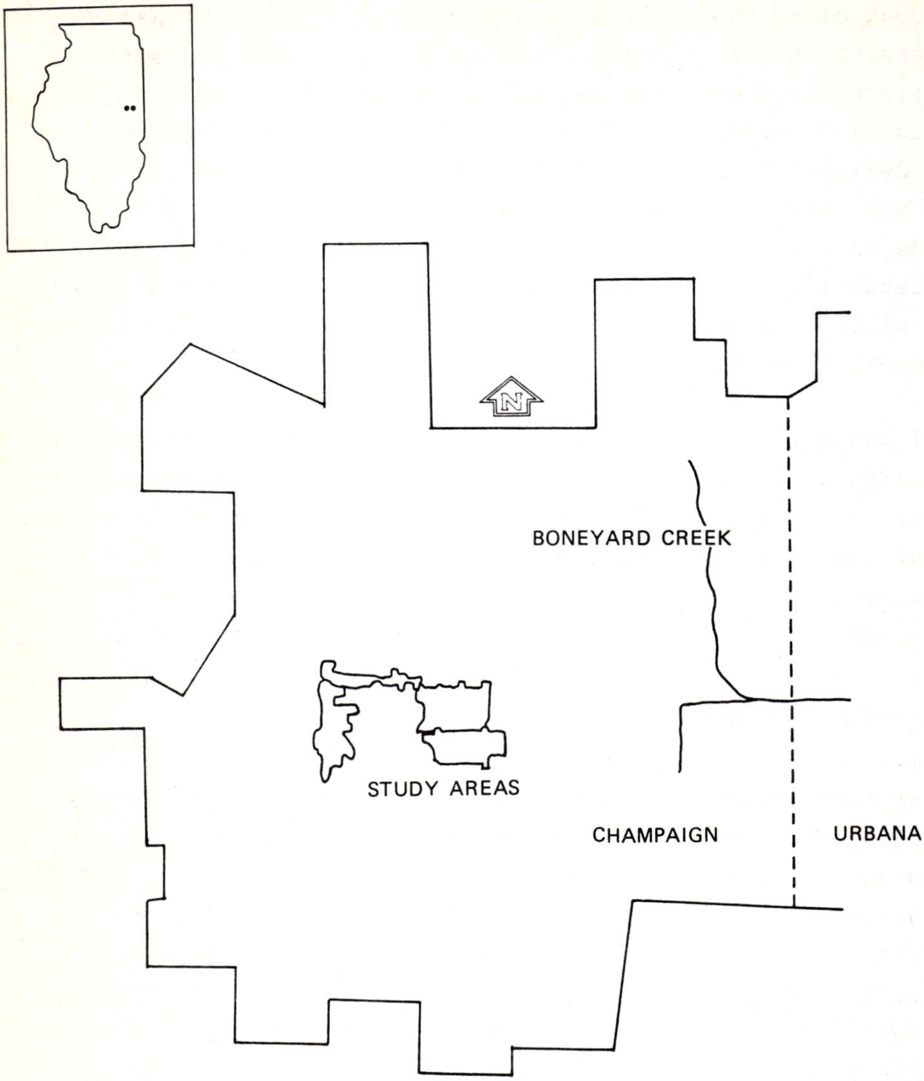

Figure 1. Champaign NURP study area - general location

acting as a control basin and the other as the experimental basin, the pairs had to be as similar as possible. Pairs of selected basins were to be similar in size, topography, drainage density, soil type, vegetative cover, land use, age and degree of development, total impervious area, street type and condition, traffic pattern and volume, and parking. They were to be close geographically and to have no major construction planned over the life of the study. The manholes selected as the sampling sites were to have single pipes in and out, with the same diameter (larger than 15 inch), with no change in flow direction, and with no other sewer or inlet flows entering. They were also to have proper configuration and general condition suitable for installation of flow metering and automatic sampling equipment. Finally, extension of electric power and telephone service to the sites had to be feasible. Clearly, it is nearly impossible to meet all of these requirements. The requirement for matched pairs of basins added immeasurably to the difficulty in finding suitable sites, but it was a key element of the study objectives and therefore had highest priority.

After the criteria are clearly documented and understood, potential sites are located in the office using the best available topographic maps, drainage maps, and aerial photography. In the Champaign study, 40 potential basins were located in this process. The field reconnaissance which follows should emphasize surface characteristics. The main concern of this inspection should be the condition of street surfaces, curbs, and gutters. At the same time, confirmation of map study results such as uniformity of land use and verification of surface grades, drainage divides, and location of inlets and manholes can take place. In Champaign, 30 of the potential sites were eliminated in the field inspection. Two additional potential sites were added during the field inspection leaving 12 sites for final investigation.

The final step was examination of the manholes at the potential sites to determine their suitability for runoff monitoring. This included a check of the sizes, composition, condition, and alignment of the pipes entering and leaving the manhole; and evaluation of any interfering flows from other sewers or inlets; and an assessment of the difficulty of extending electric power and telephone service to the site.

Figure 2 shows the final site selection and illustrates an important point. The instrumentation package should be considered as a system. If the package is to include rainfall or air monitoring or any variable that has some aerial distribution, the instruments of the system should function as a network. By keeping the basins in the same proximity, the instruments that are designed to provide information at a point are also providing network information and redundancy in case of failure at one point.

Data System Selection

One might begin to think of their favorite brand of stage recorder or automatic water sampler at this point, but a larger decision needs to be made first. This relates to the overall type of data recording and instrument control to be utilized in the data system design. Three general types of systems are identified below. These include a distributed system, a data logger, and a centrally-controlled system.

Distributed Systems

At one end of the system design spectrum are the fully independent or stand alone devices such as recording raingages, stage recorders, and automatic samplers. These devices tend to rely on their own on-board clocks and some type of recorder. The recorder may utilize a pen trace on a circular or strip chart, a punched paper tape, or magnetic medium. Newer stage measuring devices are available with built in rating curves so that discharge is recorded rather than stage. This can save considerable time in the office if the flow control device in use supports a standard and stable stage discharge relationship. Stand alone devices tend to be reliable since most have been around for some time. They also tend to be well built and can withstand the rigors of field use.

A distributed system offers maximum flexibility in terms of location of equipment. The devices do not need to be connected to each other or to a central system and most have their own source

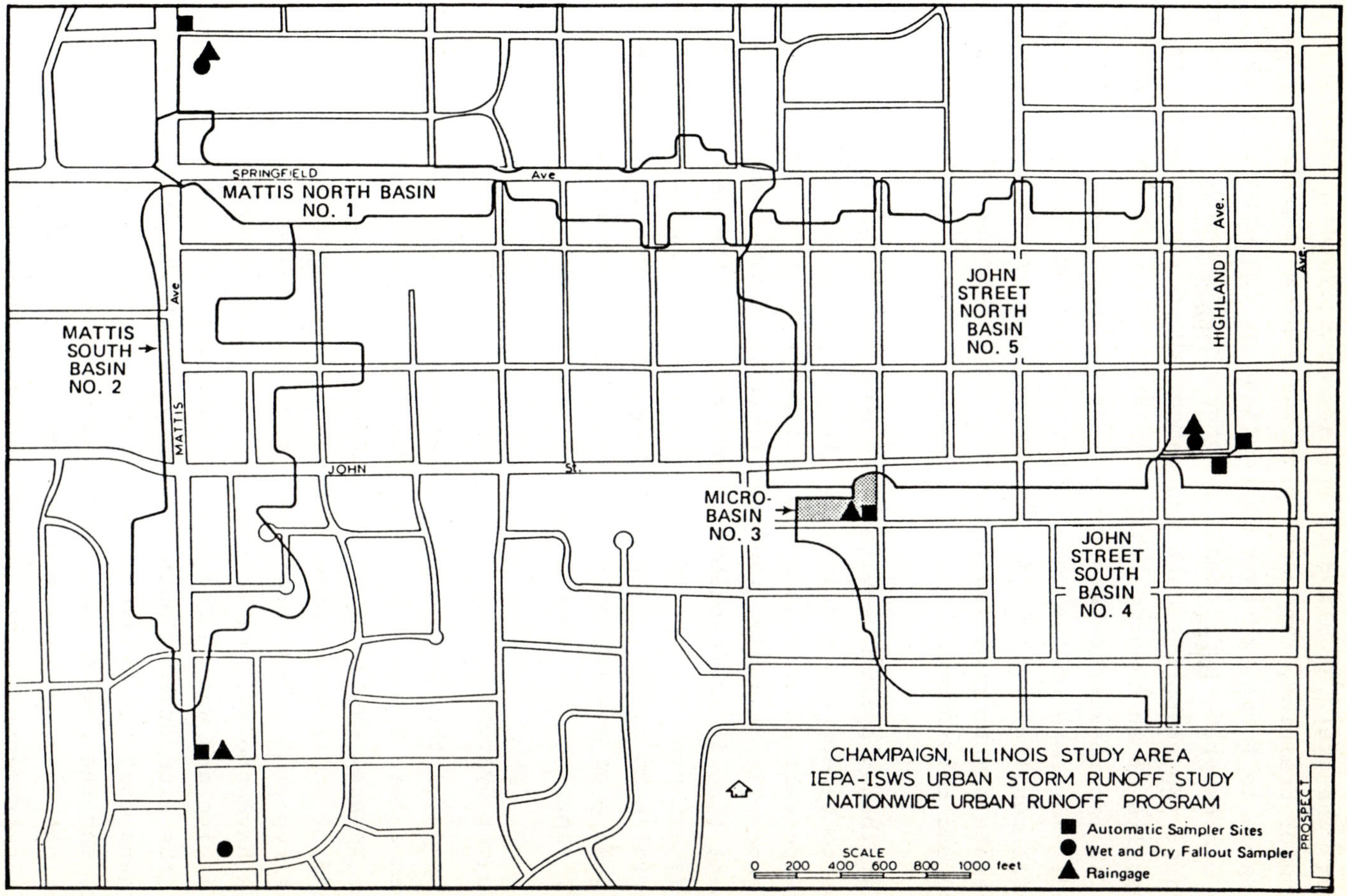

Figure 2. Champaign NURP study area - basin boundaries and instrument locations

of power. The staff required to operate this system needs no electronics training. The data from the system is available in human readable format so that technicians or students in the office can generate hydrographs or other relationships with relative ease. No elaborate data management scheme is required to have results of an event quickly available for analysis or review. The importance of this consideration should not be underestimated. For some of the more automated systems, the generation of useful results may depend on a systems analyst or programmer to generate new or modified programs. Considerable time can be lost during this process.

The distributed system requires more manpower than the other systems to be described below, but the training of the staff is not as specialized as with other systems. The instrumentation, while generally reliable, requires frequent field inspection to make sure that it is functioning and ready for the next event. The greatest problem with this system is its reliance on separate clocks for each device. Urban studies require accurate timing of the sequence of events and the relationship between rainfall, runoff, sampling, or other functions.

Data Logger Systems

A variation of the distributed system utilizes a device generally known as a data logger. Depending on the sophistication of the logger, several channels are available which will record data from end of the line sensors such as tipping bucket raingages or various types of stage recorders. Micro-processor controlled loggers have the ability to send an initiation signal to a sampling device and record the time of sampling. In most cases, the end of the line devices must be hard wired to the data logger. This means that the devices must be in close proximity to the logger, and the logger must be located in the field. It is likely that one data logger would be required for each basin or cluster of instruments in the project. A desirable location for a raingage may have to be altered, for example, if it is to feed into a logger some distance away. The data logger has the advantage of offering a common clock for each device it controls or for which it stores data. Staff training requirements for this system are more stringent than for the distributed system. This is particularly true for the office staff since the magnetic media produced by the logger will require computer interpretation before any useful information is available. Some loggers offer standard data reduction programs for micro-computers, but it is unlikely that these will serve the needs of all studies. Again the requirement for and reliance on an experienced programmer should not be underestimated.

Centrally-Controlled System

Finally, one could envision a centralized system that relies on a micro-computer located at the project headquarters that communicates with the end of the line devices by leased phone lines, radio, or even satellite. In this system, the end of the line devices are hard-wired to a remote interface which contains a modem or other communications device. The interface is similar in function to a data logger except that it transfers information directly to the central computer. The interface also relays messages from the central computer to samplers or similar end of the line devices that are under control of the computer or computer operator. This system offers the most flexibility in programming samplers to operate in relation to real time information being relayed to the computer from the various end of the line sensors. It is also the most reliant on a trained programmer to generate useful results.

A major advantage is that all raingages, depth sensors, samplers, or other end of the line devices are operated from the same clock. Synchronization of data is automatic and precise. Data are recorded directly into magnetic storage, eliminating any chart reading operations. Only this type of system will allow regular status checks of instruments on a 24 hour a day basis. Depth sensors, for example, can record depth values on the system printer every hour so that the operator in the office can tell if they are functional. Other status switches can be used to signal power off conditions or even vandalism. If event simulation is part of the study, the simulation can be compared with the observed event as it occurs. Simulated results can even be used to trigger sampling sequences. Perhaps the major disadvantage of this kind of system is the need for highly trained staff to set up, maintain, and operate the equipment.

The Champaign study employed a system similar to the one just described. Figure 2 introduced earlier shows the location of raingages, and automatic samplers included in the system. There also existed a stage sensing device at each sampler station. A remote interface was installed at each of the four clusters of instruments. The remote interface in this case was a Recon II manufactured by Sangamo Weston. It had the capability of receiving hard wired signals with at least 8 separate addresses of the following types:

1. Status/Alarm: 8 Status/Alarm inputs for relay closures.
2. Analog: 6 points, 0-5 volt, 0-4 milliamp (ma), or 4-20 ma, with 8 bit coding accuracy providing 0.5% accuracy.
3. Control: 4 two-state or 8 unitary controls, contact closure rated at least 200 ma and 30 volts for 200 milliseconds.
4. Pulse Accumulator: accepts one tipping bucket raingage signal and provides accumulation of up to 255 pulses before reset with two registers to prevent overflow.

The system is shown schematically in figure 3. The bubblers indicated on the figure were the stage indicating device being used with the system. The system was designed prior to the common availability of micro-computers and was configured around a Heath H-11 system. Experience with operation of this system for a two year period indicated that the electronics were the most reliable part of the system. Communications caused problems for some periods of time due to erratic conditions on the telephone lines. By far the most frequent problems were related to the mechanical devices at the end of the line. This was particularly true of the bubblers, and to some extent, the automatic samplers.

The designer of a new data collection system should consider a centrally controlled system as an option. The electronics of the systems are very reliable and the flexibility in sampling schemes has the potential for producing the maximum amount of information

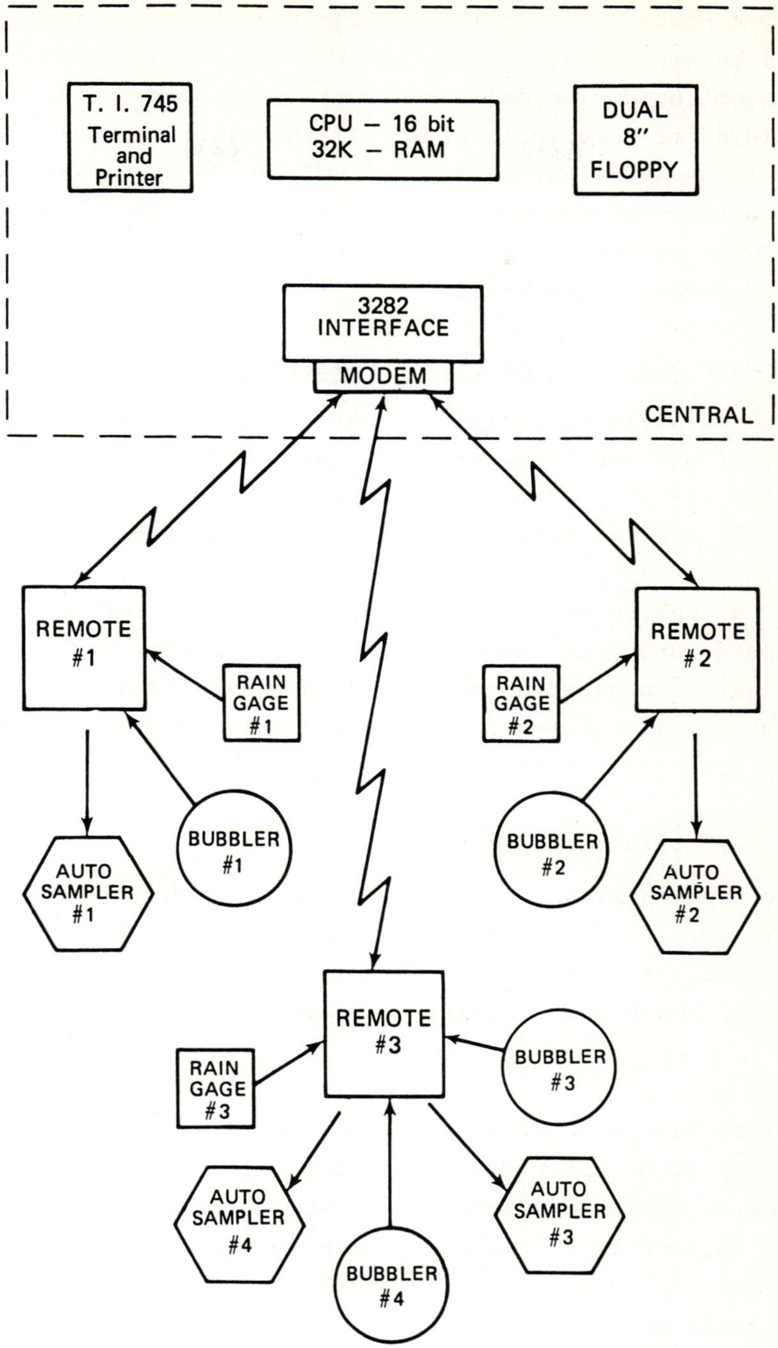

Figure 3. Schematic diagram of telemetry system

for minimum costs. The ability to monitor the condition of instruments between events also increases the probability of successful sampling during desired events. The amount of time spent in the field for routine instrument maintenance may be reduced, but this is an area that should not be slighted regardless of the type of system employed. The potential for faster and more thorough analyses of the data exists, but is dependent on an experienced programmer on the staff.

Selection and Installation of Instrumentation

Proper installation of equipment begins with the site selection process. If equipment requirements have not been considered during this stage, good installations may not be possible. Safety of the staff and security of equipment must be a major consideration in the installation phase. The instruments must also be placed in an environment where they can function reliably. In the Champaign study the remote stations, bubblers, and automatic samplers were installed above ground in insulated shelter houses. This reduced temperatures in summer and extended the sampling season in the winter months.

Discharge Determination Devices

Often the most difficult problem associated with urban runoff studies is reliable monitoring of flow (3). This is a two part problem usually requiring the installation of a device to generate critical depth and a device to measure and record water level.

Many devices have been developed over the years, that are based on a weir or other restriction to flow. Although theoretical rating curves can be developed for these devices, it is always desirable to perform flow measurements in place. The Palmer Bowlus Flume (4) is available commercially in many pipe sizes and has a reliable theoretical rating if approach conditions are good and there is no chance for backwater. A convenient instrument for use in the urban environment is the electromagnetic water current meter. This device generates a magnetic field parallel to its longitudinal axis. As water, a conductor moves through the magnetic field in the plane normal to the longitudinal axis, it produces a voltage in the sensor proportional to its velocity. Information is available from Marsh-McBirney Incorporated, Gaithersburg, MD.

Figure 4 illustrates the installation of an asymmetric flume described by Wenzel (5). This flume restricts the flow on one side of the pipe thus leaving the invert free from obstruction. It also has the capability of functioning during either free surface or pressure flow. In this particular installation, the pipe was significantly out of round thus complicating the theoretical rating. Dye dilution tests were made to confirm the resulting rating, but the pressure flow rating could not be tested and was discounted in the study. Theoretical rating curves should always be verified by field measurements.

Figure 5 shows a simpler installation where the free overfall of a pipe into the manhole was used as the flow control. A bubbler was installed 3 diameters upstream from the overfall where normal depth was assumed to occur. The float switch shown in this manhole was used to indicate submergence of the overfall.

The device used to measure depth or water level in the Champaign study was a Sigmamotor LMS-300 level recorder. It operates on 110 volt AC, has its own compressor and has an accuracy of 1% in an operating range of 0 to 1 meter. The signal generated by this device is a 4-20 ma current proportional to the pressure required to force a bubble of air through an orifice located at the invert of the storm sewer. That pressure is in turn proportional to the depth of water over the orifice. These bubblers are convenient to use in the restricted area of underground drainage, but require nearly daily maintenance to insure reliable performance.

Automatic Sampler

Several samplers are available commercially with the capability of performing discrete or composite sampling. The sampler must be able to withdraw a sample of water from the storm sewer on command from the remote station or some other device and store the sample of water in a refrigerator until it can be picked up

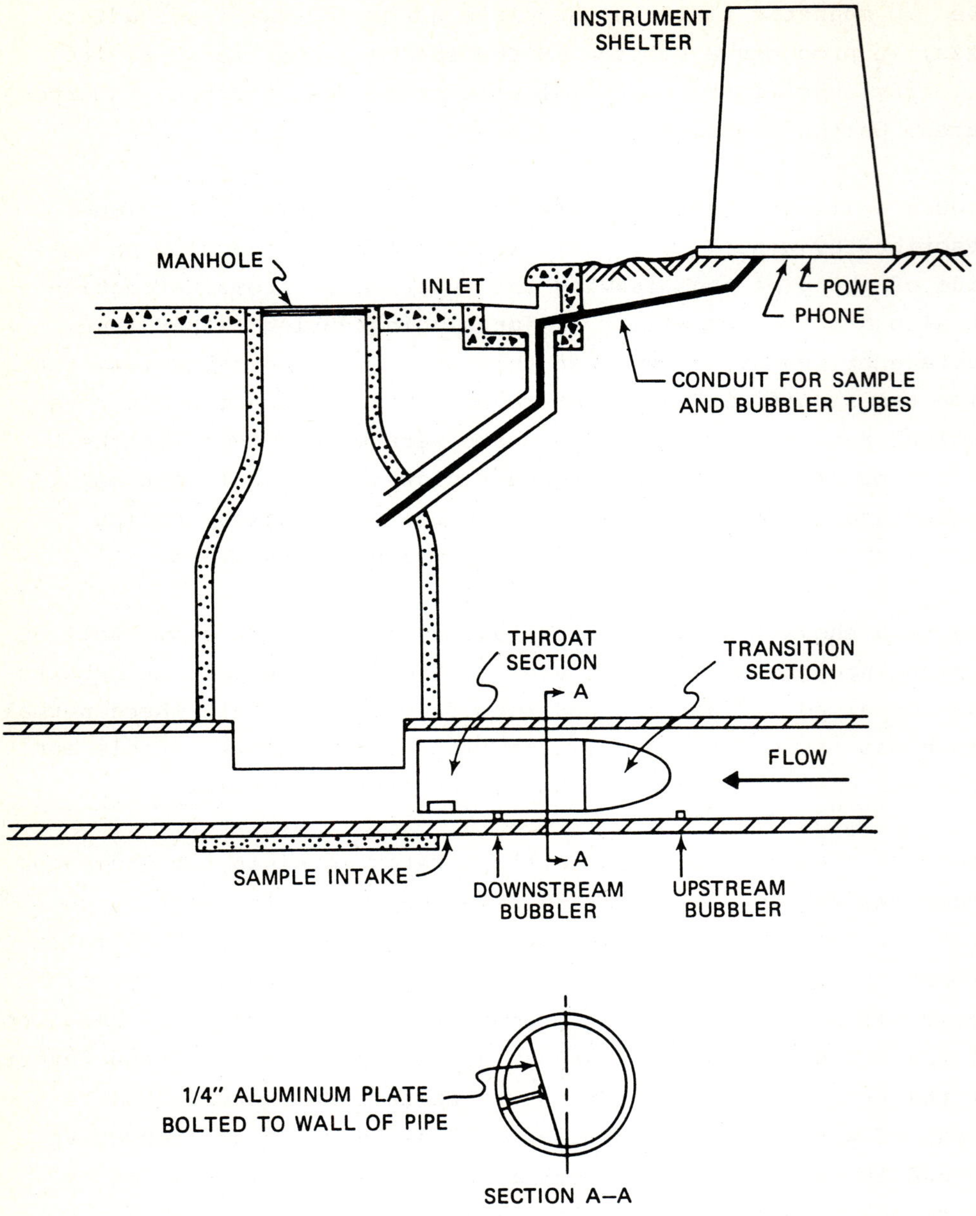

Figure 4. Underground installation with asymmetric flume

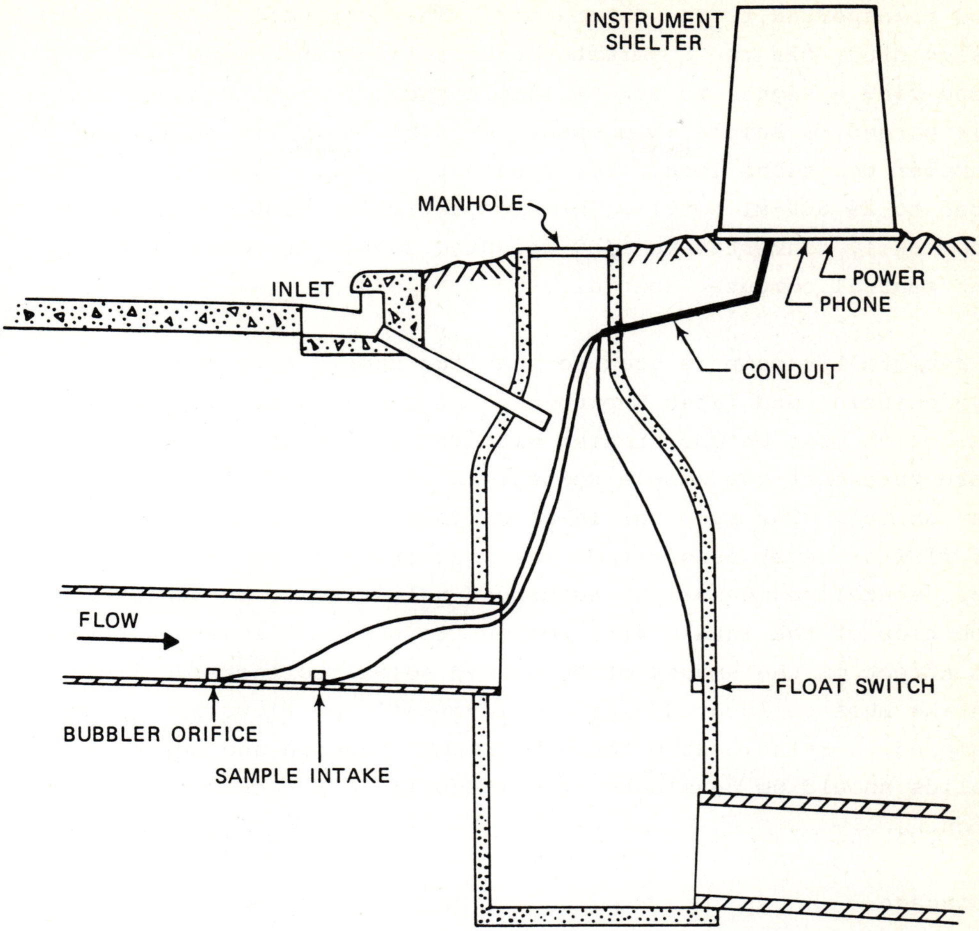

Figure 5. Underground installation with free overfall

and transported to the laboratory. The unit used in the Champaign study was the Sigmamotor 6301 refrigerated sampler. Upon receiving a signal to take a sample the 3/8 inch suction line is air purged, a sample is pumped, the line is purged again, and the sampler positions itself for the next sample. Samples were limited to 24 500-ml bottles in this particular sampler, although it was easily converted to flow-weighted composite sampling using the central computer control.

A peristaltic pump is used so that the sample only contacts the Tygon tubing and latex tubing used in the suction line. Sampler selection must be coordinated with the analytical laboratory to make sure that the sample collection and preservation criteria can be met. The pump should be capable of generating a velocity of flow at least as great as that occurring in the main flow if a representative measure of suspended solids is to be achieved. Location of the intake with respect to the pipe invert also has an affect on the amount of suspended solids collected. Since the intake must be located near the invert if low flows are to be sampled, a relationship between intake location and suspended solids should be developed experimentally for each sampling location.

Raingage

Weather Measure P-105 tipping bucket raingages were used in the Champaign network. The 8 inch diameter collector funnels the precipitation to a dual cup that tips with each 0.01 inch of precipitation. This type of gage is not as accurate for high rates of rainfall as the weighing bucket type, but is much more suited to an automated network. Sampling of flow was initiated by the first tip of a raingage and an increase in stage at the sampling point. The variability of rainfall in areas subject to thunderstorms is so great that even small urban basins should have at least two gages. If more than one basin is to be instrumented in the same area, an effort should be made to distribute the raingages in such a way as to form a network covering the study area. It is often difficult to find adequate exposure for a raingage in a heavily landscaped urban area. The area above the gage should be free from vegetation or other obstructions within a 45 degree angle in all directions. Some of the gages in the Champaign study were mounted on a 2.5 meter pole to increase exposure and reduce vandalism. This is not a desirable practice, but one of the compromises required in the urban area.

Equipment Shelter

The shelters used in Champaign were designed for electrical switching equipment but worked very well for the urban study. The shelters were about 1.3 meters square and 1.5 meters tall. They were light green in color and caused minimum distraction in the residential neighborhoods where they were installed. They were Western Power Products Model 42-2. A typical installation included underground wiring and piping for both aesthetic and security reasons.

Wet-Dry Fallout

These devices were manufactured and loaned to the project by ISWS, but are commercially available from Aerochem Metrics, Bushell, Florida. Two plastic buckets are installed on a frame about one meter above ground. A lid covers one bucket and exposes the other to dry fallout. A sensor on the lid detects rain and the lid moves to cover the dry fallout bucket and expose the other bucket to catch a rainfall sample. After rainfall ceases, the lid again moves and exposes the dry fallout bucket once more. At each fallout sampling location, a gross accumulation bucket was also left exposed to collect a combined wet and dry sample for comparison purposes. High volume air samplers were also installed on the upwind and downwind side of high traffic streets to determine the degree to which street dirt was resuspended.

Air Quality Samplers

The "Hi Vol" high volume atmospheric particulate sampler remains the standard device for determining total suspended particulates. Two versions of the sampler are available. The basic unit collects particles in the range of 1 to 80 microns and the modified unit collects particles in the 1 to 10 micron range. The units function by using a vacuum motor to pull large amounts of air over specified time through a glass fiber filter. Strips of filter are subsequently analyzed for quantity and quality of particulates. Experience with air sampling in the Champaign study are described by Gatz (6). For the collection of other atmospheric parameters the reader is referred to standard texts on the subject (7).

Summary

An attempt has been made to generalize a broad subject using the experience gained from an existing data collection system to illustrate potential problems and give guidelines for similar systems. Site selection was chosen as the beginning point for the design of a data collection system because it is often overlooked and this later leads to data collection problems. The site criteria must begin with items related to project objectives, but must also include the limitations or restrictions imposed by available staff, equipment and funds. The first decision related to hardware selection must be the degree to which the system is to be automated or computerized. It is clear that there are benefits to be gained from highly automated and computerized systems, but staff trained in electronics and programming are required to take advantage of these benefits.

Monitoring results need to be made available to the professional staff as soon as possible so that deficiencies in the system can be corrected early in the project. This means that a data management system, graphics outputs, and data summaries must be developed prior to collection of data if an automated system is to work efficiently. For the same reason, laboratory results must also be made available as quickly as possible - particularly in the early part of the study. If a flexible data collection system has been designed and if results are made available in a convenient format and analyzed quickly, the data collection program can be modified throughout the project to reflect changing needs as some questions are answered and others are raised.

References

1. Terstriep, Michael L., G. Michael Bender, and Douglas C. Noel, Nationwide Urban Runoff Project, Champaign, Illinois: Evaluation of the Effectiveness of Municipal Street Sweeping

in the Control of Urban Storm Runoff Pollution, Illinois State Water Survey, CR300, 1982.

2. Athayde, D.N., P.E. Shelley, E.D. Driscoll, D. Gaboury, and G. Boyd, Results of the Nationwide Urban Runoff Program, Volume 1, USEPA, 1983.

3. Kirkpatrick, G.A. and P.E. Shelley, Sewer Flow Measurement -A State of the Art Assessment, Environmental Protection Technology Series EPA-600/2-75-027, 1975.

4. Grant, D.M., Open Channel Flow Measurement Handbook, Instrument Specialties Company (ISCO), Lincoln, Nebraska, 1979.

5. Wenzel, H.G., Jr., Development of a Meter for Measurement of Sewer Flow, Research Report No. 74 of the Water Resources Center, University of Illinois at Urbana-Champaign, 1973.

6. Gatz, Don, Susan Wiley, and Lih-Ching Chu, Characteristics of Urban and Rural Inhalable Particulates, Illinois State Water Survey, CR 308, 1983.

7. World Meterological Instrument and Observing Practices, 4th Edition, World Meteorological Organization, WMO-No. 8.PP.3, Geneva Switzerland, 1973.

DATA EVALUATION IN FIELD STUDIES OF URBAN RUNOFF QUALITY: ASPECTS OF ASSESSING THE MEASUREMENT INTERVAL

ir. Frans H.M. van de Ven
IJsselmeerpolders Development Authority
P.O. Box 600
8200 AP LELYSTAD
Netherlands

Introduction

The objectives of systems for data collection on urban runoff quality can be basically divided in two groups. The first deals with solving a problem and the second wiih controlling a process. Both groups are shown in Figure 1.

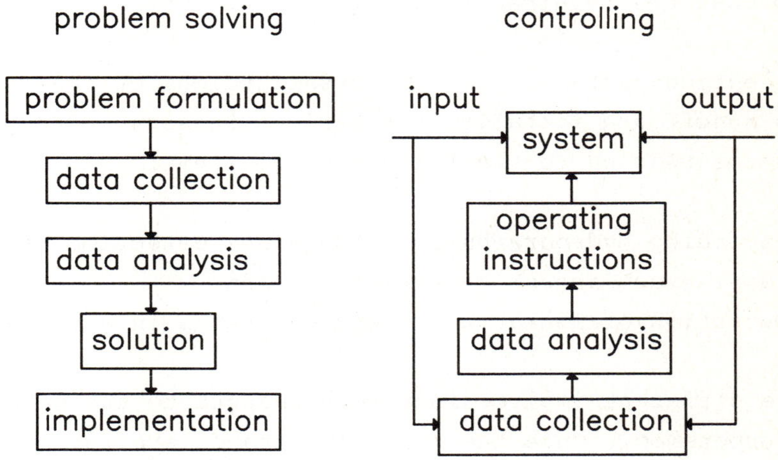

Figure 1. Data collection in a problem-solving framework.

NATO ASI Series, Vol. G10
Urban Runoff Pollution.
Edited by H.C. Torno, J. Marsalek, and M. Desbordes

The solving of a problem as an objective is among others found in research projects and in the work of consultants. Objectives dealing with control are not only limited to industrial plants, but they are also encountered in environmental operations, e. g. in regulation of pollutant discharges designed to avoid exceedance of threshold concentrations in the receiving water body.

The required accuracy for data collection depends highly on the study objectives and possibilities with respect to the accuracy of the results. The formulation and quantification of the required accuracy is a very difficult but essential part of setting up data collection programmes. Limited finances, manpower, equipment and managerial problems, as well as the characteristics of the variables to be investigated, may make the study goals impossible to reach. Yet, a clear formulation of accuracy requirements forms the basis for a data collection system that produces the required results.

Data collection systems used for investigations of the quality of urban runoff are installed for various reasons. Generally, these can be reduced to the following two items:

- Recording hydrographs/pollutographs: detection of level exceedances; determining loads.
- Detection of linear and stepwise trends in a variable.

Both have different implications with respect to the data collection programme. Once the accuracy demands are formulated, a minimum measurement interval must be applied to obtain the required results. The variability of, and the correlation between, the measurement results are important controlling factors in assessing that interval.

In urban runoff quality studies the measurement problems are generally limited to point measurements at an outfall. Occasionally the quality is sampled at different sites along a watercourse, but regional measurement networks are seldom used, except for measurements of precipitation quantity and quality.

For this reason, this paper is limited to the data collection strategy for one-site measurements; the problem of network density is beyond the scope of this paper.

The main feature of a one-site data collection system is the measurement interval T_s between two consecutive observations or its reciprocal, the measurement frequency f_s. Operational networks frequently distinguish between sampling frequency and measurement frequency. Principally because of the associated costs, the measurement frequency of numerous variables may be smaller than the frequency of field site visits and collection of samples for laboratory chemical analyses (3).

In this paper the relationship between the measurement interval T_s, and the errors introduced by non-continuous measurement, is elucidated and the consequences regarding the objectives of the data collection programme are given.

First, the accuracy of the determination of hydrographs/pollutographs is discussed and, secondly, the problem of trend detection is presented. The first part therefore deals with such topics as sampling distortion and interpolation error, while The second one deals with the detectability of certain changes.

Sampling Distortion

As we know from the Fourier transformation, almost every continuous signal of a variable can be decomposed into an infinite series of sines and cosines. Every element of that series is defined by its amplitude, frequency and phase shift. The changes in a variable are however not registered continuously but they are recorded at some discrete time interval T_s. This has considerable implications for the frequencies that are still detectable.

According to the Shannon theorem, a certain frequency f can be observed only when $f_s > 2f$. It is observed correctly only if no frequencies larger than $1/2\ f_s$ are present in the continuous signal. Otherwise the spectrum is distorted, because the real spectrum F(f) is repeated every $n.f_s$ with n = (.... -2, -1, 0, 1, 2, ...) as indicated in Figure 2. The resulting frequency spectrum is the sum of these spectra.

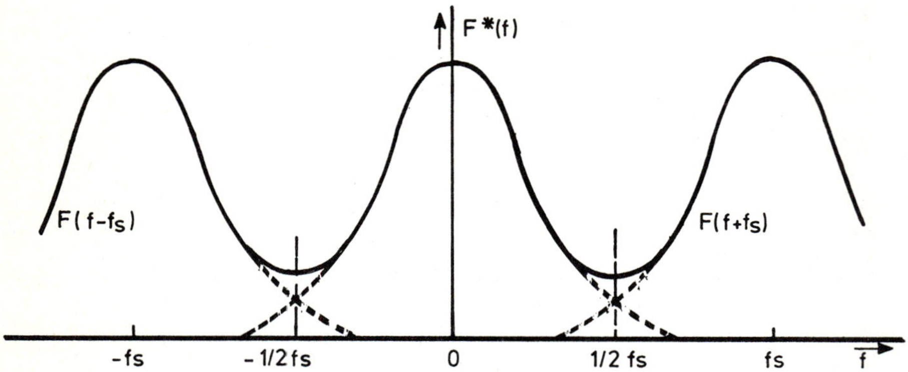

Figure 2. The spectrum of a signal F(f) measured with a frequency f_s.

This leads to the conclusion that the frequency of measurement must be chosen in such a way that not too much information has frequencies higher than 1/2 f_s, in order properly detect the continuous behaviour of a variable. The sampling acts as a high frequency (low pass) filter.

Interpolation

The measurement of the continuous behaviour of a variable f(t) results in a series of observed values with equidistant temporal sample spacing. In calculations, these values are normally interpolated. To obtain an appreciation of combined errors that are caused by the measurement and interpolation, the behaviour of the error function e(t), shown in Figure 3, is investigated. f't is the series of observed values and V(t) is the interpolation filter.

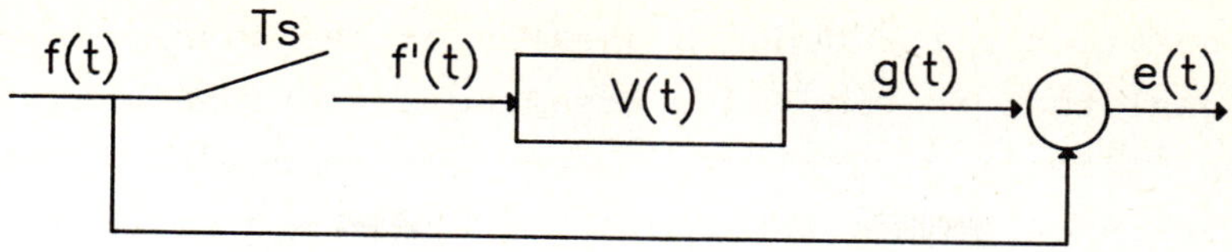

Figure 3. Sampling and interpolation system

The functioning of three commonly used interpolation filters is indicated in Figure 4. The filters b and c are called predictive, because at time $t = -1/2\ T_S$ and $t = -\ T_S$ respectively, the next value is already known. These filters can be applied only in existing time series, not in continuous processes for control purposes.

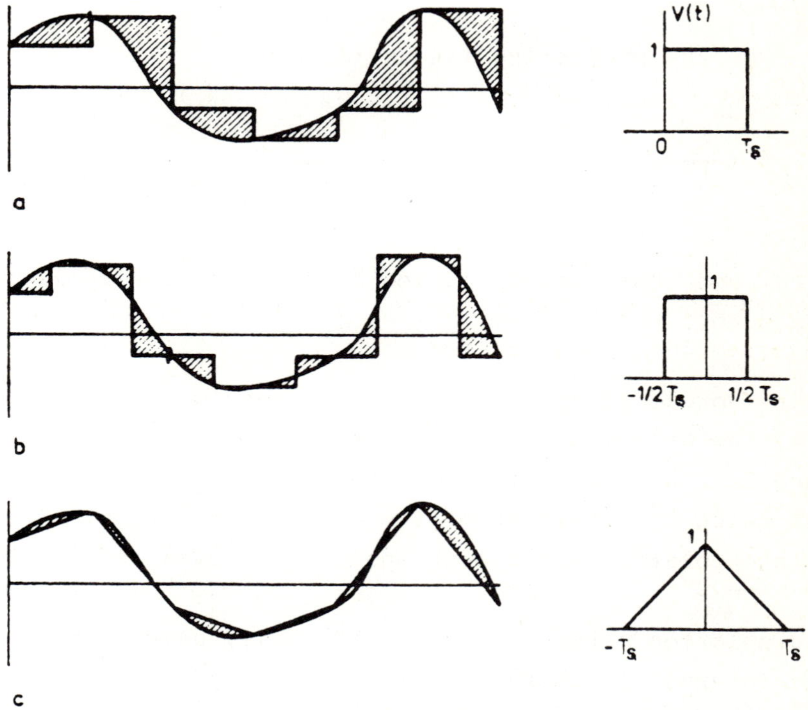

Figure 4. Three commonly used interpolators V(t) and their effect in reconstructing the continuous behaviour of a variable.

a: Non-predictive zero-order hold filter;

b: Predictive zero-order hold filter;

c: Predictive linear filter

The interpolation error depends greatly on the characteristics of the signal that is sampled. Assuming a stationary system with known variance of f(t), σ_f^2, and autocorrelation function $\rho(t)$, the variance of the interpolation error σ_e^2 is (4):

a. For the non-predictive zero-order hold filter a:

$$\sigma_e^2 = 2\,\sigma_f^2 - \frac{2\,\sigma_f^2}{T_s} \int_0^{T_s} \rho(t)dt$$

b. For the predictive zero-order hold filter b:

$$\sigma_e^2 = 2\,\sigma_f^2 - \frac{2\,\sigma^2}{T_s} \int_{-1/2\,T_s}^{1/2\,T_s} \rho(t)dt$$

C. For the predictive linear filter c:

$$\sigma_e^2 = \frac{5}{3}\,\sigma_f^2 + \frac{1}{3}\,\sigma_f^2\,\rho(T_s) - \frac{4\,\sigma_f^2}{T_s} \int_0^{T_s} [1-\frac{t}{T_s}]\,\rho(t)dt$$

It can be seen that the expectated mean of the interpolation error is zero.

For a first-order (linear reservoir) process with a time constant T_1, the behaviour of the interpolation error is indicated in Figure 5. The upper limit of the interpolation error variance is twice the variance of the original function. The linear interpolator is the best one in this case, as expected.

The interpolation error can be considerable, unless the measurement interval is chosen small enough. In practice T_s is often chosen as 0.1 T_1 or even smaller.

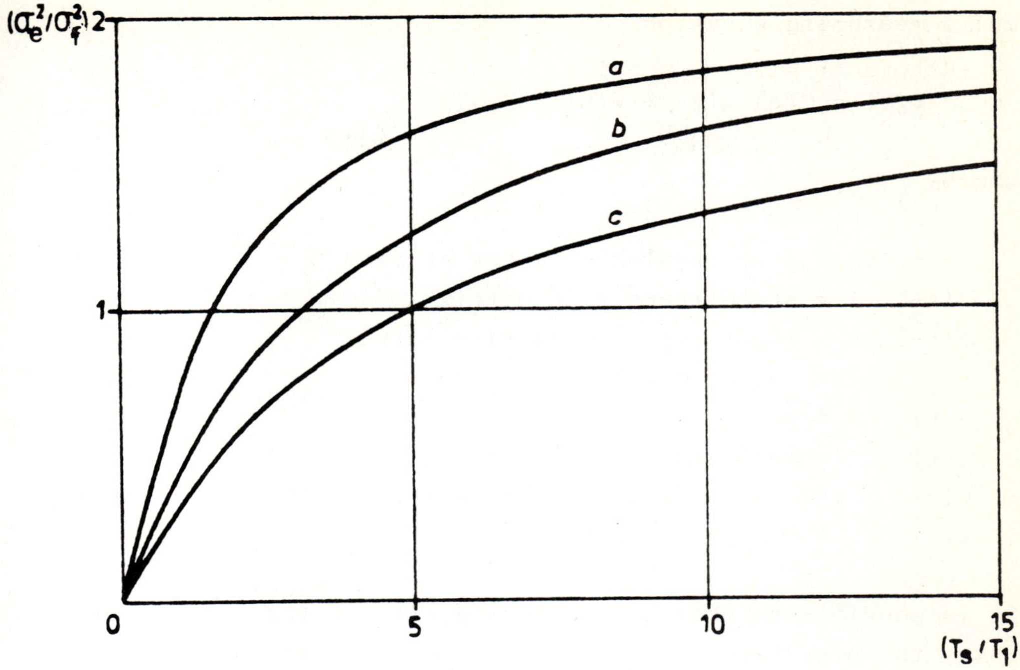

Figure 5. The interpolation error versus the measurement interval for a first-order process.

Kalman Filter

Until now, we have discussed the interpolation problem as if we knew nothing about the system that governs the behaviour of a variable. The interpolation can certainly be improved if we are able to add some a-priori knowledge to the interpolation process. Moreover, we know that the measurements, without exception are subject to measurement error. Depending on the measurement method, this error can be considerable. To include this error in the interpolation would be another improvement in dealing with the interpolation problem. The Kalman-Bucy filter includes both features. After replacing differentials by differences, the filter can be described by the equation:

$$x(t) = A(t-1,t)x(t-1) + B(t-1,t)u(t) + v(t)$$

and a measuring equation:

$$y(t) = C(t)\ x(t) + w(t)$$

where:

$x(t)$ = state variable vector at time t
$A(t-1,t)$ = state transition matrix
$B(t-1,t)$ = input state transition matrix
$u(t)$ = input vector
$v(t)$ = model error
$y(t)$ = measurement
$C(t)$ = state-measurement transition matrix
$w(t)$ = measuring error

It is possible to obtain an estimate $\hat{x}(t)$ of the state $x(t)$ using the measurements $y(to), \ldots\ldots, y(t)$. This estimate is given by the following set of equations (5):

$$x^*(t) = A(t-1,t)\hat{x}(t-1) + B(t-1,t)u(t)$$
$$P^*(t) = A(t-1,t)\ \tilde{P}(t-1)\ A^T(t-1,t) + Q_1(t-1)$$
$$K(t) = P^*(t)\ C^T(t)\ [C(t)P^*(t)\ C^T(t) + R_1(t)]^{-1}$$
$$\tilde{P}(t) = [1 - K(t)\ C(t)]\ P^*(t)$$
$$\hat{x}(t) = x^*(t) + K(t)[y(t) - C(t)\ x^*(t)]$$

where:

$x^*(t)$ = predicted state estimate
$\hat{x}(t)$ = corrected (optimal) state estimate
$P^*, \tilde{P}$ = covariance matrices of state estimates
A^T = transponed matrix A
$Q_1(t)$ = covariance matrix of model errors
$K(t)$ = Kalman gain
$R_1(t)$ = covariance matrix of measurement errors

The behaviour of every variable can be studied by going through a time series and by solving this set of equations for each timestep.

The Kalman filter functions as a predictor-corrector method with respect to the state variables in the vector x(t). This vector can be composed of a number of constituent concentrations. The predicted value $x^*(t)$ is corrected with the help of the measurement y(t) and gives a new estimate $\hat{x}(t)$. The best extimate of the actual value at time t is, therefore, given by C $\hat{x}(t)$ and differs from y(t) because of the measurement inaccuracy.

The a-priori knowledge about the process is modelled by the relations in the model equations, plus a description of the input u(t). The behaviour of the variable u(t) must be well known.

The main feature of the Kalman filter, however, is that a covariance matrix of the state estimates is updated every time step. The diagonal elements p_{ii} of this matrix $\tilde{P}$ give the variance of the estimation (prediction) errors comparable to σ_e^2 for the interpolation error. Now suppose that a process continues a number of timesteps without a new measurement being made. Because a new observation is missing, the calculation scheme is continued, assuming $\hat{x}(t) = x^*(t)$. During these time steps the variances in the matrix P* increase. As soon as an observation y(t) becomes available, the p_{ii}'s decrease instantaneously (see $\tilde{P}$).

In order to get sufficiently good measurements, a maximum can be assigned to the value of p_{ii}. Before p_{ii} exceeds this $p_{ii,max}$, a new measurement must be made. The time between the old and new measurement is T_s. The measurement frequency must be chosen sufficiently high to avoid the exceedance of $p_{ii,max}$. On the other hand, the p_{ii}'s show a minimum. Less variance cannot be achieved with the prevalent model. The sources of variance are the measurement errors and the model errors. Increasing the measurement frequency does not influence the variance of the measurement errors, because the measurement technique and accuracy remain the same. The variance due to the insufficiency of the model is reflected in the covariance matrix Q_1; the better the model, the smaller the variance in Q_1.

Let us consider a one-dimensional example. Given a first-order process:

$$T_1 \frac{dx}{dt} + x = f(t)$$

Replacing differentials with differences and reorganizing gives:

$$x(t) = (1 - \frac{\Delta t}{T_1})\ x(t - \Delta t) + \frac{\Delta t}{T_1} f(t)$$

where Δt is the time step of the finite difference. This equation is the first equation of the Kalman filter when $x(t) = x^*(t)$ = predicted state variable, and $x(t-\Delta t) = \hat{x}(t-\Delta t) = x^*(t-\Delta t)$ in the case that no new measurement $y(t-\Delta t)$ is available. The second equation of the filter, the covariance of the predicted state estimate, is given by:

$$P^*(t) = [1 + (\frac{\Delta t}{T_1})^2 - 2 \frac{\Delta t}{T_1}]\ P^*(t-\Delta t) + Q_1$$

If a new measurement then becomes available, $\tilde{P}(t-\Delta t) = P^*(t-\Delta t)$ because $K(t)=0$. As stated earlier, Δt should be less than or equal to $0.1\ T_1$ in order to get a reasonable description of the discretized process. Assuming $\Delta t = 0.1\ T_1$, the expression for $P^*(t)$ becomes:

$$P^*(t) = 0.81\ P^*(t-\Delta t) + Q_1$$

Assuming that $P^*(0) = 1 = Q_1$, the behaviour of $P^*(t)$ can be calculated when no new measurement y becomes available. This behaviour is shown in Figure 6.

Now let us assume that $P^*(t)$ drops to the level $P^*(0)$ when a measurement becomes available for the corrector (which is not necessarily true). In this case, the measurement interval can be determined by setting a limit to $P^*(t)$. For example, if we require that $P^*(t) < 4$, then $T_s \leq 6\,\Delta t$.

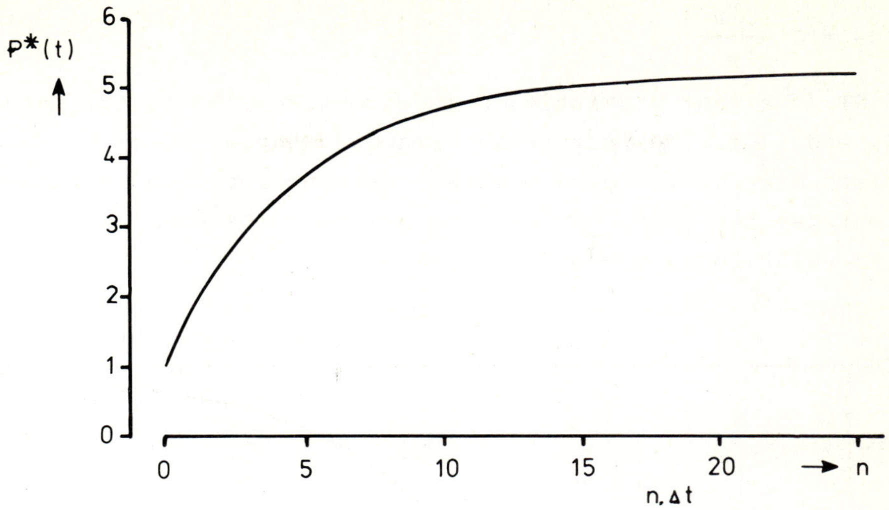

Figure 6. The development of the variance of the prediction x*(t) with increasing t (=n Δt), and without a new measurement being made.

Trend Detection

Another objective for routine water quality measurements can be trend detection. Many surface water samples are taken more or less regularly in order to investigate ihe presence of a gradual change in a water qualiiy variable. Lettenmaier (1) discussed the detectability of such a trend in relation to the characteristics of the variable and to the frequency of measurement. Among others, this method is used to optimize the routine sampling programmes for Dutch surface waters (2).

To investigate the presence of a trend, the data series is subjected to a statistical test. The power, P, of such a test is defined as the chance that a trend that is hidden in the data is actually present. Now let:

$$N_T = C \frac{T_r}{\sigma}$$

where NT is a test statistic, C is a constant, T_r is the total trend, and σ is the (corrected) standard deviation of the estimated mean of the data series. Lettenmaier found a relation between the power of the test and N_T. The basic shape of this relation is given in Figure 7.

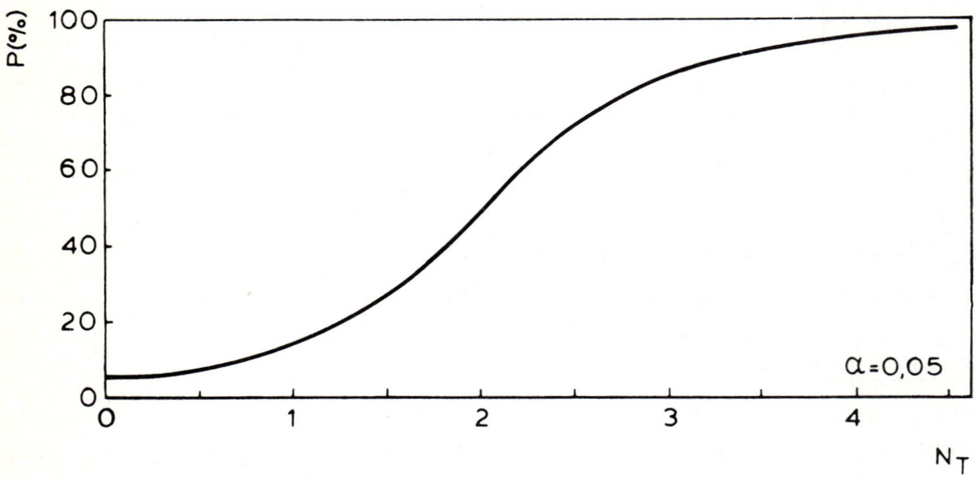

Figure 7. The relation between the power of the test for trend detection and the test statistic N_T.

The test used for a step trend is the Mann-Witney test. For linear trends the Spearman test is used. Neither test requires assumptions on the probability distribution of the data.

For independent data the standard deviation of the mean is:

$$\sigma = \frac{\sigma_f}{\sqrt{N}}$$

Where σ_f is the standard deviation of the variable f(t) and N is the number of observations. If the temporal length of the data series is $T = N.T_s$, N_T can be written as:

$$NT = \frac{C.T_r.\sqrt{T}}{\sigma_f \sqrt{T_s}} = \frac{C.T_r.\sqrt{N}}{\sigma_f}$$

The relation between the power of the test and value of T_s is now fixed for given values of T, Tr and σ_f.

The data, however, are only seldom independent. In most cases they show some autocorrelation, that is indicated by the autocorrelation function $\rho(t)$. In that case the real number of measurements N must be replaced by the effective number N^1:

$$N^1 = \frac{N}{1+2 \sum_{i=1}^{N-1} (1 - \frac{i}{N}).\ \rho(iT_s)}$$

To avoid extensive calculations, Schilperoort et. al. (2) assumed a Markov model for the interdependency, so that $\rho(iT_s) = [\ \rho(Ts)]^i. = \rho_1{}^i$.

In Figure 8, the resulting relations between T_s and the power P are indicated for various values of ρ_1 and T_r, while the other variables were kept constant.

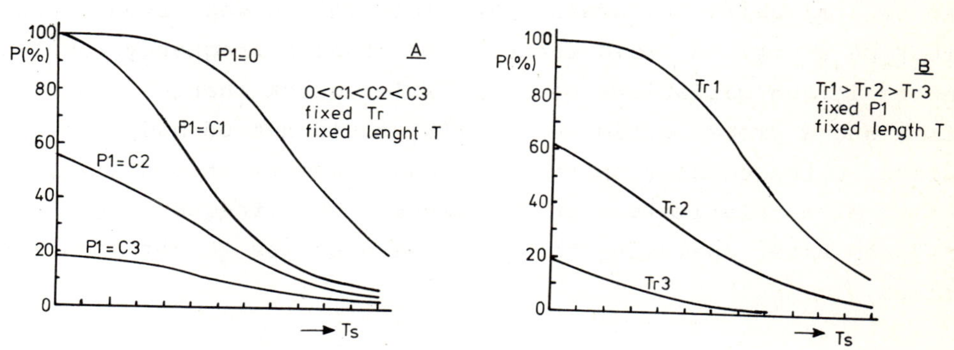

Figure 8. The relation of the measurement interval Ts and the power of the test for trend detection;

a. various autocorrelations b. various trend magnitudes

As can be seen in Figure 8, the correlation strongly influences the detectability of a certain trend. In some cases, the power (the chance to detect a present trend) is less than 50% even if continuous measurements are made. The same is true if the trend is weak. The relationship between the measurement interval and the detection of trends is clearly stated in Figure 8. We can conclude that the Lettermaier theory provides us a good tool for the choice of the frequency of measurement.

Discussion

Depending on the objectives of the measurements, the relation between the frequency of measurement and the error, standard deviation or uncertainty can be calculated in order to establish the measurement frequency. Good estimates, however, of $\rho(t)$, σ_f, R and Q are indispensable. For these estimates, though, we already need data, so that we enter a vicious circle when we start measurements. The less we know about the behaviour of the variables and the processes, the more frequently measurements must be made in order to obtain good estimates and to collect information on the process that can be used as a-priori knowledge for the interpolation.

To get estimates of $\rho(t)$, σ_f, R and Q, many measurements have to be available. The measurement interval between these data must be short, to avoid the cut off of high frequency components in the variations of a variable; quick changes would get lost and a proper estimate of $\rho(t)$ could not be made. This leads to the conclusion that the measurements should always start at a relatively high frequency. This frequency can be reduced after analyzing the first data series to get good estimates.

For the first "educated" guess of the sampling frequency, assumptions are normally made for the autocorrelation and the variance. These assumptions are taken on the safe side.

When the objectives of the project change, the frequency of measurement must be re-assessed using the new accuracy requirements. This flexibility in the data collection strategy with respect to the measurement interval is an absolute necessity.

Summary

Data are collected in field studies of urban runoff quality to find hydrographs/pollutographs or trend changes. Assessing the frequency of measurements is one of the problems to be addressed when such a study is initiated. Data have to be evaluated in order to deduce the frequency of measurements from the allowed variability and/or uncertainty in the objectives. This paper gives a number of calculational approaches and considerations. In the behaviour of a variable, components with frequencies higher than half the frequency of measurement cannot be detected.

Depending on the type of interpolation between observations, the interpolation error can be calculated from the statistical features of the behaviour of a variable. If the system that is controlling that behaviour is known, and if an estimation of the system and measurement errors can be made, a Kalman filter can be best applied for the interpolation. The variance of the estimated value of a variable in the Kalman filter can be used to assess the frequency of measurements.

For trend detection, a statistical test is applied to the data series. The power of such a test depends on the trend magnitude, the variability in the behaviour of the variable and the effective number of data available for the test. The frequency of measurements is related to this effective number and therefore to the power of the test.

For applying these theories, estimates of the statistical features of the variable have to be available. Starting the observations with a relatively high frequency of measurements gives the best estimates for these features, and allows for a reduction of the frequency later in the data collection programme without harming the study objectives. A change in the objectives of the study requires a renewed assessment of the frequency of measurements and generally leads to a change in that frequency. The data collection system must allow for such a change.

References

1. Lettenmaier, D.P. Detection of trends in water quality data from records with dependent observations. Water Resour. Res. 12 (5): 1037-1046; 1976

2. Schilperoort, T., S. Groot, B.G.N. Van de Wetering, F. Dijkman. Optimizing the sampling frequency in routine water quality programmes (in Dutch). H_2O.15 (5); 1982

3. Steele, T.D. Strategies for water quality monitoring. In: J.C. Rodda (ed). Facets of hydrology Volume II. John Wiley & Sons Ltd. 1985

4. Van der Grinten, P.M.E.M., J.M.H. Lenoir. Statistical process control (in Dutch). Prisma Technica 50, Het Spectrum. 1973.

5. Van der Kloet, P., H. de Niet, A.G. Posthumus. Determination of a measuring network along rivers. In: J.A. Cole (ed). Advances in Hydrometry (proceedings of the Exeter Symposium). IAHS Publ. no. 134. 1982

6. Van de Ven, F.H.M. Aspects of the sampling frequency and network density of the network for groundwater level observations (in Dutch). Inst. for Geotechnics, Delft. 1978

7. Van Geer, F.C. An equation-based theoretical approach to network design for groundwater levels using Kalman filters. In: Improvements of methods of long term prediction of variations in groundwater resources and regimes due to human aciivity (proceedings of the Exeter Symposium). IAHS Publ. no. 136. 1982

DISCUSSION - FIELD STUDIES OF URBAN RUNOFF QUALITY

In the discussion which followed the presentation of the papers, the following points were raised:

1. There was a general conclusion that we are still unable to accurately predict pollutant concentrations on a site-specific basis. It is also not possible to use "national" average values.

2. Dr. Geiger stated that, in his opinion, we do not have enough data available to support detailed runoff quality modeling. This view was not shared by all participants.

3. Dr. James pointed out that the cost of data collection and data handling equipment is coming down rapidly. He was, at the same time, concerned that we do not have enough information about the physical situation in the catchment, such as ground cover, percent imperviousness, types of structures, drainage configuration, and particularly if and how these factors change between data sets. He felt that our attention should now focus on better spatial and temporal resolution, and the relationship between urban runoff pollutant loads and receiving water quality impacts.

4. Mr. Ellis pointed out that in reality it is the costs which control the frequency of our sampling. Decisions are made on an economic rather than scientific basis.

5. Dr. Roesner emphasized that we still cannot "count the dead bodies" - we are still unable to show conclusively that urban runoff pollution translates into measurable environmental impacts. Furthermore, until we are able to do this, it will be difficult to argue that more research funds are needed. Dr. Schilling noted that we have counted fish kills, but that it is very difficult to sort out and rank the causes of such incidents. He also notes that in Germany (as indeed in the United States) after virtually all wastewater is treated in secondary treatment facilities, there have still been only slight improvements in receiving water quality. The reason for this is urban runoff pollution and agricultural washoff.

6. Dr. Huber pointed out that the time constants in typical receiving waters are much longer than the time constants in the runoff processes, and that we need to reflect this in our sampling and modeling programs.

7. Mr. van de Ven argued that his trend analysis technique was a very useful tool in establishing large-scale (regional) sampling networks. Mr. Falk, on the other hand, felt that such a requirement was simply not realistic. Dr. Geiger stated that if one accepted Mr. van de Ven's thesis the sampling interval should be about 1/10 of the time constant of the linear reservoir. This would result in about a 5-minute sampling interval for the Munich-Harlaching catchment. However, observations showed that a 1-minute sampling interval was necessary early in the storms,

decreasing to perhaps 20-30 minutes later. It would appear that this is an area warranting a good deal of further study.

8. Mr. Jacquet pointed out that Mr. van de Ven's method does not take into account systematic measurement error.

9. It was further agreed that political constraints when implementing engineering designs often are of greater impact than modeling accuracy.

DETERMINISTIC MODELING OF URBAN RUNOFF QUALITY

Wayne C. Huber
Dept. of Environmental Engineering Sciences
University of Florida
Gainesville, Florida 32611

Introduction

Modeling Objectives

Urban runoff quantity and quality problems are both historic and current. Cities haved long assumed the responsibility of control of stormwater flooding and treatment of point sources of wastewater (e.g., municipal sewage). But within the past two decades new responsibilities have arisen with the recognition of the severe pollution potential of urban nonpoint runoff, especially from combined sewer overflows (CSOs) and to a lesser degree, from stormwater runoff (51,53). With recognition of the quality problems has come the need for adequate analytical tools for their analysis. This need coupled with the advent of modern computer technology has led to many complex techniques for analysis of urban runoff quality.

This paper is primarily a review of the literature; it is not possible in limited space to provide full technical details about all methods that will be described. The focus is on deterministic modeling of quality processes. The distinction between other techniques such as regression or statistical is not always clear, however, and there will necessarily be some overlap in the discussion that follows. Furthermore, most quality processes are driven by quantity processes. That is, it is usually necessary to predict hydrographs before predicting "pollutographs" (a plot of concentration versus time). Hence, there will be brief discussions of rainfall-runoff-routing considerations in the paper as well. But the focus will be on physically-based simulation modeling of urban runoff quality processes with few details on the hydrology or on primarily statistical (e.g., regression) methods. However, the opposite of "statistical" does not neces-

NATO ASI Series, Vol. G10
Urban Runoff Pollution.
Edited by H. C. Torno, J. Marsalek, and M. Desbordes

sarily imply "deterministic," which can only be used loosely in reference to prediction of urban runoff quality, since such predictions are notoriously difficult to perform accurately and subject to large errors unless there are adequate site-specific data to calibrate and verify them.

Models may be used to address many of the objectives of urban runoff quality analysis, including:

1. Characterize the urban runoff.
2. Provide input to receiving water analysis.
3. Determine effects, sizes and combinations of control options.
4. Perform frequency analysis on quality parameters.
5. Provide input to cost-benefit analyses.

Objectives 1 and 2 characterize the magnitude of the problem, whereas objectives 3, 4 and 5 are related to the analysis and solution of the problem. Computer models allow some types of analysis, such as frequency analysis, to be performed that could rarely be performed otherwise since periods of measurement are seldom long enough. It should always be borne in mind, however, that use of measured data is usually preferable to use of simulated data, particulary for objectives 1 and 2 in which accurate concentration values are needed. In general, models are not good substitutes for good field sampling programs.

It is generally accepted that combined sewer overflows usually (but not invariably) cause receiving water problems that can be quantified and documented, e.g., low dissolved oxygen, lake eutrophication, beach closures, fish kills, impaired aesthetics, etc. (51, 177) but the negative impacts of stormwater runoff are not nearly so clear (53,66). For instance, the extensive U.S. Environmental Protection Agency (EPA) Nationwide Urban Runoff Program (NURP) found very few instances of impaired beneficial use of receiving waters (in the sense of violation of a standard or other criteria mentioned above) in spite of the fact that the stormwater itself may have high concentrations of pollutants (45,161). Most stormwater-induced negative receiving water impacts result from temporarily high bacteria levels that may cause beach closings. The question then arises, is it necessary to be concerned about stormwater quality at all? If the answer is negative, then there is no reason to model or be concerned about deterministic methods for prediction of stormwater quality. Indeed, almost all storm drainage facilities are designed on the basis of quantity control. In most instances, it is doubtful that knowledge of the pollutant concentrations or loads in the stormwater would affect the design!

As another example, stormwater quality regulations of the State of Florida may be satisfied by providing for retention of the runoff from the first inch (25 mm) of rainfall, or the first 0.5 inch (13 mm) of runoff for areas less than 100 acres (40 ha). In such cases there is clearly no benefit in conducting a water quality modeling exercise that may be both complex and useless. As receiving water quality standards become more conservative and environmentally sensitive, stormwater quality will assume a more important roll in drainage design (53). And parameters such as sediment affect the hydraulic characteristics of the drainage system at the very least. The moral is simply to avoid engaging in an involved stormwater quality modeling study unless the results are expected to make a difference and affect the design. Suit the analysis to the task.

Data Requirements

There is an unavoidable and critical need for site-specific quality (and ancillary quantity) data for calibration and verification of urban runoff quality simulation models. In fact, the models are almost worthless without such data whenever accurate concentrations and loads are required, as in studies of receiving water impacts. There is somewhat more flexibility when only relative values are required, for instance, to determine the relative effects of control options. But in any application in which absolute values are needed, all models require site-specific data since there is such a large variability in such data, even locally (45,77). Generalized data from other studies are suitable for rough estimates or "first cut" parameter estimates, at best. To reiterate, <u>predictions of absolute values of concentrations and loads will not be credible without local, site-specific data for calibration and verification</u>. This is in contrast to prediction of hydrographs which may be reasonably accurate early in the calibration procedure just on the basis of initial parameter estimates.

Required Temporal and Spatial Detail

In most studies of urban runoff quality, the overall objective is related to control of receiving water quality, which may be highly influenced by urban nonpoint source inputs. Receiving water concentrations, then, are the final indicator of the impacts of urban runoff and associated control measures. Fortunately, from the analysis point of view, in most cases of practical interest receiving waters respond only slowly to pollutant inputs, as indicated in Table 1 (42,81). The implication is that it is seldom necessary to consider more than just total pollutant loads (from storms) for receiving water analysis. Detailed time histories (pollutographs) are usually not required since such short time increment variations are strongly damped by the receiving water response. This is fortunate because it is easier to predict total storm loads than it is to predict detailed pollutographs during a storm. The latter might be necessary only for analysis, say, of bacteria in a small stream.

On the other hand, short time increment pollutographs may be required for studies of control options, the effectiveness of which is dependent on the temporal sequence of loads and concentrations. For instance, is there a "first flush" (high concentration at the beginning of the storm) to capture? At what time during the storm do regulators release overflows from combined sewers and what are the resulting loads? When should control devices be activated for real-time control? Answers to these questions may indeed depend upon prediction of an accurate time history of concentrations during the storm as well as upon prediction of total loads. But in many applications, the impact of controls on receiving water quality can still be determined simply through their effect on total loads.

The point is to match the level of analysis to the task. If detailed time histories are not required for further analysis,

Table 1. Required Temporal Detail for Receiving Water Analysis. After Driscoll (42) and Hydroscience (81)

Type of Receiving Water	Key Constituents	Response Time
Lakes, Bays	Nutrients	Weeks - Years
Estuaries	Nutrients, DO(?)	Days - Weeks
Large Rivers	DO, Nitrogen	Days
Streams	DO, Nitrogen Bacteria	Hours - Days Hours
Ponds	DO, Nutrients	Hours - Weeks
Beaches	Bacteria	Hours

then do not take the effort to predict them. Rather, concentrate on prediction of total loads from storm events, a much easier task for which many more supporting data are available.

Similarly, unnecessary spatial detail can often be avoided for quality modeling when only the concentrations and loads at the catchment outfall are desired. Only as much detail is needed in a simulation as is required to answer subsequent questions about control options or to ensure a realistic simulation. For instance, the detailed drainage network may not need to be simulated unless it affects scour or overflow timing in a combined sewer. A highly lumped catchment schematization is often sufficient for prediction of receiving water impacts. On the other hand, a pipe cannot be simulated unless there is an inflow at the upstream end, and some degree of spatial resolution is necessary if the effect of moving or "kinematic" storms is to be simulated (86).

Continuous Versus Single-Event Simulation

Various levels of detail in the modeling effort can be applied to meet the various objectives listed above. In most cases, following initial formulation of project goals and objectives, it is desirable to establish an initial rough or "first cut" estimate of the magnitude of the problem using various screening or "Level I" procedures. A few of these are mentioned later since most fall in the context of regression or statistical procedures.

Following the screening estimates, the analysis may proceed in a planning or preliminary design phase for which continuous simulation is very well suited. Continuous simulation is characterized by long simulation times (e.g., several years), relatively long time steps (e.g., one hour), and a relatively coarse catchment schematization (e.g., use of a highly lumped system with low spatial resolution and little of the drainage network simulated). Rainfall input to the simulation usually consists of a long-term historical record, such as the 25-year hourly record available for most U.S. weather stations from the U.S. National Weather Service. Single event simulation usually has just the opposite characteristics: short simulation times (e.g., several hours), short time steps (e.g., a few minutes or less), and a more detailed catchment schematization. Rainfall input usually consists of a "design event" of some type, the choice of which may be controversial (3,12,13,79,84,107,111,125,169). Adams and Howard (3) make an especially strong argument against the use of design storms. However, as pointed out below, a preliminary continuous simulation often permits an informed selection of design events for subsequent detailed analysis.

The output from a continuous simulation may be analyzed statistically to produce frequency analyses and other comprehensive summaries of long-term impacts of the urban runoff. For instance, the return period of various quantity and quality characteristics may be calculated, such as for runoff volumes, pollutant loads, peak concentrations, durations, peak runoff rates, interevent times, etc. Such an analysis is performed after separation of the continuous runoff and quality time series into an assumed series of independent events, usually by specifying a minimum number of intervening dry-weather hours or minimum interevent time (67).

As an example, consider the results of a hypothetical two year continuous simulation presented in Table 2. Rankings for the various parameters are shown to differ depending on the parameters. That is, the largest rainfall event (by depth) is the second largest flow volume event, the fifth largest peak flow event, the fourth largest suspended solids load event, the sixth largest BOD load event, and finally, the largest nitrate load event. The only corresponding ranking for the largest rainfall depth is with the largest nitrate load! In general, each parameter has its own unique sequence of event rankings, although it is true that the highest values for all parameters tend to be found from the same "cluster" of storms. Even this effect would diminish if a longer simulation time were used. The unique sequences result from the different effect of antecedent conditions on different parameters. For example, wet antecedent conditions favor high runoff volumes but low quality loads. These results

Table 2. Ranking of Storm Events by Quantity and Quality

33rd St. Drainage Area, Lake Calhoun, Minneapolis
Two-year Continuous SWMM Simulation, 1960-61
Minimum Interevent Time = 6 hours

Date of Storm Corresponding to Indicated Parameter (yr/mo/day)

Rank	Total Rain	Total Flow	Peak Flow	SS Load	BOD Load	NO3 Load
1	60/09/07	61/08/18	60/08/25	61/05/14	61/10/10	60/09/07
2	61/08/18	60/09/07	61/08/18	61/10/10	61/05/14	61/04/14
3	61/10/28	61/10/28	61/07/30	61/03/26	61/03/26	61/15/14
4	60/08/25	60/08/25	61/10/28	60/09/07	61/10/28	61/03/26
5	61/10/10	61/10/10	60/09/07	61/10/28	61/08/18	61/08/18
6	61/05/14	61/03/26	60/06/28	61/08/18	60/09/07	61/10/28
7	61/03/26	61/05/14	61/09/08	60/06/28	60/08/05	61/10/10
8	60/06/16	60/06/16	61/10/10	60/08/05	60/06/16	60/05/20

make it abundantly clear that it is incorrect to equate return periods of runoff design parameters with those of conditional rainfall frequencies, as is so often done through the use of intensity-duration-frequency (IDF) curves.

The frequency analysis on the parameter of interest is of tremendous usefulness for deciding upon design levels. There is no need to tie the analysis to rainfall statistics using IDF curves, and a selection of "design events" may be chosen from the historical record on the basis of the frequency analysis for use in subsequent more detailed single event simulations, avoiding the use of synthetic design storms. The engineer thus has a much more firm statistical basis for the discussion of, say, a "5-year design." Since the continuous simulation typically uses a coarse catchment schematization to save on computer time, more detailed single event simulations are usually necessary for final designs.

The virtues of continuous simulation in urban hydrology are discussed in general by Linsley and Crawford (103) and Huber et al. (79). Applications to urban hydrology began with the STORM model in 1974 (141) and are now becoming more prevalent (58,62,84,139). Continuous simulation is only possible, of course, using computer models, whereas single event simulation is possible in some cases simply as a desk-top procedure, using hand calculators. Although usually applied only to quantity parameters, the method of derived probability distributions has also been applied to urban hydrology, e.g., 2,31,74,81,179. The resulting analytical solutions can be very useful for preliminary analysis.

Background of Urban Water Quality Modeling

In the United States, the first comprehensive studies conducted to document the quality and effects of combined sewers were carried out in the 1960's in several cities; Field and Struzeski (51) and Field (49,50) present an extensive bibliography of the history of research and development related to the problem of control of combined sewer overflows, and Lager et al. (98) present additional technical information. Concurrently, several studies were conducted in Europe and elsewhere on combined and separate sewers; see, for example, various proceedings editted by Helliwell (69), Yen (176) and Balmer et al. (15). International activities have also been reported in technical reports prepared as part of the International Hydrological Program (IHP) and published by the American Society of Civil Engineers (ASCE) Urban Water Resources Research Program; these are summarized by McPherson and Zuidema (112). More recent summaries may be found in proceedings editted by Delleur and Torno (35). A comprehensive review of the international literature on urban hydrology in general is beyond the scope of this paper.

Characterization of stormwater runoff quality followed the earlier CSO studies by a few years; efforts that are related to modeling are reviewed by Huber et al. (79). Notable among the early studies is that of Sartor and Boyd (144), who reported on data from ten U.S. cities, followed ten years later by the massive EPA Nationwide Urban Runoff Program (NURP) during which over a million water quality data points were analyzed for over 30 U.S. cities (45). A study that was highly influential to modeling was conducted by the American Public Works Association (APWA) in Chicago in 1969 (8), and was one of the earliest to measure surface accumulations of "dust and dirt," a ubiquitous parameter in buildup-washoff formulations. In total, there are hundreds of urban runoff studies in which water quality samples have been taken and reported in the literature, all of which have enhanced present modeling capabilities.

Deterministic (or physically-based) formulations of water quality models for urban runoff evolved from the need for analytical tools with which to evaluate the emerging CSO and stormwater runoff problems. The late 1960's saw the development of various receiving water quality models and watershed models, which naturally led to extensions into urban hydrology. Although most researchers had at least applied elementary statistics and regression analysis to their data, perhaps the first attempt to simulate water quality from a first principles (although really heuristic) approach was the development of the Storm Water Management Model (SWMM) during 1969-71 (79,114,142), sponsored by the EPA. The continuous simulation Storage, Treatment, Overflow, Runoff Model, STORM, (79,141) appeared just a few years later, sponsored by the U.S. Army Corps of Engineers, Hydrologic Engineering Center (HEC). These were followed by several others; useful summaries that include descriptions of urban water quality models are given by Brandstetter (26), Huber and Heaney (76), Kohlhaas (95) and EPA (44). Support for EPA's water quality models is described by Barnwell (16). In addition, several texts discuss water quality modeling, particularly those by Overton and Meadows (124), Kibler (92), Whipple et al. (170) and Hall (65).

A summary is given in Table 3 of six operational urban water quality models. Operational implies 1) a user's manual and documentation, 2) use by other than just the model developer, and 3) continued support, usually by a government agency. The reader should note that there are many more water quality models described in the literature that are probably not operational in the sense just described, and a multitude of purely hydrologic models, i.e., models that simulate just the rainfall-runoff process (76). Table 3 is presented with the knowledge that it is incomplete, especially for listings of models not developed in the U.S. and Canada, but is at least representative of models that can be distributed to new users with sufficient documentation and user support such that they can be used successfully.

With the exception of the QQS model, all the water quality models listed in Table 3 originated in the U.S. Although there are many literature citations to ongoing water quality modeling work in Europe, Japan and elsewhere, there does not appear to be an accepted operational "standard" set of quality models as there is in the U.S. and Canada, although the NIVA model in Norway might be an exception (101). (Most urban water quality modeling in Canada relys on improved versions of U.S. models. Cooperation is enhanced by the Storm and Water Quality Modelers Users' Group, jointly sponsored by the EPA and the Ontario Ministry of the Environment.) This is in contrast to urban runoff quantity, for which there are several operational models for various countries

Table 3. Six Operational Urban Runoff Quality Models

Model	Sponsoring Agency	Year Originated	No. Pollutants	Simulation Type*	References
DR3M-QUAL	USGS	1982	4	C,SE	7
FHWA	FHWA	1981	13	SE	37,38
HSPF	EPA	1976	10	C,SE	17,91
QQS	EPA	1976	2	C,SE	60,61
STORM	HEC	1974	6	C	80,141
SWMM	EPA	1971	10	C,SE	68,79,114 115,142

*C = Continuous simulation
SE = Single event simulation

in Europe (35). There is a tendency to build a "new and improved" version of existing models or otherwise to adapt existing models to new situations. For instance, there are many examples in the literature of calibrations and useful improvements to SWMM and STORM (e.g., 133,134,140,147).

In summary, there are several operational urban water quality models that are deterministic in the sense of predicting water quality by means other than regression or simpler assumptions. These methods will be the subject of the remainder of this paper.

Predictive Methods

Introduction

Several methods are routinely utilized for quality prediction; these will be introduced here, with more details on some of them presented later. Methods related to source generation include 1) constant concentration, 2) regression equations and loading functions, 3) rating curves (concentration or load as a function of flow rate or volume, respectively), 4) buildup-washoff, 5) land surface erosion, and scour and deposition in sewers, and 6) other miscellaneous sources. This paper will highlight the buildup-washoff idea with some attention given to the other methods and related topics such as pollutant transport.

Constant Concentration

This is the simplest of the many methods and lends itself well to use with an unaltered water quantity model since loads can easily be generated as the product of concentration times runoff volume (e.g., 90). In fact, the method is probably most powerful when used with a very sophisticated hydraulic model for accurate simulation of CSOs and stormwater discharges, such as the Extran Block of the SWMM model (142). In other words, an accurate hydraulic analysis may sometimes be more important for estimation of quality effects than use of a more sophisticated but possibly hydraulically unrealistic water quality model.

The principal question is, of course, what concentration to use? The multitude of urban runoff studies alluded to earlier gives little help for clear delineations on the basis of land use, location, hydrology, etc. (45), although an interesting result of the NURP studies was the overwhelming log-normality of the event mean concentrations. The problem is reduced to the need for site-specific data, such as flow-weighted mean concentrations, for each study area. However, this is a necessity for virtually any method used for urban water quality prediction. Ranges of concentrations that may be encountered may be found in many references (e.g., 73,77,92,98,106,170) as well as in companion papers at this conference.

Use of a constant concentration cannot account for variations within a storm or between storms (unless a basis is established for variation between storms). Thus, the method is best suited for generation of loads to receiving waters, and less well suited for studying control options, such as detention ponds, that may be sensitive to intrastorm variations in concentration.

A powerful extension of the constant concentration idea utilizes the frequency distribution of the event mean concentrations (e.g., log-normal) and requires the standard deviation in addition to the mean of the data (45,81). When coupled with similar receiving water statistics, derived frequency distributions of receiving water concentrations can be obtained (40).

Regression and Loading Equations

Many studies in which water quality has been measured have attempted to relate the effluent data to catchment, demographic and hydrologic characteristics, usually through regression analysis. Since such statistical or empirical results will be extensively reviewed elsewhere in this conference, no attempt will be made here for a thorough survey. As examples, however, see the works of McElroy et al. (110), Miller et al. (117), Hemain (70), Kobriger et al. (94), Brown (29), Desbordes and Servat (36) and Hogland et al. (73). A recent review of regression formulations for highway runoff quality is provided by Miracle (118). Techniques such as these can be used for first-cut estimates of urban runoff quality, with the usual provisos about applicability of regression analysis.

Rating Curves

These are a special case of regression analysis in which concentration or event load is regressed against flow rate or runoff volume, usually in a power function form, with no other independent variables used. Based on formulations commonly used in sediment transport analysis (63,164), this is a very convenient form for data analysis and use by models. It will be discussed more later.

Buildup and Washoff

This is certainly the most common physically-based formulation used in urban runoff quality models. "Buildup" is a term that represents all of the complex spectrum of dry-weather processes that occur between storms, including deposition, wind erosion, street cleaning, etc. The idea is simply that all such processes lead to an accumulation of solids and perhaps other pollutants which are then "washed off" during storm events. The distinction between dry- and wet-weather processes is not always clear, especially for continuous modeling, as will be discussed extensively by James and Boregowda (83) in a companion paper. For example, buildup on pervious surfaces is certainly less obvious, where erosion and solution mechanisms induced by rainfall and runoff may play the dominant roll. Nonetheless, many physically-based empirical formulations have been developed to describe the buildup and washoff processes for modeling purposes, and these will be described in this paper.

Erosion, Scour, Deposition

Rainfall and runoff over pervious surfaces generate erosion of solids through dissipation of rainfall energy and entrainment due to the boundary shear stress of the overland flow. Similar entrainment (scour) and deposition processes occur in the drainage network and sewer system, with the potential to contribute significant sediment and associated pollutants from combined sewers as part of the "first flush" of material deposited during low dry-weather flows. Erosion from barren soil in urban areas can be significant, and erosion control is often an integral part of construction activities (64,68,170). Unfortunately, the general area of sediment transport is still one of the least quantifiable in the field of hydrology and hydraulics; limited information will be given later.

Other Sources

Other source processes to be discussed include the contribution to runoff quality of rainfall and snow quality, contributions of catchbasins (or gully pots) and dry-weather flow, and adsorption of pollutants onto sediment particles.

Dry-Weather Processes -- Buildup

Introduction

Since a companion paper by James and Boregowda (83) will consider this topic in detail, only a brief summary of common modeling options will be discussed herein. As used in most modeling studies to date, the idea of "buildup" incorporates all dry-weather processes that either generate or remove pollutants from impervious surfaces. Sources include (98): atmospheric fallout, vegetation and leaf litter, land surface erosion, spills, pavement degredation, automobile emissions and decay, anti-skid and deicing compounds, snow accumulation, construction, etc. Dry-weather removals may occur due to street sweeping, wind erosion, extraneous flows of water, etc. Although efforts have been made to quantify various components of the net buildup process, particulary those that are related to traffic (24), most data consist of end-of-pipe concentrations, lending themselves only to gross calibrations. Hence, virtually all models have eventually returned to purely empirical formulations, in which buildup is only a function of time (dry-weather days before a storm), although factors such as street sweeping can be dealt with explicitly. Unfortunately, since virtually no physics are included in such formulations, they are just as subject to the need for site-specific data as are any regression equations.

Available Buildup Data

As part of the influential APWA study in Chicago (8), street surface accumulation of "dust and dirt" -- defined as anything passing through a 0.25-inch mesh screen -- was measured by sweeping with brooms and vacuum cleaners. The accumulations were measured during dry weather for different land uses and curb lengths, and the data were normalized in terms of pounds of dust and dirt per dry day per 100-feet of curb or gutter, with the implication that the buildup was a linear function of dry days. Following the APWA study, similar buildup studies were reported by AVCO (14) for Tulsa, Oklahoma, by Sartor and Boyd (144) and Pitt and Amy (129) for ten U.S. cities, and by Shaheen (145) for highways in the Washington, D.C. area. More recently, Pitt (128) reports on extensive data gathered both on the street surface and in runoff for San Jose, California. A drawback of the earlier studies is that street surface accumulations and stormwater concentrations were seldom measured simultaneously; hence, it is difficult to draw direct conclusions on their interrelationships. Summaries of such data are provided by Amy et al. (11) and most comprehensively, by Manning et al. (106), in which many problems and facets of sampling and measurements are also discussed. For instance, some data were obtained by sweeping, others by flushing; the particle size characteristics and degree of removal from the street surface differ for each method.

The Chicago APWA study also analyzed the dust and dirt for the constituents BOD5, COD, total coliforms, total nitrogen and total phosphate and presented the results in terms of mass of pollutant as a fraction of dust and dirt, e.g., milligrams pollutant per gram dust and dirt or parts per thousand by mass. This procedure was followed by most of the other studies cited above; the best summary of all buildup and pollutant fraction data is given by Manning et al. (106). The pollutant fractions (or "potency factors" as they are sometimes called) thus represent one method of calculating the buildup of specific pollutants if the buildup of dust and dirt is known. Alternatively, the accumulations of the pollutants can be measured directly, bypassing the intermediate dust and dirt parameter. Models may include one or both options (79). Linear buildup and pollutant fractions were used in the original SWMM model (114) and in the STORM model (80,141); both were heavily influenced by the APWA Chicago study (8). Although many of the early studies cited above normalized the buildup data in terms of mass per length of curb, more recent studies tend to present the data normalized by area, i.e., mass per area, which avoids knotty problems such as determination of the equivalent curb length of a parking lot or park.

Alternative Buildup Formulations

Most data from studies cited above imply linear buildup since they are given in units such as lb/curb-mile per day. However, there is ample evidence that buildup can be nonlinear; Sartor and Boyd's (144) data are most often cited as examples (Figure 1) in which surface accumulations tend to approach assymptotic values with elapsed time since cleaning or storm washoff, with additional nonlinear relationships also reported by Terstriep et al. (158) and Pitt (128). Ammon (10) has summarized several such studies and determined that the buildup relationships generally fall into one of four functional forms: linear, power function, exponential, or Michaelis-Menton:

Linear:

$$P = a\,t \qquad [1]$$

where P = mass of pollutant on surface,
t = time since last cleaning or storm runoff, and
a = coefficient.

Power:

$$P = a\,t^{b} \qquad [2]$$

where b = exponent.

Exponential:

$$P = P_L(1 - e^{-bt}) \qquad [3]$$

where P_L = limiting (assymptotic) surface load.

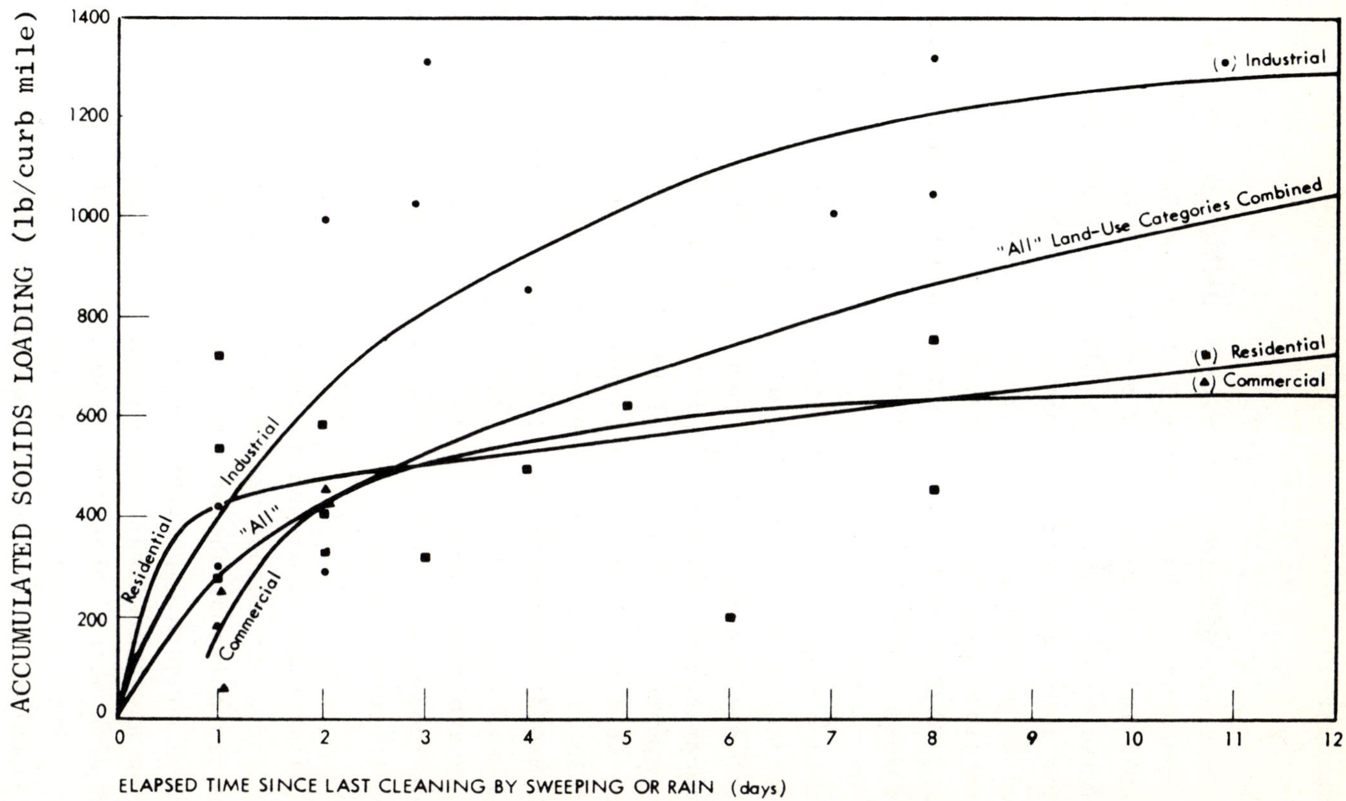

Figure 1. Buildup data reported by Sartor and Boyd (144).

Michaelis-Menton:

$$P = P_L\ t/(a + t) \qquad [4]$$

The coefficients in equations 1-4 take on different meanings and units depending on their context. Thus, coefficient a is a rate parameter in equations 1 and 2 and a "half saturation constant" in equation 4. Units will obviously vary and be appropriate to the units used for time and mass. The equations can also be used to predict mass per curb length or mass per area. Elapsed time since last storm, t, may be ambiguous and must be defined, e.g., number of hours since last storm of 0.05 inches (1.3 mm) or more (117).

The four buildup equations are illustrated in Figure 2 for hypothetical dust and dirt, DD, buildup with purely arbitrary numerical values chosen for the coefficients (79). Although the linear and power equations do not include an upper bound, one can arbitrarily be imposed. At least one of the four equations should certainly be able to fit reported literature or sampled data. Of the six models listed in Table 3, HSPF and STORM use linear buildup, DR3M-QUAL uses exponential, FHWA uses Michaelis-Menton, and SWMM gives the user the choice of any of the four methods (including an arbitrary upper limit). The QQS model uses a slightly more complex polynomial function of time that results in an assymptotic formulation.

For single event simulation the choice of functional form is unimportant since any method can be adjusted to give the desired mass at the beginning of the storm. For continuous simulation the choice may be more important, at least in deciding whether or not to impose an upper limit or assymptote. In this context, the interaction with storm interevent time is relevant, as sketched in Figure 3. Small interevent times (or street cleaning intervals) mean there is more dependence on buildup rate, whereas long interevent times mean there is more dependence on buildup limit. Parameter estimation efforts may be directed accordingly. Interestingly, some regression studies (117) have shown little dependence of storm runoff loads on preceding dry days, implying

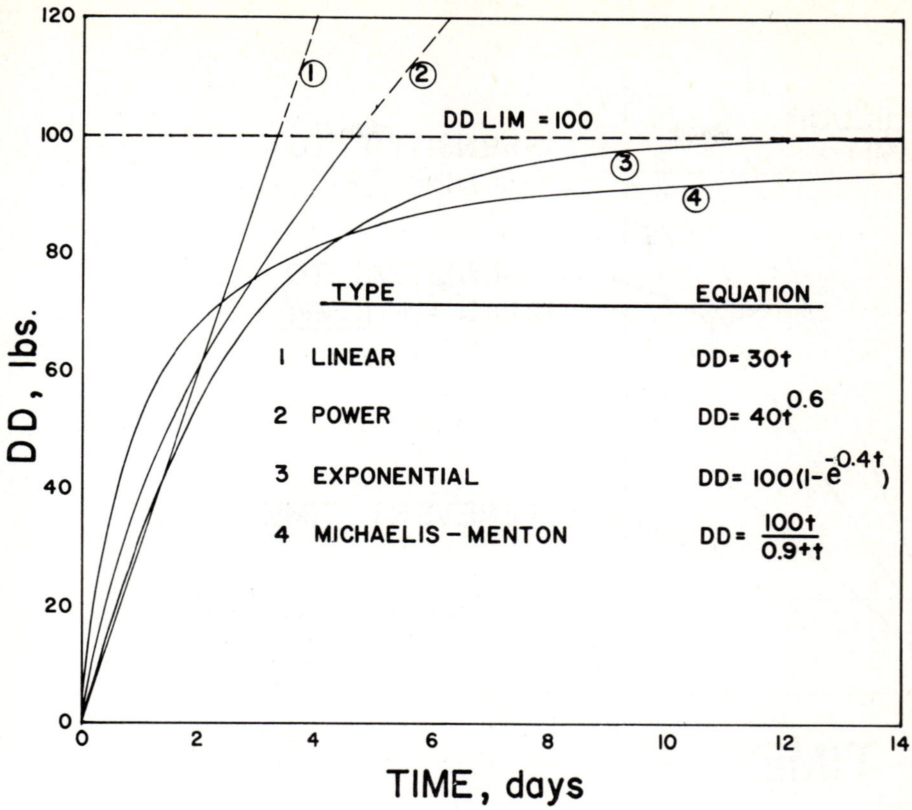

Figure 2. Hypothetical buildup rates according to alternative buildup formulations, using arbitrary parameter values (Ref. 79). DD = "dust and dirt." DDLIM = limiting surface load, = P_L of equations 3 and 4.

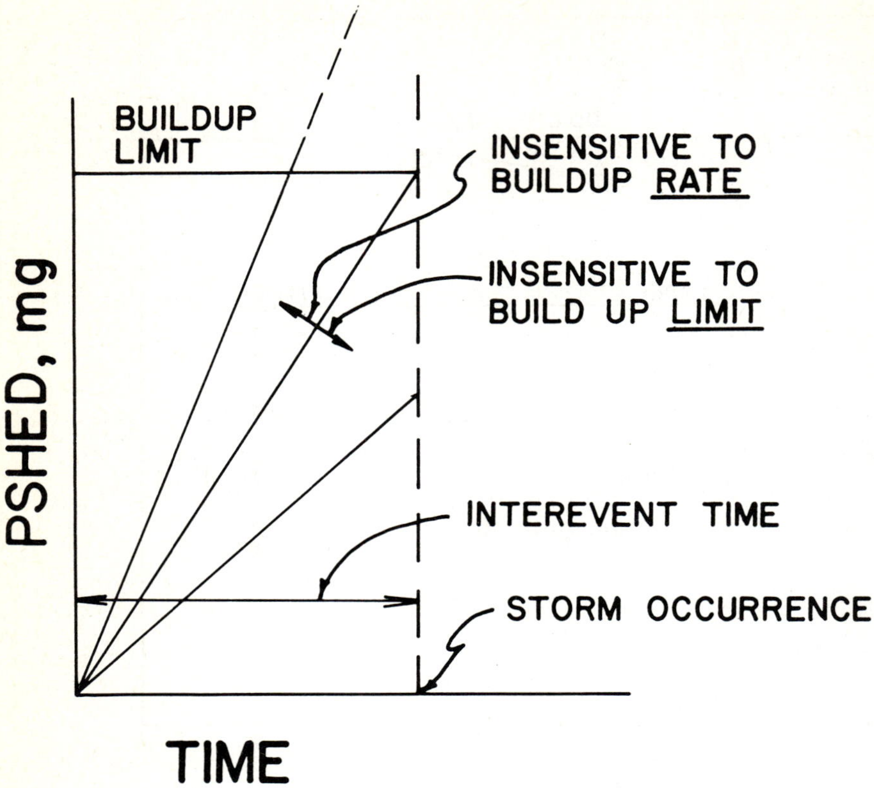

Figure 3. Interaction of buildup parameters with storm interevent time (Ref. 79). PSHED = mass of pollutant on surface.

dependence only on the buildup limit (and not on the functional form). This appears to be especially true for highways in which surface material tends to remain at a constant level in a state of suspension due to eddies created by traffic.

When snowmelt is simulated, buildup may be used to account for the application of deicing chemicals. The amount applied varies according to depth of snowfall and local practice. Loading rates and other guidelines are discussed, for example, by Field et al. (52), Richardson et al. (137), Proctor and Redfern and J.F. MacLaren (133,134), Waller and Coulter (168), and Krukar and Cook (97). When simulated by the empirical equations described above, buildup should only be permitted when snow is on the surface (79).

Buildup (and related) methods such have these have not escaped criticism in the literature. Whipple et al. (171) criticized the linear buildup assumption in the original SWMM model. Jewell et al. (88) emphasized the need for calibration of whichever functional form is chosen, and Sonnen (153) has highlighted the problems and weaknesses with quality modeling formulations and also presented guidelines for needed research. Alley and Smith (6) choose an exponential buildup equation and provide parameter estimation procedures.

Dry-Weather Removals

The primary removal mechanism considered in most models is that due to street cleaning, usually by mechanized sweeping. The usual assumption is that cleaning removes a fraction of accumulated pollutants when it occurs; an extensive table of removal efficiencies as a function of pollutant and street cleaner type is given by Pitt (128). If, in fact, street sweeping does remove a specified fraction of material on each occurence, then a time history of accumulation will occur such as sketched in Figure 4 (79), in which an equilibrium mass will eventually result when the buildup during the interval between cleanings just equals the fraction removed during a cleaning. However, such values may be misleading (e.g., removal of 40 % of the surface BOD by one pass

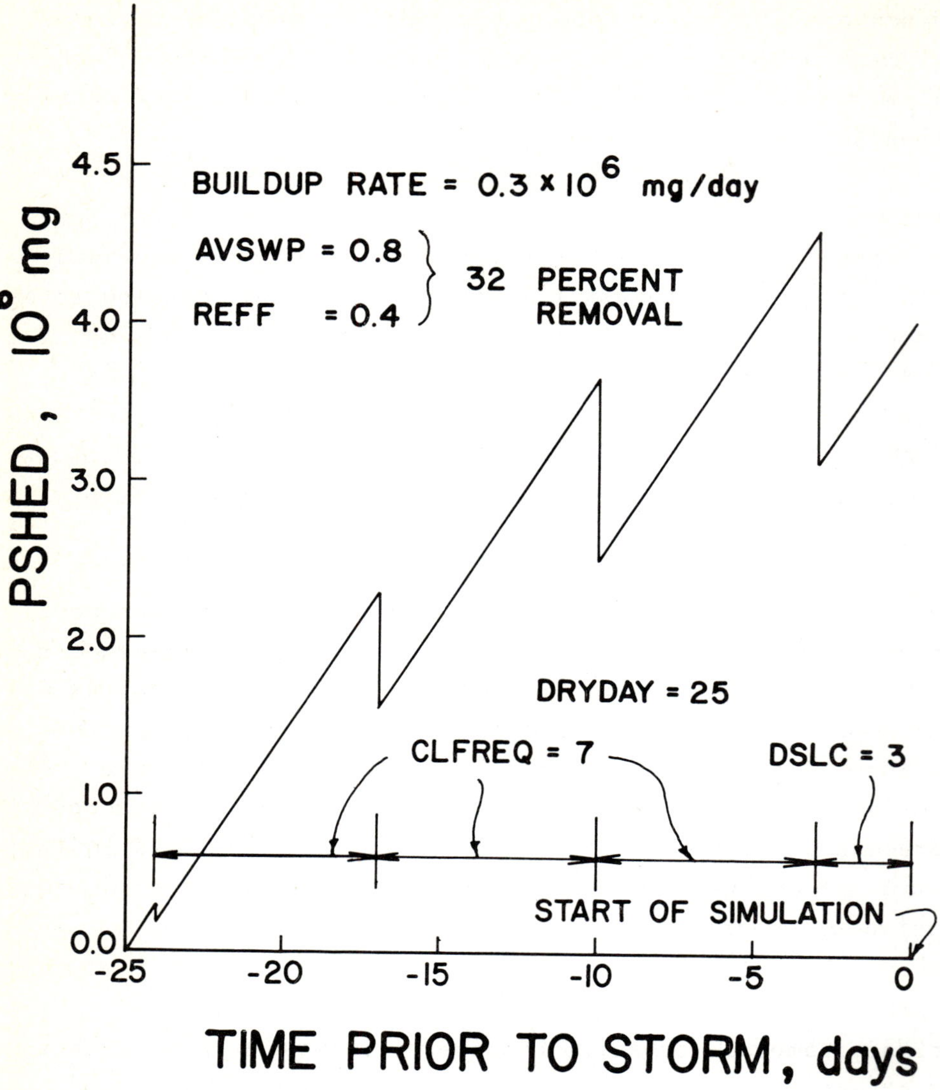

Figure 4. Hypothetical time history of pollutant mass on surface, with linear buildup and street sweeping (Ref. 79). AVSWP = availability factor, REFF = street sweeping efficiency, PSHED = mass of pollutant on surface, CLFREQ = street cleaning interval, DRYDAY = days since last storm, DSLC = days since last cleaning.

of a mechanical sweeper), since the overall effectiveness for stormwater pollution control is highly dependent on the cleaning frequency and on how much of the surface area is actually cleaned. The NURP studies (45) concluded that street sweeping was generally ineffective as a technique for improving the quality of urban runoff on the basis of ten study sites subjected to detailed analysis. Some additional effectiveness might accrue if streets are cleaned prior to the beginning of the rainy season after a long dry season or as part of cleanup after a winter season that has involved distribution of sand and deicing chemicals. Of course, these results do not detract from the usual aesthetic purposes of street cleaning.

Summary

Most operational urban water quality models generate surface loads by an empirical buildup mechanism based on fitting of functional forms to observed surface accumulation data. Although there are attempts to include causal mechanisms in the buildup process (24), there are relatively few data to support such relationships, and the entire process might more easily be considered a calibration exercise, in which the buildup parameters for whichever functional form is chosen will be calibrated as part of the comparison of predicted and measured runoff quality.

Washoff

Introduction

Washoff is the process of erosion or solution of constituents from a subcatchment surface during a period of runoff. If the water depth is more than a few millimeters, processes of erosion may be described by sediment transport theory in which the mass flow rate of sediment is proportional to flow and bottom shear stress, and a critical shear stress can be used to determine incipient motion of a particle resting on the bottom of a stream channel (63,164). Such mechanisms might apply in street gutters and larger channels, although no operational urban runoff model incorporates the available physics for these mechanisms into its pollutant generation routines. For thin overland flow, rainfall energy can also cause particle detachment and motion. This effect is incorporated into some predictive methods for erosion from pervious areas (e.g., 91,173) and might apply also to washoff from impervious surfaces, although the effect of a limited supply (buildup) of erodable material must also be considered for this latter case.

Ammon (10) reviews several theoretical approaches for urban runoff washoff and concludes that although the sediment transport based theory is attractive, it is often insufficient in practice because of lack of data for parameter (e.g., critical shear stress) evaluation, sensitivity to time step and discretization, uncertain cross-sectional geometry of the flowing water, and because simpler methods usually work as well (still with some theoretical basis) and are usually able to duplicate observed washoff phenomena. It might be added that simpler formulations are consistent with the overall state of the art of urban runoff quality prediction, which, it is emphasized, is crude and always requires calibration regardless of the level of sophistication of the component mechanisms. Although concepts of sediment transport theory have been applied to scour and deposition processes in the sewer and drainage network, almost all modeling activities for washoff from impervious surfaces have focused on improvements to the various empirical formulations outlined below. Generation of pollutants from pervious surfaces may also incorporate some additional theory as described later when discussing erosion.

Empirical Washoff Formulations

The basis for many of the empirical formulations used by models is the results of Sartor and Boyd (144), part of which are reproduced in Figure 5. These results were obtained using a portable rainfall generator with jets 4 to 8 ft above the asphalt or concrete surface to simulate heavy to moderate rainfall. The results are remarkable in that there is an unmistakable tendency toward an exponential, or first-order decay washoff process. Nakamura (120,121) shows similar experimental results in his plots showing linear relationships between the log of remaining mass of pollutant on the surface versus cumulative runoff volume (equivalent to elapsed time when the runoff rate is constant).

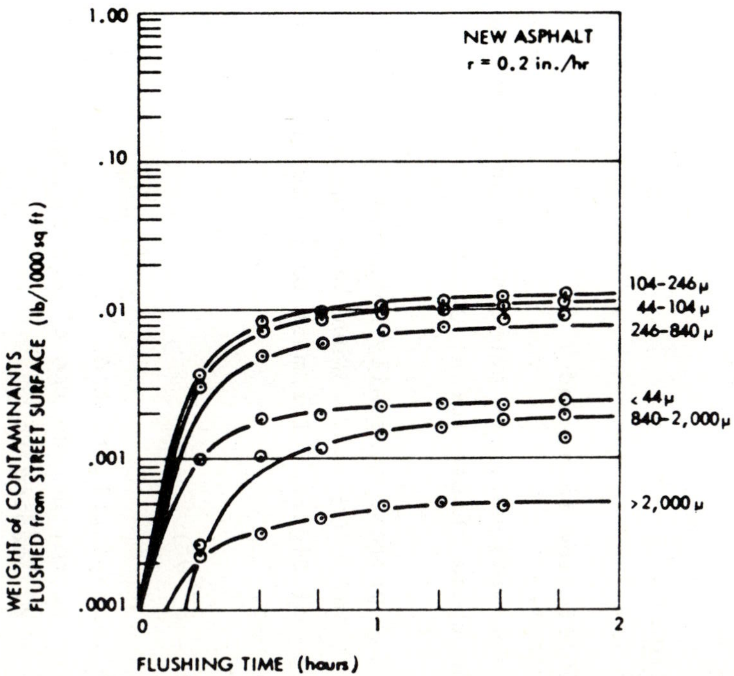

a.

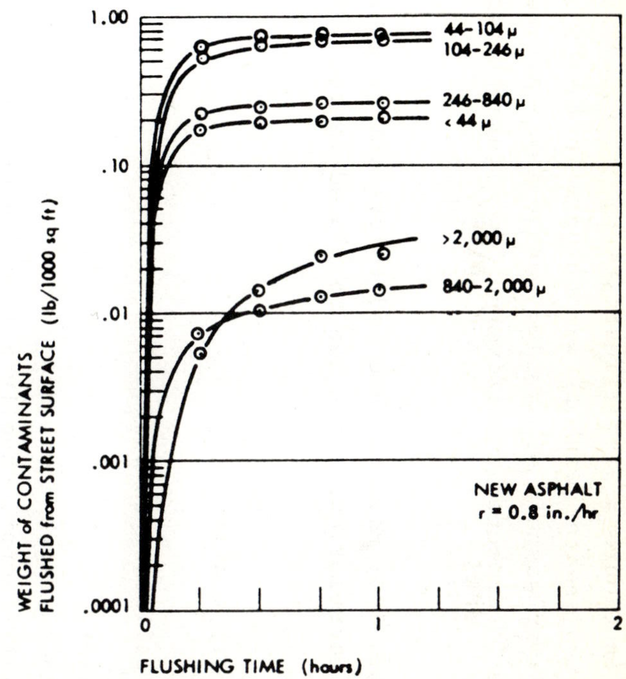

b.

Figure 5. Results of washoff experiments by Sartor and Boyd (144).
a. Runoff rate = 0.2 in/hr. b. Runoff rate = 0.8 in/hr.

The results shown in Figure 5 also illustrate the effect of "rainfall" intensity (greater washoff with higher intensity) and particle size (greater washoff of smaller particles). Although both results would be expected intuitively, few figures illustrate the effects so well.

The Sartor and Boyd (144) results have led to variations on the first order decay formulation. Most can be placed in the general form,

$$dP/dt = -K_w \, f_1(r) \; P \; AV(r) \tag{5}$$

where P = mass of pollutant on surface,
t = time,
K_w = coefficient,
r = runoff rate, depth/time, and
$AV(r)$ = availablity factor, usually a function of r.

The functional forms for f_1 and AV are to be discussed, and the rate dP/dt is the mass rate of washoff (mass/time) as a function of time. Washoff concentration can be obtained by dividing dP/dt by the runoff flow rate,

$$C = \frac{dP/dt}{Q} = \frac{dP/dt}{A\ r} \tag{6}$$

where $Q = A\ r$ = runoff flow rate, and
A = surface area.

There may be need for various units conversions in equations 5 and 6, depending on those chosen for A, Q and r.

The first order decay process is easily obtained from equation 5 by letting $f_1(r) = 1$ and $AV(r) = 1$. When integrated, the familiar exponential form of Figure 5 for mass of pollutant washed off, $P_o - P$, is obtained,

$$P_o - P = P_o \; (1 - e^{-K_w t}) \tag{7}$$

where P_o = initial mass of pollutant on surface.

An obvious extension is to include the effect of runoff rate explicity as $f_1(r) = r$, leading to

$$P_o - P = P_o(1 - e^{-K_w rt}) = P_o(1 - e^{-K_w R}) \qquad [8]$$

where $R = \int r\,dt$ = cumulative runoff depth.
It may readily be shown (89) that equation 8 may be applied on a time-step basis with variable runoff rate, r, which is equivalent to using the form with cumulative runoff, R. This first order assumption was implemented in the original 1971 SWMM model (114) prior to the experimental justification published later by Sartor and Boyd (144) in 1972. (The first order washoff formulation was developed by Mr. A.J. Burdoin, a consultant to the SWMM project.) Another assumption that has persisted in the literature is that 0.5 inches (13 mm) of cumulative runoff, R, will result in 90 % washoff (P/P_o = 0.1), yielding K_w = 4.6/in, which is only a mathematical, not an observed, truth. Several of the assumptions related to this first order washoff formulation are discussed critically by Sonnen (153).

The formulation of equation 5 is not completely without physical basis since sediment transport theory predicts dependence of scour upon flow rate (63,164), but it is basically an empirical formulation. A primary difficulty with the idea that the washoff rate is linearly proportional to runoff rate (culminating in equation 8) is that concentrations can only decrease with time during the storm event. If the division indicated in equation 6 is carried out with $f_1(r) = r$ and AV = 1, there results

$$C = K_w\ r\ P\ /\ A\ r = K_w\ P\ /\ A \qquad [9]$$

in which concentration must decrease continuously since P decreases continuously. Even in the original SWMM model this was recognized as a problem since the there was difficulty in calibrating the quality portion to observed runoff data, which often showed increases in concentration during a storm event as runoff rate increased (115). This led to the use of "availability factors," usually of the form

$$AV = a + b\ r^c \qquad [10]$$

where a, b and c = coefficients dependent on the pollutant. Availability factors of the form of equation 10 were incorporated into the original and Version II of the SWMM model (78,114), the STORM model (80,141), the DR3M-QUAL model (7) and the FHWA model (37). If the exponent c in equation 10 is greater than zero, concentrations during a storm will increase with runoff rate. The choice of the four parameters a, b, c and K_w, will lead to various forms of the pollutograph, which can be analyzed both in the form of cumulative load, P_o-P, (mass) washed off as a function of time or cumulative storm volume, R (4,5,43), or as instantaneous load, dP/dt, as a function of runoff rate, r (79).

Although the four parameters can be used to obtain most pollutograph shapes, and calibration guidelines are available (4), it is hard to justify the need for so many washoff parameters in light of the available data base. This led to a simpler formulation in the Version III SWMM model (79) of the form,

$$dP/dt = K_w \, r^c \, P \qquad [11]$$

where c = exponent obtained from calibration.
Equation 11 is equivalent to the use of an availability factor with a = 0 and b = 1 and results in only two calibration parameters, K_w and c. The DR3M-QUAL model (7) also permits a form of this idea but fixes the value of c = 2. Clearly equation 11 also permits concentrations to increase as r increases as long as c > 1. Concentrations that <u>decrease</u> as flow increases, such as might occur for dissolved constituents, may also be simulated by setting c < 1. The first formulation of equation 11 known to the writer was in an unpublished report to the Boston District Corps of Engineers in 1974, authorship unknown.

The exponential washoff formulation in its various forms has been discussed extensively in the literature. Ammon (10) discusses washoff formulations in general and evaluates K_w values for Sartor and Boyd's (144) data (with washoff linearly proportional to runoff rate), with values of from 5.7 to 9.3 per inch for a runoff rate of 0.2 in/hr, and 2.9 to 4.2 per inch for a runoff rate of 0.8 in/hr. Interestingly, no clear dependence on particle size was found. Values of K_w for individual pollutants (not sediment, as in the Sartor and Boyd data) might be several orders of magnitude less, as reported by Sueishi et al. (157). Sonnen (153) estimates values of K_w from sediment transport theory ranging from 0.052 to 6.6 per inch, increasing as particle diameter decreases, rainfall intensity decreases, and as catchment area decreases, pointing out that 4.6 per inch is relatively large compared to his calculated values. Nakamura (120,121) conducted new washoff experiments that verified the exponential form and also showed dependence of the washoff coefficient on roughness, slope, runoff rate and cumulative runoff volume. The dependence varies with the pollutant. Hogland et al. (73) used a form of equation 5 in which the washoff load rate is proportional to a polynomial in r. Price and Mance (132) generate sediment washoff using a complex function of rainfall energy, critical shear stress and overland flow rate. Boregowda (24) and Shivalingaiah and James (146) combine removal functions based on overland flow (a power function), rainfall energy, and vehicles during the storm into an empirical equation for removal at each time step.

The exponential washoff equation is not the only method used in operational models. HSPF (91) uses a sediment-related algorithm taken from the ARM model (41) and earlier versions (32) in which a sediment transport capacity is first computed as as power function of the overland flow rate, r. If the capacity to transport is greater than the quantity stored on the surface, all solids are removed. Otherwise, the removal is equated to the capacity just computed. This method is thus very similar to the rating curve idea to be discussed later. Other pollutants are then computed as a fraction of the sediment by multiplication by potency factors. The method thus involves two power function coefficients, plus potency factors for each pollutant, perhaps lending itself more easily to calibration than does an exponential washoff equation with separate parameters for each pollutant.

The QQS model (60) is the only one of the six listed in Table 3 to use rainfall instead of runoff to compute washoff from impervious surfaces. An impulse response similar to an instantaneous unit hydrograph for runoff is required for each pollutant. The convolution with "effective rainfall" is then carried out on a time step basis with the inclusion of two more functions, the first representing the diurnal variation of pollutant washoff due to land use, and the second representing the decrease of washoff with increasing rainfall duration (the exponential washoff idea). The response function for pollutants might be based on instantaneous unit sediment graphs (172); the other two functions must be based on intuition and experience with the model. Calibration might be more difficult than the other models discussed since functions must be fit, not just coefficients. However, agreement between QQS predictions and measured quality data may be seen to be as good as for other of the operational models discussed (60).

Observed Pollutograph Variations

It was mentioned above that one method of characterizing the nature of the pollutograph within a storm is to plot load (mass/time) versus runoff rate (depth/time) or flow rate (volume/time). Alternatively, cumulative load (mass) could be plotted versus cumulative runoff depth (4,5). Examples of the former may be seen in Figure 6 for a primarily solids constituent (total suspended solids), a primarily dissolved constitutent (nitrite plus nitrate), and a "mixed" constituent (BOD). The curves show hysteresis and are not smooth because the runoff hydrograph (not shown) is not unimodal and increases at different times during the storm. The hysteresis is caused simply by the fact that concentrations are higher at the beginning of the runoff than at the end, corresponding to the idea that the load should be proportional to the amount of material on the surface at each time step (the first order washoff idea).

Can a model reproduce such shapes (if necessary!)? As an example, hypothetical SWMM simulations are shown in Figure 7, using equation 11 (79). The shape is very flexible, and hysteresis can clearly be enhanced or damped depending on the choice of parame-

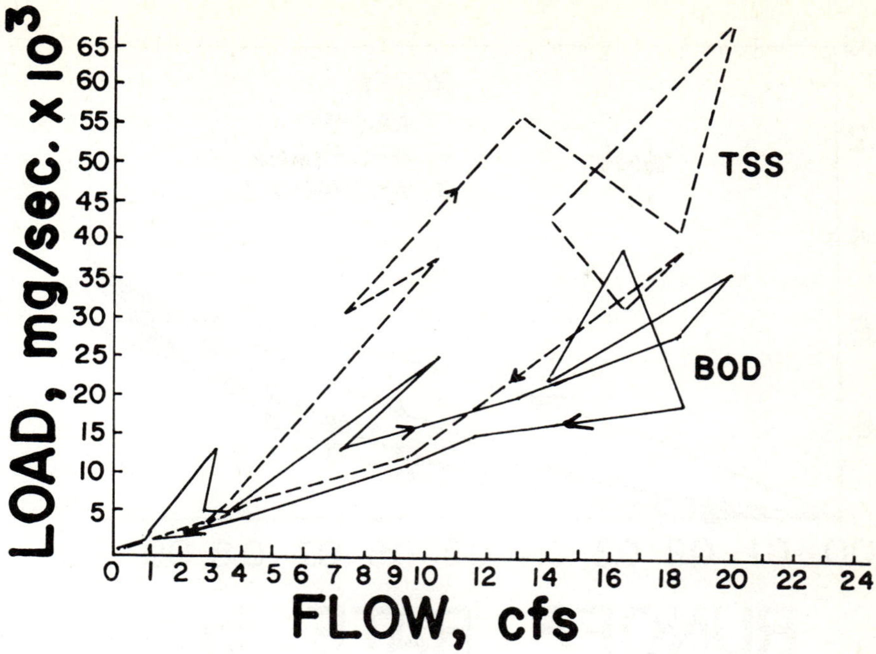

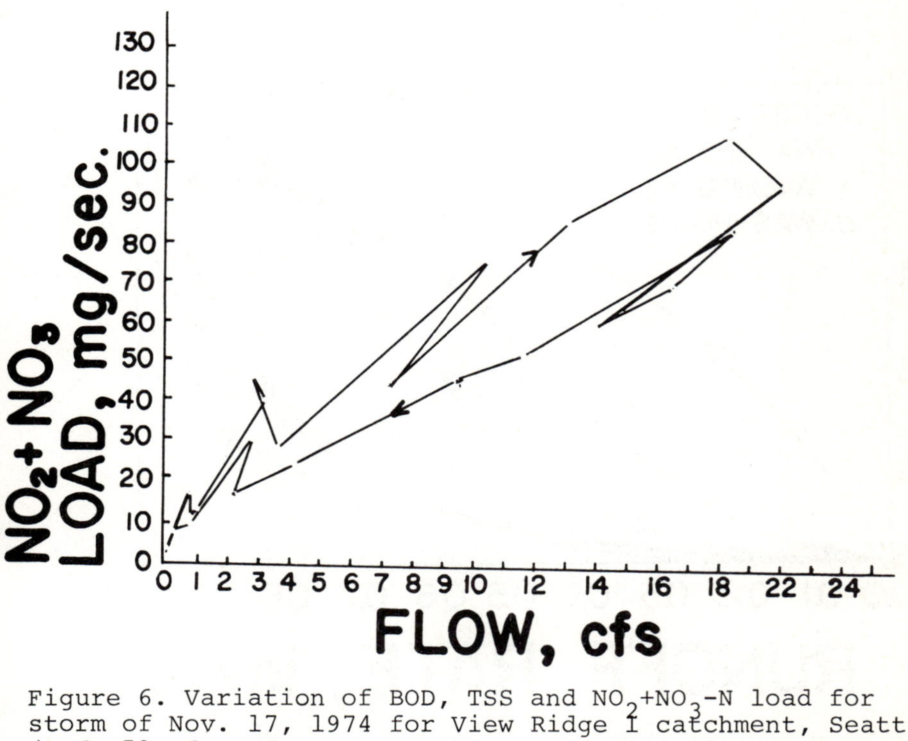

Figure 6. Variation of BOD, TSS and NO_2+NO_3-N load for storm of Nov. 17, 1974 for View Ridge I catchment, Seattle (Ref. 79, data from Ref. 77). Connected points trace time history in clockwise cycle.

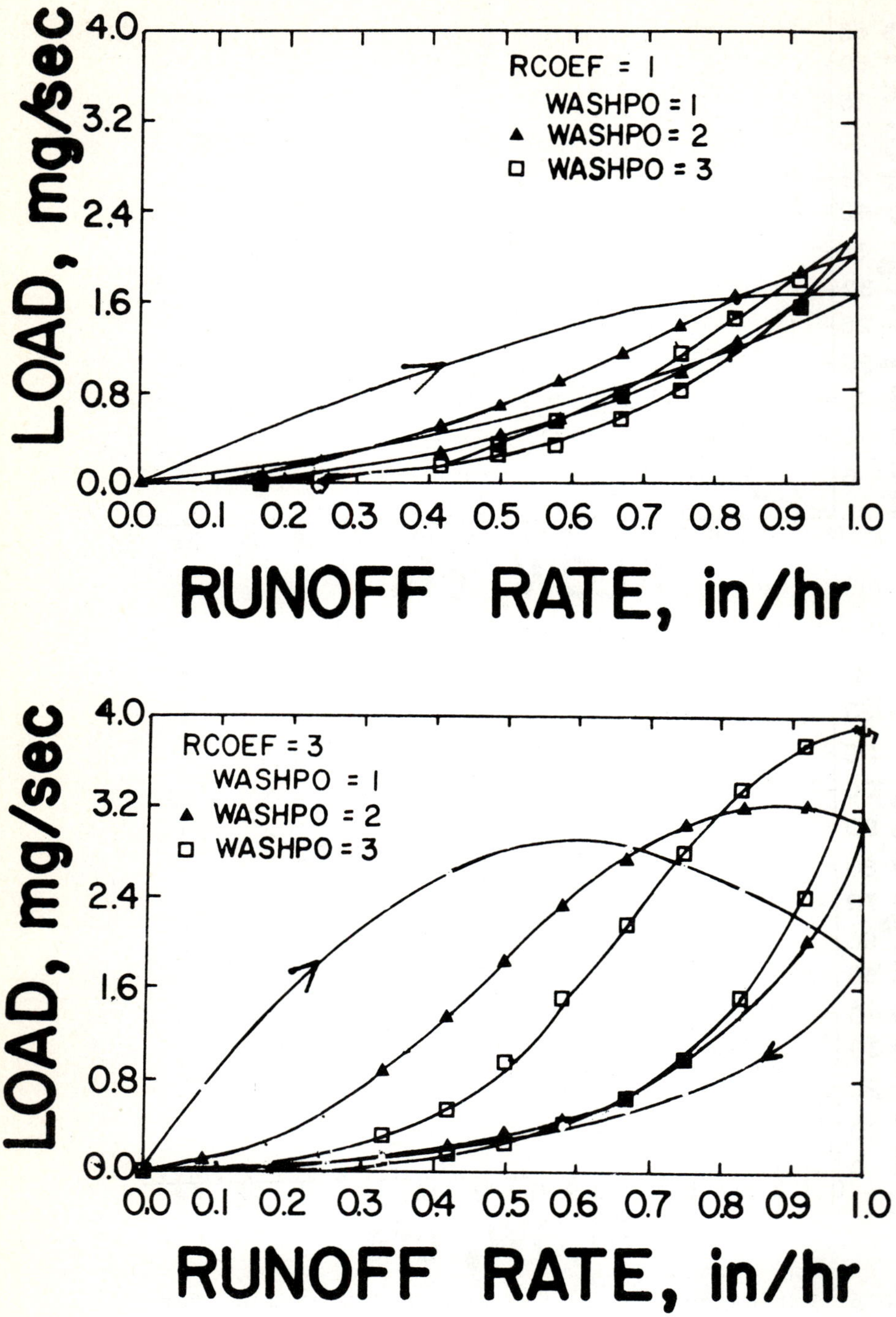

Figure 7. Hypothetical time history of load generated by SWMM with a triangular runoff hydrograph (Ref. 79). RCOEF = coefficient K_W in equation 11. WASHPO = exponent c in equation 11.

ters. In this example the curves are smooth since they are a response to a simple, symmetric triangular runoff hydrograph (not shown) which peaks at 1.0 in/hr at 60 min. When applied to an irregular realistic hydrograph, the irregularities shown in Figure 4 may also be simulated.

Summary

Washoff formulations tend to be empirical in nature and are strongly related to early experimental work performed by Sartor and Boyd (144). Most, but not all, operational urban water quality models use a form of the exponential or first order washoff equation, with enough parameters to permit fitting to observed data. The need for calibration of washoff parameters arises primarily from the desire to simulate pollutograph variations during a storm. As discussed early in this paper, this may or may not be necessary in practice. If only total storm event loads are needed, then considerable calibration effort may be avoided, since subtle temporal variations can be ignored. An alternative to the exponential washoff idea that is easily adapted to storm event calibration and use of existing data is the rating curve methodology, discussed next.

Rating Curves

Basis

A rating curve is simply a regression of load rate against flow rate, total load against total runoff, or concentration against flow rate or runoff volume. Almost all such regressions are power functions of the form

$$F = a\, Q^{b} \qquad [12]$$

or

$$M = a\, V^{b} \qquad [13]$$

where F = load rate, mass/time,
Q = flow rate, volume/time,
M = total load, mass,
V = runoff volume, and
a,b = regression coefficients (different for equations 12 and 13).

The power function (or log-log) relationship is chosen as one that commonly gives a good fit to the data (75,96,149,156,160, 166), although other relationships (e.g., linear) are occasionally found in multiple regression studies (19,36,70,117,131). The power function is based on empirical sediment transport relationships, in which the functional form of equation 12 is especially evident in streams (63,151,164).

Equations for instantaneous concentration, F/Q, or event mean concentration, M/V, can easily be derived by dividing equation 12 or 13 by Q or V, respectively. However, the total load, M, is often computed by multiplying a flow-weighted concentration, C, by the runoff volume. This can lead to spurious correlation (21), in which higher correlation coefficients may be obtained when regressing M = C V versus V than are justified from regressions of C versus V. For this reason Miller et al. (117) did not include runoff volume in their list of independent variables for regression analysis. Nonetheless, this is routinely done in urban runoff quality studies for which it may be argued (149,166) that in spite of the spurious correlation, a regression of M versus V gives a better estimate of M than does a regression of C versus V, followed by multiplication by V (to obtain M from the regression equation). Benson (21) also supports the idea that there is nothing wrong, per se, with using a relationship between M and V such as equation 12 as long as it is not used to infer a relationship between C and V (by dividing both sides by V). If there is no relationship between C and V then one would expect a coefficient $b \cong 1$, implying a constant concentration. For example, for the model of equation 13, Huber (75) reports values of the exponent b ranging from 0.71 to 1.41. Indeed, many concentration data are scattered in this way (39,166).

Intrastorm Data

Variations of pollutographs during a storm may sometimes be accounted for using the rating curve formulation of equation 12. Consider the data of Figure 6 plotted on log-log scales in Figure 8 (79). Although the narrowness of the curves is somewhat deceptive because of the several log cycles used on the ordinate, the

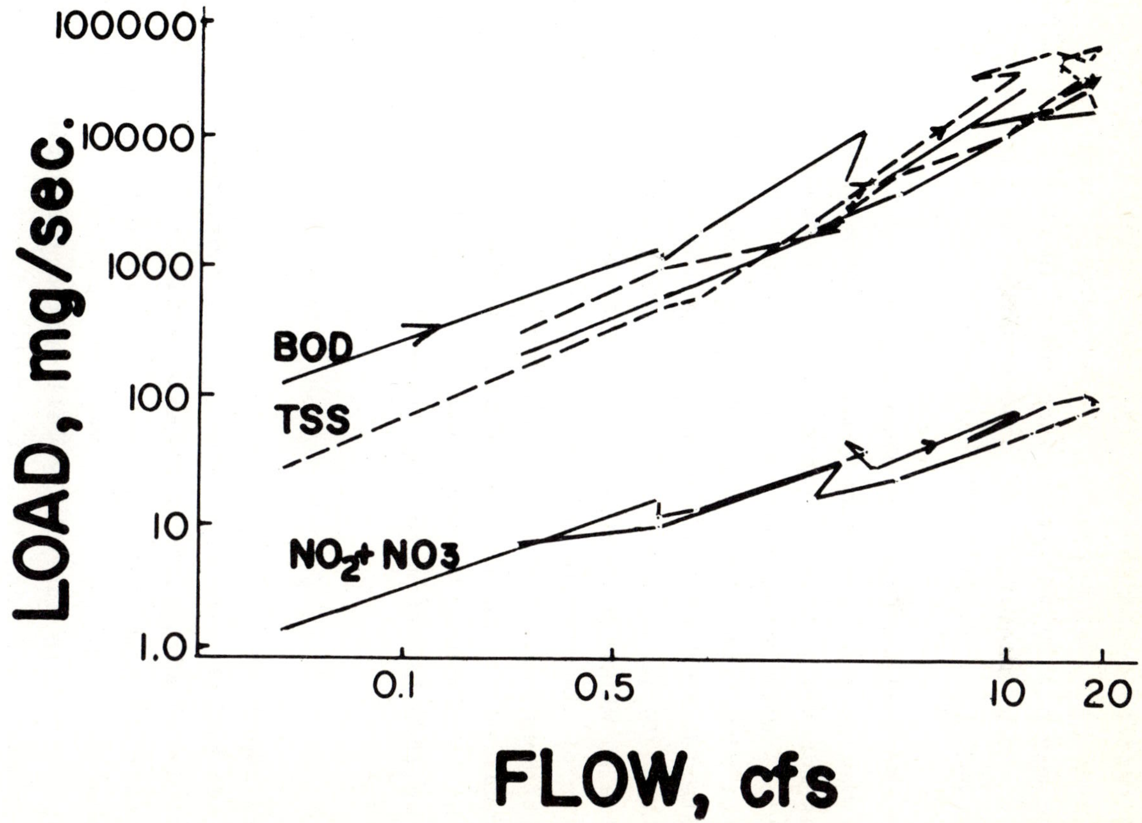

Figure 8. Log-log plot of load versus flow for same data as shown in Figure 6 (Ref. 79). Connected points trace time history in clockwise cycle.

curves still tend to collapse to a line, particularly the curve for nitrite plus nitrate, primarily a dissolved parameter. A straight line may easily be fit to such data, in this case yielding a slope b = 0.65 implying that concentration is inversely related to flow, i.e., proportional to $Q^{-0.35}$.

The HSPF washoff methodology described earlier is essentially a rating curve but only for sediment (91). The SWMM model permits rating curves for any pollutant instead of the exponential washoff formulation, with or without accompanying buildup (79). If buildup is not simulated then there is no upper bound on the amount of material that can be generated from the surface. The DR3M-QUAL model (7) permits a form of a sediment rating curve for pervious areas in which total storm event sediment yield is a power function of the product of total storm volume and peak flow and is apportioned over time steps on the basis of the ratio of the incremental volume during a time step to the sum of the squares of the incremental volumes over all time steps. Other pollutants can then be estimated as linear functions of sediment concentration.

Summary

Rating curves represent a convenient formulation for washoff prediction that bypass several of the parameter estimation problems associated with buildup and exponential washoff. They are purely empirical and must rely on local data for the required regression analyses, but again, <u>all</u> methods must rely on local data. Rating curves are especially well suited for estimation of total storm loads since many data are analyzed on this way. Even intrastorm variations may be modeled using rating curves of the form of equation 12 if hysteresis is small in plots of load rate versus flow rate (e.g., Figures 6 and 8).

Washoff from pervious areas deals more directly with the subject of erosion, but is no less difficult to predict, as will be discussed next.

Land Surface Erosion

Importance

A surprisingly large portion of an urban area can be subject to erosion due to construction activities, unpaved roads and unmaintained lots (64,170). To this could be added undeveloped portions of an overall urban area that may or may not be subject to erosion on the basis of their land cover, such as agricultural enclaves and parks. Finally, erosion is often evident along the banks of depressed urban stream channels. (Scour and deposition within such channels will be discussed in the next section.)

Aside from the loss of soil, erosion causes sediment deposits at points in the drainage network with low velocities and can diminish the hydraulic capacity of the drainage system. Scour of such deposits can cause quantity and quality problems further downstream. And other pollutants are often adsorbed onto sediment particles and may be computed as a fraction of the sediment concentration, as will be discussed in more detail later.

Theory

Many of the same comments apply to erosion and sediment prediction as were made at the beginning of the previous discussion of washoff. Erosion during overland flow is due both to dissipation of rainfall energy and the tractive force of the bed shear stress due to the flow. Although the full power of sediment transport theory may be brought to bear upon the problem, most formulations for watersheds are empirical in nature and are often reduced to simple formulations involving rainfall intensity and/or overland flow rate (164); occasionally an urban runoff model will attempt to incorporate both effects simultaneously (24,91,132). Most, however, deal with sediment erosion separately, at least as far as consideration of rainfall energy is concerned, e.g., SWMM (68,79), STORM (80), DR3M-QUAL (7).

Although it is possible to develop complex representations of the erosion process, most of the urban models are based on simpler formulations developed for non-urban applications (32,41,150). Examples of such formulations will be discussed below as applied in urban runoff quality models.

Semiempirical Methods

Omitting many theoretical justifications and relevant literature citations, land surface erosion models may consist of two components: detachment and transport by rainfall, and detachment and transport by overland flow. In its simplest form, rainfall detachment is a power function of rainfall or rainfall excess intensity,

$$S_r = k_1 \, i^{b_1} \qquad [14]$$

where S_r = detached sediment, mass or mass/area,
i = rainfall intensity,
b_1 = exponent, and
k_1 = coefficient incorporating many factors.

A typical value for the exponent b_1 is 2.0 (41,150). The coefficient k_1 is the product of any grouping of the following factors, depending on the particular model: 1) units conversion, 2) erodability index, 3) topography factor, 4) land use factor, 5) availability or cover factor, 6) management factor, and 7) factors depending on measured sediment yield from the watershed. The HSPF model (91) formulation is based on the earlier work of Crawford and Donigian (32) and uses factors 1, 2, 5 and 6. The Universal Soil Loss Equation, USLE (173,174), is included in SWMM (68,79) and STORM (80) and uses factors 1, 2, 3, 5 and 6 in addition to a different functional form for rainfall energy (the product of intensity times its logarithm). Equation 14 is usually applied on a time step basis, yielding a potential flux of detached sediment.

Sediment transport capacity may be a function of rainfall also (150) but is taken as a function of overland flow in the HSPF model and not considered in SWMM or STORM (all detached sediment is assumed to be transported off the watershed). The power function for this relationship is

$$F_c = k_2 \, Q^{b_2} \qquad [15]$$

where F_c = sediment transport capacity, mass/time,
Q = overland flow rate,
b_2 = exponent, and
k_2 = coefficient.

The exponent b_2 ranges between 1.6 and 2.0 (41); Solomon and Gupta (150) use a value of 5/3. The coefficient, k_2, must be obtained by calibration.

Detachment (scour) by overland flow is yet a third power function,

$$S_o = k_3 Q^{b_3} \quad [16]$$

where S_o = detached (scoured) mass,
b_3 = exponent, and
k_3 = coefficient.

A value of 2/3 is used by Solomon and Gupta (150) for the exponent b_3. Again, k_3 must be found by calibration, and the equation is applied on a time step basis.

Within HSPF, if $F_c > S_i + S_o$ then all detached sediment is transported off the watershed. If the inequality is not satisfied, then the model removes sediment at the rate F_c. Variations on this option are explained in the HSPF User's Manual (91). There is assumed to be an unlimited supply of sediment.

Use of equations 14, 15 and 16 is typical of simplified sediment erosion models and is probably more reasonable than use of the USLE on a storm event basis. The USLE (173,174) was developed for annual sediment yield. Although it does predict erosion as a function of rainfall intensity and is perhaps calibratable, its use on a time step basis in SWMM and STORM is certainly beyond the range of its development and probably questionable. The sediment erosion method used by the DR3M-QUAL model has been described earlier in conjunction with rating curves. In the four models that explicitly model sediment erosion, all other pollutants resulting from the erosion are computed as a fraction of sediment concentration. These fractions (potency factors) will be discussed later.

Summary

Erosion algorithms in urban water quality models are based on similar algorithms developed for non-urban applications, particulary agricultural applications. Functionally, erosion is proportional to power functions of rainfall intensity, overland flow rate or both, with a large array of factors to be incorporated into the preceding coefficents. Calibration is an inevitable requirement, for which there is some help in the literature (32,41,150,151).

Sediment scour and deposition in the drainage network itself lend themselves more to a "proper" theoretical analysis, although in the end they are hardly less tractable, as will be seen in the next section.

Scour and Deposition in Sewers and Channels

Importance

Deposition of solid material during dry-weather flow in combined sewers and subsequent scour during wet-weather flow has long been assumed to form a significant contribution of solids to combined sewer overflows. Field and Struzeski (51) document several examples, to which might be added numerous other case studies and measurements (e.g.,28,59,102,126,127,155). This is often, but not always, evidenced by the presence of a "first flush," in which high concentrations appear early in the storm due to the scour of deposited sediment. Storm sewers subject to infiltration and base flow, and sanitary sewers with cross connections and infiltration can also experience dry-weather deposition, leading to scour during storm runoff events, although the pollution problem may not be as severe as for combined sewers. In addition, natural channels that serve as storm drains are subject to scour during wet weather and may cause sedimentation problems at the very least.

It is in fact the hydraulic problems caused by sedimentation (deposition) in sewers that motivated the first analytical work in this area and the development of design guidelines such as self-cleansing velocities (9). In reality, self-cleansing velocities are complex functions of pipe hydraulics and sediment transport characteristics and may not remain constant for the same pipe (1,175). The problem is further complicated by the very nature of the sediment found in combined and other sewers which tends to be a highly cohesive mixture of organic and inorganic material (126) that does not behave in the same manner as sand or other particles used to develop sediment transport theory. Although sediment transport theory can certainly be successfully applied to prediction of sediment flux in streams and to design of stable channels (63,164), it is probably true that good experimental data are still lacking to be able to adequately predict scour and deposition of the highly heterogeneous material found in sewers. However, the problem has had considerable study, especially in relation to effects on pipe friction (28,154,162) and solids estimation and control (126,127,155).

Ackers (1) presents a potentially very useful application of theory and sewer data for prediction of interactions of flow and sediment discharges, water and sediment depths and sediment concentration. For instance, it is possible to use his results to estimate the depth of sediment that may accumulate in a sewer with the attendant reduction in cross sectional area and need for higher hydraulic gradients. However, his method was not installed formally as part of an urban water quality model at the time of publication. This paper will focus on methods currently used in models presented earlier.

Applied Theory

Perhaps the most ambitious attempt at modeling of scour and deposition in sewers has been the work of Sonnen (152), in which he simulated sediment transport processes including bedload, suspended load and washload using established relationships summarized by Graf (63) and Vanoni (164) that incorporate critical shear stress, friction and roughness interrelationships, and settling theory. Up to ten settling velocity ranges (as characterized by particle size and specific gravity) can be simulated in order to represent the heterogeneous nature of the sediment. The sediment transport routines were added to the quality portion of the Extran Block of SWMM (142) which in the end was unfortunate because the quality portion was eventually dropped from Extran on the basis that the state of the art of urban quality modeling did not justify linkage with such a sophisticated quantity model. Thus the routines added by Sonnen (152) for scour and deposition got eliminated in the transition. He was able to illustrate first flush and other reasonable results with his model, however. If resurrected it would represent a more theoretically based approach for modeling within sewer systems than any routines currently used in operational models.

Of the six operational models listed in Table 3, only SWMM has a scour-deposition routine for the drainage network, located in the Transport Block (79). This simpler formulation came after Sonnen's (152) work and uses the Shields criterion (63,164) to estimate the diameter of the particle subject to incipient motion, d_c, given an average specific gravity of the sediment mixture. Scour and deposition will only be simulated if the specific gravity is entered with a value greater than 1.0. For each such pollutant a particle size distribution of up to four ranges is entered. The model then assumes that particles in the flow with diameter greater than d_c will move to the bed and remain, and particles in the bed with diameters less than d_c will be entrained with the flow. The amount (mass) of particles moving to and from the bed is computed knowing the particle size distribution and the mass of particles in suspension and in the bed. The methodology thus maintains for each conduit a time history of the maximum particle diamter in suspension (i.e., in motion), and the minimum particle diameter in the bed without routing the different settling velocity ranges independently. The model has the desirable characteristic that smaller particles tend to be in the flow and larger in the bed, and upstream deposition will mean less deposition downstream since the dividing diameter is routed (by flow weighting) downstream. Individual pollutants can be more or less "soluble" depending on the particle size distribution input for them.

As implemented, the scour-deposition routine in SWMM can be calibrated (e.g., by adjusting the specific gravity and particle size distribution) but is only a weak theoretical representation since Shields' criterion has never been verified using solids taken from sewers. There is also no effect of deposition on bed geometry or roughness; flow calculations are conducted as if there were no sediment and the sewer conduit was clean. In contrast, Acker's (1) calculations show how easily a pipe can become more than half full of sediment if concentrations are high. Perhaps the best that can be said is that the routine behaves logically, conserves mass, can be calibrated, and does not have to be used!

In the context of usefulness, it should be mentioned that a common way of modeling deposition and first flush effects is simply to increase the "surface" loadings by altering the buildup parameters. Given effluent data for calibration this method may be satisfactory. A causal model is still required, however, for simulation of control measures such as sewer flushing.

Summary

Scour and deposition are important in urban drainage systems from both a hydraulic and water quality aspect. Although empirical procedures are available for estimation of dry-weather solids buildup and wet-weather discharge, the most theoretically based model of the process (152) is no longer available. Alternatives (79) may produce reasonable results but must rely more heavily on calibration than on first principles from sediment transport theory.

Pollutant Fractions

Background

Several of the operational water quality models permit calculation of the concentration of a desired pollutant by equating it to a fraction of the concentration of another pollutant, i.e., $C_1 = f\ C_2$, where f = pollutant fraction (or potency factor or partition coefficient). This has long been associated with non-urban models in which it is customary to first compute sediment or "solids" discharge, and then compute other pollutants as fractions of the sediment concentration under the assumption that they are adsorbed onto the sediment particles (41). This has also been the practice in urban water quality modeling because the original SWMM model (114) was influenced so strongly by the Chicago APWA study (8) in which other pollutants were measured as a fraction of "dust and dirt." In addition, the samples were filtered prior to analysis, thus measuring the soluble portion only. Hence, although BOD5, say, could be calculated as a fraction of dust and dirt, the Chicago values could not account for the insoluble fraction. During calibration of the original SWMM using combined sewer data from San Francisco and Cincinnati, it was necessary to add an arbitrary 5 % of suspended solids to BOD5 to account for the insoluble portion of the BOD and achieve a reasonable calibration in combined sewers. Thus, two types of pollutant fractions may be identified: 1) calculation of a pollutant concentration as a fraction of "dust and dirt" or sediment or "solids" concentration, and 2) calculation (or addition to) a pollutant concentration as a fraction of another pollutant concentration, e.g., as a fraction of suspended solids. Data for the two types arise from different sources, as will be discussed.

Fractions of suspended solids to be added to BOD5 and other pollutants were incorporated into both SWMM and STORM. These could cause very deceptive calibrations to the unwary user, since if suspended solids concentrations were high, even 5 % of their value could dominate the BOD calculation (87), especially when it was not obvious whether or not such fractions could be altered or set to zero. The solution, of course, is simply to permit the user to input his/her own values for such fractions, or ignore them altogether. This is done to varying degrees in most models.

Modeling Options

Among the six operational models discussed, STORM (80) and SWMM (79) permit individual pollutants to be calculated for buildup purposes as a fraction of dust and dirt. Dust and dirt can be an actual sediment parameter or hypothetical. In Chicago, it consisted of all material passing through a 0.25-inch mesh screen. The fractions to be input are arbitrary (i.e., supplied by the user) in both models, but in STORM the pollutants are pre-specified (suspended solids, settleable solids, BOD, nitrogen, orthophosphate, coliforms), whereas in SWMM they may consist of up to any ten constituents. Pollutant buildup in SWMM may optionally be accomplished for each pollutant individually, without needing to use dust and dirt and the pollutant fractions. The advantage of the fractions is that there are considerable data available (discussed below). The HSPF model (91) has essentially the same flexibility as SWMM, for up to ten arbitrary pollutants. In the FHWA model (37), the user may choose to simulate up to 13 of a pre-specified list of pollutants, all of which are calculated as a fraction of suspended solids. Unfortunately, the fractions can only be changed by altering the Block Data subroutine.

The DR3M-QUAL model (7) computes impervious area buildup individually for up to four constituents. However, for pervious area runoff, a linear function (including the constant) of "sediment" concentration is calculated for each. The QQS model (60) does not use dust and dirt buildup; rather, individual parameters are used for up to two pollutants.

The STORM model arbitrarily adds fractions of suspended and settleable solids to the other pollutants. This is inflexible and can cause calibration difficulties as already mentioned. In SWMM an arbitrary matrix of pollutant fractions may be entered, in which any pollutant can be computed as the sum of fractions of other pollutants, if desired. As just described, SWMM, HSPF and DR3M-QUAL can compute any pollutant as a fraction of sediment or "solids," where the sediment is calculated by an erosion algorithm. The use of the various forms of pollutant fractions seems to work most satisfactorily for constituents that are transported primarily in particulate form and somewhat less so for constituents that are primarily in a dissolved state. Modeling is usually simplified if the pollutant is treated as the total of its dissolved and particulate forms, although some studies do report concentrations by particle size (106,129,144,145).

Data Sources

The best summary of pollutant mass as a fraction of dust and dirt mass is provided by Manning et al. (106). These pollutant fractions may be used directly in models such as SWMM, STORM and HSPF or indirectly to develop buildup parameters for individual pollutants.

For purposes of calibration, fairly arbitrary values of fractions of suspended and settleable solids were incorporated into early versions of SWMM and STORM. In an effort to evaluate potency factors for various constituents in both urban and agricultural runoff, Zison (178) examined available data and developed regression relationships as a function of suspended solids and other paramenters. (A recent update [25] of Zison's work emphasizes rate constants more than potency factors.) His only urban catchments were three from Seattle (77), for which several water quality and storm event parameters were available. Unfortunately, statistically meaningful results could only be obtained using log-transformed data, and simple fractions of the type required for potency factors are seldom reported. Zison (178) acknowledged this and suggested that model modifications might be made or piecewise-linear approximations be made to the power function relationships. Several other studies also report linear relationships among pollutants (70,94,104,135), although there is a minor difficulty in using many of the results that include an intercept (constant) term in the regression since since only the DR3M-QUAL model permits the input of this term directly, and then only for computing other pollutants as a linear function of sediment. The presence of an intercept is difficult to interpret in any event since it implies a constant or base level concentration regardless of the value of the independent variable.

Summary

Pollutant fractions or potency factors may be used to compute concentrations as a fraction of dust and dirt, sediment or "solids" concentrations, and/or to compute or add to pollutant concentrations as a fraction of the concentration of other pollutants. Many data are available, although not always in the desired functional form. It is perhaps regrettable that so much attention has been devoted to regression and so little to more basic sience in the evaluation of these fractions. For example, Sonnen (153) points out that it may be possible to apply basic chemical principles such as the Freundlich isotherm to describe adsorption phenomena. (This is available as an option in HSPF for pesticide simulation.)

Other Quality Sources

Precipitation

There is now considerable public awareness of the fact that precipitation is by no means "pure" and does not have characteristics of distilled water. Although low pH (acid rain) is the best known parameter, many other substances can also be found in precipitation, including organics, solids, nutrients, metals and pesticides. Nutrient concentrations, particularly nitrogen parameters, are likely to be higher in rainfall than in runoff (93,108), and Randall et al. (136) reported higher rainfall loadings than runoff loadings for residential and pasture land uses, indicating uptake by the land surface. Runoff loadings were higher than rainfall for highly tilled agricultural land and for a commercial shopping center. In a study in Knoxville, Tennessee, Betson (22) found precipitation to contribute a large fraction of measured runoff quality loads for several parameters. Rainfall concentration and loading data are now widely available (22,27,71,93,106,108,123,136,163) for use in urban quality models.

The SWMM (79) and DR3M-QUAL (7) models allow modeling of precipitation quality explicitly, through use of a constant concentration. These loads are then added to loads generated by other sources. In other models, precipitation quality can only be simulated indirectly, such as by adjustment of parameters to achieve a constant concentration. Randall et al. (136) point out that rainfall concentration is inversely related to cumulative depth during a storm, with a tendency for event precipitation loads to vary only over a small range. Thus, if intrastorm variations are important, rainfall contributions might be simulated by buildup-washoff. Another approach is to simulate the precipitation scavenging process directly from its basic physics. A modification to SWMM using this idea is presented by Boregowda (24).

Snow quality is more complex because it acquires pollutants by uptake from surface sources as well as by atmospheric scavenging. For this reason it is customary to simulate snow quality by means of the buildup options. Proctor and Redfern and James F. MacLaren (134,135) provide a review of snow quality data. Simulation of snowmelt quality is not straight forward, however, since concentrations can be much higher in snow packs than in runoff, especially for particulates and heavy metals. Such constituents may be simply deposited beneath the snow pack during ordinary melt sequences and not reach the drainage system.

Catchbasins

Catchbasins or gully pots are found at stormwater inlets in many cities. They were originally intended to provide pollutant removal by the trapping of coarse debris and solids, thus preventing sewer clogging and siltation, and also intended to prevent emanation of odors from combined sewers by providing a water seal. The volume of catchbasin sumps varies considerably, ranging from 2.8 to 78 ft^3 (0.08 to 2.2 m^3). This volume is typically reduced by a large quantity of solids trapped in the sump.

An extensive study of catchbasins by Lager at al. (99) indicated that the treatment effect of catchbasins was minimal because of infrequent maintenance; the median cleaning frequency in 1973 was once per year. As a result, they are more likely to represent a source rather than a sink of pollutants, as material is flushed from the sumps during storms; they may also contribute to the first flush effect. A study by Fletcher and Pratt (55) indicated that mean contributions to total storm loads could be as high as 10 to 30 %. However, experience with SWMM simulations of catchbasins indicates that their contribution to total loads is negligible in most cases; results are insensitive to their presence (79).

SWMM (79) is the only urban quality model to simulated catchbasins explicitly. It is done simply by assuming a pollutant load present at the beginning of a storm, which is then flushed in proportion to the volume of runoff passing through the catchbasin, leading to an exponential reduction in pollutant mass in the catchbasin. This is also the technique used by Fletcher and Pratt (55). Other models may easily include catchbasin contributions simply by increasing surface buildup quantities.

Dry-Weather Flow

In combined sewers, dry-weather flow represents a significant fraction of the total flow on an annual basis, although on an annual basis only 3 to 5 % of it is lost from the system during overflows (51). Baseflow may also occur in storm sewers due to infiltration or seepage from banks into unlined channels and be of low quality (20). Even during overflows, the quality of the CSO is influenced more by scour of deposited sediment and by the quality of the entering storm water than by the effect of mixing of sanitary and storm water (51). Nonetheless, it may be necessary to simulate the dry-weather flow if only to model the system hydraulics adequately, and inclusion of quality is not difficult.

Dry-weather flow quality values are best obtained from treatment plant or other sampling records. The base flow in the drainage network may then be modeled with a constant concentration. Treatment plant records can also be used to infer wet-weather quantities and overflows (100,116). An embellishment is to allow hourly and daily correction factors, in which average values of flow and concentration may be multiplied by factors to correct for diurnal and weekly variations (143,161,167).

Among the operational models, SWMM, STORM and QQS allow the specification of baseflow explicitly, with average (constant) concentrations. Flows and quality are subject to hourly and daily correction factors. Baseflow and quality may be entered as a time series in HSPF and are not an option in the DR3M-QUAL or FHWA models. Both SWMM and STORM also allow estimation of dry-weather flow quantity and quality on the basis of regression relationships with demographic factors, but these would rarely be an adequate substitute for site-specific data.

Summary

Sources of pollutants in storm and combined sewage include surface accumulation, erosion, scour, precipitation, catchbasins, and dry-weather flow. Mathematical formulations are available to simulate all sources, incorporating various degrees of the actual physics and chemistry of the indicated processes, and all rely heavily on site-specific data for their accuracy. The next section describes the transport of pollutants through the drainage network once they have been generated.

Pollutant Transport in the Drainage Network

Importance

During a storm event a pollutant spends relatively little time in the sewer system -- usually less than an hour for typical sewer system lengths and velocities. Hence, pollutant routing does not usually receive the attention that flow routing does. This is probably justified in light of preceding discussions since it is usually more important to simulate the total quantities correctly than their intrastorm variations, with the exceptions noted earlier concerning evaluation of controls and real-time operation. Nonetheless, pollutants do not move instantaneously from inlets to the drainage system outlet, and if the drainage system is simulated, methods must be used to provide for pollutant routing. Although the theory is well understood for streams and channels (e.g., 54), the complex hydraulic phenomena found at some structures in a sewer system can still make the task challenging.

Theory

The fundamental transport processes of advection and diffusion may be combined in the advective-diffusion equation,

$$\partial C/\partial t + U\,\partial C/\partial x = D\,\partial^2 C/\partial x^2 +/- S \qquad [17]$$

where C = concentration,
t = time,
x = distance along the conduit,
U = average velocity,
D = longitudinal dispersion coefficient, and
S = term representing all sources and sinks, including decay.

The complete equation is not ordinarily applied in an urban drainage network, although its solution for an instantaneous plane source is the basis for dilution (time of travel) methods for velocity measurements in channels (72). When the right hand side of equation 17 is set to zero, it corresponds to pure advection (plug flow) in which pollutant concentration distributions are routed downstream with no change in shape, a common modeling choice. The other extreme is represented by omitting the two middle terms (the advective and diffusive terms), corresponding to a completely mixed "tank" or continuously stirred tank reactor, CSTR, also a common modeling choice when applied to conduits in series. The "middle ground" may be represented by omitting only the diffusive term (third term), in which pollutants are subject to advection plus the action of sources and sinks. However, since residence times in the drainage system are so brief, biological and chemical decay and transformations are usually omitted from models of channels and pipes, leaving scour and deposition as the only source-sink processes to consider; this has been discussed earlier. Of course, decay and other source-sink processes are key elements in the simulation of storage in the network. Bowie et al. (25) present recent information on rate constants.

Modeling Implementations

Pure advection or plug flow is probably the method of choice for routing since it best preserves time lags in the pollutographs. Both the DR3M-QUAL (7) and the QQS (60) models simulate plug flow using Lagrangian techniques that avoid a numerical solution of the left hand side of equation 17. Rather, a bookkeeping system is established for each pipe whereby the total mass of a pollutant in a pipe consists of one or more plugs of material advected into the pipe during previous time steps. The outflow of pollutant from the pipe during a time step is the sum of the mass contained in enough plugs to satisfy the required outflow volume, under a "first in - first out" criterion. There is some "numerical mixing" that occurs during this summation, but overall it is minimal, particularly if the time step is approximately equal to the travel time in the pipe. (The time step must not exceed the travel time in any pipe, however.) Since pipe or channel lengths vary throughout the network, the time step is limited by the shortest travel time, and longer pipes will experience more numerical mixing. However, Lagrangian schemes of this type are often invoked specifically to minimize numerical diffusion that may accompany finite difference solutions of the pure advection equation (33,54). Plug flow routines may also be used to simulate storage unit routing (7,79,148).

Pollutant routing in the SWMM (79) and the FHWA (37) models is accomplished by complete mixing within each channel or pipe. The mixed outflow becomes the inflow to downstream segments, and is easily computed without an extensive bookkeeping procedure. On the other hand, it is subject to considerable artificial mixing, since pollutants entering during a time step are mixed throughout the entire length of the segment, although the increase in concentration at the downstream end will usually be gradual. Thus, plug flow routing can reproduce much more rapid changes in the pollutograph than can routing by complete mixing. However, complete mixing lends itself more readily to simulation of complex hydraulic structures, for which only a mixing volume is required instead of an estimate of travel time.

The HSPF model (91) basically performs pollutant routing by complete mixing with modifications to account for reaches in which the travel time may be less than the time step. All models perform complete mixing at junctions and inflows. The STORM (80,141) routing is the simplest of any: it does none since the drainage network is not simulated. Time variations in the outflow might still be reasonable if actual travel times are expected to be much less than the time step, which is usually one hour.

Summary

Routing of quality parameters through the drainage network has not received as much attention in modeling as has the generation of the parameters from the various sources dicussed earlier. This is usually justified on the basis of the short travel times and residence times within the drainage system. For prediction of total event pollutant loads the routing is perhaps unimportant, but if the detailed temporal variations of the pollutograph are desired or if scour and deposition within the drainage network must be simulated, then the routing method assumes greater importance. Although routing by complete mixing is somewhat more versatile in terms of representing more of the complex hydraulic structures found in a sewer system, plug flow routines are expected to be preferable for simulation of sharp concentration changes characteristic of transport in the well-defined channels of a drainage network.

Storage Treatment Processes

Introduction

Modeling of storage and treatment processes as applied to storm and combined sewage is such an involved topic that space prohibits an extensive dicsussion in this paper. Instead, there will only be room for mention of processes to be considered and literature citations.

Storage-treatment facilities might be classified as detention versus non-detention, structural versus non-structural (e.g., a tank versus a roadside grassed swale), and high-rate versus conventional sewage treatment. The prevalence of detention facilities in urban areas, especially in new developments, indicates the need for models that can simulate removal processes that may occur during storage for quantity control; this will be the emphasis of this section. A survey of modeling activities for conventional waste treatment (and other relevant topics) is given by James (82). Modeling guidelines and data often accompany other descriptive and design information about high-rate treatment processes unique to sewer systems and urban runoff.

See the reviews by Field (49,50) for references.

Modeling Detention Facilities

Flow routing in detention facilities is easily accomplished by conventional reservoir routing techniques such as the Puls method or nonlinear reservoirs (165). The resulting inflow and outflow hydrographs and storage volumes are then used by quality routing routines that may employ pure advection (7,79,148) complete mixing (79,113) or a more theoretically exact, but complex solution of the advective-diffusion equation (47,113). Quality routing generally follows the same principles as outlined in the previous section on routing through the drainage network, although complications can arise when there are multiple pathways out of a device, particularly for sludge removal (79,122).

The more challenging problem is usually to simulate the removal processes. These removals can occur through physical means, such as sedimentation, filtration or other solids removal processes, or through biological or chemical actions. Physical processes may often be characterized in terms of their action on particles; hence a measure of treatability, such as the distribution of settling velocities, is essential. Sedimentation can then be simulated through work originating with Camp (30) and updated by several investigators for ultimate implementation in models such as SWMM (79,122) and DR3M-QUAL (7). For this purpose information on the particle size and specific gravity distributions is required (or more directly, the settling velocity distribution). These can be input as distributions (7) or as discrete ranges (79,152). Settled material forms sludge. This is optionally routed out of the storage unit in SWMM and not accounted for in DR3M-QUAL. Neither model accounts for scour of deposited material such as might occur in a detention pond subjected to an extreme hydrologic event. When particle sizes are simulated, it is also straight forward to simulate removal by screening devices (79).

Treatment by any of the three processes, physical, chemical or biological, can also be simulated through the use of empirical removal functions, such as the classical exponential form of removal as a function of detention time (46). Most mathematical removal functions of this type may be incorporated into the SWMM model through the use of generalized removal functions. Removal can be functionally be related to concentration or removal of any pollutant, or to inflow or detention time.

HSPF (91) can simulate detention processes to some degree by treating a storage unit as a stream or reservoir segment, for which extensive algorithms are included. STORM (79,141) simulates storage without any removal, and treatment strictly as a percent removal. QQS (60) simulates some of the particle settling mechanisms but apparently through the use of default values for settling characteristics. The FHWA model (37) simulates removal by complete mixing with first order decay, as can the SWMM, HSPF and QQS models.

The critical information in any such modeling exercise is the treatability of the influent to the detention facility. Pollutants that are in particulate form or adsorbed onto solids are relatively easy to remove by physical treatment, such as sedimentation. Dissolved pollutants are typically much more difficult to remove in a high flow setting such as for storm and combined sewage. Indeed, there is some evidence that "wet" detention ponds can increase the quantity of nitrogen released in the effluent (48).

Information Sources

Abundant information is now available on almost every aspect of detention facilities. Poertner (130) has prepared a manual with much practical design, construction, maintenance and analysis information. Proceedings edited by DeGroot (34) provide similar coverage, along with considerable data and examples. And quality data and performance functions (e.g., statistical analysis) are available from too many sources to provide a comprehensive list (e.g., 18,45,48,109,119).

Other Modeling Considerations

Computer Requirements

All of the models discussed in this paper originated on large or "main frame" computers, and all continue to be available for such machines (16). However, this is no longer a routine requirement with the current explosion in microcomputer population and capabilities. For example, SWMM (56,85) and STORM (23) have been adapted for microcomputers, and other quality models probably could be if the effort was taken to make minor modifications and recompile the Fortran code. Many rainfall-runoff models are advertised for microcomputers. The implication is that access to urban water quality models will increase as they are converted to microcomputers, with an accompanying elevation of the level of analysis that can be attempted by engineers on a routine basis. Even continuous simulation is possible when sufficient disk storage is available for the input precipitation time series (23,138).

Advancing computer technology will also lead further into the realm of expert systems approaches to modeling in which the user is guided along the subtle pathway of parameter estimation. One attempt has already been made along these lines for the HSPF model (57). A key element of such software is simply to point the user to the various sources of data, and to help the user to document his/her parameter estimates.

Modeling Limitations

The caveat that has been emphasized repeatedly in this paper is that urban runoff quality modeling cannot provide meaningful absolute values of concentrations and loads without adequate, site-specific monitoring data for calibration and verification. Even relative magnitudes of variables will be suspect without some evidence from field sampling that, say, street cleaning really does reduce pollutant loads.

Aside from the needs for data, models can by no means perform all the tasks that a study might need. None of the models simulate physics or chemistry from first principles, except in a cursory manner for particle settling, for instance. In particular, the processes of buildup and washoff are represented in models almost wholly in an empirical fashion. Although erosion estimates have some basis in theory, they still must rely upon calibration for parameter estimates. As long as this is borne in mind, the models can be used successfully, given adequate data. In some cases, simple formulations lead to easier parameter estimation, since the number of parameters may be far fewer and easier to evaluate (e.g., an intercept and slope for a rating curve) than for physically-based parameters.

Pollutant routing through complex hydraulic structures is probably not adequately simulated by any model, particularly when sediment pathways are strongly guided by secondary currents. This is an area in which physical models are still widely used. Of the six operational models discussed, only HSPF presently simulates subsurface quality processes, at the expense of an entirely new tier of required soil and groundwater quality parameters. This transport pathway may become increasingly important as the population of detention ponds and other land surface "treatment" facilities continues to rise.

The economics of urban runoff quality control is usually evaluated after the modeling exercise. However, the ability of continuous simulation to produce a frequency analysis on any desired parameter is a tremendous boon to estimation of costs, benefits and reliability.

Case Studies

Perhaps the most critical need at present is simply better publication of user experience in the form of case studies with operational models. Although every user's manual indicates information about case studies, these are often in the form of obscure reports. Users should be encouraged to publish their modeling experience in accessible journals, and sponsoring agencies should encourage publication of case studies more often. Probably the best source of such information at present are proceedings of the the Storm and Water Quality Users' Group Meetings and other reports published by the EPA Center for Water Quality Modeling (16), in which may be found many case studies using most of the models mentioned in this paper, plus several more. Several examples of analysis, data collection, and many very practical considerations dealing with implementation of controls are presented by Lynard et al. (105).

Case studies are most conspicuously absent from this paper and must remain so due to considerations of length. However, references for every model discussed refer to case studies and model applications.

Conclusions

Deterministic models of urban runoff quality are much beyond their infancy. However, certain modeling formulations have had a remarkable longevity, particulary the buildup-washoff concept that dates from 1971 (114). Although there have been passing forays into sediment transport theory (152), there have been relatively few breakthroughs in terms of better implmentation of the physics of most of the processes involved, or even into more causal empirical relationships (24). The passage of time has instead added a massive database for use in model calibration, verification and improvements (e.g., 45,59,77). It has been observed, however, that because most monitoring programs only sample at the "end of the pipe," few of these data relate to the fundamental physics of the simulated processes (153). Instead, estimation of empirical parameters may now be done with somewhat more confidence.

An array of operational water quality models now exists for the engineering community in the sense of documentation, user support and experience; six have been discussed extensively in this paper (DR3M-QUAL, FHWA, HSPF, QQS, STORM, SWMM). Without exception, these models are capable of accurate simulation of hydrographs and pollutographs, given adequate data for calibration. By the same token, none of the models can be assumed to accurately predict absolute magnitudes of concentrations and loads without such data, although they may be very useful for comparative studies and estimation of relative effects. Caveat emptor!

Acknowledgements

The writer is grateful for support in modeling and other activities in urban hydrology from several projects sponsored by the U.S. Environmental Protection Agency. The many consultations with colleagues and students at the University of Florida and elsewhere are also gratefully acknowledged, particularly Dr. James P. Heaney.

References

1. Ackers, P. Urban Drainage: The Effects of Sediment on Performance and Design Criteria. Proceedings of the International Conference on Urban Storm Drainage, University of Southampton. Pentech Press. London. 535-545; 1978 April.

2. Adams, B.J. and J.B. Bontje. Microcomputer Applications of Analytical Models for Urban Stormwater Management. Proceedings of the Conference on Emerging Computer Techniques in Stormwater and Flood Management, Niagra-On-The-Lake, Ontario. American Society of Civil Engineers, New York. 138-162; 1983 November.

3. Adams, B.J. and C.D.D. Howard. The Pathology of Design Storms. Publication 85-03, University of Toronto, Dept. of Civil Engineering, Toronto, Ontario. 1985 January.

4. Alley, W.M. Estimation of Impervious-Area Washoff Parameters. Water Resources Research. 17(4):1161-1166; 1981 August.

5. Alley, W.M., Ellis, F.W. and R.C. Sutherland. Toward A More Deterministic Urban Runoff Quality Model. Proceedings of the International Symposium of Urban Storm Runoff, University of Kentucky, Lexington. 171-182; 1980 July.

6. Alley, W.M. and P.E. Smith. Estimation of Accumulation Parameters for Urban Runoff Quality Modeling. Water Resources Research. 17(6):1657-1664; 1981 December.

7. Alley, W.M. and P.E. Smith. Multi-Event Urban Runoff Quality Model. U.S. Geological Survey Open File Report 82-764, Reston, VA. 1982.

8. American Public Works Association. Water Pollution Aspects of Urban Runoff. Federal Water Pollution Control Administration, Contract WP-20-15, Washington, DC. 1969.

9. American Society of Civil Engineers and Water Pollution Control Federation. Gravity Sanitary Sewer Design and Construction. ASCE Manuals and Reports on Engineering Practice No. 60, New York. 1982.

10. Ammon, D.C. Urban Stormwater Pollutant Buildup and Washoff Relationships. Master of Engineering Thesis, University of Florida, Dept. of Environmental Engineering Sciences, Gainesville. 1979.

11. Amy, G., Pitt. R., Singh, R., Bradford, W.L. and LaGraff, M.B. Water Quality Management Planning for Urban Runoff. EPA-440/9-75-004 (NTIS PB-241689), Environmental Protection Agency, Washington, DC, 1974 December.

12. Arnell, V. Analysis of Rainfall Data for Use in Design of Storm Sewer Systems. Proceedings of the International Conference on Urban Storm Drainage, University of Southampton. Pentech Press, London. 71-86; 1978 April.

13. Arnell, V. Rainfall Data for the Design of Sewer Pipe Systems. Report Series A:8, Chalmers University of Technology, Dept. of Hydraulics, Goteborg, Sweden. 1982.

14. AVCO Economic Systems Corp. Storm Water Pollution from Urban Land Activity. EPA Report 11034FKL07/70 (NTIS PB-195281), Environmental Protection Agency, Washington, DC. 1970 July.

15. Balmer, P., Malmqvist, P-A. and A. Sjoberg. Eds. Proceedings of the Third International Conference on Urban Storm Drainage, Chalmers University, Goteborg, Sweden. 4 Vol. 1984 June.

16. Barnwell, T.O., Jr. EPA's Center for Water Quality Modeling. Proceedings of the Third International Conference on Urban Storm Drainage, Chalmers University, Goteborg, Sweden. 2:463-466; 1984 June.

17. Barnwell, T.O., Jr. and J.L. Kittle. Hydrologic Simulation Program - Fortran, Development, Maintenance and Applications. Proceedings of the Third International Conference on Urban Storm Drainage, Chalmers University, Goteborg, Sweden. 2:463-466; 1984 June.

18. Bedient, P.B. Evaluation of Effects of Stormwater Detention in Urban Areas. Research Project Completion Report, Project No. B-244-TEX, (NTIS PB82-256736), Rice University, Dept. of Environmental Science and Engineering, Houston, TX. 1982 May.

19. Bedient, P.B., Harned, D.A. and Characklis, W.G. Stormwater Analysis and Prediction in Houston. Journal of the Environmental Engineering Division, ASCE, 104(EE6):1087-1100; 1978 December.

20. Bedient, P.B., Lambert, J.L. and P. Machado. Low Flow and Stormwater Quality in Urban Channels. Journal of the Environmental Engineering Division, ASCE, 106(EE6):421-436; 1980 April.

21. Bensen, M.A. Spurious Correlation in Hydraulics and Hydrology. Journal of the Hydraulics Division, ASCE, 91(HY4):35-42; 1965 July.

22. Betson, R.P. Bulk Precipitation and Streamflow Quality Relationships in an Urban Area. Water Resources Research. 14(6):1165-

1169; 1978 December.

23. Bontje, J.B., Ballantyne, I.K. and B.J. Adams. User Interfacing Techniques for Interactive Hydrologic Models on Microcomputers. Proceedings Conference on Stormwater and Water Quality Management Modeling, Burlington, Ontario, Report R128, Computational Hydraulics, McMaster University, Hamilton, Ontario. 75-89; 1984 September.

24. Boregowda, S. Modeling Stormwater Pollutants in Hamilton, Canada. Ph.D. Thesis, McMaster University, Hamilton, Ontario. 1984 September.

25. Bowie, G.L., Mills W.B., Porcella, D.B., Campbell, C.L., Pagenkopf, J.R., Rupp, G.L., Johnson, K.M., Chan, P.W.H., Gherini, S.A. and C.E. Chamberlin. Rates, Constants, and Kinetics Formulations in Surface Water Quality Modeling (Second Edition). EPA/660/3-85/040, Environmental Protection Agency, Athens, GA. 1985 June.

26. Brandstetter, A.B. Assessment of Mathematical Models for Storm and Combined Sewer Management. EPA-600/2-76-175a (NTIS PB-259597), Environmental Protection Agency, Cincinnati, OH. 1977 August.

27. Brezonik, P.L. Nutrients and Other Biologically Active Substances in Atmospheric Precipitation. Proceedings Symposium on Atmospheric Contribution to the Chemistry of Lake Waters, International Association of Great Lakes Research. 166-186; 1975 September.

28. Broeker, H.W. Impact of Depositions on Sewer Operation. Proceedings of the Third International Conference on Urban Storm Drainage, Chalmers University, Goteborg, Sweden. 3:879-888; 1984 June.

29. Brown, R.G. Relationship Between Quantity and Quality of Storm Runoff and Various Watershed Characteristics in Minnesota, USA. Proceedings of the Third International Conference on Urban Storm Drainage, Chalmers University, Goteborg, Sweden. 3:791-800; 1984 June.

30. Camp, T.L. Sedimentation and the Design of Settling Tanks. Transactions ASCE, 111:895-936; 1946.

31. Chan, S. and R.L. Bras. Urban Storm Water Management: Distribution of Flood Volumes. Water Resources Research. 15(2):371-382; 1979 April.

32. Crawford, J.H. and A.S. Donigian, Jr. Pesticide Transport and Runoff Model for Agricultural Lands. EPA-660/2-74-013, Environmental Protection Agency, Washington, DC. 1973 December.

33. Cunge, J.A., Holly, F.M. Jr. and A. Verwey. Practical Aspects of Computational River Hydraulics. Pittman Publishing Inc., Boston, MA. 1980.

34. DeGroot, W. Ed. Stormwater Detention Facilities. Proceedings of the Conference, Henniker, NH. American Society of Civil Engineers, New York. 1982 August.

35. Delleur, J.W. and H.C. Torno. Eds. Urban Hydrology. Proceedings of the International Symposium, Baltimore, MD. American Society of Civil Engineers, New York. 1983 May.

36. Desbordes, M. and E. Servat. Solids in Urban runoff, Statistical Analysis of French Experimental Data. Proceedings of the Third International Conference on Urban Storm Drainage, Chalmers University, Goteborg, Sweden. 3:947-956; 1984 June.

37. Dever, R.J., Jr., Roesner, L.A. and J.A. Aldrich. Urban Highway Storm Drainage Model Vol. 4, Surface Runoff Program User's Manual and Documentation. FHWA/RD-83/044, Federal Highway Administration, Washington, DC. 1983 December.

38. Dever, R.J., Jr., Roesner, L.A. and D-C. Woo. Development and Application of a Dynamic Urban Highway Drainage model. Urban Stormwater Hydraulics and Hydrology, Proceedings of the Second International Conference on Urban Storm Drainage, Urbana, IL. Water Resources Publications, Littleton, CO. 229-235; 1981 June.

39. Diniz, E.V. and W.H. Espey, Jr. Maximum Utilization of Water Resources in a Planned Community, Application of the Storm Water Management Model, Volume I. EPA-600/2-79-050c, Environmental Protection Agency, Cincinnati, OH. 1979 July.

40. Di Toro, D.M. Probability Model of Stream Quality Due to Runoff. Journal of Environmental Engineering, ASCE. 110(3):607-628; 1984 June.

41. Donigian, A.S., Jr. and H.H. Davis, Jr. User's Manual for Agricultural Runoff Management (ARM) Model. EPA-600/3-78-080, Environmental Protection Agency, Athens, GA. 1978 August.

42. Driscoll, E.D. in Benefit Analysis for Combined Sewer Overflow Control. Seminar Publication. EPA-625/4-79-013, Environmental Protection Agency, Cincinnati, OH. 1979 April.

43. Ellis, F. W. and R.C. Sutherland. An Approach to Urban Pollutant Washoff Modeling. Proceedings International Symposium on Urban Storm Runoff, University of Kentucky, Lexington. 325-340. 1979 July.

44. Environmental Protection Agency. EPA Environmental Data Base and Model Directory. Information Clearinghouse (PM 211A), Environmental Protection Agency, Washington, DC. 1983 July.

45. Environmental Protection Agency. Results of the Nationwide Urban Runoff Program, Volume I Final Report. NTIS PB84-185552, Environmental Protection Agency, Washington, DC. 1983 December.

46. Fair, M.F., Geyer, L.C. and D.A. Okun. Water and Wastewater Engineering. John Wiley and Sons, New York. 1968.

47. Ferrara, R.A. and D.R.F. Harleman. Hydraulic Modeling for Waste Stabilization Ponds. Journal of the Environmental Engineering Division, ASCE. 107(EE4):817-830; 1981 August.

48. Ferrara, R.A. and P. Witkowski. Stormwater Quality Characteristics in Detention Basins. Journal of Environmental Engineering, ASCE. 109(2):428-447; 1983 April.

49. Field, R. The USEPA Office of Research and Development's View of Combined Sewer Overflow Control. Proceedings of the Third International Conference on Urban Storm Drainage, Chalmers University, Goteborg, Sweden. 4:1333-1356; 1984 June.

50. Field, R. Urban Runoff: Pollution Sources, Control, and Treatment. Water Resources Bulletin. 21(2):197-206; 1985 April.

51. Field, R. and E.J. Struzeski, Jr. Management and Control of Combined Sewer Overflows. Journal Water Pollution Control Federation. 44(7):1393-1415; 1972 July.

52. Field, R., Struzeski, E.J., Jr., Masters, H.E. and A.N. Tafuri. Water Pollution and Associated Effects from Street Salting. EPA-R2-73-257 (NTIS PB-222795), Environmental Protection Agency, Cincinnati, OH. 1973 May.

53. Field, R. and R. Turkeltaub. Urban Runoff Impacts: Program Overview. Journal of the Environmental Engineering Division, ASCE, 107(EE1):83-100; 1981 February.

54. Fischer, H.B., List, E.J., Koh, R.C.Y., Imberger, J. and N.H. Brooks. Mixing in Inland and Coastal Waters. Academic Press, New York. 1979.

55. Fletcher, I.J. and C.J. Pratt. Mathematical Simulation of Pollutant Contributions to Urban Runoff from Roadside Gully Ponds. Urban Stormwater Quality, Management and Planning, Proceedings of the Second International Conference on Urban Storm Drainage, Urbana, IL. Water Resources Publications, Littleton, CO. 116-124; 1981 June.

56. Gancarz, D.H. and W.C. Huber. The SWMM Storage/Treatment Block for the IBM-PC. Proceedings Conference on Stormwater and Water Quality Management Modeling, Burlington, Ontario, Report R128, Computational Hydraulics, McMaster University, Hamilton, Ontario. 53-58; 1984 September.

57. Gaschnig, J. Development of a Knowledge-Based Expert System for Water Resource Problems. Final Report, SRI Project 1619, Stanford Research Institute, Menlo Park, CA. 1981 August.

58. Geiger, W.F. Urban Runoff Pollution Derived from Long-Time Simulation. Proceedings National Symposium on Urban Hydrology and Sediment Control, University of Kentucky, Lexington. 259-270; 1975 July.

59. Geiger, W.F. Characteristics of Combined Sewer Runoff. Proceedings of the Third International Conference on Urban Storm

Drainage, Chalmers University, Goteborg, Sweden. 3:851-860; 1984 June.

60. Geiger, W.F. and H.R. Dorsch. Quantity-Quality Simulation (QQS): A Detailed Continuous Planning Model for Urban Runoff Control, Volume I, Model Description, Testing, and Applications. EPA-600/2-80-011 (NTIS PB80-190507), Environmental Protection Agency, Cincinnati, OH. 1980 March.

61. Geiger, W.F. and H.R. Dorsch. Quantity-Quality Simulation (QQS): A Detailed Continuous Planning Model for Urban Runoff Control, Volume II, User's Manual. EPA-600/2-80-116 (NTIS PB80-221872), Environmental Protection Agency, Cincinnati, OH. 1980 August.

62. Geiger, W.F., LaBella, S.A. and G.C. McDonald. Overflow Abatement Alternatives Selected by Combining Continuous and Single Event Simulations. Proceedings National Symposium on Urban Hydrology, Hydraulics and Sediment Control, University of Kentucky, Lexington. 71-79; 1976 July.

63. Graf, W.H. Hydraulics of Sediment Transport. McGraw-Hill Book Co., New York. 1971.

64. Guy, H.P. Sediment Problems in Urban Areas. U.S. Geological Survey, Circular 601-E, Reston, VA. 1970.

65. Hall, M.J. Urban Hydrology. Elsevier Applied Science Publishers. New York. 1984.

66. Heaney, J.P. and W.C. Huber. Nationwide Assessment of Urban Runoff Impact on Receiving Water Quality. Water Resources Bulletin. 20(1):35-42; 1984 February.

67. Heaney, J.P., Huber, W.C., Medina, M.A., Jr., Murphy, M.P., Nix, S.J. and S.M. Hasan. Nationwide Evaluation of Combined Sewer Overflows and Urban Stormwater Discharges, Vol. II: Cost Assessment and Impacts. EPA-600/2-064b (NTIS PB-266005), Environmental Protection Agency, Cincinnati, OH. 1977 March.

68. Heaney, J.P., Huber, W.C., Sheikh, H., Medina, M.A., Doyle, J.R., Peltz, W.A. and Darling, J.E. Urban Stormwater Management Modeling and Decision Making. EPA-670/2-75-022 (NTIS PB-242290), Environmental Protection Agency, Cincinnati, OH. 1975 May.

69. Helliwell, P.R. Ed. Urban Storm Drainage. Proceedings of the International Conference on Urban Storm Drainage, University of Southampton. Pentech Press, London. 1978 April.

70. Hemain, J.C. Statistical Analysis of Runoff Quality Data from French and U.S. Catchments. Urban Stormwater Quality, Management and Planning, Proceedings of the Second International Conference on Urban Storm Drainage, Urbana, IL. Water Resources Publications, Littleton, CO. 133-140; 1981 June.

71. Hendry, C.D. and P.L. Brezonik. Chemistry of Precipitation at Gainesville, Florida. Environmental Science and Technology.

14(7):843-849; 1980 July.

72. Herschy, R.W. Ed. Hydrometry, Principles and Practices. John Wiley and Sons, New York. 1978.

73. Hogland, W., Berndtsson, R. and Larson, M. Estimation of Quality and Pollution Load of Combined Sewer Overflow Discharge. Proceedings of the Third International Conference on Urban Storm Drainage, Chalmers University, Goteborg, Sweden. 3:841-850; 1984 June.

74. Howard, C.D.D. Theory of Storage and Treatment Plant Overflows. Journal of the Environmental Engineering Division, ASCE, 102(EE4):709-722; 1976 August.

75. Huber, W.C. Urban Wasteload Generation by Multiple Regression Analysis of Nationwide Urban Runoff Data. in Workshop on Verification of Water Quality Models, R.V. Thomann and T.O. Barnwell, Eds. EPA-600/9-80-016 (NTIS PB80-186539), Environmental Protection Agency, Athens, GA. 167-175; 1980 April.

76. Huber, W.C. and J.P. Heaney. Analyzing Residuals Discharge and Generation from Urban and Non-Urban Land Surfaces. in Analyzing Natural Systems, Analysis for Regional Residuals - Environmental Quality Management, D.J. Basta and B.T. Bower, Eds. Resources for the Future, Washington, DC. (also available as EPA-600/3-83-046, NTIS PB83-223321). Chapter 3, 121-243; 1982 June.

77. Huber, W.C., Heaney, J.P., Aggidis, D.A., Dickinson, R.E. and R.W. Wallace. Urban Rainfall-Runoff-Quality Data Base. EPA-600/2-81-238 (NTIS PB82-221094), Environmental Protection Agency, Cincinnati, OH. 1981 October.

78. Huber, W.C., Heaney, J.P., Medina, M.A., Peltz, W.A., Sheikh, H. and G.F. Smith. Storm Water Management Model User's Manual, Version II. EPA-670/2-75-017 (NTIS PB-257809), Environmental Protection Agency, Cincinnati, OH. 1975 March.

79. Huber, W.C., Heaney, J.P., Nix, S.J., Dickinson, R.E. and D.J. Polmann. Storm Water Management Model User's Manual, Version III. EPA-600/2-84-109a (NTIS PB84-198423), Environmental Protection Agency, Cincinnati, OH. (also available from Dept. of Environmental Engineering Sciences, University of Florida, Gainesville). 1981 November.

80. Hydrologic Engineering Center. Storage, Treatment, Overflow, Runoff Model, STORM, User's Manual. Generalized Computer Program 723-S8-L7520, Corps of Engineers, Davis, CA. 1977 August.

81. Hydroscience, Inc. A Statistical Method for Assessment of Urban Stormwater Loads - Impacts - Controls. EPA-440/3-79-023, Environmental Protection Agency, Washington, DC. 1979 January.

82. James, A. Ed. Mathematical Models in Water Pollution Control. John Wiley and Sons, New York. 1978.

83. James, W. and S. Boregowda. Continuous Mass-Balance of Pollu-

tant Build-up Processes. NATO Workshop on Urban Runoff Pollution, Montpellier, France. 1985 August.

84. James, W. and M.A. Robinson. Continuous Models Essential for Detention Design. Proceedings of the Conference on Stormwater Detention Facilities, Henniker, NH. American Society of Civil Engineers, New York. 163-175; 1982 August.

85. James, W. and M.A. Robinson. PCSWMM3 - Version 3 of the Executive, Runoff and Extended Transport Blocks, Adapted for the IBM-PC. Proceedings Conference on Stormwater and Water Quality Management Modeling, Burlington, Ontario, Report R128, Computational Hydraulics, McMaster University, Hamilton, Ontario. 39-52; 1984 September.

86. James, W. and Z. Shtifter. Implications of Storm Dynamics on Design Storm Inputs. Proceedings Stormwater and Water Quality Management Modeling and SWMM User's Group Meeting, Report CHI-R81, Computational Hydraulics, McMaster University, Hamilton, Ontario. 55-78; 1981 September.

87. Jewell, T.K. and D.D. Adrian. SWMM Storm Water Pollutant Washoff Functions. Journal of the Environmental Engineering Division, ASCE, 104(EE5):1036-1040; 1978 October.

88. Jewell, T.K., Adrian, D.D. and F.A. DiGiano. Urban Stormwater Pollutant Loadings. Publication No. 113, Water Resources Research Center, University of Massachusetts, Amherst. 1980 May.

89. Jewell, T.K., Adrian, D.D. and D.W. Hosmer. Analysis of Stormwater Pollutant Washoff Estimation Techniques. Proceedings International Symposium on Urban Storm Runoff, University of Kentucky, Lexington. 135-141; 1980 July.

90. Johansen, N.B., Linde-Jensen, J.J. and P. Harremoes. Computing Combined System Overflow Based on Historical Rain Series. Proceedings of the Third International Conference on Urban Storm Drainage, Chalmers University, Goteborg, Sweden. 3:909-918; 1984 June.

91. Johanson, R.C., Imhoff, J.C. and H.H. Davis, Jr. User's Manual for Hydrological Simulation Program - Fortran (HSPF). EPA-600/9-80-015, Environmental Protection Agency, Athens, GA. 1980 April.

92. Kibler, D.F. Ed. Urban Stormwater Hydrology. Water Resources Monograph 7, American Geophysical Union, Washington, DC. 1982.

93. Kluesener, J.W. and G.F. Lee. Nutrient Loading from a Separate Storm Sewer in Madison, Wisconsin. Journal Water Pollution Control Federation. 46(5):920-936; 1974 May.

94. Kobriger, N.P., Meinholz, T.L., Gupta, M.K. and R.W. Agnew. Constituents of Highway Runoff, Volume III, Predictive Procedure for Determining Pollution Characteristics in Highway Runoff. FHWA/RD-81/044 (NTIS PB81-241911), Federal Highway Administration, Washington, DC. 1981 February.

95. Kohlhaas, C.A. Ed. Compilation of Water Resources Computer Program Abstracts. U.S. Committee on Irrigation, Drainage and Flood Control, Denver, CO. 1982.

96. Konno, H., Nonomura, S. and T. Ohshima. Sediment Discharge on Land-Grading Areas in Kohoku, Japan. Urban Stormwater Quality, Management and Planning, Proceedings of the Second International Conference on Urban Storm Drainage, Urbana, IL. Water Resources Publications, Littleton, CO. 209-217; 1981 June.

97. Krukar, M. and J.C. Cook. Optimization and Testing of Highway Materials to Mitigate Ice Adhesion. EPA-600/2-78-035, Environmental Protection Agency, Cincinnati, OH. 1978 March.

98. Lager, J.A., Smith, W.G., Lynard, W.G., Finn, R.G. and E.J. Finnemore. Urban Stormwater Management and Technology: Update and User's Guide. EPA-600/8-77-014 (NTIS PB-275264), Environmental Protection Agency, Cincinnati, OH. 1977 September.

99. Lager, J.A., Smith, W.G. and G. Tchobanoglous. Catchbasin Technology Overview and Assessment. EPA-600/2-77-051 (NTIS PB-270092), Environmental Protection Agency, Cincinnati, OH. 1977 May.

100. Laukkanen, R. Indirect Estimation of Pollutant Loads on the Basis of Treatment Plant Data Records. Proceedings of the Third International Conference on Urban Storm Drainage, Chalmers University, Goteborg, Sweden. 3:929-938; 1984 June.

101. Lindholm, O.G. Modelling of Sewerage Systems. in Mathematical Models in Water Pollution Control, A. James, Ed. John Wiley and Sons, New York. Chapter 10, 227-246; 1978.

102. Lindholm, O.G. Dry Weather Deposition in Combined Sewers. Proceedings of the Third International Conference on Urban Storm Drainage, Chalmers University, Goteborg, Sweden. 3:861-868; 1984 June.

103. Linsley, R.K. and N.H. Crawford. Continous Simulation Models in Urban Hydrology. Geophysical Research Letters. 1(1):59-62; 1974 May.

104. Little, L.M., Horner, R.R. and B.W. Mar. Assessment of Pollutant Loadings and Concentrations in Highway Stormwater Runoff. FHWA Project WA-RD-39.12.1, Report No. 17, University of Washington, Dept. of Civil Engineering, Seattle. 1983 April.

105. Lynard, W.G., Finnemore, E.J., Loop, J.A. and R.M. Finn. Urban Stormwater Management and Technology: Case Histories. EPA-600/8-80-035 (NTIS PB81-107153), Environmental Protection Agency, Cincinnati, OH. 1980 August.

106. Manning, M.J., Sullivan, R.H. and T.M. Kipp. Nationwide Evaluation of Combined Sewer Overflows and Urban Stormwater Discharges - Vol. III: Characterization of Discharges. EPA-600/2-77-064c (NTIS PB-272107), Environmental Protection Agency, Cincin-

nati, OH. 1977 August.

107. Marsalek, J. Synthesized and Historical Storms for Urban Drainage Design. Proceedings of the International Conference on Urban Storm Drainage, University of Southampton. Pentech Press, London. 87-99; 1978 April.

108. Mattraw, H.C., Jr. and C.B. Sherwood. Quality of Storm Water Runoff from a Residential Area, Broward County, Florida. Journal of Research, U.S. Geological Survey. 5(6):823-834; 1977 November-December.

109. McCuen, R.H. Water Quality Trap Efficiency of Storm Water Management Basins. Water Resources Bulletin. 16(1):15-21; 1980 February.

110. McElroy, A.D., Chiu, S.Y., Nebgen, J.W., Aleti, A. and F.W. Bennett. Loading Functions for Assessment of Water Pollution from Non-Point Sources. EPA-600/2-76-151 (NTIS PB-253325), Environmental Protection Agency, Washington, DC. 1976 May.

111. McPherson, M.B. Urban Runoff Control Planning. EPA-600/9-78-035, Environmental Protection Agency, Washington, DC. 1978 October.

112. McPherson, M.B. and F.C. Zuidema. Urban Hydrological Modeling and Catchment Research: International Summary. ASCE Urban Water Resources Research Program Technical Memorandum No. IHP-13 (NTIS PB-280754), American Society of Civil Engineers, New York. 1977 November.

113. Medina, M.A., Jr., Huber, W.C. and J.P. Heaney. Modeling Stormwater Storage/Treatment Transients: Theory. Journal of the Environmental Engineering Division, ASCE, 107(EE4):781-797; 1981 August.

114. Metcalf and Eddy, Inc., University of Florida and Water Resources Engineers, Inc. Storm Water Management Model, Volume I - Final Report. EPA Report 11024DOC07/71 (NTIS PB-203289), Environmental Protection Agency, Washington, DC. 1971 July.

115. Metcalf and Eddy, Inc., University of Florida and Water Resources Engineers, Inc. Storm Water Management Model, Volume II - Verification and Testing. EPA Report 11024DOC08/71 (NTIS PB-203290), Environmental Protection Agency, Washington, DC. 1971 August.

116. Meuller, J.A. and D.M. Di Toro. Combined Sewer Overflow Characteristics from Treatment Plant Data. EPA-600/2-83-049 (NTIS PB83-224543), Environmental Protection Agency, Cincinnati, OH. 1983 June.

117. Miller, R.A., Mattraw, H.C., Jr. and M.E. Jennings. Statistical Modeling of Urban Storm Water Processes, Broward County, Florida. Proceedings International Symposium on Urban Storm Water Management, University of Kentucky, Lexington. 269-273; 1978 July.

118. Miracle, D.L. Review of Existing Methods for Predicting Pollutant Loads from Highway Runoff. Master of Engineering Technical Report (unpublished), University of Florida, Dept. of Environmental Engineering Sciences, Gainesville. 1985.

119. Murrey, M.P.H. and J.J. Ganczarczyk. Storage for Storm Water Quality Control - Meadowvale Test Site Study. Research Report No. 63, Canada-Ontario Agreement on Great Lakes Water Quality, Environmental Protection Service, Environment Canada, Ottawa, Ontario. 1977.

120. Nakamura, E. Factors Affecting Stormwater Quality Decay Coefficient. Proceedings of the Third International Conference on Urban Storm Drainage, Chalmers University, Gotéborg, Sweden. 3:979-988; 1984 June.

121. Nakamura, E. Factors Affecting the Removal Rate of Street Surface Contaminants by Overland Flow. Journal of Research, Public Works Research Institute, Ministry of Construction, Japan. 24; 1984 November.

122. Nix, S.J., Heaney, J.P. and W.C. Huber. Revised SWMM Storage/Treatment Block. Proceedings Storm Water Management Model Users Group Meeting, Annapolis, Maryland, EPA-600/9-79-003 (NTIS PB-290742/6BE), Environmental Protection Agency, Washington, DC. 1978 November.

123. Novotny, V. and B.W. Kincaid. Acidity of Urban Precipitation and its Buffering During Overland Flow. Urban Stormwater Quality, Management and Planning, Proceedings of the Second International Conference on Urban Storm Drainage, Urbana, IL. Water Resources Publications, Littleton, CO. 1-9; 1981 June.

124. Overton, D.E. and M.E. Meadows. Stormwater Modeling. Academic Press, New York. 1976.

125. Patry, G. and M.B. McPherson. Eds. The Design Storm Concept. EP80-R-8, GREMU-79/02, Civil Engineering Dept. Ecole Polytechnique de Montreal, Montreal, Quebec. 1979 December.

126. Pisano, W.C., Aronson, G.L., Queiroz, C.S., Blanc, F.C. and J.C. O'Shaughnessy. Dry-Weather Deposition and Flushing for Combined Sewer Overflow Pollution Control. EPA-600/2-79-133 (NTIS PB80-118524), Environmental Protection Agency, Cincinnati, Oh. 1979 August.

127. Pisano, W.C. and C.S. Queiroz. Procedures for Estimating Dry Weather Pollutant Deposition in Sewerage Systems. EPA-600/2-77-120 (NTIS PB-270695), Environmental Protection Agency, Cincinnati, OH. 1977 July.

128. Pitt, R. Demonstration of Non-Point Pollution Abatement Through Improved Street Cleaning Practices. EPA-600/2-79/161 (NTIS PB80-108988), Environmental Protection Agency, Cincinnati, OH. 1979 August.

129. Pitt, R. and G. Amy. Toxic Materials Analysis of Street Surface Contaminants. EPA-R2-73-283 (NTIS PB-224677), Environmental Protection Agency, Washington, DC. 1973 August.

130. Poertner, H.G. Ed. Urban Stormwater Management. Special Report No. 49, American Public Works Association, Chicago, IL. 1981.

131. Pratt, C.J. and J.R.W. Adams. Sediment Washoff into Roadside Gullies. Urban Stormwater Quality, Management and Planning, Proceedings of the Second International Conference on Urban Storm Drainage, Urbana, IL. Water Resources Publications, Littleton, CO. 174-183; 1981 June.

132. Price, R.K. and G. Mance. A Suspended Solids Model for Storm Water Runoff. Proceedings of the International Conference on Urban Storm Drainage, University of Southampton. Pentech Press, London. 546-555; 1978 April.

133. Proctor and Redfern, Ltd. and J.F. MacLaren, Ltd. Storm Water Management Model Study - Vol. I, Final Report. Research Report No. 47, Canada-Ontario Agreement on Great Lakes Water Quality, Environmental Protection Service, Environment Canada, Ottawa, Ontario. 1976 September.

134. Proctor and Redfern, Ltd. and J.F. MacLaren, Ltd. Storm Water Management Model Study - Vol. II, Technical Background. Research Report No. 48, Canada-Ontario Agreement on Great Lakes Water Quality, Environmental Protection Service, Environment Canada, Ottawa, Ontario. 1976 September.

135. Racin, J.A., Howell, R.B., Winters, G.R. and E.C. Shirley. Estimating Highway Runoff Quality. Report FHWA/CA/TL-82/11, California Department of Transportation, Sacramento. 1982 September.

136. Randall, C.W., Grizzard, T.J., Helsel, D.R. and D.M. Griffin, Jr. Comparison of Pollutant Mass Loads in Precipitation and Runoff in Urban Areas. Urban Stormwater Quality, Management and Planning, Proceedings of the Second International Conference on Urban Storm Drainage, Urbana, IL. Water Resources Publications, Littleton, CO. 29-38; 1981 June.

137. Richardson, D.L., Terry, R.C., Metzger, J.B., Carroll, R.J. and A.D. Little. Manual for Deicing Chemicals: Application Practices. EPA-670/2-74-045 (NTIS PB-236152), Environmental Protection Agency, Cincinnati, OH. 1974 December.

138. Robinson, M.A. and W. James. Continuous Variable Resolution Stormwater Modelling on a Microcomputer. Proceedings Conference on Stormwater and Water Quality Management Modeling, Burlington, Ontario, Report R128, Computational Hydraulics, McMaster University, Hamilton, Ontario. 59-73; 1984 September.

139. Robinson, M.A. and W. James. Chedoke Creek Flood Storage Computed by Continous SWMM and Dynamic Storms. Report No. R124, Computational Hydraulics, McMaster University, Hamilton, Ontario. 75-89; 1984 September.

140. Roesner, L.A., Giguere, P.R. and L.C. Davis. User's Manual for the Storm Runoff Quality Model RUNQUAL. Report to Southeast Michigan Council of Governments, Water Resources Engineers, Springfield, VA. 1977 July.

141. Roesner, L.A., Nichandros, H.M., Shubinski, R.P., Feldman, A.D., Abbott, J.W. and A.O. Friedland. A Model for Evaluating Runoff-Quality in Metropolitan Master Planning. ASCE Urban Water Resources Research Program Technical Memorandum No. 23 (NTIS PB-234312), American Society of Civil Engineers, New York. 1974 April.

142. Roesner, L.A., Shubinski, R.P. and J.A. Aldrich. Storm Waer Management Model User's Manual Version III: Addendum I, EXTRAN. EPA-600/2-84-109b (NTIS PB84-198431), Environmental Protection Agency, Cincinnati, OH. 1981 November.

143. Roy F. Weston, Inc. Hourly Diurnal Flow Variations in Publicly-Owned Wastewater Treatment Facilities. EPA-600/2-81-218 (NTIS PB82-107954), Environmental Protection Agency, Cincinnati, OH. 1981 September.

144. Sartor, J.D. and G.B. Boyd. Water Pollution Aspects of Street Surface Contaminants. EPA-R2-72-081 (NTIS PB-214408), Environmental Protection Agency, Washington, DC. 1972 November.

145. Shaheen, D.G. Contributions of Urban Roadway Usage to Water Pollution. EPA-600/2-75-004 (NTIS PB-245854), Environmental Protection Agency, Washington, DC. 1975 April.

146. Shivalingaiah, B. and W. James. Algorithms for Buildup and Washoff and Routing Pollutants in Urban Runoff. Proceedings of the Third International Conference on Urban Storm Drainage, Chalmers University, Goteborg, Sweden. 4:1445-1456; 1984 June.

147. Shubinski, R.P., Knepp, A.J. and C.R. Bristol. Computer Program Documentation for the Continuous Storm Runoff Model SEM-STORM. Report to Southeast Michigan Council of Governments, Water Resources Engineers, Springfield, VA. 1977 July.

148. Smith, G.F. Adaptation of the EPA Storm Water Management Model for Use in Preliminary Planning for Control of Urban Storm Runoff. Master of Engineering Thesis, University of Florida, Dept. of Environmental Engineering Sciences, Gainesville. 1975.

149. Smolenyak, K.J. Urban Wet-Weather Pollutant Loadings. Master of Engineering Thesis, University of Florida, Dept. of Environmental Engineering Sciences, Gainesville. 1979.

150. Solomon, S.I. and S.K. Gupta. Distributed Numerical Model for Estimating Runoff and Sediment Discharge of Ungaged Rivers 2. Model Development. Water Resources Research. 13(3):619-629; 1977 June.

151. Solomon, S.I. and S.K. Gupta. Distributed Numerical Model for Estimating Runoff and Sediment Discharge of Ungaged Rivers 3.

Comparison with Other Simple Techniques. Water Resources Research. 13(3):631-636; 1977 June.

152. Sonnen, M.B. Abatement of Deposition and Scour in Sewers. EPA-600/2-77-212 (NTIS PB-276585), Environmental Protection Agency, Cincinnati, OH. 1977 November.

153. Sonnen, M.B. Urban Runoff Quality: Information Needs. Journal of the Technical Councils, ASCE. 106(TC1):29-40; 1980 August.

154. Stephensen, D. Effect of Sediment on Flow in Part Full Pipes. Urban Stormwater Quality, Management and Planning, Proceedings of the Second International Conference on Urban Storm Drainage, Urbana, IL. Water Resources Publications, Littleton, CO. 166-173; 1981 June.

155. Stotz, G. and K. Krauth. Factors Affecting First Flushes in Combined Sewers. Proceedings of the Third International Conference on Urban Storm Drainage, Chalmers University, Goteborg, Sweden. 3:869-878; 1984 June.

156. Sueishi, T., Morioka, T., Hasegawa, Y. and K. Hohji. Long-Term Effects of Suspended Solids in Urban Stormwater. Urban Stormwater Quality, Management and Planning, Proceedings of the Second International Conference on Urban Storm Drainage, Urbana, IL. Water Resources Publications, Littleton, CO. 191-200; 1981 June.

157. Sueishi, T., Morioka, T. and S. Kawamoto. Comparison Among Models for Stormwater and Pollutant Runoff in View of their Reliability, Sensitivity, and Applicability. Proceedings of the Third International Conference on Urban Storm Drainage, Chalmers University, Goteborg, Sweden. 2:605-614; 1984 June.

158. Terstriep, M.L., Bender, G.M. and D.J. Benoit. Buildup, Strength, and Washoff of Urban Pollutants. Preprint 3439, ASCE Convention and Exposition, Chicago, IL. 1978 October.

159. Torno, H.C. The EPA Nationwide Urban Runoff Program (NURP). Proceedings of the Third International Conference on Urban Storm Drainage, Chalmers University, Goteborg, Sweden. 4:1465-1474; 1984 June.

160. Tucker, C.G.J. and G.H. Mortimer. The Generation of Suspended Solids Loads in Urban Stormwater. Proceedings of the International Conference on Urban Storm Drainage, University of Southampton. Pentech Press, London. 695-704; 1978 April.

161. Tucker, L.S. Sewage Flow Variations in Individual Homes. ASCE Combined Sewer Separation Project, Technical Memorandum No. 2. American Society of Civil Engineers, New York. 1967 February.

162. Urcikan, P. Frictional Losses in Pipe Flow of Storm Waters Having Various Concentrations of Sediments. Proceedings of the Third International Conference on Urban Storm Drainage, Chalmers University, Goteborg, Sweden. 4:1475-1486; 1984 June.

163. Uttormark, P.D., Chapin, J.D. and K.M. Green. Estimating Nutrient Loadings of Lakes from Non-Point Sources. EPA-660/3-74-020, Environmental Protection Agency, Washington, DC. 1974 August.

164. Vanoni, V.A. Ed. Sedimentation Engineering. American Society of Civil Engineers, New York. 1975.

165. Viessman, W., Jr., Knapp, J.W., Lewis, G.L. and T.E. Harbaugh. Introduction to Hydrology. Harper and Row, New York. Second Edition. 1977.

166. Wallace, R.C. Statistical Modeling of Water Quality Parameters in Urban Runoff. Master of Engineering Technical Report (unpublished), University of Florida, Dept. of Environmental Engineering Sciences, Gainesville. 1980.

167. Waller, D.H. Peak Flows of Sewage from Individual Houses. ASCE Combined Sewer Separation Project, Technical Memorandum No. 9. American Society of Civil Engineers, New York. 1968 January.

168. Waller, D.H. and W.A. Coulter. Winter Runoff from an Urban Catchment. Research Report No. 41, Canada-Ontario Agreement on Great Lakes Water Quality, Environmental Protection Service, Environment Canada, Ottawa, Ontario. 1977.

169. Wenzel, H.G., Jr. and M.L. Voorhees. An Evaluation of the Urban Design Storm Concept. Research Report 164, Water Resources Center, University of Illinois, Urbana-Champaign. 1981 August.

170. Whipple, W., Grigg, N.S., Grizzard, T., Randall, C.W., Shubinski, R.P. and L.S. Tucker. Stormwater Management in Urbanizing Areas. Prentice-Hall, Inc., Englewood Cliffs, NJ. 1983.

171. Whipple, W., Jr., Hunter, J.V. and S.L. Yu. Effects of Storm Frequency on Pollution from Urban Runoff. Journal Water Pollution Control Federation. 49(11):2243-2248; 1977 November.

172. Williams, J.R. A Sediment Graph Model Based on an Instantaneous Unit Sediment Graph. Water Resources Research. 14(4):659-664; 1978 August.

173. Wischmeier, W.H. and D.D. Smith. Rainfall Energy and Its Relationship to Soil Loss. Transactions American Geophysical Union. 39(2):285-291; 1958 April.

174. Wischmeier, W.H. and D.D. Smith. Predicting Rainfall-Erosion Losses from Cropland East of the Rocky Mountains. Agricultural Handbook 282, U.S. Dept. of Agriculture, Washington, DC. 1965.

175. Yao, K.M. Sewer Design Based on Critical Shear Stress. Journal of the Environmental Engineering Division, ASCE, 100(EE2):507-520; 1978 April.

176. Yen, B.C. Ed. Urban Stormwater Quality, Management and Planning. Proceedings of the Second International Conference on Urban Storm Drainage, Urbana, IL. Water Resources Publications,

Littleton, CO. 1981 June.

177. Yousef, A.Y., Wanielista, M.P., McLellon, W.M. and J.S. Taylor. Eds. Urban Stormwater and Combined Sewer Overflow Impact on Receiving Water Bodies, Proceedings of the National Conference, Orlando, Florida, November 1979. EPA-600/9-80-056 (NTIS PB81-155426), Environmental Protection Agency, Cincinnati, OH. 1980 December.

178. Zison, S.W. Sediment-Pollutant Relationships in Runoff from Selected Agricultural, Suburban, and Urban Watersheds, A Statistical Correlation Study. EPA-600/3-80-022, Environmental Protection Agency, Athens, GA. 1980 January.

179. Zukovs, G. Development of a Probabilistic Runoff Storage/Treatment Model. Master of Engineering Thesis, University of Toronto, Dept. of Civil Engineering, Toronto, Ontario. 1983.

CONTINUOUS MASS-BALANCE OF POLLUTANT BUILD-UP PROCESSES

William James and Shivalingaiah Boregowda
Civil Engineering Department, McMaster University
Hamilton, Canada L8S 4L7

Introduction

The authors discuss the concept of "pollutant build-up" from the perspective of continuous rather than event modelling. Pollutant sources are described, including anthropogenic activities, such as atmospheric deposition and scavenging, automobiles, and pavement degradation, as well as natural processes. Redistribution and removal processes occurring during dry periods are also described (including biological decay, intentional removals and wind or eddy-induced transport).

Several earlier studies are reviewed in terms of land-use correlations, pollutant fractions or potency factors of dust and dirt. Fitted procedures commonly used in available, operational models are summarised. Several case studies are cited. Daily mass balances are treated in more detail. Procedures for parameter estimation including sensitivity analysis are given.

It is argued that the traditional idea that build-up (addition) is discretely distinct from washoff (removal) should be modified. A better approach simply encompasses all contributory dry-period and wet-period processes (removal, addition, and distribution), that are appropriate to continuous modelling. In this way, weather situations that are not clearly dry or wet can be appropriately modelled, and we avoid having to label redistribution processes as "build-up" or "wash-off".

NATO ASI Series, Vol. G10
Urban Runoff Pollution.
Edited by H. C. Torno, J. Marsalek, and M. Desbordes

There are many and diverse sources of pollutants, man-made and natural. Generally, pollutants derived from natural sources can be related to swampgas, salt sprays from the sea, dust picked up by winds, smoke from naturally lit fires, terpenes, resins and pollens from forests, fog, photochemical ozone, nitrogen oxide and other oxidants, gases, vapours, and particulates derived from volcanoes, geysers and fissures and climatic factors. Most of these natural pollutants depend on the geographical location of the city under consideration. We have not presented algorithms for these natural processes.

In urban situations, industries, automobiles, house heating, sanitary or combined sewer overflows, and gardens (fertilizers, insecticides, and pesticides) contribute significant quantities of pollutants to surface waters and the atmosphere. Rain falling on an urban area gains pollutants from the air, dusty roofs, littered and dirty streets, sidewalks, traffic by-products (gasoline and oil drippings, tire residuals, brake wear, vehicular exhaust), metallic corrosion, hazardous material spills and from chemicals applied for fertilization, control of ice, rodents, insects and weeds. Many of these contaminants are redistributed by winds and other activities (such as snow removal).

Single event simulation requires all initial conditions such as soil moisture and start-up pollutant loadings on the surface and in the atmosphere etc. to be specified for each wet event. Most of these initial estimates are subject to error, to a degree depending on the experience of the modeller. On the other hand, continuous simulation requires only one set of initial data, usually well ahead of any critical set of design episodes. The start-up error is then not usually significant.

Pollutant build-up and redistribution over an impervious area is a continuous process both in wet and dry periods. Pollutants savenged from atmosphere by rain cannot be termed "build-up" or "wash-off". Because of the complex interaction of the dry, "inter-event", predominantly pollutant-accumulating period and the subsequent wet, washoff-producing period, continuous modeling is to be preferred for quality models.

Background Review

Various human activities in urbanised areas increase the pollutant load accumulating on the surface and in the atmosphere. Many of these pollutant generating activities are more intense on impervious areas than on pervious areas. Impervious areas contribute higher BOD, COD, toxic metals, nutrients, oil and grease etc. However, solids eroded from unprotected and bare lands, and nutrients from pervious farm lands cannot be neglected.

Most of the pollutant quantities correlate reasonably well with the mass of dust and dirt (DD) accumulated over dry periods. Therefore, DD has been taken as a reference for estimating concentrations of various pollutants.

Due to the inadequacy of the generally available urban stormwater quality data base, the deposition and removal processes are not completely or even adequately understood. Some urban non-point-source pollution prediction models assume linear accumulation rates, but there is general consensus that the DD accumulation rate should be a decreasing function of time. A wide variety of accumulation relationships have appeared in the storm water literature[(1)]. The accumulated DD loading rate has been related to the elapsed time since street sweeping, or time since one half inch rainfall, whichever comes last[(2)]. The equations developed for different land uses with their correlation coefficients were summarized by Sartor and Boyd[(2)] and the observed data and fitted curves plotted. The poor correlations show that land use classifications and antecedent time period may not adequately depict the accumulation process.

Geomorphologic and climatic variables may also be important independent variables.

McCuen and Sutherland[3] used a pattern search technique to fit equations to Sartor and Boyd's approximate curves and their best-fit equations were tabulated.

The San Jose study[4] showed that, immediately after street cleaning, the street DD loading increases rapidly but levels off after several days.

Nunno[5] found that the non-linear buildup of a pollutant (BOD) between storms has a limit, regardless of the length of the dry period. The best fits between the initial pollutant loads and the antecedent dry days supported exponential and Michaelis-Menton relationships.

Shaheen[6] examined samples collected from roadways over various deposition periods. These data revealed that the deposition rate of non-traffic related pollutants is linear with time while the accumulation levels off substantially on the pavements after several days due to the displacement of the material onto areas adjacent to the road ways. For traffic related loadings the net accumulation rate is given by:

$$\frac{dL}{dT} = K_d - K_1 L$$

where: L = the pollutant loading, kg/km

T = total traffic, axles

K_d = desposition rate per axle, kg/axle-km.

K_1 = loss rate per axle.

On integration, the above equation leads to

$$L = L_m\ (1-e^{-K_1 T})$$

where: $L_m = K_d/K_1$ = maximum pollutant loading, kg/km.

This equation can be transformed into an accumulation equation by assuming $t=T/N$, where t is accumulation period in days and N is average daily traffic in axles/day. In this case:

$$L = L_m\ (1-e^{-Kt})$$

where: $K = K_1 N$

For the high and slow speed lanes of North Capitol Street, Washington, D.C., Shaheen estimated:

K_d = 0.00288 lb/axle-mile (0.0008kg/axle-km),
K_1 = 0.00001 to 0.00003 per axle and
N = 40,000 axle/d

this leads to limits for L of 96 and 288 lb/mile (28 and 82 kg/km) and for K of 0.4 and 1.2 d^{-1}

Terstriep et al.[7] sampled total solids accumulated on the streets of Urbana-Champaign, Illinois, sequentially during a 10 day period following a runoff event. The total solids increased during the dry period for all three sites. The accumulation rate was quite high in commercial streets compared to the two residential streets.

In a 57.6 acre residential catchment study in Burlington, Ontario, Marsalek[8] found the antecedent dry period to be the most important variable in a linear regression model for suspended solids concentration.

Mance and Harman[9] found that for most pollutants studied, between 60 to 90 percent of the variance in the total volume discharged could be explained by multiple regression analysis using antecedent dry period, and magnitude of the previous runoff as independent variables. However, the latter two variables only accounted for a maximum of six percent of the variance. In other words the characteristics of the latest event were primarily responsible for determining the mass of pollutants discharged. This would seem to indicate an ample DD supply; one that is not time constrained. Waller[10] also observed that the antecedent dry period had little statistical effect on the quality of runoff in Halifax.

In an extensive statistical data analysis of a 47 acre residential catchment and a 39 acre transportation catchment in Broward County, Florida, Miller et al.(11) found the antecedent dry period to be a significant independent variable. Peak discharge was the most important variable in the regression analysis.

Novotny and Goodrich-Mahoney(12) derived an equation for DD accumulation near the curb:

$$D = \frac{A}{B}\left(1 - e^{-Bt}\right)$$

where:

- $A = A_f(W/2) - 5.02\,R_d - 6.29\,P_a + 1.15\,T_d$
- $B = 0.0166\,(e^{-0.0884H})\,(T_s + W_s)$
- D = dust and dirt (suspended solids) accumulated near the curb (g/m).
- W = street width (m).
- A_f = atmospheric fallout (g/m^2-d).
- R_d = residential density (dwellings/ha).
- P_a = percent open area near the site,
- T_d = traffic density (1000 axles/d).
- H = curb height (cm).
- T_s = traffic speed (km/h) and
- W_s = wind speed (km/h).

Ammon(1) supported the relationship developed by Meta System after reanalyzing the data from APWA(13), Sartor and Boyd(2) and Pitt(4):

$$L = \frac{-a}{1+bt} + \frac{ac}{b^2}\ln(1+bt)e^{-ct} + c(1+bt)\,e^{-ct}$$

$$+ c^2(1+bt)^2 e^{-ct} + \frac{a}{b} + c + \frac{c^2}{4}$$

where:

- L = accumulated loading, mg
- t = accumulation period, d
- and a,b,c = constants (typical values are not given).

It was found that climate and average daily traffic have a larger effect on loadings than land use. Loadings were low, both at very high and very low traffic counts, and higher at intermediate levels. This suggests that both deposition and scour from traffic-related activities are important factors.

A simplified equation for calculating the buildup rate on street pavements was developed by Gupta et al[14]:

$$Ts = 0.002\ N^{0.89}$$

where: Ts = total solids (kg/km-d)

N = average daily traffic (vehicles/d).

After reviewing the existing theories on DD buildup, it is evident that the buildup process is usually described by an empirical relationship. Except for the work of Novotny and Goodrich-Mahoney[12], all other equations do not consider meteorological parameters. But DD accumulation is the constant interaction of addition and removal processes due to various activities in the catchment on the one hand, and redistribution processes related to meteorological parameters on the other hand. Hence, a daily mass balance of these processes over a dry period would represent the buildup process better than would empirical relationships. A better approach involves: discretize and model all the accumulation, washoff and redistribution processes and relate them to meteorological parameters.

Existing Algorithms

Many urban runoff quantity and quality models have been developed during the last two decades. In SWMM3, options for both linear and nonlinear (exponential and Michaelis-Menton) pollutant accumulation over a dry day period are given:

$$Cf = Df\ Dm$$

$$UL = Dd\ Dm$$

$$EL = Cf\ t^{p} \qquad \text{(power linear)}$$

$$EL = UL\ (1-e^{-pt}) \qquad \text{(exponential)}$$

$$EL = ULt/(Cf + t) \qquad \text{(Michaelis Menton)}$$

where: EL = pollutant buildup, (mg or MPN)

Cf = coefficient, kg/d.

t	=	time in days
P	=	exponent, constant
UL	=	upper limit for load, kg.
Df	=	DD buildup rate, kg/100 m, curb-d. or kg/ha-d.
Dm	=	Gutter length, 100 m or subcatchment area, ha
Dd	=	Limiting buildup quantity, kg/100m or kg/ha.

In STORM[15] pollutant accumulation for each one hour time step when runoff is less than 0.0005 inch/h (0.0122 mm/h) is given by:

$$Lt = Dd\ Q\ Gt + Lt\text{-}1$$

where:

Lt	=	pollutant loading on subcatchment at time 't'(mg).
Dd	=	dust and dirt loading rate (g/d-100 m)
Q	=	fraction of dust and dirt consisting of pollutant, in mg/g
G	=	total gutter or curb length in subcatchment, 100m
t	=	time step, d.
Lt-1	=	pollutant loading on subcatchment at t-1, (mg)

The pollutant accumulation in the Hydrologic Simulation Program (HSP), Agricultural Runoff Management Model (ARM), and the Nonpoint Source Pollutant loading model (NPS) is formulated as:

$$Lt = Lt\text{-}1\ (1\text{-}Rg) + D$$

where:

Lt	=	pollutant accumulation at time t, (t/ha).
Lt-1	=	pollutant accumulation at time (t-1), (t/ha).
Rg	=	general removal fraction
D	=	pollutant deposition during time interval, (t/ha)

The general removal fraction is as follows in HSP and ARM:

$$R_g = F \frac{E}{f} R_W K_b$$

where: F = fraction of impervious area.

E = efficiency of street cleaning on the impervious area.

f = frequency of street cleaning.

R_W = fraction of pollutant removal by wind

K_b = fraction of pollutant decay by biochemical processes.

The general removal rate in the NPS Model is an input parameter and there is one value for each month.

The Quantity Quality Simulation (QQS) program(16) has an asymptotic accumulation relationship formulated as:

$$L = (1-\{(az+1) / (bz + cz + 1)\}) L_m$$

where: L = pollutant accumulation during dry period, kg/ha

z = dry period (number of 5 minute intervals).

a,b and c = empirical coefficients.

L_m = maximum pollutant loading (kg/ha).

The magnitude of the coefficients a, b and c is in the range 0.0001 to 0.0000001 and depends on the pollutant being accumulated.

Mass Balance Approach

Pollutant concentration in storm water is governed by three types of process: buildup, washoff and transport. Pollutants scavenged during rainfall, and materials previously deposited in storm sewers and catchbasins, also add to the instantaneous concentration. The mass balance for pollutants in urban storm runoff for a single storm event and for a single catchment is:

$$P_r = P_i + P_s + P_p + P_d$$

where: P_r = total mass flux of pollutants in the subcatchment pollutograph mg/s

P_i = total mass flux of accumulated pollutants washed off impervious areas, mg/s.

P_s = total mass flux of pollutants scavenged by rain during its motion through the atmosphere, mg/s.

P_p = total mass flux of pollutants washed off pervious areas, mg/s.

P_d = total mass flux of previously deposited pollutants scoured from storm sewers and catch basins, mg/s.

The accumulation process on impervious surfaces areas can be expressed:

P_a = (Atmospheric dustfall + surface input) - Removals

or

$$P_a = L_a+L_v+L_s+L_p+L_e-R_v-R_b-R_i-R_w$$

where: P_a = daily mass flux of pollutant accumulated, mg/d

L_a = daily mass flux of atmospheric dustfall, mg/d

L_v = daily mass flux of pollutants from vehicles, mg/d

L_s = daily mass flux of pollutants from special activities such as construction, demolition of structures, mg/d

L_p = daily mass flux of pollutants from population related activities such as lawn cutting, insecticides and pesticides, spray and fertilizer on lawns, mg/d

L_e = daily mass flux of pollutants from vegetation, mg/d.

R_v = daily mass flux of pollutants removed by vehicle generated eddies, mg/d

R_b = daily mass flux of pollutants removed by biological decomposition, mg/d

R_i = daily intentional removal of pollutants, e.g. street cleaning, mg/d

R_w = daily removal of pollutants by natural wind, mg/d

Each of the above processes is now discussed in detail.

Atmospheric deposition

In a highly industrialized and urbanized catchment, atmospheric contributions to runoff water quality have been found to be considerable(17,18,19). Atmospheric fallout on subcatchments depends mainly on source areas and physiographic and climatic conditions. A statistical method for predicting representative dustfall averaged over an individual subcatchment using monthly average wind velocity and monthly total precipitation as independent parameters and dustfall as a dependent parameter, was earlier developed by the authors and interfaced with SWMM3 for runoff quality prediction(20).

Vehicle input

Highway runoff pollutants important to water quality are vehicle dependent. Metals (lead, zinc, copper, chromium, and iron), nutrients (nitrogen and phosphorous) and hydrocarbons are largely contributed by motor vehicle traffic. The specific source and type of these solids depends on where the vehicle has travelled and where the pollutants were picked up. Probable common sources of these solids include parking lots, urban and industrial areas, construction sites, unpaved roads and farming areas. Traffic byproducts, gasoline and oil drippings, tire residuals, brakewear, vehicle exhaust and metallic corrosion are also major contributors to pollutant accumulation. Many studies have focussed on vehicle input and developed algorithms for the prediction of runoff quality(6). These daily quantities produced per subcatchment could be quantified:

$$L_V = NLM$$

where:
- L_V = mass of DD produced by vehicles mg/d.
- N = number of vehicle axles active each day in a subcatchment (product of total population and P assuming two axles per vehicle)
- P = population to vehicle ratio = 0.4 to 0.5
- L = mass of DD produced per vehicle per km travelled (800 mg/axle-km in Hamilton)

M = total length of road in subcatchment, km.

Population input

Litter deposits in urban areas include solid wastes deposited on surfaces, private and municipal waste collection operations, animal and bird fecal droppings, and other cultural deposits. The DD component of litter (materials less than 3.5 mm size) is regarded as having the greatest pollution potential of street surface refuse materials.

Although most of the litter is originally greater in size than that of DD, it is possible that a significant portion of the street DD originates from the mechanical fracture of litter. An American Public Works Association study(13) reported that street DD in residential areas increased as population density increases, because of additional pedestrian and roadway traffic.

Therefore, population input is formulated as:

$$L_p = P\ L$$

where: L_p = total mass of DD due to population, mg/d.

P = total population in a subcatchment

L = per capita mass of pollutants/day, (200mg/cap-d in Hamilton)

Vegetation Input

Fallen dead leaves and grass clippings in urban areas also contribute significant quantities of DD to both pervious and impervious areas. The rate of vegetation input increases substantially during the fall and depends on the density of vegetation. In addition to DD load, nutrients and other chemicals are leached to runoff water, depending upon the contact period with water(21,22). The study conducted by Prasad in the Borough of Etobicoke in the Municipality of Metropolitan Toronto on autumn leaves showed that potentially large organic loads originate from leaves in an urban area.

The amount of leaf fallout per tree per year is about 14.5 to 26 kg/yr[23]. The tree density could vary from 10-500 trees/ha depending upon the utility of land. Since most of the leaves are removed by regular street cleaning or arrested in the vegetation, only a fraction joins the runoff. Therefore an average rate of 0.3 to 6.0 kg/ha-d (assuming leaves fall over a period of 90 days and a tree density of 10-400 trees/ha) is assumed in modelling vegetation input. The vegetation input may be estimated by:

$$L_e = V\,A$$

where: L_e = Vegetation input DD, mg/d

V = Vegetation load rate to DD which varies with the landuse, and density of trees, mg/ha-d.

A = area, ha

Special activities

This component is added to the DD accumulation only in the case of construction, demolition of structures or any other pollutant producing activities in the subcatchment. This is very site-specific and could be treated as a separate subcatchment with an additional pollutant load. DD produced by special activities is estimated by:

$$L_s = S_L\,A$$

where: L_s = total mass of DD producted by special activities in mg/d.

S_L = DD loading, mg/ha-d

A = special activity area, ha.

Removals

The DD accumulated on pavements is subjected to physical, chemical and biological processes which ultimately reduces the accumulation rate almost to zero over an extended period of dry days. This may be due to biological decomposition of organic material, wind and vehicle induced eddies and regular street cleaning practices.

Biological Removal

Organic pollutants decompose into simple substances in the presence of organisms, with or without oxygen. This reduces the total quantity of pollutants available at the time of washoff. Nitrate and phosphate are exceptions to the above process as they are normally increased by the addition of decomposed end products. This removal of solids by biological decomposition may be estimated:

$$R_b = PF(1-e^{-kt})$$

where: R_b = total mass of DD removed by biological decomposition, mg/d.

P = average mass of DD available on the surface, mg/d.

F = fraction of decomposable DD

k = decay coefficient, 1/d.

t = number of days.

Vehicle removal

Vehicle generated eddies transport pollutants from the streets to ineffective impervious and pervious areas from which they may not be washed off during storm events(24). The physical transfer or loss of solids from effective to ineffective surfaces results in a net reduction of DD. This could be estimated by:

$$R_v = N\ F\ P$$

where: R_v = mass of DD removed due to vehicle generated eddies, mg/d.

N = number of vehicles in a subcatchment

P = DD available for eddy-transport, mg/d.

F = $\theta_1(1-e^{-\theta_2 S})$

θ_1 = maximum fraction of DD that could be removed

θ_2 = decay coefficient, h/km

S = vehicle speed, km/h

Wind removal

The transport of accumulated DD due to natural wind also reduces the available quantity for washoff. Dust removal is a function of wind speed, curb height and presence or absence of an adjacent pervious area which acts as a pollutant sink. The general pattern of airflow on street pavements dictates the loss rate. Natural wind induced losses may be expressed(12) as:

$$R_W = 0.0116\ (e^{-0.088H})\ W$$

where: R_W = fraction of DD lost due to wind

H = curb height, cm

W = wind speed, km/h.

Intentional Removal

Regular street cleaning enhances aesthetics and minimises pollutant loadings to the receiving water. The frequency and efficiency of cleaning depends on traffic density and the cleaning equipment respectively. Solids remaining on the pavement after street cleaning are finally washed off by rain. DD remaining on the catchment after street cleaning is given by(25):

$$R_i = \frac{L_a - R_{ma}}{L_a}$$

or

$$R_{ma} = L_a(1 - R_i)$$

where: R_{ma} = DD remaining on subcatchment surface, mg

L_a = DD present before sweeping, mg

R_i = DD removed by sweeping = $A_v\ R_f$

A_v = availability factor = $0.6\ (2.5\ P_d)^{0.2}$

P_d = population density, persons/ha.

R_f = removal efficiency.

Net accumulation

The net DD accumulation on an impervious area may then be calculated on a daily basis considering both addition and removal processes. The present authors wrote a subroutine (NEWBLD) using the algorithms discussed above (the last ten equations) for the prediction of DD accumulation over a dry day period on individual subcatchments. A provision to input the sweeping dates as time series for all the subcatchments is also included in the subroutine. The fraction of DD arested in vegetation and buildings due to the canopy effect is also added to the accumulation. NEWBLD has been inserted in our version of the RUNOFF block of SWMM3 and is called CHGQUAL herein(26).

Wet Weather Pollutant Buildup

The quantity of pollutant scavenged during a storm depends on the quantity available, intensity of rainfall, collection efficiency and the areal coverage of the storm. Pollutant scavenging rate is exponential in time(18,27). The process may be simplified for modelling purposes. For interfacing with CHGQUAL, the algorithm for aerosol scavenging was formulated as:

$$S_c = S_o\ e^{-R_a t}$$

where: S_c = quantity of aerosol remaining in the atmosphere at time t, mg

S_o = quantity of aerosol present before scavenging, mg.

R_a = rain scavenging rate constant, 1/s.

t = time, s

The quantity of aerosol present before scavenging is calculated assuming a triangular distribution of pollutant concentration within the atmospheric mixing zone. The total mass is given by:

$$S_o = AHCF$$

where: A = area of subcatchment, m^2

H = atmospheric mixing height, m

C = mean atmospheric aerosol concentration, mg/m3

F = pollutant fraction in aerosol, mg/mg

Rain scavenging rate, as calculated by Slinn[28] is:

$$R_a = \frac{1}{0.0045}\left(\frac{I}{25.4}\right)^{3/4} C_f$$

where: I = intensity of rainfall, mm/h

C_f = collection efficiency, (10^{-2} to 10^{-1} for particle size 1-5 µm and drop radius 0.1 - 1.00 mm; see Ref. 27)

The rate of pollutant scavenging is obtained by differentiating the scavenging algorithm:

$$\frac{dS_c}{dt} = S_o(-R_a)e^{-R_a t}$$

or

$$P_s = -S_c R_a$$

where: P_s = aerosol pollutant scavenging rate, mg/s

S_c = aerosol pollutant quantity remaining in the atmosphere, mg

R_a = rain scavenging rate, 1/s

The amount of gaseous pollutant absorbed into rain droplets depends on the pollutant concentration potential between the drop and the gas. The equilibrium concentration in the droplet is attained in a short period of travel and is assumed not to be rate limited. Therefore pollutant concentration is constant throughout the storm. The total pollutant quantity depends on the quantity of runoff and the pollutant equilibrium concentration. Hence, gas scavenging is written:

$$P_g = 1000\ Q\ C_q$$

where: P_g = rate of gaseous pollutant added to the runoff water, mg/s

Q = runoff rate, m3/s

C_q = equilibrium concentration, mg/1

1000 = conversion factor, m3 to litre

Canopy washoff algorithms

The leaf canopy provides an additional lodging area for atmospheric dust. The quantity of dust arrested in the canopy depends mainly on wind velocity and frequency, area density of leaves and atmospheric dust concentration. A simplified formulation is developed in this section, suitable for adaptation to stormwater management modelling. Pollutant buildup in the canopy may be modelled by assuming a linear buildup during dry weather using dustfall rates predicted by a model such as ATMDST(29) for individual subcatchments. The canopy pollutants similarly may be washed off using an exponential relation between pollutant flux and storm intensity and duration. Of course, not all pollutants washed off the canopy will reach the overland flow because of soil infiltration and also because of filtration due to motion through vegetation. For these reasons it is often assumed that pollutants from canopy washoff are not significant. Algorithms for simulating the canopy contribution may easily be developed as follows:

$$P_c = D\ t\ A\ E_f\ F$$

where:

- P_c = total pollutant quantity on the canopy before the rain, mg
- D = atmospheric dust, mg/eff.h-m^2
- t = dry days, d
- A = total canopy area, m^2
- F = fraction of pollutants in atm. dust, mg/mg
- E_f = number of effective hours, (24 x wind frequency)

The quantity of pollutant remaining after wash off in the canopy is given by:

$$P_w = P_c\ e^{-C_1 I t/3600}$$

where:

- P_w = quantity of pollutant remaining in the canopy after wash off at time t, mg
- P_c = quantity of pollutant available initially on the canopy, mg
- C_1 = canopy washoff constant (0.4), 1/mm
- I = rain intensity, mm/h

t = time, s

Washoff rate in mg/s can be calculated by differentiating the previous equation:

$$P_r = \frac{-P_w}{3600} I C_1$$

The concentration of pollutant in mg/1 within the canopy is written as:

$$C_c = \frac{P_r}{IA} 3600$$

and pollutant flux to runoff water is given by:

$$F_c = C_c\, q$$

where: F_c = pollutant flux in canopy, mg/s

C_c = pollutant concentration in canopy, mg/1

q = quantity of flow joining runoff water, 1/s

A = area of canopy, m^2

P_r = $d(P_w)/dt$, mg/s

Continuous Mass Balance

Although a wide variety of pollutant accumulation relationships have been proposed by many investigators, depending upon their field data, a relationship suitable for all catchments is yet to be derived. In view of this problem, subroutine NEWBLD was developed as a substitute to the BUILD routine of SWMM3 RUNOFF. NEWBLD is based on individual buildup processes. The ten equations discussed above are used in the NEWBLD subroutines. NEWBLD accounts for daily addition and removal processes depending on the prevailing meteorological and geological conditions of an individual subcatchment. The ATMDST routine predicts atmospheric dustfall which is then transferred to NEWBLD. The new buildup algorithm does not follow any conventional accumulation rate curve even though it is non-linear in nature. The FORTRAN code is given in the CHGQUAL User's Manual(29).

Sensitivity tests

DD buildup rate decreases over a period of dry days and becomes almost zero, attaining steady state. Both addition and removal processes are governed by wind velocity and source concentration. Even if the pollutant input from all sources is constant, the dispersion and advection processs, mainly controlled by meteorological factors, distribute pollutants onto the subcatchments. Therefore a sensitivity test on the buildup algorithm was performed for three different land uses using an actual data set for the Chedoke Creek Catchment in Hamilton. The analysis showed that buildup is sensitive to wind velocity and vehicle speed. Since meteorological parameters fluctuate widely, random values of wind velocity and frequency and average values of remaining parameters were used to generate results that were very similar to the DD buildup data observed by various researchers. The effect of street cleaning frequency on DD buildup was also studied.

Application of Continuous Mass-Balance Method

A mathematical model is simply a quantitative expression of a process or phenomenon one is observing, analyzing, or predicting. Since no process can be completely observed, any mathematical expression of a process will involve some element of uncertainity. Hence, any mathematical model formulated to represent a process or phenomenon will be conceptual to some extent and the reliability of the model will be based upon the extent to which it can be or has been verified. Model verification in turn depends on the data available to test the model scientifically and the resources available to perform the scientific tests. A reliable data base simplifies the problem of model building and testing. After ensuring the availability of good data, a series of code verification, sensitivity, calibration, and validation tests can be designed to produce sufficient information to answer a wide range of questions about the model's performance[30].

Data Acquisition

Hamilton's urban drainage system comprises an area of approximtely 120 km^2 which drains through 20 major combined sewer outfalls to Hamilton Harbour and Coote's Paradise. Among the three major drainage areas of Hamilton, the Chedoke Creek Catchment was chosen for examination in this study.

This drainage system is 26.8 sq.km. in area, and about 46% of the area is served by separate sewers. The predominant land use is low to medium density single family residential housing. The Chedoke Creek outfall channel receives runoff from 93% of the area and ultimately drains to Coote's Paradise[31].

In the late 1970's and in early 1980's, a study was initiated to investigate the pollutional potential due to runoff. The Computational Hydraulics Group (CHG) at McMaster University set up a dense hydrometeorological data acquisition network in the test catchment. Simultaneously, a side weir overflow model[32], and continuous and interactive storm water models were developed and tested for Hamilton.

Since the storm cells are dynamic, stationary rain data are too unrealistic to use in urban runoff studies[33,34]. A higher density of rain recording stations was necessary to track the storm direction and intensity. An inexpensive, higher precision and resolution, reliable and automatic system for rainfall data was developed[35,36].

Runoff samples were collected at three points in the Chedoke Creek Catchment at approximately 5 minute intervals during storm events. The runoff water samples were analysed for BOD, suspended solids, total nitrogen, total phosphorous and total coliform organisms.

Information pertaining to street cleaning, population and vehicle density, was obtained from the Hamilton City Hall. The CHG archive provided all the data needed for model development.

Calibration

The CHGQUAL was calibrated against the Chedoke Creek data base. Data collected on 6 storms during summer 1980 on flow, BOD, SS, total nitrogen and total phosphorous at 2 stations were used for calibration. The hyetographs used in the calibration were recorded at 3 rain stations[31]. Since storms are dynamic, the hyetograph for an individual subcatchment is not the same as the observed hyetographs at the rain recording stations. Identifying the individual hyetographs for all the subcatchments was done by careful study of the three observed hyetographs and fixing the direction and approximate speed of the storm. The observed peak flow and time to peak were also used as guidelines to fit the hyetograph to each individual subcatchment. The minor adjustment for the flow and pollutant concentration was done by calibration. Since the model is very sensitive to most input data, the storm starting time was lagged to account for storm travelling time, and pollutant load rates and meteorological data were adjusted to match the predicted results with the observed data. The pollutographs produced by CHGQUAL agreed well with observed data when compared to SWMM3 results. Finally, the results predicted by both CHGQUAL and SWMM3 models were compared; the statistics of both best fit lines of CHGQUAL and SWMM3 were compared with the ideal line. The results indicated that CHGQUAL predicts quality better than does SWMM3, at least for the conditions pertaining to Hamilton at that time.

Validation

Whatever the model sophistication, it is hard to match observed data exactly because of the complexity of the real process. The data used for validation were collected in the summers of 1981 to 1983. Pollutant parameters used for validation were BOD, SS, total nitrogen, and total phosphorous. The validated results were good with respect to time to peak and peak concentration.

Discussion and Conclusions

Current deterministic models usually compute storm water runoff quality in sequential steps. Estimation of DD accumulation during dry weather is the first stage and estimation of washoff of the DD during runoff events is the second stage. Routing is a third stage, if pollutant concentration is to be computed at the downstream end of the storm sewer network. Pollutant sedimentation, dispersion and transport in the receiving water is the final stage. Since the processes of accumulation, redistribution, washoff, transport and dispersal are complex in nature, a general model for the complete process suitable for application to all catchments has yet to be developed. This contribution should be seen as a small step in that direction.

The data collected by researchers on DD and pollutant buildup show linear and non-linear accumulations, with a high degree of scatter. The scatter represents variations due to unexplained processes. Because of this, using a fitted equation to estimate DD for a specific episode of dry weather and a subsequent storm may not relate to the actual processes involved at that time in that place. For example, high winds transport DD from effective to ineffective areas; unreported street sweeping prior to a storm considerably reduces DD; thermal inversions may significantly increase fallout. Clearly a simple linear or exponential equation using the number of drydays as the only independent variable is not sufficient to depict complex buildup processes. For these reasons, this review focussed on overcoming the shortcomings of current methods of computing DD buildup by developing process oriented buildup algorithms, based on daily mass balances of DD or a pollutant. Several FORTRAN codes were written to interface with the SWMM3 package. The NEWBUILD routine in CHGQUAL calculates buildup on all the subcatchments on an individual basis; it is not based on the five land use classifications. This gives the user a capability of modelling management strategies for reducing buildup.

In addition to impervious area washoff, (a) aerosol and gas scavenging, (b) canopy washoff and (c) pervious area erosion add to runoff water quality during the storm. Most models currently treat both aerosol and gas scavenging together as a constant concentration throughout the simulation. Aerosol scavenging varies exponentially with the intensity of rainfall. It also varies with particle radius and collection efficiency. Gas scavenging remains almost constant throughout the simulation because it is not rate limited and attains equilibrium concentration within a small distance of travel. These two processes are modelled separately in CHGQUAL.

The vegetation canopy in Ontario accumulates more dustfall than the land surface because of the large surface area of leaves and air filtration through the canopy. Pollutants washed off the canopy during a storm joins the runoff only after satisfying depression storage and exceeding infiltration. Of the total quantity of pollutant washed off from the canopy only a portion joins the runoff. The remainder either infiltrates or is arrested in the grass filtration blanket. Consequently the net canopy effect on pollutant runoff may not be significant under normal conditions when the grass filtration blanket is absent or thin. On the other hand, when the location of the catchment is very close to dust prone areas, the canopy effect may be significant. In Ontario, the forest canopy is widespread, and so this phenomena was also modelled in CHGQUAL.

Most of the computed pollutographs calibrated well against observed pollutographs by adjusting model input parameters rather than model constants. This confirms the necessity for careful sensitivity analysis. CHGQUAL was validated against a 1981-83 summer data set using average calibrated model parameters. The validated results were good with respect to time to peak and peak concentrations.

Conclusions

The art of modelling urban stormwater quality is considered to be in need of a larger data base and better knowledge of pollutant buildup, redistribution, washoff, transport and dispersal. The attempt made in this review to better describe the basic physics of processes involved in storm water quality is a start towards a complete process oriented model. The statistical comparison between SWMM3 and CHGQUAL for both calibration and validation showed that CHGQUAL probably represents an improvement. However the tests were carried out on a very small data set, so it is too soon to make any generalised statement on the predictibility of CHGQUAL. Since CHGQUAL is process oriented, more capabilities for investigating sophisticated management techniques based on the physical processes are now possible.

Of course, the algorithms used here are by no means the final solution to the problem of storm water quality modelling. But an attempt has been made to identify the processes, to model them separately and to use them in an overall model for management of runoff water quality. Until a sufficiently large data base is available which will allow further description of the processes involved in urban runoff water quality, models such as CHGQUAL should continue to be used with caution.

References

1. Ammon, C.D., "Urban Stormwater Pollutant Buildup and Washoff Relationships", M. Engineering Thesis, University of Florida, 1979.

2. Sartor, J.D. and Boyd, G.B., Water Pollution Aspects of Street Surface Contaminants, Environmental Protection Technology Series, EPA-R2-72-081, U.S. Environmental Protection Agency, Cincinnati, 1972.

3. McCuen, R.H. and Sutherland, R., "Flood Runoff from Urban Areas", Technical Report No. 33, Water Resources Research Center, University of Maryland, College Park, Maryland, 1975.

4. Pitt, R., The Potential of Street Cleaning in Reducing Non-point Pollution", Urban Stormwater Management Workshop Proceedings, EPA-600/9-78-017, U.S. Environmental Agency, Cincinnati, pp. 91-119, 1978.

5. Nunno, T.J., "Application and Testing of Stormwater Modeling Technology", M.S. Project, Department of Civil Engineering, University of Massachusets, Amherst, 1978.

6. Shaheen, D.G., "Contributions of Urban Roadway Usage to Water Pollution", Environmental Protection Technology Series, EPA-600/2-75-004, U.S. Environmental Protection Agency, Cincinnati, 1975.

7. Terstriep, M.L., Bender, G.M. and Benoit, D.J., "Buildup, Strength and Washoff of Urban Pollutants", Preprint 3439, Presented at American Society of Civil Engineers Convention, Chicago, 1978.

8. Marsalek, J., "Stormwater Quality in the Malvern Catchment", Presented at the Stormwater Management Model Users Group Meeting, Toronto, Ontario, 1976.

9. Mance, G. and Harman, M., "The Quality of Urban Stormwater Runoff", Urban Storm Drainage, Proceedings of International Conference on Urban Storm Drainage, Southampton, England, 1978.

10. Waller, D.H., "Factors that Influence Variations in the Composition of Urban Surface Runoff", Water Pollution Research in Canada, No. 7, pp. 68-95, 1972.

11. Miller, R.A., Mattraw, H.C. and Jennings, M.E., "Statistical Modelling of Urban Storm Water Processes, Broward County, Florida", Proceedings, International Symposium on Urban Storm Water Management, University of Kentucky, Lexington, pp. 269-273, 1978.

12. Novotny, V. and Goodrich-Mahoney, J., "Comparative Assessment of Pollution Loadings from Non-point Sources in Urban Land Use", Progress in Water Technology, Vol. 10, Nos. 5/6, pp. 775-785, 1978.

13. American Public Works Association, "Water Pollution Aspects of Urban Runoff", Federal Water Pollution Control Administration, Report WP-20-15, U.S. Department of the Interior, Washington, D.C. 1969.

14. Gupta, M.K., Agnew, R.W. and Kobriger, N.P., "State-of-the-Art Report on Highway Runoff Constituents, Envirex Inc. for FHWA Contract DOT-FH-8600, 1978.

15. HEC (Hydrologic Engineering Center), US Army, Corps of Engineers, "Storage, Treatment, Overflow, Runoff Model (STORM)", Computer Program 723-58-12520, Davis, California, 1976.

16. Geiger, W.F. and Dorsch, H.R., "Quantity-Quality Simulation (QQS): A Detailed Continuous Planning Model Description, Testing and Applications", EPA-600/2- 80-011, Environmental Protection Agency, Cincinnati, OH, 1980.

17. Shivalingaiah, B. and James, W., "Atmospheric Pollution in Relation to Storm Water Quality Modelling; Literature Review for an Industrial City", Storm Water Quality Model User's Group Meeting, Washington, D.C. EPA-600 9-82-015, pp. 274-295, 1982.

18. Randall, C.W., Helsel, D., Grizzard, T. and Hoehn, R., "The Impact of Atmospheric Contaminants on Storm Water Quality in an Urban Area", Prog. Wat. Tech., Vol. 10, Nos. 5/6, pp. 417-431, 1978.

19. Novotny, V., "Acidity of Urban Precipitation and its Buffering During Overland Flow", Second International Conference on Urban Storm Drainage, Urbana, Illinois, U.S.A., June 1981.

20. James, W. and Shivalingaiah, B., "Stormwater Quality and Atmospheric Pollution Modelling in an Industrial City", 6th Canadian Hydrotechnical Conference, Ottawa, pp. 707-722, June 1983.

21. Prasad, D., Henry, J.G. and Kovacko, R., "Pollution Potential of Autumn Leaves in Urban Runoff", International Symposium on Urban Storm Runoff, Lexington, Kentucky, pp. 197-201, 1980.

22. Chen, W.C., Hudson, J.M.R., Gherini, A.S., Dean, J.D. and Goldstein, A.R., "Acid Rain Model: Canopy Module", J. ASCE, Vol. 109, No. 3, pp. 585-603, 1983.

23. Heaney, J.P. and Huber, W.C., "Storm Water Management Model, Refinements, Testing and Decision Making",

Department of Enviromental Engineering and Sciences, University of Florida, 1973.

24. Sehmel, G.A., "Particle Resuspension from an Asphalt Road Caused by Car and Truck Traffic", Atmospheric Environmental Pergamon Press, Vol. 7, pp. 291-309, 1973.

25. Huber, W.C., Heaney, T.P., Nix, S.J., Dickenson, R.E. and Polmann, D.J., "Storm Water Management Model User's Manual, Version III", U.S. Environmental Protection Agency, Cincinnati, OH, 1981.

26. Shivalingaiah, B., "Modelling Storm Water Pollutants in Hamilton, Canada", Ph.D. Thesis, McMaster University, Hamilton, Canada, pp. 1-252, 1984.

27. Slinn, W.G.N., "Some Approximation for the Wet and Dry Removal of Particles and Gases from the Atmosphere", Water, Air, Soil Pollution, 7, pp 515-543, 1977.

28. Slinn, W.G.N., "Precipitation Scavenging: Some Problems, Approximate Solutions and Suggestions for Future Research", Precipitation Scavenging ERDA 41 Symposium Series, CONF-741003, pp. 1-60, 1974.

29. Shivalingaiah, B. and James, W., "CHGQUAL User's Manual", CHG Publications, Hamilton, 1984.

30. James, W. and Robinson, M.A., "Standard Terms of Reference to Ensure Satisfactory Computer-Based Urban Drainage Design Studies", Canadian Journal of Civil Engineering, Vol. 8, No. 3, pp. 294-303, March 1981.

31. James, W. and Henry, D., "Stormwater Management Model for Hamilton Harbour, Summer Field Program 1981 (Development of a Stormwater Overflow Model for Pollutional Loadings to Hamilton's Receiving Water, Summer Field Program and Data for Period May to October 1981)". Report to the Ontario Ministry of the Environment. CHI Publicatons, (about 300 pp.) March 1982.

32. James, W. and Mitri, H., "Modelling Sideweir Diversion Structures for Stormwater Management", Canadian Journal of Civil Engineering, Vol. 9, No. 2, pp. 197-205, June 1982.

33. Shtifter, Z., "A Kinematic Storm Model for Urban Drainage Studies", M.Eng. Thesis, McMaster University, May 1980.

34. Scheckenberger, R., "Dynamic Spatially Variable Rain Models for Storm Water Management", M.Eng. Thesis, McMaster University, 1983.

35. James, W., Haro, H., Robinson, M.A. and Kitai, R., "Hydrologic Data Acquisition: Innovative, High

Resolution, Pogrammable Instrumentation for Storm Water Management", Proceedings of the Conference on Storm Water and Water Quality Management Modelling, U.S. EPA, Washington, D.C., EPA 600/9-82-015, pp. 128-151, 1982.

36. Haro, H., Kitai, R. and James, W., "Precipitation Instrumentation Package for Sampling of Rainfall", Institute of Electrical and Electronics Engineers (Transactions on Instrumentation and Measurement), Vol. IM32, No. 3, pp. 423-429, September 1983.

37. Thomann, V.R., "Verification of Water Quality Models", J. ASCE, Environmental Engineering Division, vol. 108, pp. 923-940, 1982.

DISCUSSION - DETERMINISTIC MODELING OF RUNOFF PROCESSES

In the discussion which followed the presentation of the papers, the following points were made:

1. Mr. Metaxatos expressed the opinion that the community's bad early experience with the Storm Water Management Model (SWMM) had led to the abandonment of models. Dr. James and Mr. Barnwell, among others, strongly stated that in their experience the opposite was true - that models were being increasingly used, particularly for quantitative assessment of runoff. They also noted that early experience was in many cases very negative, but that this was due as much to inexperience in model use as to flaws in the models themselves. Dr. Huber noted that early reliance on model "default" values was a major source of these difficulties.

2. Dr. James noted that other possible sources of modeling data in the average municipality, such as air monitoring and street maintenance organizations, will become increasingly important as more of their data become available in automated form.

3. Mr. Metaxatos questioned whether or not we were able to relate land uses to pollutant accumulations. Dr. Huber replied that we are not able to do so, nor does he expect that we will be able to do so in the future, because "land use" is too broad a term to characterize the pollutant-generating capacity of an area.

4. Dr. Roesner explained that, in the U.S., at least, one of the reasons that models are not more frequently used in water quality analyses was that the water quality objectives do not require it - either because the time scales are too long, or because many States have arbitrary rules which require, for instance, the capture of the first inch of runoff.

5. Dr. Medina expressed the opinion that legal challenges will force fairly sophisticated model use, and that data to support these models is still going to be a problem.

6. Dr. Ferrara pointed out that when evaluating whether or not models are "good" it is important to remember that there are three uses of models.

   a. To help organize and visualize cause and effect relationships in the catchment and receiving water.

   b. To assist in making ordinal comparisons of alternatives.

   c. Detailed analysis and design.

   For the first two, at least, a high level of model accuracy is not too important.

7. Dr. Geiger asked what was the preferred method of establishing inter-event times for rainfall or runoff. Dr. Huber replied that he used two methods:

   a. The use of auto-correlations - inte-revent time is defined when it drops below significance level.

b. Choose inter-event time statistically such that the standard deviation equals the mean.

Dr. Huber (and others) further noted that it is important to relate inter-event times to runoff, rather than rainfall, events, and that this is a compelling argument for the use of continuous simulation.

8. Dr. Gonenc expressed the opinion, which was supported by others, that models are already sufficiently well developed, and that our emphasis now should be on how they are practically applied, including improvement of our understanding of quantity/quality relationships.

9. Mr. Ellis stated that he was not too optimistic about our ability to improve the predictability of accumulation and transport processes, with the possible exception of solids. He further noted that hysteresis (in both directions, though clockwise is prevalent) is evident in many systems, and increases the inefficiency of using lumped-parameter inputs. Dr. Ferrara, on the other hand, pointed out that the hysteresis hypothesis is not necessarily valid. It could be a linear function if the rainfall driving the calculations were different. Mr. Ellis agreed.

10. Mr. B. N. Jacobsen asked if data on measured mass-removals from street-sweeping machines had been used for comparison with model simulations. Dr. James replied that, in Hamilton, the data were very hard to get from the street maintenance people. Dr. Huber also referred to the Pitt (1979) study cited in his paper which showed the difficulties of obtaining quantitative street-sweeper performance data.

11. Mr. Urbonas pointed out that, as an end-user of research results (as a member of the staff of an urban drainage district), his requirements were much less detailed than those of research groups. Mr. Weatherbe supported this, pointing out that he has worked with very complex receiving water systems and found that if one tries to evaluate the causes of "major" changes over long time periods the inputs must be aggregated and process modeling simplified.

STATISTICALLY BASED MODELLING OF URBAN RUNOFF QUALITY: STATE OF THE ART

J.C. Hémain
Laboratoire Central des Ponts et Chaussées
B.P. 19, 44340 Bouguenais
France

Introduction

Models or modelling techniques may be divided into two main groups:

- Deterministic models which comprise causal relationships between various variables controlling the processes modelled, and
- Stochastic models which comprise relationships between probability levels of the variables included in the processes.

Considering the first group, other classifications may be introduced, for example:

- Physically-based models, derived from theoretical approaches (fluid mechanics, system analysis, etc.). These have been described in ref. (9).
- Statistically-based models, mainly developed from statistical analysis of experimental data. They may be purely empirical, but can also take advantage of statistical theories.

Such a classification has been proposed by many authors (1). In each group, one could find other classes, according to the model objectives (planning, design, management), model inputs (single event model or continuous simulation model), and the time interval to be considered (minutes, hours, or longer).

The division between the physically-based and statistically-based models is not well defined. For example, parameters of a physically-based model can be fitted to data using statistical

NATO ASI Series, Vol. G10
Urban Runoff Pollution.
Edited by H. C. Torno, J. Marsalek, and M. Desbordes

techniques such as in the case of the availability factor A in the SWMM model:

$$M = A \times P \, [1 - e^{-KQ\Delta t}] \qquad (1)$$

$$A = a + bQ^{c} \quad (A \leq 1)$$

where M is the total suspended solids (TSS) mass which is transported during a given time interval Δt, Q is the runoff discharge during this time interval, P is the TSS mass on the catchment surface at the beginning of the time interval, and a, b, c and K are empirical constants. In the same manner, purely statistically-based relationships could be checked by a physically-based analysis to test their physical reliability. Moreover, both of the two main approaches, deterministic and stochastic, could be used in the same model as shown in refs. (2) and (3).

<u>Statistically-Based Modelling: A Reliable Approach</u>

The statistical approach presents many advantages. First of all, it allows one to check or to choose the explanatory variables (or parameters) which have the greatest influence on runoff quality. Factors affecting runoff quality are very numerous and could not be classified a priori in the order of importance. Considering the existing models, there are probably tens of factors affecting runoff quality. Numerous statistical analyses performed on reliable data (4-6) give fairly variable or even contradictory results, from one catchment to another. For example, some authors (7) do not agree with the influence of the antecedent dry weather duration on runoff pollution, even though it appears as one of the most important parameters in physically-based models. Similarly, land use, another important parameter of physically-based models for pollution load estimates, is not always statistically significant, even when considering separate storm sewer systems (8).

Moreover, the statistical analysis makes it possible to check relationships among various pollution parameters (or indicators), or to show similar behaviour of some parameters. Also, statistical analysis should reduce the amount of necessary expensive data required to develop a new model or to calibrate a given one. For example, it was observed that the ratio between COD and BOD_5 for residential land runoff generally lies in the range from 6 to 10. Such a statistical result is used in the SWMM for the "potency factor" default values (pollutant ratios in "dust and dirt") which are 40 mg/g for COD and 5 mg/g for BOD_5.

A synthesis approach used in statistically-based models is of primary interest. These models are composed of relationships between input variables (rainfall and/or runoff), output variables (instantaneous or mean values of pollutant discharges and/or loads), and the parameters describing the drainage system, independently of the various components of the whole runoff pollution process. As shown in refs. (9) and (10), these components are numerous and often complex: pollutants have various origins and forms of production, they are spread over the catchment areas by wind and traffic, their nature may be changing with time due to chemical or biochemical transformations, local practices such as street cleaning can modify the processes, and further changes take place in the sewers (deposits) etc. All the components of the runoff pollution process cannot be simulated at present because the data required are generally not available. Most of the available data have been collected at catchment outlets and are not suitable for the modelling of upstream runoff pollution processes. They can be used only in a lumped model approach which may be in agreement with the statistical approach.

Statistical modelling can also benefit from the statistical and sampling theories, as far as the data make this possible. Various relationships can be checked using many well-known statistical tests which have been generally computerized. However, it will be often more difficult to develop a statistically-based model than to fit an available physically-based model to the available data. When developing a statistically-based model, one should remember that the data are the most important part of the work (11): the experimental design should be first carefully studied, depending on the phenomenon investigated, and the measurements should be done precisely according to the well-known standards. This follows from the fact that a statistically-based model generally cannot give more information than the data contain. This last remark also gives the limitations of statistically-based models. They should not be transposed from one catchment to another, cannot give information related to catchment changes, and do not yield accurate results for events outside of the observed events range.

Finally, the use of given statistically-based models should be limited to well instrumented catchments. They may be excellent tools for operational purposes, such as pollution control in drainage systems, including real-time control. In the last case, as the length of the data series continuously increases, the model can be calibrated periodically for the observed events. Recognizing the importance of data for the development of a statistically-based model, one must also remember that data are also very important for the calibration of physically-based models.

<u>Various Statistically-Based Tools</u>

Statistically-based models of runoff pollution may be developed for planning, design or operational purposes.

The first level reflects approximate long-term planning operations designed to compare various urbanization scenarios and/or to roughly estimate the pollution loadings from non-point as well as point sources of pollution (e.g., domestic wastes). For example, such computations may include estimates of pollutant annual loads which were simply obtained by multiplying a mean pollutant concentration by the annual runoff volume (which may be given by the annual rainfall and catchment characteristics) or by using some more complex general formulae. Such "desktop" models use mean pollution parameter values that can be found in the literature. The user must choose those values which seem pertinent to the catchment studied and follow from such criteria as land use, population density, climatic conditions, etc. Examples of such approaches are the general formulae established in 1977 by the University of Florida from the available U.S. data (12) (see Table 1). Such formulae are required to simulate the effectiveness of a given control facility. Some research has been done in France to develop similar formulae from French experimental catchment data.

The second modelling level concerns the estimate of pollutant loads during a storm event. For this purpose, statistically-based models use simple or multiple regression equations in linear or non-linear forms. In these equations, a given pollution variable Y_i (the concentration or a loading of a i-th pollutant) is related to one or more variables X_j of the storm event: the total rainfall depth, rainfall duration, runoff volume, peak runoff, dry weather duration, etc. The general forms are:

$$Y_i = a + b\,X_1 + c\,X_2 \qquad (2)$$

or $$Y_i = a'\,X_1^{b'}\,X_2^{c'} \qquad (3)$$

Simple regression equations in which Y_i is generally a power function of the runoff flow rate or volume are usually called "rating curves". These equations must be fitted using sample data from the studied catchment. There may be as many equations as the number of catchments or data samples. These models can be used in studies of runoff pollution impact on receiving waters. Because the receiving water quality is not necessarily sensitive to pollutant load changes during short time intervals, total event loads may be sufficient for such studies (13).

Table 1. General Equations for Prediction of Annual Pollutant Loadings in the U.S.A. (12).

Separate sewer networks: $M_s = \alpha(i,j)*P*F(PD)*\gamma$

Combined sewer networks: $M_c = \beta(i,j)*P*F(PD)*\gamma$

where M is the annual loading of pollutant j from land use i (kg/ha/year)

P is the annual rainfall (mm/year)

PD is the population density (per./ha)

γ is the street sweeping efficiency, defined as

$$\gamma = \begin{cases} Ns/20 & \text{for } 0 < Ns \leq 20 \text{ days} \\ 1.0 & \text{for } Ns > 20 \text{ days} \end{cases}$$

where Ns = Number of days between sweepings

F(PD) = Function of population density depending on land use as follows:

residential, $F(PD) = 0.142 + 0.134 * PD^{0.54}$

commercial and industrial, $F(PD) = 1.0$

others, $F(PD) = 0.142$

The parameters $\alpha(i,j)$ and $\beta(i,j)$ are listed in the table below.

Parameter	Land Use i	Pollutant j			
		BOD_5	TSS	Total PO_4	Total N
α (separate)	1-Residential	0.0353	0.721	0.00149	0.00579
	2-Commercial	0.141	0.981	0.00335	0.0131
	3-Industrial	0.0535	1.29	0.00312	0 0122
	4-Other	0.0050	0.119	0.00044	0.00267
β (combined)	1-Residential	0.145	2.97	0.00614	0.0239
	2-Commercial	0.583	4.06	0.00138	0.0539
	3-Industrial	0.221	5.30	0.0129	0.0504
	4-Others	0.0206	0.491	0.00182	0.0111

Equations (2) and (3) are generally fitted to the data considering that these data are mutually independent. However, the simulation of continuous events should be preferable and should lead to better results. In that case, the conventional "rating curves" or other regression equations can be linked to conceptual models of pollution accumulation during dry weather periods. This has been done by Servat (3) for French data (14). Some recent studies seem to show that "rating curves" can lead to good results without the use of pollution accumulation models.

The most sophisticated level is related to the pollutograph modelling, i.e., the modelling of pollution concentration or load changes with time. The main objective of these models is to check the efficiency of such pollution control facilities as detention basins, sewer overflow device regulators, etc. Such models can also be used for operational purposes. Pollutographs are described by regression equations between pollution indicators Y_i, and runoff variables, such as runoff discharge Q(t) or runoff volume at time t, and/or time. Such equations are frequently written in the following form:

$$Y_i(t) = a\ Q\ (t)^b \qquad (4)$$

The above form is used in the Non Point Source Pollutant Loading Model NPS, developed in 1976 by Hydrocomp. Inc. for the U.S. Environmental Protection Agency (15). Another form is:

$$Y_i(t) = a'\ Q^{b'}(t)\ t^c \qquad (5)$$

Equations (4) and (5) are coupled with hydrological or hydraulic models and with mass transport models that describe runoff pollution changes within the sewer system (taking into account solid deposits, for example). One should notice that such "rating curves" are an optional procedure in the SWMM Level III model which should be considered as a primarily physically-based model.

As stated earlier, another approach can be used which is purely stochastic in nature and is concerned with the study of probability distributions of runoff pollution parameters. For this approach the independence of the observed variables is necessary. It could be well adapted to the study of combined sewer overflows. Pollutant loads to receiving waters are given in terms of probability. Sewer overflow facilities, detention basins or treatment plants should be designed according to probability levels of exceedance of some given pollution parameters. However, such an analysis needs long term records of runoff pollution data which are rare. Sometimes the probability distribution of a given pollution parameter can be estimated from long time rainfall series (2). Similarly, a deterministic rainfall-runoff-pollution model can be used to simulate pollution data, the probability distribution of which is then studied. This is done, for example, in the STORM model (1).

Review of Significant Past Studies and Results

It is impossible to review all the studies and results here. Rather, the better known studies that constitute basic references will be noted.

The first such a study was conducted by Weibel et al. in 1963 (16) in a residential experimental catchment (11 ha) in Cincinatti (U.S.A.). The study data were subsequently used for calibration of the Battelle Wastewater Management Model (1). Many other models have been tested using these data.

1969, the American Public Works Association (APWA) conducted important study of runoff pollution in 18 catchments in ago (17). The main results were the estimates of daily ulation rates of dust and dirt for various land use, and tes of various pollutant loads in dust and dirt. These are given in Table 2. They are still considered as reference data which were used as default values in known models as SWMM and STORM.

Table 2. Pollutant Accumulation Rates on Urban Catchments in Chicago (17)

Land Use	Dust and Dirt (DD) Accumulation (kg/day/curb km)	Pollutant rate (mg/g of DD)					
		TSS	BOD 5	COD	PO4	N	Coliforms*
Residential single family housing	10.4	1000	5.0	40	.05	0.48	1.3
Residential Multi-family housing	34.2	1000	3.6	40	.05	0.61	2.7
Commercial	49.1	1000	7.7	39	.07	0.41	1.7
Industrial	68.5	1000	3.0	40	.03	0.43	1.0

* 10^6 x Most Probable Number/g

In 1970, the AVCO System Corporation performed a similar study in 15 experimental watersheds in Tulsa (USA) and tried to find reliable relationships using many tools of statistical analysis: statistical tests, simple and multiple regression analysis, principal component analysis, etc. (18). Unfortunately, the results were not statistically significant.

The Sartor and Boyd's study (19) of runoff pollution in eight U.S. cities should be considered as the beginning of a new approach to runoff pollution studies. Many important results were obtained, among which the most important ones have influenced model builders in the following:

- accumulation and wash-off of pollutants should be modelled separately,
- land use and the antecedent dry weather period are the most important factors controlling runoff pollution. However, the stormwater quality is not very sensitive to that last factor as evidenced by the correlation coefficients obtained by the authors for data in Figure 1.

Pollution wash-off should be modelled as:

$$M = P\,[1 - e^{-KQt}] \tag{6}$$

where M is the removed mass, t the lapsed time since the beginning of the event, P the initial amount of pollutants on the catchment, K is a constant, and Q is the runoff discharge at time t. Such a formula is still used in many models, such as SWMM, STORM, HSP, LANDRUN, etc. (1).

In 1977, the first effective urban runoff quality data base was created with the financial support of the U.S. EPA. Initially data were collected from 25 experimental catchments in 11 U.S. cities. New data have been added since then and the results of classical statistical analysis of these data have been published (20). The base has been used by many researchers in order to find explanatory variables of the runoff pollution process, and to test or develop various models (see refs. (4), (5), for example).

Since 1977, large volumes of data have been collected in many countries, sometimes through national programmes. The important Nationwide Urban Runoff Program (NURP) managed by the U.S. EPA must be mentioned here. It involved 28 local planning projects and data were collected in over 80 experimental catchments. A final report was recently published (21). In France a national programme was established in 1980-1981. Although only four catchments were monitored, almost all the runoff events were sampled. Consequently, four rather long rainfall-runoff-pollution time series are now available and these series represent a fairly rare data base. The final report will be published shortly (22). Various analyses of runoff pollution data, collected all over the world since the beginning of the 70's, have led to an increasing knowledge of the runoff pollution processes. Most of the progress has been made in runoff pollution characterization. Pollution sources and magnitudes are known better, especially in comparison with wastewater pollution (23) (24) (25). A literature review on

Figure 1. Solids Accumulation Rates (19).

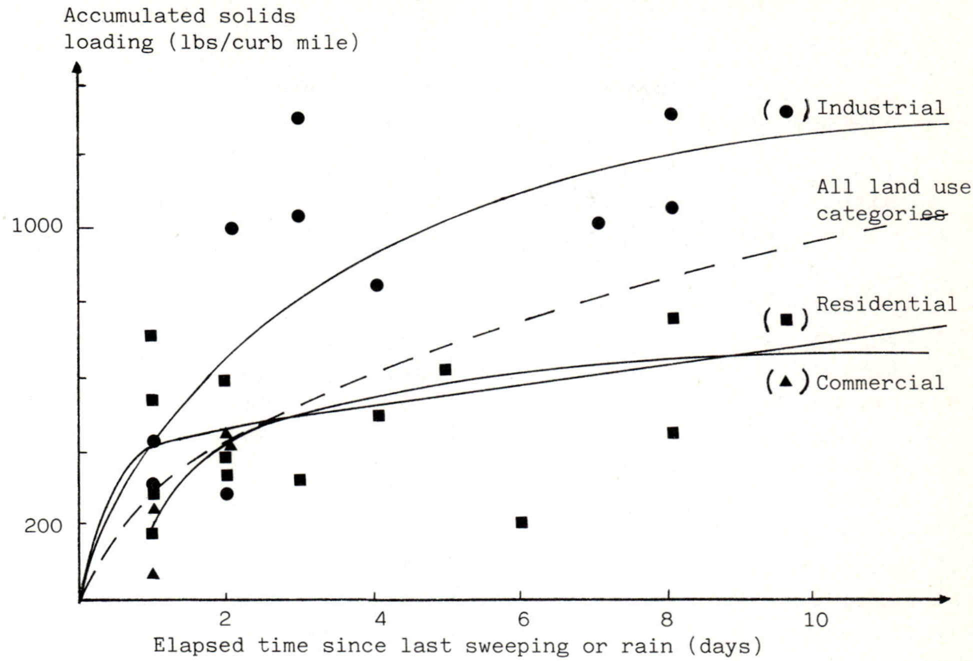

this question is presented elsewhere in these proceedings (26). The modelling of runoff pollution has not improved much in spite of many research studies (1). Even for the simplest "models" (pollutant annual loads, for example) it is not easy to find the right reference values, because the results vary widely from one author to another, in spite of efforts to standardize the measurements, data collection and data analysis procedures (4).

Table 3 gives statistical parameters of COD concentrations for 29 residential catchments (more than 10 values for each catchment) which were tested during the NURP. The coefficient of variation (COV), from one event to another, varied from 0.39 to 1.38, and the median value varied from 30 mg/l to 145 mg/l, from one site to another. There are significant differences between "the data recommended for planning level purposes by U.S. EPA" (21) and the mean annual concentrations for three French residential catchments (see Table 4). Except for lead, French runoff pollution seems to be higher than that in the

Table 3. N.U.R.P. Site-Mean Values of COD Event-mean Concentrations for Residential Land Use (21).

Site	Area (Acre)	Pop Dens. (per/acre)	Imp. %	No. of Obs	COD Event Mean Concentration (mg/l)			
					Mean	COV	Median	90% Conf. Limits
CO1-Big Dry Cr	33	19	41	16	129	.75	105	79-139
CO1-Cherry	57	24	38	14	122	.66	102	77-136
CO1-116/Claude	167	14	24	15	137	.74	103	76-139
DC1-La Keridge	68	21	33	44	60	.66	50	43-58
DC1-Stratton	8	-	-	31	51	.45	45	39-53
IL1-John N	54	18	19	31	126	.80	98	79-122
KS1-Overton	58	8	38	14	162	.67	135	101-180
MD1-Bolton Hill	14	30	51	19	218	1.38	128	85-193
MD1-Homeland	23	9	29	13	172	.73	139	101-192
MD1-Mt. Wash.	17	12	29	20	168	.85	128	96-170
MD1-Res Hill	10	55	76	13	177	.85	135	94-194
WA1-Surrey	95	9	29	118	48	.54	42	39-46
WI1-Burbank	63	15	50	27	39	.79	30	24-38
WI1-Hastings	33	17	51	23	41	.55	36	30-44
FLI-Young Apts	9	-	6	12	73	.96	52	34-79
TX1-Hart	378	9	40	11	82	.83	63	42-94
WI1-Lincoln	36	18	57	16	91	.95	66	46-94
TN1-R2	89	4	13	11	45	.39	42	34-52
DC1-Westleigh	41	3	21	39	51	.46	46	41-52
KS1-IC-92 nd	63	-	37	11	176	.98	126	80-197
IL1-John S.	39	18	18	29	111	.80	87	69-108
TN1-R1	69	11	33	11	120	.96	87	56-135
WA1-Lake Hills	102	12	37	127	44	.54	38	36-41
IL1-Mattis S.	28	22	37	30	180	.72	146	119-178
FL1-Charter Hdg	42	-	16	12	55	.64	47	35-64
DC1-Fairidge	19	-	34	48	51	.46	47	42-52
IL2-Comb Inlets	524	8	17	24	138	.90	102	78-134
NC1-1023	324	6	27	34	90	.97	64	51-82
DC2-Stedwick	27	15	34	45	45	.60	39	34-45

Table 4. Water Quality Characteristics of Urban Runoff

Constituent	Event-Mean Concentration (mg/l)	
	Recommended value for median urban sites in USA	Mean value estimates for French catchments
TSS	100	200 to450
BOD_5	9	12 to 38
COD	65	80 to200
Tot. P	0.33	0.82 to1.80
TKN	1.50	3.30 to6.11
Tot. Pb	0.14	0.085 to0.16
Tot. Zn	0.16	0.29 to0.33

U.S.A., considering average values and the sampling methods used to collect data. Thus, to use a "desktop" model is not as easy as it would seem and, in this case, limited measurements in agreement with the study objectives would not make the whole approach more comprehensive.

One finds the same uncertainties regarding the influences of catchment rainfall and runoff variables. In other words, no single formulation of a regression equation adequately predicts stormwater quality for all pollutants and basin conditions. This has been shown by Hémain (4) and Jewell (5). Table 5 (4) gives the variables best correlated with the TSS and COD mean concentrations and total loads for ten catchments. Also shown is the correlation coefficient R (linear or nonlinear) and an estimate of the relationships reliability (low, fair or good) based on various statistical criteria. All the rainfall and runoff parameters are listed at least once. However, the same variable may have opposite effects from one catchment to another and for the same pollution indicator. Unfortunately, good correlations are very few. Similarly, the study of antecedent dry weather period effects leads to similar variable results for various catchments and pollution parameters.

Table 5. Explanatory Variables Best Correlated with Total Event Loads M and Event-Mean Concentrations C (4).

Land Use	Catchments	No of Obs	MTSS	MCOD	CTSS	CCOD
Single family Resi-	Seattle View Ridge One	16	QM (R =.538) Low		-	
dential	Broward County Res.	18		-		H (R = -.731) Fair
Multi-family Resi-dential	Dade County	11	IM (R =.794) Low	IM (R =.819) Fair	IM (R=.676) Low	-
Commer-	Seattle Southcenter	12	H(R=.881 Fair		QM (R = -.627) Fair	
cial	Broward County Com.	23	DR (R=0.681) Fair	DR (R=0.741) Fair	Q (R = -.460) Low	Q (R = -.528) Low
Indus-trial	Seattle South Seattle	16	I(R=.569) Low		-	
	Nice Barron de Berre	16	V(R=.884) Good	IM (R=.875) Fair	IM (R=.751) Fair	IM (R= .651) Low
	Velizy Ouest	22	QM (R=.917) Good	QM (R=.812) Fair	DP (R=-.451 Low	DR (R= -.567 Low
	Les Ulis Sud	22	Q(R=.763) Fair	V(R=.598) Low	DP (R= -.777) Fair	DP (R= -.833) Fair
	Broward County Transport-ation	27		V(R=.837) Good		IM (R= -.554) Low

R: Correlation Coefficient -: No Significant Change

Independent variables: H = Rainfall Volume; I = average rainfall intensity; IM = maximum 5-min. rainfall intensity; DP = rainfall duration; V = runoff volume; Q = average runoff discharge; QM = runoff peak; DR = runoff duration

Jewell and Adrian (5) conducted a similar study for nine catchments included in the Urban-Rainfall-Runoff-Quality Data Base of the U.S. EPA (four of them were also studied by Hémain (4)). These authors paid special attention to multiple regression equations (Stepwise technique). The results showed the same variability as those given earlier.

Regarding the influences of different variables on mean event concentrations or total event loads, Hémain (4) listed, in the decreasing order of importance, the maximum 5-min. rainfall intensity or the peak runoff (not studied by Jewell and Adrian (5)), and then the runoff volume or the mean runoff discharge. In both studies (4,5), the antecedent dry weather duration did not seem to be a significant variable. Moreover, Hémain indicated that the linear regression equations were more suitable, while Jewell and Adrian found that log-log models produced the greatest number of best fits.

Many other studies produced similar results. For example, Melanen and Laukkanen (6) found for seven catchments in Finland that only 18% to 37% of pollutant mean concentration variances can be explained by simple linear regression equations. According to these authors, the most significant variable seems to be the maximum 5-min. rainfall intensity. Hémain (22) and Servat (27) found similar results for four French catchments.

So, the literature review revealed a large number of various models for runoff pollution. As an example, some of the best regression equations for French catchment data are given in Table 6. At this time, it is not possible to identify the "best model". Moreover, as shown in Table 7 (5), the same equation yields better or worse fit for various catchments. One would think that the changes in the R coefficient could be explained by catchment parameters, such as land use or climatic conditions (mean antecedent dry weather duration, storm duration and frequency, etc.). Such hypotheses were confirmed by Servat (3). However, as far as we know, any given runoff

Table 6. Some of the Best Statistical Relations Fitted to French Data

Catchment	Regression Equation	No. of Observa-tions	Correl-ation Coeffi-cient
Velizy Ouest	MTSS = .77 QM-70	22	.92
Aix Zup	MTSS = 1.42 QM+0.69 V-58.8	51	.95
Les Ulis Nord	MCOD =.6 QM+15 DTS+.37 DR-37.4	69	.89
Les Ulis Nord	MBOD5 = 4.5 H+3.7 DTS-4.3	59	.86
Maurepas	CTSS = 3.42 IM+22.9 DTS+68.2	89	.70
Les Ulis Sud	CCOD = 2300 $DP^{-.563}$	22	.83
Nice Barron de Berre	CCOD = 3.11 IM+8.25 DTS+3.6	16	.88

Explanatory variables: DTS = antecedent dry period, for others see Table 5.

Units used: M = kg, C = mg/l; QM = m^3/s; V = m^3; DTS = days; DR and DP = hours; H = mm; IM = mm/h.

Table 7. Model Performance in Various Catchments: Storm Event Loads (5) Expressed as MTSS (Kg/ha) = a DTS^b[days] DP^c[min.] Q^d[cm/h]

Basin	No.of Cases	a	b	c	d	Multiple R^2	F
Overall	118	213	0.481	0.313	1.20	0.63	65.6
Lincoln,Neb #2	8	519	0.634	0.798	1.35	0.86	8.30
Lincoln,Neb #3	10	526	0.109	0.905	1.48	0.87	13.8
Seattle,Wash.#1	19	3.47	0.498	0.969	0.906	0.65	9.16
Seattle,Wash.#3	21	55.4	0.251	0.622	1.12	0.57	7.64
Seattle,Wash.#4	17	76.9	0.235	0.487	1.23	0.76	14.0
Durham, N.C.	33	421	-	1.03	1.15	0.83	72.0

$$F = \frac{R^2/k}{(1-R^2)\ (N-k-1)}$$ where N = number of cases and k = 3

(number of independent variables in the equation). The computed value of F may be compared with tabulated values of the F distribution to test the null hypothesis (R = 0).

pollution model which was established elsewhere but calibrated against local data, should not generally give better results than a statistically-based model developed from these data only.

Pollutograph models are less numerous, probably because of limitations of the usually available data. The model used in the Battelle package (28) can be summarized as

$$M(t) = a + b\,Q(t) + c\,V(t) \tag{7}$$

where M(t) is the pollutant mass, Q(t) is the runoff discharge, and V(t) is the cumulated runoff volume at time t lapsed since the event beginning. Another simple equation was used in the NPS package (4) in the following form:

$$M(t) = a\,Q^b(t)$$

The Wells (1) model is based on the Markov process:

$$\ln M(t) = a + b \times \ln M(t-1) + ct + d\,Q(t) \tag{8}$$

The Wells model produced a good fit for the West Lafayette catchment, as shown in Figure 2.

For 62 pollutographs recorded during the French runoff pollution research programme, the above models were found suitable (29). The results given by the general form

$$M(t) = a\,Q^b\,(t)\;t^c$$

are listed in Table 8 for COD (Jewell and Adrian). Once more, one should notice that the c parameter varies from 0.055 to 0.615. Moreoever, the exponents vary from one pollutant to another.

When testing 15 different models, Jewell and Adrian (5) have shown that equation 5 or any equation including t, Q(t) and

Figure 2. Example of Results Given by Wells Model (Eq. 8) Fitted to the Data Collected in West Lafayette, U.S.A. (1)

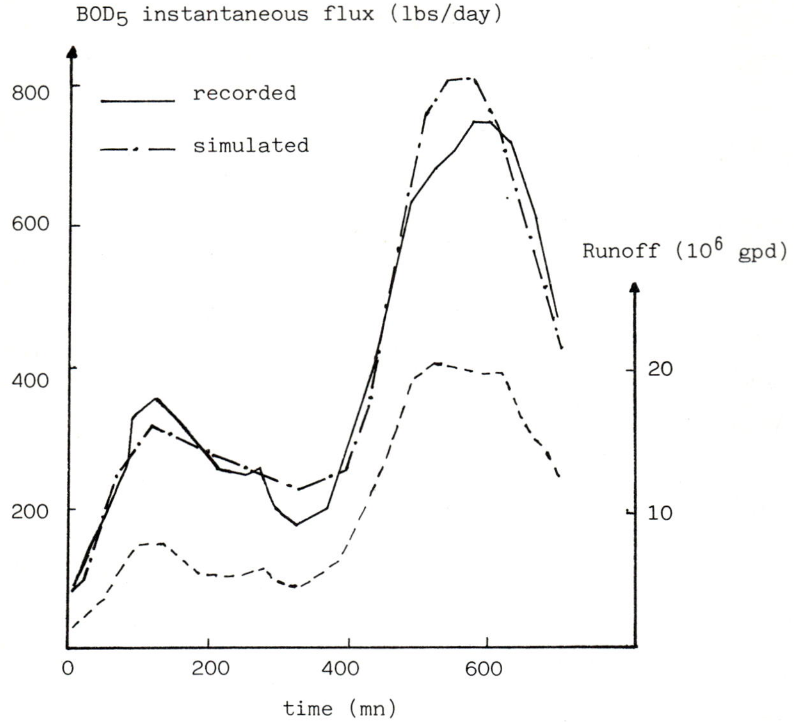

V(t) only (i.e., the models 1, 2, and 6 to 9 in Table 9), never produced the best fit of data, except in the case of the Durham data. Even more surprising is the fact that model number 8, i.e.,

$$M(t) = a + b \ln Q(t) + c \ln(t) + d \ln V(t) \tag{9}$$

never produced the best fit, while model number 14, which differs from model number 8 by a simple climatic factor (almost constant for a given catchment), often produced the best fit. Moreover, looking at the stepwise results, Q(t) is always the first selected variable, and t and V(t) are the second or third selected variable for 60% of all events. It follows that these variables are probably of great importance for runoff pollutograph modelling. Deviations from this conclusion could be

Table 8. Model Performance in Various Catchments: Instantaneous Flux (5) Expressed as $C\ COD = a\ Q^b\ (t)\ t^c$.

Basin	Cases	a	b	c	Multiple R^2	F
Overall	1741	0.544	1.06	-0.0629	0.78	3043
LN1	104	1.10	0.999	-0.178	0.76	303
LN2	108	2.28	1.26	-0.185	0.85	304
LN3	142	2.53	1.02	-0.282	0.72	180
BROW	226	0.0704	0.614	-0.359	0.66	216
SEA 1	103	0.542	1.08	-0.0555	0.83	237
SEA 3	122	0.361	0.963	-0.107	0.80	240
SEA 4	126	6.35	1.09	-0.615	0.82	274
DUR	467	3.47	1.11	-0.275	0.90	2099

explained by measurement procedures, various flow conditions in sewers, etc. This indicates that the experimental design must be very carefully checked in order to allow a reliable comparison of various data banks.

Stochastic modelling of such runoff pollution parameters as event-mean concentrations or loads may be of interest in the study of combined sewer overflows (CSO). Some such models have been established (24) as for example those presented by Van der Heyden and Lijklema (2). The frequency analysis of runoff pollution parameters is done in a conventional way, combining the probability distributions of various hydrological variables involved in the runoff pollution process: the rainfall depth

Table 9. Models That Fit Best for Each Basin, Instantaneous Flux (5)

Basin	Model Number			Compilation of Models for Basins	
	Suspended Solids	COD	Total-P	Model Number	Number of Best Fits
Overall	14	14	14	1	0
LN1	12 or 13	14	-	2	0
LN2	12	3 or 12	-	3	2
LN3	14	12	-	4	0
WND	14	-	-	5	0
BROW	-	14	14	6	2
SEA1	12 or 15	3 or 12	13	7	2
SEA3	12	12	12	8	0
SEA4	14	12 or 14	12	9	0
DUR	7,11,13 or 15	6,10,12 or 14	6,7,10,11,12 13,14 or 15	10	2
				11	2
				12	13
				13	4
				14	12
				15	3

or mean rainfall intensity, rainfall duration, etc. It is assumed that the variables are random and statistically independent. If not, one should take into account correlations between the variables. It has been found that mean concentrations of pollutants in runoff are not highly correlated with hydrological parameters, except perhaps for TSS and the maximum rainfall intensity. In combined sewers, the influences of sanitary wastes result in less variable concentrations and less dependence on hydrological variables. In some models, CSO concentrations are assumed to be equal to mean observed values (30). From a more general point of view, the log-normal distribution should fit well the CSO concentrations data. The

Figure 3. Maurepas Catchment: Two Parameter Lognormal Distribution Fitted to TSS Concentrations (Maximum Likelihood) (31)

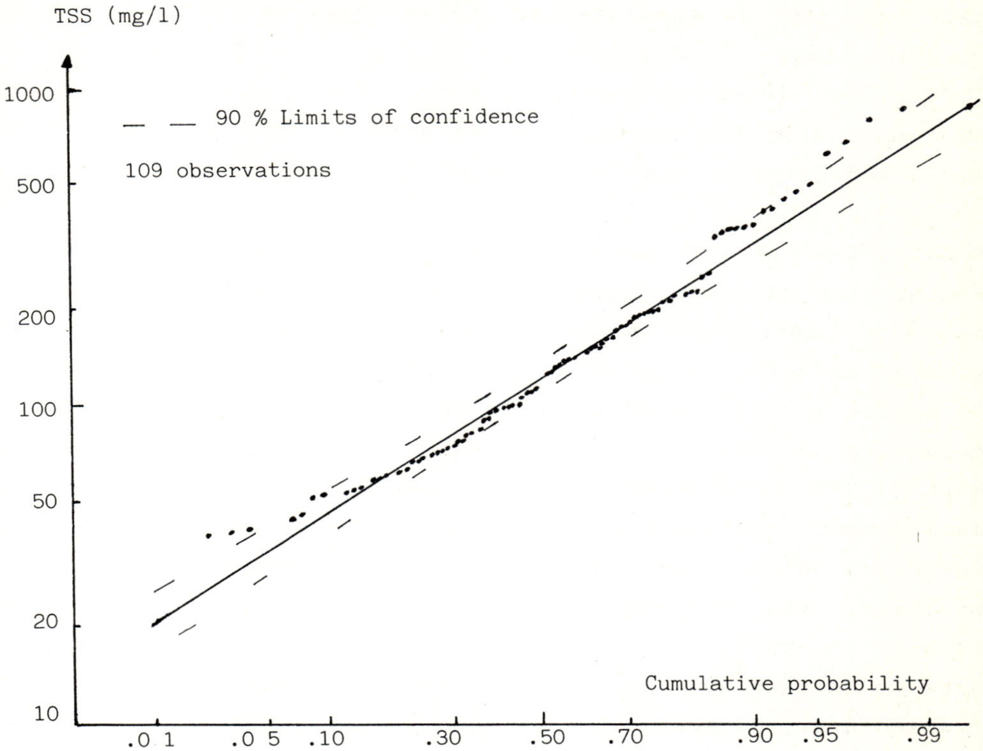

NURP study has shown such results (21), particularly for TSS, total phosphorus, total Pb, and total Zn, while COD concentrations were not log-normally distributed. A similar analysis of French data shows that the log-normal distribution should be suitable for mean concentrations of runoff pollutants (31). Figure 3 gives an example of TSS distribution for the Maurepas catchment. However, it must be emphasized that the reliability of such a distribution is mainly dependent on the data series length according to the sampling theory. That is to say that the above mentioned models should be applied to hydrological data series and runoff quality data series having similar lengths. This represents the main difficulty at present and probably will continue to be a problem for a long time since runoff pollution measurements are quite expensive and conducted over short periods.

Recommended Modelling Methodology

This subject was discussed many times during international meetings, such as the International Conferences on Urban Storm Drainage held in Southampton, U.K. (1978), Urbana, U.S.A., (1981), and Göteborg, Sweden (1984). Thus only a summary of the most significant findings related to the statistically-based modelling of runoff pollution is presented here.

First, if it is true that "no particular model is consistently better than others in predicting stormwater pollutant runoff", it seems preferable to collect data for the catchment to be studied in order to build a specific model for that catchment. Of course special attention must be paid to the data collection depending on the study objectives. The experimental design must be the first step in this approach and should define the measurement methodology: pollution parameters, number of measuring points, time interval, sampling methods and devices, analysis, etc. The experimental part is of primary importance because the model accuracy will depend on the data reliability. Data collection and analysis are discussed elsewhere in these proceedings. However, it must be emphasized that flow measurements should be made with the highest degree of accuracy possible in good engineering practice.

Regarding data analysis, it should be remembered that simple data plots may give interesting qualitative information. Before making any regression analyses, one should study all possible correlations among the variables. The correlation matrix often gives elements for a final choice in the model formulation. Mean or instantaneous concentrations may be preferred to pollutant discharges or loads. The last mentioned variables are directly proportional to runoff variables and so one should beware of spurious correlations. Moreover, no model form should be excluded a priori, even the simplest one as the simple linear regression.

Finally, the selection of a given model should be based on the analysis of the physical meaning of the equations, which have been established (according to some logical concepts of the runoff pollution process), and on such well established statistical criteria as the correlation coefficient, confidence limits, parameter sensitivity to sample sizes, probability distribution testing, etc. Should several models yield similar results, one should retain the simplest model with the smallest number of variables and the easiest to use model, with reference to the usually available data. A model with rainfall parameters should be simpler to use than another one with runoff parameters, because rainfall data are usually available, while runoff data are generally computed by a rainfall-runoff model. Model users must remember that the quality of output of a stormwater management model can be no better than the quality of the input data. Once established, a given statistically-based model should not be used outside of its experimental field. This is the main limitation of statistical models.

Points to Highlight

Data reliability is of primary importance for statistically-based models and, therefore, the data must be collected for a rather long period. Because the rainfall data series are usually fairly long, the reliability of the rainfall data corresponding to runoff pollution data can be tested using a long time series. The above reliability problem is even more important in frequency analysis of runoff pollution parameters (as in CSO impact studies). In that case, the population distribution of the parameters is generally unknown.

Another point should be addressed: Is it necessary or not to model separately pollutant accumulation during dry weather periods and pollutant washoff during storm events? If it is necessary, this should not be done by applying statistically-based models to the available data. It is possible to assume a priori that the pollutant mass washed off from a given catchment during a storm event should depend on the available pollutant mass spread over the catchment at the beginning of a runoff event. In that case, two separate models must be used for continuous simulations. Statistically-based models should not be used for this purpose. The model presented by Servat (3) is based on a concept that pollutant accumulation is described by a conceptual equation while the pollutant wash-off is given by a multiple regression equation in which the available pollutant mass, given by the conceptual equation, is an independent variable. However, Servat's results show that such a modelling approach does not give any better results than a simpler statistically-based model which includes conventional variables only. Moreover, it was suggested (10) that the traditional concept of the build-up (accumulation) being discretely distinct from the wash-off (removal) should be modified. This subject should be discussed further because it may change the future modelling approaches.

References

1. Hémain, J.C. Etude bibliographique des modèles de qualité des eaux drainées par les réseaux d'assainissement urbain. Laboratoire d'Hydrologie Mathématique Rapport 8/79, Montpellier, 246 pp, 1979.

2. Van der Heijden R., Lijklema L., Aalderink, R.H. A statistical methodology for the assessment of water quality effects of storm water discharges. NATO workshop on Urban Runoff Pollution, Montpellier, France, August 25-30, 1985.

3. Servat, E. TSS, BOD5, COD accumulation and transport over urban catchment surface: A modelling approach. NATO workshop on Urban Runoff Pollution, Montpellier, France, August 25-30, 1985.

4. Hémain, J.C. Statistical analysis of runoff quality data from French and U.S. catchments. Proceedings of the 2nd International Conference on Urban Storm Drainage, Urbana, Illinois, U.S.A., June 15-19, 1981.

5. Jewell, Thomas K. and Adrian, Donald D. Statistical analysis to derive improved stormwater quality models. Journal WPCF, 54(5): 489-499, May 1982.

6. Melanen, Matti J. and Laukkanen, Risto H. Urban runoff quality in Finland and its dependence on some hydrological parameters. IAWPR/Pergammon Press Ltd. Prog. Wat. Tech 12: 1073-1083. 1980.

7. Whipple, W. Jr., Hunter, J.V. and Yu, S.L. Effects of storm frequency on pollutant from urban runoff. Journal WPCF, 49 (11):2243-2248, November 1977.

8. Deutsch, J.C. et al. Caractérisation de la pollution des eaux de ruissellement par temps de pluie. Société Hydrotechnique de France. XVIIe journées de l'Hydraulique, Nantes, September 14-16, 1982, 8 pp.

9. Huber, Wayne C. Deterministic modelling of urban runoff quality. Review paper. NATO workshop on Urban Runoff Pollution, Montpellier, France, August 25-30, 1985.

10. James, W. Pollutant build-up processes as a net dry-weather effect. NATO workshop on Urban Runoff Pollution, Montpellier, France, August 25-30, 1985.

11. Desbordes, M. and Hémain, J.C. Problèmes posés par la modélisation de la qualité des eaux pluviales urbaines. Société Hydrotechnique de France. XVIIe Journées de l'Hydraulique, Nantes, September 14-16, 1982 7pp.

12. Field, R., Tafury, A.N. and Masters, H.E. Urban runoff pollution control technology overview. US EPA Environmental Protection Technology Series, report EPA 600/2-77-047, March 1977, 91 pp.

13. Huber, Wayne C. Session A9: Special model studies. General discussion. Proceedings of the 3rd International Conference on Urban Storm Drainage. Göteborg,, Sweden, June 4-8, 1984.

14. Deutsch, J.C. and Hémain, J.C. Main results of the French national programme of urban runoff quality measurement. Proceedings of the 3rd International Conference on Urban Storm Drainage, Göteborg, Sweden, June 4-8, 1984.

15. Donigian, A.S. Jr, and Crawford, N.H. Modelling non-point pollution from the land surface. US EPA Ecological Research Series. Report EPA 600/3-76-083, July 1976, 279 pp.

16. Weibel, S.R. et al. Urban land runoff as a factor in stream pollution. Journal WPCF. 36:914-924, 1964.

17. American Public Works Association. Water pollution aspects of urban runoff. U.S. Department of the Interior. Federal Wat. Pollut. Control Administration, Contract No. WA 66-23, WP20-15, January 1969, 271 pp.

18. AVCO Economic Systems Inc. Storm water pollution from urban land activitiy. FWQA Contract No. 14-12-187. Publ. No. 11034 FKL, April 1970, 342 pp.

19. Sartor, J.D. and Boyd, G.B. Water pollution aspects of street surface contaminants. US EPA Environmental Protection Technology Series, report EPA/R2-72-081, November 1972, 236 pp.

20. Huber, Wayne C. Urban rainfall-runoff-quality data base. Update with statistical analysis. US EPA Special Reports Series, Report EPA 600/8-79-004, 1979, 255 pp.

21. US EPA. Water Planning Division. Results of the nationwide urban runoff program - volume I - final report. US EPA, Washington, D.C., December 1983, 186 pp.

22. Servat, E. Mesure de la pollution du ruissellement pluvial urbain. Campagne expérimentale française. Rapport de synthèse. Service Technique de l'Urbanisme, Paris, 1985, 80 pp.

23. Yen, B.C., Editor. Urban stormwater quality, management and planning. Proceedings of the 2nd International Conference on Urban Storm Drainage. Urbana-Illinois, USA, June 15-19, 1981.

24. Balmer, P., Malmqvist, P and Sjöberg, A., Editors. Proceedings of the 3rd international conference on urban storm drainage. Göteborg, Sweden, June 4-8, 1984.

25. Novotny, V. and Chesters, G. Handbook of nonpoint pollution. Van Nostrand Rainhold Compagny, Toronto.

26. Ellis, Bryan J. Pollutional aspects of urban runoff. Review paper, NATO Workshop on Urban Runoff Pollution. Montpellier, France, August 25-30, 1985.

27. Desbordes, M and Servat, E. Solids in urban runoff. Statistical analysis of French experimental data. Proceeding of the 3rd International Conference on Urban Storm Drainage, Göteborg, Sweden, June 4-8, 1984.

28. Arnett, R.C. et al. Diurnal flow and quality patterns in a combined sewerage system. Proceedings of the 46th Annual Conference of WPCF. Cleveland, Ohio, USA, September 1973.

29. Hémain, J.C. Groupe de travail "Pollution des Eaux de Ruissellement Urbaines". Mesure de la pollution du ruissellement pluvial urbain. STU, Paris, 1985.

30 Loganathan, G.V. and Delleur, J.W. Effects of urbanization on frequencies of overflows and pollutant loadings from storm sewer overflows: a derived distribution approach. Water Resources Research, 20 (7):857-865, July 1984.

31 Chamayou, G. Etude statistique de la distribution des teneurs moyennes en divers indicateurs de pollution des eaux de ruissellement pluvial sur quatre bassins versants urbains. Draft report. Laboratoire d'Hydrologie Mathématique, Montpellier, 1985, 30 pp.

A STATISTICAL METHODOLOGY FOR THE ASSESSMENT OF WATER QUALITY EFFECTS OF STORM WATER DISCHARGES

Rene T.J.M. van der Heijden, Lambertus Lijklema* and
R. Hans Aalderink*
*Department of Water Pollution Control, Agricultural University,
De Dreijen 12
6703 BC Wageningen
The Netherlands.

Introduction, Background

This paper presents a simple approach for the prediction of the probability of oxygen depletion in a stagnant water receiving combined sewer overflows. Before embarking upon this methodology it is appropriate to elucidate shortly the Dutch conditions with respect to sewerage and water quality and the framework in which this study was executed.

The majority of the sewer systems in the Netherlands is of the combined type; a recent nationwide inventory reports 90%. There is a tendency towards more separate systems; since 1975 35% of the newly constructed systems were separate. Most systems have outlets discharging into stagnant or semi-stagnant receiving waters (86%), the majority of which has a small surface area. Such waters comprise polder ditches, canals, ornamental ponds in parks, moats etc. Traditionally the combined sewer systems have been designed to restrict the overflow frequency in the range of 5-10 per year, without much specific insight in the local water quality effects In its most elementary form the design was based upon the so-called "dot-graph" in which the volume (in mm) and duration (in hr) of all rain events in a period of 37 years are plotted; see Figure 1.

NATO ASI Series, Vol. G10
Urban Runoff Pollution.
Edited by H.C. Torno, J. Marsalek, and M. Desbordes

This graph also shows the storage capacity of the system and the excess pumping capacity (pumping capacity of treatment plant minus averaged dry weather flow).

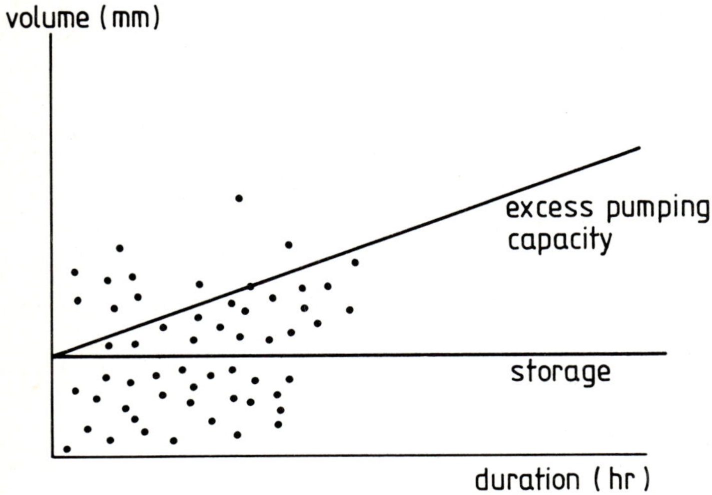

Figure 1. "Dot-graph" of rain events and interception in sewerage system.

The number of dots above the upper line divided by the number of years is the overflow frequency. Storage is generally in the range of 6-10 mm, excess pumping capacity from 0.5-1.0 $mm.hr^{-1}$.

In the period 1970-1985 the sewage treatment capacity in the Netherlands has been extended from about 5 to 25 million population equivalents and before long virtually all domestic sewage will be treated. This has caused the question of the effects of sewer overflows on the receiving waters to become topical. One of the results is a nationwide study including aspects such as rainfall statistics, relations between system characteristics and discharged load of pollutants, deposition and erosion of solids in sewers, retention facilities and effects on receiving waters.

Although much of the experience from abroad is very useful, the Dutch conditions are at variance with most other countries regarding the sewage systems (slight slope, high storage), the receiving waters (stagnant, shallow) and also with respect to climatic factors. Hence a part of the expertise should be acquired nationally. The involvement of our Department in the study of the water quality effects has raised the universal question whether a deterministic approach or a statistical approach should be pursued in predicting effects. Typical examples in literature of both approaches are given in (3) and (5). Data collection on a specific site, including measurement of rainfall, sampling of overflow volumes, duration and quality and monitoring of water quality in the receiving pond in principle allows a comparison of both methodologies. Presently the emphasis is on the deterministic description of the relation: overflow - water quality but this paper contains a preliminary statistical analysis of the observations and an exploration of a method to predict the distribution of oxygen concentrations in a receiving water.

Data Base

For the study data on 239 rain events, collected in the village Loenen during the period 28-6-1981 to 24-7-1983, were available. The data were fitted to a gamma distribution:

$$f(X) = \frac{1}{\beta^{\alpha} \Gamma(\alpha)} X^{\alpha-1} \exp - \frac{X}{\beta} \qquad (1)$$

Results are presented in Table I.

Table I. Statistics of Rainfall Data in Loenen

Variable	Volume	Duration	Intensity	Dry period
Units	mm	hr	mm/hr	hr
Average	5.34	5.98	1.38	72.46
Stand.Dev.	5.36	5.32	1.52	121.86
Coeff. of Variation	1.0045	0.8886	1.1020	1.6871
α	0.9926	1.2663	0.8243	0.3536
β	5.3801	4.7232	1.6742	204.93
Gamma (α)	1.0043	0.9032	1.1380	2.5192

The time series is short and the parameter estimation for the gamma function on the basis of moments is certainly not the best, but here the emphasis is on the illustration of the methodology and not on the accuracy. Some of the fitted and observed (cumulative) probability density functions are shown in Figure 2. The definition of an event as independent, considering the length of the previous dry period, deserves special attention. Selection of very short intervals as the minimum dry period introduces autocorrelation in volumes and causes difficulties in a proper handling of overlapping effects. On the other hand long intervals will smooth peak discharges and will lead to underestimation of the number of overflows. In certain studies the dry weather period has been forced to fit an exponential distribution in order to use some of the attractive properties of this distribution (coefficient of variation = 1). This can be achieved by variation of the minimum duration of a dry period. It is felt by the authors that the intrinsic dynamic properties of the system (autocorrelation in rainfall, time needed to fill and to empty the sewerage system plus storage facilities) deserve priority in guiding the selection over an advantage in the handling of the analytical instruments. In the data for table I the minimum dry weather period chosen was 4 hr.

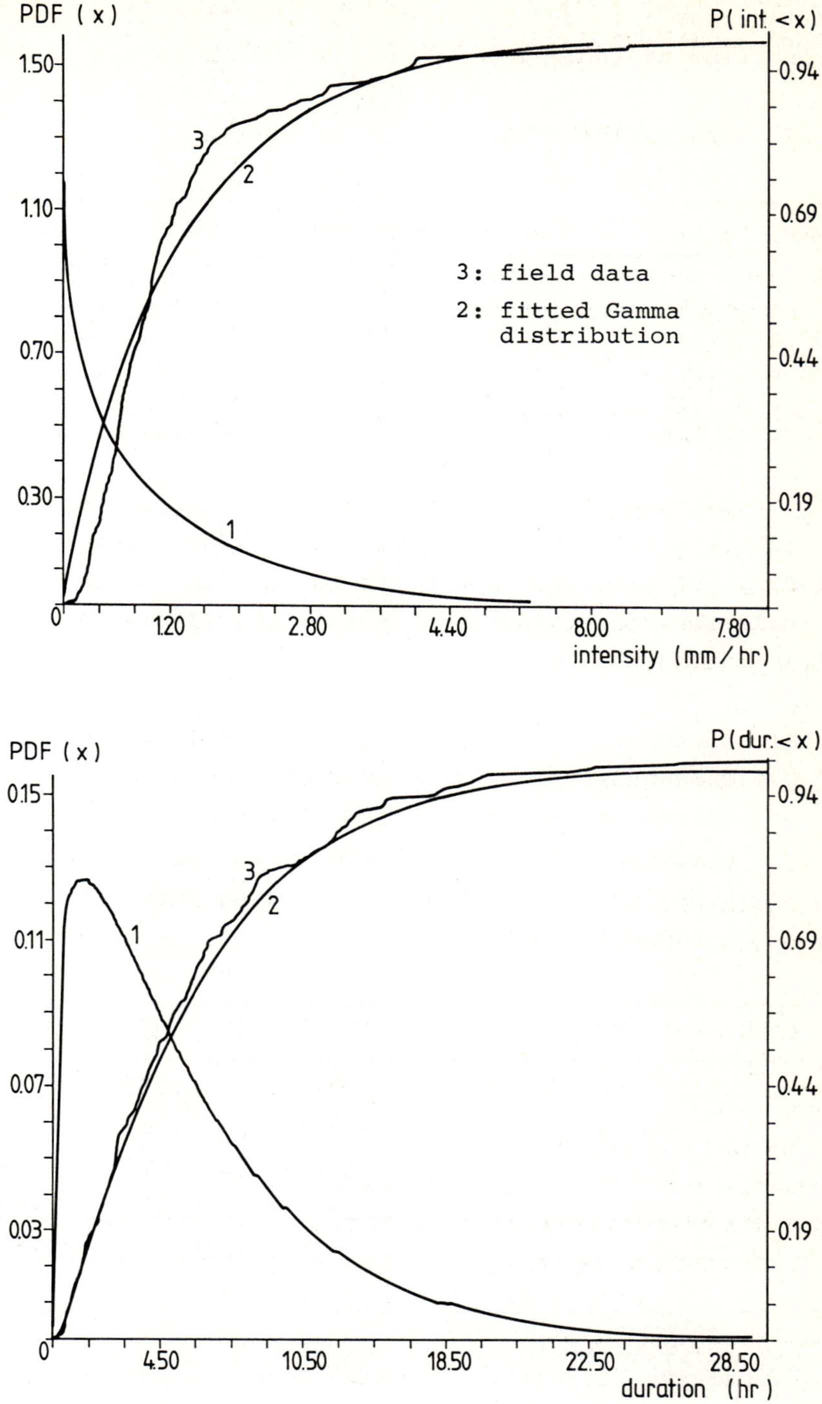

Figure 2. Probability Density Functions (PDF) and Distribution P

The data on quality used in this study comprise 26 overflows. The properties of their log values are summarized in table II.

Table II. Statistics of BOD in CSO at Loenen

Variable	tot. BOD	BOD after settling
Units	g/m^3	g/m^3
Average (log)	3.665	2.893
Stand.Dev. (log)	0.654	0.612
Coeff. of Variation (log)	0.178	0.211

Further handling of the data for statistical assessment of the water quality effects of overflows requires a proper handling of existing correlations. Again, the data base is small, but the correlation coefficients shown in table III mainly serve the illustration of a method.

Table III. Correlation Coefficients of Rainfall and Overflow Characteristics. D = Duration of Storms, V = Volume of Rain Event, I = Intensity, D.P. = Dry Period between Events, V.O. = Volume of Overflow, t. BOD = total BOD in Overflowing Water and BOD-s = BOD after Settling.

Variable	V	I	D.P.	V.O.	t. BOD	BOD-s
D	0.615	-0.339	-0.019	-0.052	-0.099	-0.102
V		0.158	0.079	0.767	-0.062	-0.151
I			0.135	-0.102	0.385	0.189
D.P.				-0.188	0.417	0.843
V.O.					-0.260	-0.438
t. BOD						0.755

The high positive correlation between Duration and Volume of rainfall prohibits their simultaneous use in a model. The negative correlation between Intensity and Duration will be discussed later.

Another high correlation coefficient is between Dry Period and BOD but because the dry period is hardly correlated with the other stochastic variables driving the system this is not a problem. Total BOD is correlated with Intensity, after settling the correlation is less. This suggests that the suspended fraction of BOD is transported preferentially at high intensities together with a more or less independent contribution of the colloidal-soluble BOD. In assessing water quality effects we will use the soluble BOD as a variable, the suspended material will settle quickly in the receiving water. Its effect on the oxygen balance will be smoothed over a longer period and can be incorporated in the Sediment Oxygen Demand (SOD), which is considered to be constant.

Model

The data base discussed in the previous section was used as a specimen for the calculation of the probability of violation of oxygen standards in a pond. The motive was a conflict on the necessity to construct an expensive detention facility. A simple water balance yields the volume of the overflow:

$$\underline{VO} = (\underline{V} - SS)A * R - S-(\underline{D} * P * CP) + DWF * \underline{D} \qquad (2)$$

in which

VO	= Volume of overflowing Water	m^3
V	= Volume of Rain	mm
SS	= Storage on Surface	mm
A	= Impervious Area	m^2
R	= Runoff Coefficient	-
S	= Storage in Sewerage System	mm
D	= Duration of Storm	hr
P	= Pumping Capacity	m^3/hr
CP	= Coefficient for effective Pumping	-
DWF	= Dry Weather Flow	m^3/hr

In equation (2) the underlined variables are stochastic, the others are treated as constants although there is a certain variability. The surface storage SS is not always available, especially during the winter when evaporation from pools, gutters etc. is slow, overlap with previous events resulting in lower values may occur. The value is low anyway. Also the Runoff-Coefficient R varies in reality. During wet periods pervious areas can contribute to some extent to run-off and for impervious surfaces similar considerations as discussed for surface storage apply. Although we may define these factors independently we cannot fully disentangle their individual contributions to overflow from field data. In principle R can be dealt with as a stochastic variable, even if field data would show a correlation with other variables this seems possible, but such data are not available at this time.

The storage S in the sewers (plus eventual retention basin) is not always available at the beginning of a storm. Di Toro and Small (2) have presented a methodology for the calculation of the effective storage as a function of mean run-off volume V, maximum storage S, pumping rate P and mean dry period. The results show little loss of effective volume when the product of pumping rate and mean dry period surpasses the mean run-off volume by a factor of two or more. At lower values the loss of effectiveness is most pronounced in large basins where large storms can be accumulated. Along these lines the long term effective storage can be calculated and substituted in equation (2) instead of S. It should be noted however that the analysis by Di Toro and Small produces long term average values and not a distribution. Also the analysis was based on the assumption of exponential distributions for runoff volume, duration and dry periods. This has been commented upon already. The use of long term average values results in the loss of information on interdependency of individual variables. The problem is complex and apparently also serious. The available data set shows that 26 overflows result from 239 storm events, a probability of 0.11. However, 9 overflows were preceded by a short dry weather period (< 13 hr) out of a total of 54 short periods; a probability of 0.17.

The criterion of 13 hr corresponds with the theoretical time needed to pump the filled system empty. So the probability of an overflow seems to be substantially higher after a short dry period. This can be attributed to the combined effects of loss of storage in the system, reduced infiltration in the pervious part of the catchment and a decrease in surface storage. The first factor dominates probably. Table III shows a negative correlation between dry period and volume of the overflow. In this study we used a constant value for the storage. In principle the technique of partition of the data in subclasses, discussed later for the relationship duration-intensity, might be applied for the dry period effect as well. Presently this is not feasible due to the small data set.

The pumping capacity P is not always employed to its full extent. In the initial phase of an event the pump(s) will be disengaged until a certain level has been reached. Also the capacity itself is a function of the filling of the system. In this study the coefficient CP was set to one.

The dry weather flow DWF varies roughly between 1 and 10% of the daily volume. This variation is smoothed when the duration of a storm increases. So especially thunder-showers may coincide with high DWF. If necessary division in day and night events can be made but use of the daily average probably introduces little error.

Equation (2) can be re-written by substitution of $\underline{V} = \underline{D} * \underline{I}$. This results in:

$$\underline{VO} = \underline{D}\,(\underline{I} * A * R - P * CP + DWF) - SS * A * R - S \qquad (3)$$

In equation (3) each stochastic variable appears only once and thus an ideal interdependency is removed. Further aspects of multiplication of stochastic variables will be discussed in a subsequent section.

Dutch studies and studies elsewhere, for instance Mueller and Anderson (4), show that the variation of pollutant concentration in overflows cannot be related unequivocally to physical phenomena in the transport chain. The limited information summarized in Table III seems to confirm this for soluble BOD. Hence the discharged load L can be approached by

$$\underline{L} = \underline{VO} * \underline{c} \tag{4}$$

where $\underline{c}$ is the (independent) stochastic concentration. In this study we used a log normal distribution for $\underline{c}$.

In the specific site studied the overflow was discharged in a pond of more or less rectangular shape; see Figure 3.

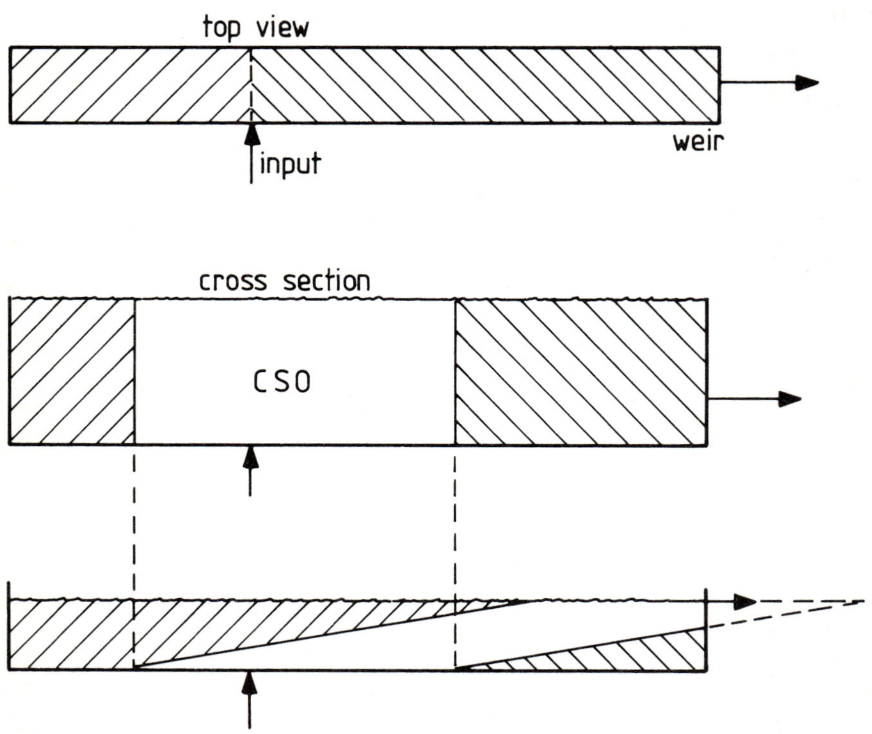

Figure 3. Definition of interception of CSO in receiving pond

A discharge results in a rising water level and actuates the weir at the far end of the otherwise stagnant pool. Depending on the volume of the overflow, a fraction of this volume will be lost from the pond when the water level has returned to the mark of the weir. This process can be envisaged as outlined in Figure 3. As the time constant for deoxygenation and reaeration is much longer than the typical time for an event, we may assume that the deoxygenation and reaeration producing the oxygen sag curve start after the overflow is finished. After levelling out with a linear velocity profile, the water will mix due to wind and an oxygen sag will develop. The BOD-s and oxygen deficit (OD) equations used are well known:

$$\frac{dC}{dt} = -K_{11}C \tag{5}$$

and

$$\frac{d(OD)}{dt} = K_{12}C - K_{22}(OD) + SOD \tag{6}$$

in which K_{11} is the BOD decay coefficient, K_{12} the deoxygenation rate and K_{22} the reaeration rate. An analytical expression for the maximum deficit can be obtained readily. The initial deficit used for the initial condition in equation (6) was set to the steady state value $SOD{:}K_{22}$. In reality this value maybe too low because the overflowing water is well aerated. Before presenting results the operations with stochastic variables should be addressed.

Operations with stochastic variables

The equations (3) and (4) lead to subtraction, addition and multiplication of a stochastic variable and a constant and to multiplication of two stochastic variables. Subtraction or addition of a constant is equivalent to shifting the abscissa. In the cumulative distribution function the appearance of negative values is no problem because the curve is simply cut for $X < 0$, but the probability density function cannot be used in numerical procedures when negative values occur. Multiplication with a scalar amounts to multiplication of the abscissa and, in order to maintain a unit surface area for the probability density function, also division of the ordinate by the same number. Multiplication of stochastic variables $\underline{a}$ and $\underline{b}$ with probability density functions $f_a(X)$, $f_b(X)$ and cumulative distributions $F_a(X)$, $F_b(X)$ will yield a stochastic variable $\underline{z}$ with associated $f_z(X)$ and $F_z(X)$. Because

$$F_z(Z) = E\,(\underline{z} < Z) = E(\underline{a} * \underline{b} < Z) \tag{7}$$

all combinations (A,B) with a product < Z can be found by letting A take all values from zero to infinity and B from zero to Z:A, hence:

$$F_z(Z) = {}_0\int^{\infty} [f_a(A) * {}_0\int^{Z/A} f_b(B)dB]\, dA$$

$$= {}_0\int^{\infty} f_a(A) * F_b(\tfrac{Z}{A})\, dA \tag{8}$$

Sometimes analytical solutions can be used when the distribution fits for instance an exponential form. An example of such an analytical approach has been presented by Di Toro (1). He calculated the distribution of concentrations in a river after mixing of runoff and upstream river water. The two contributing flows and concentrations were assumed to be log-normally distributed. One approach assumed that the resulting concentration is also distributed log-normally and the moments are then calculated from the moments of the constituent variables, neglecting correlations between flows and concomitant concentrations. A second approach accounted for these correlations by direct evaluation of the integral defining the probability that the contributing flows and concentrations result in a concentration exceeding a dummy value. The latter method is comparable to the approach defined by equation (8) and yielded better results than the first, where the resulting distribution was forced in the trammels of a chosen distribution.

Equation (8) prevents this, but for analytical solutions the constituent terms should still be characterised by an analytical formulation for their distribution, including parameter values such as means and standard deviations or variance. A problem that we feel here is that each probability density function has its own variance, mean etc., but the reverse is not true. Hence a correct value for mean and variance does not necessarily imply a correct representation of the distribution as well. Especially in situations such as encountered in overflows, for instance defined by equation (3), the tail of the distribution should be represented correctly and there is no real confidence that analytical solutions are the best approach here.

The integral of equation (8) can also be evaluated numerically. This is somewhat time consuming because for each value of Z of interest the integral should be evaluated, but there is no need to assume any a priori distribution for $\underline{a}$, $\underline{b}$ or $\underline{z}$; a table of values is sufficient. The product in equation (4) could be calculated in this way if data are available. The major problem that remains is the handling of correlations. Figure 4 illustrates this. Curve 1 is the distribution of the volume of the precipitation as observed but fitted to the gamma distribution. Curve 4 is the product simulation using the fitted curves for intensity and duration, Figure 2, by substitution in equation (8). This product heavily overestimates the frequency of high volumes. The main reason is that the negative correlation between intensity and duration is not considered. Hence the number of combinations of high intensity and long duration is overestimated. The correlation is due to seasonal variations; the high intensity thunder storms in the summer soon blow over, whereas the winter rains may last longer but generally do not reach high intensities. In order to improve the product simulation the probability density functions for duration were classified in categories.

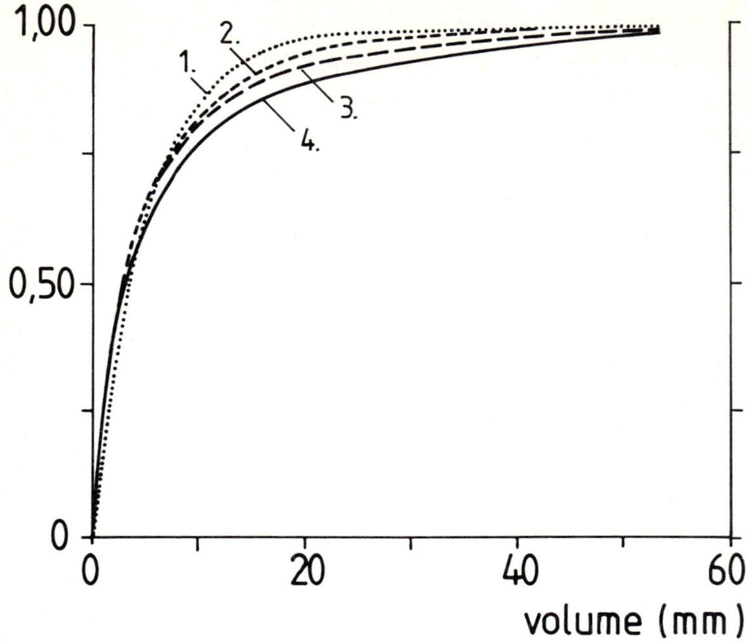

Figure 4. Distribution of Volume of Rain calculated by different procedures; see text.

In Figure 5 curve 3 represents the probability for all events, curves 1 and 2 are the density functions for intensities higher than 2.50 and 0.87 mm/hr and curves 4 and 5 represent the intensities below these values. The differences in position of the curves will be clear. Returning to Figure 4, the curve 3 is the result of the sum of the integrations for the subclasses divided around the insity of 0.87 mm/hr and curve 2 the same with 2.50 mm/hr as pivot. Where prediction of extreme volumes is important it might be wise to select a higher value than 2.50 mm/hr for the discrimination of the subclasses. It will be clear that subclassification is a powerful way of approaching reality. As rain data are sufficiently available this can be done readily for duration and intensity.

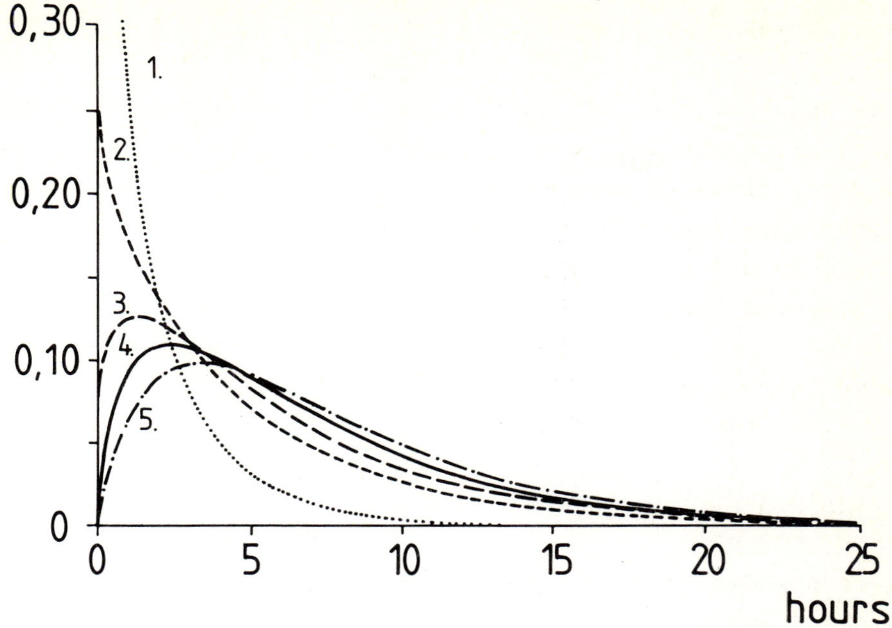

Figure 5. Probability Density Functions for Duration of Rainfall sub-classified for different Intensities.

It has been discussed already that the duration of dry periods apparently has a strong influence on the frequency of overflows. Table III does not reveal significant correlations of the dry period with duration, intensity or volume, but this may change when we would consider the data for short dry periods separately. In order to account for the effect that after a short dry period not always the full interceptor storage capacity is available, an estimate should be made of the effective value of S in equations (2) and (3); see also the previous discussion of Di Toro and Small's paper. The evalauation of equation (3) can be separated in two classes; one category involves all events with dry periods greater than the emptying time for a filled interceptor, for which equation (3) is unchanged. For the short dry periods subclass the right hand side should include a term for the probability of left-over runoff from the previous storm or:

$\underline{D} * \underline{I} - \underline{DP} * P * CP$ (9)

where ideally D and I values should be used evaluated for all storms preceding short duration dry periods DP and again subdivided further to account for the correlation between D and I. The value for this new term equation (9) should lie between zero and S, but because DP has a minimum value (say 4 hr) it seems better to use a maximum value of $\underline{D} * \underline{I} = S$. This way of accounting for overlapping effects of storms has not yet been elaborated.

Results

Figure 6 shows loads discharged into the receiving pond (curves 3 and 4) and loads remaining after a fraction of the discharged load is washed away over the weir as outlined in Figure 2; curves 1 and 2.

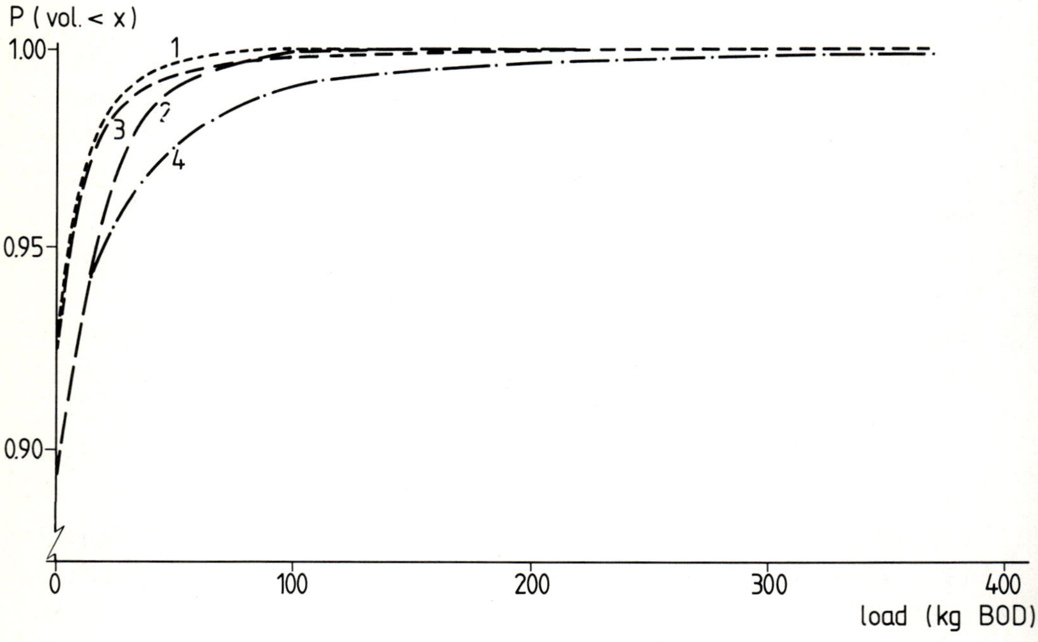

Figure 6. Distribution of discharged load, curves 3 and 4, and retained load in receiving Pond, curves 1 and 2.

Curves 1 and 3 used subclasses for intensities above and below 2.50 mm/hr, the other curves were based on 0.87 mm/hr as transition value. Curves 1 and 2 are steeper because extreme remaining loads are highly improbable because in that case a large volume is involved and much water leaves the pond immediately. The loads remaining in the pond cause an oxygen depletion; the distribution of the minimum values is shown in Figure 7. The negative O_2 concentrations are due to the calculation in terms of deficit, where the value may surpass the saturation value. The maximum oxygen concentration of 6.3 g/m^3 is lower than the saturation concentration as a consequence of the sediment oxygen demand.

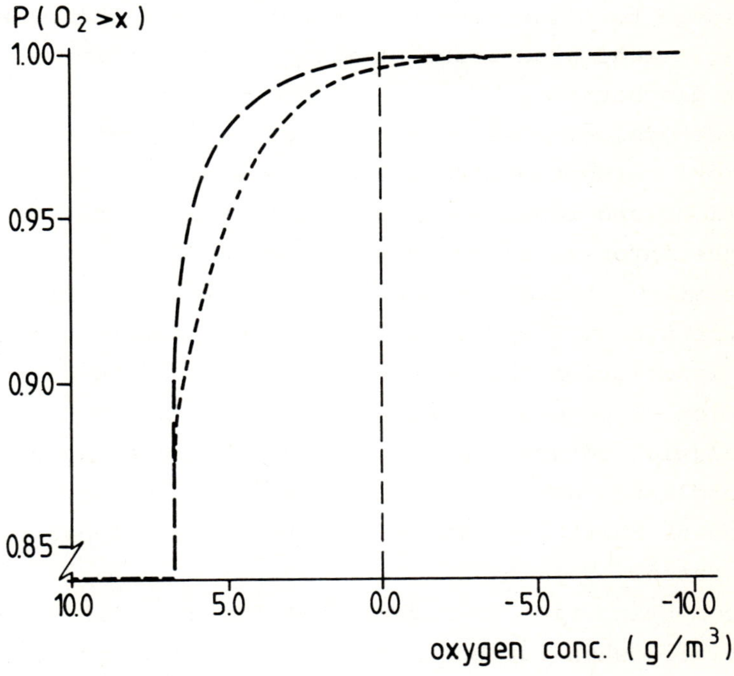

Figure 7. Distribution of Oxygen Concentrations in receiving Water

The higher curve is related to the intensities divided at 2.50 m/hr, the lower curve applies to 0.87 mm/hr. Using the (better) first curve we can calculate for instance the frequency of oxygen concentrations below 1.3 g/m³. The probability is 1 - 0.9973 = 0.0027; with an average of 113 storms per year this means a frequency of 0.3 or once in 3-4 years. The effects of measures such as increased storage S or pumping capacity P or reduced imperviousnes affecting A, can be evaluated readily.

Conclusions and Summary

- A simple runoff - overflow - water quality model has been presented to illustrate the potentials and the problems of a statistical prediction of water quality effects of storm water discharges.
- The methodology presented here has the advantage that no a priori assumptions regarding the shape of the forcing functions and resulting distributions are required.
- All the information in the forcing functions is preserved and carried through the calculations into the resulting distributions. The result obtained ultimately contains more information than other studies, yielding only a fraction of a total or similar quantities.
- Correlation between variables can be handled by division in subclasses as shown for intensity and duration. The effect of short dry periods on storage still has to be elaborated.
- The computing time is higher than for analytical approaches, especially when further subdivisions are introduced.

References

(1) Di Toro, Dominic M. Probability model of stream quality due to runoff. Journal of the Environmental Engineering Division, ASCE, 110 (3): 607-628; 1984.

(2) Di Toro, Dominic M.; Small, Mitchell J. Stormwater interception and storage. Journal of the Environmental Engineering Division, ASCE, 105 (1): 43-54; 1979.

(3) Driscoll, E.D. and Associates, Inc. Combined sewer overflow analysis handbook. Volume I: Procedures and example analysis. Prepared for EPA, contract no. 68-01-6148. July 1981.

(4) Mueller, J.A.; Anderson, A.R. Combined sewer overflow quality from treatment plant data. J. Water Poll. Control. Fed. 51 (5): 958-973; 1979.

(5) Medina, M.A.; Huber, W.C.; Heany, J.P.; Field, R. River quality model for urban stormwater impacts. J. of the Water Resources Planning and Management Div. ASCE 107 (1): 263-280; 1981.

TSS, BOD5 AND COD ACCUMULATION AND TRANSPORT OVER URBAN CATCHMENT SURFACES: A MODELLING APPROACH

Eric Servat
Laboratoire d'Hydrologie Mathematique
Universite des Sciences et Techniques du Languedoc
34060 Montpellier Cedex
France

Introduction

There is an increasing awareness of pollution problems in urban sewer management. This is the reason why, in France, a national experimental program on urban runoff pollution has been started (3). In the study reported in this paper, the data which have been collected for more than one year from the four experimental catchments (Maurepas and Les Ulis near Paris, Aix-Zup and Aix-Nord in Aix-en-Provence) have been used. We propose here to develop a modeling approach for the production-accumulation mechanisms and for the surface transport of TSS, BOD_5 and COD. For TSS, we will summarize the main results and refer the reader to already published papers (5). Several rainfall-runoff pollution modeling objectives can be theoretically defined. Nevertheless, the nature and accuracy of the available data will set certain model limits. Within the defined framework of the French experimental catchment program, monitoring was carried out at the outfalls of various watersheds. The raw data were based on several samples collected during rainfall events and thus sampling can be viewed as a spatially-varying phenomenon that can be characterized by a pollutant mean concentration value. The modeling approach we have developed is directly derived from the objectives of the national monitoring program to determine estimates of transported pollutant mass over a long duration (about one year) and for a "space scale" which relates to small catchments (less than 50 ha).

NATO ASI Series, Vol. G10
Urban Runoff Pollution.
Edited by H. C. Torno, J. Marsalek, and M. Desbordes

Empirical modeling of the rainfall-runoff process has been shown to possess certain limits (5). Moreover, it appears that the determination of statistical relationsips would be unduly influenced by high magnitude events for which the probability of occurrence is quite low.

For these reasons a conceptual procedure is proposed which is compatible with the representativeness of the sampling program and with the trends arising from the data analysis (5,6).

TSS Accumulation and Transport Modelling Approach

This section summarizes the main results for the modeling approach developed earlier (2,5).

A linear TSS accumulation model, based on the following hypotheses, was selected :

- A constant production rate within a given time interval.
- An initial mass of pollutant (stock) close to the final one on the surface of the catchment when considered over a long time period.
- The transported mass during an event i is less than or equal to the available one, i.e., $Ei \leq Mdi$.

In such cases, the accumulated mass is taken to be proportional to the dry weather period:

$$Ai = Pr \times DTSi \qquad \text{(Eq. 1)}$$

in which :

Ai = accumulated mass (kg) over catchment surface during dry weather period DTSi (in time interval units) which separates rainfall events i and i-1.

Pr = TSS production rate during a time interval (kg/time interval unit).

It is then possible to write (5):

$$Pr = \frac{\sum_{i=1}^{n} Ei}{\sum_{i=1}^{n} DTSi} \qquad \text{(Eq. 2)}$$

Pr, as computed by Equation 2, was checked such that the constraint Ei $\leq$ Mdi was appropriate for each event.

The computed daily production rate for the catchments under consideration are:

- 1.7 kg ha^{-1} d^{-1} in Aix-Zup
- 2.6 kg ha^{-1} d^{-1} in Maurepas
- 3.1 kg ha^{-1} d^{-1} in Les Ulis

(we could not model masses in Aix-Nord because of unresolved problems in volume estimates).

Examination of the chronology of sampled rainfall events can determine successive residual masses over the catchment surfaces within the limits of an undetermined constant. Indeed, we have assumed that a full surface "wash up" has been reached once in order to be able to compute Pr (see Fig. 1, Maurepas example). Given that the previous hypotheses was verified, the proposed linear accumulation model was adopted in preference to others (e.g. asymptotic accumulation, power function accumulation, parabolic accumulation with an upper limited stock, different from the chosen criteria)(5). With regard to TSS transport, the objective of the modeling procedure was to reproduce the transported mass during each event and also the total transported mass within the observed event series. Three control variables were chosen for this step:

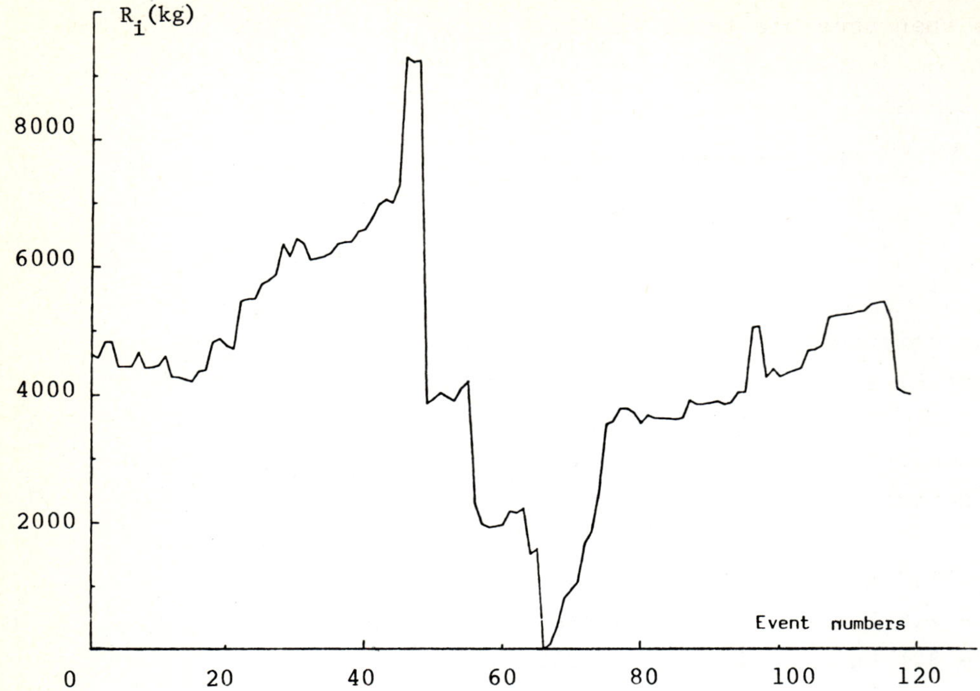

Figure 1 - Residual mass over catchment surface (Ri) after each rainfall event.

Catchment: Maurepas
(Linear accumulation model)

- Available mass, Md (kg)
- Maximum intensity within a five-minute time interval, Imax 5 ($mm\ h^{-1}$)
- Runoff Volume, VR (m^3)

The chosen model is:

$$E = K.Md^{\alpha}\ Imax5^{\beta}\ VR^{\gamma} \qquad \text{(Eq. 3)}$$

in which:

E = transported mass during any event (kg)

Md, Imax5, VR = defined above

K, α , β, γ, = parameters peculiar to each catchment.

Model parameters were identified using Rosenbrock's method (4), based on the minimization of the sum of the square of the deviations between observed and computed values, subject to the two following constraints:

- The transported mass cannot be negative.
- The transported mass cannot be greater than the available one.

Table 1 summarizes the values of K, α, β and γ.

Table 1 - K, α, β, γ values for each catchment

E = K Md Imax5 VR

	AIX-ZUP	LES ULIS	MAUREPAS
K	0.697	0.996	0.412
α	0.324	0.161	0.165
β	0.636	0.818	0.921
γ	0.307	0.421	0.382

In each case the different parameters have comparable values. This is interesting and positive in so far as one one of the aims is to obtain a transport law that could be generalized, at least in terms of its form.

Results expressed as deviations and criterion functions are given in Table 2. The reproduction of the total observed transported mass shows reasonable accuracy as an absolute value. For events with high transported masses, the difference between observed and computed mass varies between ± 10% and ± 30%. This is quite good, since it embraces most of the particle masses which are discharged to the receiving waters.

Table 2 - Results for the three experimental catchments: Total Suspended Solids - TSS

$$E = K\ Md^{\alpha}\ \ Imax5^{\beta}\ \ VR^{\gamma}$$

	AIX-ZUP	LES ULIS	MAUREPAS
Average carried mass (kg)	225.56	417.70	187.25
Average quadratic deviation as a % of the average carried mass	14.1%	10.7%	10.6%
Criterion function	50.981	92.757	20.408
Observed and computed total mass deviation (%)	- 5.8%	+ 5.3%	- 13.0%
Deviation-calibration (%)	- 6.1%	1.4%	12.0%
Deviation-verification (%)	- 5.6%	9.8%	- 33.6%

Rainfall intensity within a five minute time interval is the lowest time boundary used to determine rainfall aggressivity during the French national program. The use of other rain gauges with very low integration times should supply data on rainfall aggressivity which is even more representative. Additional climatic variables such as wind speed, humidity, etc... would further improve the results.

BOD5 and COD Accumulation and Transport Modeling Approach

As a first step, a statistical analysis of the observed mean concentrations and masses was performed (6). The results are similar to those we could get for TSS. Concerning the observed BOD_5 and COD masses, which are the parameters we are interested in, the explanatory variables seem to be Imax5 and VR, and to a lesser degree the dry weather period between two rainfall events (DTS). The transport model will be partially built up from these first results.

Accumulation Modelling Approach

Four pollutant accumulation models were tested including an asymptotic model, a power function, a parabolic model with an upper limited stock, and a linear model.

The asymptotic model:

$$M(t)= M(to)\ e^{-K(t-to)} + \frac{Pr}{K}(1 - e^{-K(t-to)}) \qquad \text{(Eq. 4)}$$

in which:

- $M(t)$ = accumulated mass in time t (kg).
- K = part of the accumulated particles removed during a time interval.
- Pr = pollutant production rate during a time interval (kg/time interval unit).

However, this model was found to produce daily rates much too high, for BOD_5 and COD, except if the asymptote is reached after a two month period (K = 0.025). Such a model is not appropriate for a phenomenon for which the time scale is on the order of a "few days". The power function model can be written:

$$M(t) + a\ t^n \qquad \text{(Eq 5)}$$

in which :

- $M(t)$ = accumulated mass in time t (kg).
- a = function of the pollutant production rate during a time interval, Pr.

and $n < 1$.

To compute a, we can assume that for tc = x days, we have:

$$\left.\frac{d\,M(t)}{dt}\right|_{t = tc} = 0.05\ Pr \qquad \text{(Eq.6)}$$

It is then possible to compute a and Pr for different values of n. Plausible values of daily production rates of pollutants (BOD_5 and COD) were reached for low n (n = 0.2) and high tc (tc = 60 days) (see Table 3).

Table 3 - BOD_5 and COD Daily Production Rates

($Kg\ ha^{-1}\ d^{-1}$)
n = 0.2 tc = 60 days

	AIX-ZUP	LES ULIS	MAUREPAS
BOD_5	.3	.2	.5
COD	8.6	1.5	3.7

However, as for the previous scheme, these results are not representative of the studied phenomenon. To get such values of Pr, if Equation 6 is confirmed with tc = 10 days, n should be about 10^{-3} or 10^{-4}, which has no meaning.

For the parabolic model with an upper limited stock:

$$M(t) = -\alpha \frac{t^2}{2} + Pr\ t \qquad \text{(Eq. 7)}$$

where:

M(t) = accumulated mass in time t (kg)
α = f (Pr)

It has also been assumed that for tc = x days, Equation 6 is verified. The upper limit of the stock is reached when:

$$\left.\frac{dM(t)}{dt} = 0\right|_{t = t\,lim} \qquad \text{(Eq. 8)}$$

As in the two previous modeling approaches, we have computed Pr for several hypotheses. Results do not agree with the "time-scale" of the phenomenon in so far as they seem to be plausible for t values about 60 days for BOD_5 and for COD, as given by Equation 6.

As a result, the linear accumulation model was chosen. Conceptually, it may not be the most satisfactory model, given that such a model does not consider the degradation phenomenon that should occur with BOD_5 and COD. However, the measuring procedures may not be accurate enough to allow a modeling approach that integrates pollutant degradation.

Given the hypotheses developed for TSS:

- A constant production rate within a time interval.
- An initial stock close to the final one on the surface catchment when considered over a long time observation.
- The transported mass during an event i less or equal to the available mass ($Ei \leq Mdi$).

It is then possible to write Equation 2 for BOD_5 and COD:

$$Pr = \frac{\sum_{i=1}^{n} Ei}{\sum_{i=1}^{n} DTSi}$$

Ei = transported mass during a rainfall event i

DTSi = the dry weather period which separates rainfall events i and i-1.

The computed daily production rates are in Table 4.

Table 4 - BOD_5 and COD Daily Production Rates, Linear Accumulation Model

	Pr (kg ha^{-1} d^{-1})		
	AIX-ZUP	LES ULIS	MAUREPAS
BOD_5	0.22	0.26	0.16
COD	1.20	1.25	1.10

Results agree with previous hypotheses except for Les Ulis, where the final stock is very different from the initial one, considering BOD_5 as well as COD. This could be due to a measuring procedure failure but this cannot be confirmed. This is the reason why two transport modeling approaches have been tested. The first one considers the available mass as a variable of the model, the second does not (since none of the accumulation models tested is fully satisfying).

Transport Modelling Approach

As mentioned before, two approaches were tested:

- The first one, given by Equation 3, is similar to the TSS model and considers the three same control variables:
    - Available mass, Md (kg)
    - Maximum intensity within a five minute time interval, Imax5 mm/h)
    - Runoff volume, VR (m^3)

- The second one considers a two-level pollution stock on the catchment surface: a first level which is usually requested, and a second one, called "deep level", which is sometimes requested but which cannot be a bounding factor. In this case, two control variables are sufficient, Imax5 and VR. The proposed equation is given by:

$$E = K' \; \mathrm{Imax5}^{\beta'} \; \mathrm{VR}^{\gamma'} \qquad \text{(Eq. 9)}$$

For BOD_5 and COD, parameter optimization was carried out using Rosenbrock's method (4). In the case of the first approach, two types of constraints were considered for each event:

- The transported mass cannot be negative.
- The transported mass cannot be greater than the available one.

For the second modeling approach, only the first constraint applies.

$$* \; E = K \; \mathrm{Md}^{\alpha} \; \mathrm{Imax5}^{\beta} \; \mathrm{VR}^{\gamma}$$

K, α, β and γ values are given in Table 5. Results expressed as deviations and criterion functions are in Tables 6 and 7 for BOD_5 and COD, respectively.

Table 5 - K, α, β, γ Values for each Catchment

E = K Md Imax5 VR

	AIX-ZUP		LES ULIS		MAUREPAS	
	BOD5	COD	BOD5	COD	BOD5	COD
K	1.255	0.795	0.555	0.820	0.361	0.122
α	-0.069	0.313	0.126	0.175	0.223	0.630
β	0.989	0.780	0.217	0.400	0.500	0.807
γ	0.227	0.226	0.451	0.506	0.291	0.142

Table 6 - BOD_5: Results for the Three Experimental Catchments

$$E = K\ Md^{\alpha}\ Imax5^{\beta}\ VR^{\gamma}$$

	AIX-ZUP	LES ULIS	MAUREPAS
Average carried mass (kg)	37.9	41.8	13.1
Average quadratic deviation as a % of the average carried mass	13.1%	7.9%	13.7%
Criterion function	1.208	0.528	0.145
Observed and computed total mass deviation (%)	10.4%	-6.1%	57.6%
Deviation calibration (%)	5.8%	2.5%	38.5%
Deviation-verification	26.8%	-14.7%	73.4%

Table 7 - COD: Results for the Three Experimental Catchments

$$E = K\ Md^{\alpha}\ Imax5^{\beta}\ VR^{\gamma}$$

	AIX-ZUP	LES ULIS	MAUREPAS
Average carried mass (kg)	213.8	205.2	85.5
Average quadratic deviation as a % of the average carried mass	9.8%	7.6%	12.0%
Criterion function	21.607	11.738	4.699
Observed and computed total mass deviation (%)	-5.2%	6.5%	73.4%
Deviation-calibration (%)	-1.0%	3.0%	26.6%
Deviation-verification (%)	19.8%	11.5%	137.6%

Strictly speaking, this model should not have been used with Les Ulis data because one of the assumptions of the accumulation model was not met. The results are not as good as they were with TSS, especially those on the Maurepas catchment. There are large differences between calibration and verification results. Therefore, if the reproduction of the total observed transported mass is quite accurate (excluding Maurepas), we cannot say that the proposed modelling approach is a fully satisfying one.

$$* \; E = K' \mathrm{Imax5}^{\beta'} \; VR^{\gamma'}$$

K', β' and γ' values are given in Table 8. Results expressed as deviations and criterion functions are in Tables 9 and 10 for BOD_5 and COD, respectively.

Table 8 - K', β', and γ' Values for each Catchment

$$E = K' \; \mathrm{Imax5}^{\beta'} \; VR^{\gamma'}$$

	AIX-ZUP		LES ULIS		MAUREPAS	
	BOD5	COD	BOD5	COD	BOD5	COD
K'	0.994	1.252	1.214	1.875	0.158	0.745
β'	1.085	0.793	0.216	0.398	0.462	0.739
γ'	0.148	0.464	0.459	0.579	0.534	0.487

Deviations between observed and computed masses are similar to those of the previous model. The reproduction of the total observed transported mass is not as good as it was with TSS, since differences between calibration and verification results are not at all negligible. However, results are of the same order as those obtained previously. The investigation shows that for the events with high transported masses, the differ-

Table 9 - BOD_5: Results for the Three Experimental Catchments

$$E = K' \, Imax5^{\beta'} \, VR^{\gamma'}$$

	AIX-ZUP	LES ULIS	MAUREPAS
Average carried mass (kg)	37.9	41.8	13.1
Average quadratic deviation as a % of the average carried mass	12.6%	7.8%	11.5%
Criterion function	1.121	0.520	0.103
Observed and computed total mass deviation (%)	-1.3%	-3.1%	25.1%
Deviation-calibration (%)	-0.25%	2.4%	13.0%
Deviation-verification (%)	-5.4%	-8.7%	35.0%

Table 10 - COD: Results for the Three Experimental Catchments

$$E = K' \, Imax5^{\beta'} \, VR^{\gamma'}$$

	AIX-ZUP	LES ULIS	MAUREPAS
Average carried mass (kg)	213.8	205.2	85.5
Average quadratic deviation as a % of the average carried mass	9.1%	7.4%	10.9%
Criterion function	18.502	11.150	3.948
Observed and computed total mass deviation (%)	-8.2%	10.0%	36.0%
Deviation-calibration (%)	-12.4%	-1.7%	-0.4%
Deviation-verification (%)	6.4%	27.0%	87.5%

ence between observed and computed masses is usually between $\pm$ 10% and $\pm$ 30%, which is interesting in so far as it embraces most of the particle masses discharged to the receiving waters.

For the same level of accuracy, it seems more appropriate to use the second model, because it is not necessary to compute available masses of pollutant. For TSS, we think that the inclusion of climatic variables such as wind speed, humidity, etc... would further improve the results.

Likewise, more rain gauges, having very low integration times, could have shown much higher intensities with shorter time intervals (30 seconds or 1 minute) that would be even more representative of rainfall aggressivity. In urban areas, rainfall aggressivity must be considered as the most important index to explain the transport of pollutants, since once they have been entrained from the ground surface they are almost certain to be flushed by the runoff.

Several other formulations were not adopted (6), including:

$$E = K' \, Md \, (1 - e^{-\beta' \, Imax5}) \, VR^{\gamma'}$$

$$E = K' \, (1 - e^{-\beta' \, Imax5}) \, VR^{\gamma'}$$

$$E = K \, Md^{\alpha'} \left(1 - \frac{1}{Imax5^{\beta'} \, VR^{\gamma'}}\right)$$

$$E = K' \, Imax5^{\beta} + \delta \, DR + \gamma \, VR + \alpha$$

in which:

DR = runoff duration (days).

They were not adopted for several reasons:

- A mean quadratic deviation which varies too much from one catchment to another.

- The criterion function shows high values.
- Differences between calibration and verification are high.

Conclusions

Using data from a national experimental measurement program, the modeling objective was to reproduce the total TSS, BOD_5 and COD loads for selected urban catchments. A two-step approach involving accumulation and transport was first performed. This approach led to good results with TSS. A linear accumulation model was chosen, which depends on a constant daily production rate and on the assumption that over a long time period the total mass produced will be removed. Simulation and rainfall-runoff TSS transport was achieved using a three-variable model (available mass, rainfall intensity within a five minute time interval and runoff volume). The results are good ($\pm$ 5%) in so far as over a long time period the total transported mass can be reproduced by the following model:

$$E = K\ Md^{\alpha}\ \ Imax5^{\beta}\ \ VR^{\gamma}$$

The same two-step approach did not lead to the same level of accuracy for BOD_5 and COD, given that catchments could not satisfy the assumptions in the accumulation models tested.

A one-step approach was then tried to simulateon rainfall-runoff BOD_5 and COD transport. In this case, only two control variables were retained. This model is conceptually different because it assumes that available mass is not a limiting factor. Results are good enough ($\pm$ 10%) so that over a long time period, the total transported masses of BOD_5 and COD can be reproduced using the following relationship:

$$E = K'\ Imax5^{\beta'}\ VR^{\gamma'}$$

The results, however, are not as good as those obtained for TSS.

For the Maurepas catchment, BOD_5 and COD computed balances are always overestimated, irrespective of the approach used. No explanation could be given for the discrepancy. More generally, for each pollutant and for each of the retained approaches, with respect to small events the reproduction of the observed masses is not very good, because the relative influence of one or another of the variables is not well known. For large-scale events, the level of accuracy (from $\pm$ 10% to $\pm$ 30%) seems to be very acceptable.

The modeling of pollutant accumulation and transport phenomena presented here is undoubtedly subject to improvement. Such an improvement, however, would require the acquisition of more and different data such as wind speed, humidity, etc.., or a methodology and a measurement protocol better suited to pollutant sampling. Likewise, other modelling objectives might be be developed. Such formulations would necessarily provide alternative views of the modeled phenomena, since they would depend on shorter time intervals. Such formulations could include pollutogram reconstruction, real-time management, and the prediction of receiving water discharge.

References

1. Chatfield, C. and Collins, A.J. Introduction to multivariate analysis. Chapman and Hall ed., London. 1980.

2. Desbordes, M. and Servat, E. Solids in urban runoff. Statistical analysis of French experimental data. Third International Congress on Urban Storm Drainage, Goteborg, Sweden, 4-8 June, 1984.

3. Hemain, J.C. Mesure de la pollution du ruissellement pluvial urbain. Rapport 1 : Suivi et bilan de la campagne experimentale. Laboratoire d'Hydrologie Mathematique, USTL, Montpellier, France. Note LHM 21/83. 1983.

4. Rosenbrock. An automatic method for finding the greatest or least value of a function. Computer Journal, vol. 3, p. 175, 1960.

5. Servat, E. Contribution a l'etude des matieres en suspension du ruissellement pluvial a l'echelle d'un petit bassin versant urbain. These de Docteur-Ingenieur, 29 juin 1984, USTL, Montpellier, France, 182 pages + 189 p. annexes. 1984.

6. Servat, E. Etude de la DB05 et de la DCO du ruissellement pluvial urbain. Essai de modelisation. Laboratoire d'Hydrologie Mathematique, USTL, Montpellier, France, Note LHM 11/85. (a paraitre). 1985.

DISCUSSION - STATISTICALLY-BASED MODELING OF URBAN RUNOFF PROCESSES

In the discussions which followed the presentation of papers, the following points were raised:

1. Dr. Geiger questioned the independence of volume and concentration in Dr. Servat's models. He also asked whether he got the same correlations with high-intensity and low-intensity storms. Dr. Servat replied that he had not looked at these questions.

2. Dr. Lijklema commented that receiving water analyses, particularly those involving sequential reactions and interdependence, were difficult and better done using deterministic models.

3. Dr. Schilling pointed out that the water resources literature was full of references to models like the ARMAX and related models, and asked why they were not used in urban runoff problems. Dr. Patry stated that one reason was that the method was not applicable if one doesn't have uniform time series.

4. Dr. Patry also discussed the problem that, in model calibration and verification, any data "points" may, in fact, be subject to great systematic error, and that it does not make too much sense to fit model results to a set of points. He further explained the usefulness of doing probability distributions of model output, using either closed-form solutions (derived probability distributions) or Monte-Carlo

analyses. These techniques are of great assistance when one cannot identify the magnitude or source of errors.

5. Mr. Jacquet asked how Dr. Lijklema had determined the k-rates for BOD_5 - the reply was that they used data that they had for surface waters.

6. Dr. James asked Dr. Hemain about the effects on his analyses of system operation in periods of no rain - the reply was that they were not significant. Dr. Desbordes reinforced this, stating that dry-weather contributions were extremely difficult to quantify, and do not seem to be too important, at least in France.

CONVENTIONAL POLLUTANT IMPACTS ON RECEIVING WATERS

Thorkild Hvitved-Jacobsen
Environmental Engineering Laboratory
University of Aalborg, Sohngaardsholmsvej 57
DK-9000 Aalborg, Denmark

Introduction

In recent years increasing concern has been devoted to urban runoff pollutant impacts on receiving waters. The following are among the most important reasons for this concern:

- Urban areas make up an important part of man's environment today. Furthermore, an increasing degree of urbanization and activities in connection with this are characteristic features of changes going on in societies all over the world. Therefore, the pollutants discharged with the runoff water from urban areas are an important factor in the contamination of the receiving waters and are supposed to remain so in the future.
- Increasing control of discharges from point sources like municipal sewage treatment plants and industrial waste outlets has increased the relative importance of the urban wet weather sources.

We are faced with two main problems from urban stormwater runoff (SWR) and combined sewer overflows (CSO): flooding and the effects of pollutants on receiving water quality. This review paper focuses on the effects of conventional pollutants, defined as biodegradable organic matter, nutrients (nitrogen and phosphorus), settleable solids and bacteria.

The effects of these pollutants on surface waters are well known.

NATO ASI Series, Vol. G10
Urban Runoff Pollution.
Edited by H. C. Torno, J. Marsalek, and M. Desbordes

Therefore, we will emphasize the characteristics of SWR and CSO compared with continuous point discharges. The key element of these characteristics is the load-effect relationship. In order to achieve certain beneficial uses of waters this relationship is a fundamental requirement for the choice and evaluation of water quality criteria.

General Aspects of Pollution from Urban Runoff

Basic Characteristics

The discharges from SWR and CSO are intermittent and therefore basically different from continuous discharges from municipal sewage treatment plants. Even though the pathway of pollutants from the drainage area implies that urban sources for both SWR and CSO are non-point, they usually enter receiving waters through well defined outlets. Therefore, considering the impacts on the receiving systems, SWR and CSO must be defined as intermittent point discharges with overland flow as an exception.

The level of concentration of the pollutants, flow characteristics, duration of the single runoff event and number of events per year are characteristics that influence the impact on the receiving water.

With respect to receiving water impacts, it is interesting to compare the concentrations of the pollutants from the intermittent point discharges with typical values of municipal sewage after secondary treatment, figure 1. The reason for this comparison is that secondary treatment still seems to be the most common treatment process for municipal wastewater.

In spite of site-to-site variations, figure 1 shows that the concentrations of the pollutants in urban runoff may be of potential importance for the deterioration of surface waters. While it is true that the concentrations of nitrogen and phosphorus in urban runoff are below the level of municipal sewage after secondary treatment, it is well known that secondary treated wastewater may cause eutrophication (3).

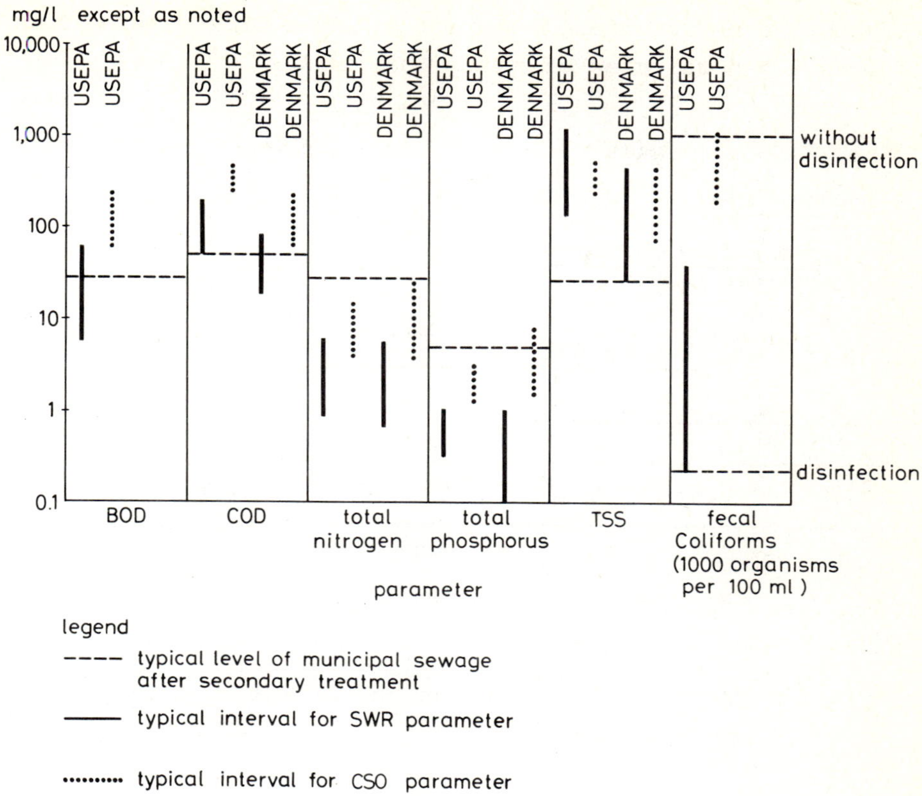

Figure 1 Representative discharge quality in SWR and CSO compared with typical municipal sewage after secondary treatment (1, 2).

The load of pollutants that enters receiving waters depends on both concentration and flow. The variability of these parameters both during a single runoff event and between events make receiving water responses complicated when compared with the impacts of continuous discharges. An appropriate simplification of these relationships and a statistical analysis seem to be the only realistic way to solve the load-effect problem (4, 5).

Types of Impacts

The time scale of urban runoff water quality problems is an essential factor to be considered. Figure 2, after Driscoll and Mancini (6), gives a general view of these time scales.

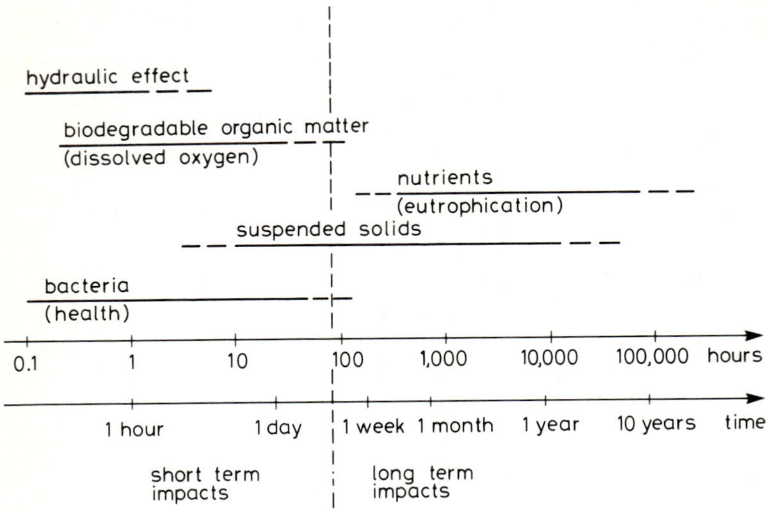

Figure 2 Time scales for urban runoff water quality problems.

In the following discussion a short term impact is defined as having its main effect within the first 2-3 days after a discharge event, whereas a long term impact causes a water quality problem beyond this scale of time. Although this definition is simplified, it is helpful in the assessment of load-effect relationships-short term impacts are related to single events and the long term impacts are related to seasonal or annual loads. As shown in figure 2, short term impacts are exemplified by the discharge of biodegradable organic matter and bacteria whereas nutrients give rise to a long term impact.

Impacts on Dissolved Oxygen

The General Problem

Dissolved oxygen (DO) depletion in receiving waters is a well known impact of sewage disposal. The classic example is the DO depletion in streams but low DO concentrations in lakes and coastal waters may also appear as a result of excessive BOD loads. In the following discussion, DO depletion processes in streams will be emphasized.

As shown in figure 1, the BOD concentration in combined sewage is roughly 5 times the concentration in stormwater runoff. Therefore, it is not surprising that CSO rather than SWR is the major reason for DO depletion.

In case of intermittent point discharges the following two phenomena are the most important concerning DO depletion:

- Degradation of discharged organic matter which exerts an oxygen demand on the water body (7, 8).
- Sediment oxygen demand caused by the scouring effect of urban runoff outfalls (9, 10).

Biodegradation of Discharged Organic Matter in Streams

Based on general knowledge of processes in rivers it appears reasonable to make a crude distinction between the following removal mechanisms for organic matter from the water phase (7):

- Degradation in the water phase.
- Extraction to the sediments, i.e. absorption and adsorption processes.
- Sedimentation.

These removal mechanisms are linked to the following two different effects on the DO consumption in the river (11).

- An immediate consumption caused by degradation of the soluble BOD fraction in the water body and by a direct absorption of organic matter by benthic organisms, e.g. bacteria and fungi. In case of a CSO discharge these processes take place in the polluted plug moving down the river.
- A delayed consumption. The removal of the organic matter from the water phase takes place without comsumption of DO. The organic matter becomes fixed to stationary objects, i.e. stones, plants and other bottom materials. In this fixed position the organic matter is degraded and a delayed oxygen

demand is exerted on the water passing by. In case of a CSO discharge this depletion will take place after the polluted plug has passed by.

The linkage between the above mentioned removal and DO consuming processes is shown in figure 3.

It is important to stress that these processes and their interaction take place as a result of both continuous and intermittent discharges. During a steady state situation little attention is given to which part of the discharged organic matter is degraded with an immediate and which with a delayed DO consumption. However, during a CSO event it becomes very important: The delayed oxygen demand will not be exerted on the volume of water from which it was discharged but in the water that subsequently flows past the sediments, e.g. after the polluted plug has passed the river reach in question. Rates of adsorption and sedimentation in a river system is in the order of 1.5-2.0 m/d (11), compared with a rate of deoxygenation in the water phase of 0.2-0.4 d^{-1}. Therefore, it is not surprising that the delayed DO consumption associated with freshly adsorbed or settled wastes from CSO play an important role, especially in small river systems near outfalls.

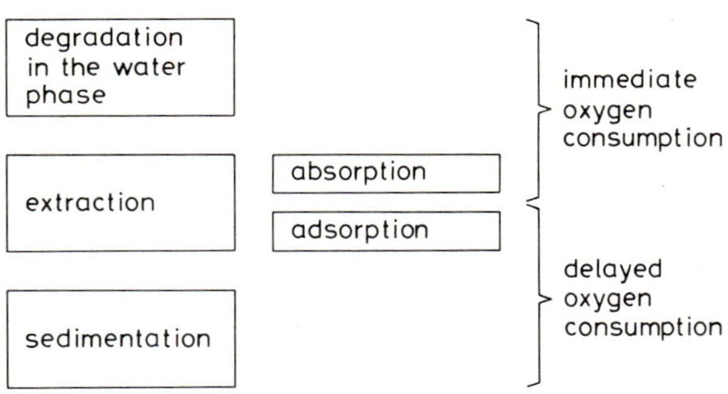

Figure 3 Terminology for processes that remove organic matter from the water phase and exert either an immediate or a delayed oxygen consumption on the river water.

Figure 4 illustrates this concept. The oxygen sag curve is shown during a CSO discharge and at different times after the event. The time variation of the oxygen concentration at two typical stations downstream is also shown. There is reason to believe that the critical deficit caused by the discharge event will be reached significantly upstream from the point where the critical concentration appears during passage of the polluted plug.

Figure 5 is a further illustration of the immediate and delayed effects of CSO discharges on the DO concentration. The figure shows the DO concentration in a small Danish river 4 km downstream of a CSO outfall during 5 consecutive days and nights. The average baseflow of the river is 150 ℓ/s and the period includes 2 overflow events. The immediate as well as the delayed effects are readily observed.

In order to calculate the effect of CSO loads on rivers it is important to investigate whether the immediate or the delayed DO consumption is the dominating process. It is not easy to indicate which characteristic of the receiving river and the overflow event will be important for the type of DO consumption. However, observations as well as analyses seem to indicate that the immediate consumption is the dominating process in larger rivers whereas small rivers with a flow below 0.5 m^3/s often show a DO depletion due to a delayed consumption.

A simple approach to a solution to the oxygen sag involving immediate DO consumption is the well known Streeter-Phelps equation. The solution to the oxygen sag based on delayed DO consumption can be regarded as an extension of the Streeter-Phelps equation involving the removal process from the water phase to the bottom and a following delayed degradation of organic matter and oxygen consumption at the bottom (11, 5). This extension results in the following equations expressing the oxygen concentration in a river subjected to a CSO load:

$$C = \bar{C} + \Delta C \frac{\beta}{\Delta\beta} - \frac{L'_o}{\frac{K_2}{K'} - 1} \left(e^{-K' \cdot t_h} - e^{-K_2 \cdot t_h} \right)$$

where $K' = k/h$

$$L'_o = K_4 \cdot \frac{P}{Q_{ov} + Q_b} e^{-K_4 \cdot t}$$

Terms are defined in list of symbols.

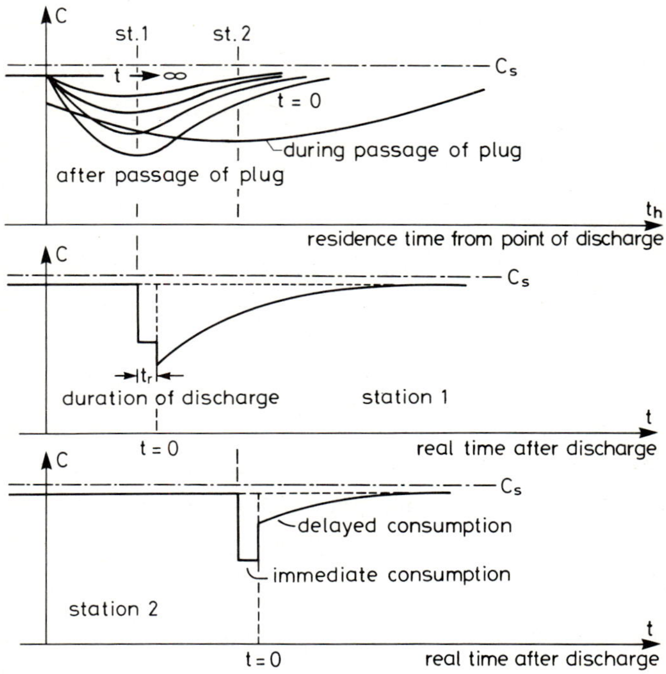

Figure 4 Illustration of oxygen sag curves during and after a CSO discharge of short duration. The immediate and the delayed oxygen demand give rise to very different observations. See text.

As an illustration of the use of these equations figure 6 shows the resulting DO profile in a river as a function of both residence time from the point of discharge (t_h) and real time after the discharge (t).

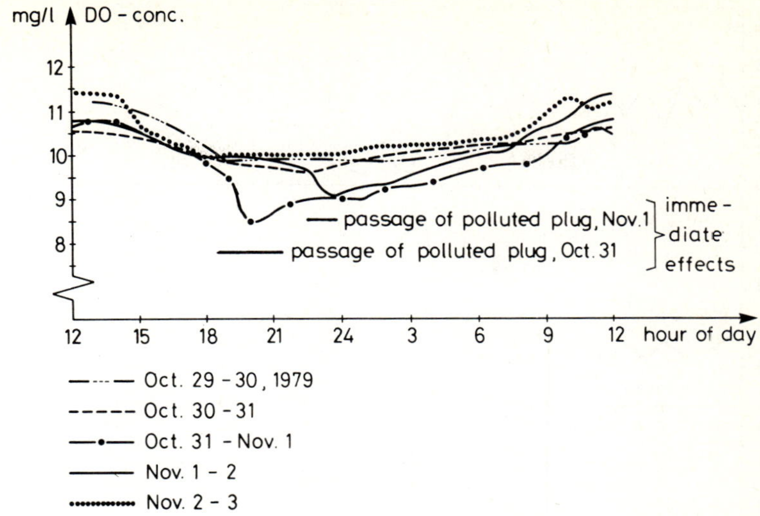

Figure 5 DO concentrations in a small Danish river during 5 consecutive days and nights. Immediate as well as delayed DO consumption is observed. The dry weather DO concentration in the night shows the background level. See text.

Many investigators have reported on the effect of CSO discharges on the DO concentrations in rivers (12, 13, 14, 15). However, the fundamental processes that cause the observed effect have not been reported. Figure 7 shows a CSO event from the river Tame, UK, where the drop in DO concentration coincides with the increase in BOD and flow (12). There are indications that the immediate and not the delayed effect on the river quality is the dominating process.

Scouring and Resuspension of Sediments

Several investigators have observed that scouring and resuspension of bottom sediments in receiving waters, especially rivers, may cause a rapid DO decline (12, 9, 10). Organic matter and other materials that have accumulated in the sediments may be subject to scouring and resuspension due to increased flow during a runoff event. An increased amount of easy biodegradable organics

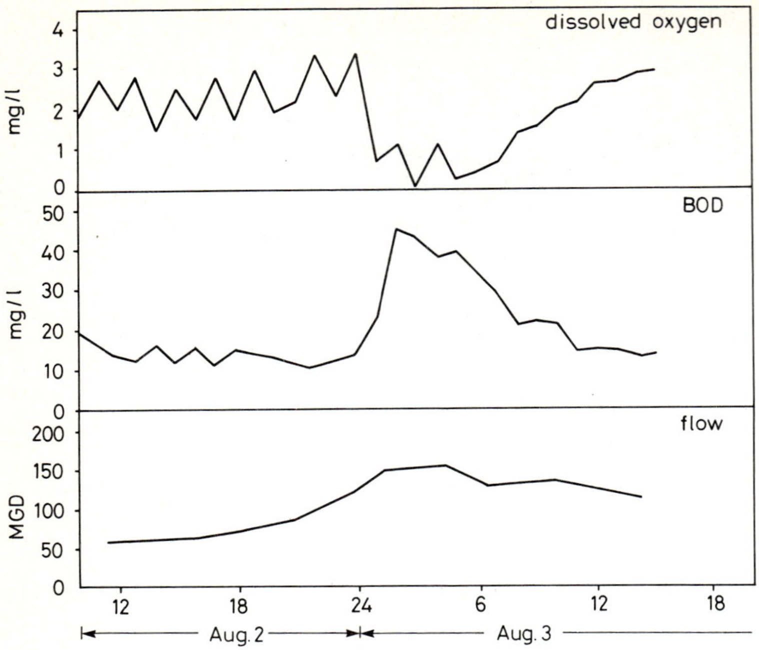

Figure 7 The effect of a storm discharge on the river Tame, UK. See text.

Table 1 Bench scale determinations of sediment oxygen demand (SOD) under undisturbed and disturbed conditions. The study area is located in the Milwaukee Metropolitan area, USA (10).

Location	Undisturbed SOD* ($g \cdot m^{-2} \cdot d^{-1}$)	Disturbed SOD* ($g \cdot m^{-2} \cdot d^{-1}$)
Kinnickinnic River	0.65	430
Mooring Basin (inner harbor)	1.40	1,370
Menomonee River	1.70	270
Milwaukee River	2.10	800
Milwaukee River	1.40	360
Milwaukee River	0.33	66

* 20 ^{o}C

and reduced species such as hydrogen sulfide and ferrous iron are brought to the surface from deeper anoxic layers. These sludges will now exert an oxygen demand on the river water flowing over the sediments. Both immediate and delayed oxygen consumption have been reported (16, 10).

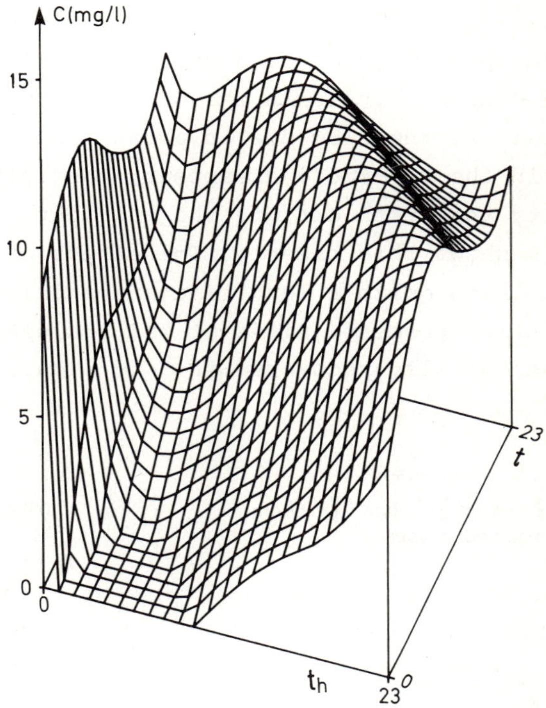

Figure 6 Resulting DO profile from a discharge of 127 kg COD to a river with a baseflow $Q_b = 50\ \ell \cdot s^{-1}$.

In order to determine the significance of the sediments on the DO balance, experiments with both undisturbed and disturbed sediments were conducted by Meinholz (10). The observed values of sediment oxygen demand (SOD) shown in table 1 clearly demonstrate the importance of scouring.

In order to quantify the differences between conditions with and without the scour mechanism, model runs have been conducted and compared with observed river data (figure 8). Comparison of these results shows the marked effect that the disturbed sediment has on the DO concentration in the river.

Diurnal DO-variations and Urban Runoff

One of the characteristics of urban runoff river impacts is related to the diurnal DO fluctuations.

In a study of several sites in the US, Keefer et al. (14) report that although sites with severe problems due to urban runoff varied considerably, all demonstrated remarkably similar hourly data records. The period prior to a storm event is characterized by fairly large diurnal cycles in the DO level, with 4 to 5 mg/ℓ variation typical. As the flow due to urban runoff increases, the diurnal cycles in the DO record disappear, possibly because increased turbidity and depth cut off the sunlight to the aquatic plants. Figure 9 shows this clearly. The effect of the storm flow on the DO level lasts quite a long time and a delayed effect seems to be the dominating process. Sites at which a single hydrograph peak was examined usually recovered in 3 to 5 days.

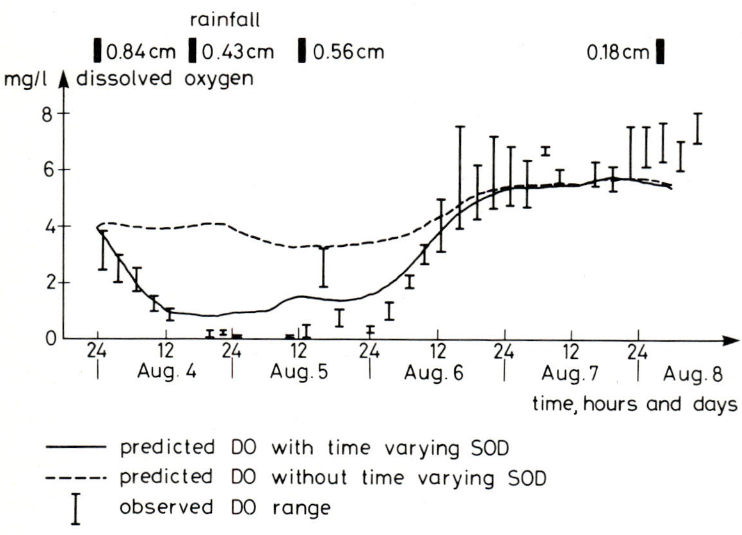

Figure 8 DO profiles showing the impact on river quality due to the scouring effect of urban runoff. Milwaukee River, Wisconsin, USA (10).

Lester (12) reports that discharges occurring during the daytime when dry weather flows of sewage are at their peak, both with respect to the rate of flow and concentration, will have greater impact than corresponding discharges during the night. This may be correct if the immediate and not the delayed DO consumption is the dominating process. Furthermore, it is probably only true in larger rivers. In small rivers with large plant photosynthesis during the day and a corresponding high respiration rate, the diurnal oxygen fluctuations may cause the more serious runoff effect on the DO concentration to take place during the night.

Eutrophication and Ecological Effects of Urban Runoff

General Aspects

The influx of nutrients, especially nitrogen and phosphorus, into receiving waters may fertilize or stimulate the growth of aquatic weeds or algae. Water quality problems associated with excessive nutrient loadings, i.e. eutrophication, manifests itself mainly by oxygen depletion and sometimes by benthic anoxic conditions, fish-kills and aesthetic problems. Furthermore, the eutrophication problem often involves substantial changes in the delicate ecological balance of the water system. Eutrophication is a problem mainly associated with nutrient loadings to lakes, estuaries and the enclosed coastal zone. To a certain extent rivers may also be affected but only the traditional problem will be emphasized here.

As shown in figure 2 the discharge of nutrients gives rise to a long term impact due to the accumulation in the receiving water. Therefore, it is the total discharge of nutrients during a period of time, e.g. one year, and not the impact related to the single event that influences the eutrophication level.

Eutrophication and the different parameters that influence the general phenomenon will not be discussed in this paper. Emphasis will be given to the characteristics of eutrophication related to urban runoff.

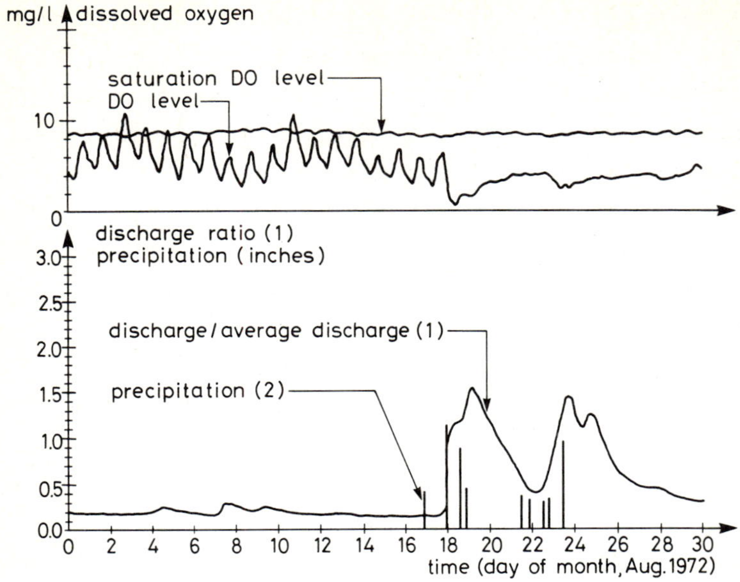

Figure 9 The effect of urban runoff on the diurnal DO cycles and DO level. Scioto River, Ohio, USA (14).

Quantity and Quality Aspects of Nutrients from Urban Runoff

As shown in figure 1 both CSO and SWR contribute to the loading of nitrogen and phosphorus on receiving water systems. However, the yearly areal loading from the catchment and not the concentration of the runoff water is the important characteristic parameter for the impact on urban lakes and similar water systems. In order to compare urban and rural contributions figure 10 gives an overview based on typical export coefficients for Danish urban and rural areas. The curves show that, contrary to nitrogen, the contribution of phosphorus might be important even though the urban area is relatively small. Furthermore, due to the fact that a great part of lakes are phosphorus limited, this nutrient seems to be the more important one to consider in urban runoff.

The availability of the nutrients plays an important role for the growth of both rooted vegetation and algae. Even though it is well known that processes taking place in the water system may convert less available particulate nutrient forms to soluble forms, a certain part of the total nutrient amount seems to be in a non-

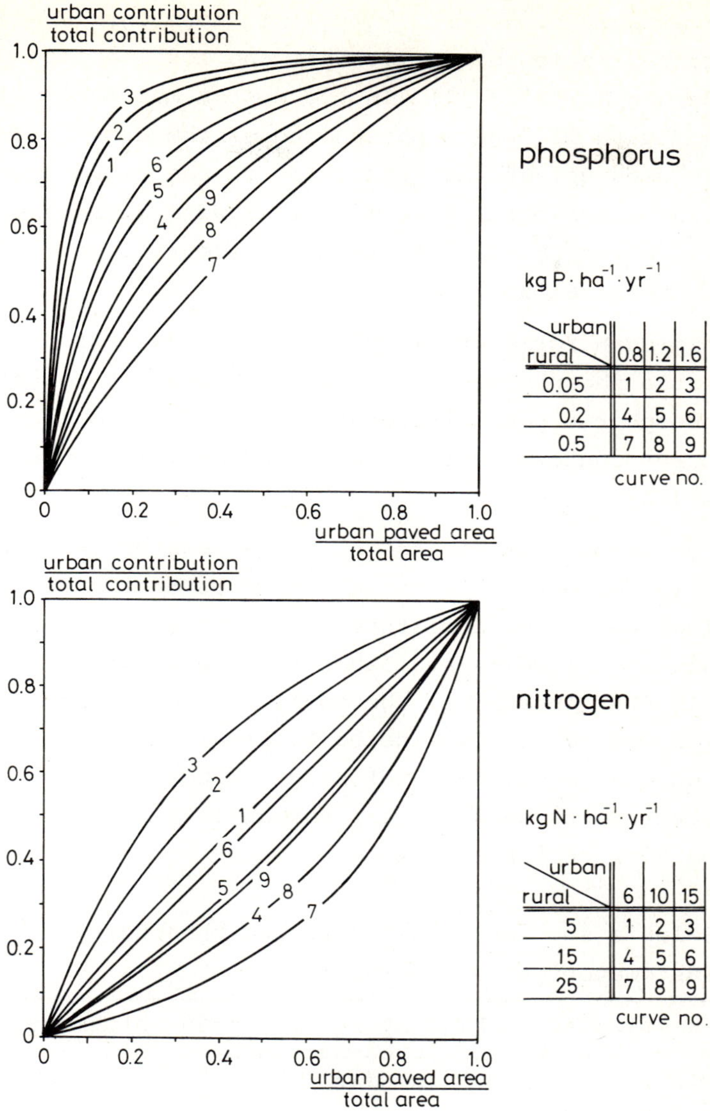

Figure 10 The distribution between urban and rural areal contributions to nutrient discharges. The values are based on typical levels for urban and rural areal discharges in Denmark (low, mean and high) as shown in the matrices.

usable form for primary producers. Figure 11 shows corresponding values of the concentration of particulate and total phosphorus in an estuary during 1979-81 (17). In the months March through May, no. 3-5 in figure 11, where phosphorus as the growth limiting nutrient is used up, the maximum value of the ratio (part. P/tot. P) is 0.75. Since part. P is identical with algae associated phosphorus, the ratio indicates that an essential part of this nutrient is in a non-available form.

The available nitrogen forms are mainly ammonia and nitrate. However, the available phosphorus is generally equivalent to the soluble ortho-phosphorus plus a fraction (0.2 or 0.3) of the particulate phosphorus content during the period of maximum production of biomass (18).

The percentage of soluble nutrients in raw wastewater and urban runoff is shown in table 2. This table indicates that nutrients in the discharge from CSO, i.e. a mixture between raw wastewater and runoff water, are more readily available for the growth of primary producers than in SWR. Therefore, the total nutrient content in CSO compared with SWR may be considered to result in the more immediate impact on the receiving water body. Values for the percentage of soluble nutrients show a great variability and only extensive time series give reliable data.

Table 2 Soluble species of nitrogen and phosphorus in raw wastewater and urban runoff.

	raw wastewater (typical Danish levels)	highway runoff, n=69 (Hvitved-Jacobsen et al. (19))	stormwater runoff, annual average. (Kluesener and Lee (20))
N (%)	60-65	15-25	23
P (%)	60-65	30-40	51

A main effect of accelerated productivity is the increase of algae photosynthesis per unit area. As a part of an intensive investigation, Harper et al. (21) studied the impact of SWR on algal productivity in the phosphorus limited urban lake Eola, Florida, USA, where urban runoff was found to be the only source of nutrients. Based on bioassay experiments it appears from figure 12 that the mixing of lake water with various proportions of untreated storm-

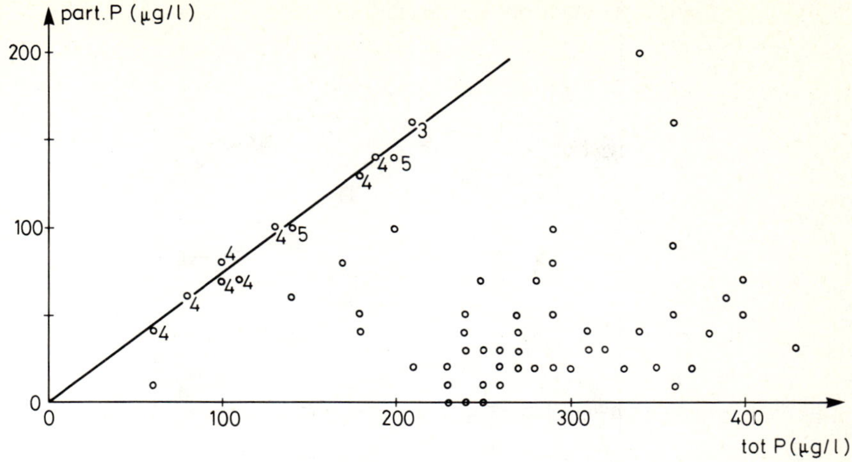

Figure 11 Corresponding values of particulate and total phosphorus in an estuary, Mariager Fjord, Denmark, during 1979-81 (17). Data indicate that an essential part of the phosphorus in the water phase is in a non-available form. See text.

water increased algae cell dry weight more than 5 times the control tests. However, when the stormwater was coagulated and the phosphorus content reduced, there was no sizable increase in algal productivity. The availability of phosphorus in the runoff water is obvious.

Lake Eola is probably one of our best investigated urban lakes. Figure 13 gives an overview of the SWR impact on the lake water quality.

Accumulation of Nutrients in Lakes Receiving SWR and CSO

Surface water bodies with long detention times such as lakes and estuaries tend to concentrate nutrients in the bottom sediments. This accumulation is increased by biological and chemical processes taking place in the water phase as well as in the sediments. Nutrients accumulated in the sediments may be returned to the water phase and become available to plant growth. Generally, these processes have been intensively studied all over the world and they

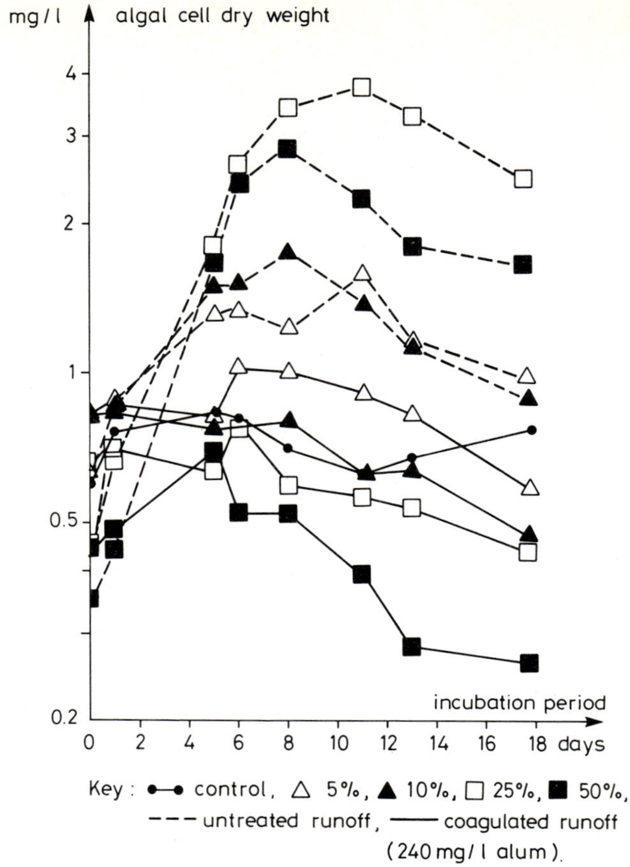

Figure 12 Response of indigenous algal species in Lake Eola to various concentrations of stormwater runoff and coagulated runoff water (21).

will not be discussed. Here, the main purpose is to show what takes place in heavily loaded urban lakes and ponds. Table 3 gives an overview on the retention of nutrients in such systems.

The data from table 3 shows that, in spite of the fact that the specific surface loading is high, an important part of the nutrients might still be expected to be retained in the system. With phosphorus as an example, this high retention in the sediments results in a diminished fertilizing effect of the nutrient loading. This concept is illustrated by the well known simplified input-output phosphorus models, e.g. the phosphorus loading graph proposed by Larsen-Mercier (24) as shown in figure 14. Furthermore, the retention of the discharged nutrients tends to protect the receiving systems with input from the lakes and ponds in question.

Table 3 Characteristics, loadings and retention of nutrients by bottom sediments for lakes and ponds receiving SWR.

	Lake Tueholm and Vallensbaek, urban pond. (Jacobsen (22)) Denmark	Lake Eola, urban lake. (Yousef et al. (23)) USA	Maitland West Pond, urban pond. (Hvitved-Jacobsen et al. (19)) USA
Surface area (ha)	35	11	1.2
Mean depth (m)	1.2	3.0	1.7
Retained (%)			
N	32	83	10+88*
P	69	86	99
Specific surface load, L ($g \cdot m^{-2} \cdot yr^{-1}$)			
N	112	21	31
P	5.2	2.5	5

* removed by denitrification

Bacterial Contamination of Receiving Waters

General Aspects

The main beneficial uses of water that are influenced by bacterial contamination from urban sources may be grouped as follows:

- Public water supply
- Recreation and aesthetics
- Fishing and shell fishing
- Irrigation, agricultural water supply

Depending on the importance of those uses, the urban dry weather flow from wastewater treatment plants may be treated in order to observe given health standards. Especially where high priority is given to the above mentioned beneficial uses, the urban wet weather sources for discharge of pathogens may impact the receiving water system to an unacceptable degree.

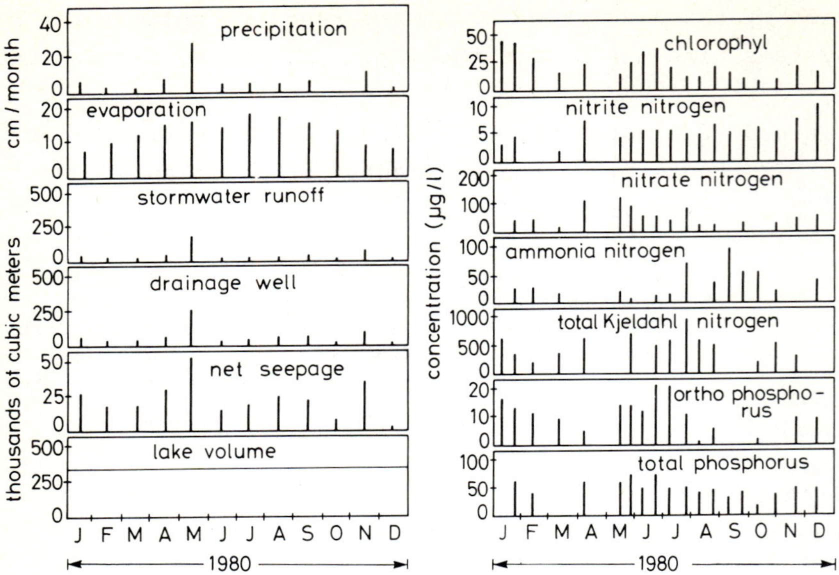

Figure 13 Hydrologic parameters for Lake Eola drainage basin and concentration of quality parameters in Lake Eola.

A relatively high density of patogens may be expected in CSO, figure 1. Also, sewer leaks and septic tank leachate may give rise to disease causing viruses and bacteria in SWR (25). However, reports on wet weather bacterial contamination of waters seem to blame CSO rather than SWR (26, 27, 28).

Based on data for a total of 156 storm events from 17 different sites it is concluded that coliform counts in urban runoff during the warmer periods of the year are approximately one order of magnitude greater than those in urban runoff during colder periods (29).

The indication of pathogenic microorganisms in waters is a general and difficult problem (30). It must be recognized that the total coliform, the fecal coliform and the fecal streptococci concentrations are widely used as indicator organisms. This is also the case in urban runoff studies on bacterial contamination of receiving waters.

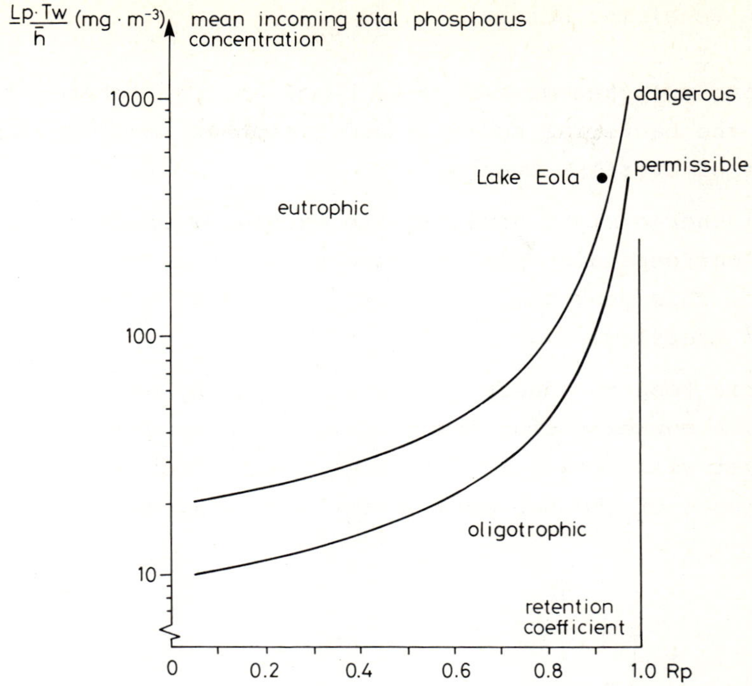

Lp ~ annual areal phosphorus load, specific surface load, ($mg \cdot m^{-2} \cdot yr^{-1}$).

Tw ~ hydraulic residence time, (yr^{-1}).

$\bar{h}$ ~ mean water depth, (m).

Rp ~ fraction of influent phosphorus retained in the sediments.

Figure 14 Annual total phosphorus load and mean depth versus the retention of phosphorus according to the Larsen-Mercier model. See text.

As shown in figure 2 bacterial contamination of water is considered a short term impact problem. The die-off of bacteria depends on a long list of general and local parameters which will not be discussed here. Both scientific investigations and practical observations show that, during a short time, a typical density of bacteria in a receiving water system resulting from a CSO discharge is reduced to a level which would be acceptable for recreational purposes like bathing. Figure 15 shows a typical die-off pattern for enteric bacteria in stormwater (31).

Bacteria in Receiving Waters

The following two phenomena seem to be of special interest with respect to the bacterial behavior and impact on receiving waters as a result of overflow events:

- CSO discharge might give rise to excessive loads of bacteria and a corresponding peak of concentration in the receiving system. This peak will last for a period of time after cease of the overflow.
- Bacteria tend to adsorp on suspended particles in the runoff water. A corresponding distribution between what will settle and what will remain in the water phase affects the disappearance of the bacteria in the receiving system.

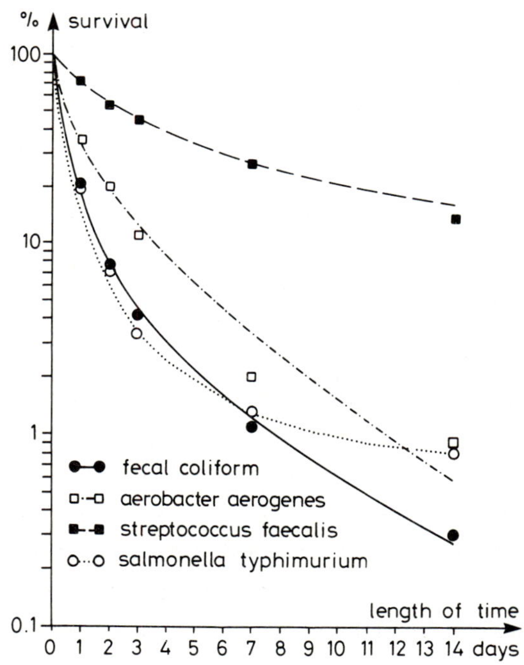

Figure 15 Persistence of selected enteric bacteria in stormwater stored at 20 °C.

Observed peaks of bacteria in receiving waters as a result of CSO are often reported (10, 32). The classic example is the inexpedient discharge of combined sewage near popular beaches (27, 28). Based on investigations carried out at a Swedish beach near Stockholm in the bathing season (July through September) in 1978, Bennerstedt and Brattberg (28) have reported on the coliform bacteria peaks, table 4. Samplings were carried out the day after the storms. High peaks even with low return periods are readily observed.

Figure 16 is a further example showing the impact of a CSO overflow discharge on the water quality of Genesee River, Ontario, Canada (32). The overflow event was caused by a rainfall of 9 mm over 8 hours. The figure illustrates that, although the region where CSO discharges take place shows relatively moderate average bacterial concentrations, high peaks are observed in the reach of the river in question.

Table 4 Measured levels (meanvalues) for total coliform bacteria in Klara Sjö, lake Mälaren, Sweden (28).

date of sampling	volume of rain (mm)	return period (months)	total coliform bacteria (per 100 ml)
Aug. 14	20	9	1,500
Sep. 3	16	10	1,000
July 6	17	7	500
Sep. 24	5	1	440
Sep. 10	3	1	210
July 11	3	1	190

As previously mentioned, it is important to identify the distribution of bacteria in connection with suspended, settleable solids contrasted with those bacteria that might be transported over longer distances in the water phase. Referring to figure 1, the concentration of suspended solids is relatively high in both stormwater and combined sewage. Therefore, the distribution of the bacteria in urban runoff discharges might be different from what is observed in the outlets from sewage treatment plants.

Schilinger and Gannon (33) have made investigations on bacterial adsorption on suspended particles in urban stormwater. They report that, because fecal coliforms showed lower tendency to adsorp than other bacterial species, they may not be the optimum group for assessing health risks. Furthermore, the high suspended solids concentration in urban runoff might result in a relatively high removal rate of pathogens to the bottom sediments. The sediment environment may prolong the survival of those bacteria because of the richness of nutrients and protection from the destructive action of sunlight.

The transportation of the overflow water in the receiving water depends on local hydraulic characteristics. As an example, Mancini (27) reports that only discharge of coliforms within 4 to 6 miles of the beaches required control and treatment in the New York harbor, USA. The dilution factor and the die-off in the water phase are main factors to be considered.

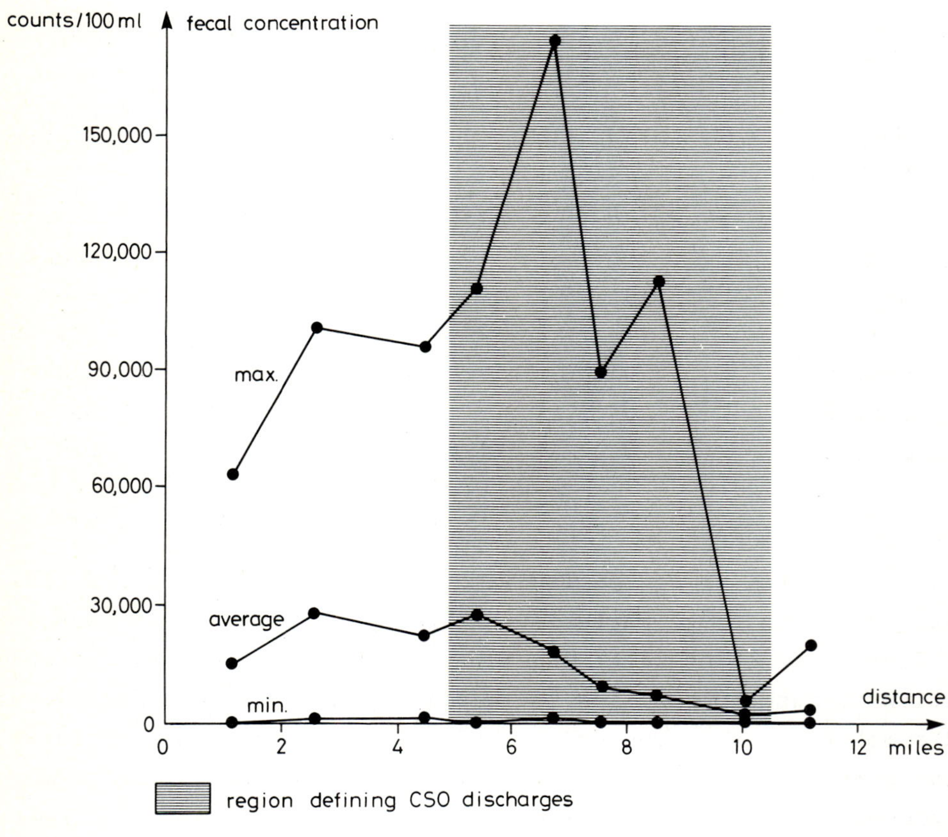

Figure 16 Minimum, average and maximum fecal coliform concentrations during a wet weather survey. Genesee River, Ontario, Canada.

Water Quality Criteria for Intermittent Point Discharges

From an engineering point of view all water pollution and protection problems have their basis in the beneficial uses made of given waters. In our effort to protect man's environment, the water quality standard and criterion play an important role:

- Water quality standard.
  A standard is a politically decided and legally set goal for a particular water system, which expresses desired uses or states.

- Water quality criterion.
  A criterion represents ideally a constituent concentration or level which results in a certain degree of environmental effect upon which scientific judgement may be based. For practical purposes a criterion means a designated concentration of a constituent that, when not exceeded, will protect an organism, an organism community or a prescribed water use or quality with an adequate degree of safety.

This definition of the water quality criterion makes the following a basic and important statement: The effects which are observed in a receiving water system for a given pollutant and pollutant concentration must be in accordance with the details of the corresponding water quality criterion. From this paper we have seen that abrupt discharges like CSO and SWR give rise to effects that are different form what will be observed for continuous discharges. It is important to notice that this is the case particularly for short term impacts related to single events. Therefore, the water quality criterion for the single event discharge of e.g. biodegradable organic matter and bacteria must be given special attention, whereas a water quality criterion for the discharge of nutrients is equivalent to what is well known for continuous discharges.

Water quality criteria are published by various agencies. In general, these criteria, which are related to continuous discharges, are similar but have changed as knowledge of the environmental effects of various pollutants have been acquired. As an example the Danish quality criterion for DO in a trout river is a daily median and minimum DO-value of 9 $mg \cdot l^{-1}$ and 6 $mg \cdot l^{-1}$, respectively.

Based on what has previously been discussed, it must be expected that a criterion based on the above given formulation is violated if CSO to the river takes place.

In recent years several papers have advocated the use of relative frequency curves and percentage time of occurrence as the proper basis for the evaluation of the performance of CSO (34). Figure 17 exemplifies this concept based on DO depletion.

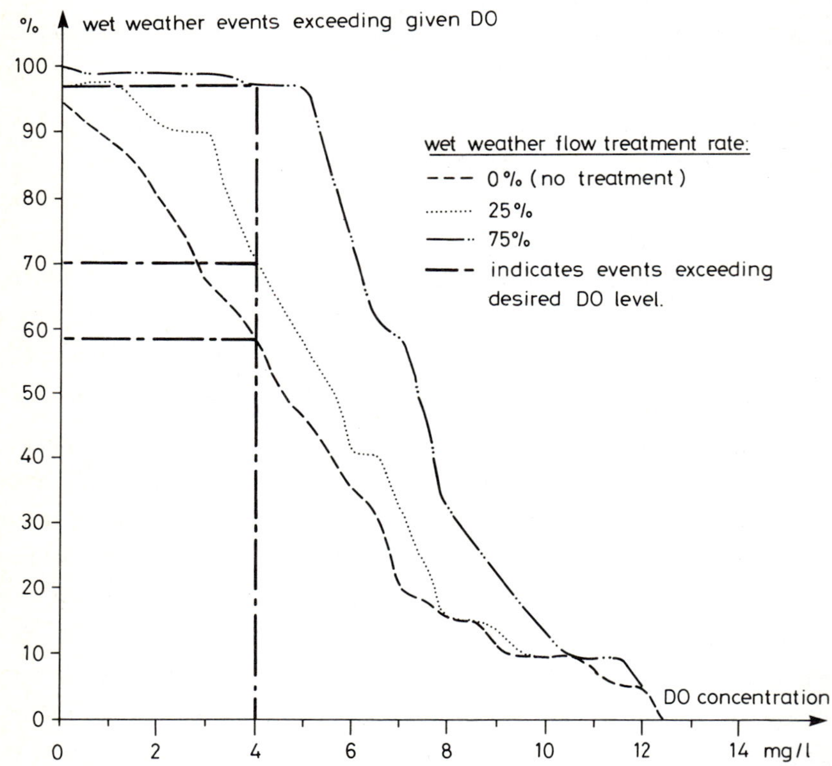

Figure 17 Example of cumulative frequency curves for minimum DO.

A fish-kill caused by depletion due to CSO will have exterminated the fish population for as long a time as it takes to build up a new population or to reestablish it by migration. The concern is: How infrequently does the event occur, or how frequently can it be alloved? This is judged by extreme event statistics, not through percentage time of occurrence.

It is the statistical concept of the approach described here that DO depletion caused by CSO has to be evaluated based on annual exceedance statistics - not on percentage time of occurrence (35). Figure 18 shows this concept. The criterion selected is that half the fish population may be killed at the DO concentration and duration indicated for the rarest events (from 8 to 16 year return period). The criterion is given for two durations of exposure time, 1 and 12 hours, for the fish population. Figure 19 shows the new criterion recommended by the Danish Water Pollution Control Committee. Three levels of water quality are given corresponding to habitats for spawing trout fish, trout fish and carp fish.

It is interesting to notice that a recent publication from USEPA (29) have used exceedance statistics on the impact of heavy metals.

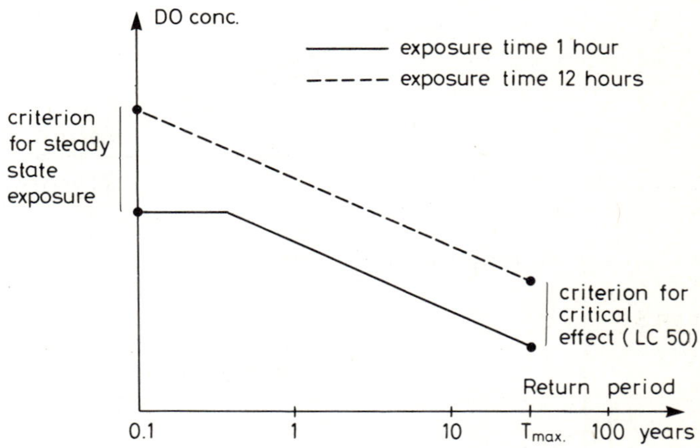

Figure 18 The principle of a water quality criterion (dissolved oxygen) as to the impact of CSO on a fish population.

The extreme event concept that has here been exemplified by a DO criterion is general. Similar criteria may be worked out for other short term impacts based on intermittent point discharges (36).

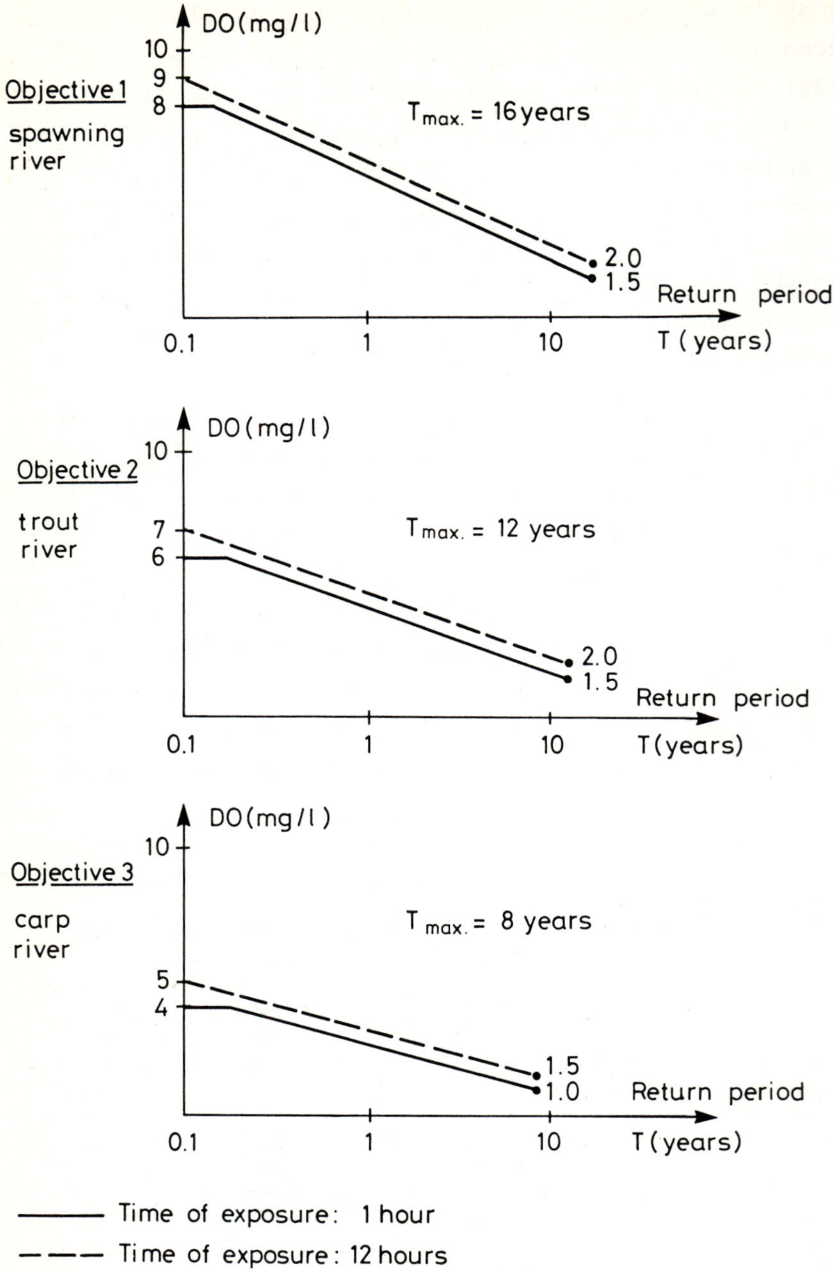

Figure 19 Recommended water quality criterion for extreme event statistics on required DO in rivers as affected by CSO.

Concluding Remarks

This paper has focused on the conventional pollutant impacts of intermittent urban point discharges on receiving waters. Emphasis has been given to those impacts that characterize the differences between intermittent discharges and continuous discharges.

There are several reasons why intermittent point discharges result in impacts different form the continuous discharges. These reasons are grouped and characterized as follows:

- The nature of the load and the pollutants give rise to both short term and long term impacts. Furthermore, the short term impacts may result in immediate (during discharge) and delayed (after discharge) effects.
- Overflow and runoff may cause increased flow, resulting in scour and resuspension that may influence water quality.
- Increased concentration of pollutants in the receiving water system may exceed the level of acceptable water quality. Furthermore, increased pollutant levels may accelerate processes with a negative influence on the water quality.
- The receiving water system is not necessarily adapted to the intermittent impact in question, i.e. natural selfpurification may be depressed.
- The rare single events may cause the critical effects.

In a US nationwide assessment of urban stormwater pollution impacts on receiving water bodies, Heaney et al. (37) stressed the fact that, due to the lack of documentation and clear definition of urban stormwater impacts, the significance of these impacts could not be assessed. It is my impression that, since those words were written, we have been walking too slowly in the right direction. Too many investigations have been devoted to determine the runoff quality compared with the impact studies of the discharged pollutants. It is my hope that we will see changes in this trend. Furthermore, we must free ourselves from site specific investigations that have a main purpose of recognizing the magnitude of the impacts and emphasize studies where the fundamental phenomena are more strongly stressed.

List of Symbols

C	oxygen concentration ($mg \cdot l^{-1}$)
$\bar{C}$	mean daily oxygen concentration in the river ($mg \cdot l^{-1}$)
C_s	saturated oxygen concentration ($mg \cdot l^{-1}$)
h	water depth (m)
$\bar{h}$	mean water depth (m)
k	1' order removal rate for adsorption or sedimentation ($m \cdot d^{-1}$)
K_2	reaeration constant (d^{-1})
K_4	1' order deoxygenation constant for organic matter fixed to stationary objects (d^{-1})
L_p	annual areal phosphorus load, specific surface load ($mg \cdot m^{-2} \cdot yr^{-1}$)
P	total load of organic matter during one overflow event, expressed as COD (g)
Q_b	baseflow in the river ($m^3 \cdot s^{-1}$)
Q_{ov}	flow passing the overflow weir ($m^3 \cdot s^{-1}$)
R_p	fraction of influent phsophorus retained in the sediments
t	real time after the discharge (d)
t_h	residence time from point of discharge (d)
t_r	duration of discharge (d)
T_w	hydraulic residence time (yr^{-1})
β, $\Delta\beta$	constants giving the scale of the daily DO fluctuations (38)
ΔC	maximum daily fluctuation in oxygen concentration ($mg \cdot l^{-1}$)

References

1. Lager, A., W.G. Smith, W.G. Lunard, R.M. Finn and E.J. Finnemore (1977), Urban Stormwater Management and Technology: Update and User's Guide, USEPA report, EPA 600/8-77-014, September 1977.

2. Hvitved-Jacobsen, T. (1985), Pollution from Urban Runoff in Denmark, report submitted to the European Water Pollution Control Association, Urban Runoff Quality Committee; Department of Civil Engineering, University of Aalborg, Denmark, February 1985.

3. OECD (Organization for Economic Cooperation and Development) (1980), Monitoring of Inland Waters; OECD eutrophication programme, the Nordic project; Nordforsk (the Nordic Cooperation Organization for Applied Research) environmental sciences publication no. 1980:2, Helsingfors, Finland.

4. DiToro, D.M. (1980), Statistics of Receiving Water Response to Runoff. In Y.A. Yousef, M.P. Wanielista, W.M. McLellon and J.S. Taylor (eds.), Urban Stormwater and Combined Sewer Overflow Impact on Receiving Water Bodies, proceedings of the National Conference, Orlando, Florida, USA, Nov. 26-28, 1979. US Environmental Protection Agency, Cincinnati, Ohio, USA, USEPA-600/9-80-056, pp. 437-465.

5. Johansen, N.B., T. Hvitved-Jacobsen and P. Harremoës (1984), Simulation of the impact of combined sewer overflows on rivers, proceedings of the eighth Nordic Hydrological Conference on Hydrology and Water Quality, Nyborg, Denmark, Aug. 6-8, 1984; Nordic Hydrological Programme, report no. 5, vol. 2, the National Agency of Environmental Protection, Copenhagen, Denmark, pp. 365-376.

6. Driscoll, E.D. and J.L. Mancini (1978), Assessment of Benefits Resulting from Control of Combined Sewer Overflows, presented at US Environmental Protection Agency Technology Transfer Seminars on Combined Sewer Overflow Assessment and Control Procedures.

7. Harremoës, P. (1982), Immediate and Delayed Oxygen Depletion in Rivers, Water Research, vol. 16, pp. 1093-1098.

8. Hvitved-Jacobsen, T. (1982), The Impact of Combined Sewer Overflows on the Dissolved Oxygen Concentration of a River, Water Research, vol. 16, pp. 1099-1105.

9. Kreutzberger, W.A., R.A. Race, T.L. Meinholz, M. Harper and J. Ibach (1980), Impact of Sediments on dissolved oxygen concentrations following combined sewer overflows, Journal Water Pollution Control Federation, vol. 52, no. 1, pp. 192-201.

10. Meinholz, T.L. (1980), Analysis of Receiving Stream Impacts on the Milwaukee River. In Y.A. Yousef, M.P. Wanielista, W.M. McLellon and J.S. Taylor (eds), Urban Stormwater and Combined Sewer Overflow Impact on Receiving Water Bodies, proceedings of the National Conference, Orlando, Florida, USA, Nov.

26-28, 1979. US Environmental Protection Agency, Cincinnati, Ohio, USA, USEPA-600/9-80-056, pp. 168-196.

11. Hvitved-Jacobsen, T. and P. Harremoës (1982), Impact of combined sewer overflows on dissolved oxygen in receiving streams. In B.C. Yen (Ed.), Urban Stormwater Quality, Management and Planning. Proceedings of the Second International Conference on Urban Storm Drainage, Urbana, Illinois, USA, June 15-19, 1981. Water Resources Publications, Littleton, Colorado, USA, pp. 226-235.

12. Lester, W.F. (1967), Effect of storm overflows on river quality, Symposium on Storm Sewage Overflows, London, UK.

13. Caspers, H. (1978), The effect of storm drain outflow on the ecosystem of the Alster lake and its canals in the centre of Hamburg. In P.R. Helliwell (Ed.), Urban Storm Drainage, proceedings of the First International Conference on Urban Storm Drainage, Southampton, UK, April 1978. Pentech Press, London, UK, pp. 670-678.

14. Keefer, T.N, R.K. Simons and R.S. McQuivey (1980), Dissolved Oxygen Impact from Urban Storm Runoff. In Y.A. Yousef, M.P. Wanielista, W.M. McLellon and J.S. Taylor (eds.), Urban Stormwater and Combined Sewer Overflow Impact on Receiving Water Bodies, proceedings of the National Conference, Orlando, Florida, USA, Nov. 26-28, 1979. US Environmental Protection Agency, Cincinnati, Ohio, USA, USEPA-600/9-80-056, pp. 223-244.

15. Lavallé, P., P. Lessard and J.-P. Villeneuve (1984), Water Quality Variations in Running Waters due to Combined Sewer Overflowing. Evaluation of Negative Influence. In P. Balmér, P.A. Malmqvist and A. Sjöberg (Eds.), The Urban Geohydrology Research Group, Chalmers University of Technology, Göteborg, Sweden, Proceedings of the Third International Conference on Urban Storm Drainage, volume 3, pp. 761-769.

16. Ellis, B. (1976), Sediments and water quality of urban storm water, Water Services, vol. 80, no. 970, pp. 730-734.

17. Hvitved-Jacobsen, T. (1982), Phosphorus availability in coastal areas - loads, effects and control (in Danish), The Danish Society of Sanitary Engineers, Copenhagen, May 25, 1982.

18. Wanielista, M.P, Y.A. Yousef and C.L. Cassagnol (1981), Stormwater Management Workshop, Department of Civil Engineering and Environmental Sciences, University of Central Florida, Orlando, Florida, USA.

19. Hvitved-Jacobsen, T., Y.A. Yousef, M.P. Wanielista and D.B. Pearce (1984), Fate of Phosphorus and nitrogen in ponds receiving highway runoff, International Symposium on Highway Pollution, London, UK, Sep. 6-9, 1983, The Science of the Total Environment, 33, pp. 259-270.

20. Kluesener, J.W. and G.F. Lee (1974), Nutrient Loading from a Separate Storm Sewer in Madison, Wisconsin, Journal Water Pollution Control Federation, vol. 46, pp. 920-936.

21. Harper, H.H., Y.A. Yousef and M.P. Wanielista (1980), Productivity responses of Lake Eola to urban runoff. In Y.A. Yousef, M.P. Wanielista, W.M. McLellon and J.S. Taylor (eds.), Urban Stormwater and Combined Sewer Overflow Impact on Receiving Water Bodies, proceedings of the National Conference, Orlando, Florida, USA, Nov. 26-28, 1979. US Environmental Protection Agency, Cincinnati, Ohio, USA, USEPA-600/9-80-056, pp. 341-370.

22. Jacobsen, B.N. (1984), Effects of Stormwater discharges to small lakes. In P. Balmér, P.-A. Malmqvist and A. Sjöberg (Eds), The Urban Geohydrology Research Group, Chalmers University of Technology, Göteborg, Sweden. Proceedings of the Third International Conference on Urban Storm Drainage, volume 4, Göteborg, Sweden, June 4-8, 1984, pp. 1367-1376.

23. Yousef, Y.A., M.P. Wanielista, R.P. Traver, H.H. Harper and T. Walsh (1982), Impact of Stormwater Runoff on Lake Eola Water Quality. In B.C. Yen (Ed.), Urban Stormwater Quality, Management and Planning. Proceedings of the Second International Conference on Urban Storm Drainage, Urbana, Illinois, USA, June 15-19, 1981. Water Resources Publications, Littleton, Colorado, USA, pp. 236-245.

24. Larsen, D.P. and H.T. Mercier (1975), Lake phosphorus loading graphs: An alternative, US Environmental Protection Agency, National Eutrophication Survey, Working Paper no. 174, Washington, D.C.: USEPA.

25. Mikalsen, K.T. (1984), Assessment of Water Quality Changes Resulting from Urbanization, Agriculture and commercial Forestry in the State of Georgia, USA. In P. Balmér, P.-A. Malmqvist and A. Sjöberg (Eds.), The Urban Geohydrology Research Group, Chalmers University of Technology, Göteborg, Sweden. Proceedings of the Third International Conference on Urban Storm Drainage, vol. 3, Göteborg, Sweden, June 4-8, 1984, pp. 801-810.

26. Benzie, W.J. and R.J. Courchaine (1966), Discharges from Separate Storm Sewers and Combined Sewers, Journal Water Pollution Control Federation, vol. 38, pp. 410-421.

27. Mancini, J.L. (1980), Impacts of Intermittent Urban Discharges on Receiving Waters. In Y.A. Yousef, M.P. Wanielista, W.M. McLellon and J.S. Taylor (eds.), Urban Stormwater and Combined Sewer Overflow Impact on Receiving Water Bodies, proceedings of the National Conference, Orlando, Florida, USA, Nov. 26-28, 1979. US Environmental Protection Agency, Cincinnati, Ohio, USA, USEPA-600/9-80-056, pp. 261-280.

28. Bennerstedt, K. and G. Brattberg (1984), Effects on Lake Mälaren from Combined Sewer Overflow. In P. Balmér, P.-A. Malmqvist and A. Sjöberg (Eds.), The Urban Geohydrology Research Group, Chalmers University of Technology, Göteborg, Sweden. Proceedings of the Third International Conference on Urban Storm Drainage, volume 3, Göteborg, Sweden, June 4-8, 1984, pp. 771-779.

29. US Environmental Protection Agency (1983), Final Report of the Nationwide Urban Runoff Program, USEPA Water Planning Division, vol. 1, Sep. 30, 1983.

30. Bonde, G.J. (1976), Bacterial Indicators of Sewage Pollution. In Coastal Pollution Control, WHO training course held under the auspices of DANIDA, Jutland, Denmark, vol. II, pp. 359-379.

31. Geldreich, E.E., L.C. Best, B.A. Kenner and D.J. Van Donsel (1968), The Bacteriological Aspects of Stormwater Pollution, Journal Water Pollution Control Federation, vol. 40, pp. 1861-1872.

32. Murphy, C.B., G.J. Welter, D.A. MacArthur and R.P. Canale (1980), The Use of Receiving Water Quality Models in Urban Runoff Pollution Abatement: Application to Marginal Benefit-Marginal Cost Analysis. In Y.A. Yousef, M.P. Wanielista, W.M. McLellon and J.S. Taylor (eds.), Urban Stormwater and Combined Sewer Overflow Impact on Receiving Water Bodies, proceedings of the National Conference, Orlando, Florida, USA, Nov. 26-28, 1979. US Environmental Protection Agency, Cincinnati, Ohio, USA, USEPA-600/9-80-056, pp. 524-556.

33. Schillinger, J.E. and J.J. Gannon (1985), Bacterial adsorption and suspended particles in urban stormwater, Journal Water Pollution Control Federation, vol. 57, no. 5, pp. 384-389.

34. Medina, M.A. (1979), Level III: Receiving Water Quality Modeling for Urban Stormwater Management, US Environmental Protection Agency, USEPA 600/2-79-100, August 1979.

35. Hvitved-Jacobsen, T. (1985), Pollution of Rivers from Overflow Structures (in Danish), Publication no. 22 from the Danish Water Pollution Control Committee, Denmark.

36. Hvitved-Jacobsen, T. (1984), Discussion of Papers on the Effects of Urban Storm Systems on Receiving Waters. In P. Balmér, P.-A. Malmqvist and A. Sjöberg (Eds.), The Urban Geohydrology Research Group, Chalmers University of Technology, Göteborg, Sweden. Proceedings of the Third International Conference on Urban Storm Drainage, volume 4, Göteborg, Sweden, June 4-8, 1984, pp. 1592-1600.

37. Heaney, J.P., W.C. Huber and M.E. Lehman (1980), Nationwide Assessment of Urban Stormwater Pollution Impacts on Receiving Water Bodies. In Y.A. Yousef, M.P. Wanielista, W.M. McLellon and J.S. Taylor (eds.), Urban Stormwater and Combined Sewer Overflow Impact on Receiving Water Bodies, proceedings of the National Conference, Orlando, Florida, USA, Nov. 26-28, 1979. US Environmental Protection Agency, Cincinnati, Ohio, USA, USEPA 600/9-80-056, pp. 407-436.

38. Simonsen, J. and P. Harremoës (1978), Oxygen and pH Fluctuations in Rivers, Water Research, vol 12, pp. 477-489.

MEASURED CSO CONTRIBUTION TO RIVER QUALITY DETERIORATION AND METHODOLOGIC APPROACH FOR NEGATIVE INFLUENCE EVALUATION

Jean-Pierre Villeneuve
Institut national de la recherche scientifique, INRS-Eau
2 700, rue Einstein
Sainte-Foy (Québec)
Canada G1V 4C7

Pierre Lavallée
Institut national de la recherche scientifique, INRS-Eau
Ministère de l'Environnement du Québec
Programme d'assainissement des eaux
3 900, rue Marly
Sainte-Foy (Québec)
Canada G1X 4E4

1. Assessment of the situation

It has been observed that, with conventionally designed interceptors, combined sewer overflows (CSOS) occur almost every time a rain event generates runoff. For example, in the City of Québec (Québec, Canada), the maximum rainfall that does not generate runoff is equal to 2.0mm (1). Although, the reality of such overflows is acknowledged and it is possible to control them through better management (real-time control, offline-system storage), the necessity of such interventions has not yet been demonstrated.

Mr. Richard Field, director of the "Storm and combined sewer program", a U.S. EPA division, explained in Göteborg, Sweden (2) that delays in bringing forward such quality interventions for CSO events are caused mainly by our inability to explain and demonstrate the impacts of intermittent overflows on the receiving waters.

Until 15 years ago, overflows were not considered a potential source of contamination of the receiving waters. Actually, CSO is a result of the design of interception systems, which is based on two hypotheses that appear increasingly impossible to verify.

NATO ASI Series, Vol. G10
Urban Runoff Pollution.
Edited by H.C. Torno, J. Marsalek, and M. Desbordes

Firstly, it was considered that, when flows in sewers caused by runoff started to increase, almost all of the contaminants were flushed to the treatment plant and, secondly, that the wastewater discharged to the receiving waters was very dilute. Thus, this water was not supposed to create environmental problems. Moreover, introducing diluted wastewater into a biological treatment process was, in view of its low strength, assumed harmful to the operation of the biological reactor. All these elements, firmly anchored in the traditional design of interceptor systems, have delayed the understanding of CSO impacts on the receiving environment. Recent research has led to a better characterization of this type of wastewater and has given a momentum to the research on receiving water deterioration due to CSOS.

The representation of wastewater quality during a rainfall period is complex. Investigations of the specifics of each catchment and each rain event made the general characterization of this type of contamination difficult. Nonhomogeneity of samples collected in various research studies made it impossible to explain the differences observed from one study to another. Actually, only very recent studies toward characterizing the water quality variations in CSOS and urban runoff water allow a better conceptualization of surface and sewer deposits accumulation and washoff.

For example, some recent research results (3, 4, 5) have shown that pollutant loads resulting from the discharge evacuation of urban runoff are considerable, even during frequent rain events. From these facts, the evaluation of the negative influence of the CSOS on the receiving waters should be revised, by stressing how much contamination is caused by CSOS. This request must take into account the processes involved in the contamination of the receiving environment and establish the required accuracy for the evaluation of the loads. The evolution of knowledge on the characterization of CSO wastewater has allowed better studies and, implicitely, a better understanding of contaminant accumulation and washout in the urban environment. The modeling of the phenomena resulting from urban runoff must continue and follow exact methods. However, an intermediate phase based on integrating CSO wastewater with the general process of water quality improvement must be implemented by the representation, for each drainage system, of the consequences related to the discharges.

The purpose of this paper is to provide a simple methodological approach for the evaluation of the negative influence of CSO wastewater on rivers. Such a methodology should define: (i) the characterization of CSO wastewater; (ii) the longitudinal and lateral diffusion of wastewater discharges in flowing water; (iii) the description of general evaluation concepts for the consequences of this type of discharge.

From the description of the impacts of such discharges on the receiving environment and in relation to a major program for characterizing CSO wastewater, we will attempt to define the required level of intervention to decrease the potential impact of intermittent CSO discharges on the receiving environment.

2. Observed deterioration of the receiving waters

The major element in demonstrating the impact on the receiving waters is the heavy load of contaminants flushed into the receiving waters during CSO events. Characterization studies of urban runoff and CSO wastewater have supported the notion of the "ecological disaster" by showing that several major rain events (the worst five), during a given year, could be responsible for 25 to 40% of the annual shock load discharged to the receiving waters by this type of wastewater (6). Following the observation of great variations in the water quality in the St. Charles River (Québec, Canada) during rain periods in the summer of 1982, we have looked for the elements that could explain those variations. Since the sanitary sewage had already been intercepted for 10 years, the only remaining source of contamination was the CSO. For this reason, we have concentrated our efforts on the impacts of such discharges on the quality of the receiving waters.

2.1 Accumulation

As shown in Lavallée et al. (7), the river pollutant load export balance is negative during rain events. For all the events measured, the amounts of contaminants leaving the segment studied are lower than 35% of the total loads discharged by CSO. The loads were measured on 65% of the drained area, and evaluated by modeling for the remaining land with the following equation:

$$L_{ij} = \bar{C}_{ij'} \int_{t_0}^{t_f} Q_{ij}\, dt \qquad (1)$$

where:

t_f : time of the end of runoff period;

t_0 : time of the beginning of the runoff period;

L_{ij}: loads flushed from a catchment j during an event i;

Q_{ij}: flow modeled by SWMM for a catchment j during an event i;

$\bar{C}_{ij}$: mean concentration of pollutants for an event i, transposed from a catchment j', representative of j.

The criteria for pairing catchments j and j' were presented by Lessard and Lavallée (5). For computation purposes, the hypothesis of a differentiation between the loads generated by various types of land use had been accepted.

From the results obtained during the rain event of June 29, 1982 for the St. Charles River (fig. 2.1), we observed that the concentrations measured at the upstream station were lower than the concentrations measured at the intermediate stations, as if there was an input of more than 10,000 kilograms of suspended solids into the river between the head and tail stations (fig. 2.2). This observation can be explained by low flow velocities and the resulting poor capacity for solids transportation.

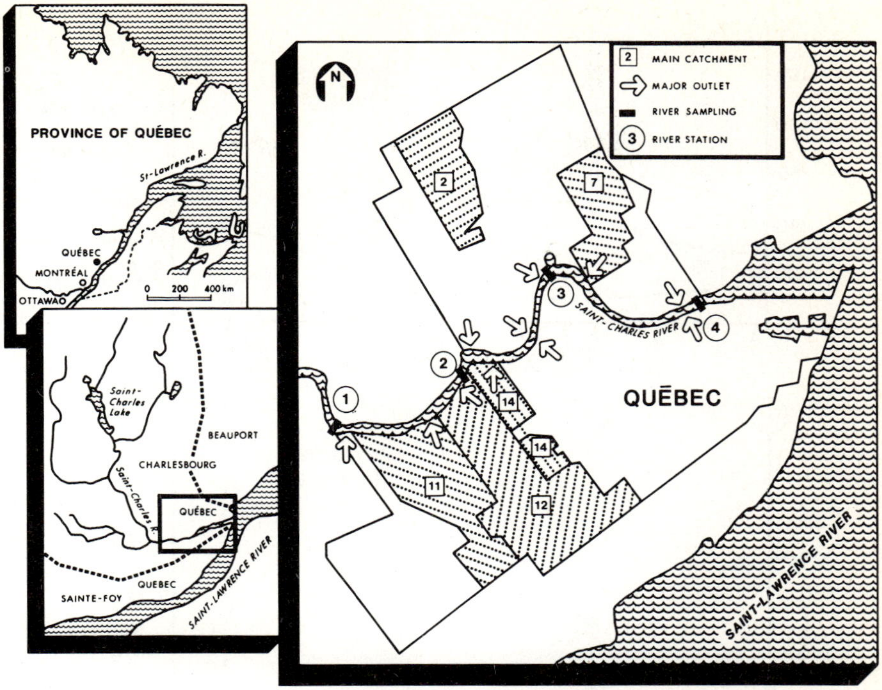

Figure 2.1: Studied Area

In view of the amounts of contaminants flushed during the rainfall by CSOs, a major accumulation process can be anticipated. Two aspects must be taken into account: the physical location of stormwater outlets and the characteristics of urban runoff wastewater. These two aspects emphasize the scope of the accumulation.

The first aspect to be considered deals with the location of the stormwater outlets (including CSOs). They are located along the shore, in places where flow velocity and water depth are low. As a consequence, diffusion is limited. The second aspect deals with suspended particles: the loads flushed into the receiving waters are easily settled, which limits their transport over long distances. A study by Klemetson (8) has showed this increased settleability of materials transported by combined sewers during rain events.

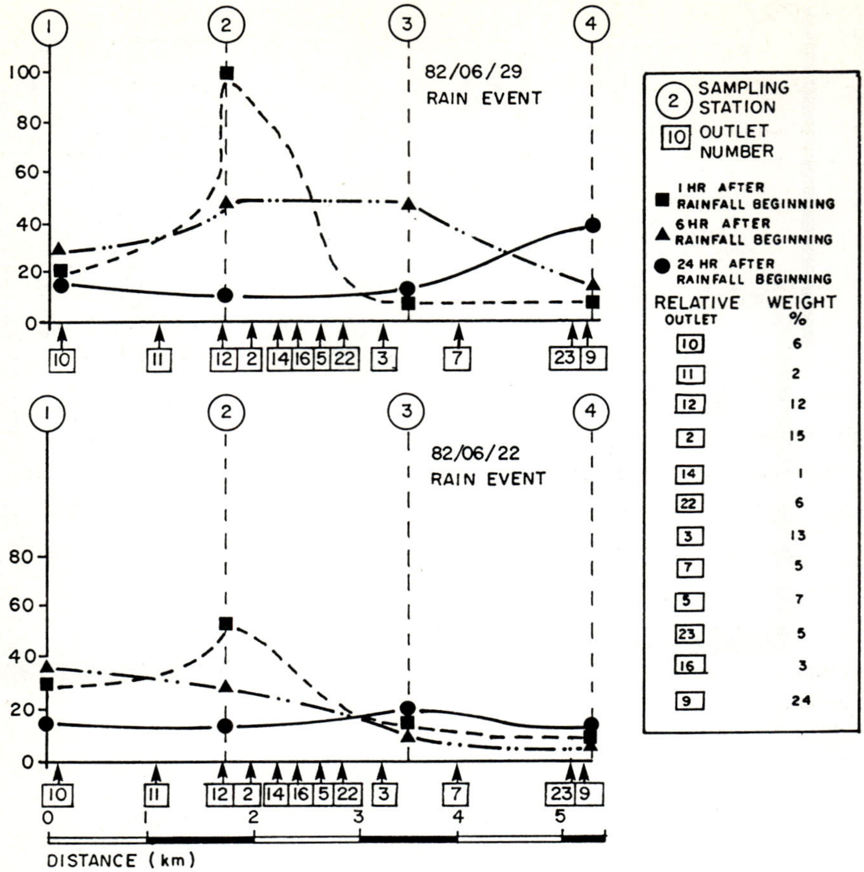

Figure 2.2: Spatial water quality variation during and after a rain event: Suspended solids

Regarding the location of stormwater outlets, our observations tend to demonstrate that lateral dispersion is very limited. The resulting dilution and the transport capacity of the river are thus decreased. A major accumulation of contaminants was also observed very near the outlets, following a lowering of the water level in the St. Charles River. These findings, presented in tables 2.1 and 2.2, show the high level of contaminants in the sediments near the overflow outlets.

The metal and organic matter contents were analyzed. The sediments taken in the whole downstream section of the river show high levels of organic matter contents, characteristic of the levels found in decomposing sludges. Downstream from the CSO zone, the sediments are highly contaminated. The concentrations of heavy metals are 10 to 15 times higher (45 mg Cu/kg; 117 mg Pb/kg; 750 mg Zn/kg) than those measured upstream. Near the overflow sites, less than 15-20 metres from the outfalls, the concentrations are as high as 980 mg Cu/kg, 1150 mg Pb/kg and 2260 mg Zn/kg.

Table 2.1: River bottom sediments quality downstream from outfalls

STATION		C_{org} [1]	N_{org} [1]	Cu [2]	Pb [2]	Zn [2]
Site	3 A	3,74	0,19	76,0	210	320
	B	2,47	0,15	62,4	194	260
	C	2,47	0,13	74,0	262	300
Site	4 A	2,36	0,13	585	810	1620
	B	1,36	0,07	890	1150	2360
	C	1,08	0,06	320	620	975

Note: The comma "," is used as a decimal point

1 %

2 mg/kg

3 A, B, C are sampling stations located 10, 12,5 and 15 meters downstream from the outfall

Table 2.2: Longitudinal river bottom sediments quality variations. Means of three sampling runs

STATION[3]		C_{org}[1]	N_{org}[1]	Cu[2]	Pb[2]	Zn[2]
Ustream	A	0,14	0,02	3,0	7,4	48,4
	B	0,26	0,03	6,4	18,9	64,7
	C	0,42	0,05	14,0	61,4	81,9
1,0 km	A	1,03	0,02	3,7	6,9	53,9
	B	0,25	0,02	3,3	6,4	57,9
	C	1,63	0,07	13,5	88,8	164
1,6 km	A	0,37	0,03	6,5	12,8	64,7
	B	0,41	0,03	6,5	18,1	77,0
	C	1,56	0,04	5,2	17,9	67,7
2,0 km	A	1,26	0,08	9,9	20,8	115
	B	1,25	0,09	8,5	15,3	99,2
	C	0,49	0,02	3,1	28,1	56,1
3,0 km	A	4,62	0,29	22,8	55,8	276
	B	3,87	0,20	23,9	65,5	307
	C	4,22	0,22	26,9	82,1	394
3,5 km	A	4,37	0,24	30,6	85,6	368
	B	4,02	0,25	27,8	79,5	340
	C	4,27	0,26	31,7	81,3	371
4,0 km	A	4,28	0,25	28,0	76,3	395
	B	3,93	0,19	25,9	80,3	383
	C	4,39	0,36	26,7	83,4	485
5,0 km	A	6,22	0,34	46,1	178	755
Downstream	B	5,21	0,25	45,6	154	688
	C	5,04	0,26	48,0	179	741

Note: The comma "," is used as a decimal point

1 %

2 mg/kg

3 A, B, C are sampling stations on a transversal section

For the downstream section of the river, the consequences of accumulations are of three kinds:

i) odors
ii) toxicity of the sediments and of the water-sediment interface
iii) vegetal hyper-productivity

In these three cases, the dissolved oxygen (DO) level in the mass of water is the key element of the problem. The sediment oxygen demand (SOD) has been evaluated as varying between 4.0 and 7.0 g/m^2/day. As presented by Krenkel and Novotny (9), this SOD level corresponds to the sludge near a municipal outlet. A consequence of such a SOD level is the reduction of the DO level in the water, particularly at the water-sediment interface, down to less than 2 mg/l. This level, considered as the anoxic threshold, is responsible for changes in the redox potential of the sediments. Below the 100 mev level, metals and phosphorus are released by the reduction of iron and manganese oxydes. Large amounts of heavy metals are made bioavailable. Because of the bioaccumulation (10,11), those toxic materials can filter into the food chain and be responsible for lethal and sub-lethal phenomena. This release of phosphorus is responsible of a growth of algae. Such phenomena were observed in the St. Charles River during the summers of 1978, 1979 and 1981.

It is important to note that the inconveniences to the population are the odors, generated by the metabolic wastes of heterotrophic, anaerobic and optional aerobic organisms when the DO conditions make aerobic mineralization impossible. The above three deterioration factors (odors, toxicity, vegetal hyper-productivity) occur in dry weather, between two rain events, thus lengthening the deterioration period beyond the duration of the rain event.

2.2 Instant variation in water quality

The accumulations of contaminants on the bottom sediments of a river are considerable and the environmental nuisances can be

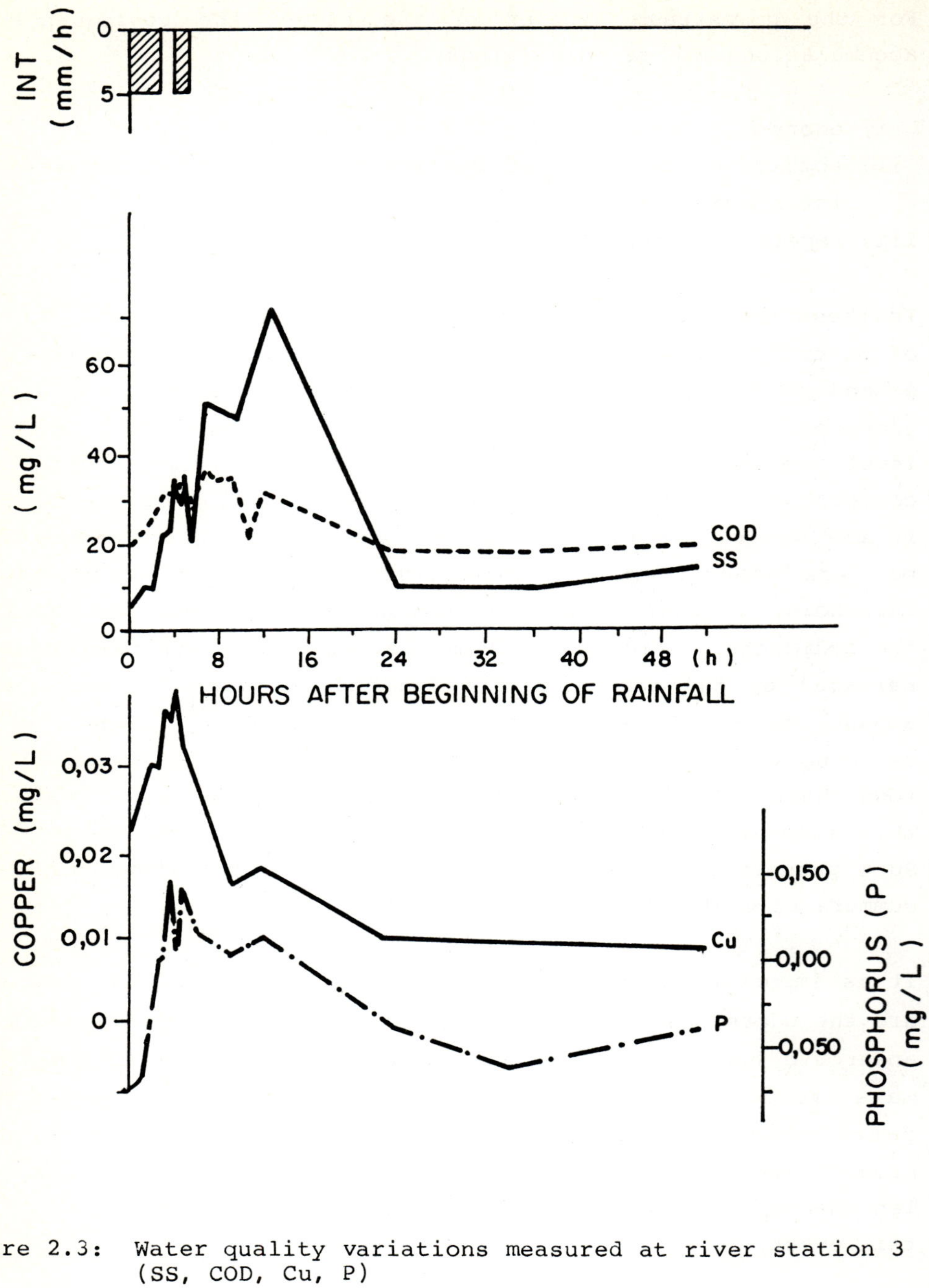

Figure 2.3: Water quality variations measured at river station 3 (SS, COD, Cu, P)

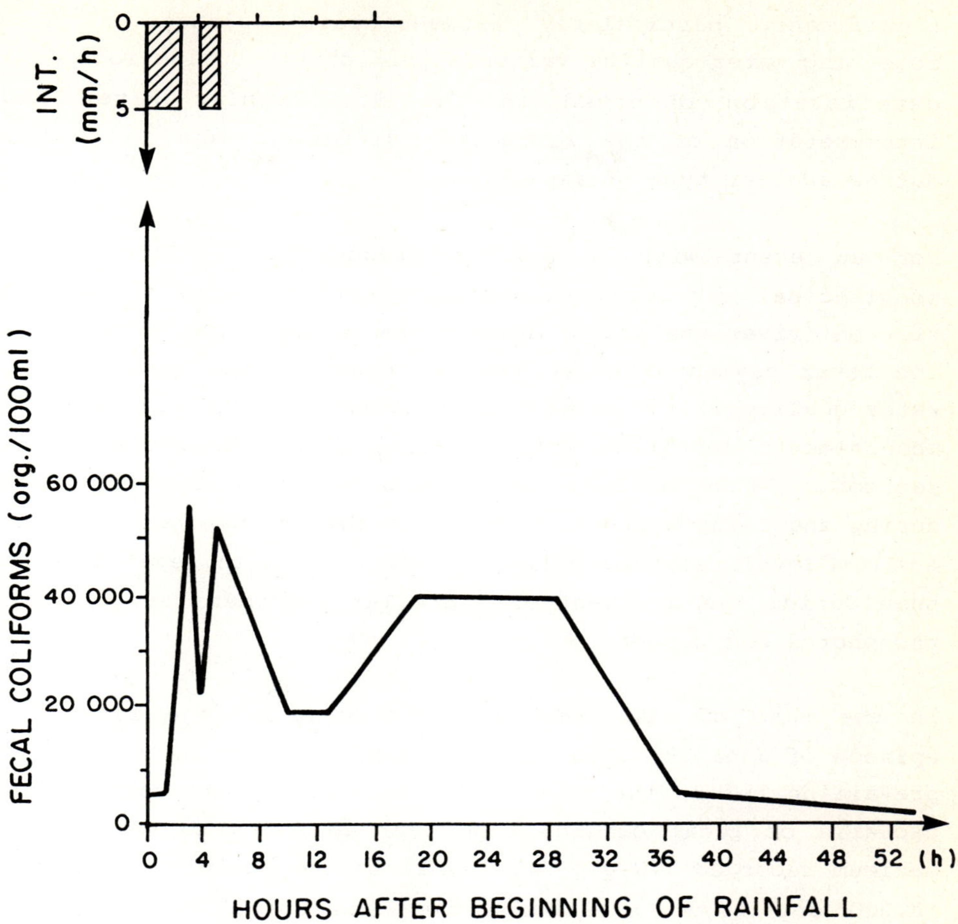

Figure 2.4: Water quality variations measured at river station 3 (fecal coliforms)

significant, particularly between rain episodes. Moreover, receiving water quality varies significantly during rainfall. The deterioration observed in the St. Charles River and the interpretation of the wastewater diffusion make it possible to define another type of impact.

For an event with a monthly frequency (7), fig. 2.2 shows longitudinal variations in water quality. At a given reference station (river station 2) located downstream from three outlets in the river segment studied, we can observe time variations in the water quality of the receiving environment (fig. 2.3, 2.4). These measurements integrate water quality along the whole transversal section. Water quality levels are very significantly different during the rainy weather compared to the dry weather. For SS, the maximum level recorded during rainfall is up to seven times higher than during the dry weather. Similar phenomena are observed for phosphorus and copper.

In the case of COD, the level measured in the river for the episode of June 29, 1982 is higher, by a factor of 2, than the one prevailing before the rain. The bacteriological counts (coliform bacteria of fecal origin) took over 48 hours to return from the maximum recorded (70,000 org/100ml) to the basic dry weather level (2,000 org/100 ml), although the mean travel time between this station and the farthest upstream outlet is less than 12 hours. It can be added that some pathogenic organisms are capable, more easily than fecal coliforms, of being adsorbed to the setleable materials (12) that are likely to accumulate near the outlets. The health hazard would therefore be greater.

However, if we look at the areas immediately downstream from each overflow outlet, we find even higher concentrations. In order to show the critical conditions prevailing near the outlets, we present the measured results of the overflow wastewater diffused in a very large river, the St. Lawrence River, which has a mean flow higher than 12,000 m^3/s. This example allows an effective visualization of the variations in the quality of the receiving waters.

The longitudinal and lateral diffusion of wastewater was measured with chemical observation techniques described in Lavallée et al. (13). Our purpose was to demonstrate:

- the low dispersion of wastewater;
- the consequent minimum dilution;
- the high concentrations of contaminants near the banks and immediately downstream from the outlets.

It was observed, for an outlet located on occurring the shore of the St. Lawrence River, (the maximum flow 4.0 m^3/sec, occurring twice/summer season), that the dilution factors were as low as 0.02 to 0.03, 1,200 metres away from the outlet (fig. 2.5). Figure 2.6 shows the longitudinal evolution of the dilution rate in this particular case. It must be noted that the flow velocities in the river vary between 0.25 and 0.45 m/sec. and that the depth is about 3 metres over the dispersion area. The longitudinal evolution of the minimum dilution rate (fd_{min}) with distance (x) was represented by the following equation:

$$fd_{min} = 0.63\ e^{-0.0024x} \qquad (2)$$

Our attempts have also demonstrated that the areas with the highest concentrations were very close to the shore. Figure 2.7 shows the lateral distribution of concentrations, 500 and 1,000 metres from the discharge point. The low lateral dispersion enhances accumulations near the shore and the resulting very high concentrations. High concentrations were measured in the St. Lawrence River, 500 metres downstream from the outlet, less than 10 metres from the shore: 65 ppb zinc, 30 ppb lead and 12 ppb copper. The bacteriological contamination level measured at the same site was 75,000 org/100 ml (coliform bacteria of fecal origin). The evaluation of the impact of intermittent discharges requires that the frequency of rain events be identified. For a given rainfall depth (h), the maximum intensity (I_{max_5}) and duration characteristics, the rainfall event can be associated with a class of events of a known exceedance probability (7). These classes are presented in table 2.3.

3. Environmental consequences of instant quality changes

The changes in the bacteriological quality of water in the receiving waters after an overflow are significant and the

Table 2.3: Characteristic rain events

CLASS (1)	OCCURRENCE	MEDIAN VALUES			ACTUAL EVENT		
		(2)	(3)	(4)	(2)	(3)	(4)
4- 6	once/week	4,6- 5,1	90-100	3,5- 8,0	4,8	155	7,2
6-10	once/week	7,1- 8,1	135-185	6,6-13,3	8,1	135	15,0
10-16	once/summer	11,5-12,8	95-155	20,0-29,0	12,0	150	25,4

1. Based on total rainfall groups
2. Precipitation (mm)
3. Duration (min.)
4. 5 minute (mm/h) maximum rainfall intensity

consequences can be directly evaluated. Water uses involving body contact are affected by this type of contamination. The persistence of contamination beyond the overflow period depends on the travel time and the possible areas of stagnation. The case measured in the St. Charles River is possibly an extreme because of the river hydrodynamic conditions, but the limited diffusion of overflow wastewater, when discharged near the shore, is instrumental in increasing the travel time and high frequency of health hazards. In addition to these hazards, we must also take into account the aesthetic deterioration of the river (floating debris, turbidity, odors). The consequences of instant changes in water quality are even more significant for the environment because the contamination occurences are highly frequent. On a statistical basis, about 30 CSO episodes occur between May 15 and September 15. Consequently, the average dry weather period

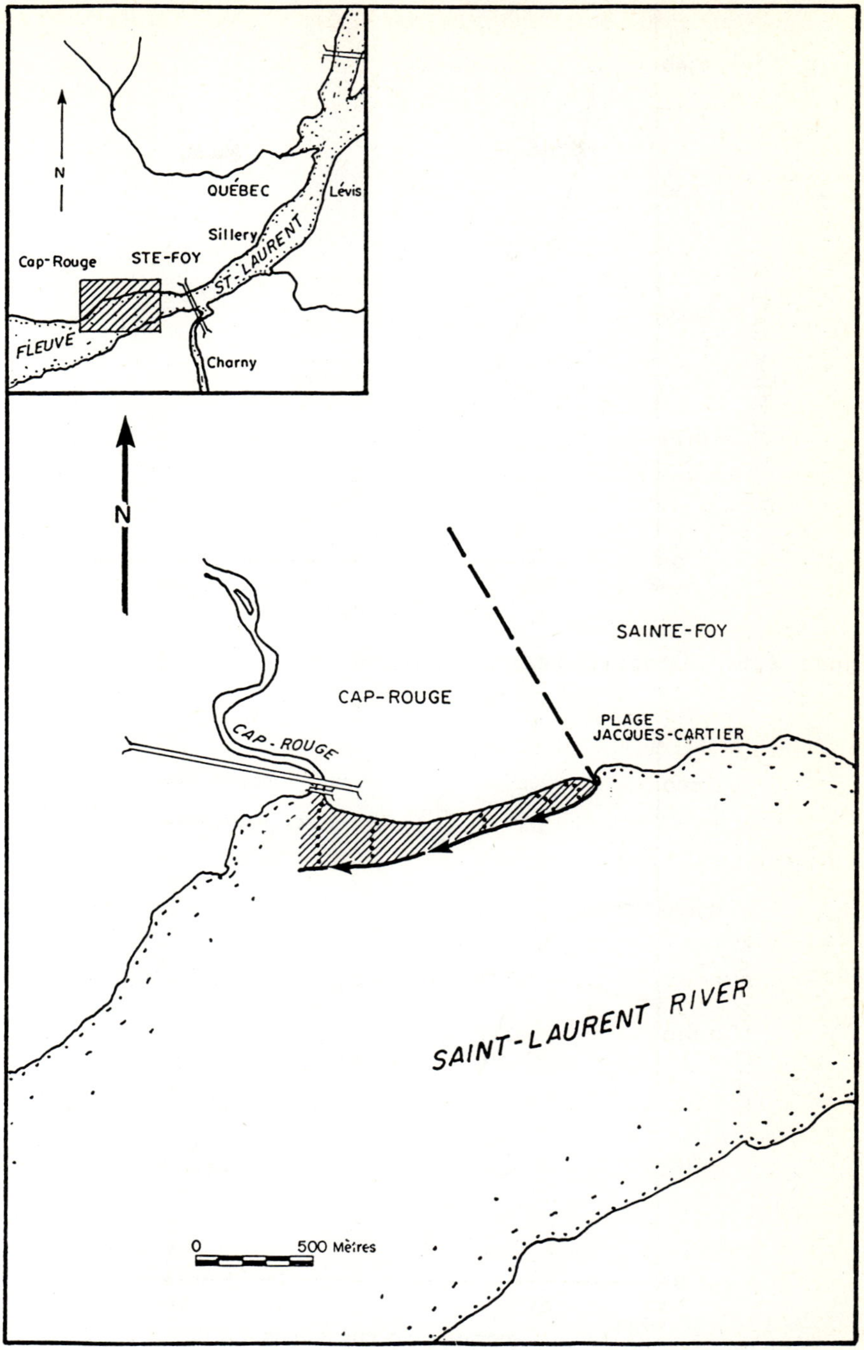

Figure 2.5: Stream diffusion pattern of a CSO outlet discharging into the St. Lawrence River

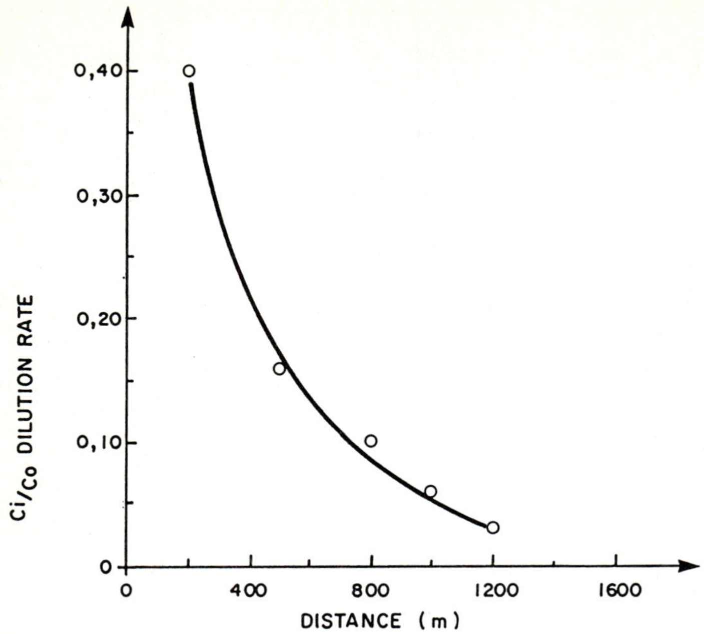

Figure 2.6: Longitudinal evolution of the dilution rate

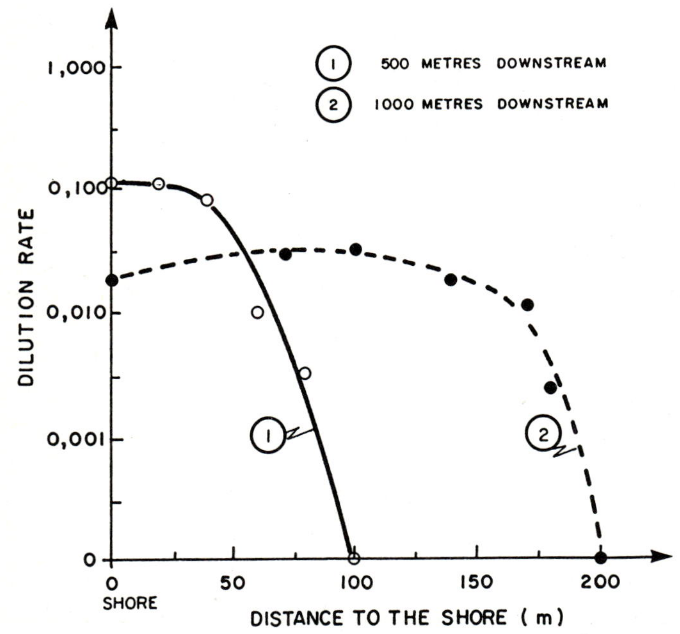

Figure 2.7: Lateral evolution of dilution rate

between the events is 100 hours. As we have observed before, the persistence of an unacceptable level of deterioration could be as long as 48 hours (the flow of contaminants is not synchronized with the displacement of water mass). The period during which the physical, chemical and bacteriological quality would be acceptable is therefore very short.

The study of biological phenomena would show that the dry weather period equilibrium in the natural environment is even shorter. The toxicity of runoff water has not been irrefutably demonstrated (14). However, heavy metal concentrations measured in the receiving waters downstream from the overflow structures are presumed toxic if reference is made to the toxicity thresholds identified as having adverse effects (table 3.1). A consequence of the episodic presence of contaminants could be sub-lethal effects on some aquatic organisms(10). Specialists in ecotoxicology have demonstrated the negative effects on green algae when exposed to copper concentrations between 5 ug/l and 10 ug/l (15). Even if the exposure to polluting substances is not continuous, the recovery period (return to initial metabolic conditions) is rather long. The physiological evolution can be represented by a drastic perturbation followed by recovery (fig. 3.1). More specific experiments would be required in order to delineate the exact period of recovery. But, knowing that copper concentrations as high as 40 ug/l have been measured in the St. Charles River during rainy periods (maximum 125 ug/l), the hypothesis of major metabolic perturbation in living organisms is reinforced.

If the recovery period varies between few hours and few days, it is also possible that the duration of the physiological recovery period is longer than the period between two rain events(5).

4. Characterization of CSO wastewater

To predict receiving water impacts due to CSO we need the wastewater characterization. Catchments sampling operations were done simultaneously with the sampling in the St. Charles River. As described in Lessard and Lavallée (16) and Lavallée and Lessard

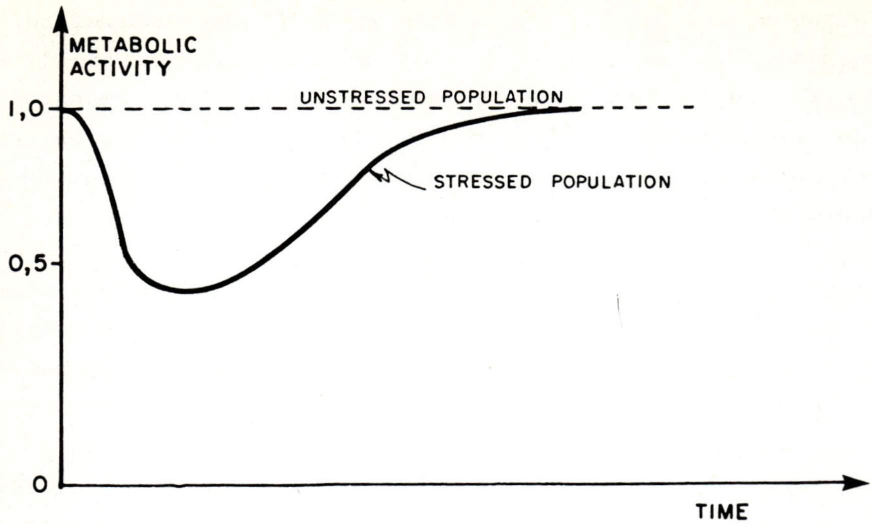

Figure 3.1: Metabolic activity fluctuation of organism under stress

Table 3.1: Comparison of water quality criteria and measured water quality levels in the river during a rain period (the June 29 event)

PARAMETER	CRITERIA (mg/l)	MEASURED MEAN CONCENTRATIONS
SS (mg/l)	11	40
COD (mg/l)	11	31
P (mg/l)	0,033	0,125
Cu (mg/l)	0,005	0,03
Pb (mg/l)	0,025	0,075
Zn (mg/l)	0,030	0,100
Fecal coliforms org./100 ml	1000	50000

(1) sampling operations took place in five major drainage catchments in the City of Québec (Québec, Canada) in 1982, and in one catchment with strictly residential land use in Sainte-Foy (Québec, Canada) during the summer of 1984. Our initial purpose was to demonstrate differences caused by various land use. The 1984 sampling focused on a catchment with a very low urbanization level (single-family housing development) in order to compare it with the catchments sampled in 1982, which were more representative of more densely populated urban centres (in terms of commercial and industrial facilities, apartment housing, high density traffic, etc.).

Research by Hemain (6) has shown that catchments with strictly residential land use and some apartment housing ($>$ 200 pers./ha) generated every year more than twice the pollutant loads (SS, COD, BOD_5) of a single-family housing area ($<$ 100 pers./ha) for the same conditions of rain, imperviousness and drainage area.

The study of regional climatic conditions seems to show that climatic variations have no influence on the loads produced for the same annual runoff volume(17).

Investigations of land use, which were made for representative catchments, have not identified other influences than those described earlier (17).

The results of our study on CSO wastewater quality variations allow the possibility of verifying some hypotheses regarding the influence of land use on the amounts of contaminants discharged into the receiving waters.

4.1 Pollutograph shapes

Figures 4.1 to 4.4 present hydrographs and pollutographs for six different meteorological events. The shapes of the pollutographs and their synchronization with the peaks of hydrographs are quite similar during individual events, regardless of the shape of the hyetograph.

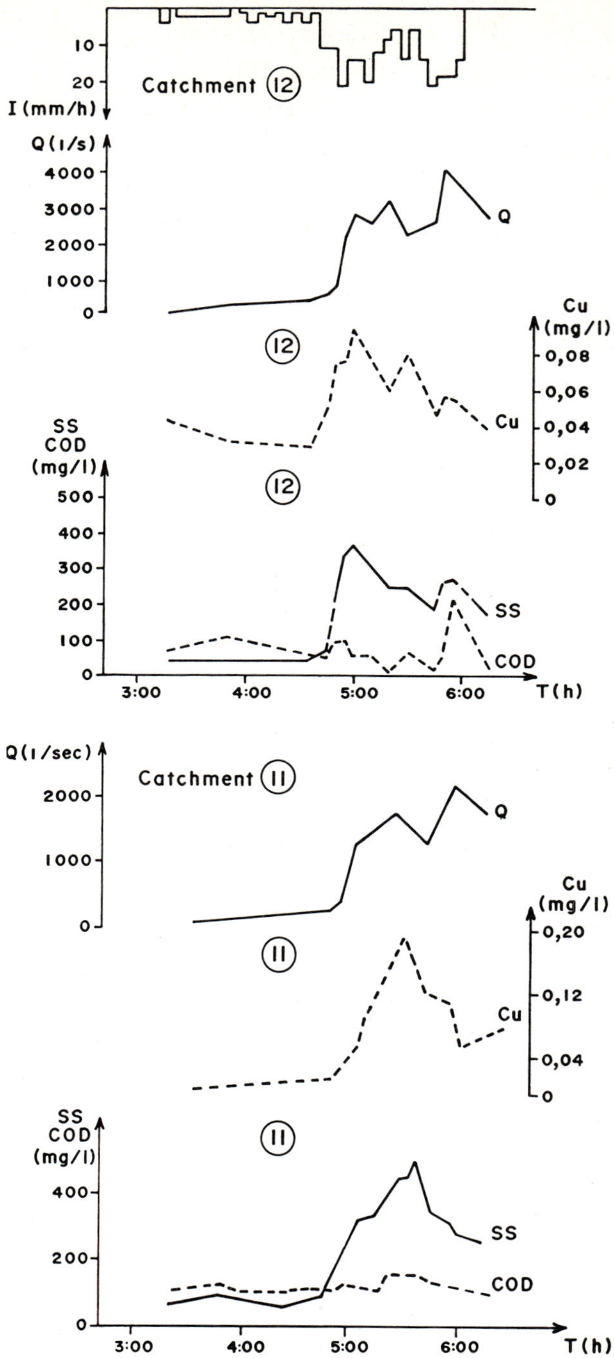

Figure 4.1: Wastewater quality variation measured in two different adjoining catchment during the same rain event (June 16, 1982).

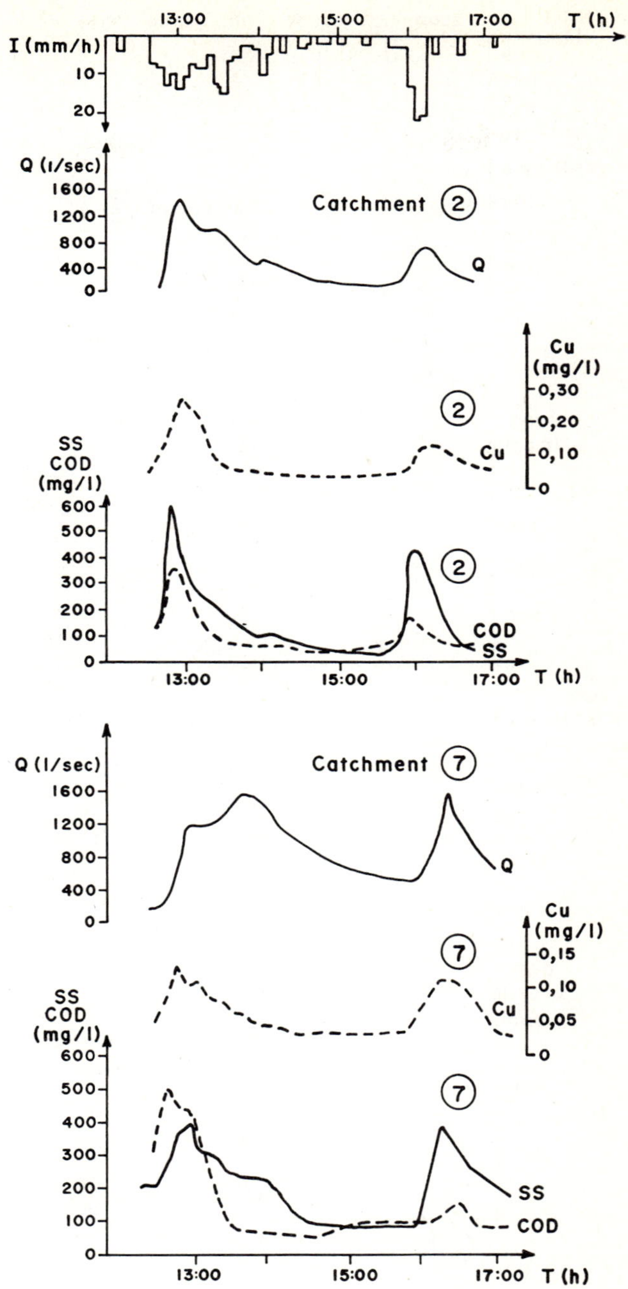

Figure 4.2: Water quality varations during successive rain events within the same precipitation area (June 29, 1982).

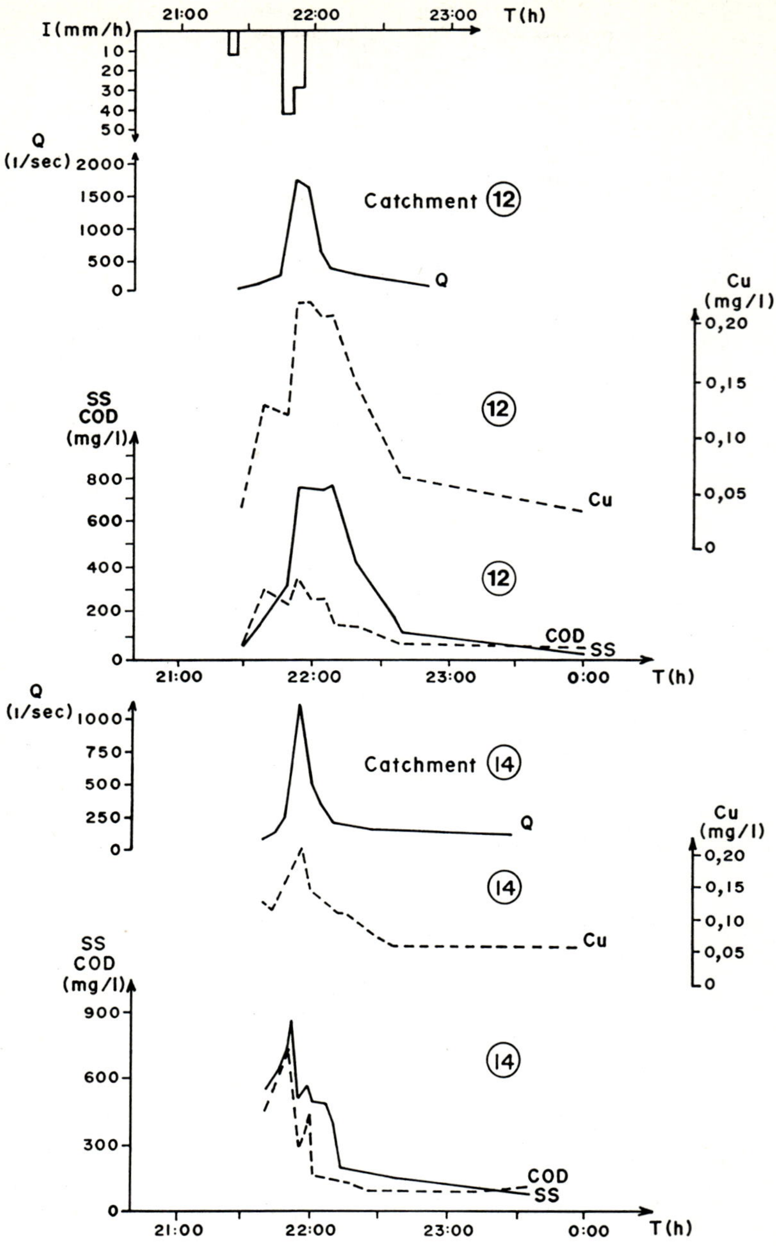

Figure 4.3: Hydrographs and pollutographs for a large catchment 12 (> 150 ha) and a small catchment 14 (< 40 ha) within the same precipitation area.

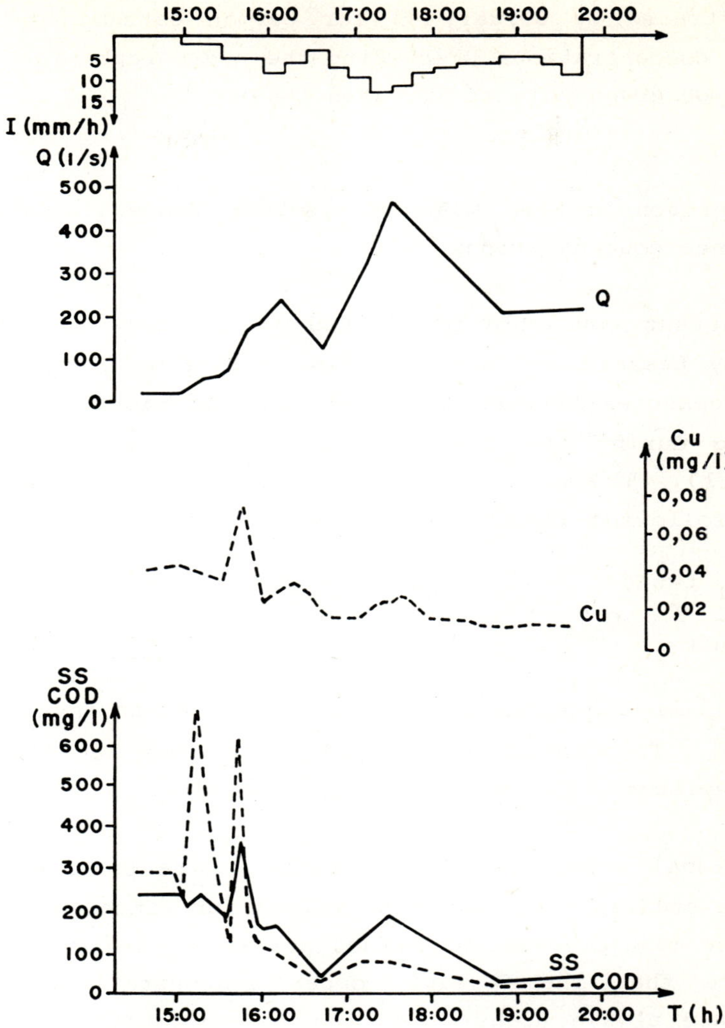

Figure 4.4: Hydrograph and pollutographs observed in catchment No. 3 on June 19, 1984

As demonstrated by Geiger (4) for German catchments, the rapid growth of concentrations preceeding the hydrograph peak (the first flush) is observed only in isolated cases.

4.2 Comparison between mean wet-weather concentrations and dry weather concentrations

Mean pollutant concentrations found in overflow wastewater were defined by Lessard and Lavallée (16). This activity showed only the discrepancies between the mean concentrations and the values prevailing during the dry weather immediately before the rain (table 4.1). The mean values ($\bar{x}$) in Table 4.1 have been computed with the following summation:

$$\bar{x}_i = \frac{1}{n} \sum_{t=1}^{t=n} C_{mit} \qquad (3)$$

where C_{mit} is the concentration measured at an instant t for an outlet i. The summation is made from the beginning of the overflow episode until the end of the overflow.

For individual events, small differences were observed among the catchments studied. By comparing values for catchments 11, 12 and 14 with the mean dry-weather values prevailing before the rain, we can notice that the SS and copper concentrations during the overflow are always higher than the concentrations prevailing in dry weather before the event.

4.3 Standardization of measured concentrations for large catchments

Figures 4.5 to 4.8 show probabilities of exceedance of contaminant concentrations measured during CSO events. This way of presenting the results has already been used by Marsalek (18). The aggregated results obtained for an exclusively commercial catchment were classified in this manner. Our purpose is to define the exceedance probabililty of a given value,

Table 4.1: Comparison of measured mean concentrations during overflow periods and dry period pollutant concentrations measured just before the start of rainfall.

RAIN EVENT	POLLUTANT	SAMPLED CATCHMENT			
		11	12	14	MEAN CONCENTRATIONS
1	SS	253	195	278	214 ± 65
	COD	61	55	48	45 ± 14
	CU	0,075	0,047	0,047	0,051 ± 0,019
2		-	-	-	-
		-	-	-	-
		-	-	-	-
3	SS	154	115	101	126 ± 20
	COD	57	84	66	130 ± 85
	CU	0,085	0,092	0,046	0,082 ± 0,027
4	SS	284	234	218	237 ± 27
	COD	105	151	200	160 ± 36
	CU	0,105	0,077	0,084	0,084 ± 0,016
5	SS	490	232	207	249 ± 80
	COD	82	86	95	87 ± 7
	CU	0,095	0,083	0,076	0,075 ± 0,015
6	SS	730	452	470	522 ± 123
	COD	242	304	239	254 ± 38
	CU	0,202	0,155	0,148	0,146 ± 0,37
7	SS	209	82	79	109 ± 53
	COD	78	98	91	84 ± 10
	CU	0,081	0,047	0,044	0,053 ± 0,015
8	SS	117	118	113	128 ± 17
	COD	186	172	145	163 ± 18
	CU	0,071	0,052	0,043	0,043 ± 0,010
9	SS	76	75	65	76 ± 6
	COD	32	80	47	71 ± 29
	CU	0,048	0,0239	0,023	0,038 ± 0,008
Dry weather period	SS	43 ± 23	124 ± 118	56 ± 21	
	COD	121 ± 79	124 ± 88	164 ± 127	
	CU	0,039 ± 0,014	0,065 ± 0,034	0,039 ± 0,006	

[1] Overflow volume was not significant

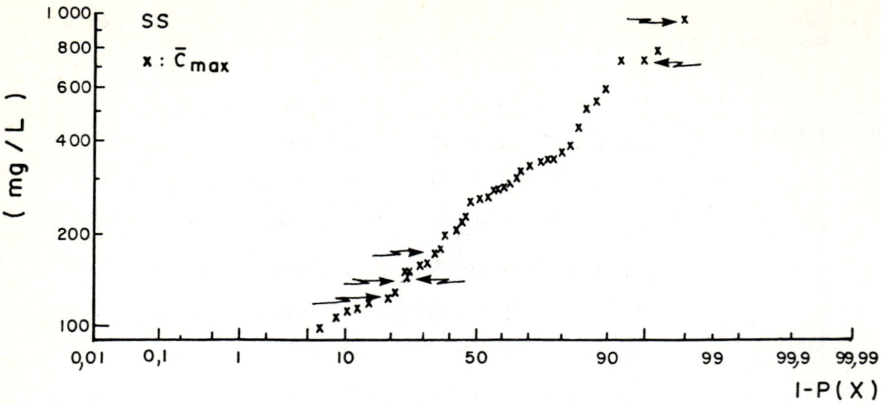

Figure 4.5: Probabilistic exceedance: SS

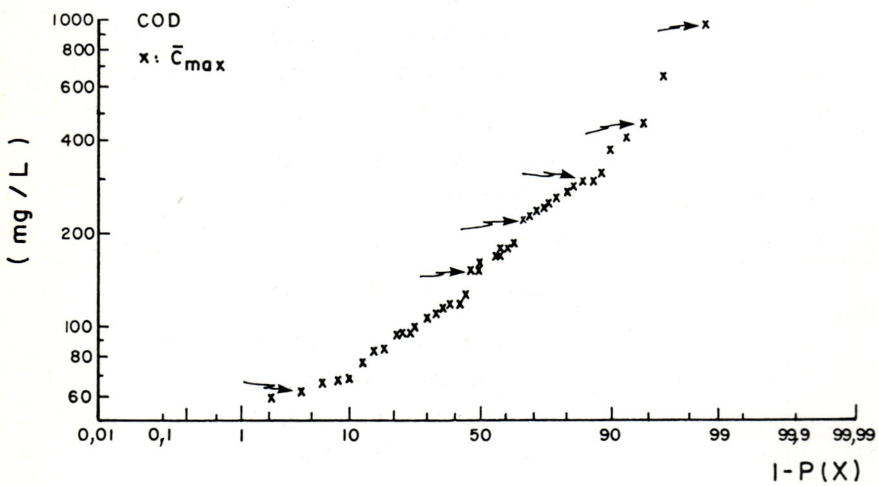

Figure 4.6: Probabilistic exceedance: COD

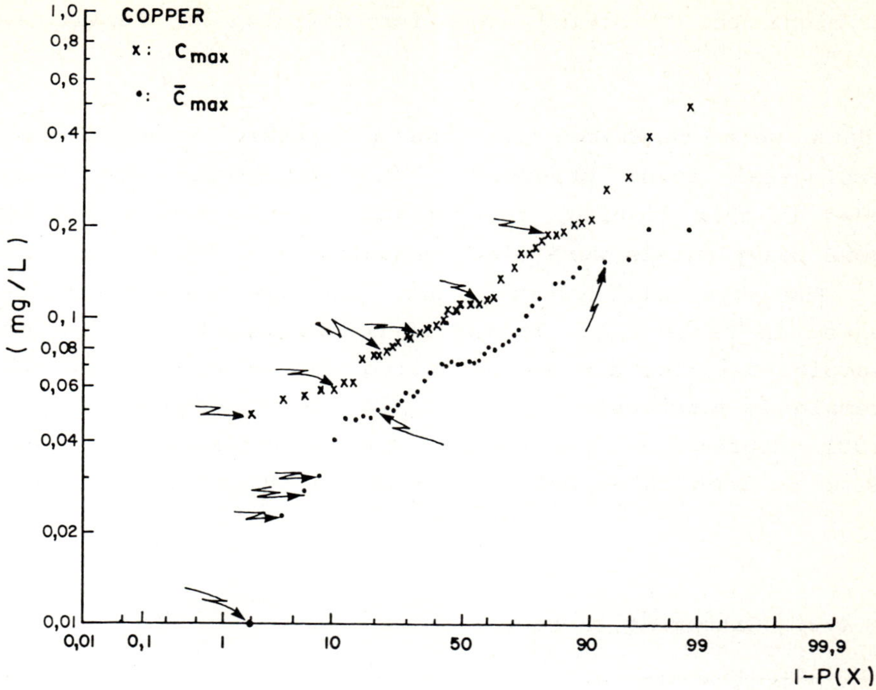

Figure 4.7: Probabilistic exceedance: Copper

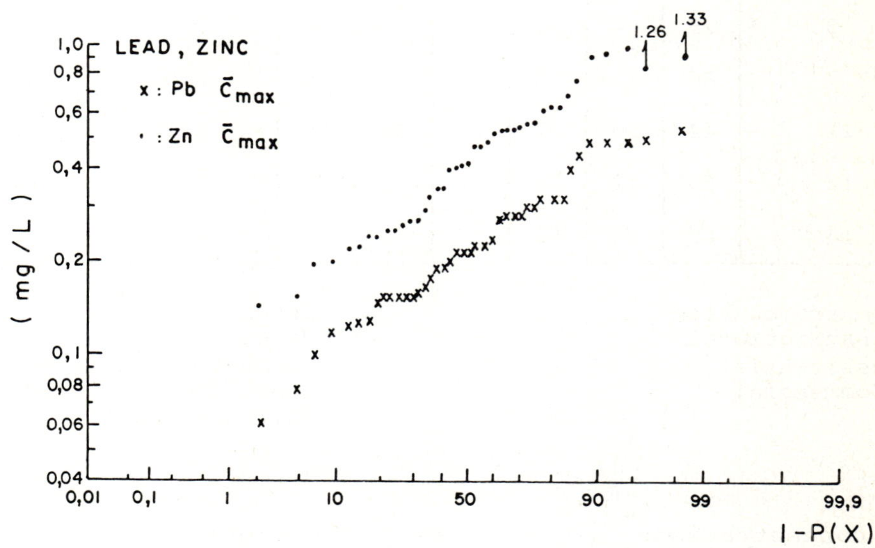

Figure 4.8: Probabilistic exceedance: Lead, Zinc

which might be a working tool for defining an "environmental hazard.".

The data were tabulated for the catchments measured and the meteorological event observed. Six catchments are therefore included in this listing, for the COD, SS and copper parameters. Two more heavy metals were also evaluated, but for five catchments only. The physical, hydraulic and land use characteristics are presented in Table 4.2. It should be mentioned that catchment 3 was sampled only for six events during the summer of 1984 and that the remaining catchments (2, 7, 11, 12 and 14) were sampled during the 1982 program for nine events. The methodologies applied were the same in both cases and are presented in Lessard and Lavallée (16).

Table 4.2: Catchment characteristics

CATCHMENT (NUMBER)*		LAND USE (%)					AREA (km^2)	7 (min)
	1	2	3	4	5	6		
2	62	10	60	23	3	4	0,70	20
3	29	4	89	3	--	4	0,70	45
7	86	12	77	2	--	9	0,61	25
11	44	10	27	3	48	12	1,02	40
12	53	12	53	12	3	21	1,81	25
14	69	29	52	8	--	11	0,32	12

1. Imperviousness
2. Institutional
3. Residential
4. Commercial
5. Industrial
6. Parks
7. Concentration time

* Identification used during studies

The concentrations plotted are the weighted mean maximum concentrations ($\bar{c}_{max}$) and, for one parameter, the instant maximum concentration ($\bar{c}_{max}$). The first variable represents the concentration of a contaminant i associated with each litre of water overflown during the time period (Δt) starting 15 minutes before the maximum flow (t_{max} - 15) and ending 15 minutes after this maximum (t_{max} + 15) regardless of the synchronization between the peaks of the hydrograph and the pollutograph. The concentration $\bar{c}_{max}$ is evaluated according to the following equation:

$$\bar{c}_{max} = \frac{1}{n} \sum_{i=1}^{n} c_i * \frac{Q_i}{Q_{moy}} \qquad (4)$$

where i = 1 corresponds to t_{max}-15, n = 30/Δt, c_i is the concentration measured during the time t_i lapsed between t_{max}-15 and t_{max}+15, Q_i is the flow recorded for the same time t_i and Q_{moy} is the mean flow measured during that period. Generally, samples were taken every five minutes, thus seven values were tabulated. The 30 minute period (from t_{max}-15 to t_{max}+15) was taken into account for full consideration of the process of evaluating the impacts on the receiving environment. It was observed during our sampling program on the St. Charles River (7) that water quality variations exceeding 350% of pollutant concentrations during thirty minute periods were measured.

All the values of $\bar{c}_{max}$ were then classified according to their exceedance probability . In the occurrence of a grouping of values caused by a particular phenomenon , a discontinuity should be observed in the classification. The analysis of the COD and SS results shows that the data are classified without induced groupings. In order to represent this , we have identified in the graphs the values obtained for catchment 3, the least urbanized in the group. The values computed for this catchment can be found at the highest level as well as at the lowest third of the values classified (fig. 4.5). There seems to be more continuity for COD, since in this case the values are evenly distributed over the whole curve (fig. 4.6). The results for copper are very different (fig. 4.7). All the events measured for catchment 3, except one (I_{max_5} = 5 mm/h), are classified in the lowest fifth of the curve with obvious discontinuity in the progression. It is to be noted that the classification of C_{max} did not show the same level of discontinuity.

Bringing forth this continuous condition of organization in the concentrations measured seems to indicate that the specific influence of the land use type is not systematic or structured, in the case of large drainage catchments, if the resulting impact on the receiving environment (Δt = 30 minutes) is taken into account. The results of Servat (18) and Novotny et al. (19) and the explanations for the accumulation and transport of contaminants they have inferred from those results can support our own results.

Novotny et al. (19) reported that accumulation of contaminants according to the time elapsed since the latest episode having generated runoff (D_{ts}) varies and may be positive or negative. Thus it appears that D_{ts} is not a key factor in the maximum load flush. Hemain (6) and Servat (19) have demonstrated that, for suspended solids, D_{ts} could be a minor factor in explaining the maximum concentrations measured. It can be expected then that two consecutive events can generate the same $\bar{C}_{max}$ concentrations even if the dry weather period D_{ts} is very short.

If on one hand, land use has influence on the annual loads produced in the comparison between residential type drainage catchments (6, 17), then on the other hand, some physical factors dependent on land use may explain the effectiveness of washoff (20). For example, the intensity of circulation could explain the composition of loads flushed into the river. If we attempt to reinforce these two explanatory hypotheses, we can draw some general conclusions from the studies of Novotny et al. (20), Servat (19) and Hemain (6):

- for catchments with exclusively residential land use, there may be variations in the volumes of the annual loads produced (BOD_5, COD, SS) according to the population density;

- automobile traffic is one determining factor in the accumulation of some contaminants, including heavy metals;

- surface wind drag (height < 10 m) may help to displace contaminants accumulated on street surfaces;

- maximum instant concentrations (C_{max}) are explained mainly by I_{max_5} for homogeneous residential catchments.

The validity of these general conclusions inferred from our observations has been verified. On the basis of our results, it appears logical to believe that for blocks of combined land use (residential, commercial, institutional and industrial), characteristic for urban areas excluding peripheries and suburbs, $\bar{C}_{max}$ values are of the same order for the same meteorological event. This has been particularly verified for large drainage catchments (> 100 ha) where compensating phenomena between land use types can help in grouping the $\bar{C}_{max}$ values. Major differences exist for various rain events. The variations observed in the $\bar{C}_{max}$ values (mg/l) are: SS, 80-1000; COD, 60-950; copper, 0.05-0.55; lead, 0.06-0.55; zinc, 0.14-1.33.

We have observed these great differences in $\bar{C}_{max}$ among rain events without elaborating on them. These differences between the loads washed off a catchment j during an event y_1 and those of an event y_2 have been studied by a number of authors. Hemain (6, 21) has brought forth the variables that can explain such fluctuations. The characteristics of the rainfall events studied, such as I_{max_5} (the maximum 5-minute intensity), I_{moy} (mean intensity), duration, the time elapsed since the preceeding runoff episode (D_{ts}), the total rainfall depth (h_t) were tested by Servat (19) and Desbordes and Servat (3) for the SS parameter.

I_{max_5} appears to explane some variations in the maximum concentrations in wet weather. However, the modeling of the maximum concentrations produced during an event y_i is not as effective as would be required. The statistical processing of the data collected during the recent sampling programs can lead, in the short term, to a greater reliability of interpretation (3).

Our preliminary approach, then, is to propose that managers translate the sensitivity of the environment (low, moderate and high) into $\bar{C}_{max}$ with an exceedance probability. Thus, for an environment described as sensitive to copper contamination, a manager will be able to evaluate the impact of CSO by considering the values exceeded only in 20% (1 - P(x) = 80%) of cases: 0.134 mg (Cu)/1.

5.0 Procedures for evaluating intervention requirements in wastewater quality improvement

A number of tools for evaluating the excess amounts of contaminants discharged into the receiving waters are available to managers. Starting with quality objectives or according to the consequences for the water quality in rivers, the tolerable amounts of sanitary wastewater are defined. Some adjustments must be made to these methods of evaluation in order to take into account non-point urban pollution. In defining intervention requirements, the water quality improvement programme of the government of Québec uses two methodologies for evaluating the impacts of a pollution source. For these methodologies, we will present only the basic principles.

5.1 Problems related to accumulation

The first method consists in evaluating the decreased dissolved oxygen level for some problem segments of rivers. One of the available tools, the QUAL-II model (22), can take into account the sediment oxygen demand to reduce DO in water. In several models of this type, the hypothesis of steady conditions makes the consideration of an intermittent source of contamination more difficult. It is possible to bypass this problem by an increase in sediment oxygen demand (SOD), when computing weekly pollutant loads due to CSO. As a result, the evaluation of the dissolved oxygen level in the problem segment is more realistic. Then, the loads which have to be treated (point and non-point sources) in order to maintain a minimum level of dissolved oxygen in the segment are identified.

5.2 Exceedance of water quality standards

The second evaluation method more widely used in Québec, because dissolved oxygen problems are not common to all river system, is based on maintaining the water quality standards in the segment immediately downstream from an overflow. From the empirical definition of a tolerative mixed zone, the manager must establish whether a source of pollution is likely to generate concentrations of contaminants higher than the maximum tolerable concentrations immediately downstream from the tolerative mixed zone. Such a zone has been defined, for the Québec region, as an area with a maximum length of 300 metres and a maximum width of 50 metres and covering 50% or less of the width of the receiving river(23). All these conditions must be met simultaneously. The evaluation therefore demands verification of the following inequality:

$$C_n - C_a - C_{rj} > 0 \tag{5}$$

where C_n is the quality standard to be kept, C_a is the mean concentration upstream from the overflow site, and C_{rj} is the concentration resulting from an overflow j immediately downstream from the tolerative mixed zone. Taking into account the dispersion of the wastewater flushed into the receiving environment, it is possible to evaluate Crj (13) as follows:

$$Crj = C_{zj}\ fd \tag{6}$$

where C_{zj} is the concentration of a contaminant z flushed by an outlet j and fd is the dilution rate evaluated at the end of the tolerative mixed zone. This dilution rate (Ci/C_o) corresponds to the ratio of the measured concentration of a substance $C_i(Z)$ 300 metres downstream from an overflow, and the concentration of the same substance in the wastewater before the mixing (C_0) which can be either measured or evaluated through modeling (13). The dilution rate depends on the flow rejected (Q_e) and on the morphological and hydrodynamic conditions of the receiving waters. For fixed morphological and hydrodynamic conditions of the receiving waters we have verified that fd varies in proportion to the variation of the outlet flow. According to equations (5) and (6), a decrease in C_{zj} or Q_e(increase of fd) would then make it possible to return to balanced conditions.

5.3. Application to the case of the tunnel in Ste-Foy

This example was selected to demonstrate the particular influence of CSOS on major rivers. The data in this example were acquired to evaluate solutions aimed at decreasing the amounts of pollutants discharged into a receiving river. For that purpose, the dispersion of the wastewater discharged at the outflow was also measured. The receiving waters near the overflow are rich in aquatic flora and fauna. In view of the environmental consequences of a shock load at this site, the environment is considered moderately sensitive. The water quality objectives for the site can be translated into criteria.

In case of copper, which is an element that has significant effects on aquatic life, there should be no more than 0.005 mg (Cu)/1300 metres downstream from the outflow, in order to ensure an aquatic life equilibrium. Since the concentration in the receiving waters upstream from the outflow is equal to 0.002 mg/l, the input to the receiving waters, diluted and dispersed, 300 metres downstream, should not exceed 0.003 mg (Cu)/l (C'_{rj}). Such an exercise can be repeated for any pollutant considered critical to the receiving environment.

The data necessary for defining the level of intervention have been identified as follows:

- upstream concentration (C_a), 0.002 mg (Cu)/l;

- quality standard (C_n), 0.005 mg (Cu)/l;

- minimum dilution rate is estimated 300 metres from the discharge site following equation (3) as .307 (sedimentation, also a cause of deterioration, is considered nonexistent for the purpose of evaluating intervention requirements);

- in view of the moderate risk level, the concentration at the outflow corresponds to a 50% exceedance probability: $\bar{C}_{max}$ is equal to 0.074 mg (Cu)/l.

By solving equation (6), it is established that the resulting concentration 300 metres from the discharge site is equal to 0.023 mg (Cu)/l. By introducing this value into equation (5), it is observed that the quality objective is not reached since the concentration, 300 m from the discharge site, is exceeded by 0.020 mg (Cu)/l. Even as far as 500 metres from the outflow, the objective concentration is exceeded in more than 60% of the transversal section. This means that excessive polluting loads have been discharged into the river.

The overload phenomena can be corrected by modulating the flow discharged at the outflow; as a consequence of this operation, the dilution is proportionally increased; the reduction of the $\bar{C}_{max}$ concentration by treatment reduction is also part of the process. The correction rate (W) is evaluated by:

$$\frac{C_{rj}}{C'_{rj}} = W \qquad (7)$$

In this particular example, W is equal to 7.67. Intervention aimed at decreasing the loads discharged by a 7.67 factor would maintain the balance of the environment either by retaining the volumes discharged or by decreasing concentrations at the outflow. It must be noted that, given the tolerable damage frequency, it has been established that exceedance of C'_{rj} could be tolerated for events with a frequency less than once per the summer season (from May 15 to September 15).

6. Conclusion

Intervention requirements for water quality improvement must be established for all point and non-point sources. As already identified by Field and Turkeltaub (24), environmental benefits from major investments in municipal wastewater treatment are very low without interventions for urban non-point sources. After reduction of point municipal and industrial discharges, the discrepancy between the dry and wet weather situations will be even greater.

The recent studies considering the reality of accumulation, washoff and transport of the contaminants present in CSO water allow better characterization of this type of wastewater and of its impact on the receiving environment. Finally, our research leads to the following conclusions:

- following accumulation of large amounts of contaminants, river sediments, downstream from CSO outlets, are ten to fifty times more contaminated than those upstream;

- in-stream pollutant concentrations downstream from outflow sites reach, during rain periods, the levels 2 to 7 times higher than the dry weather levels;

- the period of recovery to initial conditions in the cases of the St. Charles and St. Lawrence Rivers stretched to 48 hours after the rainfall (physiological recovery of biological organisms should be even longer);

- classification by increasing values of mean weighted maximum concentrations ($\bar{C}_{max}$) for the SS and COD parameters does not bring forward a difference according to land use, in the case of large drainage catchments.

Our approach aims to provide tools and methods for an immediate integration of CSOS with the sources "accounted for" by water quality managers. We are convinced that the development of more effective tools for forecasting the loads discharged will be achieved; but it is imperative that in the meantime develope a methodological approach to evaluate their impact.

High-frequency rain events are a major source of problems for the receiving waters. But even if major events are responsible for more than one third of the annual loads discharged into the river, the more frequent episodes are responsible for drastic changes in water quality. The time required to return to initial balance conditions, being longer than the period between two runoff episodes, supports the idea of persistent deterioration between rain events. Moreover, because of their number, they are, according to other studies, responsible for more than 60% of the annual loads produced by stormwater runoff.

The planned interventions, however, are more realistic if two aspects mentioned in this study are considered and used as a basis for urban discharge controls:

- overflow loads are more easily settleable than the sanitary wastewater;

- the runoff volumes for frequent events are smaller than those from exceptional events.

On such a basis, storage, or treatment facilities specially designed for decreasing the loads discharged to rivers in wet weather will be smaller and will be used more effectively for all events. The control of exceptional events and the economic consequences of erecting large works, rarely used to full capacity, should not be considered as short term priorities.

Remarks

The methodology, results and discussion included in this paper were elaborated after Pierre Lavallée's, Ph. D. Thesis.

Acknowledgments

The authors thank Quebec City and Ste-Foy City authorities for allowing the presentation of some results obtained during studies of their sewer network.

References

1 LAVALLÉE, P. and P. LESSARD. Importance de la contamination des débordements de réseau unitaire. Sciences et Techniques de l'eau, 17(4): 353-357: 1984.

2 FIELD, R.. CSO-Techniques in the U.S.A. State of the art. Proceedings of the Third International Conference on Urban Storm Drainage, Goteborg, Sweden, 4: 1333-1356: 1984.

3 DESBORDES, M. and E. SERVAT, Solids in urban runoff. Statistical Analysis of French Experimental Data. Proceedings of the Third International Conference on Urban Storm Drainage, Goteborg, Sweden, 3: 947-957: 1984.

4 GEIGER, W. Characteristics of combined sewer runoff. Proceedings of the Third International Conference on Urban Storm Drainge, Goteborg, Sweden, 3: 851-860: 1984.

5 LESSARD, P. and P. LAVALLÉE. Combined sewer overflows: The Québec city experience. Proceedings of the Third International Conference on Urban Storm Drainage, Goteborg, Sweden, 3: 831-840: 1984.

6 HEMAIN, J.C. Mesure de la pollution du ruissellement pluvial urbain. Laboratoire d'hydrologie mathématique, Université des sciences et techniques du Languedoc, Montpellier, 3: 74: 1983.

7 LAVALLÉE, P., LESSARD, P. and J.P. VILLENEUVE. Water quality variations in running waters due to combined sewer overflowing. Estimations of negative influence. Proceedings of the Third International Conference on Urban Storm Drainage, Goteborg, Sweden, 3: 761-769: 1984.

8 KLEMETSON, S.L. Factors affecting stream transport of combined sewer overflow sediments. Journal of Water Pollution Control Federation, 57(5): 390-397: 1985.

9 KRENKEL, P.A. and V. NOVOTNY. River water quality model construction. In: Modelling of rivers, HSIEH Winshen ed., John Wiley and Sons, Toronto: 1979.

10 LAVALLÉE, P., COUTURE, P. and J.C. AUCLAIR. Approche conceptuelle à l'aide de bio-essais in situ pour améliorer la fiabilité de l'interprétation écotoxicologique de variables physico- chimiques dans le cas de problème en assainissement de l'eau. Sciences et techniques de l'eau, 52(3): 261-265: 1984.

11 LAVALLÉE, P. Les impacts de la pollution diffuse urbaine. M.Sc. Thesis, INRS-Eau, 108 p.: 1980.

12 SHILLINGER, J.E. and J.J. GANNON. Bacterial adsorbtion and suspended particles in urban stormwater. Journal of Water Pollution Control Federation, 57(5): 384-389: 1985.

13 LAVALLÉE, P., SHOIRY, P. and J.P. VILLENEUVE. Propagation longitudinale et latérale des eaux usées déversées en cours d'eau. Sciences et techniques de l'eau, 17(4): 361-366: 1984.

14 COUTURE, P. Contribution des bio-essais avec algues pour l'étude des impacts environnementaux en eaux douces. Doctorat 3° cycle Thesis, Université Paul-Sabatier, Toulouse, 128 p: 1983.

15 THOMAS, W.H. and D.L.P. SEIBERT. Effects of copper on the dominance and the diversity of algae: controlled ecosystem pollution experiment. Bull. Mar. Sc., 27: 23-33: 1977.

16 LESSARD, P. and P. LAVALLÉE. Caractérisation des eaux usées de débordements de réseau unitaire (accepted for publication Can. Journal of Civil Engineering, 27 p.): 1985.

17 DEUTSCH, J.C. and J.C. HEMAIN. Main results of the French Quality Measurement. Proceedings of the Third International Conference on Urban Storm Drainage, Goteborg, Sweden, 3: 939-946: 1984.

18 MARSALEK, J. Caractérisation du ruissellement de surface issu d'une zone urbaine commerciale. Sciences et Techniques de l'eau, 17(2): 163-167: 1984.

19 SERVAT, E. Contribution à l'étude des matières en suspension du ruissellement pluvial à l'échelle d'un petit bassin versant urbain. Docteur- Ingénieur Thesis, Laboratoire d'hydrologie mathématique, Université des sciences et techniques du Languedoc, Montpellier, 182 p., 7 annexes: 1984.

20 NOVOTNY, V., SUNG, H.M., BANNERMAN, R. and K. BARM. Estimating non point poolution from small urban watersheds. Journal of Water Pollution Control Federation, 57(4): 339 et ss: 1984.

21 HEMAIN, J.C. Approche méthodologique pour l'étude de la pollution du ruissellement pluvial urbain. Docteur-Ingénieur Thesis, Laboratoire d'hydrologie mathématique, Université des sciences et techniques du Languedoc, Montpellier, 128 p., 2 annexes: 1980.

22 ROESNER, L.A., MONSER, J.P. and D.E. EVENSON. Computer program documentation for the stream quality model QUAL-II. EPA, System Development Branch: 1977.

23 GOUIN, D. La détermination des objectifs de traitement dans un projet d'assainissement. Sciences et Techniques de l'eau, 17(4): 383-388: 1984.

24 FIELD, R. et R. TURKELTAUB. Urban runoff receiving water impacts: Program overview. J. Env. Eng. Div. Proc., American Society of Civil Engineering, 107(EE1): 1981.

DISCUSSION - CONVENTIONAL POLLUTANT IMPACTS ON RECEIVING WATERS

Dr. Hvitved-Jacobsen was prevented from attending due to illness, and his paper was summarized by Mr. Barnwell. In the discussion which followed the presentation of the papers, the following points were raised:

1. There was considerable discussion of the 8-16 year DO-based water quality criteria outlined by Dr. Hvitved-Jacobsen. Mr. van de Ven noted that available time series data on DO may not be adequate to establish such a standard. Mr. B. N. Jacobsen stated that for now it is more important to enforce the principle outlined in Dr. Hvitved-Jacobsen's criteria regardless of whether background data were available. Dr. Lijklema pointed out that bacterial effects occur over much longer time intervals, and viral effects even longer when considering survival in sediments.

2. Mr. B. N. Jacobsen asked Mr. Lavallee whether it was possible, based on his studies, to describe the relationship between sediment oxygen demand and combined sewer overflows, noting that this demand is very important in modeling studies. Mr. Lavallee replied that this is very difficult to do, because of the problem of disturbance of existing deposits. He also stated that they did, however, have further work going on in this area.

3. Dr. Patry asked Mr. Lavallee about whether or not he had taken into account the scour of bottom deposits in the river. Mr. Lavallee replied that these were not a significant factor.

4. Mr. B. N. Jacobsen asked if it was possible to distinguish between sediment oxygen demand and the oxygen demand from combined sewage. Mr. Lavallee responded that their design was based on model studies which did not consider sediment oxygen demand.

5. Dr. James stated that in Toronto, the approach to relief of combined sewer overflows has been to concentrate on block-by-block analysis of their system to provide local relief, and to exclude storm water from the combined sewer system. They are also, in some cases, considering storing sanitary sewage during storm events so that combined system, in effect, becomes a storm sewer. Mr. Lavallee responded that it is of course best to control pollution at the source, but that this is not now possible.

6. Dr. Gonenc asked how the concentrations of pollutants in the stream could be higher near the stream banks, and Mr. Lavallee replied that this is due to the restriction of in-stream diffusion by the river bank.

7. Mr. Jacquet asked why Quebec City was chosen for the study. Mr. Lavallee replied that this was the only major city in Quebec with complete interception of its combined sewage.

8. Dr. Medina asked whether or not Quebec City's system was operational. The reply was that it was partially operational (one treatment device out of the nine that are planned).

9. Mr. Weatherbe asked about the data which indicated that the worst case for pollution in their study occurred on the flood tide. Mr. Lavallee responded that they were unable to collect data on the effects during tidal ebb because no storms occurred during those periods.

TOXIC POLLUTANTS: IMPACT AND FATE IN RECEIVING WATERS

Raymond A. Ferrara
Department of Civil Engineering
Lafayette College
Easton, Pennsylvania 18042
USA

Introduction

Evaluation of the impacts to receiving water quality as a consequence of uncontrolled releases of toxic pollutants to the environment is a relatively new science having formally developed within the last decade or two. The magnitude and severity of the problem have, however, spawned a concentrated effort resulting in significant advances in the state of our technology for evaluating these impacts. Such an evaluation relies upon consideration of conservation of mass in one form or another, whether that be a mathematical model (i.e. a set of equations) or a physical model (i.e. a laboratory or field scale experiment). The distribution and fate of a toxic chemical in the aquatic environment is controlled by two broadly defined processes, namely transport or those processes which tend to move a toxic chemical between neighboring regions of space, and transformation or those processes which tend to chemically, physically, or biologically alter the form of a toxic chemical. Typical transport processes include advection and diffusion, whereas typical transformation processes include biodegradation and photolysis among others. The various transport and transformation processes which determine the distribution and fate of toxic chemicals and

NATO ASI Series, Vol. G10
Urban Runoff Pollution.
Edited by H.C. Torno, J. Marsalek, and M. Desbordes

and hence their impact in receiving waters, are reviewed later in this paper. The principles of conservation of mass are utilized to provide a general framework within which these processes are integrated. However, first it is important to discuss the characteristics of toxic pollutants which distinguish them from the more conventional measures of pollution such as biochemical oxygen demand, phosphorus, coliform bacteria and others.

Characteristics of Toxic Pollutants

In a more general sense, "toxic" pollutants are actually a subset of the class of "hazardous" pollutants. A chemical is considered hazardous if it can be identified as one or more of the following - ignitable, corrosive, reactive, toxic. All four of these characteristics are likely to result in acute, immediate impacts to receiving waters when contaminants are released to the environment in sufficient quantities. Those hazardous substances which are toxic may also cause chronic impacts due to long-term low level exposure. Classically questions of toxicity have indeed dealt with evaluation of the immediate lethal impact associated with exposure to a chemical or some other toxic agent. However, the need for toxicity evaluation of the sub-lethal nature of many chemicals in the environment has been recognized. Carcinogenic, mutagenic, and teratogenic impacts throughout all levels of the biosphere have been attributed to low level exposure to toxic chemicals.

The characteristics of toxic pollutants are best illustrated by comparison with conventional pollutants (17). The most glaring difference is the sheer magnitude in numbers. Classically problems of receiving water quality have had to deal with only a limited number of natural pollutants such as biochemical oxygen demand, total phosphorus, and total dissolved solids. Manufactured toxic pollutants number in the thousands with many more being synthesized daily. Adverse water quality impacts for conventional pollutants become manifest at concentrations on the order of parts per million (1 ppm = 1 mg/l), whereas for toxics, impacts occur at concentrations on the order of parts per billion (1 ppb = 1 μg/l) or parts per trillion (1 ppt = 1 ng/l). Hence, the pollutant loading becomes significant possibly six orders of magnitude sooner for toxics. Transformation processes and reaction rates for conventional pollutants are fairly well defined, fairly rapid for those that do react, and generally produce harmless end-products. Reaction rates for many toxic pollutants are poorly defined and relatively slow. Furthermore, the end product of a toxic pollutant's degradation may be more harmful than the original pollutant. Finally, many of the toxic pollutants are adsorbed to solids in the water column and accumulate in the sediments of waterbodies. As a result their residence time in a system may be much larger than for conventional pollutants. Certain properties of these toxics also result in bioaccumulation as discussed below, a problem which does not exist with conventional pollutants.

Appendix I provides a list of the 129 Priority Toxic Water Pollutants prepared by the United States Environmental Protection Agency (USEPA). Toxicity may be related to either acute or chronic effects to the indigenous biota or to humans who ingest contaminated water or organisms residing within contaminated water. The water quality criteria reported in Appendix I have been abstracted from the USEPA Treatability Manual (29), an exhaustive five volume reference developed to provide data and information on treatability of industrial waste streams. Other sources (17,30) exist which may be used to supplement the information provided in Appendix I. The criteria provided are divided into three main categories -

Freshwater Aquatic Life, Saltwater Aquatic Life, and Human Health.

Criteria for Human Health Impacts

The criteria for human health consist of two sub-categories -- (1) the water quality criteria identified for the protection of human health from ingestion of contaminated water and contaminated aquatic organisms, and (2) the water quality criteria identified for the protection of human health from ingestion of contaminated aquatic organisms only. For any individual chemical, the criterion for the latter is always greater than or equal to the criterion for the former. For certain chemicals, the human health criteria in each of the sub-categories are identified in terms of a single value for the protection of human health from the "Toxic" properties of the chemical. One "Toxic" value is identified for ingestion of contaminated water and organisms, and another for ingestion of contaminated organisms only. For other chemicals, to provide maximum protection of human health from potential carcinogenic effects, the ambient water concentration and hence the criterion should be zero (based on the non-threshold assumption for the chemical). However, since a zero level may not be attainable at the present time, water concentration levels which may result in incremental increases of cancer risk over a human's lifetime are estimated. Criteria are identified for incremental cancer risks of 10^{-5} (i.e. one in one hundred thousand), 10^{-6} (i.e. one in one million) and 10^{-7} (i.e. one in ten million). Note that for any particular chemical, the criteria state that an order of magnitude decrease in cancer risk corresponds to an order of magnitude decrease in the water quality criteria, or conversely cancer risk is linearly related to water concentration according to the following:

$$CR = \alpha C \tag{1}$$

where CR = cancer risk

α = a constant

C = water concentration

Any consistent set of units can be used in the above equation. When considering a particular chemical, α for ingestion of contaminated organisms only will be less than or equal to α for ingestion of contaminated water and organisms. It is interesting to note that for many of the pesticides (e.g. Toxaphene), the criteria for corresponding cancer risks do not differ much whether considering ingestion of water and organisms or organisms alone, whereas for many of the other chemicals there is generally a substantial difference.

For certain chemicals, the USEPA has promulgated standards for public potable water supplies. These standards are reported in Table 1 as maximum contaminant levels (MCL), and are promulgated for protection of the public health (31). Although not a list of "toxic" pollutants per se, most of the contaminants in Table 1 can also be found in Appendix I, and the impacts of these chemicals to human health are indeed toxic. It is beyond the scope of this paper to review the toxic effects of all of the chemicals listed in Appendix I and Table 1. Those effects include impairment of internal organ functions, nervous system disorders, blood cell malfunction, and others in addition to suspected carcinogenic, teratogenic and mutagenic effects. Consequently, one impact to receiving waters as a result of the presence of toxic chemicals is the limitation of the waterbody's use as a potable water supply.

Table 1. USEPA Primary Drinking Water Standards (31)

Contaminant		MCL (mg/l)
Inorganics:		
Arsenic		0.05
Barium		1.00
Cadmium		0.010
Chromium		0.05
Lead		0.05
Mercury		0.002
Nitrate (as N)		10.00
Selenium		0.01
Silver		0.05
Fluoride		
Annual average of maximum daily air temperatures of community in which water system is situated		
^{o}F	^{o}C	
53.7 and below	12.0 and below	2.4
53.8-58.3	12.1-14.6	2.2
58.4-63.8	14.7-17.6	2.0
63.9-70.6	17.7-21.4	1.8
70.7-79.2	21.5-26.2	1.6
79.3-90.5	26.2-32.5	1.4
Organics:		
Chlorinated hydrocarbons:		
Endrin (1,2,3,4,10,10-hexachloro-6,7-epoxy-1,4,4a,5,6,7,8,8a-octo-hydro-1,4-endo,endo-5,8-dimethanonaphthalene)		0.0002
Lindane (1,2,3,4,5,6-hexachlorocyclohexane, gamma isomer)		0.004
Methoxychlor (1,1,1-trichloro-2,2-bis {p-methoxyphenyl}ethane)		0.1
Toxaphene ($C_{10}H_{10}Cl_8$-technical chlorinated camphene, 67-69% chlorine)		0.005
Chlorophenoxys:		
2,4-D (2,4-dichlorophenoxyacetic acid)		0.1
2,4,5-TP silvex (2,4,5-trichlorophenoxypropionic acid)		0.01
Total Trihalomethanes		0.10

Criteria for Aquatic Life Effects

The impacts to receiving waters are of course not limited to human health effects from direct consumption of contaminated water. As noted above, consumption of aquatic organisms residing in contaminated waters is also responsible for human health impacts. The process is simply the last step in a sequence of transfers of contaminants from prey to predator throughout the biosphere. The original source of the contaminant for the food chain is the contaminant's presence in the waterbody enabling movement of a chemical into an organism from ingestion of contaminated water. The term bioaccumulation more broadly refers to the accumulation of a chemical within an organism due to the ingestion of contaminated food as well as water. The degree of bioaccumulation varies from one organism to the next and may result in the phenomenon of biomagnification, namely the sequential increase in concentration of a chemical within organisms in going sequentially higher in the food web. The concentration of a chemical within an aquatic organism is often expressed in relation to the chemical's concentration in the surrounding water by means of the bioconcentration factor, β.

$$\beta_i = \frac{C_{oi}}{C_w} \qquad (2)$$

where β_i = bioconcentration factor of organism i

C_{oi} = concentration of chemical in organism i (e.g. μg chemical per gram of organism)

C_w = concentration of chemical in surrounding water (e.g. mg chemical per liter water)

The bioconcentration factor can be estimated by an equation of the following form (19,23).

$$\log \beta = a \log K_{ow} + b \tag{3}$$

where K_{ow} = the octanol-water partition coefficient,
a,b = constants

As evidence of the phenomenon of biomagnification, it has been generally noted that the bioconcentration factor increases with organism size, which in itself is a relative (albeit crude) indication of an organism's position in the food web (23,27).

Biomagnification as an impact resulting from the presence of toxic pollutants in receiving waters is a manifestation of a complex set of interacting biochemical and physical processes. Three characteristics functionally related to an organism's position in the food web may be responsible.

1. Metabolic rate is generally inversely proportional to organism size and weight (27). Similarly organism's rates of excretion of bioaccumulated chemicals decreases with organism size and weight. Therefore, larger organisms, namely those at the top of the food chain will cleanse themselves of ingested chemicals at a slower rate than smaller organisms, namely those at the bottom of the food chain.

2. Biomass density is generally inversely proportional to position in the food chain (19,27). Hence there is less biomass within which a chemical is concentrated at higher trophic levels.

3. Many of the toxic organic chemicals of concern are lipophilic (fat soluble) (4,19). Lipid content in organisms generally increases with organism size and age, hence making organisms at higher levels in the food chain more susceptible to bioaccumulation.

A most recent publication (16) reports a review of the literature on biomagnification of contaminants through successive trophic levels involving gill-breathing aquatic animals. The study concludes that (1) heavy metals with the exception of methylmercury do not biomagnify greatly in either freshwater or marine food webs although feeding is an important source for bioaccumulation of many toxic heavy metals, and (2) organic compounds which appear to have significant potential for biomagnification include polychlorinated biphenyls (PCBs), benzo(a)pyrene, naphthalenes, and possibly a few organochlorine insecticides such as dieldrin, endrin, kepone, and mirex.

It appears clear there is a great potential for bioaccumulation of toxic pollutants by aquatic organisms. The noted biological build-up in concentration of chemicals such as PCBs is due to a set of complex phenomena including direct uptake from the water, excretion rate, rate of transfer through the food chain, and decrease of biomass density with organism size. Thomann (27) concludes that for Lake Ontario salmon about one-third of the observed PCB concentration in the fish is due to transfer from lower levels in the food chain and about two-thirds is due to direct uptake from the water. Studies by Reinert (24) and Chadwick and Brocksen (5) conclude that the amount of dieldrin bioaccumulated in an aquatic ecosystem was directly proportional to the concentration of dieldrin in the water, and that not more than 16% of the dieldrin accumulated by the fish could have come from the fish. Neely (19) points out that transfer of chemicals through the food chain alone becomes an important link in biomagnification when an environmental interface is crossed. For example, birds that feed on fish must bioaccumulate directly through the food. For terrestrial animals, it is transfer through the food chain which is directly responsible for biomagnification.

The answer to the question of whether biomagnification occurs for many toxic chemicals in aquatic systems depends on the semantic definition of the term. A strict definition of biomagnification relies on a sequential increase in chemical concentration within organisms via transfer through the food chain. This clearly occurs in terrestrial ecosystems where food chains are more linear and highly structured than in aquatic systems, and where direct uptake of chemicals through the external medium (i.e. the physical environment) is minimal (16,19). This definition of biomagnification may also be termed food-chain concentration (20). In aquatic ecosystems, direct uptake from the external medium (i.e. the water) is significant as pointed out above, and direct transfer of chemicals through the food chain (i.e. the strict definition of biomagnification) plays a lesser role. The sequential increase of chemicals in going from one level in the food chain to another is apparent in aquatic ecosystems (Table 2). However, it is essential to realize the pathways responsible for that observed phenomenon.

Criteria for the Freshwater and Saltwater Aquatic Life in Appendix I are defined for acute and chronic effects. The criteria for acute effects represent maximum concentrations which should not be exceeded at any time in order to prevent the more immediate (i.e. acute) toxic effects to freshwater life associated with the presence of the various chemicals. The criteria for chronic effects represent temporally averaged (e.g. 24-hour average) concentrations which should not be exceeded in order to prevent those toxic effects which become manifest gradually over time. Note that as expected for any individual pollutant, the criteria for chronic effects are always less than the analogous criteria for acute effects. Criteria for all of the priority pollutants are not available. However, more values for acute than for chronic effects have been identified, likely a consequence of the increased experimental effort and uncertainty

Table 2. PCB in Lake Ontario Ecosystem
(data abstracted from Reference 27)

Ecosystem Component	PCB Concentration (ug PCB/gm wet weight organism) Average	Standard Deviation	Assumed Range of Organism Size (μm)
Net Plankton (64 μm mesh)	0.72	0.35	$10^2 \rightarrow 10^4$
Alewife, Smelt Sculpin	3.24	2.14	$10^4 \rightarrow 2.5 \times 10^5$
Coho Salmon	6.47	2.85	$2.5 \times 10^5 \rightarrow 10^6$

Table 3. Octanol-Water Partition Coefficient for Selected Toxic Chemicals
(data obtained from reference 29)

Chemical	log K_{OW}
Bis(chloromethyl) ether	-0.38
Bis(2-chloroethoxy) methane	1.26
Dimethyl pthalate	2.12
Benzidine	1.81
Pentachlorophenol	5.01
4-Nitrophenol	1.91
Benzene	2.13
Nitrobenzene	1.85
Toluene	2.69
Hexachlorobenzene	6.18
Naphthalene	3.37
PCB 1016	4.38
PCB 1260	7.14 (estimated)
Chloroform	1.97
Carbon Tetrachloride	2.64
Trichloroethylene	2.29
Tetrachloroethylene	2.88
Chlordane	2.78
Aldrin	5.17
DDT	3.98
DDE	5.69

arising from the need to extrapolate results from short term experiments to identify chronic toxicity.

Distribution of Toxic Pollutants in Receiving Waters

Analysis of the impacts of toxic pollutants in aquatic systems ultimately requires some method for describing their distribution throughout the water environment. Once accomplished, the results of this analysis may be compared with the criteria presented above, possibly calling for additional analysis to evaluate management strategies for altering the toxic pollutant's presence in terms of decreasing concentration and/or spatial distribution. Such analyses utilize (whether rigorously or not) the principle of conservation of mass which may be expressed simply as follows:

$$\frac{\Delta M}{\Delta t} = I_M - O_M + R_M + S_M \tag{4}$$

where $\frac{\Delta M}{\Delta t}$ = the change in pollutant mass in the subsystem per unit time,

I_M = the inflow of pollutant mass to the subsystem from adjacent regions of the system,

O_M = the outflow of pollutant mass from the subsystem to adjacent regions of the system,

R_M = the rate of production of pollutant mass within the subsystem due to biological or chemical reactions, and

S_M = a supply of pollutant mass to the subsystem from sources external to the system.

The terms I_M and O_M are functions of the hydrodynamics of the system. They are transport processes each of which are comprised of advection and the generally combined diffusion/dispersion mixing term. Advection describes the transport of pollutant mass in terms of the movement of the carrier fluid. Thus in aquatic systems, the advected rate of movement of the pollutant mass is the same as the rate of water movement. The diffusion/dispersion term combines molecular diffusion, turbulent diffusion and dispersion, all of which are normally expressed as proportional to the local concentration gradient transporting the pollutant in the direction of decreasing concentration. These processes are common to all pollutants, conventional and toxic, and are treated extensively in many references (e.g. 6,7,10, 21,28). It is not the objective of this paper to reiterate that literature.

The term S_M accounts for other transport processes which provide an external source (or sink) of pollutant mass for the system. Point and nonpoint sources of pollution are good examples of S_M. Such sources often add a pollutant with flow into the system and are of course particular to each system. However, there are other processes included in S_M which do not add flow to the system. These include volatilization, namely the transfer of a pollutant in gaseous form across the air-water interface, and sedimentation, namely the transfer of a pollutant in particulate form across the sediment-water interface. Both of these terms are important in determining the aquatic distribution and fate of toxic chemicals, and will be demonstrated below. Figure 1 provides a sample representation of the interacting transport processes I_M, O_M, and S_M which determine the distribution of toxic pollutants in receiving waters.

The last term in equation 4, R_M, encompasses several reaction processes including adsorption and desorption, biodegradation, hydrolysis, and photolysis. For some toxic chemicals which dissociate, application of the principles of acid-base equilibria are necessary to identify the distribution of the contaminant between its ionic and neutral form.

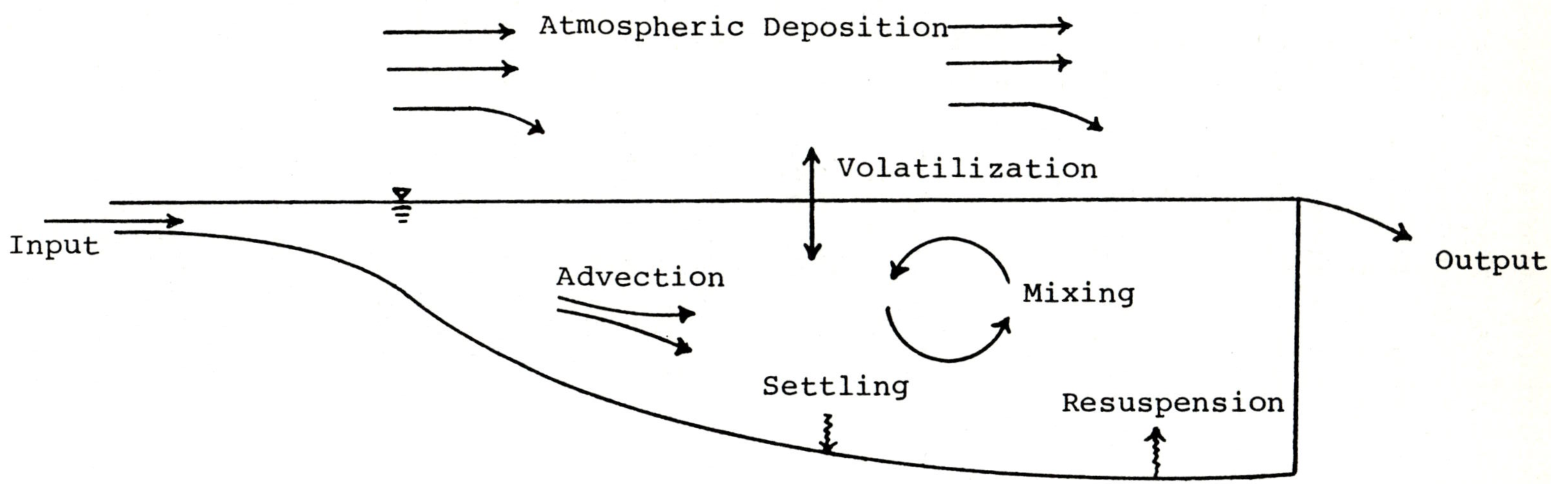

Figure 1. Transport Processes in an Aquatic System

These terms are described in detail below particularly with respect to the toxic organic chemicals. For notational convenience,

$$R_M = r_M V \quad (5)$$

$$S_M = s_M V \quad (6)$$

where V = volume of the subsystem. Hence r_M and s_M represent rate of change in concentration ($ML^{-3}T^{-1}$) rather than rate of change of mass (MT^{-1}) as denoted by the corresponding R_M and S_M. Designation for units are M=mass, L=length, T=time. For example ($ML^{-3}T^{-1}$) designates mass per volume per time, or concentration per time.

Sorption

Sorption in the context of this paper refers to chemical accumulation on the surface of a solid for example a soil particle. In many cases, the sorption process is rapid, and equilibrium between adsorbed and dissolved phases of a chemical may be assumed. Although some evidence regarding the phenomena of kinetics and hysteresis in sorption has been reported (8,13,33), a nonlinear singular sorption isotherm is often assumed.

$$X = K_f \, c^N \quad (7)$$

where X = mass contaminant adsorbed per mass solid, (MM^{-1}),

K_f = a constant,

c = dissolved contaminant concentration, (ML^{-3})

N = a constant.

For N=1, equation (7) becomes linear and K_f is referred to as K_p the partition coefficient. Values for K_p vary widely and are a function of soil type, organic carbon content, and chemical contaminant (8,13,14,15,17).

$$K_p = K_{oc} X_{oc} \tag{8}$$

where K_{oc} = K_p expressed on an organic carbon basis,

X_{oc} = organic carbon content in solids (mass fraction).

Where the particle size distribution of the solids is known

$$K_p = K_{oc} [0.2 (1-f) X^s_{oc} + f X^f_{oc}] \tag{9}$$

where f = mass fraction of fine sediments (diameter <50 μm),

X^s_{oc} = X_{oc} of coarse sediments,

X^f_{oc} = X_{oc} of fine sediments,

K_{oc} is estimated from the experimentally determined octanol-water partition coefficient, K_{ow}

$$\log K_{oc} = a \log K_{ow} + b \tag{10}$$

Note the similarity between equations (3) and (10). Values for K_{ow} for various toxic chemicals are provided in Table 3.

Under the assumption of linear sorption

$$c_T = X S + c \tag{11}$$

$$F_d = \frac{1}{1+K_p S} \tag{12}$$

$$F_a = \frac{K_p S}{K_p S+1} \tag{13}$$

$$F_a + F_d = 1 \tag{14}$$

where c_T = total (adsorbed + dissolved) contaminant concentration, (ML^{-3})

S = solids concentration, (ML^{-3})

F_a = fraction of the total contaminant in the adsorbed state,

F_d = fraction of the total contaminant in the dissolved state.

These equations may be used to distribute a contaminant between its adsorbed and dissolved states, which has implications for other reaction and transport terms.

Biodegradation

Biodegradation of toxic organic chemicals can occur via either of two mechanisms - (1) metabolism wherein organisms utilize the contaminant as a food source, and (2) cometabolism or detoxification wherein organisms are provided with another substrate as a food source but also degrade a contaminant effectively detoxifying the environment (1,2,3,22,25,26). For metabolism

$$r_M = -\frac{\mu_m}{y} B \frac{c}{K_c + c} \theta^{(T-20)} \tag{15}$$

and for co-metabolism

$$r_M = -K_B \, B \, c \frac{Q}{K_Q + Q} \theta^{(T-20)} \tag{16}$$

where μ_m = maximum growth rate constant, (T^{-1}),

y = yield coefficient, mass of B produced per mass c consumed, (MM^{-1}),

K_B = bacterial degradation rate coefficient, $(L^3M^{-1}T^{-1})$,

B = bacterial concentration, (ML^{-3}),

K_c = a constant representing c limitation, (ML^{-3}),

Q = limiting substrate concentration, (ML^{-3}),

K_Q = a constant representing Q limitation, (ML^{-3}),

θ = a constant representing a temperature factor, and

T = temperature, (^{o}C).

At any given temperature, both equations (15) and (16) may be reduced to first order reactions of the form

$$r_M = K_1 \; c \tag{17}$$

where K_1 = first order reaction rate coefficient, (T^{-1}). This is true for metabolism if (1) B is approximately constant, and (2) c is much less than K_c. For co-metabolism, equation (17) is appropriate if (1) B is approximately constant, and (2) Q is approximately constant or much greater than K_Q. Values for K_1 range from 10^{-1} to 10^{-7} day^{-1} for toxic organic chemicals (17).

Equations 15, 16, and 17 are written for a dissolved chemical. Many organic chemicals are adsorbed to solid surfaces. Equations analogous to these may also be written for the particulate forms of toxic pollutants. Whether or not chemicals in the adsorbed state are available for biodegradation is under debate (17).

Hydrolysis

Hydrolysis is the process of contaminant transformation due to reactions with water. Three types are possible - neutral, acid catalyzed, and base catalyzed (17,32). The process is expressed mathematically as follows:

$$r_M = - K_Y \; c \tag{18}$$

where $K_Y = K_n + K_a[H^+] + K_b[OH^-]$, (19)

K_n = neutral hydrolysis rate coefficient, (T^{-1}),

K_a = acid catalyzed hydrolysis rate coefficient, $(L^3M^{-1}T^{-1})$,

K_b = base catalyzed hydrolysis rate coefficient, $(L^3M^{-1}T^{-1})$,

$[H^+]$ = molar concentration of hydrogen ion, (ML^{-3}), and

$[OH^-]$ = molar concentration of hydroxide ion, (ML^{-3}).

Chemicals in an adsorbed state may be protected from acid or base catalyzed hydrolysis. Values for K_Y range from 10^{-1} to 10^{-8} day^{-1} for toxic organic chemicals (17).

Photolysis

Photolysis is the process whereby chemicals are broken down due to the absorption of the radiant energy of light. Two pathways are possible -- (1) direct photolysis wherein light is absorbed directly by the chemical, and (2) indirect (sensitized) photolysis wherein light is absorbed by intermediary compounds (toxic or non-toxic) which transfer energy to the toxic contaminants. The reaction is expressed as

$$r_M = -K_d\,c - K_s\,c = -K_{pT}\,c \tag{20}$$

where K_d = direct photolysis rate coefficient, (T^{-1}),

K_s = sensitized photolysis rate coefficient, (T^{-1}),

K_{pT} = total photolysis rate coefficient, (T^{-1}),

$= K_d + K_s$

Methods for estimating values for the photolysis rate coefficients are provided in Mills et. al. (17) largely based on the work of Zepp and co-workers (34).

Volatilization

Volatilization refers to the transfer of contaminents across the air-water interface. Henry's Law is used to express equilibrium between the gas and water phases of a chemical.

$$c_u = \frac{K_H}{R_u T} c' \tag{21}$$

where c_u = concentration of gas phase in the atmosphere, (ML^{-3}),
c' = concentration of ionically neutral form of c, (ML^{-3}),
K_H = Henry's Law constant, (atm L^3M^{-1}),
R_u = universal gas constant, (L^3 atm M^{-1} $^oK^{-1}$),
T = temperature, (^oK).

Transport of the chemical across the system boundary is proportional to the difference between the actual concentration in the water phase and the equilibrium value.

$$s_M = -\frac{K_v}{z} \left(c' - \frac{c_u R_u T}{K_H}\right) \tag{22}$$

where K_v = volatilization rate coefficient, (LT^{-1}),
z = mixed depth of waterbody, (L).

For most toxic chemicals, c_u is approximately zero. Hence,

$$s_M = -\frac{K_v}{z} c' \tag{23}$$

Application of the two-film theory of volatilization and Henry's Law gives (6,17)

$$\frac{1}{K_v} = \frac{R_u T}{K_H k_g} + \frac{1}{k_l} \quad (24)$$

where k_g = mass transfer coefficient in the gas phase across the gas film, (LT^{-1}),

k_l = mass transfer coefficient in the liquid phase across the liquid film, (LT^{-1}).

For highly soluble chemicals K_H is relatively small and $k_l >> k_g$. Then

$$K_v \simeq \frac{K_H k_g}{R_u T} \quad (25)$$

For slightly soluble chemicals, K_H is relatively large and $k_g >> k_l$ then

$$K_v \simeq k_l \quad (26)$$

Typical values for k_l are 20 cm/hr with a range of 0.4 - 40 cm/hr, and for k_g, 3000 cm/hr with a range of 400 - 4000 cm/hr. Ratios of k_l/k_g range from 0.001 (high wind mixing) to 0.01 (low wind mixing) in lakes (6).

Settling

The removal of toxic chemicals in particulate form from the water column due to settling and subsequent incorporation into a waterbody's sediments is represented by a settling velocity, v_s. In any vertically mixed section of an aquatic system, the mass flux due to settling over a given time interval Δt is

$$J_s \, \Delta t = - v_s \, \Delta t \, c_p \quad (27)$$

where J_s = rate of mass flux due to settling, $(ML^{-2}T^{-1})$,

v_s = particle settling velocity, (LT^{-1}),

c_p = concentration of particulate toxic chemicals with settling velocity v_s, (ML^{-3}).

The term $(v_s\ \Delta t)$ represents the depth of particles which would have left the section during Δt. Dividing equation (27) by Δt and the depth of the section, z, gives

$$\frac{J_s}{z} = s_M = -\frac{v_s}{z}\ c_p \tag{28}$$

For silts and sand v_s may range from 5 to greater than 700 meters per day.

Acid-Base Equilibria

The consequence of chemical ionization may be significant in that it will determine the fraction of the total contaminant mass available for other s_M or r_M processes. A common example is volatilization where only the neutral form, and not the ionized form is capable of transport across the air-water interface. Since ionization reactions are rapid, it is appropriate to assume equilibrium at any time. For a monoprotic organic acid, the appropriate equations are:

$$HA = H^+ + A^- \tag{29}$$

$$K_a = \frac{[H^+]\ [A]}{[HA]} \tag{30}$$

$$\frac{[A^-]}{C_{TA}} = \frac{1}{1 + 10^{(pK_a - pH)}} \tag{31}$$

$$\frac{[HA]}{C_{TA}} = \frac{1}{1 + 10^{(pH - pK_a)}} \tag{32}$$

and for a monoprotic organic base,

$$H_2O + B = HB^+ + OH^- \tag{33}$$

$$K_b = \frac{[HB^+][OH^-]}{[B]} \tag{34}$$

$$\frac{[HB^+]}{C_{TB}} = \frac{1}{1 + 10^{(pK_b - pOH)}} \tag{35}$$

$$\frac{[B]}{C_{TB}} = \frac{1}{1 + 10^{(pOH - pK_b)}} \tag{36}$$

where

HA and B = neutral forms of chemicals

A^- and HB^+ = ionized forms of chemicals

H^+ = hydrogen ion

OH^- = hydroxide ion

K_a and K_b = ionization constants

C_{TA} and C_{TB} = total dissolved concentration of chemicals, i.e. the sum of the neutral and ionized forms

$pH = -\log [H^+]$

$pOH = -\log [OH^-]$

$pK_a = -\log K_a$

$pK_b = -\log K_b$

and brackets, [ ], represent the concentration of a chemical form. For acids, the chemical will be primarily in its neutral form for pH less than pK_a and in its ionized form for pH greater than pK_a. For bases, the opposite is true. The concentrations of toxic organic chemicals are generally small enough that they do not affect the pH of the environment. For an organic acid, $c = C_{TA}$ and $c' = [HA]$; for an organic base, $c = C_{TB}$ and $c' = [B]$. Note then that

$$c' = \frac{1}{1 + 10^{(pH - pK_a)}}\, c = \alpha_a c \text{ for an organic acid,} \tag{37}$$

$$c' = \frac{1}{1 + 10^{(pOH - pK_b)}}\, c = \alpha_b c \text{ for an organic base,} \tag{38}$$

and equation (23) becomes

$$s_M = -\frac{K_v{}'c}{z} \tag{39}$$

where $K_v{}' = K_v\alpha_a$ for an organic acid (40)

$K_v{}' = K_v\alpha_b$ for an organic base (41)

System Model Formulation

Integration of the above processes requires some form of system model. A comprehensive workshop assessing the role that mathematical models play in predicting the concentration of chemicals in the environment is presented in reference 7. For the sake of simplicity, application herebelow will be to a fully mixed receiving water (for example, a lake). The governing equation (4) with R_M and S_M terms as described above becomes

$$\frac{d(Vc_T)}{dt} = Q_i c_{Ti} - Q_e c_T - K_1 c_T V - K_y c_T V - K_{pT} c_T V - \frac{K_v{}'}{z} c\, V - \frac{v_s}{z} c_p V + S_M \tag{42}$$

where Q_i and Q_e represent the influent and effluent flow rates, respectively, to the system, S_M represents other inputs of the chemical not associated with Q_i, and all other terms are as defined previously. Equation (42) assumes (1) that biodegradation, hydrolysis and photolysis act on the chemical in particulate and dissolved phases, (2) that only the dissolved phase is subject to volatilization, and (3) that only the particulate phase settles. If the particulate phase exists as a result of adsorption and is not available for biodegradation, acid-catalyzed and base-catalyzed hydrolysis, and photolysis, then equation (42) becomes

$$\frac{d(Vc_T)}{dt} = Q_i c_{Ti} - Q_e c_T - K_1 F_d c_T V - K_y F_d c_T V - K_n F_a c_T V$$
$$-K_{pT} F_d c_T V - \frac{K_v'}{z} F_d c_T V - \frac{v_s}{z} F_a c_T V + S_M \quad (43)$$

Solution of equations (42), (43) or some modification of either at any time first requires solution of the sorption and acid-base equilibrium equations for F_a, F_d, and K_v'. Of course any particular chemical may not be subject to all of the terms included above. Only those reactions which are of importance to the particular chemical need be included in the conservation of mass. Note that all of the terms are expressed as first order with respect to c_T and equation (43) may be written as

$$\frac{d(Vc_T)}{dt} = Q_i c_{Ti} - Q_e c_T - \gamma c_T V + S_M \quad (44)$$

where $\gamma = -[K_1 F_d + K_Y F_d + K_n F_a + K_{pT} + \frac{K_v'}{z} F_d + \frac{v_s}{z} F_a]$ (45)

Steady-state and time-variable analytical solutions for systems with constant parameters are readily available (6,28). For dynamic systems, numerical solutions may be necessary as are customarily used for conventional pollutants.

Contaminants which are adsorbed to particulate material and deposited in the sediments may be subject to a variety of processes including further burial, resuspension, desorption and diffusion back to the water column, biodegradation, and hydrolysis. Detailed treatment of sediment-water column interactions concerning toxic chemicals may be found in references 6 and 12.

Distribution of Toxic Pollutants in the Biosphere

In one sense, movement of toxic contaminants throughout the ecosystem can be viewed as analogous to movements of these same contaminants throughout the physical environment. One approach considers the aquatic food chain as a linear continuous system analogous to a linear continuous waterbody (for example, a river) (27). The food chain becomes a one-dimensional system. Transfer of contaminants along the one-dimensional food chain have governing conservation of mass equations similar in form to those for transfer of contaminants along a one-dimensional physical system. Velocity in trophic space is analogous to veloity in physical space. Length of organism in trophic space is analogous to length in physical space. Although this approach has several limitations, it is a rather elegant insight into the dynamics of contaminant transfer through the ecosystem and the physical system.

A second more common approach requires segmenting the food chain into discrete compartments (Figure 2). The number of compartments chosen is arbitrary and will depend upon the analyst's objectives and data base. Again, conservation of mass is the governing principle, and results in a set of equations of the following form.

$$\frac{d(w_i \nu_i)}{dt} = K_{ui} w_i c - K_{ei} w_i \nu_i + \alpha_{i,i-1} G_{i,i-1} w_i \nu_{i-1} \qquad (46)$$

where w_i = wet weight of average organism in compartment i, (M_i),

ν_i = mass specific contaminant concentration in compartment i, $(M_c\ M_i{}^{-1})$,

K_{ui} = uptake rate of contaminant from water for compartment i, $(L^3 T^{-1} M_i{}^{-1})$,

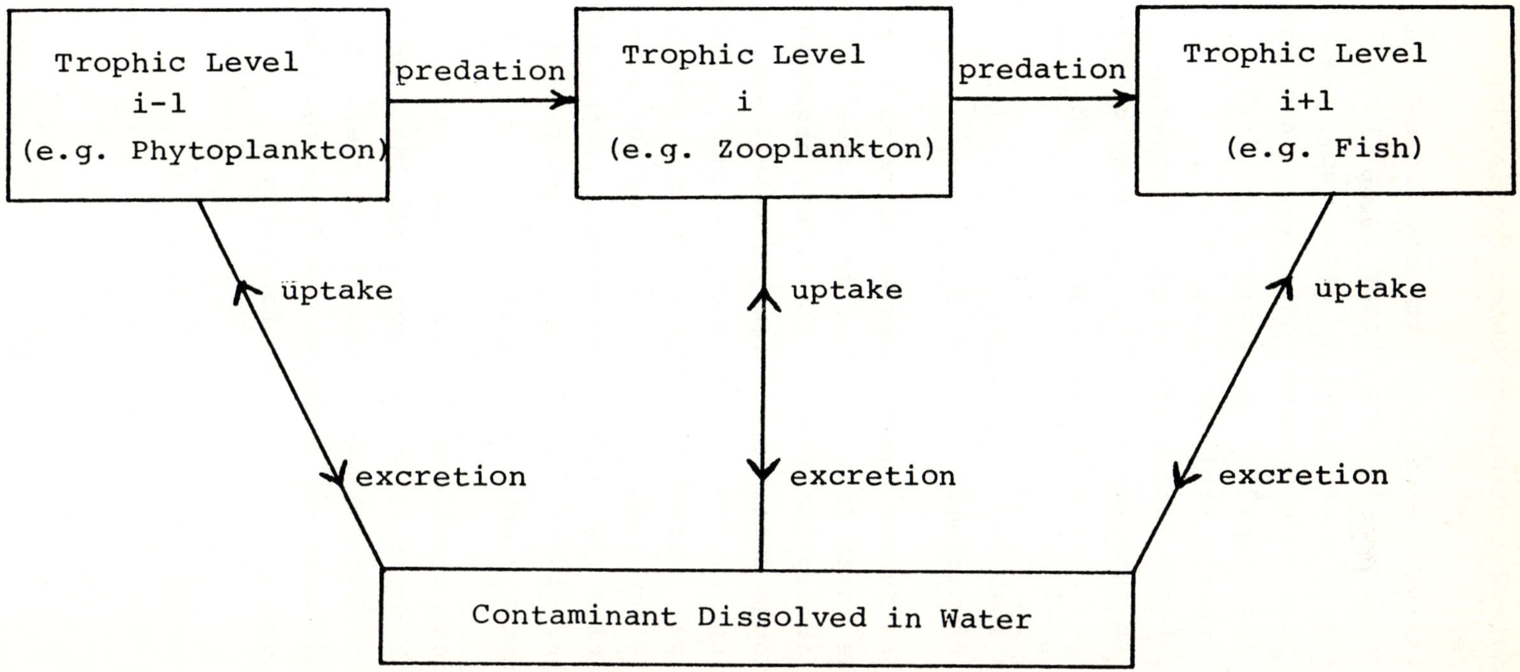

Figure 2. A Sample Compartmentalized Food Chain

c = dissolved contaminant concentration in water, $(M_c L^{-3})$,

K_{ei} = contaminant excretion rate coefficient for organism in compartment i, (T^{-1}),

$\alpha_{i,i-1}$ = contaminant assimilation efficiency; i.e. portion of grazed contaminant incorporated into body tissue of organism i, (M_c incorporated per M_c ingested),

$G_{i,i-1}$ = weight specific grazing rate of organism i on organism i-1, i.e. mass prey consumed per mass of predator per time, $(M_{i-1} M_i^{-1} T^{-1})$,

ν_{i-1} = mass specific contaminant concentration in compartment i-1, $(M_c M_{i-1}^{-1})$.

Equation (46) considers three factors - contaminant uptake directly from the water (for example, due to breathing or adsorption to and absorption through the skin), organism self-purification through normal excretion processes, and transfer of contaminants up the food chain through normal predator-prey relationships. Recall that these concepts were presented above in the discussion of bioaccumulation and biomagnifiction. Equation (46) enables evaluation of the relative importance of contaminant transfer through the food chain and direct uptake from the external medium.

Analytical solutions to equation (46) under assumptions of constant coefficients are available (6). Of course again, numerical solutions may be readily adapted. At steady-state conditions and constant coefficients, the compartmentalized food chain gives the following

$$\frac{\nu_i}{c} = \frac{K_{ui}}{K'_i} + \frac{\alpha_{i,i-1} G_{i,i-1}}{K'_i} \frac{\nu_{i-1}}{c} \tag{47}$$

where K'_i = a net decay rate coefficient expressing the decrease in the mass specific contaminant concentration due to excretion and net organism weight gain (growth minus respiration), (T^{-1}).

Note that ν_i/c is the bioconcentration factor, β_i of equation (2). The first term on the right hand side of equation (47) expresses the increase in β_i due to direct uptake from the external medium (the water), and the second term represents the increase in β_i due to transfer up the food chain.

Closure

The above presentations enable evaluation of toxic pollutant impacts to receiving water quality and aquatic life residing within contaminated waterways. The last link in the total assessment involves the impact to human health. Models for the distribution of toxic pollutants in receiving waters provide concentration distributions which may be compared with the water quality criteria of Appendix I. This is satisfactory for evaluation of the impacts to freshwater and saltwater aquatic life. Human health impacts as noted in Appendix I may also be evaluated in terms of receiving water concentrations only; however, as further noted, these impacts are a function of whether the human is ingesting contaminated water and contaminated organisms, or contaminated organisms alone. The derivation of such criteria normally requires some information regarding the relationship between the receiving water concentration distributions and the bioconcentration factor. This information, in one form or another, is derived from the principles outlined above in formulating models of the distribution of toxic pollutants throughout the aquatic system. The results are input to models of human physiological response to ingestion of toxic chemicals, often determined by extrapolation of the results of animal laboratory studies. The response determination is typically for a reference man who is one that weighs 70 kilograms and consumes 2 liters of water and 6.5 grams of aquatic organisms per day (9). Response for a reference woman and a reference child are sometimes chosen (18).

Consequently, it is clear that any assessment of the impacts of toxic pollutants to receiving waters requires a comprehensive framework integrating expertise from many disciplines. This paper has concentrated, and indeed only briefly, on those aspects related to the transport and transformation of toxic chemicals throughout aquatic systems. Such an exercise is an essential step in understanding past and predicting future contaminant concentration distributions in receiving waters. Comparison of these concentration distributions with water quality criteria enables impact assessment.

References

1. Alexander, M. Biodegradation of chemicals of environmental concern. Science. 211(9): 132-138; January 1981.

2. Alexander, M. Biodegradation of toxic chemicals in water and soil. In Dynamics, exposure and hazard assessment of toxic chemicals (R. Haque, editor). Ann Arbor Science Pub., Inc., Ann Arbor, MI. 1980, pp 179-190.

3. Baughman, G.L., D.F. Paris, and W.C. Steen. Quantitative expression of biodegrdation rate. In Biotransformation and fate of chemicals in the environment (A.W. Maki,K.L. Dickson, and J. Cairns, Jr. editors) American Society of Microbiology. Washington, D.C., 1980, pp 105-111.

4. Belton, T.J., B.E. Ruppel, K. Lockwood PCB's (Aroclor 1254) in fish tissues throughout the State of N.J.: a comprehensive survey. Office of Cancer and Toxic Substances Research, New Jersey Department of Environmental Protection, Trenton, NJ, November 1982, 36 pp.

5. Chadwick, G.F. and R.W. Brocksen, J. Wildlife Management 33:693; 1969.

6. Chapra, S.C. and K.H. Reckhow. Engineering approaches for lake management Volume 2: Mechanistic modeling. Butterworth Publishers. Woburn, MA 1983. 492 pp.

7. Dickson, K.L., A.W. Maki, J. Cairns, Jr. (editors). Modeling the fate of chemicals in the aquatic environment. Ann Arbor Science, The Butterworth Group. Ann Arbor, MI. 1982. 413 pp.

8. DiToro, D.M. and L.M. Horzempa. Reversible and resistant components of PCB adsorption - desorption isotherms. Environ. Sci. Technol. 16(9): 594-602; 1982.

9. Falco, J. (U.S. Environmental Protection Agency, Washington, D.C.) personal communication from Stanford Research Institute International. Seafood consumption data analysis final report. 1980.

10. Fischer, H.B. et al. Mixing in inland and coastal waters. Academic Press, NY. 1979, 483 pp.

11. Haque, R. (editor). Dynamics, exposure and hazard assessment of toxic chemicals. Ann Arbor Science Publishers, Ann Arbor, MI. 1980, 496 pp.

12. Jaffe, P.R. and R.A. Ferrara. Modeling sediment and water column interactions for hydrophobic pollutants. Water Research. 18(9): 1169-1174; 1984.

13. Jaffe, P.R. and R.A. Ferrara. Desorption kinetics in modeling of toxic chemicals. ASCE Journal Environmental Engineering. 109(4): 859-867; August 1983.

14. Karickhoff, S.W. Organic pollutant sorption in aquatic systems. ASCE Journal of Hydraulic Engineering 110 (6): 707-735; June 1984.

15. Karickhoff, S.W., D.S. Brown, and T.A. Scott. Sorption of hydrophobic pollutants in natural sediments. Water Res. 13: 241-248; 1979.

16. Kay,S.H. Potential for biomagnification of contaminants within marine and freshwater food webs. US Army Corps of Engineers Waterways Experiment Station. Vicksburg, MS Technical Report D-84-7, November 1984, 166 pp.

17. Mills, W.B. et al. Water quality assessment: A screening procedure for toxic and conventional pollutants Part 1. U.S. Environmental Protection Agency, Athens, GA. EPA-600/6-82-004a. September 1982, 570 pp.

18. Moghissi, A.A. et al. Methodology for environmental human exposure and health risk assessment. In Dynamics, exposure and hazard assessment of toxic chemicals (R. Haque,

ed.). Ann Arbor Science Publishers, Inc., Ann Arbor, MI. 1980, pp 471-484.

19. Neely, W.B. Chemicals in the environment - distribution, transport, fate, analysis. Marcel Dekker, Inc., NY, 1980.

20. Odum, E.P. Fundamentals of ecology. W.B. Saunders Co. Philadelphia, PA., 1971, 574 pp.

21. Orlob, G.T. (ed.) Mathematical modeling of water quality: streams, lakes, and reservoirs. International Institute for Applied Systems Analysis, John Wiley & Sons, NY, 1983, 518 pp.

22. Paris, D.F., W.C. Steen, G.L. Baughman, and J.T. Barnett, Jr. Second-order model to predict microbial degradation of organic compounds in natural waters. Applied and Environmental Microbiology. 41(3): 603-609; 1981.

23. Ray,S.M. and N.M. Trieff. Bioaccumulation of anthropogenic toxins in the ecosystem. In Environment and health (N.M. Trieff editor) Ann Arbor Science Publishers, Inc., Ann Arbor, MI 1980, pp 93-120.

24. Reinert, R.L., J. Fish Res. Board Can. 29: 1413; 1972.

25. Subba-Rao, R.V., H.E. Rubin, M. Alexander. Kinetics and extent of mineralization of organic chemicals at trace levels in freshwater and sewage. Applied and Environmental Microbiology. 43(5): 1139-1150; 1982.

26. Tabak, H.H., S.A. Quave, C.I. Mashni, E.F. Barth. Biodegradability studies with organic priority pollutant compounds. Journal Water Pollution Control Federation. 53(10): 1503-1518; 1981.

27. Thomann, R.V. Size dependent model of hazardous substances in aquatic food chain. U.S. Environmental Protection Agency. Duluth, Minnesota. EPA-600/3-28-036, April 1978, 40 pp.

28. Thomann, R.V. Systems analysis and quality management. Environmental Science Services Divisiion, Environmental Research and Applications, Inc., NY, NY. 1972, 286 pp.

29. U.S. Environmental Protection Agency. Treatability Manual Volume I. Treatability data. Office of Research and Development. Washington, D.C., EPA-600/2-82-001a. September 1981.

30. U.S. Environmental Protection Agency. Quality criteria for water. Washington, D.C., July 1976, 256 pp.

31. U.S. Environmental Protection Agency. National interim primary drinking water regulations. Federal Register, part IV, December 24, 1975.

32. Wolfe, N.L. Determinigng the role of hydrolysis in the fate of organics in natural waters. In. Dynamics, exposure and hazard assessment of toxic chemicals (R. Haque editor). Ann Arbor Science Publishers, Inc., Ann Arbor, MI. 1980, pp 163-178.

33. van Genuchten, M.T., J.M. Davidson, and P.J. Wierenga. An evaluation of kinetic and equilibrium equations for the prediction of pesticide movement through porous media. Soil Sci. Soc. Amer. Proc. 38: 29-35; 1974.

34. Zepp, R.G. Assessing the photochemistry of organic pollutants in aquatic environments. In Dynamics, exposure, and hazard assessment of toxic chemicals. (R. Haque editor) Ann Arbor Science Publishers, Inc. Ann Arbor, MI. 1980, pp 69-110.

Appendix I. Water Quality Criteria for 129 Priority Toxic Water Pollutants*

Chemical	Freshwater Aquatic Life		Saltwater Aquatic Life		Human Health: Ingestion of Contaminated Water and Organisms				Human Health: Ingestion of Contaminated Organisms Only			
	Acute	Chronic	Acute	Chronic	Toxic	Cancer Risk of 10^{-5}	Cancer Risk of 10^{-6}	Cancer Risk of 10^{-7}	Toxic	Cancer Risk of 10^{-5}	Cancer Risk of 10^{-6}	Cancer Risk of 10^{-7}
Metals and Inorganics:												
Antimony	9,000	1,600			146				45,000			
Arsenic	440	40	508			22[b]	2.2[b]	.22[b]		175[b]	17.5[b]	1.75[b]
Asbestos						300,000[r]	30,000[r]	3,000[r]				
Beryllium	130	5.3				37[b]	3.7[b]	.37[b]		641[b]	64.1[b]	6.41[b]
Cadmium	s	t	59	4.5	10							
Chromium (hexavalent)	21	.29	1,260	18	50							
Chromium (trivalent)	u	44	10,300		170				3,433			
Copper	v	5.6	23	4								
Cyanide	52	3.5	30	2.0	200							
Lead	w	x	668	25	50							
Mercury	1.7[b]	.57[b]	3.7	.025	.144				.146			
Nickel	y	z	140	7.1	13.4				100			
Selenium	260	35	410	54	10							
Silver	aa	.12	2.3		50							
Thallium	1,400	40	2,130		13				48			
Zinc	bb	47	170	58								
Ethers:												
Bis(chloromethyl) ether	238,000					.038[b]	.0038[b]	.00038[b]		18.4[b]	1.84[b]	0.184[b]
Bis(2-chloroethoxy) methane[g]	360	122										
Bis(2-chloroethyl) ether	238,000					0.3	.03	.003		13.6	1.36	.136
Bis(2-chloroisopropyl) ether	238,000				34.7				4,360			
4-Bromophenyl phenyl ether[g]	360	122										
2-Chloroethyl vinyl ether[g]	360	122										
4-Chlorophenyl phenyl ether[g]	360	122										

Appendix I. (continued)

Chemical	Freshwater Aquatic Life		Saltwater Aquatic Life		Human Health: Ingestion of Contaminated Water and Organisms				Human Health: Ingestion of Contaminated Organisms Only			
	Acute	Chronic	Acute	Chronic	Toxic	Cancer Risk of 10^{-5}	Cancer Risk of 10^{-6}	Cancer Risk of 10^{-7}	Toxic	Cancer Risk of 10^{-5}	Cancer Risk of 10^{-6}	Cancer Risk of 10^{-7}
Phthalates:												
Bis(2-ethylhexyl) phthalate	940[ℓ]	3[ℓ]	2,944[ℓ]		15,000				50,000			
Butyl benzyl phthalate	940[ℓ]	3[ℓ]	2,944[ℓ]									
Di-n-butyl phthalate	940[ℓ]	3[ℓ]	2,944[ℓ]		34,000				154,000			
Di-n-octyl phthalate	940[ℓ]	3[ℓ]	2,944[ℓ]									
Diethyl phthalate	940[ℓ]	3[ℓ]	2,944[ℓ]		350,000				$1.8x10^{6}$			
Dimethyl phthalate	940[ℓ]	3[ℓ]	2,944[ℓ]		313,000				$2.9x10^{6}$			
Nitrogen Compounds:												
Acrylonitrile	7,550	2,600[a]				0.58	0.058	0.006		6.5	0.65	0.065
Benzidine	2,500					1.2[b]	0.12[b]	0.012[b]		5.3[b]	0.53[b]	0.053[b]
3,3'-Dichlorobenzidine						.103	.0103	.00103		.204	.0204	.00204
1,2-Diphenylhydrazine	270					.422	.042	.0042		5.6	.56	.056
N-nitrosodimethylamine	5,850[k]		$3.3x10^{6}$[k]			14[b]	1.4[b]	.14[b]		160	16	1.6
N-nitrosodiphenylamine	5,850[k]		$3.3x10^{6}$[k]			49	4.9	.49		161	16.1	1.61
N-nitrosodi-n-propylamine	5,850[k]		$3.3x10^{6}$[k]									
Phenols:												
2-Chlorophenol	4,380											
2,4-Dichlorophenol	2,020	365										
2,4-Dimethylphenol	2,120											
4,6-Dinitro-o-cresol	230[j]		4,850[j]		13.4				765			
2,4-Dinitrophenol	230[j]		4,850[j]		70				14,300			
2-Nitrophenol	230[j]		4,850[j]									
4-Nitrophenol	230[j]		4,850[j]									
Parachlorometa cresol	30											
Pentachlorophenol	55	3.2	53	34								
Phenol	10,200	2,560	5,800									
2,4,6-Trichlorophenol		9.70				12	1.2	.12		36	3.6	.36

Appendix I. (continued)

Chemical	Freshwater Aquatic Life		Saltwater Aquatic Life		Human Health: Ingestion of Contaminated Water and Organisms				Human Health: Ingestion of Contaminated Organisms Only			
	Acute	Chronic	Acute	Chronic	Toxic	Cancer Risk of 10^{-5}	Cancer Risk of 10^{-6}	Cancer Risk of 10^{-7}	Toxic	Cancer Risk of 10^{-5}	Cancer Risk of 10^{-6}	Cancer Risk of 10^{-7}
Aromatics:												
Benzene	5,300		5,100	700[a]		6.6	0.66	0.066		400	40.	4.0
Chlorobenzene[c]	250	50[a]	160	129		7.2[b]	0.72[b]	0.072[b]		7.4[b]	0.74[b]	0.074[b]
1,2-Dichlorobenzene[d]	1,120	763	1,970		400				2,600			
1,3-Dichlorobenzene[d]	1,120	763	1,970		400				2,600			
1,4-Dichlorobenzene[d]	1,120	763	1,970		400				2,600			
2,4-Dinitrotoluene	330	230	590	370[a]		1.1	.11	.011		91	9.1	.91
2,6-Dinitrotoluene												
Ethylbenzene	32,000		430		1,400				3,280			
Hexachlorobenzene[c]	250	50[a]	160	129		7.2[b]	0.72[b]	0.072[b]		7.4[b]	0.74[b]	0.074[b]
Nitrobenzene	27,000		6,680									
Toluene	17,500		6,300	5,000	14,300				424,000			
1,2,4-Trichlorobenzene[c]	250	50[a]	160	129		7.2[b]	0.72[b]	0.072[b]		7.4[b]	0.74[b]	0.074[b]
Polynuclear Aromatic Hydrocarbons:												
Acenaphthene[e]	1,700	520	970	500		28	2.8	.28		311	31.1	3.11
Acenaphthylene												
Anthracene												
Benzo(a)anthracene[e]			300			28	2.8	.28		311	31.1	3.11
Benzo(a)pyrene[e]			300			28	2.8	.28		311	31.1	3.11
Benzo(k)fluoranthene[e]			300			28	2.8	.28		311	31.1	3.11
3,4-Benzofluoranthene			300			28	2.8	.28		311	31.1	3.11
Benzo(ghi)perylene[e]			300			28	2.8	.28		311	31.1	3.11
2-Chloronaphthalene[e]	1,600		7.5			28	2.8	.28		311	31.1	3.11
Chrysene												
Dibenzo(a,h)anthracene[e]			300			28	2.8	.28		311	31.1	3.11
Fluoranthene[e]	3,980		40	16	42	28	2.8	.28	54	311	31.1	3.11
Fluorene												
Indeno(1,2,3-cd)pyrene[e]			300			28	2.8	.28		311	31.1	3.11
Naphthalene[e]	2,300	620	2,350			28	2.8	.28		311	31.1	3.11
Phenanthrene												
Pyrene												

Appendix I. (continued)

Chemical	Freshwater Aquatic Life		Saltwater Aquatic Life		Human Health: Ingestion of Contaminated Water and Organisms				Human Health: Ingestion of Contaminated Organisms Only			
						Cancer Risk of				Cancer Risk of		
	Acute	Chronic	Acute	Chronic	Toxic	10^{-5}	10^{-6}	10^{-7}	Toxic	10^{-5}	10^{-6}	10^{-7}
Polychlorinated Biphenyls[cc] (PCB's):	2.0	.014	10	.030		.79	.079	.0079		.79	.079	.0079
PCB 1016												
PCB 1221												
PCB 1232												
PCB 1242												
PCB 1248												
PCB 1254												
PCB 1260												
Halogenated Hydrocarbons:												
Bromoform												
Carbon Tetrachloride	35,200		50,000			4.0	0.40	0.04		69.4	6.94	0.69
Chlorodibromomethane												
Chloroethane												
Chloroform	28,900	1,240				1.9	.19	.019		157	15.7	1.57
Dichlorobromethane[h]	11,000		12,000	6,500		1.9	.19	.019		157	15.7	1.57
Dichlorodifluoromethane[h]	11,000		12,000	6,500		1.9	.19	.019		157	15.7	1.57
1,1-Dichloroethane												
1,2-Dichloroethane	118,000	20,000	113,000			9.4	0.94	0.094		2430	243	24.3
1,1-Dichloroethylene	11,600		224,000			.33	.033	.0033		18.5	1.85	.185
1,2-Dichloropropene	23,000	5,700	10,300	3,040								
1,3-Dichloropropene	6,060	244	790		87				14,100			
Hexachlorobutadiene	90	9.3	32			4.47	.45	.045		500	50	5
Hexachlorocyclo-pentadiene	7.0	5.2	7.0									
Hexachloroethane	980	540	940			19	1.9	0.19		87.4	8.74	0.87
Methyl Bromide												
Methyl Chloride												
Methylene Chloride												

Appendix I. (continued)

Chemical	Freshwater Aquatic Life		Saltwater Aquatic Life		Human Health: Ingestion of Contaminated Water and Organisms				Human Health: Ingestion of Contaminated Organisms Only			
	Acute	Chronic	Acute	Chronic	Toxic	Cancer Risk of 10^{-5}	10^{-6}	10^{-7}	Toxic	Cancer Risk of 10^{-5}	10^{-6}	10^{-7}
Halogenated Hydrocarbons: (continued)												
1,2-Trans-Dichloroethylene	11,600[f]		224,000[f]									
Tetrachloroethylene	5,280	840	10,200	450		8	.8	.08		88.5	8.85	.885
1,1,2,2-Tetrachloroethane	9,320	2,400	9,020									
1,1,1-Trichloroethane	18,000		31,200		18,400				1.03×10^{6}			
1,1,2-Trichloroethane	18,000	9,400				6.0	0.6	0.06		418	41.8	4.18
Trichloroethylene	45,000	21,900[a]	2,000			27	2.7	.27		807	80.7	8.07
Trichlorofluoromethane[h]	11,000		12,000	6,500		1.9	.19	.019		157	15.7	1.57
Vinyl Chloride						20	2.0	.2		5246	525	52.5
Pesticides:												
Aldrin	3.0		1.3			.74[b]	.074[b]	.0074[b]		.79[b]	.079[b]	.0079[b]
alpha - BHC	100[p]		.34[p]			92[b]	9.2[b]	.92[b]		310[b]	31[b]	3.1[b]
beta - BHC	100[p]		.34[p]			163[b]	16.3[b]	1.63[b]		547[b]	54.7[b]	5.47[b]
delta - BHC	100[p]		.34[p]									
gamma - BHC	100[p]		.34[p]			186[b]	18.6[b]	1.86[b]		625[b]	62.5[b]	6.25[b]
Dieldrin	2,500	1.9	710	1.9		.71[b]	.071[b]	.0071[b]		.76[b]	.076[b]	.0076[b]
Chlordane	2.4	.0043	.04	.004		4.6[b]	.46[b]	.046[b]		4.8[b]	.48[b]	.048[b]
4,4'-DDT	1.1	.001	.13	.001		.24[b]	.024[b]	.0024[b]		.24[b]	.024[b]	.0024[b]
4,4'-DDE	1,050		14									
4,4'-DDD												
a-Endosulfan[m]	.22	.056	.034	.0087	74				159			
b-Endosulfan[m]	.22	.056	.034	.0087	74				159			
Endosulfan sulfate	.22	.056	.034	.0087	74				159			
Endrin[n]	.18	.0023	.037	.0023	1.0							
Endrin Aldehyde[n]	.18	.0023	.037	.0023	1.0							
Heptachlor[o]	.52	.0038	.053	.0036		2.78[b]	.28[b]	.028[b]		2.85[b]	.29[b]	.029[b]
Heptachlor epoxide[o]	.52	.0038	.053	.0036		2.78[b]	.28[b]	.028[b]		2.85[b]	.29[b]	.029[b]
Toxaphene	1.6	.013	.070			7.1[b]	.71[b]	.071[b]		7.3[b]	.73[b]	.073[b]

Appendix I. (continued)

Chemical	Freshwater Aquatic Life		Saltwater Aquatic Life		Human Health: Ingestion of Contaminated Water and Organisms				Human Health: Ingestion of Contaminated Organisms Only			
	Acute	Chronic	Acute	Chronic	Toxic	Cancer Risk of 10^{-5}	Cancer Risk of 10^{-6}	Cancer Risk of 10^{-7}	Toxic	Cancer Risk of 10^{-5}	Cancer Risk of 10^{-6}	Cancer Risk of 10^{-7}
Oxygenated Compounds;												
Acrolein	68	21	55		320				780			
Miscellaneous												
Isophorone	117,000		12,900		5.2				520			
2,3,7,8-Tetrachlorodibenzo-p-dioxin (TCDD)												

* Where no value is specified, existing data are insufficient to define a criterion. All values are in μg/l except where noted by superscript b or r.

[a] data not definitive; stated value based on limited data

[b] ng/l

[c] values defined for all chlorinated benzenes

[d] values defined for all dichlorobenzenes

[e] values for cancer risk defined in ng/l for all Polynuclear Aromatic Hydrocarbons. Data insufficient to define criterion for individual constituents.

[f] values defined for all dichloroethylenes

[g] values defined for all haloethers

[h] values defined for all halomethanes

[j] values defined for all nitrophenols

[k] values defined for all nitrosamines

Appendix I. (continued)

Chemical	Freshwater Aquatic Life		Saltwater Aquatic Life		Human Health							
					Ingestion of Contaminated Water and Organisms				Ingestion of Contaminated Organisms Only			
						Cancer Risk of				Cancer Risk of		
	Acute	Chronic	Acute	Chronic	Toxic	10^{-5}	10^{-6}	10^{-7}	Toxic	10^{-5}	10^{-6}	10^{-7}

[l] values defined for all phthalate esters

[m] values defined for all endosulfans

[n] values defined for all endrins

[o] values defined for all heptachlors

[r] fibers/liter

s=exp [1.05 (ln H) - 8.53] where H = hardness in mg/l as $CaCO_3$

t=exp [1.05 (ln H) - 3.73] where H = hardness in mg/l as $CaCO_3$

u=exp [1.08 (ln H) + 3.48] where H = hardness in mg/l as $CaCO_3$

v=exp [0.94 (ln H) - 1.23] where H = hardness in mg/l as $CaCO_3$

w=exp [1.22 (ln H) - 0.47] where H = hardness in mg/l as $CaCO_3$

x=exp [2.35 (ln H) - 9.48] where H = hardness in mg/l as $CaCO_3$

y=exp [0.76 (ln H) + 4.02] where H = hardness in mg/l as $CaCO_3$

z=exp [0.76 (ln H) + 1.06] where H = hardness in mg/l as $CaCO_3$

aa=exp [1.72 (ln H) - 6.52] where H = hardness in mg/l as $CaCO_3$

bb=exp [0.83 (ln H) + 1.95] where H = hardness in mg/l as $CaCO_3$

cc=values defined for all polychlorinated biphenyls

AQUATIC FATE AND TRANSPORT MODELING TECHNIOUES FOR PREDICTING ENVIRONMENTAL EXPOSURE TO ORGANIC PESTICIDES AND OTHER TOXICANTS--A COMPARATIVE STUDY

L.A. Mulkey, R.B. Ambrose, Jr., and T.O. Barnwell, Jr.
Environmental Research Laboratory, U.S. EPA
Athens, GA 30613
USA

Introduction

Evidence of potentially harmful effects of pesticides and other toxic organic chemicals on aquatic organisms has led to intensive efforts toward environmental risk assessment for existing and new chemicals. Laboratory bioassay tests, environmental damage episodes like fish kills, and information about chemical properties have all been studied with a view toward estimating or assigning risk associated with chemical use patterns, waste discharges, and related management practices. The most direct procedure for evaluating environmental risk attendant to chemical use or release is a field study in which aquatic systems are monitored for damage, causal factors delineated, and results extrapolated or generalized. In this manner, risk is allocated based upon observed damage Another popular procedure has been allocation of risk based on observation of the chemical per se (not its effects). In this case risk is allocated based only on predictions of potential damage.

These environmental observation procedures have been effectively applied to a number of toxicant-related environmental problems, most notably those associated with chlorinated hydrocarbon pesticides. Such methods are not without their shortcomings, however. In the evaluation of observed damage, one may be able to delineate the specific causal factors but the frequency of their occurrence (and hence the

NATO ASI Series, Vol. G10
Urban Runoff Pollution.
Edited by H.C. Torno, J. Marsalek, and M. Desbordes

risk or probability of their reoccurrence) cannot be discerned short of observing future episodes through numerous repeated studies. The same is true for observations of toxicant concentrations designed to infer risk or probabilities of occurrence. In both cases, observations can be made only after the chemical in question is in widespread use and discharges are identified and measured. It would seem more prudent to evaluate environmental risk of toxic organic chemicals before they are released and widely used.

Concern about pesticides and other toxic chemicals is related to effects or damage. The concept of risk reflects the likelihood of probability of such effects. Obviously, if an effect is induced, the organism affected must first have been exposed to the toxicant for a time period and intensity sufficient to inflict the damage. The study of this exposure, or exposure assessment, is defined as a quantitative evaluation of the concentration of chemical toxicants in contact with receptor populations for various environmental media as the toxicant is released, transported, and transformed among and within environmental compartments.

The source of toxic chemicals in a river basin is an important factor in the selection of techniques to analyze the problem. Toxic chemical releases may arise both from nonpoint sources such as agricultural fields from which pesticides are transported in runoff and from point sources such as chemical manufacturing plants from which chemicals are discharged in effluent. Nonpoint sources are characterized by highly variable loadings with rainfall-runoff events often dominating the timing and magnitude of chemical inputs to the receiving waters. Point sources are usually much less varied and are often thought to provide steady (not varying in time) inputs to the aquatic system. Exposure assessments for toxic chemicals in the aquatic environment must accommodate both the nonpoint or variable loading case and the point or steady loading case.

Chemical concentrations in aquatic environments are also determined by the transport and transformation processes known to occur in river basins. Chemical transport is dominated by mass movement in water or on sediment and to a lesser extent in the air by volatilization processes. As chemicals are transported by water or sediment, they are subject to a number of physical, chemical, and biological processes that can lead to either transport or transformation. In general, the influence of transport on chemical behavior is determined by the local environmental regime (flow rates, sediment concentrations, etc.); transformation processes are largely determined by chemical properties and to a lesser extent by environmental conditions.

A Mathematical Statement of the Problem

Exposure assessment as a part of risk evaluation implies the need to estimate frequency of exposure. Exposure can be viewed as the realization of the joint probability of an ambient concentration and the existence of receptor populations. The probability of ambient concentrations can be approximated by calculating their frequency of occurrence over a sufficient period of time. A convenient way to represent frequency is the cumulative distribution function (cdf). A cdf can be expressed as a graph of the concentrations (on the abcissa) exceeded for various fractions of a time period (on the ordinate). A cdf can be related to average expected return periods for different concentrations. A convenient way to quantify the ambient concentration required for environmental exposure is the cumulative frequency diagram. A cumulative frequency diagram is simply a graph of the concentration levels exceeded a given percentage of a time period. Given the cdf for a specific chemical in a specific environment, one can, for example, say that once every 10 years the concentration of a specific chemical will exceed a specified level. The concentration level can then be related to biological effects information (e.g., LC_{50} values) and a return interval for damage (or risk) results.

A cdf can be readily constructed given sufficient data in the form of time-series chemical concentrations. Rarely are any such data available, however, and when short time periods are sampled, the statistical reliability of the resulting data is always suspect. An acceptable alternative to field sampling is mathematical modeling. With modeling, the problem of generating time-series estimates for chemical concentrations remains, however. In aquatic systems, both dissolved and sorbed chemical states are of interest. Falco and Mulkey (5), Falco et al. (6), Onishi et al. (14), and Mulkey and Falco (13) have developed and summarized the governing equations for toxicant concentrations in the water column. In the following development, it is convenient to define concentration-related symbols as in Table 1. The following longitudinal one-dimensional mass-balance equations apply.

$$\frac{\partial C_w}{\partial t} = \frac{\partial}{\partial x}\left(D \frac{\partial C_w}{\partial x}\right) - \frac{\partial (C_w U)}{\partial x} + W \pm \sum R_j \qquad [1]$$

$$\frac{\partial C_s}{\partial t} = \frac{\partial}{\partial x}\left(D_s \frac{\partial C_s}{\partial x}\right) - \frac{\partial (C_s U_s)}{\partial x} + W' \pm \sum R_{sj} \qquad [2]$$

where C_w, C_s = concentration of constituent in the dissolved and sorbed states, respectively (M/L^3)

D, D_s = dispersion coefficient for dissolved and sorbed constituents (L^2/T)

U, U_s = flow velocity of water and sediment (L/T)

W, W_s' = source strengths of dissolved and sorbed constituents, ($ML^{-3}T^{-1}$)

R_j, R_{sj} = rate of production or elimination of dissolved and sorbed constituent by pathway j ($ML^{-3}T^{-1}$)

x = distance (L)

t = time (T)

Table 1. Concentration-Related Symbols Used in Mathematical Equations

Symbol	Definition	Units
C	Concentration of chemical (bed or water)	mgc/l_T
C_w	Concentration of dissolved chemical in water	mgc/l_T
C_w'	Concentration of dissolved chemical in water $C_w' = C_w/\Theta$	mgc/l_w
C_s	Concentration of sorbed chemical in water	mgc/l_T
C_s'	Concentration of sorbed chemical in water $C_s' = C_s/S$	mgc/kgs
C_p	Concentration of dissolved chemical in bed (pore water)	mgc/l_T
C_p'	Concentration of dissolved chemical in bed (pore water) $C_p' = C_p/\Theta$	mgc/l_w
C_b	Concentration of sorbed chemical in bed	mgc/l_T
C_b'	Concentration of sorbed chemical in bed $C_b' = C_b/S$	mgc/Kgs
C_2	Concentration of sediment (bed or water)	mgs/l_T
S	Concentration of sediment (bed or water) $S = C_2 * 10^{-6}$	kgs/l_T
S'	Concentration of sediment (bed or water) $S' = S/\Theta$	kgs/l_w
Θ	Porosity, or volume water per volume segment	l_w/l_T

Equations 1 and 2 include the influence of dispersion, advective transport, sources, and sinks on the change in toxicant concentration or mass. Equation 2 is applied to the sorbed state of the constituent and hence the sediment transport must be included. Sediment movement is accounted for in three terms: U_s, the sediment velocity in the x-direction; W', the sorbed pollutant (attached to sediment) sources; and ΣR_{sj}, the sink terms for the sorbed pollutant including settling and/or resuspension.

Sediment transport is an important process in hydraulics and river mechanics, but its role in pollutant transport is not fully known. Several investigators (10,18) have shown that for most organic toxicants, particularly non-polar compounds, the sediment organic matter determines the extent of sorption and hence the importance of sorbed-pollutant transport. The particle size of the sediments is influential, but Rao and Davidson (18) have shown that the organic matter content of each size fraction is the controlling factor. Organic matter content and surface area relationships suggest that silt and clay size sediment fractions are most relevant in sediment transport. One outcome of this rationale is the idea that in river systems, wash load or suspended sediment is the important pollutant transport medium rather than bed load. If bed load transport can be ignored, the problem remains to estimate suspended sediment. For washload, the stream transport capacity typically is in excess of sediment supply, and the problem is one of estimating sediment source loading--namely, watershed erosion. Thus, the key sediment-related term in equation 2 is the source term, W'.

Bed load transport *per se* may be relatively unimportant in pollutant transport, but interaction between the stream bed and the overlying water should not be ignored. Concentration gradients between the bed material and the overlying water will result in sorption and desorption and thus the stream bed can act either as a sink or a source of contaminants. For the stream bed one can write:

$$\frac{\partial C_p}{\partial t} = \frac{\partial}{\partial y} (D_b \frac{\partial C_p}{\partial y} - \frac{\partial (U_p C_p)}{\partial y} \pm \sum R_{pj} \qquad [3]$$

$$\frac{\partial C_b}{\partial t} = \pm \sum R_{bj} - \frac{V_s C_b}{L_s} + \frac{V_d C_s}{L_w} \qquad [4]$$

where y = vertical direction (L)

C_p, C_b = bed concentration of chemical in the dissolved and sorbed states, respectively (M/L^3)

D_b = vertical dispersion coefficient for dissolved constituents (L^2/T)

R_{pj}, R_{bj} = rate of production (+) or elimination (-) of dissolved and sorbed chemical by pathway j ($ML^{-3}T^{-1}$)

U_p = velocity of net pore water movement into or out of the bed (L/T)

V_d, V_s = deposition and scour velocity of sediment between bed and water column (L/T)

L_w, L_s = depth of water compartment and active bed layer (L)

Equation 3 accounts for diffusion-dispersion and pore water transport of pollutants between the bed and the overlying water column. Equation 4 accounts for sediment-bound transport by including the scour and deposition terms, V_d and V_s. Estimation usually requires a rigorous treatment of hydrodynamic forces and characterization of sediments (e.g., cohesiveness). Several approaches are available for this purpose, but many chemical exposure models either omit these processes entirely or lump them into an overall calibrated dispersion or exchange coefficient.

Equations 1-4 describe the fate and transport of dissolved and sorbed pollutants, but an interaction term must be added. Specifically, the interaction of dissolved organic toxicants with entrained sediment must be accounted for mathematically. If one assumes sorption is kinetically a first order process, then a rate equation similar to the other expressions can be written as:

$$\frac{R_s}{S} = k_s C_w' - k_d C_s' \qquad [5]$$

where R_s = net rate of chemical transfer between sorbed and dissolved state, ($ML^{-3}T^{-1}$)

k_s = rate constant for sorption, ($L_w^3 M_s^{-1} T^{-1}$)

k_d = rate constant for desorption, (T^{-1})
C_s' = concentration of sorbed pesticide, (M/M_s)
C_w' = concentration of dissolved pesticide (M/L_w^3)
M_s = mass of sediment, (M)

If one assumes $R_s = 0$ then

$$k_s C_w' = k_d C_s'$$

$$\text{or} \quad C_s' = \frac{k_s}{k_d} C_s' \qquad [6]$$

Note that equation 6 is the linear form of the Freundlich isotherm with the ratio $k_s/k_d = K_d$ being the distribution (or partition) coefficient. This coefficient is known to be a function of both the toxicant and the characteristics of the sediments onto which sorption has occurred. Most organic pollutants of current interest are non-polar, hydrophobic compounds whose K_d values correlate quite well with the organic fraction of the sediment. Rao and Davidson (18) and Karickhoff et al. (10) have developed empirical expressions for estimating appropriate values.

Both pollutants and sediment can result from rainfall-runoff or point source discharges. They are estimated by a variety of methods, including models, but are assumed as given for the procedures discussed in this paper. The problem remaining in the mathematical model formulation of aquatic fate and transport, then, is definition of the transformation rates, R_j.

The rate equation for the last terms in equations 1-4 cannot be specified in a general form because, depending on the process under consideration, many variables may be influencing the rate, leading to a multi-term and often non-linear relationship. It is convenient, however, to assume the rates are first order with respect to the pollutant concentration for given environmental conditions, leading to

$$R = \sum_{j=1}^{n} k_j C \qquad [7]$$

where R = overall rate of dissipation processes or total first-order degradation rate ($ML^{-3}T^{-1}$)

k_j = first order rate constant for the jth process, which can vary in space and time (1/T)

n = number of processes operating on chemical

Only rarely does equation 7 provide a rigorous description of the kinetics of pollutant degradation in aquatic systems, but if the influencing variables that lead to more complex expressions are reasonably constant for a specific situation then the influence can be lumped into the single value, Σk_j.

Using selected pesticides several investigators have studied individual processes (11,17,22-24). Smith et al. (20) have identified the major pathways that lead to transformation of organic pollutants in the aquatic environment. The role of these processes is receiving increasing attention and laboratory protocols for related measurements have been developed (12). The general processes identified include hydrolysis, photolysis, oxidation, and microbial degradation. Additional processes that influence the transport of pesticides and other toxic substances include volatilization and sorption. These transformation processes are reviewed in detail in another paper in this symposium.

Alternative Solutions to the Problem

To estimate the frequency of occurrence of a given concentration, one would require a long time history of measured or simulated data, typically encompassing the range of the system's natural variability. In the general case, streamflow and sediment load must also be estimated, but for this analysis it is assumed that both are known. This assumption is not unreasonable given the extent to which major river systems are gaged. Where nonpoint source loads are of major concern, simulation of flows may be preferred because the toxicant source loads and the flow (both water and sediment) originate from the same locations within the watershed. Transport processes that occur between field-sized areas and stream channels (e.g., infiltration, sediment deposition) can result in less pollutants being delivered to the stream than are delivered to the immediate field boundary. If nonpoint source loads are estimated independently from stream flow, then the usual procedure is to apply a polluant "delivery ratio" (fraction of erosion and pollutants leaving field sites that reach streams). The following alternative solution approaches assume known discharge and flows. For the nonpoint loads, delivery ratios are assumed to be unity.

Analytical Solution of Flow Equation

If equations 3 and 4 are ignored, then under certain conditions an analytical solution to equations 1 and 2 can be developed (7,13). Specifically, if dispersion is assumed to be negligible and flow velocities and sediment concentrations are set at constant values, equations 1, 2, and 5 can be solved for the pulsed loading case to yield:

$$C = C_o \exp \frac{\sum_{j=1}^{n} k_j + \sum_{j=1}^{n} k_{js} K_D S \tau}{1 + K_D S} \qquad [18]$$

where C_o = initial toxicant load or concentration (M/L^3)

k_j = pseudo-first order rate constant for the jth process for dissolved state (1/T)

k_{js} = pseudo-first order rate constant for the jth process for sorbed state (1/T)

n = total number of processes operating

K_D = distribution or partition coefficient (L^3/M)

S = suspended sediment concentration (M/L^3)

τ = travel time, x/U (T)

The analytical solution above is obviously a gross approximation of reality, but the time-varying nature of the loadings can be preserved and the influence of suspended sediment is accommodated, however simplistically. Because time variation in the loads is preserved, it follows that cdfs can be generated and an estimate of risk is possible. A major shortcoming of the approach is the inability to readily accommodate the influence of bottom sediments. A crude approximation is possible by assuming net transfers as first-order rates, but data on which to base such as empirical approach are virtually nonexistent.

Steady-State Solutions

If risk is to be inferred from time series concentration data, steady-state solutions (i.e., solutions for which $\partial c/\partial t = 0$) are of limited value in exposure assessment. Baughman and Burns (2) argue for use of steady-state equations wherever possible and, for certain boundary conditions and study objectives, this approach has enjoyed considerable success (8,19). For cases where the time variation is caused solely by flow, judicious (piecewise) application of steady-state methods at specific flow conditions may produce useful estimates of risk. In this situation, the concentration frequencies generated are determined largely by flow variations and the problem reduces to one of single-variable probabilities. If other environmental factors (e.g., pH, light, temperature, sediment) are also time varying, then their joint frequencies must be known *a priori* if steady-state models are to be applied even for selected conditions in order to generate reasonable concentration frequency distributions. Steady-state approximations (i.e., the assumption $\partial c/\partial t = 0$) of dynamic systems are most reasonable when the system response to varying inputs is more rapid than the rate of change of the varying inputs. If, however, the actual inputs that force a system toward a quasi steady-state condition change rapidly, then that steady-state condition will not be achieved. As a result, the system will continually exhibit a transient response, moving toward changing steady state targets.

The first steady-state organic pollutant model was that of Smith et al. (20). More recently, Burns, Cline, and Lassiter (3) developed the Exposure Analysis Modeling System (EXAMS), which uses second-order transformation kinetics for the fate processes described in the previous section and calculates bed sediment-water column interactions by dispersive exchange.

Dynamic Solutions

TOXIWASP (1) is a dynamic chemical and sediment model based on the Water Analysis Simulation Program (WASP) (4). WASP is a general modeling package designed for planning analysis of water quality problems in streams, lakes, reservoirs, and estuaries. Site and pollutant specific models can be formulated with this program by linking it to different water quality kinetic subroutines. WASP uses the compartment modeling approach whereby segments can be arranged in a one-, two-, or three-dimensional configuration. Pollutant transport is based on user-specified flow and dispersive mixing between segments. Although segment volumes are time-constant, all other parameters and forcing functions may vary in space and time. WASP calculates time-varying water quality parameter concentrations using an explicit, backward difference numerical solution to the mass flux form of the one dimensional advective-dispersive equation:

$$\frac{\Delta M_j}{\Delta t} = \sum_{i=1}^{n} \left[Q_i C_i + E_i A_i \left(\frac{\Delta C}{L}\right)_i\right] + W_j - KV_j \qquad [9]$$

where ΔM = change in constituent mass

ΔC = change in constituent concentration (M/L^3)

Q = water flow (L^3/T)

E = longitudinal dispersion (L^2/T)

A = cross-sectional area (L^2)

L = characteristic mixing length (L)

W = mass loading (M/T)

K = kinetic degradation or transformation rate ($ML^{-3}T^{-1}$)

V = segment volume (L^3)

subscript "i" = segment number

subscript "j" = adjacent segments

Δ = difference symbol

Because this equation calculates the mass transported through the crosssections A_i, it is termed the "mass flux" form.

TOXIWASP was created from WASP by adding a set of subroutines that calculate chemical degradation (the last term in equation 9) and that calculate special bed sediment transport processes such as sediment-water dispersive exchange, pore water percolation, and net burial-scour. The chemical degradation routines are simplified from the EXAMS kinetic structure (3).

From chemical characteristics of a compound and the environmental parameters of the system, TOXIWASP formulates a total degradation rate. This rate is based on a simple addition of the pseudo-first order rates due to hydrolysis, oxidation, biolysis, volatilization, photolysis, and other processes. Sorption onto sediments and onto biomass is calculated assuming local equilibrium, using a chemical-specific partition coefficient, and spatially varying environmental organic carbon fractions.

TOXIWASP applies equation 9 to calculate sediment and chemical concentrations for every segment, including surface water, subsurface water, surface bed, and subsurface bed. Sediment is assumed to be a conservative constituent that is advected and dispersed among water segments and that can settle to and scour from bed segments. Within the bed, sediment can be "transported" to lower layers through burial, or to upper layers through erosion. (Bed layers have constant thicknesses, their vertical location is referenced to the bed surface, which rises with deposition or falls with scour.)

As stated above, chemicals undergo a pseudo first-order decay, based on summation of several process rates, some of which are actually second-order. The effective first order decay rate can vary with time and is recalculated as often as necessary throughout a simulation. The chemical is advected and dispersed among water segments and can be transported to the bed by sorption onto sediment and subsequent settling, or by dispersive mixing between the top bed layer and the bottom water layer. Within the bed, a chemical can migrate downward through burial, or migrate upward through erosion or pore water percolation. No lateral migration of a chemical within the bed is allowed.

Also in this study, the Hydrologic Simulation Program--FORTRAN (HSPF) was operated as a dynamic organic chemical pollutant and sediment simulation model. The full HSPF program is a comprehensive package for simulation of watershed hydrology and water quality (9). From knowledge of the time history of rainfall, characteristics of the land surface and prevailing agricultural practices, HSPF predicts the time history of runoff, sediment load, and nutrient and pesticide loads. Based upon these results and information about the chemicals involved and the stream channels in the watershed, HSPF then predicts the time history of water quality and quantity throughout the watershed. These calculations are based on kinematic wave routing and an explicit, backward difference numerical solution to the advective equation:

$$\frac{\partial(AC)}{\partial t} + \frac{\partial(AuC)}{\partial x} = \frac{AW}{V} - AK \qquad [10]$$

where u = advective velocity (L/T) and other terms are as previously defined for equation 9.

The RCHRES module of HSPF, which implements equation 10, was recently modified to handle sediment and chemical interactions with the bed. These routines were simplified from SERATRA (15) to simulate a one-layer bed. Sorption onto three size fractions of sediment is calculated as a linear, reversible process.

HSPF applies equation 10 to calculate sediment and chemical concentrations for every segment and its associated bed in a one-dimensional network. The three size fractions of sediment can be advected between water segments and undergo scour from and deposition to the bed depending on stream sediment transport capacity. Sediment transport capacity may be calculated from an empirical power function of flow or the relationships of Colby or Toffaleti (21). Bed thicknesses and properties vary throughout the simulation.

A number of distinct degradation processes can be specified, including hydrolysis, oxidation, photolysis, volatilization, biodegradation, or a lumped first-order decay rate may be specified. The chemical is advected between water segments and can be transported to the bed either by sorption onto suspended sediment and subsequent settling or by direct sorption to the bed surface. The chemical can reenter the water column by desorption from the bed surface or by scour of bed sediment containing sorbed chemical.

Although HSPF has sophisticated simulation capabilities, its data management and manipulation features make it especially convenient for completing more simplified analyses such as the analytical method described previously (Eq. 18). Indeed, the otherwise tedious task of operating an analytical method with daily loads was reduced to a straightforward application of HSPS's time series management system.

Bed Sediment-Water Column Exchange Solutions

A very important, but poorly quantified, set of processes governs chemical exchange between the bed and the water column. A chemical may sorb onto suspended sediments and settle with the sediments to the bed, or sorb directly to the bed surface when water concentrations are high. When water concentrations are low, the chemical may desorb from the bed surface. Turnover of the bed surface due to either physical or biological processes accelerates this sorption-desorption process. Diffusion of pore water into overlying water will also exchange dissolved chemicals. Percolation of pore water will move dissolved chemical up through the bed into the water column; infiltration can leach chemical down through the bed.

Because these processes are poorly quantified and variable in both space and time, most chemical models now take a largely empirical approach to sediment-water exchange. HSPF and SERATRA describe the exchange as linear sorption-desorption between dissolved chemical in the overlying water and organic sediment in the bed:

$$EXCH = RSED * KT [KPB * C_w' - C_b'] \qquad [11]$$

where EXCH = net mass exchange (M/T)
RSED = upper layer bed sediment mass (M)
KT = sorption-desorption rate (1/T)
KPB = partition coefficient of the chemical onto bed sediment (L_w^3/M_s)

EXAMS and TOXIWASP describe sediment-water exchange as a dispersion process:

$$EXCH = DISPV\ (C_p' - C_w') + DSPSED*DISPV*S'\ (C_b' - C_s'\ \frac{KPB}{KPW}) \qquad [12]$$

where DISPV = volumetric fluid dispersion (L_w^3/T)

DSPSED = spatially variable sediment turnover ratio (1.0 in EXAMS)

KPW = partition coefficient of the chemical onto suspended sediment in the water column (L_w^3/M_s)

The volumetric dispersion varies spatially, and is calculated by:

$$DISPV = \frac{SCALE * D_b * A_s * FRW}{L_c} \qquad [13]$$

where D_b = vertical pore water dispersion coefficient (L^2/T)

A_s = surface area (L^2)

FRW = porosity, or volumetric fraction of pore water in the bed

L_c = characteristic mixing length (L)

SCALE = units conversion factor

Because TOXIWASP and EXAMS assume local equilibrium between dissolved and sorbed phases of a chemical in each segment, equation 12 can be rewritten:

$$EXCH = DISPV * \frac{1 + KPB * S'}{KPB} * [C_b' - KPB * C_w'] \qquad [14]$$

where DSPSED = 1.0

Equation 14 is now in the same form as equation 11. Thus, the HSPF-SERATRA sorption-desorption rate KT can be related to the EXAMS-TOXIWASP dispersion rate by equating RSED *KT to DISPV * $[\frac{1+KPB*S'}{KPB}]$ and solving for KT. If one also uses the relationship, RSED = BVOL*S'*FRW, then the solution for KT becomes,

$$KT = \frac{SCALE * D_b * A_s}{BVOL * L_c} \quad \left[\frac{1*KPB*S'}{KPB*S'}\right] \qquad [15]$$

where BVOL = the EXAMS-TOXIWASP bed volume (L^3)

SCALE = a unit conversion factor

= 2.584×10^{-4} when $D_b = m^2/hr$, $A = ft^2$, $L_c = ft$, $BVOL = ft^3 \times 10^6$

L_c = ft

BVOL = $ft^3 \times 10^6$

For chemicals with high partition coefficients, the last term in equation 15 is approximately 1.0, simplifying the expression for KT to:

$$KT = \frac{SCALE * D_b * A_s}{BVOL * L_c} \qquad [16]$$

Although the basic sediment-water exchange mechanisms are mathematically equivalent, each model has its own elaborations. The exchange rates are spatially variable in EXAMS and TOXIWASP, and these models include percolation of pore water. TOXIWASP adds spatially variable sediment turnover ratios, spatially variable net bed scour or deposition rates, and settling of suspended sediment. HSPF and SERATRA simulate three sediment size fractions. Scour-deposition and settling are calculated as dynamic processes. SERATRA adds multiple sediment layers, which can be added through deposition or subtracted through erosion. HSPF allows spatially variable sorption-desorption rates. The use of all these mathematical formulations is limited by the lack of good, thorough data necessary to calibrate and test the models on a site-specific basis.

An Example of Alternative Solutions

Application of the various models described above remains largely an untested process although various components have been evaluated for their ability to simulate field observation (19). Much can be learned, however, by applying each model to the range of problems usually encountered in many practical environmental exposure assessments. Results from each model can be evaluated for their relative differences and a better understanding of the various models and their appropriate use can be achieved.

The applications problem selected for this study involves modeling a river reach characterized in Figure 1. Figure 1 actually represents a gaged section of the Yazoo River in the State of Mississippi, USA. Flow records were available for the gages shown in Figure 1b and have been used

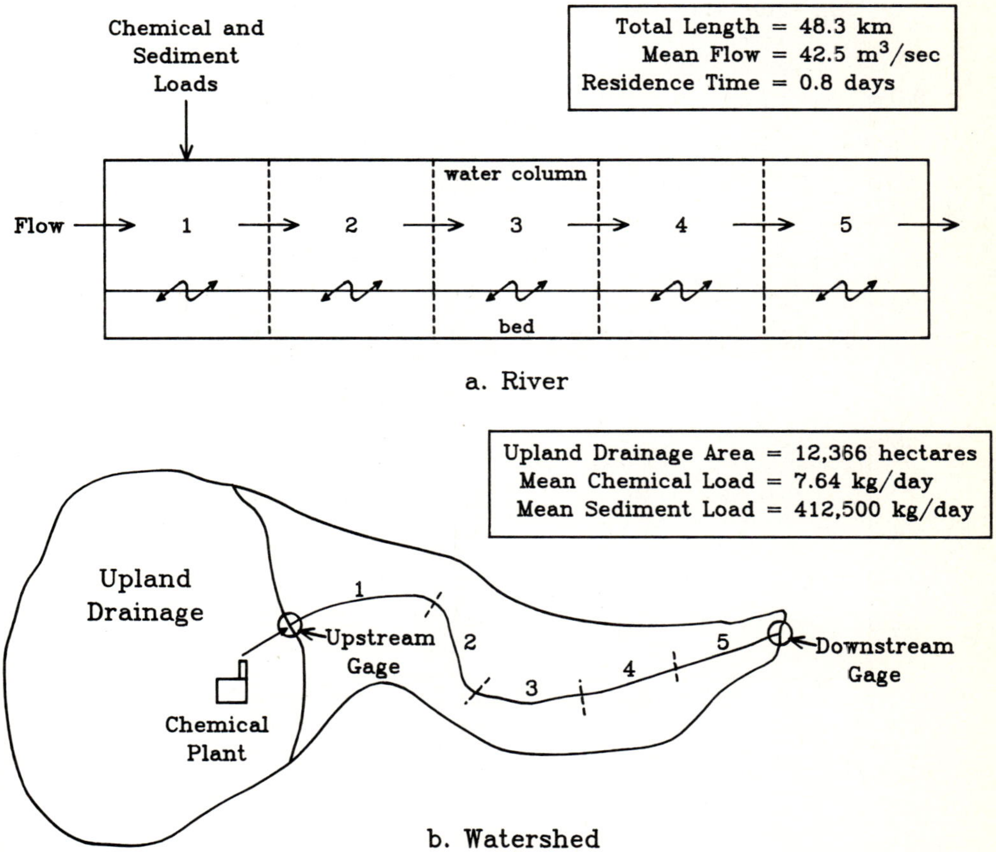

Figure 1. Overview of river environment.

in previous studies of pesticide exposure within the river basin (16). Flow is shown in Figure 2 for 1972. The chemical and sediment loads for 1972 are also shown in Figure 2 as derived (not measured) from a previous modeling study of pesticide runoff in the area. The annual-average flows and loads are also given in Figure 1.

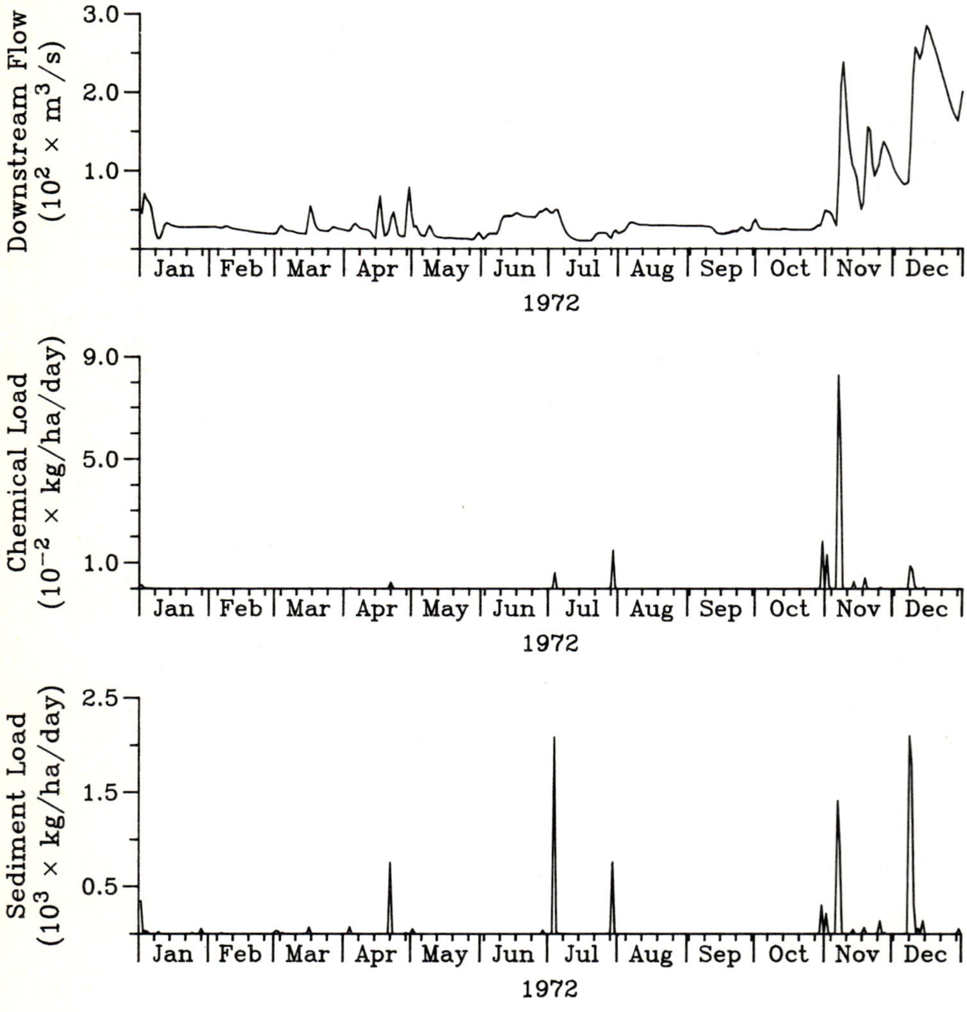

Figure 2. Unsteady loads and flow.

The environment and its characterization in Figures 1 and 2 can be combined with various loading scenarios to generate four general conditions: (1) steady river flow and steady loads represented by mean flows and input loads in Figure 1, (2) steady river flow and dynamic input loads represented by the mean flow in Figure 2 and the dynamic loads in Figure 2, (3) dynamic river flows and steady input loads represented by the flows in Figure 2 and the mean input load in Figure 1, and (4) dynamic river flows and input loads as represented in Figure 2. These four cases can be thought to represent actual real world situations or they can be perceived as possible ways to represent the same situation. For example, the conditions represented by condition 4 could be approximated by the other three conditions for purposes of analysis or modeling.

Another key set of variables that influence results are those related to the chemical properties of the input loads. Recall from previous discussion and from the description of each model that several chemical, biological, and transport processes are chemical-specific and must be uniquely characterized (parameters estimated) for each chemical. To simplify this process, however, one can specify a lumped first-order rate constant for all transformations and further specify a bulk sediment-water partition coefficient. These two parameters capture the chemical-specific information needed for the comparison study intended here. The values chosen for this study for each of two hypothetical chemicals represent a relatively non-persistent compound that sorbs weakly to sediments and a more persistent compound that sorbs much more strongly to sediment.

The range of general conditions and chemicals, and the models investigated for each are summarized in Figure 3. Tables 2-3 give the detailed environmental characteristics required for the various models. Most of these data were derived from gaging information, but some were assumed for

Case	Flow	Load	Chemical	Analytical	EXAMS	TOXIWASP	HSPF
	Conditions			Models			
1	S	S	1		*		
	S	S	2		*		
2	S	D	1	*		*	
	S	D	2	*		*	
3	D	S	1		*	*	*
	D	S	2		*	*	*
4	D	D	1			*	*
	D	D	2			*	*
5	D	H	1				*
	D	H	2				*

S = steady

D = varying daily

H = varying hourly

1 = high decay (0.2 day^{-1}), low partitioning (KOC = 200 l_w/kg_s)

2 = low decay (0.02 day^{-1}), high partitioning (KOC = 2000 l_w/kg_s)

Figure 3. Conditions investigated by models.

purposes of this study. Note that in Figure 3 five conditions rather than only the four discussed above are given. The last condition accommodates utility of an hourly varying load for comparison with the daily varying load for case 4. Recall from the discussion of each model that some processes included in one are not included in others. Most notable among these is the treatment of bed sediments. The analytical model (equation 8) does not include the influence of bed sediments. Thus, TOXIWASP used in case 2 was configured to omit bed-sediment interactions to enable a more meaningful comparison between the TOXIWASP results and those from the analytical model. Also recall from previous discussion of steady-state

Table 2

Parameter Values for River Simulations

	Segment Number				
Water Sediment Bed	1 2	3 4	5 6	7 8	9 10
Geometry					
Length m	9.66	9.66	9.66	9.66	9.66
Mean Water Volume m^3	$5.25x10^5$	$5.51x10^5$	$5.80x10^5$	$6.04x10^5$	$6.29x10^5$
Bed Volume m^3	$1.39x10^4$	$1.39x10^4$	$1.39x10^4$	$1.39x10^4$	$1.39x10^4$
Surface Area m^2	$2.28x10^5$	$2.28x10^5$	$2.28x120^5$	$2.28x10^5$	$2.28x10^5$
Mean Water Depth m	2.30	2.42	2.54	2.65	2.76
Bed Depth m	0.061	0.061	0.061	0.061	0.061
Transport					
Sediment-Water Dispersion cm^2/sec	$6.59x10^{-5}$	$6.59x10^{-5}$	$6.59x10^{-5}$	$6.59x10^{-5}$	$6.59x10^{-5}$
Mean Flow m^3/sec	35.7	39.1	42.5	6.1	9.4
Mean Residence Time days	0.17	0.16	0.16	0.15	0.15
Sediment Characteristics					
Organic Fraction, Water	0.03	0.03	0.03	0.03	0.03
Organic Fraction, Bed	0.06	0.06	0.06	0.06	0.06
Bulk Density, Bed kg/l	2.70	2.70	2.70	2.70	2.70
Bed Porosity	0.50	0.50	0.50	0.50	0.50

Table 3

Channel Properties Used in HSPF Routing

Segment No.	Flow (m^3/s)	Depth (m)	Surface Area (m**2)	Volume (**2)	Flow-Thru Time (days)
	0.1	0.4	111673.	44199.	5.12
	0.5	0.7	138774.	101247.	2.34
	1.0	0.9	152387.	144681.	1.67
	5.0	1.8	189369.	331421.	0.77
1,3,5	10.0	2.3	207945.	473599.	0.55
	50.0	4.2	258410.	1084877.	0.25
	100.0	5.5	283758.	1550283.	0.18
	500.0	10.1	352622.	3551244.	0.08
	1000.0	13.1	387211.	5074707.	0.06
	0.1	0.3	101261.	30557.	3.54
	0.5	0.5	124827.	68437.	1.58
	1.0	0.7	136597.	96852.	1.12
	5.0	1.3	168387.	216917.	0.50
7,9	10.0	1.7	184264.	306979.	0.36
	50.0	3.0	227147.	687532.	0.16
	100.0	3.9	248566.	976991.	0.11
	500.0	7.1	306413.	2179178.	0.05
	1000.0	9.2	335306.	3083959.	0.04

models the conditions for which dynamic systems could be approximated by repeated (piecewise) application of such models. Application of EXAMS to case 3 conditions was accomplished by obtaining steady-state solutions at the selected flow conditions shown in Table 4. These flows span the range of observed flows shown in Figure 2. Because only flow varied in this case and because the variations were not dramatic (flow is regulated by an upstream reservoir), one might expect a piecewise application of a steady-state model to be reasonable practice. Application to other situations where loads and flows are both dynamic is not recommended by the authors.

Investigation of the conditions and models shown in Table 1 allowed comparisons in several ways. For some cases different models were applied to the same condition allowing evaluation of differences in models. Differences in results

Table 4

Flow Frequency Analysis of River Flows

	Segment No.	Frequency						
		5%	15%	30%	50%	70%	85%	95%
Inlet gage	1	8.5	11.3	19.3	24.1	31.1	46.7	142
	3	9.4	12.5	19.9	24.8	32.5	50.7	153
	5	10.4	13.6	20.6	25.5	33.8	54.7	164
	7	11.3	14.7	21.3	26.2	35.1	58.6	176
	9	12.3	15.9	22.0	26.9	36.4	62.6	187
Outlet gage	0	13.2	17.0	22.7	27.6	37.7	66.5	198

were directly attributable to differences in the models being applied to the same condition. The range of conditions was also selected to examine modeling results under various simplifying assumptions. For example, case 4 can represent an additional assumption imposed on case 5. In other words, the hourly varying loads of case 5 were approximated by the same loads aggregated to daily varying values in case 4. Similarly, case 3 can be viewed as an approximation to case 4 followed by cases 2 and 1. Finally, each condition and model was investigated for each of two chemicals characterized as shown in Figure 2.

Estimation of environmental risk remained the objective for exposure assessment modeling in this study and, therefore, time series concentrations had to be produced by the models in order to generate cumulative probability density functions for the chemicals of interest. The concentration cdf generated from each model simulation was used as the basis for comparing results for each of the conditions and chemicals presented in Figure 3. Each cdf was derived by assigning class intervals, pooling daily average concentrations into each appropriate class interval, and accumulating the totals over the range of simulated values (minimum to maximum).

Results and Discussion

The model results for case 1 cannot be expressed as time series or in terms of cdf's because only steady-state values were estimated. For the other cases, however, time series concentrations for dissolved chemical in the water column and sorbed chemical in the bed were estimated by each model. As stated earlier, EXAMS results for case 3 are not time series *per se*, rather they are specific steady-state solutions to specific flow conditions (see Table 3) that cover the range of flows. Typical results for cases 4 and 5 for each of the two chemicals are shown in Figure 4. The short residence time for

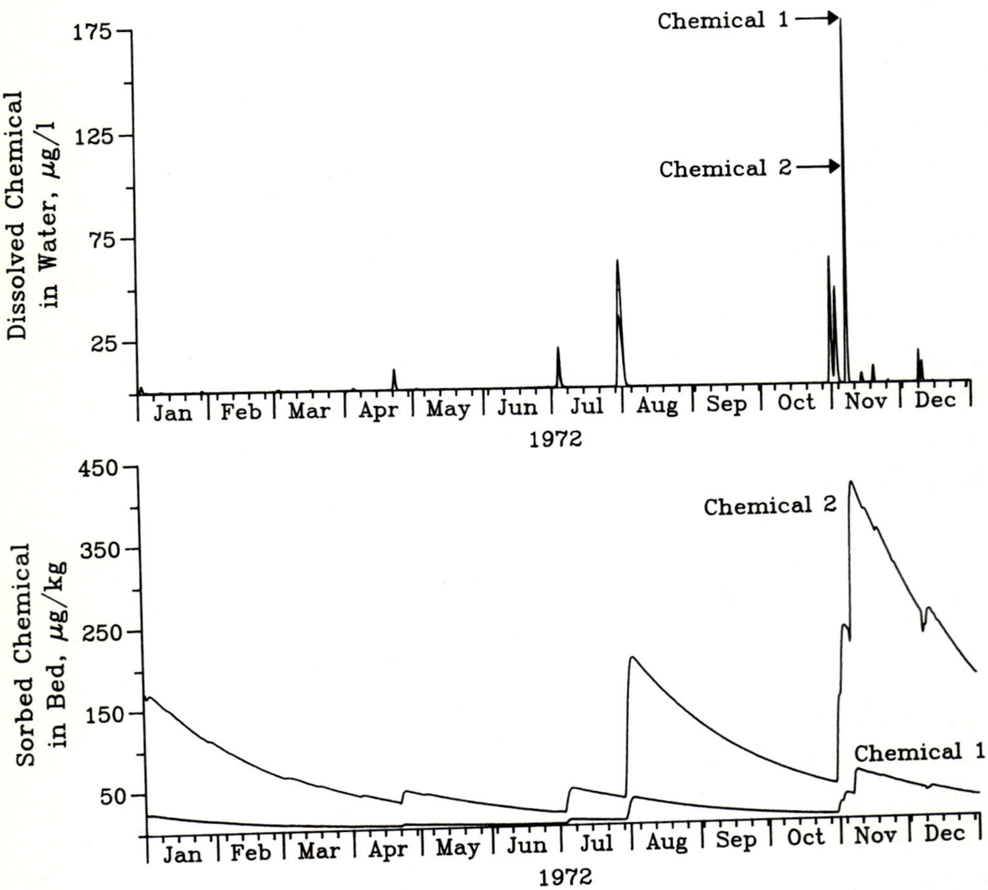

Figure 4. Downstream chemical response to unsteady loads and flow.

the river system precludes a major difference in the concentration profiles as a result of degradation but the differences due to partitioning are more pronounced. This is readily apparent from the bed sediment concentrations shown in Figure 4.

The time-series concentrations of Figure 4 are useful to examine but perhaps more insight can be gained by examining each model output as cumulative frequencies. The cdf's estimated by each model for each of the conditions outlined in Table 2 and for the flow and loading given in Figures 1 and 2 are shown in Figures 5 and 6. Semi-log representations of concentration were used only because the range of values spanned several orders of magnitude.

It is useful to take case 5 as our best estimate of the true cumulative probabability of the chemical concentrations at the downstream gage. Water concentrations below about the 75th percentile result from the desorption of chemical from the bed rather than direct chemical loading. Case 4 predicts slightly higher "chronic" concentrations because average daily events allow more chemical to sorb to the bed than average hourly events, even if the total mass loadings are the same. Higher bed storage during events leads to greater desorption between events. The differences between HSPF and TOXIWASP for case 4 are probably numerical, the result of constant time steps in the former and variable time steps in the latter.

Case 2, featuring constant flow, daily loads, and no bed sediment, mimics cases 4 and 5 above the 95th percentile where concentrations are driven by large loads. Below this percentile, concentrations for case 2 fall progressively lower, lacking desorption. The concentration of zero, not represented on this semilog graph, has a cumulative frequency of about 75%. TOXIWASP (no bed sediment included) predicts slightly higher concentrations than the analytical solution

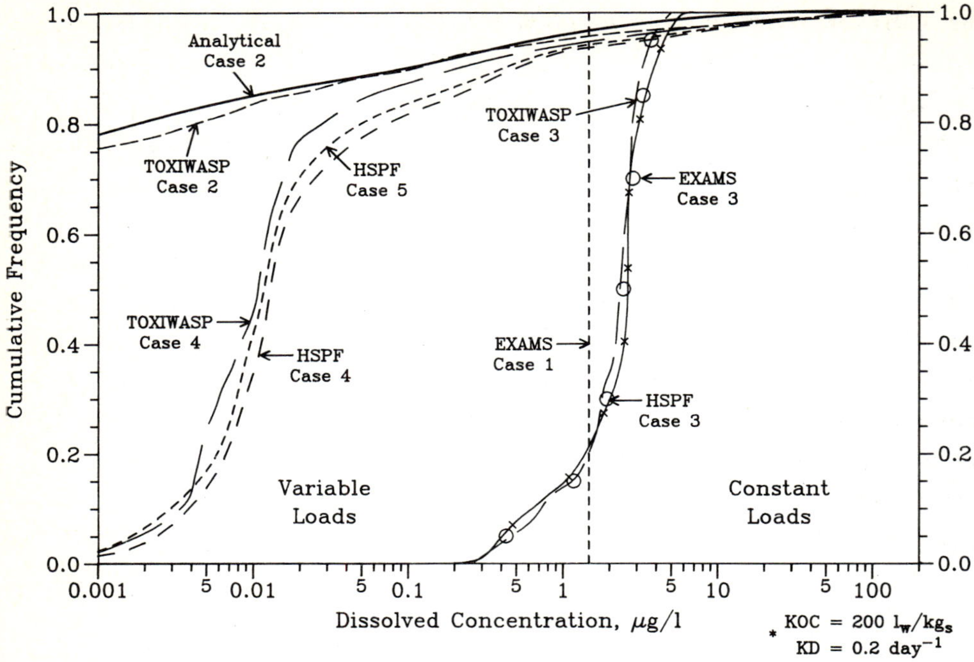

Figure 5. Dissolved chemical 1 in water.

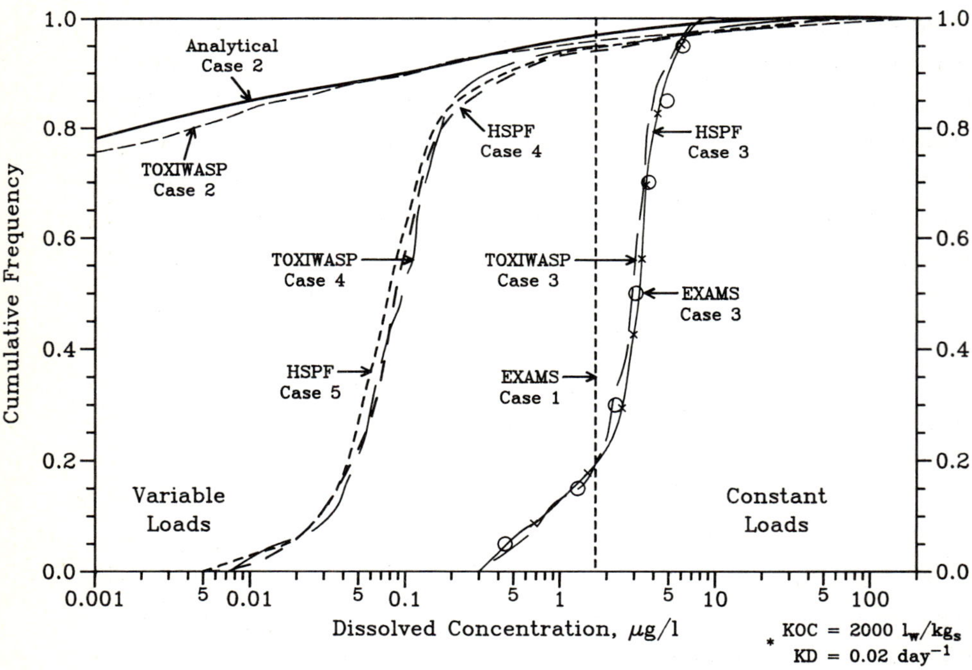

Figure 6. Dissolved chemical 2 in water.

for a given cumulative frequency because of numerical dispersion.

Substituting the constant average loading for the dynamic loading, case 3 results differ remarkably from cases 2, 4, and 5. Concentration variations are due entirely to flow variations. The case 3 results confirm that piecewise steady-state solutions at various flows (EXAMS) can approximate the dynamic system under study if certain conditions are met (e. g., steady loads, gradually varying flows). The true steady-state case 1 concentration is less than the case 3 median because the average yearly flow used was higher than the median flow. The case 3 concentration varies from the mean flow, true steady state concentration by a factor of about $\pm$ 3.

Comparing the two chemicals for each case, we find concentrations for chemical 2 always exceed those for chemical 1. This is primarily due to differences in the decay rate. Dissolved chemical 1 decays at a rapid 0.2 per day rate, whereas dissolved chemical 2 decays at a slower 0.02 per day rate. Referring to Figure 3, note that chemical 2 builds up in the bed to a much greater extent than chemical 1 because its partition coefficient is higher (2000 l/kg versus 200 l/kg). This same factor reduces the dissolved concentrations for chemical 2 during peak events.

Conclusions

Care must be taken in drawing general conclusions from specific studies. Our results should be valid for rivers similar to the moderate-sized reach studied here. Interactions between varying environmental parameters and chemical decay rates, not studied here, can add another dimension of complexity to comparátive analyses. Indeed, such conditions may yield vastly different results for some cases, especially those investigating steady-state approximations to dynamic systems. Based on this limited comparative study, several conclusions were drawn:

1. Steady-state models gave reasonable estimates of the median concentration for the constant load, variable flow situations.

2. Flow variation alone produced concentrations up to a factor of ±3 around steady state values obtained with mean annual flows.

3. Flow conditions taken from the *a priori*, known flow cumulative frequency distribution were successfully used for steady-state approximations of the chemical cumulation frequency distribution for conditions of steady load and varying flow.

4. Steady-state models applied to rivers with pulse loadings were inappropriate.

5. All approaches using non-steady loading gave similar results for the high concentration, infrequent events that could produce acute toxic effects.

6. Analysis of the high concentration, infrequent events did not require modeling interaction with the bed sediment.

7. It was necessary to simulate bed sediment if analysis of low concentrations were desired.

8. The use of daily averaged input loads rather than hourly input loads made only a minor difference in simulation results for low concentrations, and a slightly greater difference for the higher concentrations.

9. Assuming constant river volumes (TOXIWASP) rather than time-variable volumes (HSPF) resulted in no substantial differences between the respective cumulative frequency distributions.

Acknowledgements

Staff at the Athens Environmental Research Laboratory provided critical support for this project. Bruce Bartell, with Computer Science Corporation, drafted the figures by computer. Annie Smith typed timely, accurate drafts. Bob Ryans provided our technical editing. Sandra Ashe assisted in the tedious typing, copying, and other preparation chores.

Parts of this paper were first presented at a UNESCO workshop, "Comparison of Application of Mathematical Models for the Assessment of Changes in Water Quality in River Basins, Both Surface and Groundwater," held in La Coruna, Spain in 1982. The paper was subsequently substantially revised based on comments by Mr. Charles Delos of the Office of Water, U.S. EPA, and Drs. Robert Swank and Lawrence Burns, Office of Research and Development, U.S. EPA. An oral presentation of the results of this work was made at the American Geophysical Union Fall Meeting in 1983.

References

1. Ambrose, R. B., S. I. Hill, and L. A. Mulkey. User's manual for chemical transport and fate model TOXIWASP-version 1. U.S. Environmental Protection Agency, Athens, GA. EPA-600/3-83-005. 1983. 198 pp.

2. Baughman, G. L. and L. A. Burns. Transport and transformation of chemicals in the environment: A perspective. In: The handbook of environmental chemistry, O. Hutzinger, Ed. Springer-Verlag Publishing Co. 1980. pp 1-77.

3. Burns, L. A., D. M. Cline, and R. R. Lassiter. 1982. Exposure analysis modeling system (EXAMS): User manual and system documentation. U.S. Environmental Protection Agency, Athens, GA. EPA-600/3-82-023. 1982. 454 pp.

4. DiToro, D. M., J. J. Fitzpatrick, and R. V. Thomann. Water quality analysis simulation program (WASP) and model verification program (MVP)-documentation. U.S. Environmental Protection Agency, Duluth, MN. EPA-600/3-81-004. 1983.

5. Falco, J. W. and L. A. Mulkey. Modeling the effect of pesticide loading on riverine ecosystems. pp. 156-165. In: Proceedings of the conference on environmental modeling and simulation. U.S. Environmental Protection Agency, Washington, DC. EPA-600/9-76-016. 1976

6. Falco, J. W., D. L. Brockway, K. L. Sampson, H. P. Kollig, and J. R. Maudsley. Models for transport and transformation of malathion in aquatic systems. pp. 97-113. In: Water quality criteria research of the U.S. Environmental Protection Agency. U.S. Environmental Protection Agency, Corvallis, OR. EPA-600/3-76-079. 1977.

7. Falco, J. W., L. A. Mulkey, K. F. Hedden, C. N. Smith, T. O. Barnwell, J. D. Dean, R. E. Lipcsei, and M. C. Smith. Estimated degradation and transport of Dimilin in selected rivers of the southern United States. U.S. Environmental Protection Agency, Athens, GA. (Unpublished report.) 1978.

8. Falco, J. W., L. A. Mulkey, R. R. Swank, R. E. Lipcsei, and S. M. Brown. A screening procedure for assessing the transport and degradation of solid waste constituents in subsurface and surface waters. Environmental Toxicology and Chemistry, Washington, DC, Vol. 1, pp. 121-134, 1982.

9. Johanson, R. C., J. C. Imhoff, J. L. Kittle, and A. S. Donigian. Hydrological simulation program--FORTRAN (HSPF): Users manual for release 8.0. U.S. Environmental Protection Agency, Athens, GA. EPA-600/9-84-066. 1984. 774 pp.

10. Karickhoff, S. W., D. S. Brown, and T. A. Scott. Sorption of hydrophobic pollutants on natural sediments. Water Research, 13: 241-248; 1979.

11. Mackay, D. and P. J. Leinonen. Rate of evaporation of low solubility contaminants from water bodies to atmosphere. Environ. Sci. Technol. 9:1178-1180; 1975.

12. Mill, T., W. R. Mabey, P. C. Bomberger, T. W. Chou, D. G. Hendry, and J. H. Smith. Laboratory protocols for evaluating the fate of organic chemicals in air and water. U.S. Environmental Protection Agency, Athens, GA. EPA-600/3-82-022. 1982. 337 pp.

13. Mulkey, L. A. and J. W. Falco. Methodology for predicting exposure and fate of pesticides in aquatic environments. In: Agricultural management and water quality, F. W. Schaller and G. W. Bailey, Eds. Iowa State Press. 1983.

14. Onishi, Y., S. M. Brown, A. R. Olsen, M. A. Parkhurst, S. E. Wise, and W. H. Walters. Methodology for overland and instream migration and risk assessment of pesticides. U.S. Environmental Protection Agency, Athens, GA. EPA-600/3-82-024. 1982.

15. Onishi, Y. and S. E. Wise. User's manual for the instream sediment-contaminant transport model, SERATRA. U.S. Environmental Protection Agency, Athens, GA. EPA-600/3-82-055. 1982. 195 pp.

16. Onishi, Y., G. Whelan, M. A. Parkhurst, A. R. Olsen, and P. J. Gutknecht. Preliminary assessment of toxaphene migration and risk in the Yazoo River Basin, Mississippi. U.S. Environmental Protection Agency, Athens, GA. Prepared for U.S. Environmental Protection Agency, Office of Pesticide Programs, Washington, DC. (Unpublished report.) 1980.

17. Paris, D. F., D. L. Lewis, J. T. Barnett, and G. L. Baughman. Microbial degradation and accumulation of pesticides in aquatic system. U.S. Environmental Protection Agency, Washington, DC. EPA-660/3-75-007. 1975.

18. Rao, P. S. C. and J. M. Davidson. Estimation of pesticide retention and transformation parameters required in nonpoint source pollution models. In: Environmental impact of nonpoint source pollution, M. R. Overcash and J. M. Davidson, Eds. Ann Arbor Science, Ann Arbor, MI. 1980. pp. 23-67.

19. Schnoor, J. L. Fate and transport of dieldrin in Coralville Reservoir: residues in fish and water following a pesticide ban. Science, 211:840-842; 1981.

20. Smith, J. H., W. R. Mabey, N. Bohonos, B. R. Holt, S. S. Lee, T-W. Chou, D. C. Bomberger, and T. Mill. Environmental pathways of selected chemicals in freshwater systems. Part I: Background and experimental procedures. U.S. Environmental Protection Agency, Athens, GA. EPA-600/7-77-113. 1977. 97 pp.

21. Vanoni, V. A., Editor. Sedimentation engineering. Prepared by the ASCE Task Committee for the Preparation of the Manual on Sedimentation of the Sedimentation Committee of the Hydraulics Division, New York. 1975.

22. Wolfe, N. L., R. G. Zepp, G. L. Baughman, and J. A. Gordon. Kinetic investigation of malathion degradation in water. Bull. Environ. Contam. Toxicol., 13:707-713; 1975.

23. Wolfe, N. L., R. G. Zepp, G. L. Baughman, and D. M. Cline. Kinetics of chemical degradation of malathion in water. Environ. Sci. Technol., 11:88-93; 1977.

24. Zepp, R. G., N. L. Wolfe, J. A. Gordon, and G. L. Baughman. Dynamics of 2,4-D esters in surface waters: hydrolysis, photolysis, and vaporization. Environ. Sci. Technol., 9:144-150; 1975.

25. Zepp, R. G. and D. M. Cline. Rates of direct photolysis in aquatic environment. Environ. Sci. Technol., 11:359-366; 1977.

DISCUSSION - TOXIC POLLUTANT IMPACTS ON RECEIVING WATERS

In the discussion which followed the presentation of the papers, the following points were made:

1. Dr. Ferrara and Mr. Ellis discussed the important regulatory factors that might control both in-pipe and receiving water toxic pollutant levels. Dr. Ellis' main criteria were pH (since many heavy metals exhibit sensitive "adsorption-edge" affinity; e.g. a 1-1.5 unit pH change could substantially increase concentrations in the dissolved phase), the DOC/POC ratio (because organic carbon content is extremely important for mobilization and transfer from solid to soluble phases), and ionic strength (increases in winter salting activity releases large amounts of exchangeable metal to the water column). Surface area effects, on the other hand, appear to be relatively negligible and main ternary association is the pollutant-organic ligand-hydrous oxide surface.

2. Mr. Ellis also suggested that more knowledge was needed on speciation effects as total metal levels do not provide indications of toxicity in terms of bioavailability.

3. Mr. Metaxatos stated that the need for use of deterministic modeling, instead of steady-state models, for the analysis of pollutant effects from urban runoff to the receiving waters is imposed by the fact that the pollutant accumulation on the streets follows a decreasing rate of increase.

4. Dr. James asked if the sediment-water interactions model could be applied in the vicinity of an outfall. Mr. Barnwell replied that it was possible in principle, but the hydrodynamics would likely be quite different than in a stream.

5. Mr. Terstriep inquired about the availability of data on sediment-water exchange. Mr. Barnwell replied that data were limited, but that a lower bound on the exchange rate could be set by molecular diffusivity. EPA is currently sponsoring some laboratory studies to try to better define this process. Model sensitivity studies indicate that it is critical to understanding the fate and transport of sorbed pollutants.

STATE-OF-THE-ART, PHYSICALLY-BASED AND STATISTICALLY-BASED WATER QUALITY MODELING

Miguel A. Medina, Jr.
Department of Civil and Environmental Engineering
Duke University
Durham, North Carolina 27706
U.S.A.

Introduction

Urban stormwater runoff processes can be subdivided into three major subsystems: (1) overland flow, (2) storage and transport through natural or engineered drainage conduits, and (3) receiving water transport. Successful receiving water quality modeling is highly dependent upon our ability to describe mathematically pollutant accumulation, surface washoff and other transport mechanisms occurring before discharge to the receiving subsystem. Because of the random nature of hydrologic events, it is appropriate to analyze water quality effects within a probabilistic setting: this can be accomplished by long-term simulation with physically-based models, or by statistically-based (e.g., derived distribution) approaches. The probabilistic models are highly dependent on how well the distributions of the original variables can be hypothesized. The physically-based models are highly dependent on the input variable time series, which includes parameters that may have been estimated from available data by statistical procedures.

This paper reviews physically-based (Eulerian and Lagrangian reference frames) and statistically-based methods

NATO ASI Series, Vol. G10
Urban Runoff Pollution.
Edited by H. C. Torno, J. Marsalek, and M. Desbordes

for predicting water quality, including opportunities for integration of both approaches. Stochastic behavior can exist in the input signal; therefore, it can also exist in the output signal _even for a purely deterministic system_, or in the _initial conditions_ of the governing differential equations. Stochastic processes are described by their _statistical measures_ (such as the mean, covariance, higher moments) or even the distribution function or characteristic function (1). There may be randomness in the system itself, rather than just the input signal. Simplification of complex interactions and reactions introduces _uncertainty_ into our representation of physical, chemical and biological processes. Rainfall and, therefore, surface runoff (streamflow) are both inherently random events. It is appropriate to analyze water quality fluctuations within a probabilistic setting. Yet, the mathematical models need not be statistical black boxes. For example, for the past decade physically-based continuous (long-term) simulation models have been commonly applied to urban catchments to predict water quality effects in terms of the probability of occurrence of events of various magnitudes. This has been accomplished by the output time series in terms of many of the same _statistical measures used in stochastic models_.

Generally, the statistical approach attempts to calculate the probability distribution of receiving water quality concentrations and mass rates given the probability distribution of model inputs: for example, the hydrology, pollutant loadings, stream temperatures, etc. This is the classical _derived distribution_ problem. If the governing differential equations of a deterministic model are recast to account for variability in parameters such as the reaction rate coefficients and inputs, they become _random differential equations_. If a deterministic model is applied repeatedly using initial conditions, inputs, and coefficients randomly selected from their probability distribution this constitutes Monte Carlo simulation. It produces an ensemble of synthetic time series which can also be analyzed in terms of the statistical characteristics of the system response. The aim of Monte Carlo simulation is to produce a set of equally-likely events that are statistically indistinguishable from the historical data, for the length of time required or appropriate to the task.

The relationship between probabilistic water quality analysis and time variable deterministic simulation has been noted earlier (e.g.,44,15). The probabilistic analysis begins with the simplest deterministic model: a mass balance at the location where the pollutant loads can be assumed to enter the stream. Then, probability distributions are assumed or derived for each of the variables in the mass balance equation (flows and concentrations for the source and the stream). Analytical methods based on moments (mean and variance) have been presented (55,15,16) as well as a semi-analytical (numerical integration) method (15) and, Monte Carlo simulations (55) of "independent" random variables. In the Monte Carlo simulations, a value was drawn randomly from each of the distributions (river flow, discharge flow, river water quality concentration, discharge water quality concentration) and these values were subsequently used to calculate the water quality concentration downstream from the discharge. Both Warn and Brew (55) and DiToro (15,16) assumed lognormally-distributed flows and concentrations. This characterization seems justified for many streams and stormwater runoff data sets on the basis of published results of the Nationwide Urban Runoff Program (NURP) (56). However, each site-specific hydrologic time series must be checked accordingly.

Conceptually-based time variable "deterministic" model simulations attempt to predict the system response time series for each variable of interest at various locations--in the exact temporal sequence that would occur from the input time series (historical or synthetically derived). After sufficiently long simulation, the predicted time series can be evaluated, as noted above, in terms of frequency distributions of the magnitude of these events and their duration. Because many of these model parameters are difficult to estimate (infiltration rates, pollutant accumulation and washoff coefficients, reaction rates, dispersion, etc.) the deterministic label is quite optimistic. Within the past decade, Monte Carlo techniques have been more frequently combined with physically-based models to attempt to produce an output time series more representative of catchment characteristics and historical hydrologic time series. These methods can also be combined in the solution of random differential equations to estimate the distributions of reaction rate coefficients. However, for Monte Carlo methods a great amount of computer time is required to assure reasonably accurate estimates of the probability density function of system response. A unifying framework is presented in which both statistical and deterministic models can be integrated in the analysis to achieve: long-term characterization of the rainfall -runoff process and derivation of both hydrologic and water quality frequency distributions. If the level at which a particular pollutant concentration impairs the use of the receiving stream is known, then the concentration probability distribution specifies the frequency (and duration in the case of simulation of transients) with which that use (fisheries, waste allocation) will be impaired. These concepts are illustrated in Figures 1 and 2. In Figure 1, cumulative water quality frequency curves are shown, for different levels of protection upstream. In Figure 2, the frequency distribution of <u>consecutive</u> hours of violation of a stream water quality standard is depicted. If the assumptions upon which the statistical models are based are met, for a particular study site and hydrologic time series, the calculations can be easily performed on a microcomputer and the data requirements are relatively simple. These models are very useful for preliminary water resource assessment. They can be followed by a more refined approach if it is justified, as outlined in Table 1. Continuous deterministic simulation (both steady-state and transient) accounts for the actual sequencing of storm event loads. Much more complex statistical procedures (often impractical) would be required to approximate this phenomena. Deterministic models also produce a more complex time history of system response to excitation , representing not only individual storm properties but also cumulative effects of closely-spaced events. These models can be relatively expensive to execute, are data intensive and require large memory devices for data manipulation and storage. Advances in microcomputer technology are making it possible to load these large programs into memory but execution times are still long.

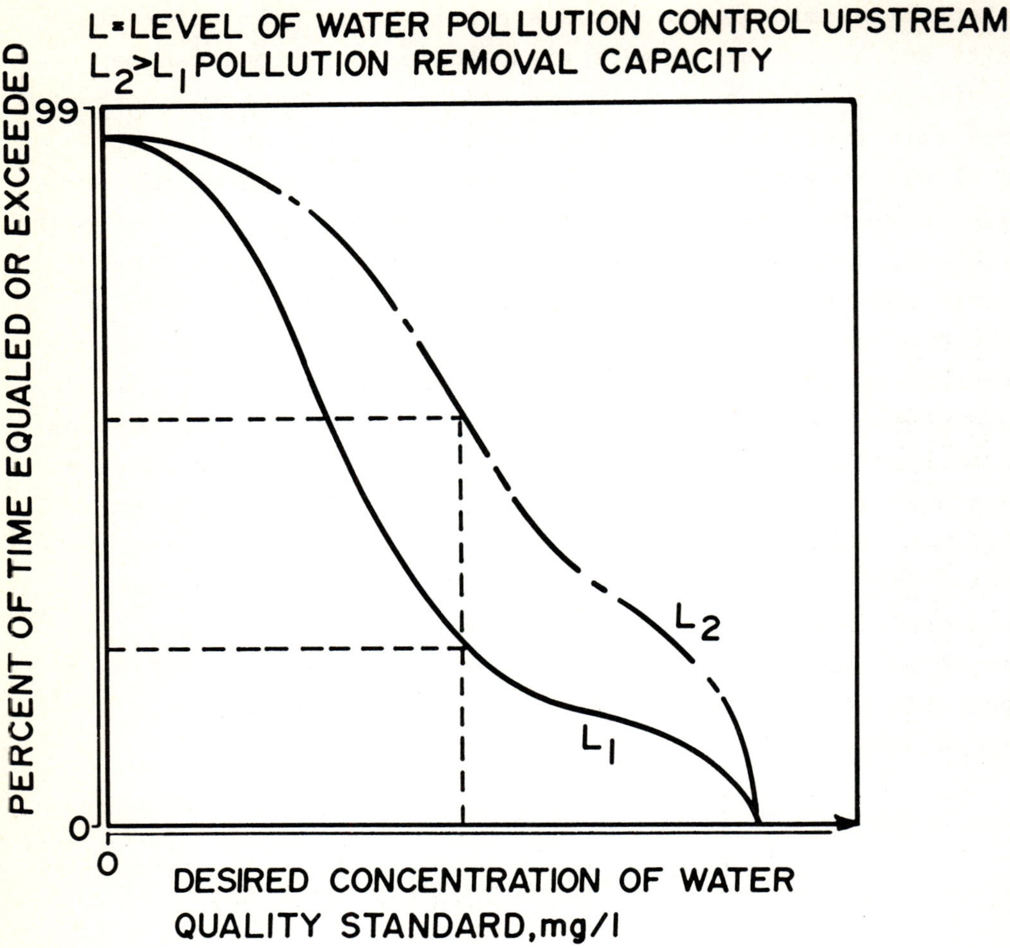

Figure 1. Cumulative Water Quality Frequency Curves for Hypothetical Watershed

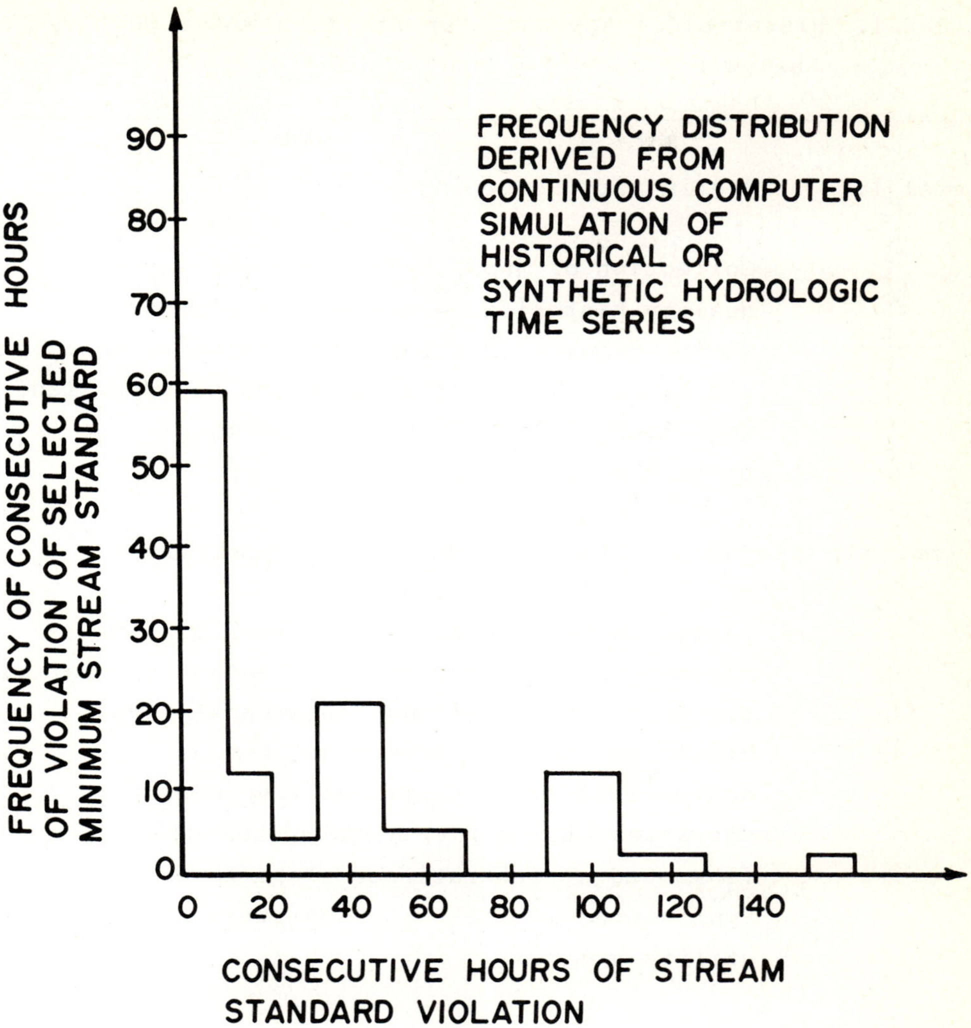

Figure 2. Frequency Distribution of Consecutive Time Periods of Violation of the Selected Stream Standard

TABLE 1. Hierarchical Approach For Instream Water Quality Analysis

---

Level I: Preliminary Screening

Collection of maps, hydrologic and water quality historical data, stream intensive survey reports, routine computation of means and extremes, inventory of point sources, land use, identification of fish species in the river reach, etc.

Level II: Statistically-Based Frequency Analysis

Identification of potential water quality impacts from derivation of pollutant cumulative frequency distributions, after analyzing historical data to determine if model assumptions (e.g., lognormality) are satisfied--for point, nonpoint and upstream flows and concentrations. This level requires application of small _microcomputer_ programs and utility modules.

Level III: Intermediate, Deterministically-Based Frequency Analysis

Steady-state, deterministic continuous simulation to derive cause/effect cumulative water quality frequency distributions, involving _mainframe_ computer or _minicomputer_. A more refined frequency analysis, requiring substantially more data than Level II, with histories of system response to many combinations of waste inputs and levels of control.

Level IV: Advanced, Deterministically-Based Frequency and Duration Analysis

Deterministic, continuous simulation of receiving water quality transients to derive both frequency and duration of water quality violations; three-dimensional computer graphics capability to view cause/effect (pollutant/water quality) relationships in time histories of system response to varying inputs and configurations of wastewater treatment unit operations, etc. At this level, it is likely that mainframe computer simulation will continue to be more convenient due to long execution times; data requirements are substantial; assessment of biological damage is possible on the basis of combined frequency and duration of water quality violations of established standards.

This paper also reviews the dependency of receiving water quality modeling upon modeling surface pollutant accumulation and washoff processes. Topics include: interfacing with surface runoff quality models; determination of rates, kinetics formulations, highly coupled reactions and feedback mechanisms; calibration and verification procedures, parameter estimation, uncertainty in prediction; and interpretation of water quality impacts in terms of water quality standards and criteria. Future research and development needs are discussed, as well as the impact of emerging micro-computer technology upon both model developers and model users. Applications are presented for urban catchments in the United States (Salem Creek, Winston-Salem, North Carolina) and Australia (Giralang, Canberra), while integrating as much as possible the physically-based and statistically-based methods, subject to data limitations.

Deterministic, Physically-based, Hydrologic and Water Quality Modeling

The majority of surface runoff and receiving water quality models are based on fixed Cartesian coordinate axes: the Eulerian spatial reference frame. However, there are some advantages to using a moving coordinate system: the Lagrangian spatial reference frame. These models are discussed in a separate section below.

Surface Runoff Quantity

Kinematic wave distributed routing rainfall-runoff models are commonly used to provide detailed simulation of storm-runoff periods with soil-moisture accounting between storms (4 , 26). Drainage basins are essentially represented as a set of linked overland-flow, channel and reservoir segments or a set of subcatchments from which overland flow drains into nodes (manholes) connected by transport elements (pipes, channels, reservoirs, etc.). Reservoir routing is usually accomplished by the modified Puls method, or an equivalent level pool routing procedure. Kinematic wave theory is used for routing flows over contributing overland-flow areas and through channels (with a linear approximation for circular, pipe segments). The solution to overland-flow problems is much simplified from the St. Venant equations to:

$$\frac{\partial y}{\partial t} + \frac{\partial q}{\partial x} = i - f \tag{1}$$

$$q = \alpha y^{m} \tag{2}$$

Any wavelike behavior must enter through the continuity equation and the approach precludes changes in surface profile due to dynamic variations. The corresponding set of equations for stream segments, or <u>channeled</u> flows, to which the only significant inputs are continuous along the stream axis and consist of rainfall, infiltration, and overland flow is

$$\frac{\partial A}{\partial t} + \frac{\partial Q}{\partial x} = q_i \tag{3}$$

$$Q = \alpha_s A^{m_s} \tag{4}$$

where Q = discharge rate, say cfs (m^3/sec)
A = flow cross-sectional area, say ft^2 (m^2)
q_i = continuous lateral inflow rate of overland flow, say cfs/ft (m^2/sec)
α_s, m_s = kinematic parameters for stream segments,
i = rainfall intensity, say ft/sec (mm/sec)
f = infiltration rate, say ft/sec (mm/sec)
y = depth of flow, say ft (m), and
α, m = kinematic parameters for overland-flow.

Typically, for turbulent flow in overland-flow segments

$$\alpha = \frac{c}{n} \sqrt{S_o} \quad \text{and} \quad m = 5/3 \tag{5}$$

where n = Manning's roughness

c = 1.49 for English units and 1.0 for metric units, and

S_o = slope.

The value of $m_s = 3/2$ is a good approximation for small streams. However, equation (4) may be replaced by the classical Manning relationship:

$$Q = \frac{c}{n} R^{2/3} A \sqrt{S_o} \tag{6}$$

where R = hydraulic mean radius.

Surface Runoff Quality

Perhaps the least deterministic aspect of water quality modeling is the procedure for computation of surface runoff quality, from assumptions made of pollutant accumulation and washoff processes. These are, generally, that: (1) the amount of pollutant which can be removed during a storm event is dependent on rainfall duration and initial quantity of pollutant available for removal; (2) no pollutants decay due to chemical changes or biological degradation during the runoff process, and (3) the amounts of pollutants percolating into the soil by infiltration are not significant. Most field data consist of measurements of stage (converted to discharge through a rating curve) and pollutant concentrations at the subcatchment outlet. These data represent the combined effects of accumulation, washoff, routing and contributions from rainfall: the pollutant loads may include contributions from different land-use types with different accumulation rates. Without local measured flow and quality data, no credibility to the predictions of surface runoff quality can be established. Nevertheless,the models could still be used to study <u>relative</u> impacts of various control alternatives upstream.

Most surface runoff quality models (5 , 26 , 27) offer either linear, exponential or Michaelis-Menton pollutant buildup formulations, or all of these as options (26). For example, the exponential decay (of solids buildup) relationship is (7):

$$P_{n,j,t} = P_{max_{n,j}} \left(1 - e^{-\gamma_{n,j} t}\right) \qquad (7)$$

where $P_{n,j,t}$ = loading of pollutant j on subcatchment n at time t, lb/acre (kg/ha),

$P_{max_{n,j}}$ = maximum (asymptotic) loading of pollutant j on subcatchment n, lb/acre (kg/ha),

$\gamma_{n,j}$ = power exponent coefficient for pollutant j on subcatchment n, day^{-1}, and

t = accumulation time, days.

The pollutant accumulation process can be related to land use by specifying a land use type for each subcatchment. Equation (7) is the most widely used relationship, as well as the exponential washoff relationship described below.

Washoff, at each time step of simulation, is made proportional to runoff rate:

$$- P_t = \frac{dP}{dt} = - a\, r^b\, P_o \qquad (8)$$

where P_t = pollutant load washed off at time t, lb/sec (kg/sec)

P_o = load available for washoff at time t, lb(kg)

r = runoff rate, in/hr (mm/hr)

a = washoff coefficient, including conversion units, and

b = a power constant.

Procedures for estimating accumulation and washoff parameters, including optimization methods are discussed by Alley (3) and Alley and Smith (2). The limitations and research needs are presented by Sonnen (48). Sediment data from streams supports the rating curve in equation (8). Rating curves can be developed independently for each pollutant, as a function of flow rate; however, spurious correlation problems arise: these are discussed later in the paper.

Eulerian Pollutant Transport

Pollutant transport processes for non-tidal river systems may be represented by the one-dimensional version of the classical advective-dispersive equation:

$$\frac{\partial C}{\partial t} + \frac{\partial}{\partial x} [E \frac{\partial C}{\partial x} - UC] \pm \Sigma S_i \qquad (9)$$

where C = concentration of water quality parameter, (pollutant), M/L^3,

t = time, T,

$-E\frac{\partial C}{\partial x}$ = mass flux due to longitudinal dispersion along the flow axis, the x direction, M/L^2T

UC = mass flux due to advection by the fluid containing the mass of pollutant, M/L^2T,

S_i = sources or sinks of the substance C, M/L^3T,

i = 1,2,3, . . ., n,

n = number of sources or sinks,

U = flow velocity, L/T, and

E = longitudinal dispersion coefficient, L^2/T,

M = mass, L = length, and T = time.

Equation (9) can be applied to shallow storage/treatment systems (detention basins, ponds), pipe segments, streams, etc. For the simplified biochemical oxygen demand-dissolved oxygen (BOD-DO) coupled reaction, for continuously discharging plane sources in the y-z plane, the governing equations can be derived from equation (9):

$$\frac{\partial L}{\partial t} + U\frac{\partial L}{\partial x} = E_x\frac{\partial^2 L}{\partial x^2} - K_{13}L \tag{10}$$

$$\frac{\partial D}{\partial t} - U\frac{\partial D}{\partial x} = E_x\frac{\partial^2 D}{\partial x^2} + K_{13}L - K_2 D \tag{11}$$

where L = BOD concentration, M/L^3

K_{13} = biochemical oxidation rate and sedimentation rate coefficient for carbonaceous BOD, 1/T,

D = DO deficit = $(C_s - C)$, M/L^3,

C_s = saturation concentration of DO at stream temperature, M/L^3 ,

C = concentration of DO in stream, M/L^3 , and

K_2 = reaeration rate coefficient, 1/T .

For steady-state analysis, equations (1Ø) and (11) become:

$$U \frac{\partial L}{\partial x} = E_x \frac{\partial^2 L}{\partial x^2} - K_{13}L \qquad (12)$$

$$U \frac{\partial D}{\partial x} = E_x \frac{\partial^2 D}{\partial x^2} + K_{13}L - K_2 D \qquad (13)$$

Solutions to these equations are presented in detail elsewhere (33 , 34) for various components of the physical system and other pollutant combinations, and are applied in continuous simulation models to derive water quality frequency curves for one of the study sites.

When the effect of nutrients on the aquatic ecosystem is of interest, a multi-constituent multi-reach receiving water quality model may be required (provided measured data for calibration and verification are available). QUAL-2E (Version 2.2) is a comprehensive and versatile model capable of simulating: algal, nitrogen, phosphorus, biochemical oxygen demand, dissolved oxygen <u>interactions</u>; algal growth rate, including algal self-shading; temperature; coliforms, and arbitrary non-conservative substances (40). This latest version is available on floppy disks for execution in microcomputers (with floating point processors). It represents a modification to QUAL-II SEMCOG (46). A summary of the differential equations solved by the earlier version of the model is presented in Table 2, with the exception of temperature. All of these equations satisfy the general form of equation (9). They constitute a system of simultaneous linear equations when expressed in an implicit finite difference form, whose coefficient matrix is tridiagonal and can be easily solved by an efficient form of Gaussian elimination (Thomas algorithm).

Table 2. Summary of Differential Equations Solved by QUAL II (46)

Conservative mineral (C)	$\frac{\partial C}{\partial t} = \frac{\partial \left(A_x E \frac{\partial C}{\partial X}\right)}{A_x \partial X} - \frac{\partial (A_x U C)}{A_x \partial X} + \frac{S_c}{A_x \Delta X}$
Algae (A)	$\frac{\partial A}{\partial t} = \frac{\partial \left(A_x E \frac{\partial A}{\partial X}\right)}{A_x \partial X} - \frac{\partial (A_x U A)}{A_x \partial X} + \frac{S_A}{A_x \Delta X} + \left(\mu_A - \rho - \frac{\sigma_1}{D_a}\right) A$
Ammonia nitrogen (NH_3)	$\frac{\partial NH_3}{\partial t} = \frac{\partial \left(A_x E \frac{\partial NH_3}{\partial X}\right)}{A_x \partial X} - \frac{\partial (A_x U NH_3)}{A_x \partial X} + \frac{S_{NH_3}}{A_x \Delta X} + \left(\alpha_1 \rho A - \beta_1 NH_3 + \frac{\sigma_3}{A_x}\right)$
Nitrite nitrogen (NO_2)	$\frac{\partial NO_2}{\partial t} = \frac{\partial \left(A_x E \frac{\partial NO_2}{\partial X}\right)}{A_x \partial X} - \frac{\partial (A_x U NO_2)}{A_x \partial X} + \frac{S_{NO_2}}{A_x \Delta X} + (\beta_1 NH_3 - \beta_2 NO_2)$
Nitrate nitrogen (NO_3)	$\frac{\partial NO_3}{\partial t} = \frac{\partial \left(A_x E \frac{\partial NO_3}{\partial X}\right)}{A_x \partial X} - \frac{\partial (A_x U NO_3)}{A_x \partial X} + \frac{S_{NO_3}}{A_x \Delta X} + (\beta_2 NO_2 - \alpha_1 \mu_A A)$

Phosphate phosphorus (P)

$$\frac{\partial P}{\partial t} = \frac{\partial\left(A_x E \frac{\partial P}{\partial X}\right)}{A_x \partial X} - \frac{\partial (A_x UP)}{A_x \partial X} + \frac{S_p}{A_x \Delta X} + \left[\alpha_2(\rho - \mu_A)A - \frac{\sigma_2}{A_x}\right]$$

Biochemical oxygen demand (BOD^c)

$$\frac{\partial BOD^c}{\partial t} = \frac{\partial\left(A_x E \frac{\partial BOD^c}{\partial X}\right)}{A_x \partial X} - \frac{\partial (A_x UBOD^c)}{A_x \partial X} + \frac{S_{BOD^c}}{A_x \Delta X} - (K^c + K_s)BOD^c$$

Benthic oxygen demand (BOD^b)

$$\frac{\partial BOD_b}{\partial t} = \frac{\partial\left(A_x E \frac{\partial BOD^b}{\partial X}\right)}{A_x \partial X} - \frac{\partial (A_x UBOD^b)}{A_x \partial X} + \frac{K_b}{A_x}$$

Dissolved oxygen (DO)

$$\frac{\partial DO}{\partial t} = \frac{\partial\left(A_x E \frac{\partial DO}{\partial X}\right)}{A_x \partial X} - \frac{\partial (A_x UDO)}{A_x \partial X} + \frac{S_{DO}}{A_x \Delta X} + [K_a(DO_x^{SAT} - DO) +$$

$$(\alpha_3\mu_A - \alpha_4\rho)A - K^c BOD^c - \frac{K_b}{A_x} - \alpha_5\beta_1 NH_3 - \alpha_6\beta_2 NO_2]$$

Coliform (F)

$$\frac{\partial F}{\partial t} = \frac{\partial\left(A_x E \frac{\partial F}{\partial X}\right)}{A_x \partial X} - \frac{\partial (A_x UF)}{A_x \partial X} + \frac{S_F}{A_x \Delta X} - K_d F$$

Radioactive material (R)

$$\frac{\partial R}{\partial t} = \frac{\partial\left(A_x E \frac{\partial R}{\partial X}\right)}{A_x \partial X} - \frac{\partial (A_x UR)}{A_x \partial X} + \frac{S_R}{A_x \Delta X} - \gamma_r R - \gamma_a R$$

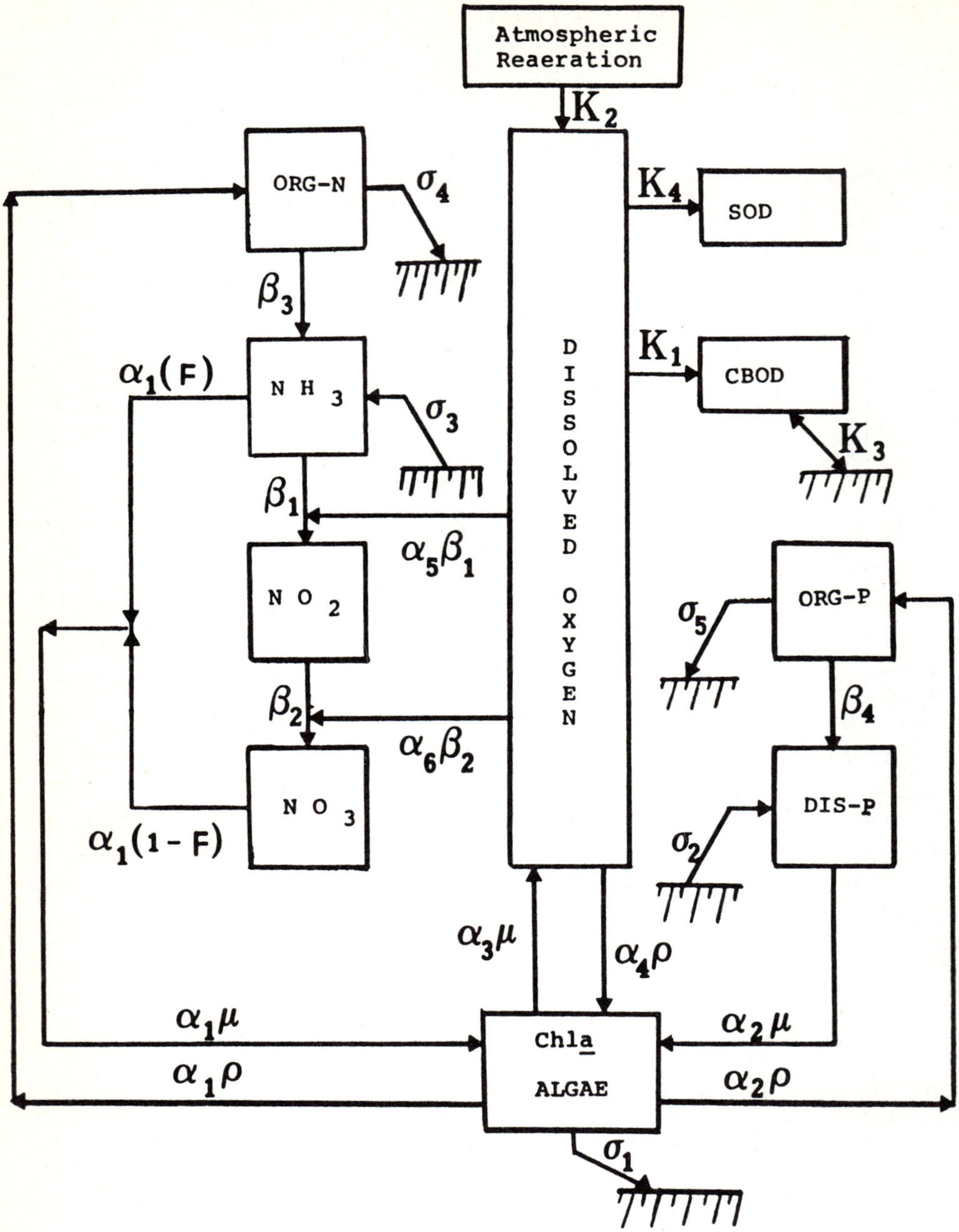

Figure 3. Conceptual Relationship Among Major Constituents Simulated by QUAL2-E (40).

For <u>completely mixed storage/treatment</u> systems

($\partial c/\partial x = 0$), equation (9) reduces to

$$\frac{dC}{dt} = \Sigma S_i \tag{14}$$

which may be rewritten, for a <u>constant volume</u> tank subjected to time-varying forcing functions of both concentration and flow rate, as (35):

$$V \frac{dc_2}{dt} = Q(t)\ [c_1(t) - C_2(t)] - K\ c_2(t)V \tag{15}$$

where V = volume of tank, L^3,
$c_2(t)$ = concentration of material in the tank and outflow, M/L^3,
$Q(t)$ = volumetric flow into and out of the tank, L^3/T, and
K = first-order decay rate constant, $1/T$.

For a well-mixed <u>variable volume</u> system, the governing equations are:

$$\frac{dV}{dt} = Q_1(t) - Q_2(t) \tag{16}$$

$$\frac{d(Vc_2)}{dt} = Q_1(t)c_1(t) - Q_2(t)c_2(t) - K_{c_2}(t)V(t) \tag{17}$$

where $Q_1(t)$ = influent fluid flow rate, L^3/T, and
$Q_2(t)$ = fluid flow rate out of the tank, L^3/T .

Advection and dispersion can also be accounted for in a variable volume storage/treatment system, using equation (16) and a modified form of equation (10). Solutions to these detention systems for non-conservative pollutants are presented elsewhere (35).

For deep, thermally stratified impoundments a complete thermal energy budget is required for water quality prediction in addition to the conservation of mass equations. In such a reservoir, all fluxes may be conceptualized as occurring along the vertical (z) axis, through horizontal slices defining discrete control volumes for which the equations of mass and heat energy conservation are written. In a review of models for stratified impoundments, Orlob (42) presents a brief chronology of model development, basic governing equations and solution techniques. For example, expressing the heat budget in terms of temperature results in:

$$\frac{\partial (V_j \theta_j)}{\partial t} = (Q_i\ \theta_i - Q_o\ \theta_o)_j + \left[\frac{1}{\rho c}\int_z^{z+\Delta z} q_{sz}\ a_z\ d_z\right]_j - Q_j \theta_j + Q_{j+1}\ \theta_{j+1} + \left(Ea\ \frac{\partial \theta}{\partial z}\right)_j - \left(Ea\ \frac{\partial \theta}{\partial z}\right)_{j+1} \quad (18)$$

where V_j = volume of the jth control element, L^3,

θ_j = temperature, ^oC,

Q_i = flow advected to the control volume in the horizontal plane, L^3/T,

Q_o = flow advected from the control volume in the horizontal plane, L^3 ,

ρ = density of water, M/L^3

c = specific heat, cal/g^{o}C

z = depth, L ,

Q_j = vertical flow, L^3/T,

a_z = horizontal area at depth z, L^2

a = horizontal area for a fixed Δz (obtained from area-elevation curves), L^2, and

E = coefficient of effective diffusion (vertical), L^2/T.

Since concentration of heat can be expressed as $\rho c \Delta \theta$, equation (18) is of the form of equation (9). The conservation of mass equation for element j is:

$$\frac{\partial (V_j C_j)}{\partial t} = - Q_j C_j + Q_{j+1} C_{j+1} + \left(Ea \frac{\partial C}{\partial z}\right)_j - \left(Ea \frac{\partial C}{\partial z}\right)_{j+1} + C_j \frac{\partial V_j}{\partial t} \pm V_j \frac{dc}{dt} \tag{19}$$

where C_j is the concentration of any quality constituent that moves with the fluid mass, and the local derivative dc/dt represents all processes other than advection and diffusion that affect the concentration in the control volume V_j (42) . Again, a set of simultaneous linear equations can be obtained by writing equation (19) for each constituent modeled. These equations can be transformed by an implicit finite difference scheme into a tridiagonal matrix solved by numerical methods.

Hinwood and Wallis (22) have classified water quality models of <u>tidal waters</u> on the basis of spatial dimensions, the reference frame (Eulerian or non-Eulerian), and the degree to which the hydrodynamic processes are included. The non-Eulerian reference frames are: The Lagrangian, in which the reference frame travels up and down the estuary with the mean water motion; and the constant upstream volume, in which the reference frame oscillates up and down the estuary during the tide cycle in such a way that a constant volume of water remains upstream of any given section (no net downstream movement over a tide cycle). In both of the non-Eulerian reference frames the <u>distance between sections expands and contracts through a tide cycle</u>. In a later article, Hinwood and Wallis (23) review basic equations, solution techniques and about 100 models of water and waste movement in tidal bays and estuaries.

Transport models for both inland and coastal waters are presented in the proceedings of a symposium on predictive ability (20), with outstanding contributions from the international community of modelers. More attention has been focused recently on the accuracy and efficiency of the <u>numerical schemes</u> for solving the advective-dispersive equation, particularly the <u>finite element</u> method, for contaminant transport problems in both surface and subsurface hydrology (53). Baca and Arnett (9) present a Galerkin finite element model, using triangular basis functions, for quasi-two-dimensional simulation of eutrophic lakes. The need for accurate numerical solution schemes may be appreciated from another perspective: in the <u>calibration</u> process, the model user may be <u>compensating for numerical simulation errors</u> in adjusting mixing coefficients and other model parameters to better reproduce observed data.

Most finite difference methods for the solution of the one-dimensional advective-dispersive equation are plagued by an artificial (numerical) diffusion in the calculation of the convective step, which is sometimes larger than the actual physical dispersion in the stream. Holly and Preissmann (24) suggest the use of cubic polynomial interpolation (rather than linear) between two spatial modes. The scheme introduces very little artificial diffusion: however, it depends on the size of the concentration distribution relative to the spatial increment (Δx), as well as the Courant number. The artificial diffusion decreases as the size of the concentration distribution increases compared to Δx .

Lagrangian Pollutant Transport

Another way to minimize numerical diffusion, which is also computationally more efficient than the simplest finite-difference scheme, is to use the Lagrangian reference frame (19). As noted above, most of the artificial (numerical) diffusion results from deficiencies in simulating the advective term in equation (9). Because the coordinate system moves at the mean flow velocity, the advection term disappears. Thus, numerical diffusion is almost entirely eliminated: the computational grid is a set of moving points which represent the centers of marked volumes of water. Longitudinal dispersion between marked volumes can be set by the analyst as considered appropriate. Although the method is straightforward, transforming the Lagrangian solution back to a fixed grid coordinate system is somewhat more involved. Jobson (29) developed an unconditionally stable model for use in upland streams and rivers, verified by use of several days of dye dilution data obtained over a 27.8-kilometer reach of the Chattahoochee River near Atlanta, Georgia.

Conceptually, in the Lagrangian framework an individual fluid parcel is followed, while accounting for all factors which tend to affect pollutant concentrations. In Figure 4, the flow segment is moving downstream with velocity U (cross-sectional mean stream velocity). Neglecting molecular diffusion, conservation of mass for a conservative substance in the element requires:

$$\frac{\partial c}{\partial t} + \frac{\partial [c(u - U)]}{\partial x} = 0 \qquad (20)$$

where c = local instantaneous value of concentration, and
u = local instantaneous velocity,

which can be defined in terms of turbulent fluctuations as

$$u = U + u_a + u' \qquad (21)$$

$$c = C + c_a + c' \qquad (22)$$

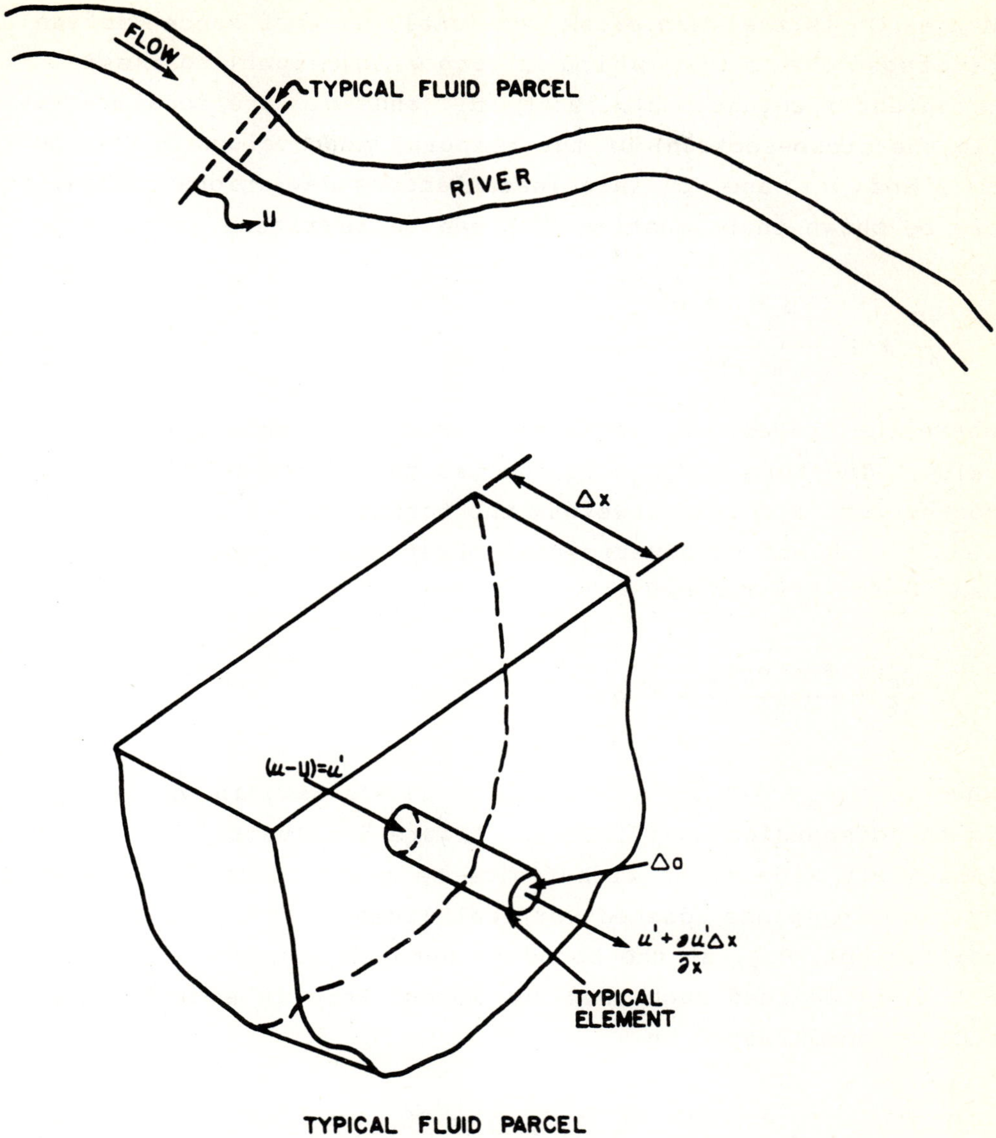

Figure 4. Typical Flow Element In A Lagrangian Transport Model (29).

where C is the mean cross-sectional value of concentration (averaged over a time which is long with respect to these turbulent fluctuations). Both u_a and c_a are local values (in the cross-section) of the temporal mean values of u and c , and u' and c' are instantaneous deviations. Then, it can be shown that equation (20) can be rewritten as (29):

$$\frac{\partial C}{\partial t} + \partial \frac{< u_a c_a + \overline{u'c'} >}{\partial x} = o \tag{23}$$

where the braces <> indicate a cross-sectionally averaged value. The term $< u_a c_a >$ designates the effect of differential convection, and far outweighs the turbulent diffusion term $<\overline{u'c'}>$. Based on experimental observations, equation (23) can be further reduced to:

$$\frac{\partial c}{\partial t} + \frac{\partial < u_a c_a >}{\partial x} = o \tag{24}$$

where $< u_a c_a > = D_x \frac{\partial c}{\partial x}$ by analogy to Fick's Law, in which D_x is the dispersion coefficient. Figure 5 illustrates an infinitely wide river with finite segments (parcels) separated by a moving plane (dashed vertical lines) (29). A mixing coefficient, DQ_k, at the boundary between segments k and k + 1 is defined such that the second term in equation (24) can be approximated by

$$\frac{\partial < u_a c_a >}{\partial x} = \frac{DQ_k (C_k - C_{k+1}) - DQ_{k-1} (C_{k-1} - C_k)}{A \, \Delta x} \tag{25}$$

where $DQ = \frac{1}{2} \int_A |u - U| \, da$

C_k = cross-sectional average concentration in segment k ,

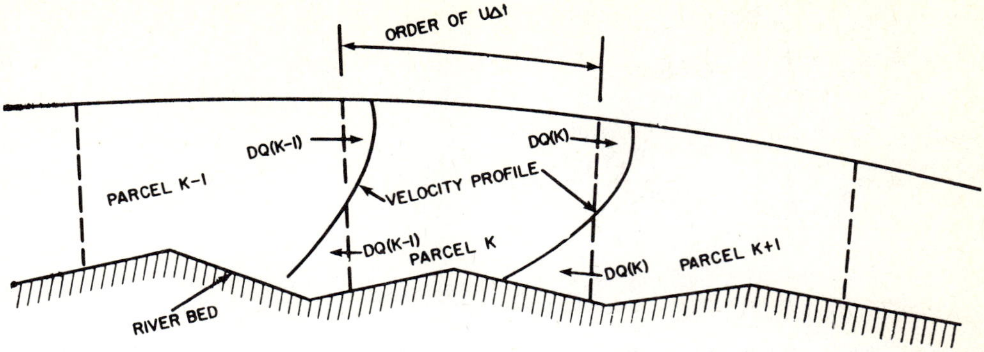

Figure 5. Longitudinal Section of River Illustrating Dispersive Process Between Fluid Parcels In Lagrangian Scheme (29).

and $A\Delta x$ is the segment volume. According to Jobson (29), an additional advantage of the Lagrangian method is the ease with which DQ can be obtained from stream gage data, as opposed to the difficulty of obtaining D_x for natural channels from dye dilution data.

The location of a particular water parcel and its concentration can be determined by integration:

$$x = x_o + \int_o^{TT} U \; d\tau \tag{26}$$

$$c = c_o - \int_o^{TT} \frac{\partial < u_a c_a >}{\partial x} \; d\tau \tag{27}$$

where TT = travel time, c_o = concentration of parcel at time zero and c is the concentration after a time lapse TT . In order to construct a solution net over space and time, equation (27) must be solved for a series of fluid parcels, and the locations of the parcels continually tracked.

Interfacing With Surface Runoff Modules

Some models provide complete simulation, from generation of surface runoff quantity and quality to transport through engineered storage and conveyance systems, to natural receiving bodies of water. It is advantageous, however, to separate the calibration process for quantity from that for quality prediction: since it is pointless to calibrate water quality parameters without correct flows. Through computer storage (memory) devices, calibrated flows can be supplied to a receiving water quality subroutine or completely separate program. Indeed, an Eulerian distributed-routing rainfall-runoff model (4) may be used to generate flows for a Lagrangian-based urban runoff quality model (5).

Discrete-Event Versus Continuous Simulation

For most hydrologic and water quality planning efforts, the probability of occurrence of events of various magnitudes and their duration is required. A model or methodology designed to simulate only discrete storm events cannot adequately provide this information. The probability attached to a particular hydrograph is not the same as that for the rainfall from which it was generated. Even when a probability level can be established for a given rainfall rate at a specified duration, it is almost impossible to assign a probability level to a storm which consists of rainfall at various rates for various durations in several possible sequences. Similarly, the probability levels computed for various water quality constituents or pollutographs are not the same as those of the hydrographs and hyetographs associated with them. Indeed, the return periods among the various water quality variables generated by the same storm event may differ. Design storms were and continue to be developed from statistical analysis of point rainfall records that include events of all types and sizes. Non-linear aspects such as antecedent moisture conditions, and both spatial and temporal variability, along with recent advances in hydrologic modeling and computer hardware should hasten the obsolescence of design storms even for water quantity applications only. More sophisticated approaches have been proposed such as combining continuous simulation with derived distribution models, and detailed screening of long-term precipitation data by computer programs to generate storm event statistics.

Rainfall Time Series Analysis

An integral part of the assessment of storm-derived pollutant loads on receiving water quality is the statistical evaluation of rainfall records. The purpose is to summarize the variables of interest (volume, duration, intensity and time between storm events) and statistically characterize the rainfall record to determine seasonal trends (8). The hourly rainfall data are summarized by storm events, each with an associated unit volume, duration, average intensity and time since the preceding storm (measured from the midpoint of the successive storms). Thus, a storm definition must be established to determine when in the hourly record a storm begins and ends. To avoid an arbitrary definition of independence, a minimum interevent time can be derived on the basis of autocorrelation analysis of the hourly rainfall of a representative year in the time series.

Statistically-Based Water Quality Models

Uncertainty in the state of water quality in a body of water arises as a result of: errors in measurement of both streamflow and pollutant concentrations; incompleteness of data (e.g., un-documented waste sources or sinks), or simply gaps in historical data; random fluctuations exhibited by natural phenomena (rainfall-runoff); and errors in the internal structure of models used in the predictive or descriptive process (e.g., conceptual errors) or errors in the estimation of model parameters. Several approaches to account for this uncertainty have been proposed, using statistically-based methods: for example, derived-distribution methods, Monte Carlo simulation, and random differential equations. These methods are reviewed, and opportunities for combining the statistical approaches with physically-based models are discussed below.

Probabilistic Models In General

Loucks and Lynn (32) proposed models to predict the probability distribution of minimum DO concentrations downstream from BOD point loads during low flows: a Markov chain was assumed adequate to describe the probability that the minimum DO was below a specified level under various stream and wasteload conditions. Thayer and Krutchkoff (50) developed a stochastic (random) model for BOD and DO concentrations in unsegmented streams. DiToro and O'Connor (14) considered the effects of two types of time variable flow on the BOD and DO concentration probability distributions in a stream. Mehta, Ahlert and Yu (37) used Autoregressive Integrated Moving Average (ARIMA) models to describe the random nature of a water quality time series that included streamflow, temperature, BOD and DO. Litwin and Joeres (30) also used a time series approach to obtain short term river DO forecasts. Moore, Dandy and deLucia (38) proposed a simple two-state (nutrients and algal biomass) eutrophication model and filtering theory to obtain an algorithm which describes the time history of variance (uncertainty) in estimates of water quality parameters. They investigated the effect of sampling on knowledge of the present trophic state of a water body, and the effect of sampling on uncertainty in future predictions of water quality. Dandy and Moore (13) combined the Streeter-Phelps equations for BOD-DO in differential form with stochastic process representations of the boundary conditions and point source inputs to the river, into a single stochastic model.

Derived Distributions

In essence, these statistical models calculate the probability distribution of receiving water pollutant concentrations, given the probability distribution of the model inputs (e.g., flows, pollutant loadings, etc.). Whereas the deterministic models attempt to predict the time histories of the output variables and perform frequency analyses on the basis of the historical series, probabilistic water quality analysis attempts to calculate the frequency distributions without computation of the exact sequence of events. Instead, the probability distributions and correlation structure of the input variables are used to compute directly the frequency distribution of the output variables. These statistical models are conceptual simplifications, which require only the statistical properties of the input time series (e.g., medians, means, coefficients of variations, cross-correlations). On the basis of exponential probability density functions for volume of runoff, interevent time, and duration of runoff -- beta density for BOD concentration and gamma density for river flow volumes during a critical period -- Loganathan and Delleur (31) derived new distributions for overflow volumes and river water quality after mixing with the untreated overflows.

The simplest mass balance that can be used to calculate river quality downstream from a pollutant discharge is:

$$C_T = \frac{Q_s C_s + Q_p C_p}{Q_s + Q_p} \qquad (28)$$

where C_T = mixed pollutant concentration in the river, downstream from the pollutant discharge location

Q_s = upstream river flow

C_s = upstream pollutant concentration

Q_p = point pollutant source flow rate, and

C_p = point pollutant source concentration.

Warn and Brew (55) proposed an analytical method based on the first two moments (mean, variance) of a two-parameter log-normal distribution, assumed to represent adequately certain ratios of the variables in equation (28) above. For nonpoint source contributions, Q_R and C_R may replace Q_p and C_p:

$$C_T = \frac{Q_s}{Q_s + Q_R} C_s + \frac{Q_R}{Q_s + Q_R} C_R \tag{29}$$

The probability model that follows assumes that Q_s, Q_R, C_s and C_R are jointly lognormally distributed. As noted by Warn and Brew (55) and as modified by DiToro (15, 16), the mass balance equation can be rewritten in the form:

$$C_T = C_R \phi + C_s (1 - \phi) \tag{30}$$

where $\phi = \frac{Q_R}{Q_R + Q_s}$ = runoff flow fraction

If ϕ and $(1 - \phi)$ were lognormaly distributed, and since C_R and C_s are also assumed to be lognormal, the products $C_R\phi$ and $C_s(1 - \phi)$ would also be lognormal. This is an approximation based on the fact that sums of lognormal random variables have been reported to have tails which are also approximately lognormal (15). Assuming independence, and if C_T is assumed lognormal, the relationships between the arithmetic moments, $\mu(C_T)$ and $\sigma^2(C_T)$, and the log mean, μ_ℓ, and log standard deviation, σ_ℓ, are (15):

$$\mu_\ell(C_T) = \ell n \left[\frac{\mu(C_T)}{1 + \nu^2(C_T)} \right] \tag{31}$$

and

$$\sigma_\ell^2(C_T) = \ell n \quad [1 + \nu^2(C_T)] \tag{32}$$

where

$$\nu(C_T) = \frac{\sigma(C_T)}{\mu(C_T)} \tag{33}$$

= coefficient of variation of C_T.

The quantiles of C_T are:

$$C_{T\alpha} = \exp\ [\mu_\ell(C_T) + z_\alpha\ \sigma_\ell(C_T)] \tag{34}$$

which is the concentration that is exceeded with probability $(1 - \alpha)$, where z_α is the standard normal α quantile.

Warn and Brew (55) suggest a numerical integration to compute the moment of ϕ; however, DiToro (15,16) notes that, since the method is approximate, numerical techniques are best reserved for the evaluation of the exact distribution of C_T. He derived the following approximate expressions:

$$\mu_\ell(\phi) = \frac{1}{2}\ [\ell n(\phi_\alpha) + \ell n(\phi_1 - \alpha)] \tag{35}$$

and

$$\sigma_\ell(\phi) = \frac{1}{2z_\alpha}\ [\ell n(\phi_\alpha) - \ell n(\phi_{1-\alpha})] \tag{36}$$

He chose $z = 1.645$ to force agreement of this straight-line approximation of the 5% and 95% quantities. Once the log mean and standard deviation are computed from equations (35) and (36), then the arithmetic moments follow from the lognormal assumption:

$$\mu(\phi) = \exp\left[\mu_{\ell}(\phi) + \frac{1}{2}\sigma_{\ell}^{2}(\phi)\right] \tag{37}$$

and

$$\nu^{2}(\phi) = \exp\left[\sigma_{\ell}^{2}(\phi)\right] - 1 \tag{38}$$

where $\sigma(\phi) = \mu(\phi)\nu(\phi)$.

These arithmetic moments of ϕ are subsequently used to compute the arithmetic moments of C_T from:

$$\mu(C_T) = \mu(C_R)\mu(\phi) + \mu(C_s)[1 - \mu(\phi)] \tag{39}$$

and

$$\begin{aligned} \sigma^{2}(C_T) &= \sigma^{2}(\phi)[\mu(C_R) - \mu(C_s)]^{2} \\ &+ \sigma^{2}(C_R)\,[\sigma^{2}(\phi) + \mu^{2}(\phi)] \\ &+ \sigma^{2}(C_s)\,[\sigma^{2}(\phi) + (1 - \mu(\phi))^{2}] \end{aligned} \tag{40}$$

where $\mu(\ )$ = mean of ( )

$\sigma^{2}(\ )$ = variance of ( ).

An alternative numerical integration (Gaussian quadrature) method which is mathematically more exact is presented by DiToro (15). Both methods are easily programmed for use with microcomputers, and results are shown to be equivalent in the application section for the study site in North Carolina (34).

Monte Carlo Simulation

Although it is treated in a separate section, Monte Carlo simulation is a combined distribution technique based on artificial (numerical) sampling. Random values are drawn from each of the distributions of the variables, initial conditions and coefficients of a deterministic model -- which is repeatedly applied until an ensemble of synthetic time series is produced that can be analyzed to define the statistical characteristics of the system response. Brutsaert (11) used a simple Streeter-Phelps BOD-DO model and triangular probability density functions to derive the cumulative probability density function of critical (minimum) DO deficit occurrence. Hornberger (25) modified the classical Streeter-Phelps model to include a term for mean daily net photosynthetic production (the effect of community metabolism on the DO balance) and evaluated the contribution of the _variance in river community_ to _overall uncertainty_ in prediction of DO concentrations. Whitehead and Young (57) also used a simplified BOD-DO model, combined with multi-variable parameter estimation techniques and Monte-Carlo simulation, to produce cumulative probability distributions of BOD and DO downstream.

Computer models based on either analytical procedures (as described above) or Monte-Carlo simulation have been operational within the Anglian Water Authority (England) in the calculation of allowable waste discharges, in terms of 95-percentile statistics (mean, standard deviation) of river flow and water quality (54).

Random Differential Equations

Although Monte Carlo simulation allows simultaneous consideration of random initial conditions, inputs and reaction rate coefficients of the deterministic model -- the computational cost is high. This may not be a significant factor in the future, with advances in micro-computer technology. An alternative approach is the solution of random differential equations. A survey of these stochastic models is provided by Padgett (43) and Finney *et al.* (18). The latter investigators numerically solved (by the alternating-direction-implicit finite difference method) the Fokker-Planck Equation (Forward Kolmogorov Diffusion Equation) to obtain the *joint probability density function* of BOD and DO. Research on Brownian motion provided the stimulus for random walk theory, random processes, random noise, spectral analysis and stochastic equations (41,49).

Rates, Constants, Kinetics Formulations, Calibration and Verification, Data Collection and Needs

State-of-the-art formulations used in surface water quality modeling have been assembled in a comprehensive report by Zison *et al* (58). Topics covered in detail include reaeration, dissolved oxygen saturation, photosynthesis, deoxygenation, benthic oxygen demand, coliform bacteria, algae and zooplankton. Some general observations were drawn worthy of further discussion:

(1) These rates, constants, and kinetics formulations should be considered an *integral part of the modeling process*, not just calibration factors. A review of Table 2 and Figure 3 should reinforce this

statement.

(2) There is no established best procedure for selection of a particular parameter and its magnitude, due to the great number of environmental factors that influence a given rate, particularly in modeling stormwater runoff quality: disagreement exists among researchers.

(3) Based on large observed differences from system to system, at least some surface water quality parameters are highly site-specific: establishment of selection guidelines is different. However, as long as the processes (physical, biological, chemical) are mechanistic (e.g., surface reaeration) rather than stochastic (e.g., turbulence) in nature, the key to generality lies in good knowledge of cause/effect relationships.

(4) Modelers should be aware of the range of appplicability of a particular formulation, and the limitations of the field measurement technique. Many data collectors never analyze their own data, nor have a clear understanding of the objectives.

(5) Increased mathematical model sophistication may be incompatible with the ability to measure additional parameters in the field, and the calibration process.

(6) A physically-based mathematical model should predict the mean after it has been properly adjusted, but model output from a calibrated model has no claims on singularity because the values used represent

only a sample from a large population.

(7) Sampling and monitoring at random may be disastrous (costly) if system response times ar not known or system behavior is not understood.

Useful reviews of data collection and instrumentation needs for urban stormwater runoff quantity and quality have been published (28). Calibration and verification procedures for water quality models have been topics of discussion for at least 60 years, since the work of Streeter and Phelps in 1925 (51). Disagreement also exists on the best procedure, but a philosophy of model confirmation may be emerging. More and better automatic calibration programs are becoming available, but through the manual procedure the analyst can obtain a great deal of insight about system behavior from resolving differences between model predictions and observed data.

For any study, the simplest feasible model should be used (47). To the extent possible, a model should be calibrated on basic physical, chemical and biological parameters (e.g., Manning's roughness, dispersion, algal growth rates, etc.) and not curve fitting parameters. Accurate water quality modeling is not possible without an adequate representation of the flow field as a prerequisite. Split-sample calibration/verification is highly recommended. As model complexity increases, the justification for using quantitative measures of verification becomes greater. A variety of simple statistical comparisons may be used (51): regression analyses, relative error, comparison of means, or the root mean square error. However, not all of the credibility of a model is embodied in verification statistics, which do not necessarily imply ability to accurately predict future water quality.

Applications

The results of applying both physically-based and statistically-based models to two urban catchments, one in the United States and the other in Australia, are presented below. For the Australian catchment, an attempt was made to evaluate in particular pollutant accumulation and washoff relationships. For the U.S. catchment, a receiving water quality analysis is presented in more detail. The modeling choices were made on the basis of data availability.

Giralang, Canberra, Australia

The Giralang urban catchment is located approximately 10 km north of the Canberra City center (39). Urbanization commenced in 1974, and by late 1976 occupation of the residential area was virtually complete: about 26 percent of the catchment is rural and the rest residential, for a total of 94.2 hectares. The generally thin layer of permeable topsoils initially absorbs rainfall quickly; however, the topsoil becomes saturated because of underlying clays, remaining saturated until drained slowly by interflow, evaporation and evapotranspiration. The climate of Canberra is continental (12): hot summers and cold winters, with altitude (580 meters) exerting a moderating influence. Rainfall is relatively uniformly distributed throughout the year: with a monthly average of about 2 in (51 mm). Stormwater flows and quality were monitored from December 31, 1976 to May 12, 1980 with some gaps in the time series. A rain gage, weir (Stevens-type recorder) and water quality samplers (stage and time activated) were located at the outlet, shown in Figure 6, which also illustrates the network of subcatchments and pipe elements used for simulation.

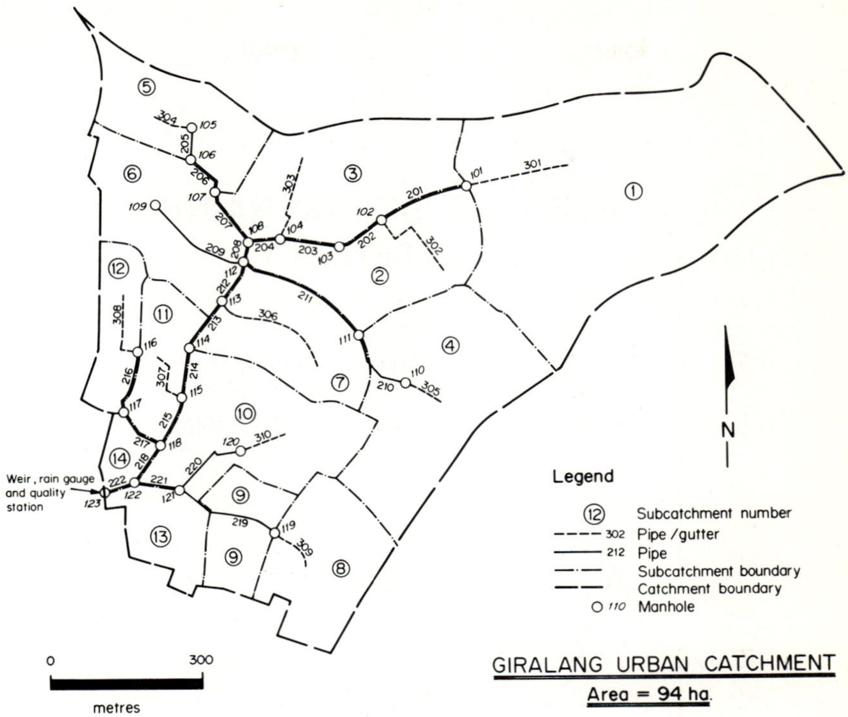

Figure 6. Giralang Urban Catchment Discretization

A detailed analysis was performed on hourly rainfall data, by grouping into storm events (using a 12-hour minimum interevent time to define independence). Then, average storm event rainfall intensity, duration, time between events and volume (depth) were computed, presented in Figures 7 through 10. Summer months are December, January and February and winter months are June, July and August. The highest rainfall intensity occurs during the summer, as well as the longest time between events (pollutant accumulation period). Storm duration is longest in October, the highest volume month, for

Canberrra. Rainfall statistics for Melbourne are also presented, for some of the storm variables, for comparison.

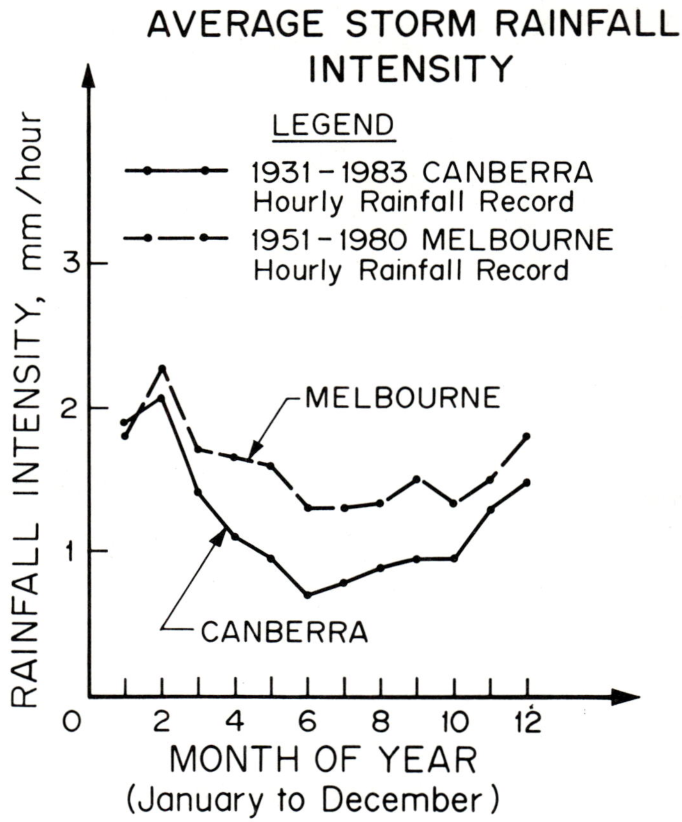

Figure 7. Average Storm Event Rainfall Intensity per Month of Year

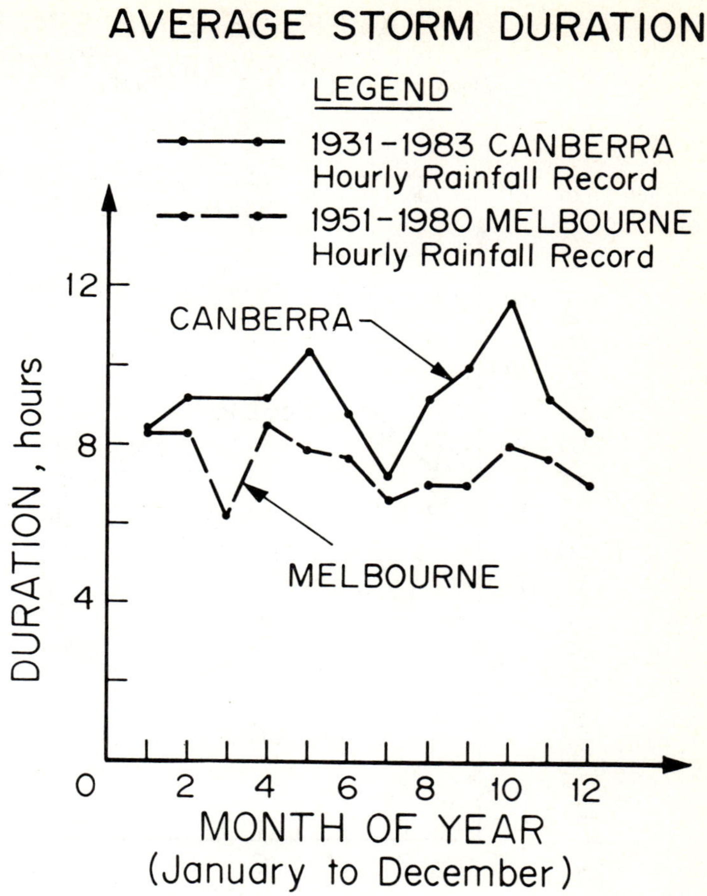

Figure 8. Average Storm Event Duration Per Month of Year

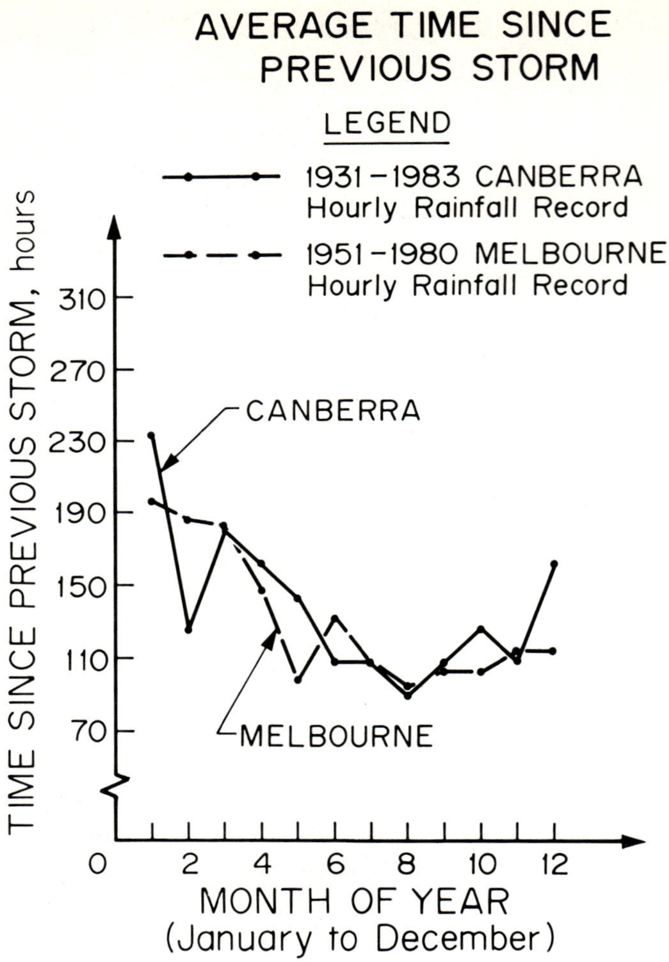

Figure 9. Average Time Since Previous Storm Event, Per Month of Year

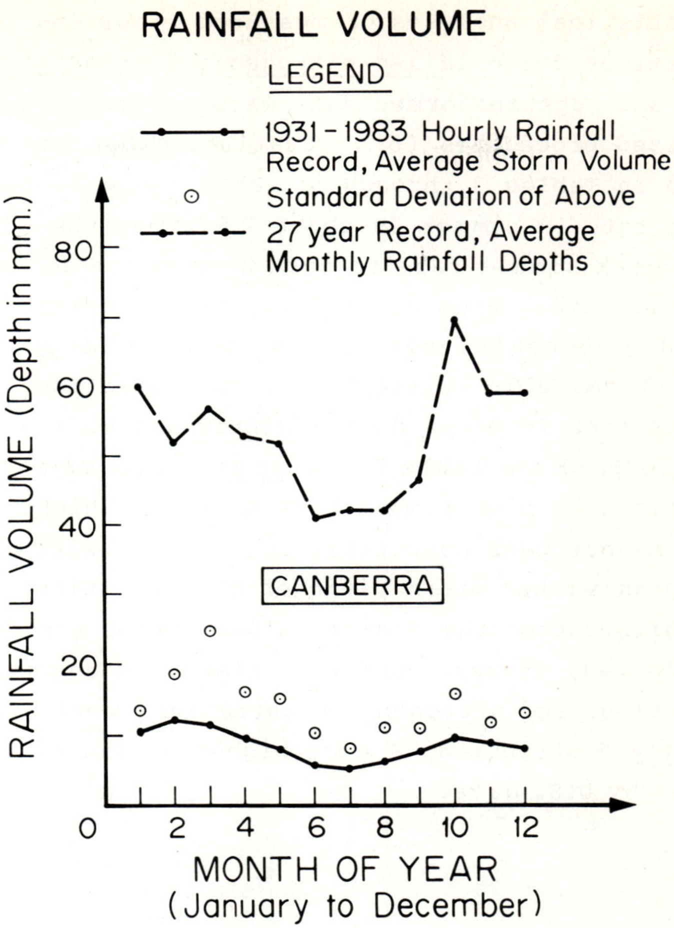

Figure 1Ø. Average Storm Event Depth (Volume) Per Month of Year

Statistical analyses of measured flows and concentrations for <u>oxidized nitrogen</u> and <u>total phosphorus</u>, for both raw and logtransformed data were performed using standardized procedures (6). Statistics for the raw data are presented in Tables 3 through 5. From Table 3, the mean discharge rate in summer is about 1.5 times the winter rate: the mean peak summer rate is more than twice as large as the mean winter rate. From bar charts, the distribution of flows was found to be positively skewed for the summer, and for all seasons. From Table 4, the higher concentrations of phosphorus tend to occur during <u>winter</u> storms (when the flows are smaller). From Table 5, the <u>summer</u> storms appear to be characterized by higher means for nitrogen, higher variance and also higher peak concentrations than in winter. The average peak winter storm nitrogen concentration is more than 3 times higher than the summer value. After the data were logtransformed, it was found that flows, phosphorus concentrations and nitrogen concentrations were approximately <u>lognormally</u> distributed -- with higher variances than those exhibited by U.S. data.

Table 3. Statistics of Measured Flows (m^3/sec)

Statistical parameters	All seasons	Summer	Winter
No. of observations	518	179	73
No. of missing obs.	-	-	-
Mean	Ø.2Ø29	Ø.2639	Ø.18Ø8
Variance	Ø.Ø938	Ø.Ø9Ø9	Ø.Ø166
Standard deviation	Ø.3Ø62	Ø.3Ø16	Ø.1288
Skewness	3.5Ø57	2.Ø391	Ø.94Ø2
Kurtosis	16.8898	5.1371	2.49Ø3
Coefficient of variation	1.5Ø89	1.1427	Ø.7123
Range	2.419Ø	1.727Ø	Ø.7Ø9Ø
Maximum	2.42ØØ	1.73ØØ	Ø.72ØØ
Minimum	Ø.ØØØ4	Ø.ØØ27	Ø.Ø1Ø4
Probability > D or < W	< Ø.Ø1	< Ø.Ø1	Ø.Ø13Ø
W: Normal/			
D: Normal	Ø.25417	Ø.19319	Ø.11756

Table 4. Statistics of Measured Total Phosphorus Concentration (mg/ℓ) as P

Statistical parameters	All seasons	Summer	Winter
No. of observations	401	155	25
No. of missing obs.	117	24	48
Mean	0.3495	0.3602	0.4866
Variance	0.0488	0.0645	0.0521
Standard Deviation	0.2209	0.2542	0.2282
Skewness	2.7161	2.8044	2.0270
Kurtosis	12.0616	12.3533	5.8725
Coefficient of variation	0.6322	0.7056	0.4691
Range	1.923	1.898	1.165
Maximum	1.96	1.96	1.275
Minimum	0.037	0.062	0.11
Probability > D or < W	< 0.01	< 0.01	0.01
W: Normal /			
D: Normal	0.14004	0.13618	0.79514

Table 5. Statistics of Measured Oxidized Nitrogen (mg/ℓ)

Statistical parameters	All seasons	Summer	Winter
No. of observations	503	177	71
No. of missing obs.	15	2	2
Mean	1.6313	1.7180	1.3301
Variance	2.9869	3.3444	0.3682
Standard deviation	1.7283	1.8287	0.6068
Skewness	3.4583	2.3106	1.0410
Kurtosis	13.6788	6.3537	0.85643
Coefficient of variation	1.0594	1.0645	0.4562
Range	10.982	10.562	2.708
Maximum	11.02	10.60	3.18
Minimum	0.038	0.038	0.47
Probability > D or < W	< 0.01	< 0.01	0.01
W: Normal/			
D: Normal	0.2062	0.21495	0.1738

Both single-event and continuous simulation were applied, calibrating first with a few discrete storm events for each of two seasons: winter and summer. Calibrations of runoff hydrographs for the storms of August 12th, 1977 and February 23rd, 1977 are presented in Figures 11 and 12, respectively. It was generally found that phosphorus pollutographs were close in shape to observed hydrographs, as shown in Figure 13 (compare to Figure 11). It has been suggested that the parameters for pollutant washoff relationships be determined on a storm by storm basis by plotting the log of total load versus the log of total flow (26). If a hysteresis effect (loop) is present, the power-exponential washoff relationship [equation (8)] should be used; otherwise, the simple rating curve method should work better. Hysteresis is exhibited in Figures 14 and 15 for both phosphorus and nitrogen loads, for the storm of August 12, 1977. for the summer storm of February 18, 1977 the plots were essentially linear for both nitrogen and phosphorus load. Although use of the rating curve method would be indicated, spurious correlation problems arise: load and flow have a common variable (loads are determined from concentrations by knowledge of the flows).

Four-year simulations of phosphorus concentration and nitrogen concentration, using SWMM (26), are presented in Figures 16 and 17. These simulations include events for which no measured data were gathered. A comparison of the statistical moments is provided in Tables 6 and 7.

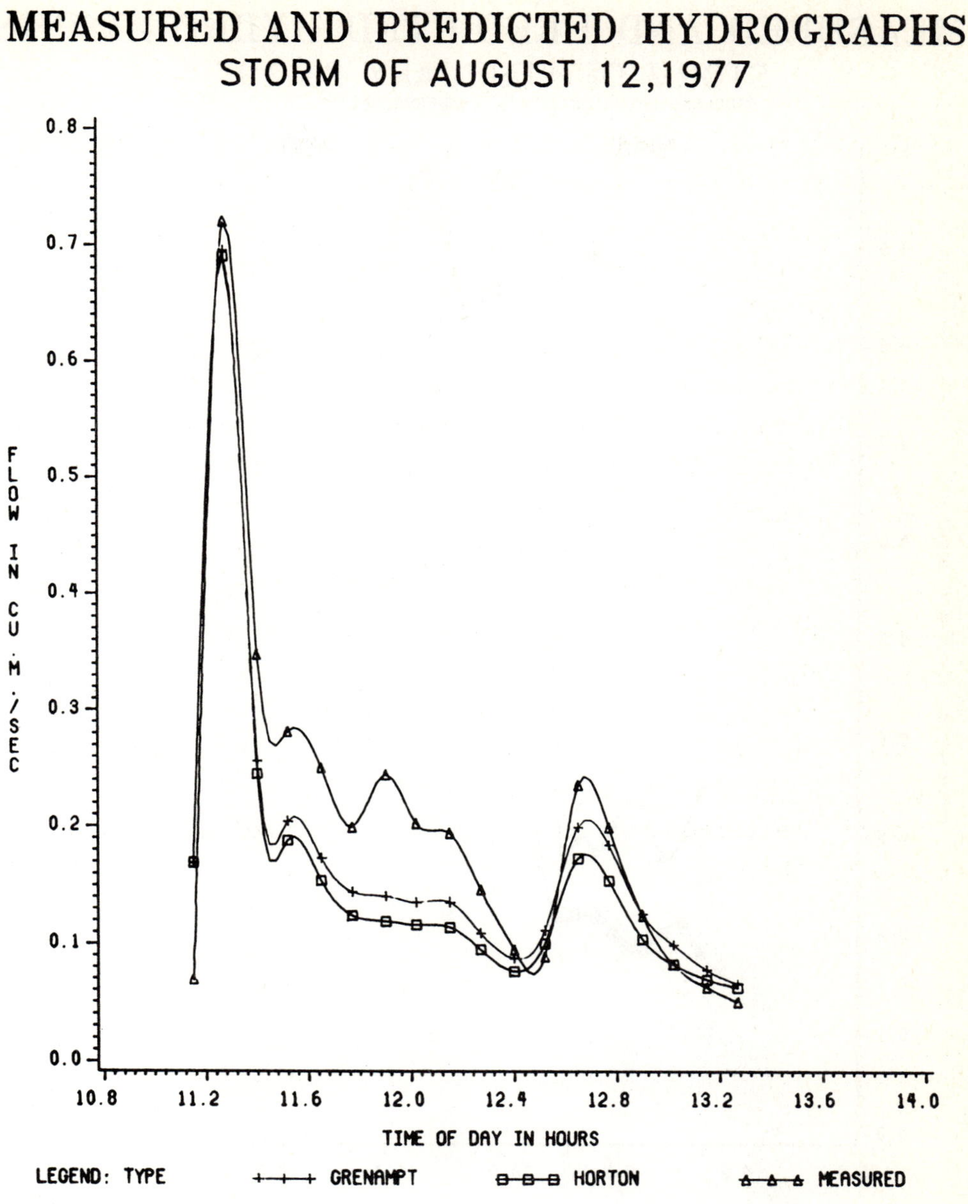

Figure 11. Measured and Predicted Hydrographs, Horton and Green-Ampt. Infiltration Models, August 12, 1977

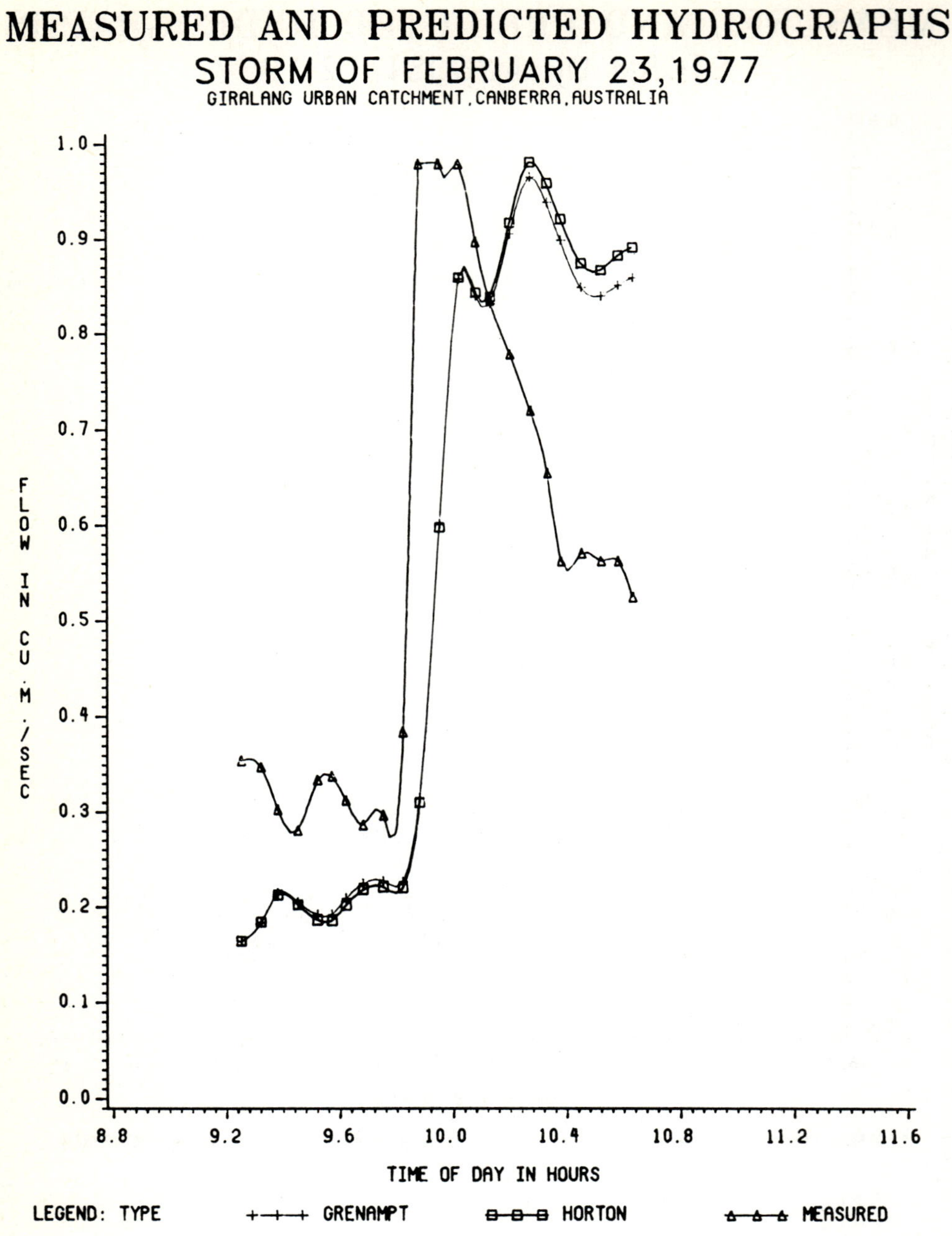

Figure 12. Measured and Predicted Hydrographs, Horton and Green-Ampt. Infiltration Models, February 23, 1977

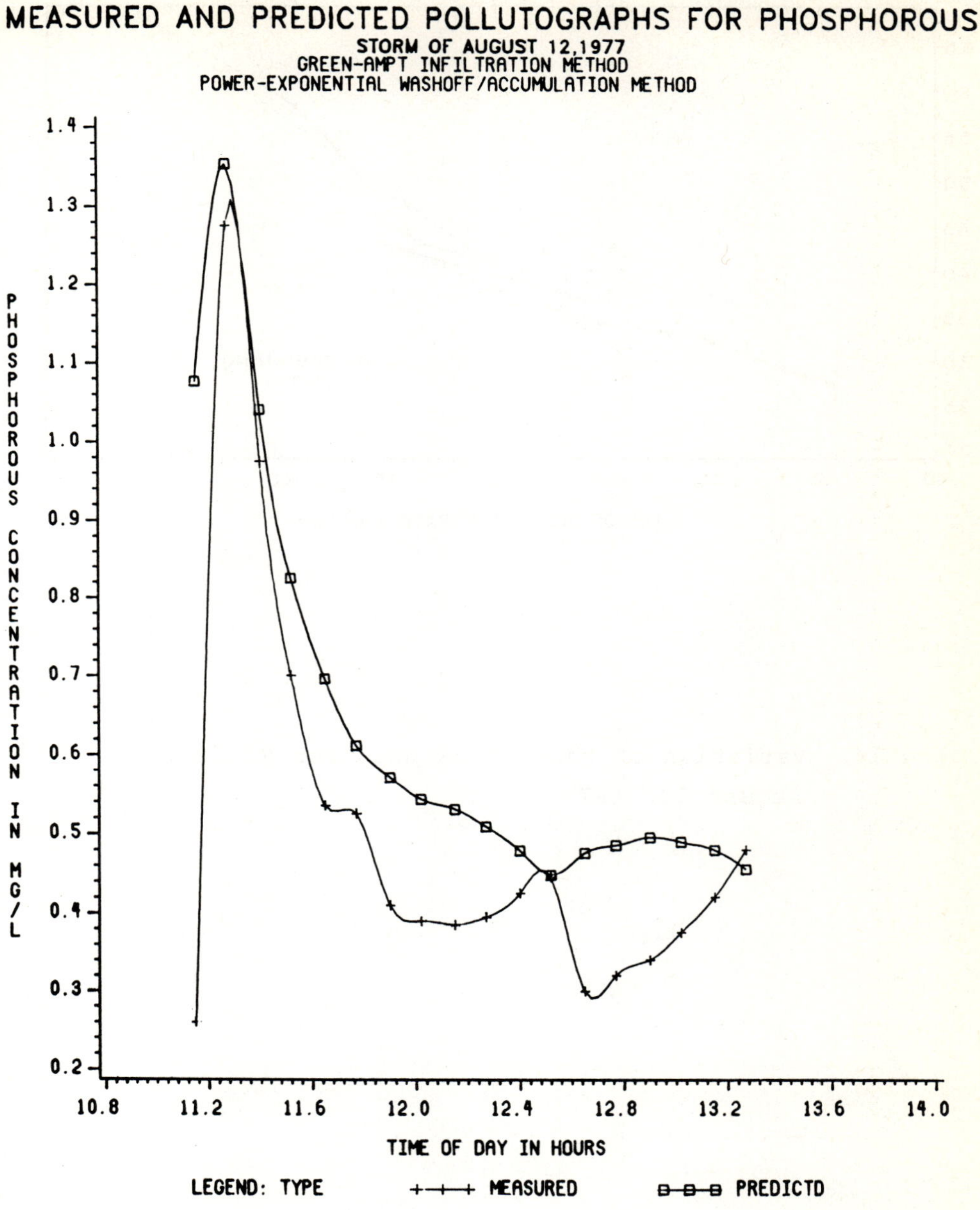

Figure 13. Measured and Predicted Phosphorus Concentration, August 12, 1977

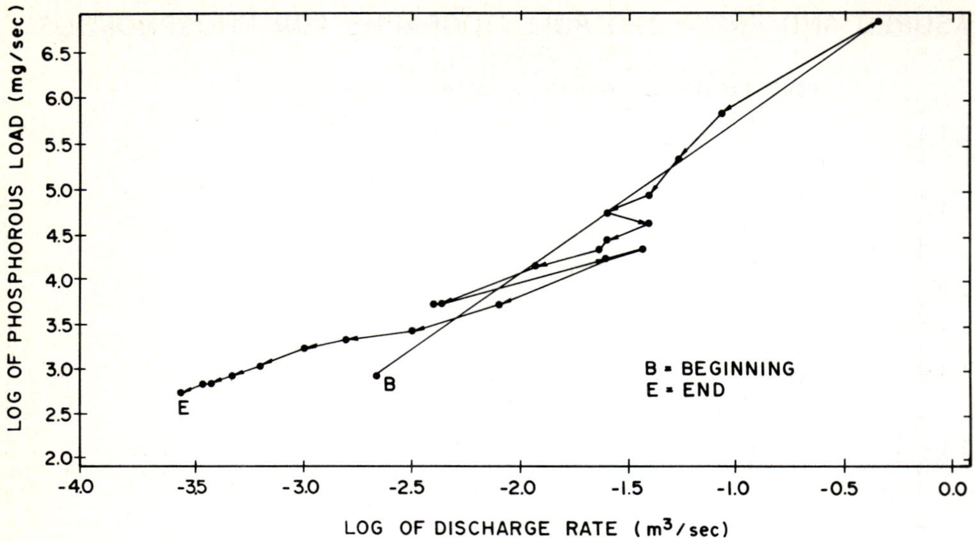

Figure 14. Variation of Phosphorus Load For Storm of August 12, 1977

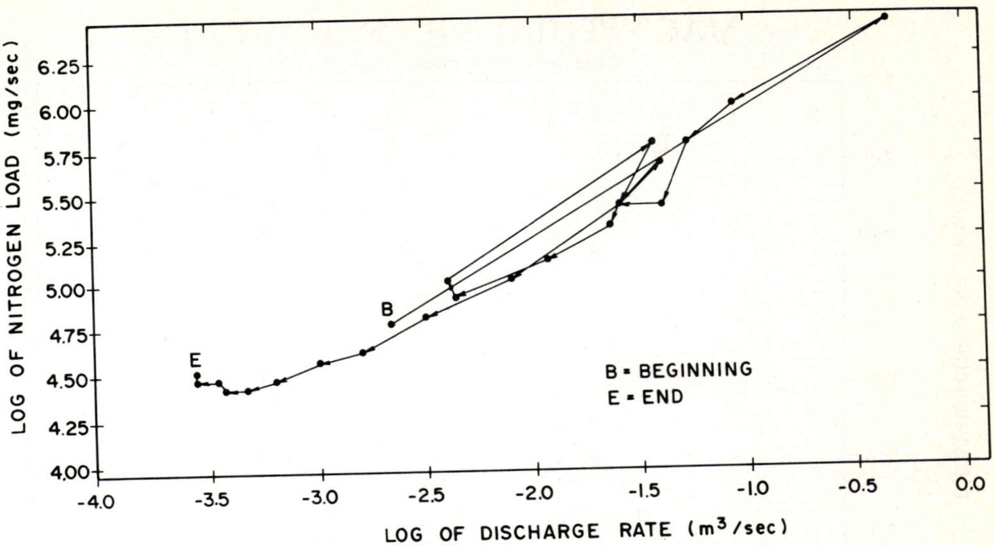

Figure 15. Variation of Nitrogen Load for Storm of August 12, 1977

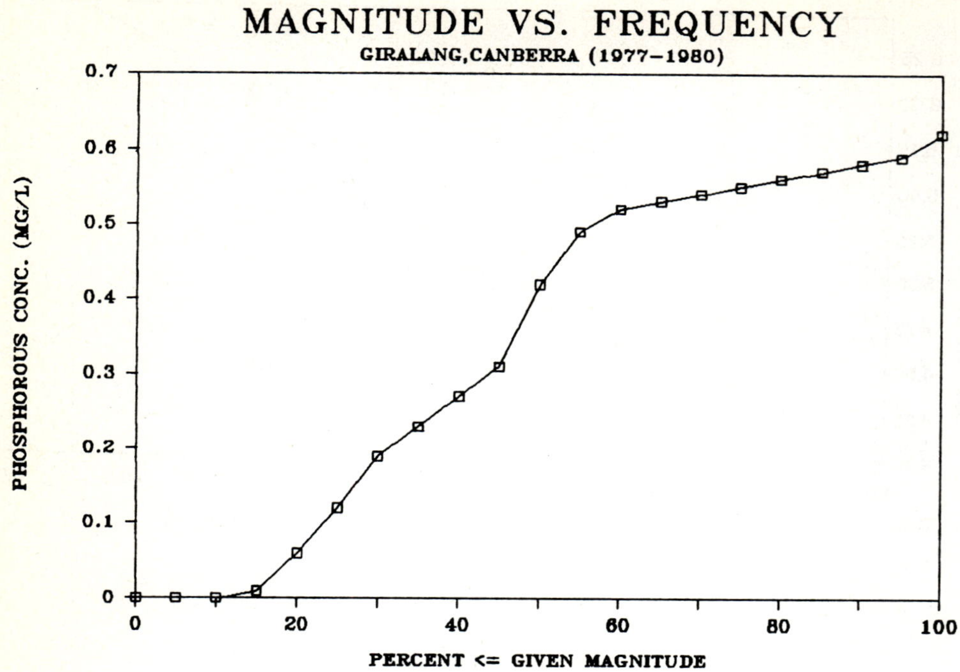

Figure 16. Continuous Simulation of Phosphorus Concentration, Giralang Urban Catchment, Canberra (1977-1980)

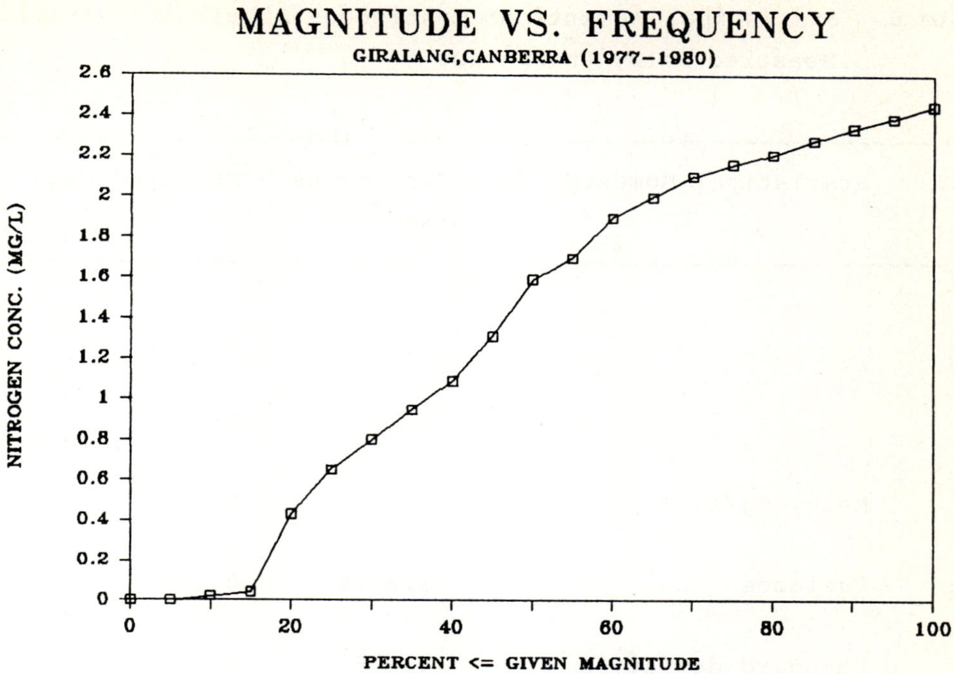

Figure 17. Continuous Simulation of Nitrogen Concentration, Giralang Urban Catchment, Canberra (1977-1980)

Table 6. Statistical Moments of Continuous Simulation Versus Measured Data, Phosphorus

Nos.	Statistical Moments	Continuous SWMM	Measured Data
			All Storms
1	Mean, mg/l	0.345	0.349
2	Variance	0.0518	0.0488
3	Standard deviation	0.228	0.221
4	Coefficient of variation	0.659	0.632
5	Coefficient of skewness	-0.392	2.7161

Table 7. Statistical Moments of Continuous Simulation Versus Measured Data, Nitrogen

Nos.	Statistical Moments	Continuous SWMM	Measured Data	
			All Storms	Winter Storms
1	Mean, mg/l	1.36	1.631	1.330
2	Variance	0.765	2.987	0.368
3	Standard deviation	0.874	1.728	0.607
4	Coefficient of variation	0.642	1.059	0.456
5	Coefficient of skewness	-0.346	3.458	1.041

Salem Creek, Winston-Salem, North Carolina, U.S.A.

Salem Creek drains most of the urban runoff from the city of Winston-Salem, North Carolina. A time series of 35 years of hourly rainfall was analyzed and grouped into storm events by defining a minimum interevent time of 12 hours of dry weather. Average storm event rainfall intensity, duration and time since the previous storm are presented in Figures 18, 19 and 20 versus each month of the year. Accurate land use

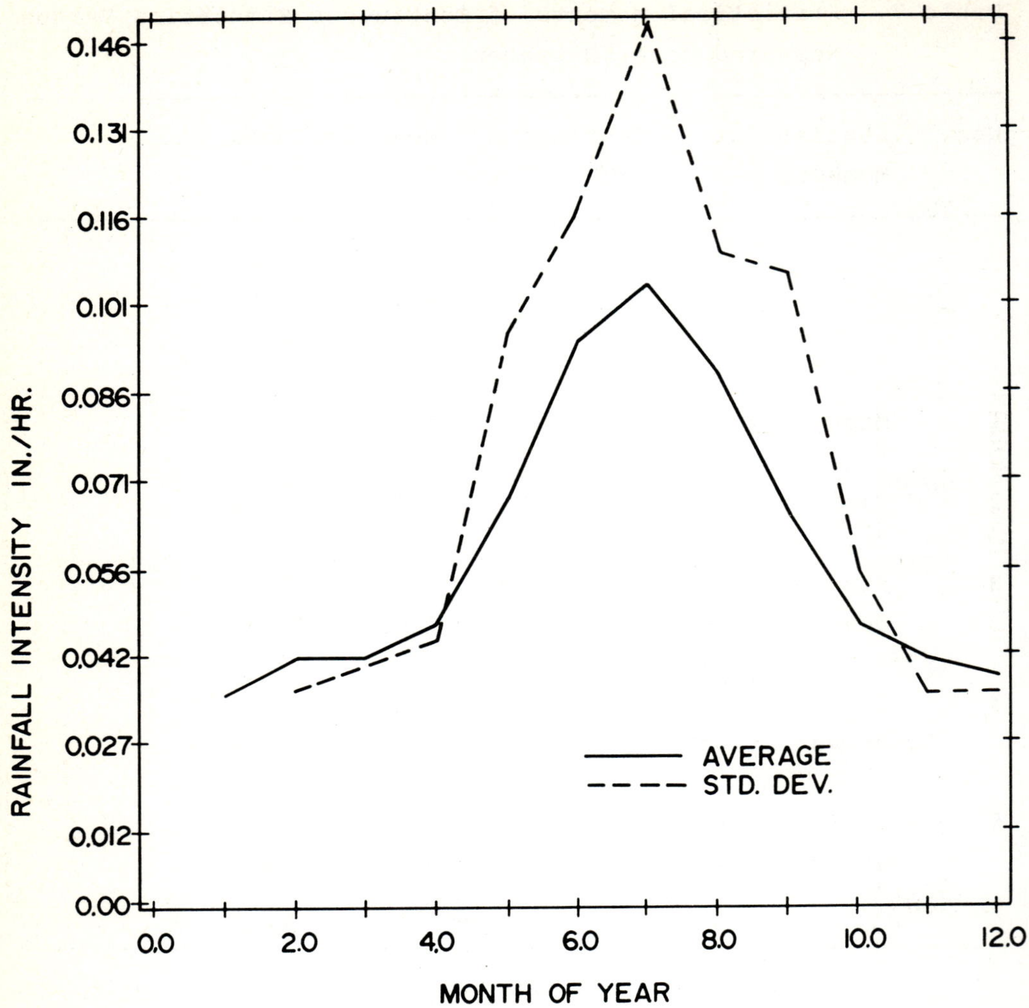

Figure 18. Storm Event Rainfall Intensity Per Month of Year, 1948-1975

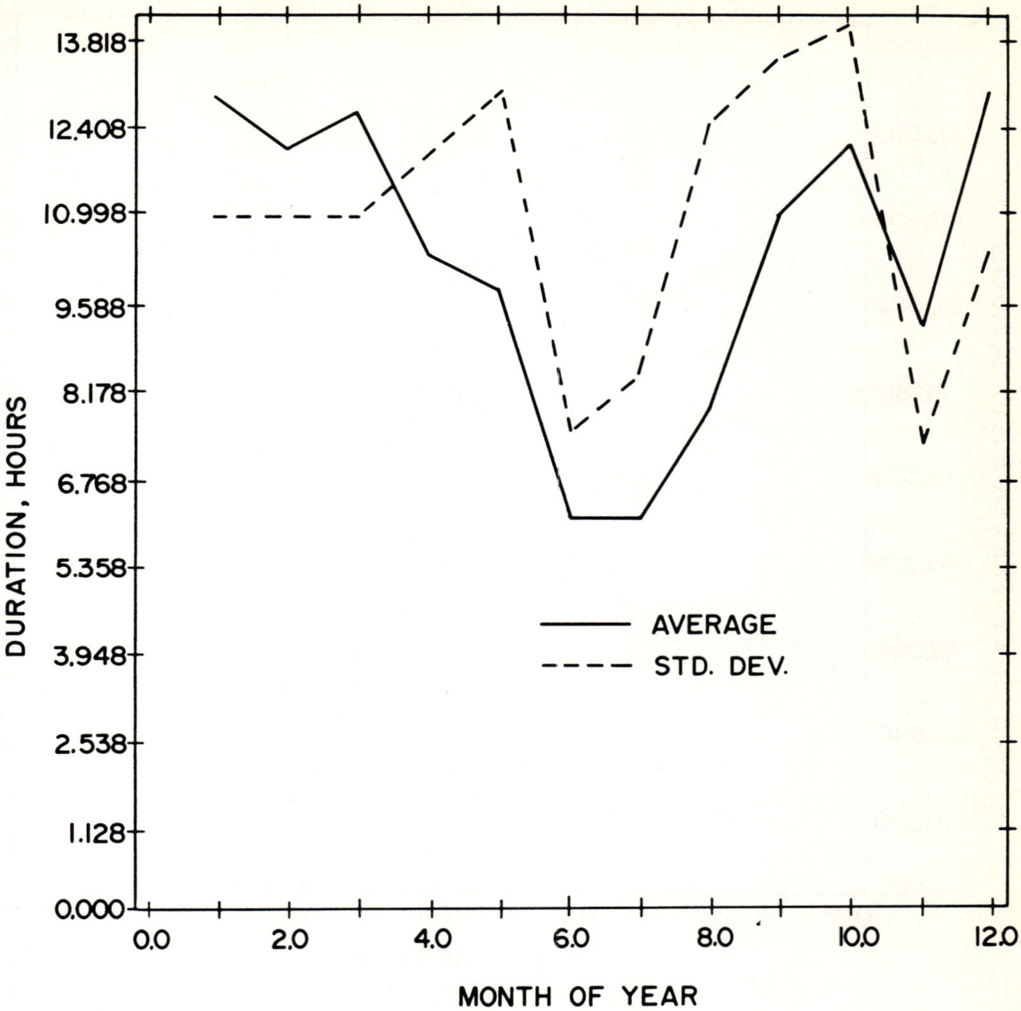

Figure 19. Storm Event Duration Per Month of Year, 1948-1975

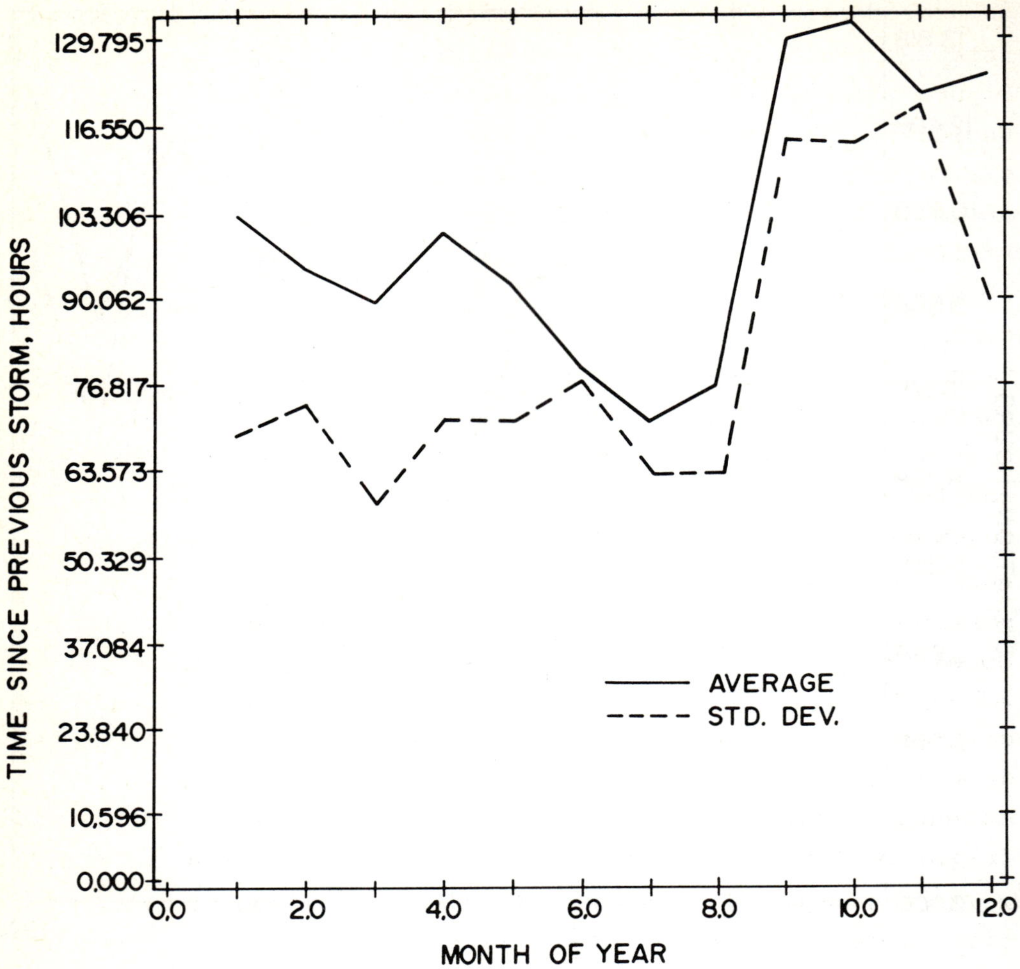

Figure 20. Time Since Previous Storm Event Per Month of Year, 1948-1975

information was obtained from the computerized Land Resources Information Service (LRIS) of the State of North Carolina (see Figure 21). Imperviousness was obtained from published U.S Geological Survey data (45). The urban catchment was discretized into overland flow and channel flow segments, as shown in Figure 22. Runoff hydrographs were calibrated and verified against measured flows, Figures 23 and 24, and key stream hydraulic and water quality parameters (e.g., velocity, depth, longitudinal dispersion, deoxygenation and reaeration rates) were adjusted to measured data from stream intensive surveys (Figure 25).

A comparison of DO cumulative frequency curves (from continuous simulation) during wet weather periods from November 1980 to August 1981 is presented in Figure 26, for both the steady-state and transient water quality models, at a selected distance downstream (34,36). The steady-state model is conservative up to a DO concentration of about 5.5 mg/l (slightly beyond the stream standard). The predicted cumulative DO frequency curves for both the continuous deterministic steady-state and the statistical (method of moments) models are very similar (see Figure 27). The frequency distribution of the duration of violations of a 5.0 mg/l DO standard is presented in Figure 28. For example, there were 8 occurrences of violation for 12 consecutive hours and one occurrence of violation for as many as 60 consecutive hours. These results could only be derived from analysis of water quality transients. The statistical derived-distribution method of moments was also applied to dry weather periods from November 1980 to August 1981, for chemical oxygen demand (COD) concentration (mg/l) and fecal coliform concentration (MPN/100ml), Figures 29 and 30. The utility of cumulative water quality frequency curves and frequency histograms of consecutive hours of violation of a stream standard may be seen from examination of Figure 31: the toxic

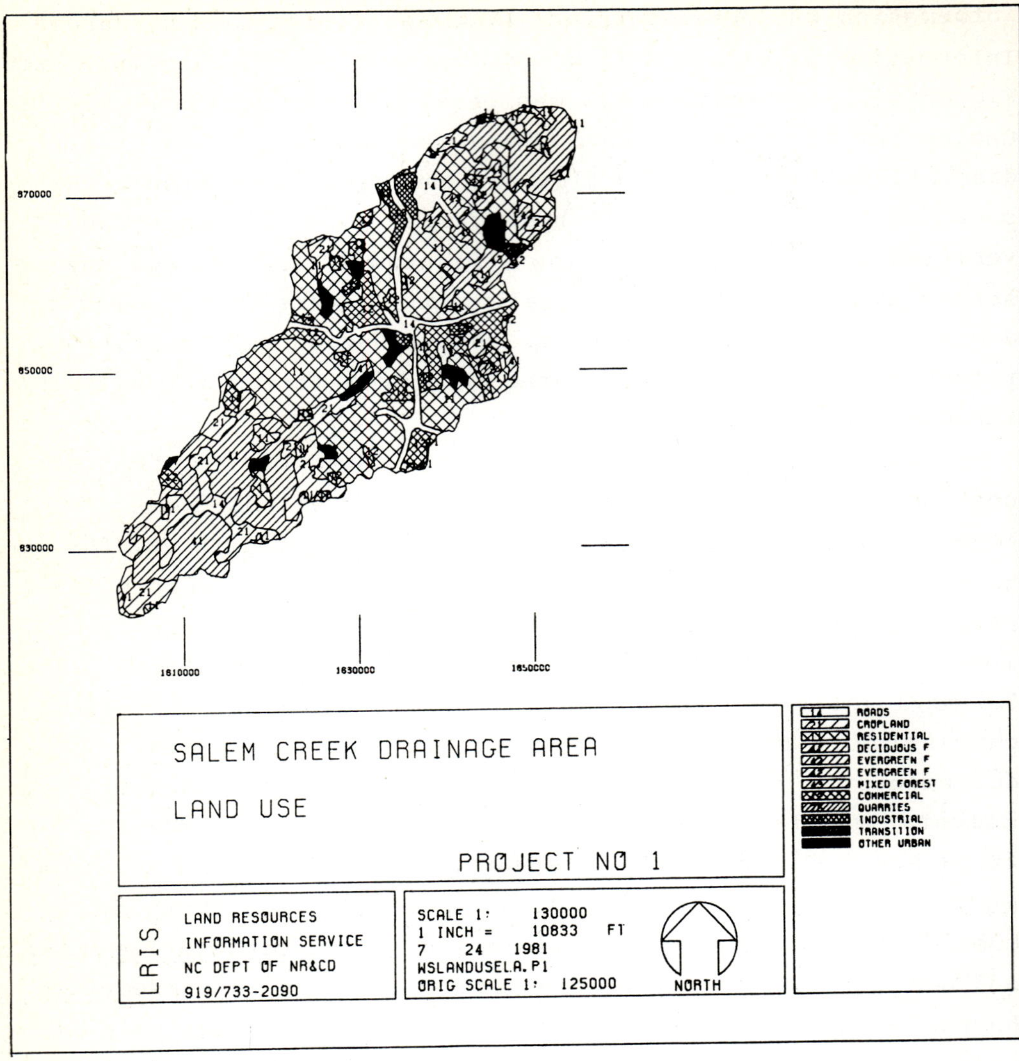

Figure 21. Land Resources Information Service Land Use Analysis, Salem Creek Drainage Basin

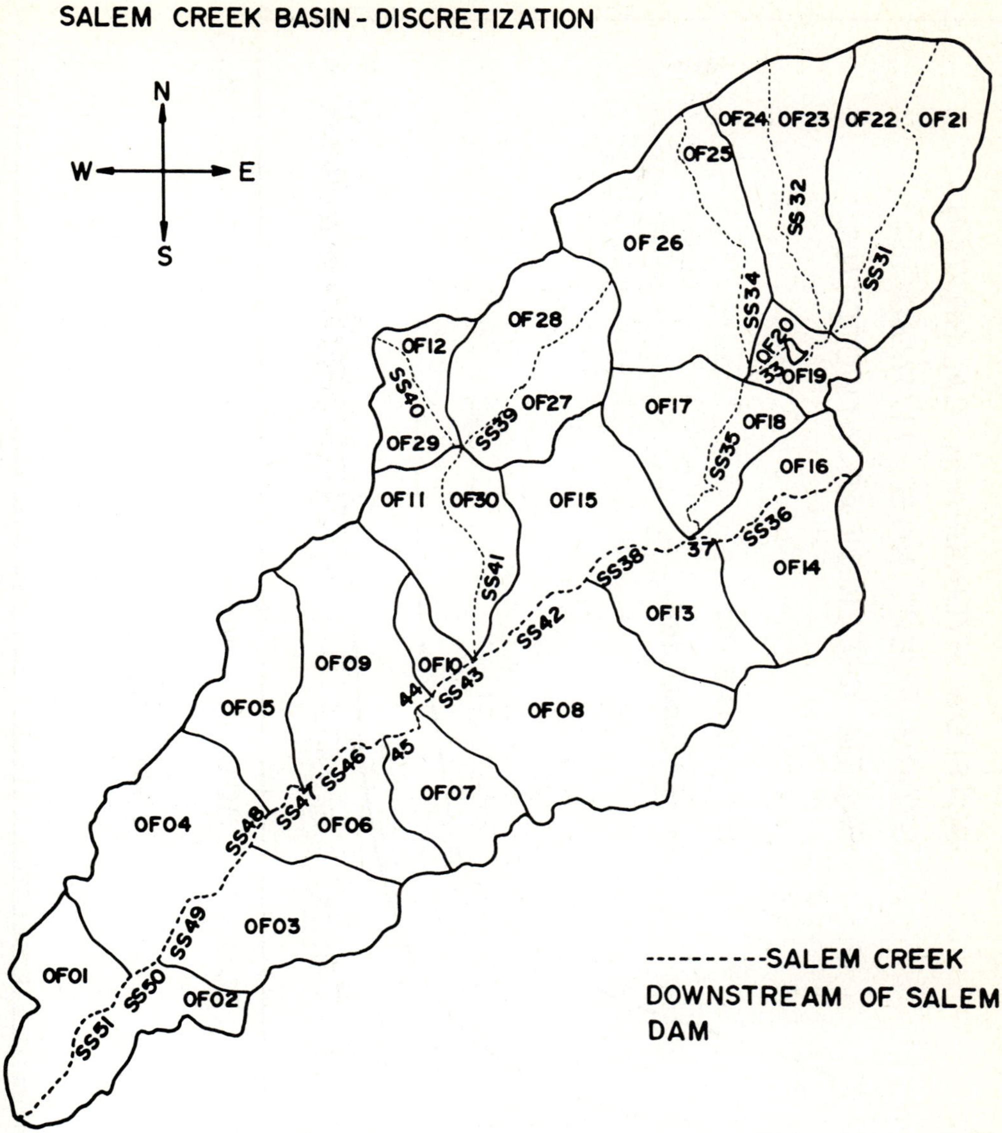

Figure 22. Discretization of Salem Creek Basin Into Overland Flow and Stream Flow Segments

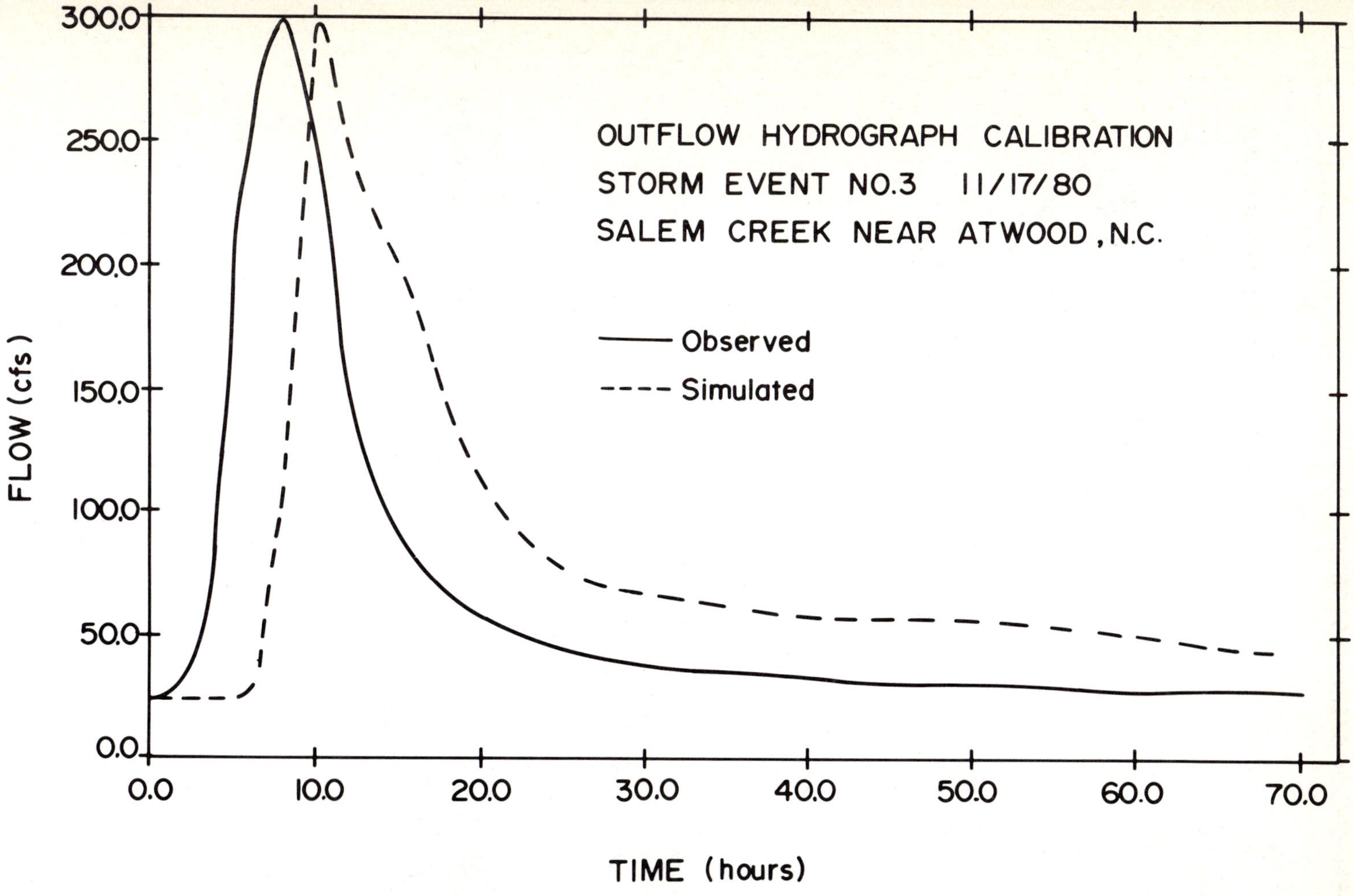

Figure 23. Salem Creek Outflow Calibration, Distributed Routing Rainfall-Runoff Model

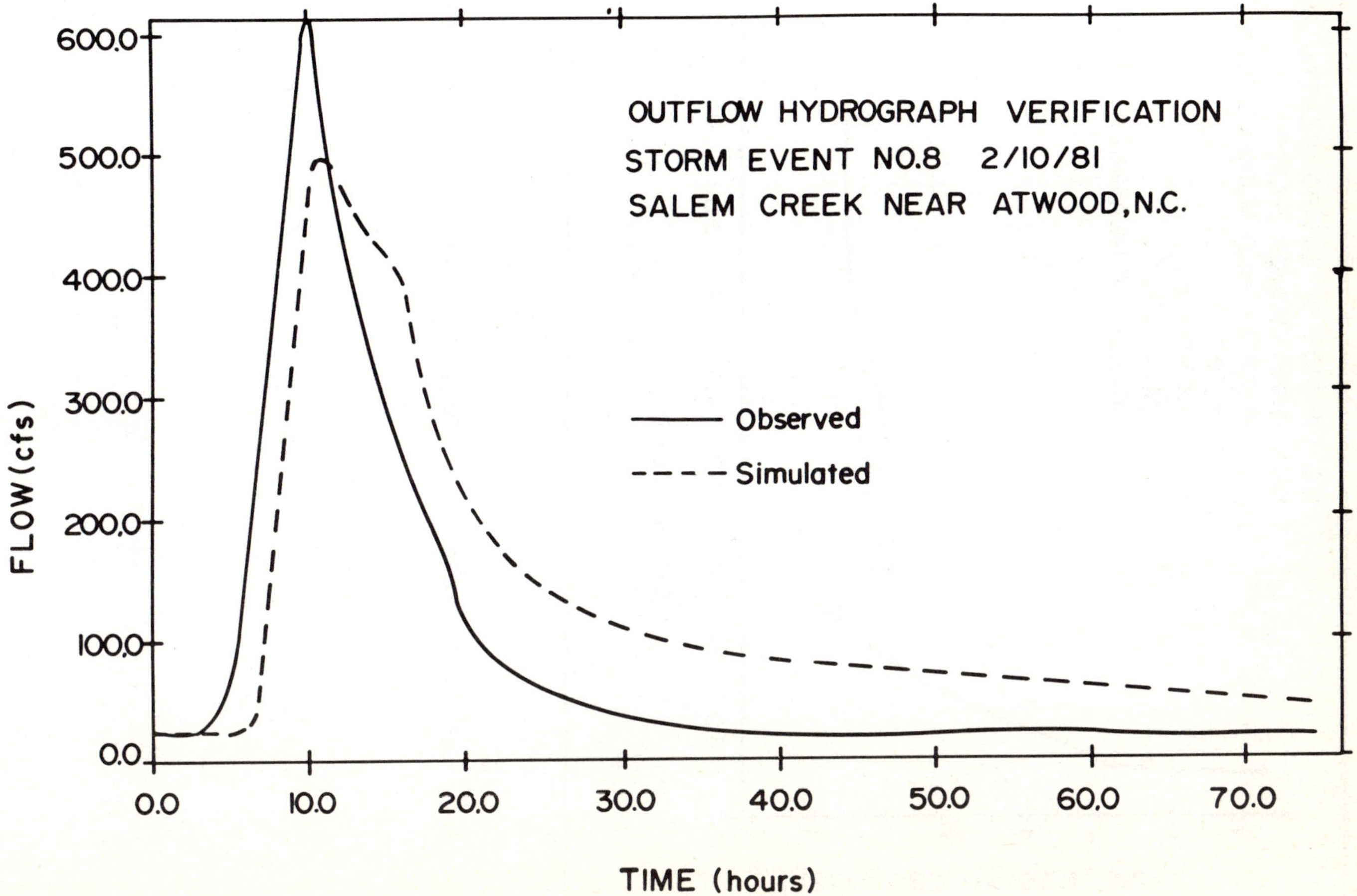

Figure 24. Salem Creek Outflow Verification, Distributed Routing Rainfall-Runoff Model

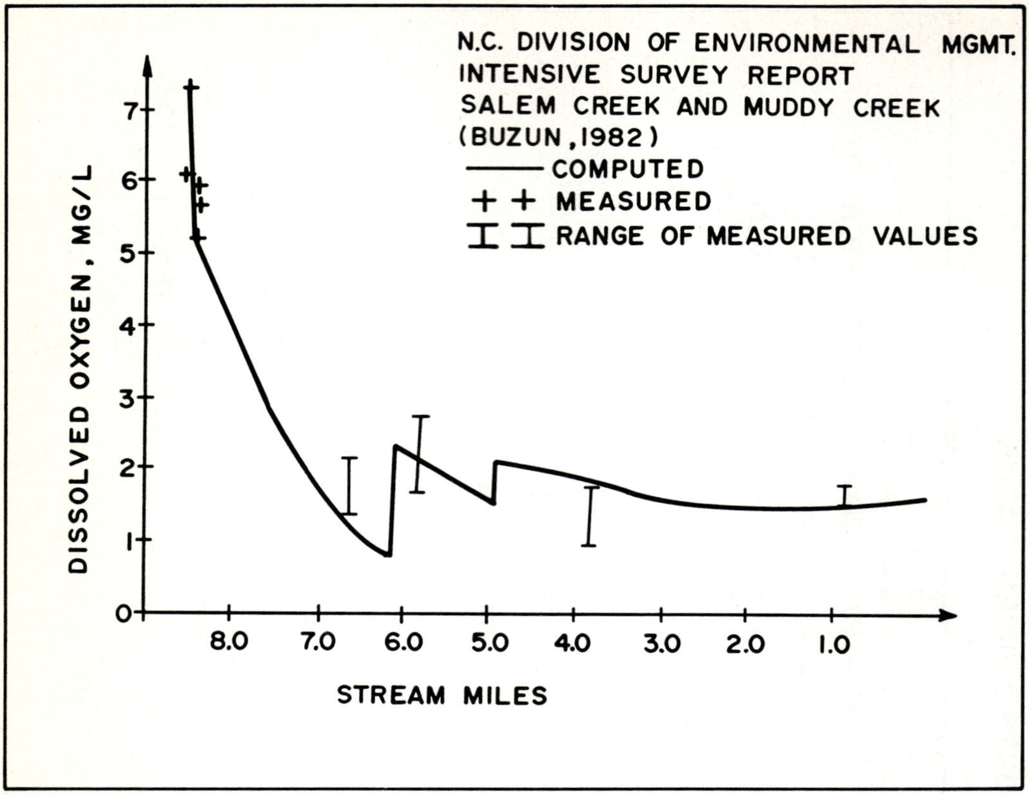

Figure 25. Calibrated Dissolved Oxygen Profile, Salem Creek

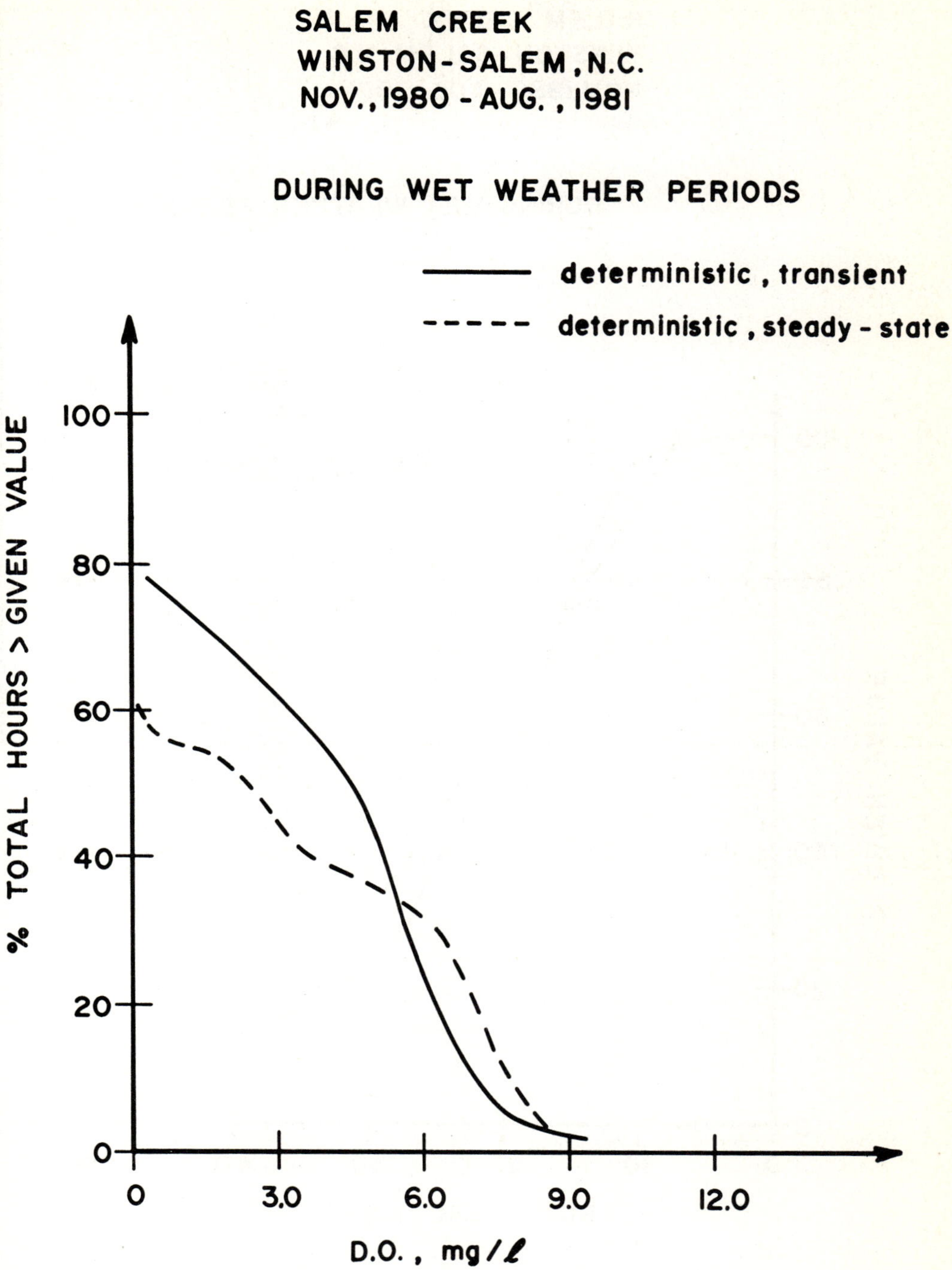

Figure 26. Continuous Physically-Based Steady-State Versus Transient Simulation of Dissolved Oxygen, Salem Creek (36)

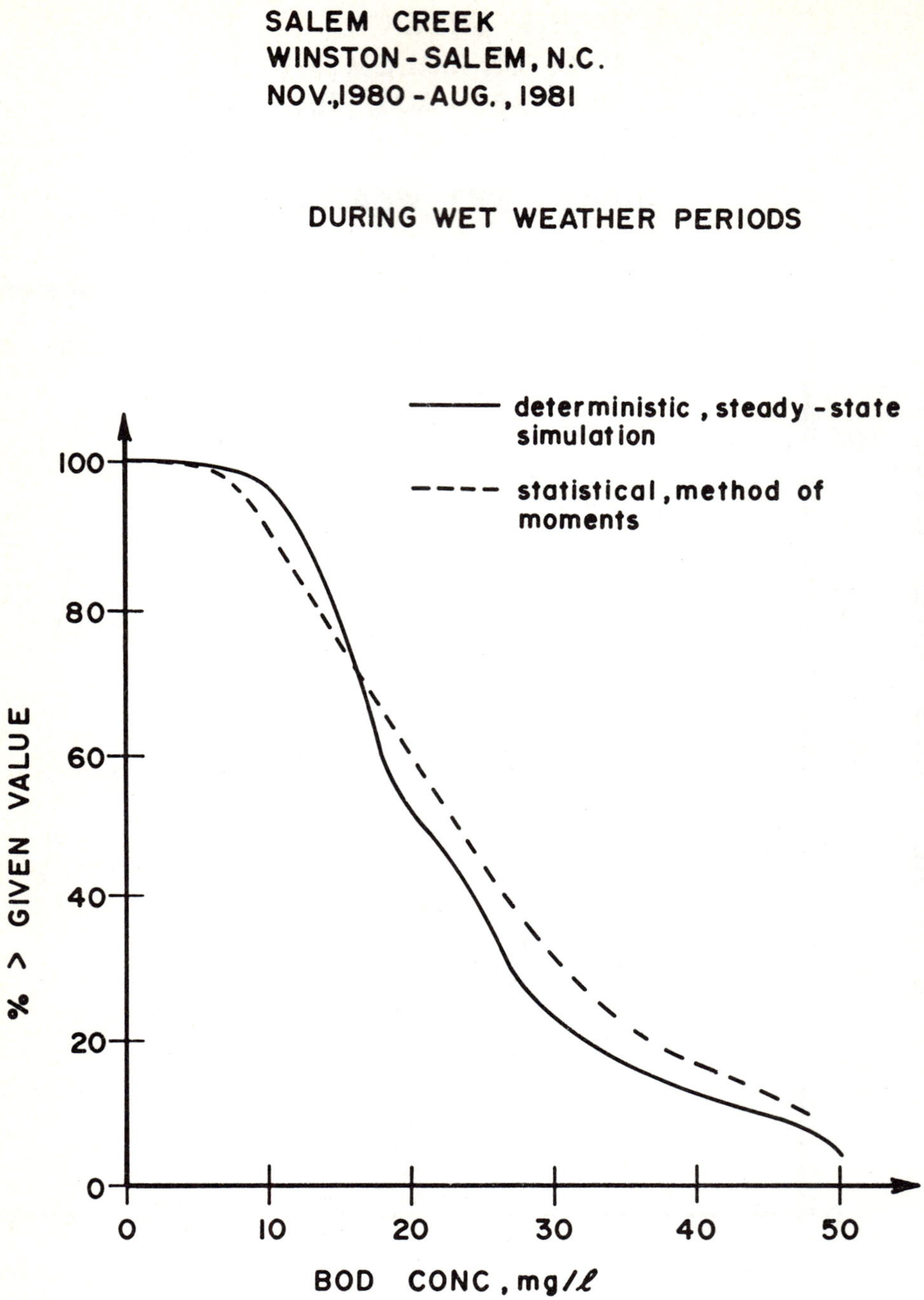

Figure 27. Physically-Based, Steady-State Versus Statistically-Based

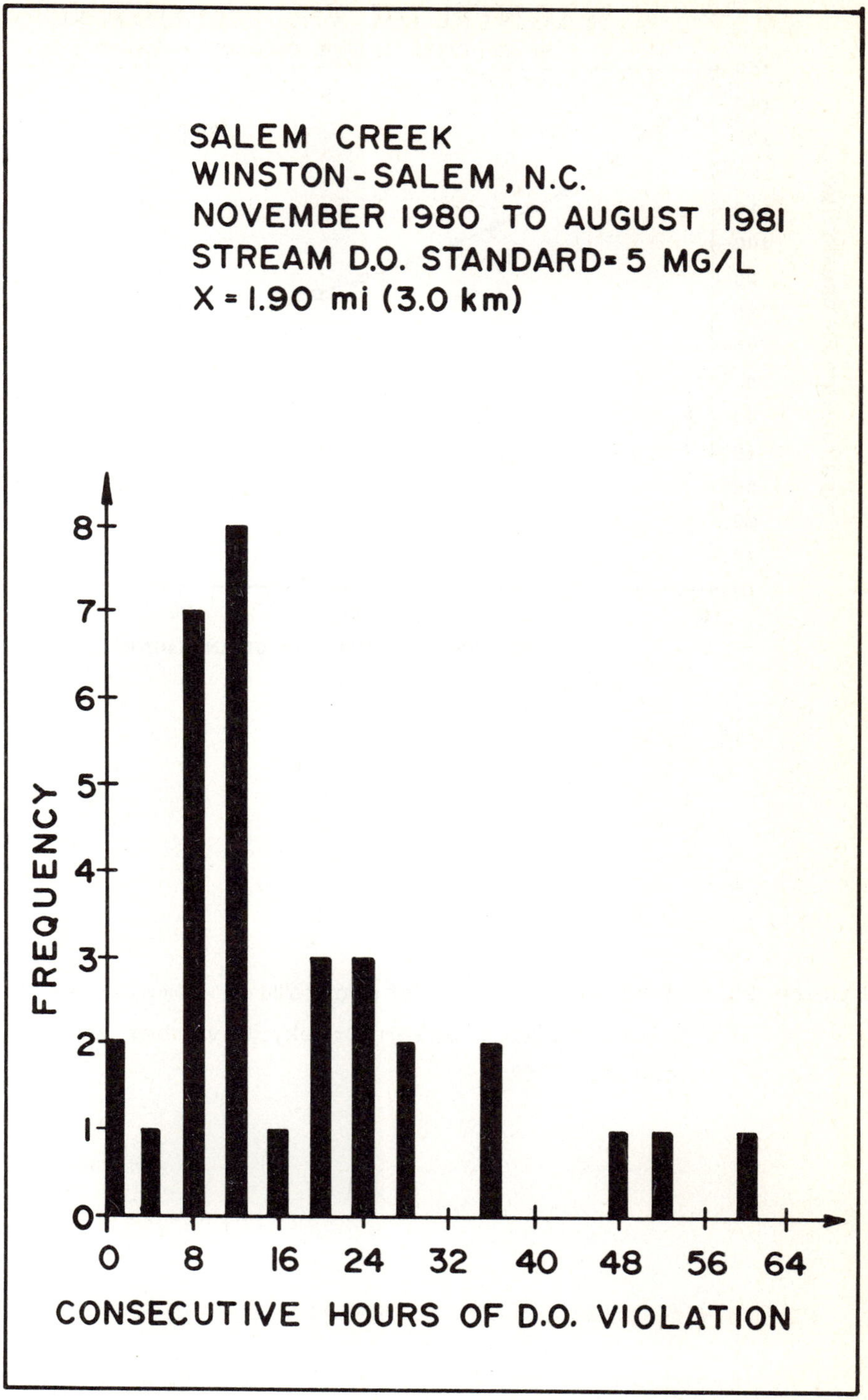

Figure 28. Frequency Distribution of Duration of Violations of Stream DO Standard, Salem Creek (36)

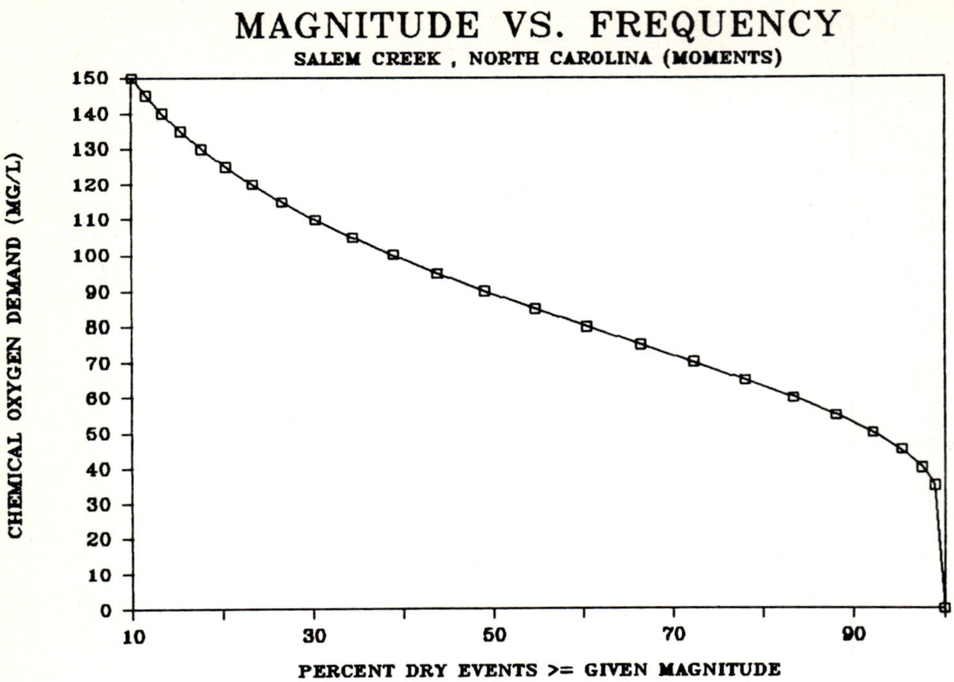

Figure 29. Predicted COD Concentration Cumulative Frequency Distribution, Salem Creek, November 1980 to August 1981

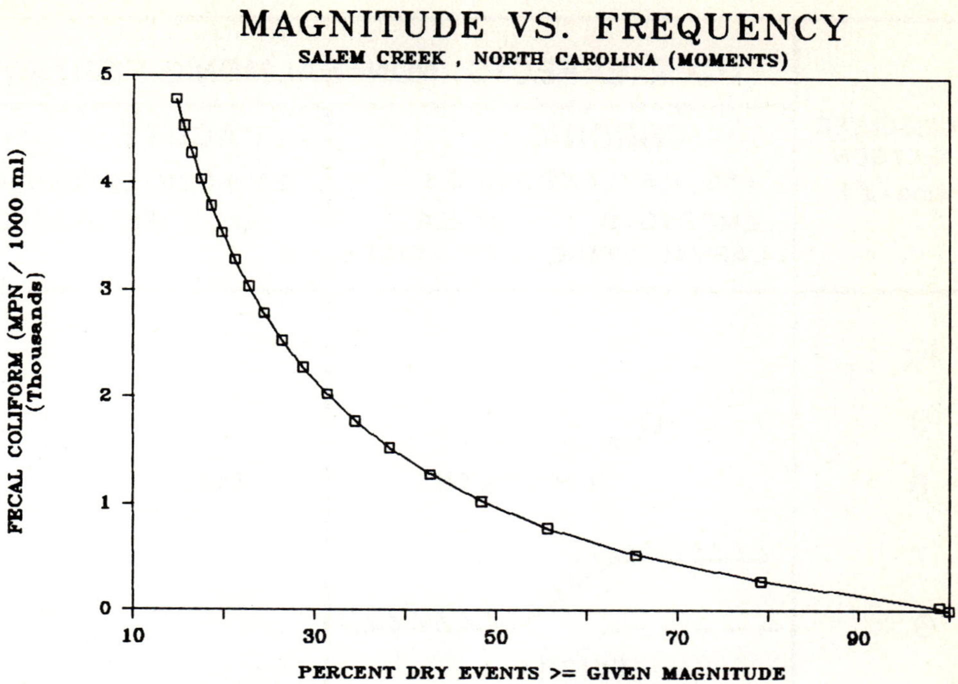

Figure 30. Predicted Fecal Coliform Concentration Cumulative Frequency Distribution, Salem Creek, Novembr 1980 to August 1981

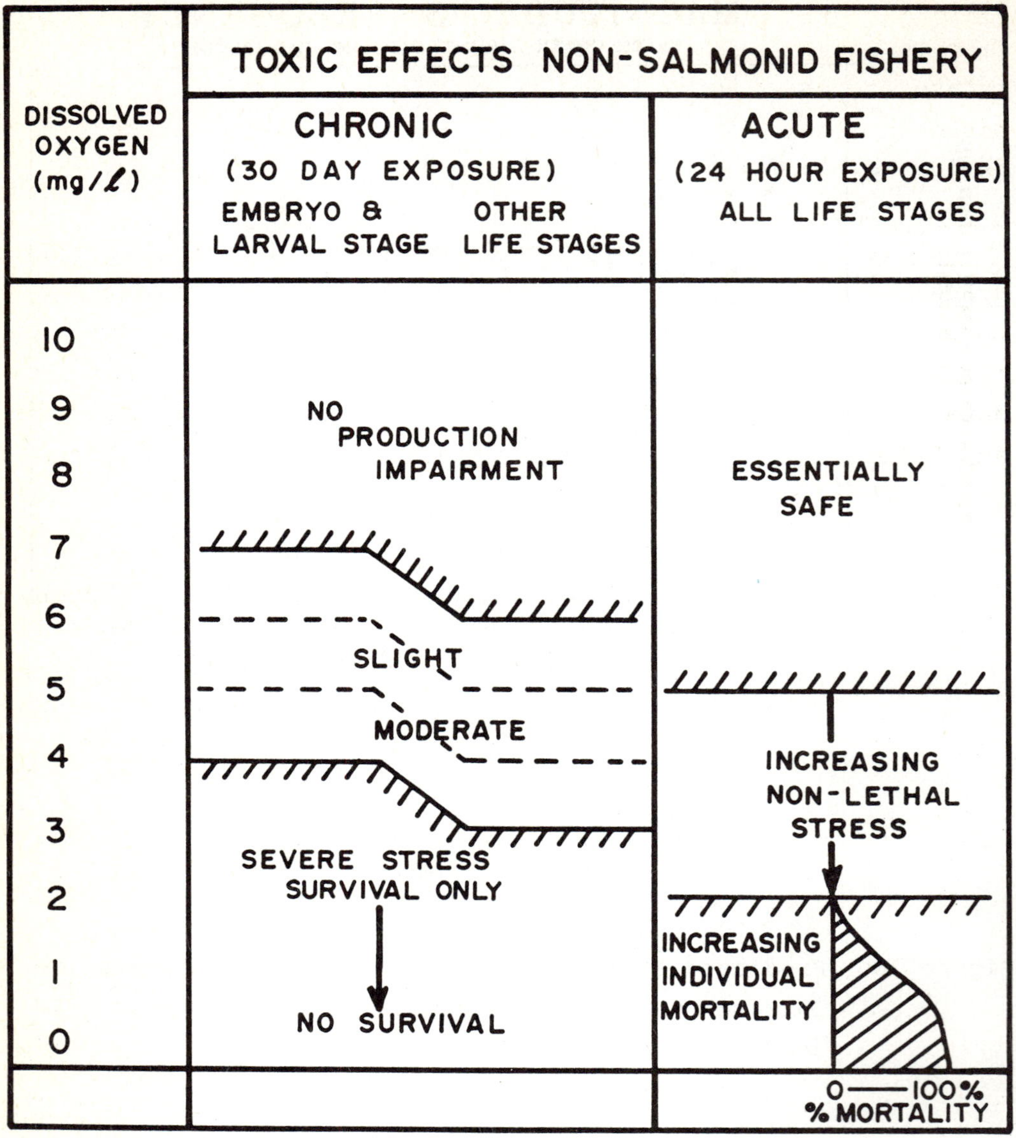

Figure 31. Effects of Chronic and Acute Exposure of Non-Salmonid Fish to Dissolved Oxygen Levels (17)

effects of both chronic and acute exposure of non-salmonid fish to various levels of DO concentration (17) can be related to predicted values. A standard for oxygen in rivers based on single event statistics (characterized by return period) has also been proposed in Denmark (21). Three-dimensional computer graphics may also prove useful to visualize storm-derived pollutant concentrations in time and space. Figure 32 illustrates the storm-related BOD concentration spikes up to 20 miles (32 km) downstream from the urban source. The graph is labeled hypothetical because data on other possible sources of BOD from several tributaries and lateral flow were not available beyond 8 miles (12.8km).

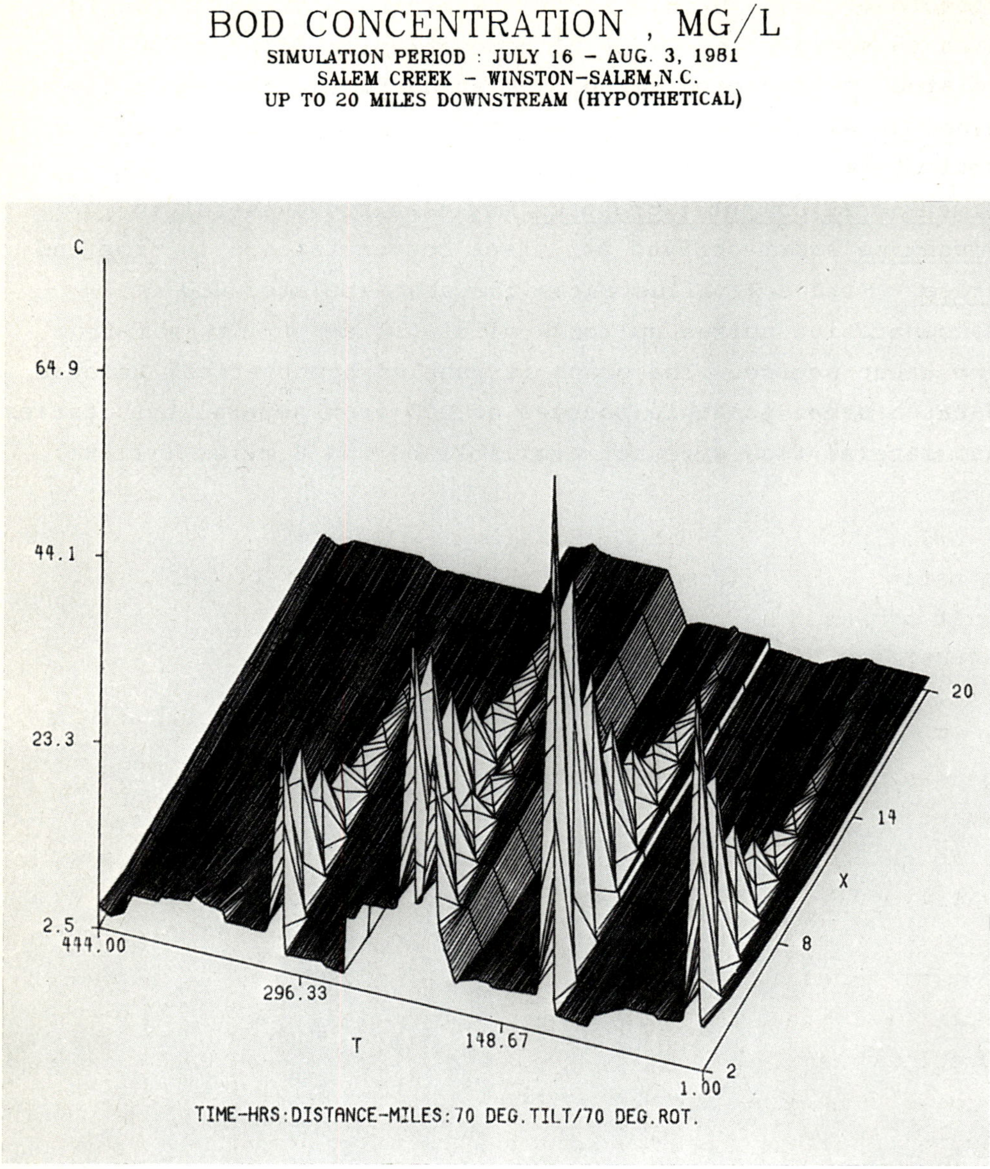

Figure 32. BOD Concentration in Time and Space, July 16-August 3, 1981 Simulation, Salem Creek

Future Directions

Advances at both extremes of the computer family; super-computers and micro-computers, will insure: (1) the survival of even larger more complex simulation models, and (2) the availability of a longer menu of models to a much larger user community. As our capacity to simulate increases, our ability to predict comes into question. Large, sophisticated models exhibit multicollinearity among the independent variables. For example, in overland flow the flow is a function of slope and depression storage; meanwhile, depression storage may be estimated as a function of slope. The inter-dependencies among water quality variables are even more numerous, complex and still poorly understood. Pollutant washoff models such as the rating curve method exhibit spurious correlation as a result of predicting pollutant load from discharge and concentration data.

While more basic research is needed to improve our understanding of pollutant accumulation and washoff processes upstream and kinetic interactions downstream, how many variables can be represented by any given model and still be properly validated? Even if all these variables could be measured, there would be uncertainty due to sampling error. Reducing uncertainty and prediction error propagation will be great challenges for the future. Given that a model has been properly validated and reasonably represents the present state of the system, the next challenge is to determine the future response of the system with confidence.

Integrating both physically-based and statistically-based methods into the overall simulation appears to provide the best opportunity to examine water quality fluctuations due to urban runoff variability. Control strategies to reduce the probability distributions of critical pollutant level durations can only be examined if the frequency distributions of consecutive violations of stream standards can be reasonably predicted. Our ability to model water quality across phases of the hydrologic cycle is quite limited in view of the difficulties in predicting within a single phase. Yet, surface/subsurface interactions must be examined in greater detail when dealing with methods of waste disposal (e.g., sanitary landfills) subject to effects from both phases of transport. Computer graphics capabilities may enhance our visualization of complex phenomena, but will not likely improve the accuracy with which the information is presented for the foreseeable future.

Acknowledgements

Research presented in this paper was partially funded by: grants from the Water Resources Research Institute of the University of North Carolina, nos. 14-34-0001-1135 and 14-34-0001-2135; the Council for International Exchange of Scholars and the Australian American Educational Foundation, who supported the author as a Visiting Fulbright Scholar with the Department of Civil Engineering, Monash University, Clayton, Victoria, Australia. The cooperation of Gary C. Codner, Russell Mein and E. M. Laurenson (Monash) and the Department of Housing and Construction (Canberra, Australia) is also gratefully acknowledged.

References

1. Adomian,G. Stochastic Systems Analysis. In: Applied Stochastic Processes, Adomian,G. (ed.). Academic Press : 1-17 ; 1980

2. Alley, W. M., Smith, P. E. Estimation of Accumulation Parameters for Urban Runoff Quality Modeling. Water Resources Research 17 (4) : 1161-1166 ; 1981 August

3. Alley, W. M. Estimation of Impervious-Area Washoff Parameters. Water Resources Research 17 (6) : 1657-1664; 1981 December

4. Alley, W. M., Smith, P. E. Distributed Routing Rainfall-Runoff Model--Version II. Open File Report 82-344. United States Geological Survey. NSTL Station. Mississippi. 1982

5. Alley, W. M, Smith P. E. Multi-Event Urban Runoff Quality Model. Open File Report 82-764. United States Geological Survey. NSTL Station, Mississippi. 1982

6. Amatya, D. M. Statistical Evaluation of Parameters In Deterministic Water Quality Models. M.S. Thesis. Duke University, Durham, North Carolina. 1985 168 pp.

7. Ammon, D. C. Urban Stormwater Pollutant Buildup and Washoff Relationships. M.S. Thesis. University of Florida. Gainesville, Florida. 1979 134 pp.

8. Areawide Assessment Procedures Manual. U. S. Environmental Protection Agency. EPA-600/9-76-014. Cincinnati, Ohio. 1976 July

9. Baca, R. G., Arnett, R. C. A Finite Element Water Quality Model For Eutrophic Lakes. In: Finite Elements in Water Resources, Gray, W. G., Pinder, G. F., Brebbia, C. A. (eds.). Pentech Press (London) : 4.125 - 4.147; 1977

10. Benton, L. P. Muddy Creek - Yadkin River Fish Kill, August 8, 1976, Forsyth County. Memorandum. Water Quality Section, Division of Environmental Management, N.C. Dept. of Natural and Economic Resources. 1976 September 24

11. Brutsaert, W. F., Water Quality Modeling by Monte Carlo Simulation. Water Resources Bulletin 11(2): 229-236; 1975 April

12. Bureau of Meteorology. Climatic Survey, Canberra and the Australian Capital Territory. Commonwealth of Australia : 68 pp. ; 1968 December

13. Dandy, G. C., Moore, S. F. Water Quality Sampling Programs in Rivers. Journal of the Environmental Engineering Division. ASCE 105(EE4): 695-712; 1979 August

14. Ditoro, D. M., O'Connor, D. J. The Distribution of Dissolved Oxygen in a Stream with Time Varying Velocity. Water Resources Research 4(3): 639-646; 1968

15. DiToro, D. M. Statistics of Receiving Water Response to Runoff. In: Proceedings of National Conference on Urban Stormwater and CSO Impact on Receiving Water Bodies. EPA-600/9-80-056, Cincinnati, Ohio. 1980

16. DiToro. D. M. Probability Model of Stream Quality Due to Runoff. Journal of Environmental Engineering. ASCE 110 (3): 607-628; 1984 June

17. Driscoll, E. D. and Associates, Inc. Combined Sewer Overflow Analysis Handbook for Use In 201 Facility Planning. U.S. Environmental Protection Agency, Washington, D.C., Vol. II; 1981

18. Finney, B. A., Bowles, D. S., Windham, M. P. Random Differential Equations In Water Quality Modeling. Utah Water Research Laboratory, Utah State University, Logan, Utah. 1979 December 41 pp.

19. Fischer, H. B. A Lagrangian Method for Predicting Pollutant Dispersion in Bolinas Lagoon, Marin County, California United States Geological Survey Professional Paper 582-B ; 1972

20. Fischer, H. B. (ed.) Transport Models for Inland Coastal Waters. Academic Press. 1981

21. Harremöes, P. Urban Storm Drainage and Water Pollution In: Urban Stormwater Quality, Management & Planning. Yen, B. C. (ed.) Water Resources Publication. Littleton, Colorado; 469-494; 1982

22. Hinwood, J. B., Wallis, I. G. Classification of Models of Tidal Waters. Journal of Hydraulics Division. ASCE 101(HY10): 1315-1331; 1975 October

23. Hinwood, J. B., Wallis, I. G. Review of Models of Tidal Waters. Journal of Hydraulics Division. ASCE 1Ø1(HY11): 14Ø5-1421; 1975 November

24. Holly, F. M., Preismann, A. Accurate Calculation of Transport In Two Dimensions. Journal of Hydraulics Division. ASCE: 1Ø3(HY11): 1259-1277; 1977 November

25. Hornberger, G. M. Uncertainty in Dissolved Oxygen Prediction Due to Variability in Algal Photosynthesis. CRES Report AS/R21. Australian National University. Canberra, Australia. 1978

26. Huber, Wayne C., *et al.* Storm Water Management Model-Version III. U. S. Environmental Protection Agency. Cincinnati, Ohio. 1983 October

27. Hydrologic Engineering Center. Storage, Treatment, Overflow, Runoff Model-STORM. Generalized computer Program 723-S8-L752Ø. U.S. Army Corps of Engineers. Davis, California. 1977 August

28. Jennings, M. E. Data Collection and Instrumentation. In: Urban Stormwater Hydrology. American Geophysical Union. Water Resources Momograph 7: 189-217; 198Ø

29. Jobson, H. E. A Practical Lagrangian Transport Model. United States Geological Survey Open-File Report 8Ø-2Ø6. NSTL Station. Mississippi. 198Ø

30. Litwin, Y. J., Joeres, E. F. Stochastic Modeling of Bi-Hourly River Dissolved Oxygen Records Monitored at a Fixed Cross-Section. In: Mathematical Models for Environmental Problems. Brebbia, C. A. (ed.) Pentech Press (London): 329-351; 1976

31. Loganathan, G. V., Delleur, J. W. Effects of Urbanization Frequencies of Overflows and Pollutant Loadings From Storm Sewer Overflows: A Derived Distribution Approach. Water Resources Research 20(7): 857-865; 1984 July

32. Loucks, D. P., Lynn, W. R. Probabilistic Models for Predicting Stream Quality. Water Resources Research 2(3): 593-605; 1966

33. Medina, M. A., Jr. Hydrologic and Water Quality Modeling for Instream Flow Strategies. WRRI Report No. 183. Raleigh, N.C. 1982

34. Medina, M. A., Jr. An Integrated Methodology for Instream Flow Strategies. WRRI Report no. 210. Raleigh, N.C. 1983

35. Medina, M. A., Jr., et al. Modeling Stormwater Storage/ Treatment Transients: Theory. Journal of Environmental Engineering Division, ASCE 107(4): 781-797; 1981 August

36. Medina, M. A., Jr., Deterministic and Statistical Water Quality Modeling. In: Computer Applications in Water Resources. Torno, H.(ed.). ASCE : 1100-1110 ; 1985

37. Mehta, B. M., Ahlert, R. C., Yu, S. L. Stochastic Variation of Water Quality of the Passaic River. Water Resources Research 11(2): 300-308; 1975

38. Moore, S. F., Dandy, G. C., deLucia, R. J. Describing Variance with a Simple Water Quality Model and Hypothetical Sampling Programs. Water Resources Research 12(4): 795-804; 1976 August

39. National Capital Development Commission. Monitoring Stormwater Flow and Water Quality in Paired Rural and Urban Catchments in the A.C.T. Technical Paper 29. Canberra City. 21 pp. ; 1980 August

40. NCASI, Modifications to the QUAL-2 Water Quality Model and User Manual For QUAL-2E Version 2.2. National Council of the Paper Industry for Air and Stream Improvement. New York, N.Y. 1985 April

41. Okubo, A. Diffusion and Ecological Problems: Mathematical Models. Springer-Verlag; 1980

42. Orlob, G. T. Models For Stratified Impoundments. In: Models For Water Quality Management, Biswas, A. K. (ed.). McGraw-Hill; 273-313

43. Padgett, W. J. A Survey of Stochastic Models For BOD and DO in Streams. In: Applied Stochastic Processes, Adomian, G. (ed.). Academic Press: 135-150; 1980

44. Papoulis, A. Probability, Random Variables and Stochastic Processes. McGraw-Hill, N.Y. 1965

45. Putnam, Arthur L. Effect of Urban Development on Floods In The Piedmont Province of North Carolina. U.S. Geological Survey Open-File Report. Raleigh, North Carolina. 1972

46. Roesner, L. A., Giguere, P. R., Evenson, D. E. User's Manual for Stream Quality Model QUAL-II. EPA 600/9-81-015; 1981 February

47. Shubinski, R. P. Use of Models as Projection Tools. In: Workshop on Verification of Water Quality Models. EPA-600/9-80-016: 62-70; 1980 April

48. Sonnen, M. B. Urban Runoff Quality: Information Needs. Journal of Technical Councils. ASCE 106(TC1): 29-40; 1980 August

49. Srinivasan, S. K., Vasudevan, R. Introduction to Random Differential Equations and Their Applications. Elsevier; 1971

50. Thayer, R. P., Krutchkoff, R. G. Stochastic Model for BOD and DO in Streams. ASCE Journal Sanitary Engineering Division 93(SA3): 59-72; 1967

51. Thomann, R. V. Measures of Verification. In: Workshop On Verification of Water Quality Models. EPA-600/9-80-016: 37-61; 1980 April

52. Thomann, R. V., Barnwell, T. O., Jr. Workshop On Verification of Water Quality Models. EPA-600/9-80-016; 1980 April

53. van Genuchten, M. Th. On the Accuracy and Efficiency of Several Numerical Schemes for Solving the Convective-Dispersive Equation. In: Finite Elements in Water Resources, Gray, W. G., Pinder, G. F., Brebbia, C. A. (eds.). Pentech Press (London): 1.71-1.90; 1977

54. Warn, A. E. The Calculation of Consent Conditions for Discharges Using Methods of Combining Distributions. Department of Operational Planning, Anglian Water Authority. Cambs, England. 1982 January

55. Warn, A. E., Brew, J. S. Mass Balance. Water Research. (14): 1427-1434; 1980

56. Water Planning Division. Results of Nationwide Urban Runoff Program. WH-554. U.S. Environmental Protection Agency. I, II, III: Washington, D. C. 1983 December

57. Whitehead , P., Young, P. Water Quality in River Systems: Monte-Carlo Analysis. Water Resources Research 15(2): 451-459; 1979 April

58. Zison, S. W., *et al.* Rates, Constants, and Kinetics Formulations in Surface Water Quality Modeling. EPA-600/3-78-105; 1978 December

EUTROPHICATION MODELLING OF SMALL URBAN LAKES

Bo Neergaard Jacobsen and Niels Nyholm
Water Quality Institute
11, Agern Allé
DK-2970 Hørsholm
Denmark

Introduction

In residential areas urban lakes may provide a high recreational value for the neighbourhood. Many urban lakes suffer from eutrophication problems leading to secondary oxygen depletion problems and an unpleasant appearance. In more severe cases primary oxygen depletion problems may occur with harmful effects on the fish population and potential mosquito and odour problems. Today, most of the dry-weather wastewater sources to such small urban lakes in Denmark have been diverted through trunk sewers to less sensitive receiving waters after treatment. This means that for small urban lakes the relative importance of stormwater discharges from combined sewer overflow (CSO) and stormwater runoff (SWR) has increased. In many cases CSO and SWR represent the entire point discharge of nutrients to such lakes.

Investigations in Birkerod Municipality

In Birkerød Municipality (situated about 20 km North of Copenhagen, population about 23,000) investigations of 19 small urban lakes receiving different loadings from CSO and SWR were conducted during 1979-84. Figure 1 shows the strategy for the investigations. Methods used for estimating loadings and for monitoring describing receiving water quality are described briefly and cause-effect relationships are discussed in this paper. Other elements in the investigation have not been commented on.

NATO ASI Series, Vol. G10
Urban Runoff Pollution.
Edited by H. C. Torno, J. Marsalek, and M. Desbordes

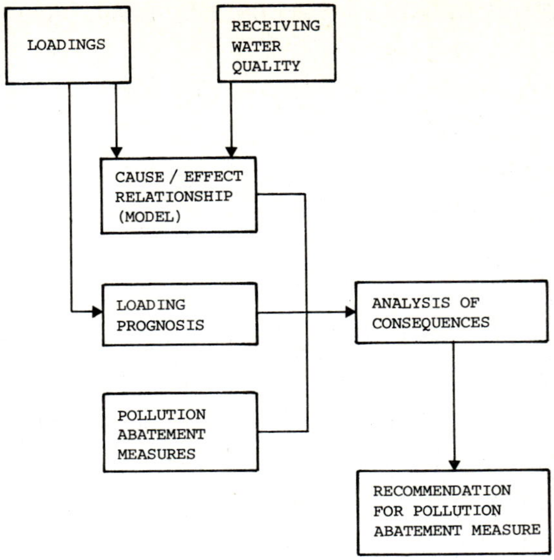

Figure 1 - General strategy for investigations in Birkerød Municipality.

Loadings

Based on field measurements conducted by Birkerød Municipality in a 2-year period 1979-81 on 4 CSO and 2 SWR discharges, estimated average concentrations of conventional pollutants have been found as shown in Table 1.

Table 1 - Estimated average concentrations (mg/l) in discharges based on field measurements (2).

	COD	BOD_5	TOTAL-N	TOTAL-P	SUSP. SOLIDS
SWR	50	6	4	0.5	50
CSO, C_R	120	23	3.2	1.2	140
Detention tank overflow, C_R	95	12	3.2	1.2	95
Dry-weather flow, C_D	470	240	33	12	230

It should be noted that concentrations for CSO and detention tank overflows refer to CR in the equation:

$$P(t) = C_D \cdot Q_D + C_R \cdot Q_R(t)$$

where: P(t) = Flow of pollution at time t after rain start (g/m3)

C_D = Concentration of the substance in dry weather flow (g/m^3)

Q_D = Dry weather flow (m^3/s)

C_R = Concentration of the substance in rain water runoff (g/m^3)

$Q_R(t)$ = Rain water runoff at time t after rain start (m^3/s).

For each of 35 CSO and overflows from detention tanks, annual and extreme loads have been calculated with the SAMBA model, based on 33 years of historic rainfall data. Principles and further details for the SAMBA model, including presentation of results, are given in References 1 and 2.

Other external loadings of N and P (total and dissolved fractions) from atmospheric outfall, groundwater inflow and upstream inflow (when appropriate) have been estimated on a monthly basis. These loadings are based on unit-area loadings found found at other surveys in the same region of the country. An overview of some data for the 19 lakes is shown in Table 2.

Receiving water quality

The investigations of receiving water quality (3) include physical/chemical measurements and biological observations at various levels of detail for the lakes.

Figure 2 shows a temperature stratification in May-June which breaks in July and is reestablished in August before finally breaking up in September. For DO there seems to be a permanent stratification from May to September, with hypolimnion values below 2 mg O_2/l. Figure 3 shows that ortho-phosphate is always present at concentrations of about 0.5 mg P/l, which indicates that the algal growth is never P-limited. The measured concentrations of nitrite + nitrate are always below 0.02 mg N/l and as shown in Figure 3, ammonia (NH) reaches a zero-level in the epilimnion during the period May-August. Elevated N and P con-

Table 2 Selected data for 19 Lakes in Birkerød Municipality

NO.	NAME	VOLUME	AREA	MAX. DEPTH	T_h	SPECIFIC SURFACE LOADING				SECCHI DEPTH
							FROM SEWER		FROM SEWER	
		$10^3\ m^3$	$10^3\ m^2$	m	year	$g\ P \cdot m^{-2} \cdot y^{-1}$	%	$g\ N \cdot m^{-2} \cdot y^{-1}$	%	m
1.1	Nordvang Sø	23.1	13.9	3.5	1.4	0.5	0	4.1	0	(0.5 - 1.1) 0.8
1.2	Hestkøbgård NN	0.9	0.6	2.0	0.04	17.7	88	288	93	
1.3	Hestkøbgård NØ	0.3	0.3	1.0	0.01	36.8	79	237	72	
1.4	Bistrup Gadekær	2.1	1.7	2.4	0.45	0.9	0	8.6	0	(0.8 - 2.0) 1.4
2.1	Birkerød Sø	245	90.3	5.8	6.6	0.25	39	3.0	9	(0.4 - 0.9) 0.7
2.2	Bistrup Kirke	3.6	2.2	3.7	0.23	9.0	93	30.1	77	(0.4 - 1.0) 0.7
3.1	Bregnerødvej	5.4	2.7	2.8	0.30	2.6	43	20.9	35	(0.5 - 1.8) 1.2
3.2	Pilegårdsparken	3.0	1.2	2.5	0.18	12.6	90	54.5	77	(0.5 - 1.3) 1.0
3.3	Trudeslund	8.6	4.4	3.8	0.37	1.8	9	16.3	8	
5.1	Vibeengen	14.8	5.2	6.1	1.6	0.6	14	6.4	4	(0.5 - 3.5) 1.4
5.2	Sækkedammen	4.2	2.9	3.0	0.26	2.2	19	13.8	8	(0.8 - 1.5) 1.1
5.3	Langedam	19.3	15.7	2.8	0.25	3.1	65	14.8	39	(0.6 - 2.2) 1.8
6.1	Rørsangervej	5.1	2.5	4.0	0.09	14.2	0.6	99	0.2	(0.6 - 1.4) 1.1
6.2	Kærvej	9.6	2.6	5.4	0.16	31	0	73	0	
6.3	Karpevænget	5.2	4.5	1.8	0.30	2.3	45	17	42	(0.5 - 1.4) 0.9
7.0	Skovhegnet	1.1	0.9	1.9	0.55	1.2	51	8.2	20	(0.5 - 1.8) 1.3
8.1	Asyldammen	6.2	3.9	4.2	0.77	1.3	9	10.6	9	(1.3 - 2.6) 2.1
8.2	Brådebæk Sø	70.9	21.0	4.5	0.60	3.5	1	25.2	1	(0.7 - 1.6) 1.1
8.3	Puemosen	71.1	18.1	8.5	0.53	4.4	23	32.2	26	(0.9 - 1.9) 1.2

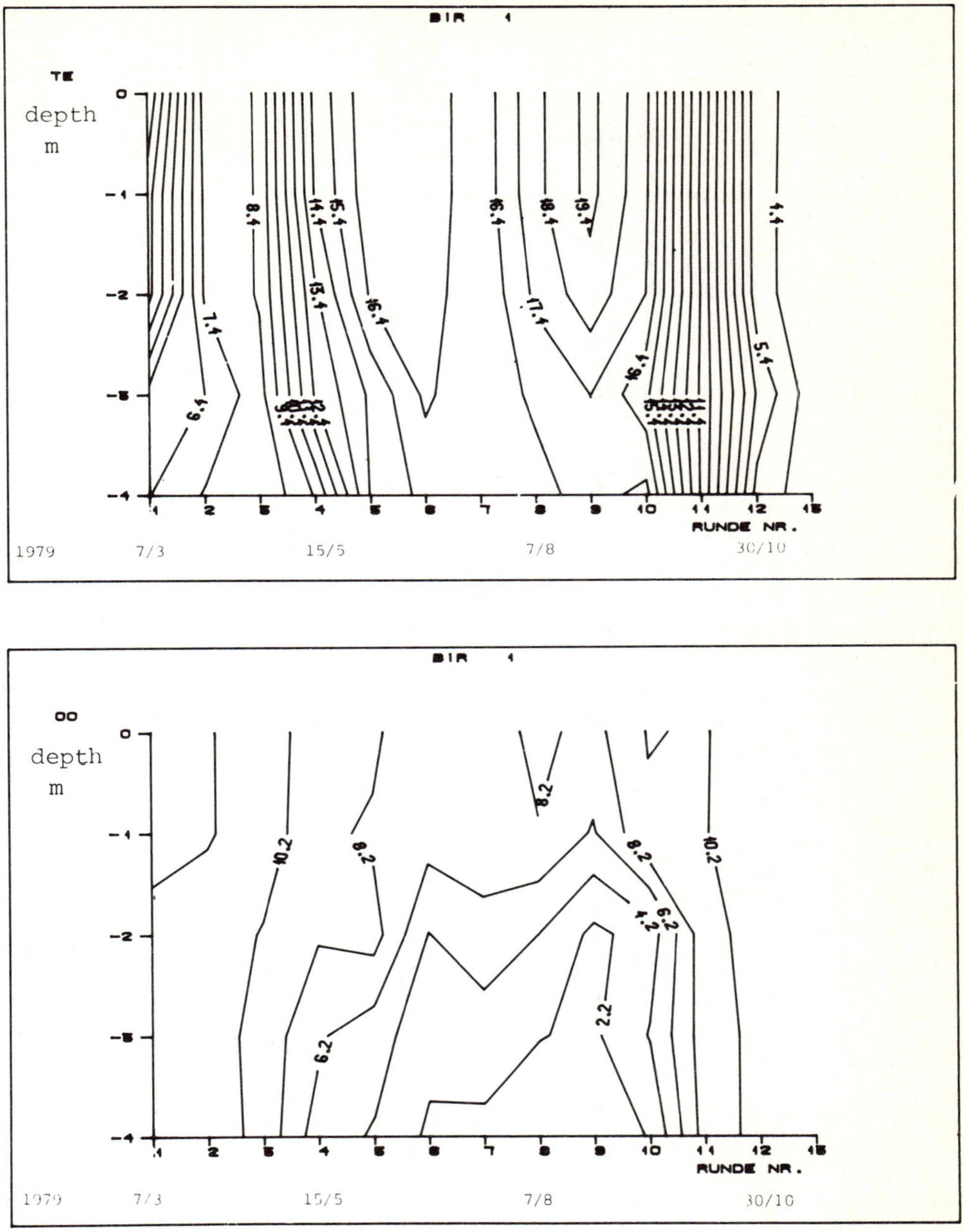

Abcissa represents time over the year
Ordinate represents water depth at station.

Figure 2 Temperature (TE in ^{O}C) and dissolved oxygen (DO in mg/l) profiles, Lake Birkerød, 1979 (3).

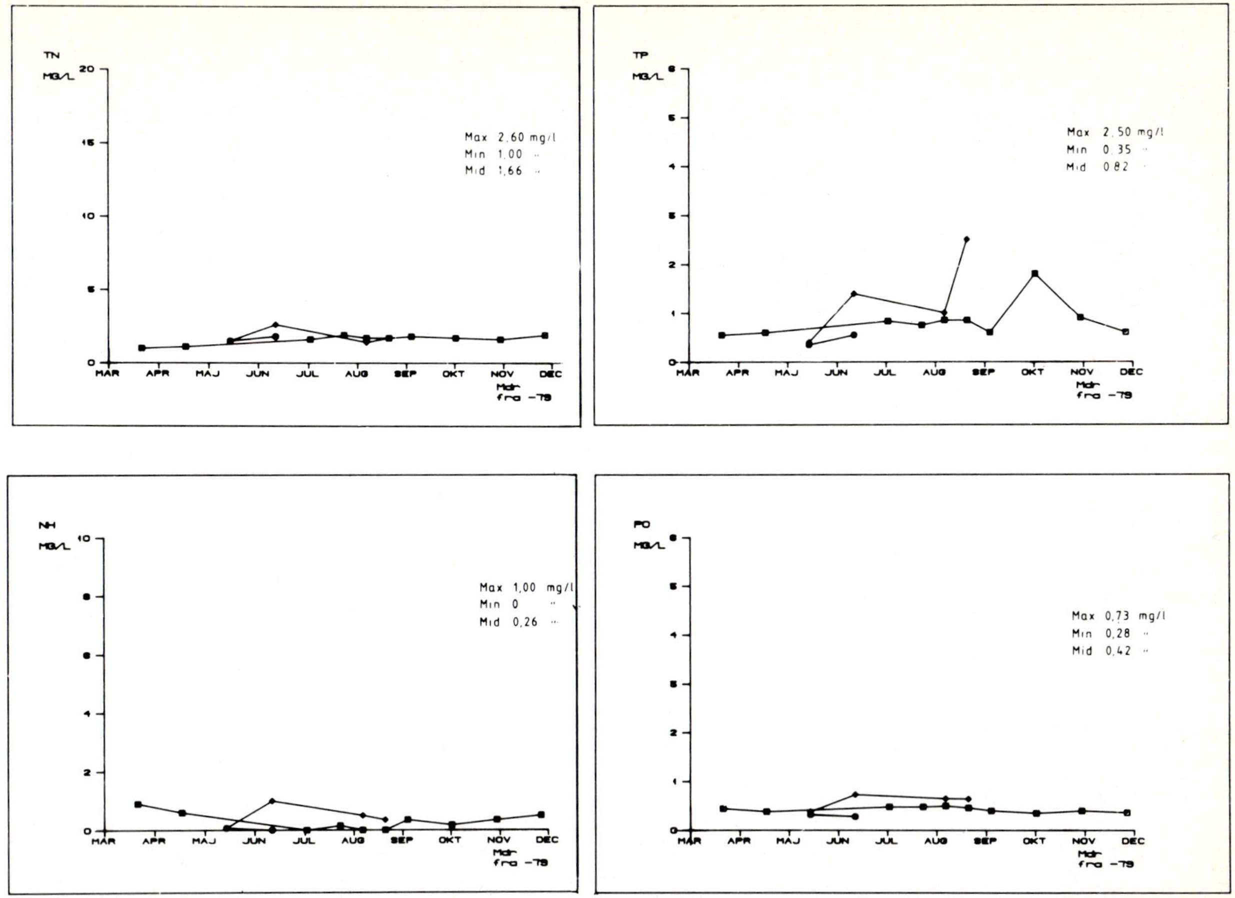

TN = Total Nitrogen (mg N/l) NH = Ammonia Nitrogen (mg N/l)
TP = Total Phosphorous (mg P/l) PO = Ortho-phosphate (mg P/l)

Figure 3 Nutrient Concentrations, Lake Birkerød, 1979 (3)

centrations in June and August refer to hypolimnion concentrations. The low concentrations of inorganic, soluble-N fractions indicate N limitation for phytoplankton primary production. It cannot be concluded from such observations, however, to what extent N is actually limiting. The intracellular nutrient concentration is the critical parameter controlling phytoplankton growth rates (5, 6). This concentration is, however, not measurable, but may be simulated using a mathematical model.

Figure 4 shows the levels and seasonal variation for phytoplankton primary production. It should be mentioned that the total yearly production of organic carbon (about 620 g C/m^2/year) is among the highest measured in Denmark. Figure 5 shows characteristics of a sample catch of the fish population obtained by electro-fishing. It appears that the fish population in Lake Birkerød is both rich and diverse in number. Furthermore, the internal balance between feeders and predators is satisfactory.

Oxygen Depletion, Empirical Models

Some of the lakes exhibit DO problems in the entire water column during long time periods. An example is Lake Langedam (No. 5.3), for which DO profiles are shown in Figure 6.

On a yearly basis the loading of organic matter from 4 CSO to Lake Langedam has been estimated to 3.4t O_2(COD), which corresponds to 1.0t C. For comparison, the primary phytoplankton production is about 3.5t C/year, which furthermore is very low compared to the external nutrient loading of Lake Langedam. The yearly external load of organic matter from CSO's, in other words, only accounts for approximately 28% of the internal primary production, and is therefore not likely to be the cause of poor DO conditions.

Using a shorter time scale such as single extreme events, the CSO's may, however, give rise to severe oxygen depletion.

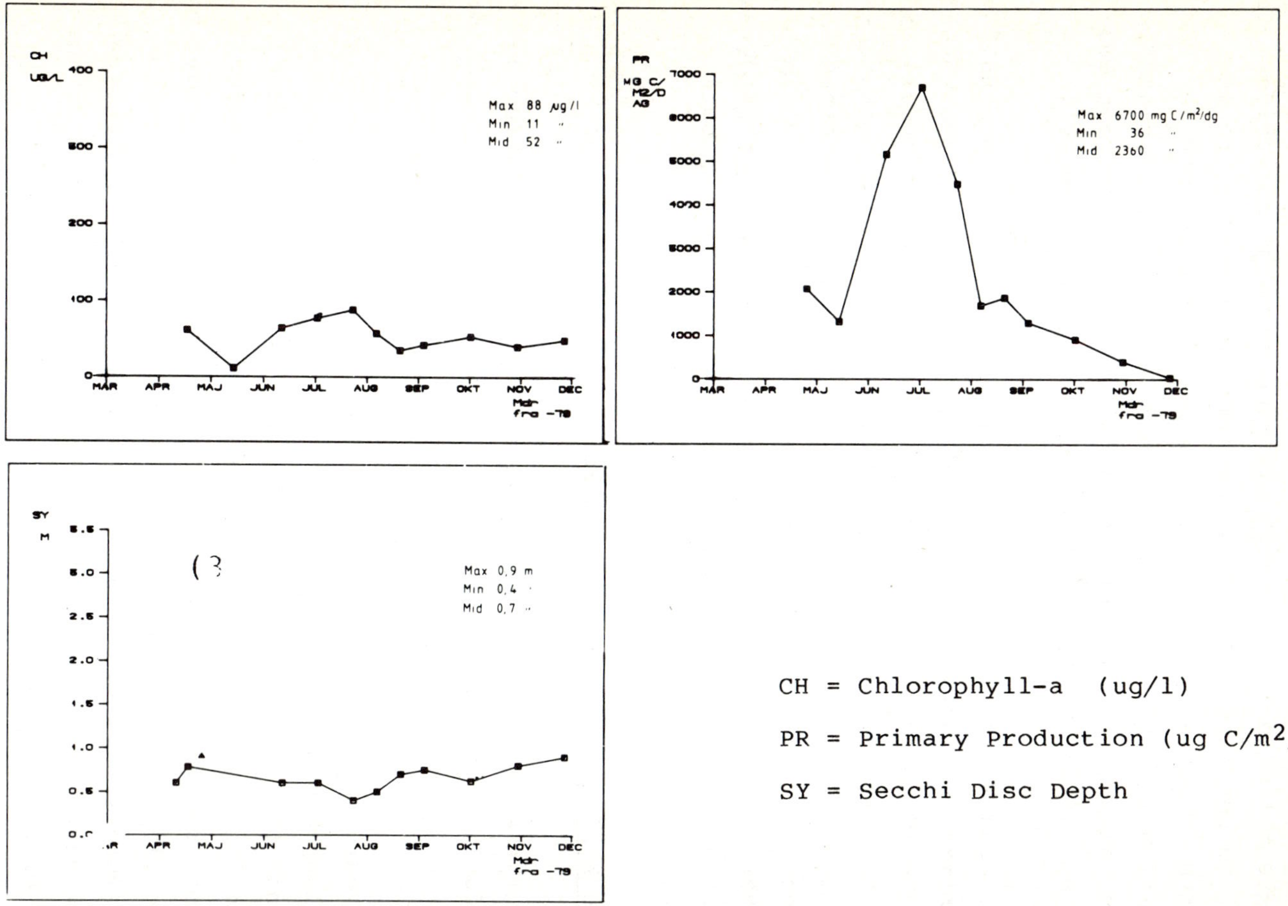

CH = Chlorophyll-a (ug/l)

PR = Primary Production (ug C/m^2/day)

SY = Secchi Disc Depth

Figure 4 Phytoplankton Growth, Lake Birkerød, 1979 (3)

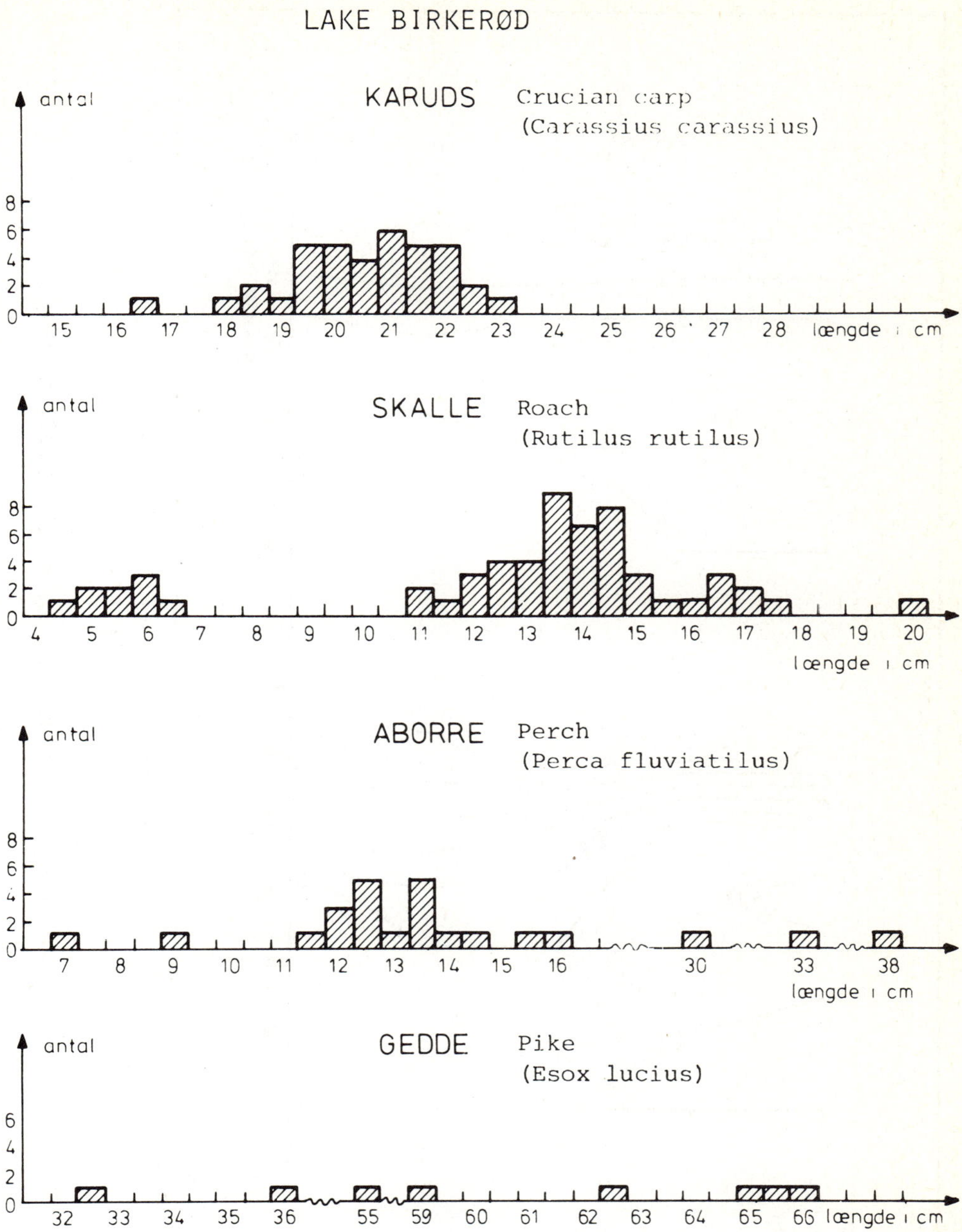

Figure 5 Length frequency diagram for dominating fish species, Lake Birkerød, 1979 (3).

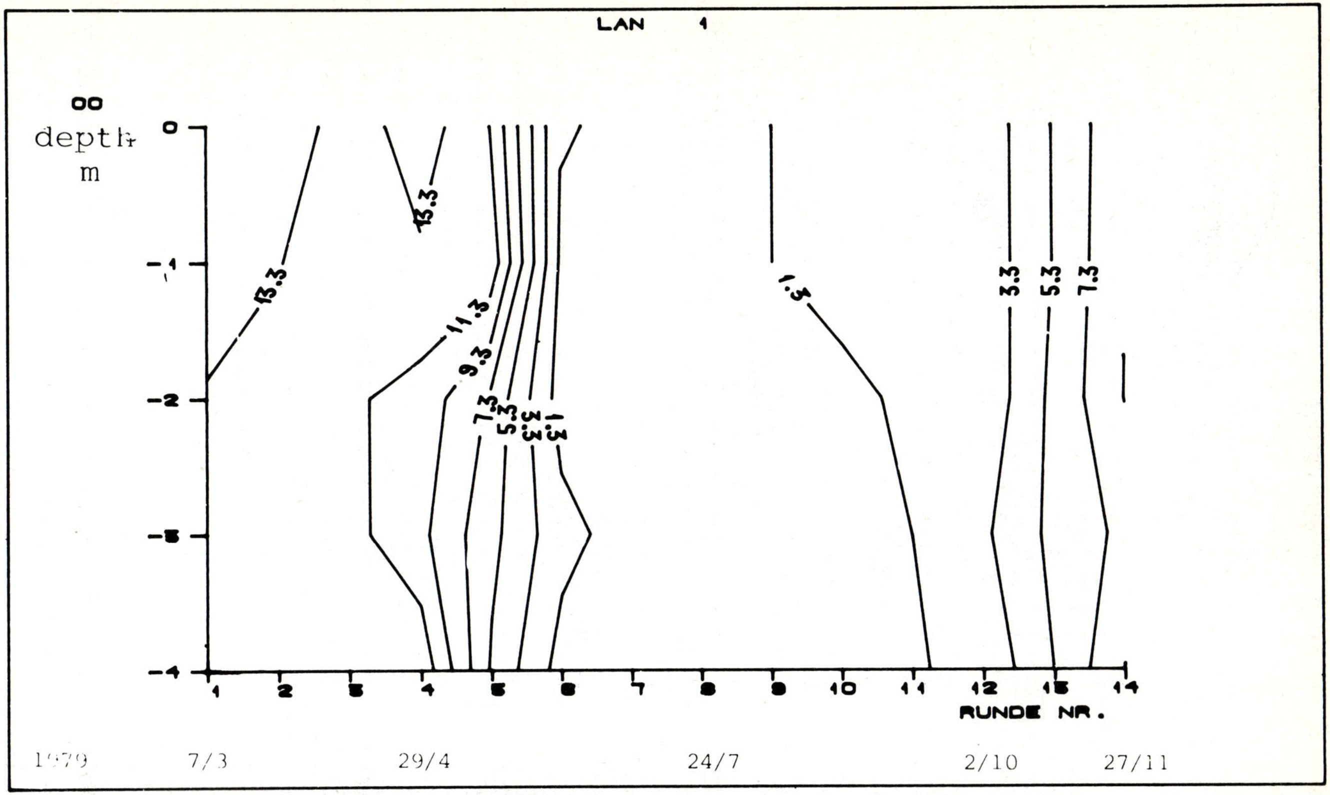

Abcissa represents time over the year
Ordinate represents water depth at station

Figure 6 DO-Profiles, Lake Langedam, 1979 (3)

Based on calculations with the SAMBA model, the total load from 4 CSO's during a storm event with a return period of 1 year (n = 1) is about 445 kg COD or about 92 kg BOD_5. As a rough estimate based on degradation rates found in other studies of CSO in Denmark (7), this will lead to an oxygen requirement of about 6 mg O_2/l for the entire lake volume during the first 24-hour period immediately following the overflows. During this short time period, surface reaeration may be neglected and phytoplankton primary production will only account for about 0.7 mg O_2/l/day. Depending on the DO-concentration before the storm event, this oxygen requirement may be detrimental to the fish population.

The very low DO concentrations observed in Lake Langedam (see Figure 6) over long time periods is believed to be the result of an unbalanced exosystem. During short periods of time extreme storm events will cause harmful effects to the fish population. As a consequence of a decreasing fish (predator) population the zooplankton population will be favored, thus resulting in increased grazing on phytoplankton leading to a low oxygen production. The lake will appear with a clear water column (high secchi disc depth), a high zooplankton biomass and a low fish biomass. With the low DO concentrations, sediment mineralization processes will take place under anaerobic conditions which may cause odour problems.

Small urban lakes will have a capacity, for receiving shock loads of organic matter. To evaluate the order of magnitude of this capacity, an empirical relationship for 15 of the lakes is shown in Figure 7.

Each data-point represents a lake. The abcissa describes a condition of the fish population evaluated by an electro-fishing survey (see Figure 5). The ordinate describes the organic load per m^3 lake volume during an extreme event (n = 1) based on SAMBA-model calculations as mentioned earlier. Data-points for 3 lakes should be placed in the indicated directions (for reasons which are not explained here) and 4 lakes with very low

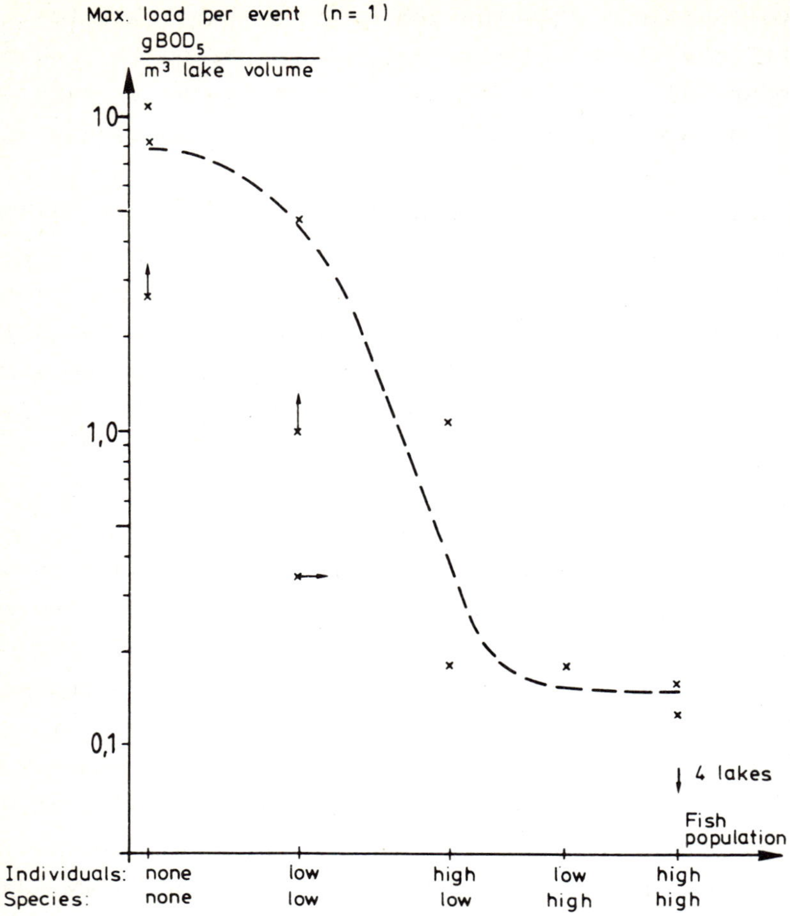

Figure 7 - Empirical relationship between extreme organic loads and the resulting fish population (4).

organic loadings are below ordinate values. The Figure can be used for the planner's first rough estimate of acceptable organic load to a lake. Above 5g BOD_5/m^3 per event (n = 1), no fish population is achieved; below 0.2g BOD_5/m^3, the organic load is not limiting for the fish population; between these limits, only O_2-tolerant species survive.

With higher lake temperatures in warmer climates the capacity for receiving organic shock-loads is expected to be lower than indicated on Figure 7.

For example, demonstrating the use of Figure 7 for Lake Langedam, it can be calculated that the present extreme BOD_5 load ($n = 1$) of about 92 kg BOD_5 per event should be decreased to 19 kg BOD_5 or 3.9 kg BOD_5 respectively if the mentioned criteria are to be met. If these reductions in extreme loads are to be accomplished by the installation of retention tanks, it will require holding volumes of about 2,000 m^3 or 3,200 m^3 respectively. Compared to the space available and the lake volume of 19,300 m^3, the use of retention tanks as the only pollution abatement measure is not regarded as realistic. In the actual situation combined measures including retention tanks and lake water aeration are recommended.

Empirical Input-Output Eutrophication Models

Hydraulic retention time T_n and external nutrient loading have an influence on the eutrophic state of a lake. Simple empirical input-output models like Vollenweider or Dillon-Riegler diagrams may be used for evaluation the expected effects of changes in the external loadings.

Figure 8 shows the position of the lakes in a Dillon-Riegler diagram. The retention coefficient, R, is referring to (input-output) + input and has been calculated by the formula (8):

$$R = 1 \cdot (1 + \sqrt{(Th)^{-1}}\ )^{-1}$$

In this type of diagram altered external hydraulic and/or P loadings may be evaluated in terms of the lake's relative position in the diagram. It should be noted that the "splitting lines" between eutrophic states are based on experiences from larger and deeper lakes abroad. These lakes will also typically have longer hydraulic retention times and fairly constant loadings so one is in fact violating the basic assumption laid down for the construction of the diagram when evaluating small urban lakes loaded with CSO and SWR.

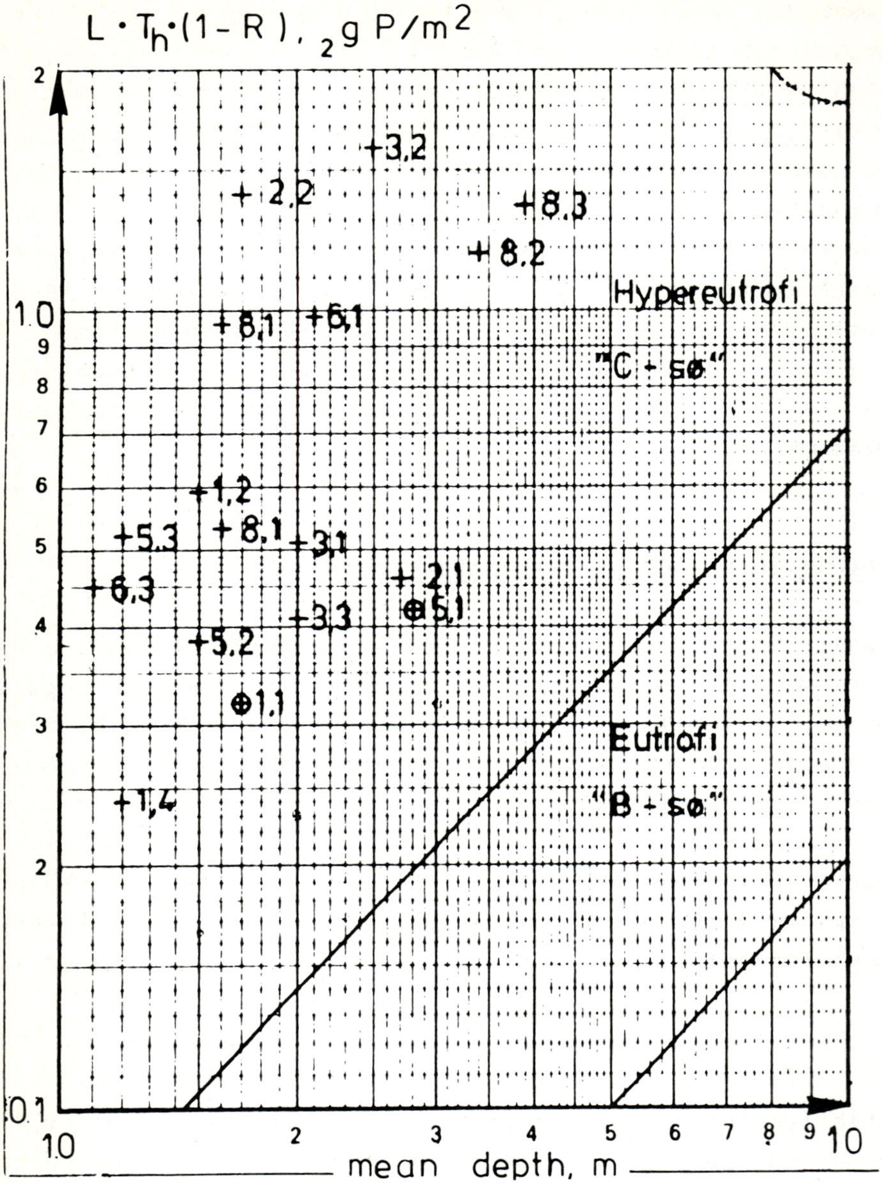

⊕ = P-limitation is indicated by model simulations
L = Specific surface loading (g P/m^2/year)
Th = Hydraulic retention time (year)
R = Retention coefficient.

Figure 8 Dillon-Riegler diagram for lakes in Birkerød Municipality. Numbers refer to Table 2.

Furthermore, most of the lakes in Birkerød Municipality seem to be N-limited rather than P-limited. A similar diagram may be constructed for N loadings, but evaluating the position of eutrophic states will in this case be even more difficult.

It is obvious that improved cause-effect relationships for evaluating the effects of altered external loadings or other pollution abatement measures in small lakes with short retention times are required.

Eutrophication Model, Dynamic Mass Balances

A more detailed evaluation of the cause-effect relationships for external loading versus state of eutrophication may be conducted using the mathematical simulation model, LAVSOE, earlier developed by the VKI.

The model consists of a set of deterministic differential equations describing an early cycle of phytoplankton growth and nutrient transformations in a lake which is assumed to be in a steady state with respect to input of nutrients from external sources. When used for predictions, the model describes a future steady state only and not the transient period following altered loadings. The model structure is shown in Figure 9.

The water body of a lake is represented in the model as a completely-mixed stirred-tank reactor. The depth is assumed to be constant, possible effects of the littoral zone are neglected, and the sediments are assumed to constitute a homogenous layer.

Sediment phosphorus and sediment nitrogen are not included in the model as actual state variables. The model describes sediment-water interactions, not on a long term basis but on a seasonal basis, and release of sediment nutrients is described as a function of temperature and sedimentation rates.

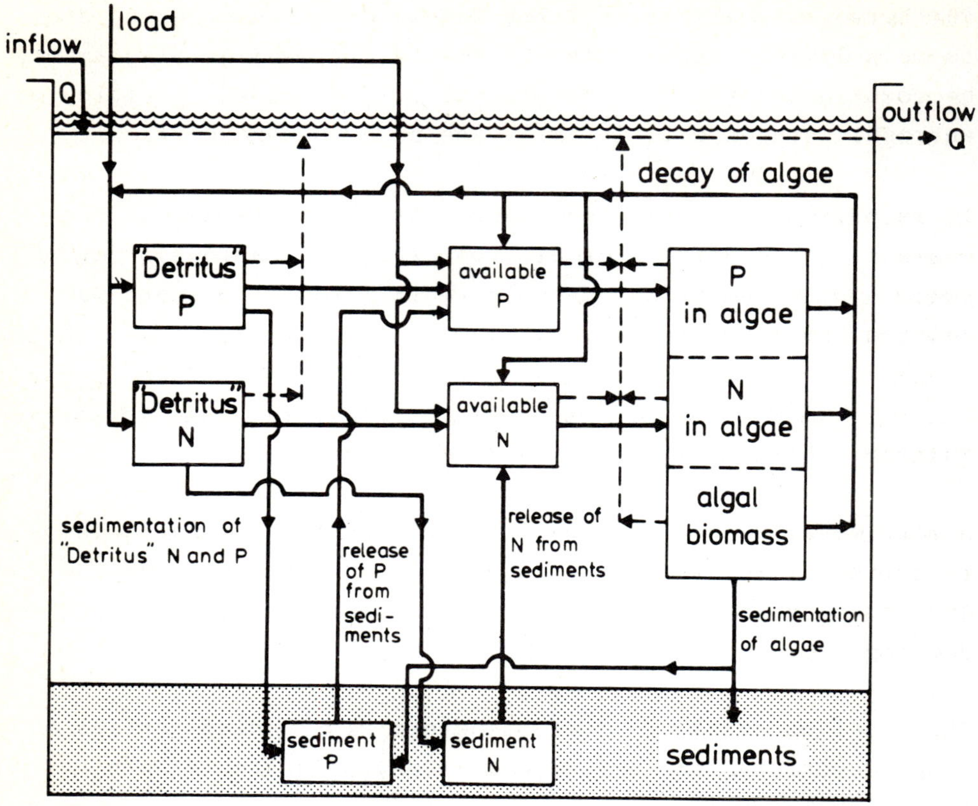

Figure 9 - Flow chart representing the LAVSOE model structure (9).

The compartments or state variables in the model are:

1. phytoplankton biomass (mg dry weight/l)
2. phosphorus in phytoplankton (mg P/l)
3. nitrogen in phytoplankton (mg N/l)
4. available phosphorus (mg P/l)
5. available nitrogen (mg N/l)
6. "detritus" phosphorus (non-available phosphorus) (mg P/l)
7. "detritus" nitrogen (non-available nitrogen (mg N/l)

The model does not include denitrification. Only total nutrient retentions are calculated. For nitrogen, this retention is a gross retention, comprising both net retention and denitrification.

No distinction is made between particulate and dissolved fractions of "detritus" phosphorus and nitrogen; these variables comprise all phosphorus and nitrogen forms not available to phytoplankton. The compartment of available phosphorus is assumed to consist of dissolved ortho-phosphate, while available nitrogen is made up of the entire pool of inorganic, dissolved nitrogen, i.e. nitrate, nitrite and ammonia.

The model describes the interactions between the compartments by the following key processes:

1. growth of phytoplankton
2. uptake of phosphorus and nitrogen by phytoplankton
3. decay of algae, including grazing by zooplankton
4. sedimentation of algae
5. sedimentation of "detritus" phosphorus and "detritus" nitrogen
6. decay of "detritus" phosphorus and "detritus" nitrogen
7. release of available nutrients from sediments.

Exogenous variables or driving variables are described as tabulated functions of the time of year. The variables include:

1. incident solar radiation
2. water temperature
3. inputs of phosphorus and nitrogen
4. water throughflow.

A detailed description of the LAVSOE model is given in (9) together with other examples of its application.

Minimal data requirements for calibration of the model include measurements during a yearly cycle of 1) primary production, 2) secchi disc depth (and preferably light transmission), and 3) concentrations of chlorophyll-a, total phosphorus, total nitrogen, dissolved ortho-phosphate and dissolved inorganic nitrogen (nitrate, nitrite and ammonia). Finally, a nutrient budget should be set up, which must be consistent with mass balances calculated from the model.

In the actual study the LAVSOE model has been calibrated and used for predicting the effects of alternative management strategies for the 4 lakes: Lake Nordvang, Lake Birkerød, Lake Langedam and Lake Vibeengen. Examples from Lake Birkerød, Lake Langedam and Lake Vibeengen are given in the following sections.

Lake Birkerød

The field measurement data shown in Figures 2-4 have been used for calibration. Table 3 shows the nutrient budget and quantification of selected processes for the year 1979. The mass balances show that the internal nutrient flux is dominant, being about one order of magnitude higher than the yearly external load.

On average, the levels of nutrient concentrations seem to be described fairly well by the simulations, although inorganic P is underestimated during the summer months. The simulated yearly variation for total N differs considerably from the measured values. The simulations show clearly N limitation but no P limitation. The relatively high values of the biological variables chlorophyll and primary production measured during the

Table 3 - Mass balances, Lake Birkerød, 1979 (4).

	TOTAL-P kg/year	TOTAL-N kg/year
Groundwater inflow	9.4	78
Atmospheric outfall	5.9	178
CSO	7.2	20
Total input	22.5	276
Total output	22.2	40
Accumulation, sediment	-1.4	223
Accumulation, water	2.6	13
Sedimentation	367	1792
Release from sediment	368	1569

(Results from simulation are shown in Figure 10.)

summer months are not reflected by the simulations and consequently secchi depth values are over-estimated.

The reason for poor simulation of yearly variations in total N may be a combination of:

1. Discrepancies between estimated and true external load of N.
2. Insufficient model description of N sedimentation/release from sediment.
3. Occurrence of blue-green algae species capable of fixating N from the atmosphere.

Because of the long hydraulic retention time, combined with the fact that stratification occasionally occurs, one should not expect a perfect simulation of measured data by the LAVSOE model. The calculated retention time for N is about 1 year but for P about 6 years. The nutrient budget shows for P practically no retention in the lake, which indicates that the lake may yet not be in equilibrium with the present P loadings but still be in a transition phase from discontinuation of sewage discharges 30 years earlier.

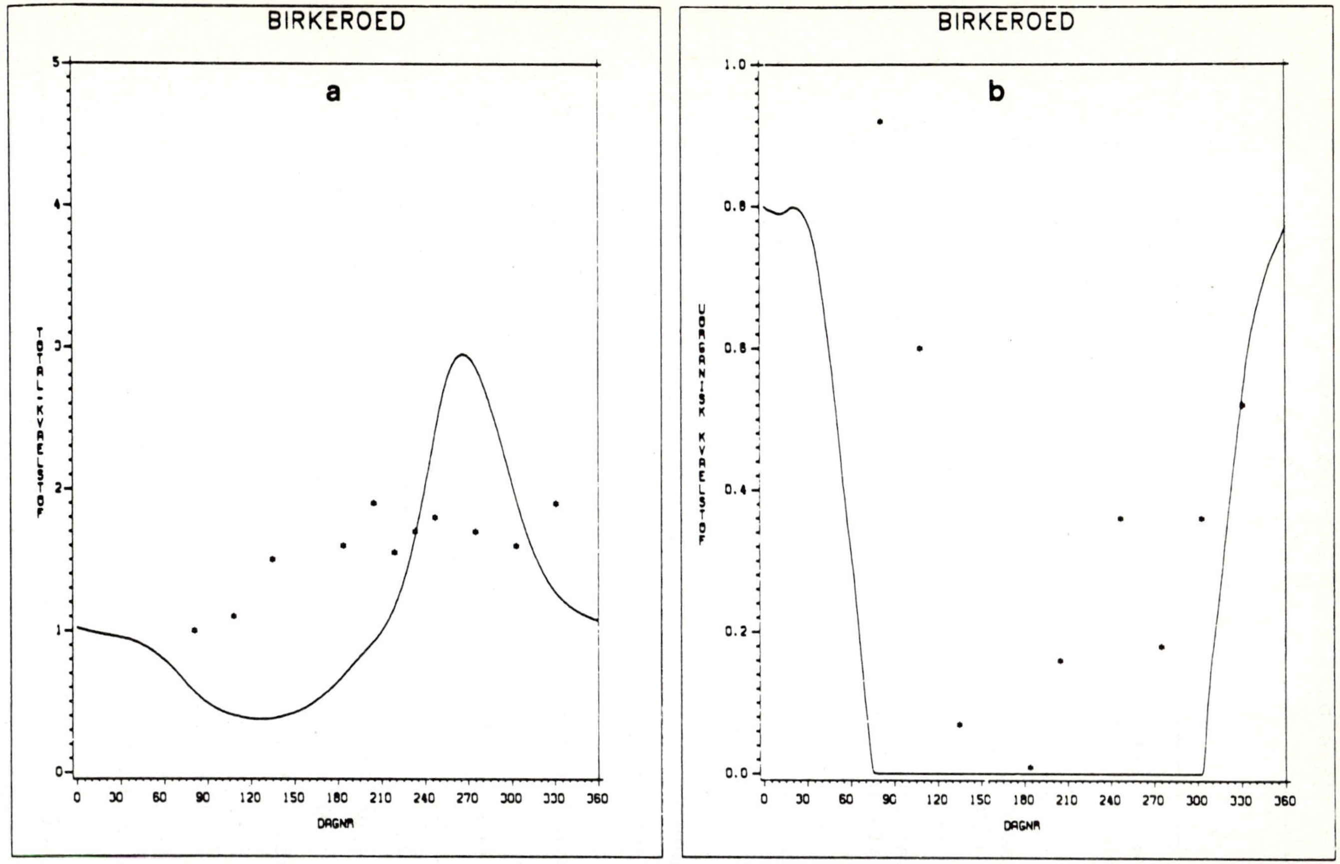

Figure 10 LAVSOE-simulation, Lake Birkerød, 1979 (4)

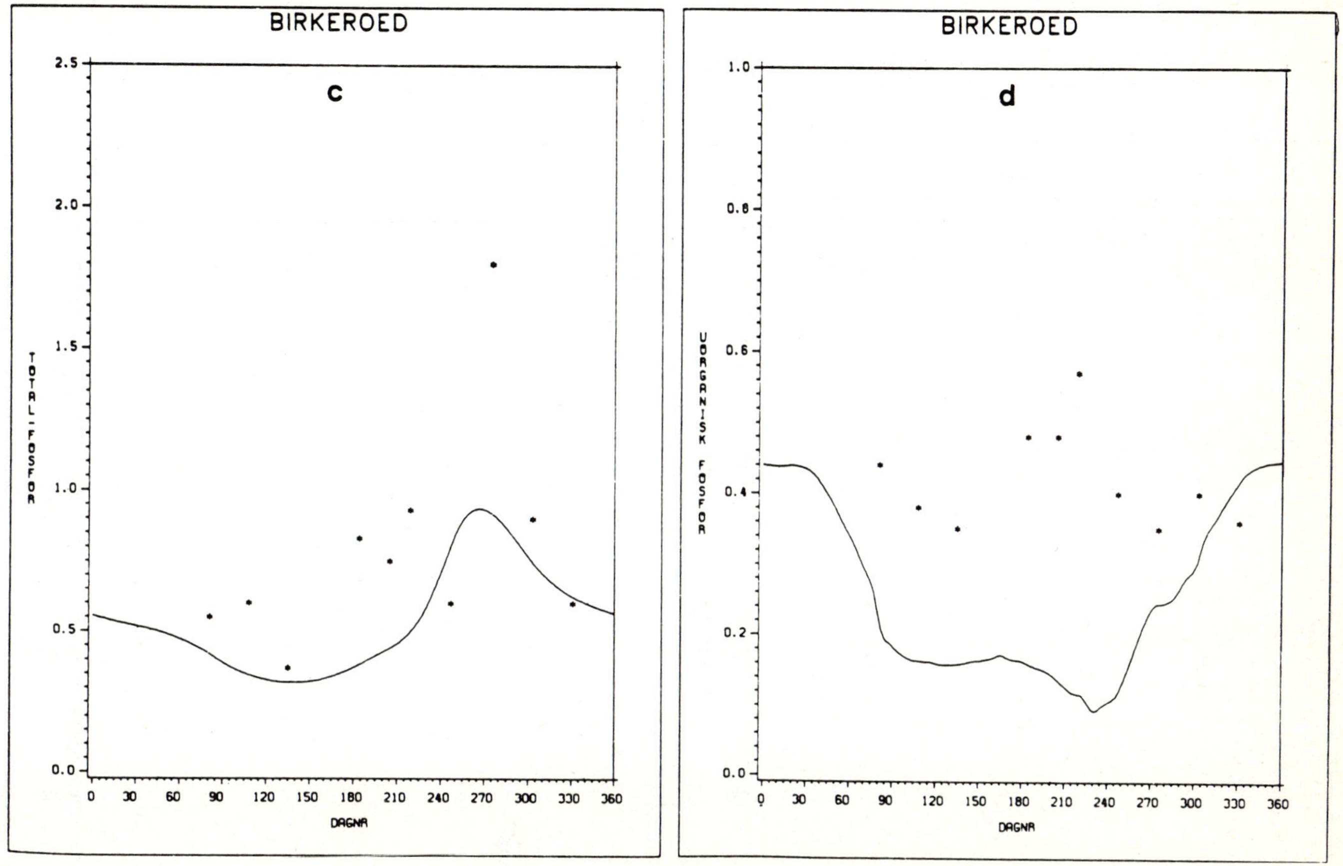

Figure 10 LAVSOE-simulation, Lake Birkerød, 1979 (4) (cont'd)

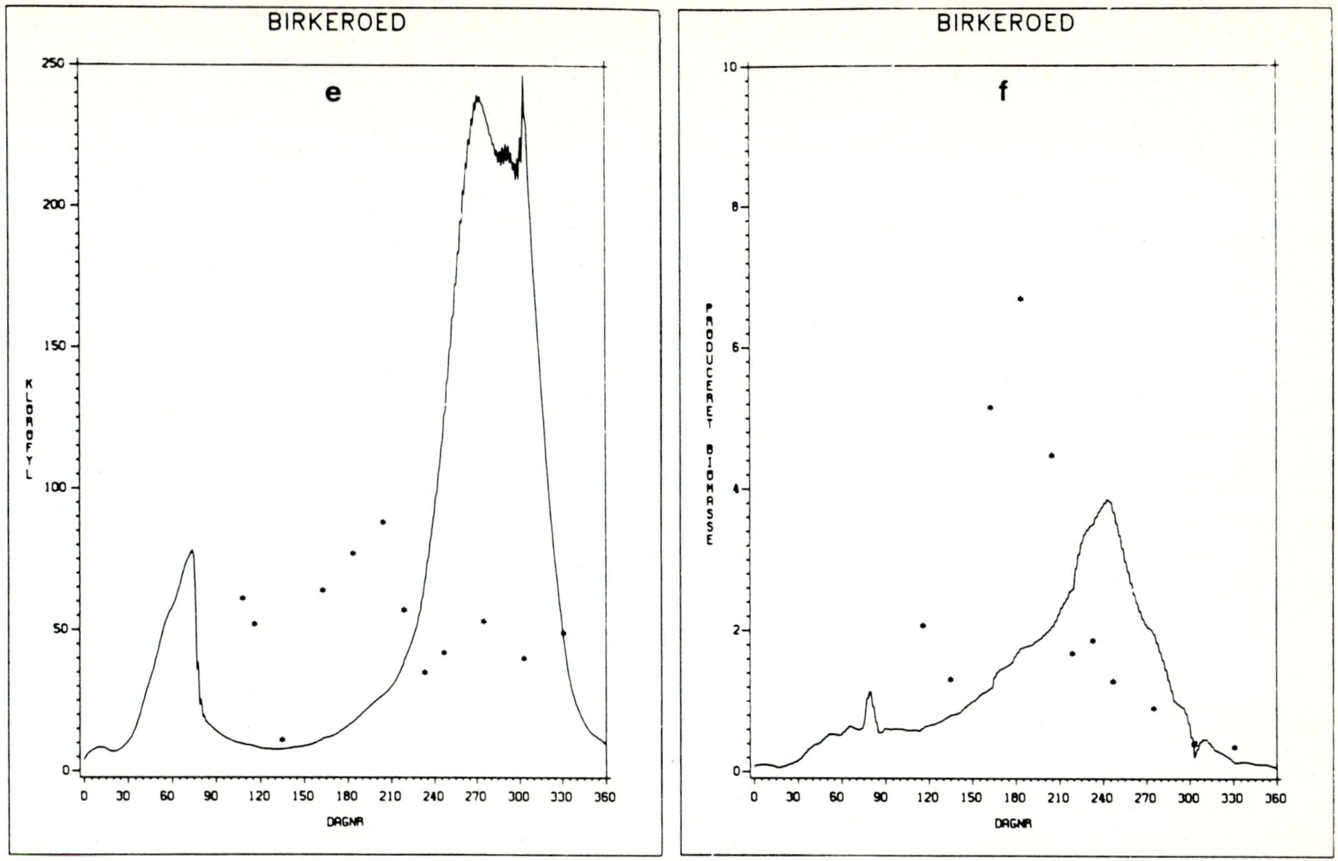

Figure 10 LAVSOE-simulation, Lake Birkerød, 1979 (4) (cont'd)

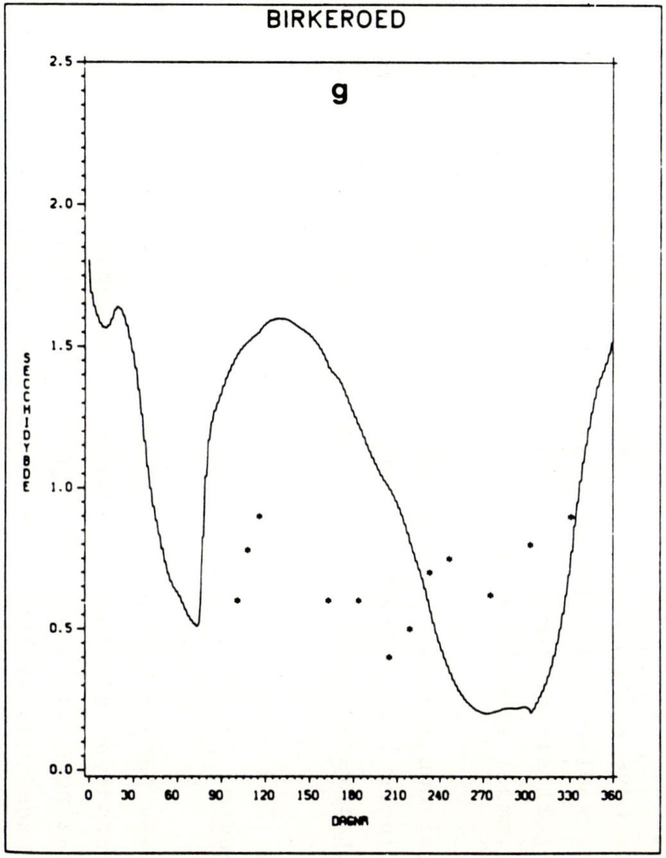

a) Total-P (mg/l)
b) Inorganic, dissolved P (mg/l)
c) Total-N (mg/l)
d) Inorganic, dissolved N (mg/l)
e) Chlorophyll (mg/l)
f) Primary production (g/m^2/day)
g) Secchi depth (m)

Figure 10 LAVSOE-simulation, Lake Birkerød, 1979 (4) (cont'd)

The model, calibrated for each lake, has been used for predicting the effect of altered external loadings. Results are shown in Table 4. For comparisons with 1979 (the calibration year), the "average year" situation and also the increased CSO-loading to average standard deviation (describing a rainy year) situation is shown.

The reason for the relatively low primary production (and high secchi depth) in 1979 compared to the average year with approximately the same yearly load is the distribution of the load over the year. In 1979 only 35% of the yearly loading with available N took place in the period May - September, whereas in the average situation these months would account for 45% of the yearly load with available N.

The various alternatives do not differ significantly from the average situation; in other words a reduction in the external loads will have a limited effect as long as the nutrient release from sediments is the dominating factor in the mass balance. In all situations the lake will be highly eutrophic with very small secchi depths.

Recommendations for lake restoration measures are therefore to reduce effects from sediment release by means of discharging hypolimnotic water or by stabilizing the sediment by chemical addition.

Lake Langedam

The nutrient budget and quantification of selected processes for Lake Langedam are shown in Table 5. In this case the external loadings play a more dominating role, being of the same order of magnitude as the internal nutrient flux.

Table 4 - Predictive simulations. Comparison of altered situations for external loading to Lake Birkerød(4).

ALT. 1: Decreased CSO-load by partly separation of sewers in catchment area.

ALT. 2: Same decrease in CSO by roof drainage infiltration.

ALT. 3: 0-discharge from CSO and SWR.

		1979	AVERAGE YEAR	AVERAGE + CSO STD.DIV.	ALT.1	ALT.2	ALT.3
External nutrient load	(kg P/year)	22.5	25	28.6	29.1	16.2	15.3
	(kg N/year)	276	283	294	362	260	257
Hydraulic retention time	(year)	6.64	6.66	6.21	4.40	8.03	8.21
Primary production	(g C/m^2/year)	420	514	511	517	512	512
Average secchi depth. Day 120-260	(m)	1.12	0.64	0.64	0.60	0.65	0.65
Minimum secchi depth	(m)	0.45	0.11	0.11	0.09	0.12	0.12
Nutrient limitation	Spring	N	(P),N	N	N	N	N
	Summer	N	N	N	N	N	N
	Fall	N	N	(N)	(N)	(N)	(N)

Table 5 - Mass Balances, Lake Langedam 1979, (4).

	TOTAL-P kg/year	TOTAL-N kg/year
Groundwater inflow	10	85
Atmospheric outfall	1.0	31
CSO	29	77
Upstream input	9.0	41
Total input	49	234
Total output	46	192
Accumulation, sediment	-2.3	28
Accumulation, water	5.2	14.3
Sedimentation	19	151
Release from sediment	21	123

The results from the simulation are shown in Figure 11.

With the relatively short hydraulic retention time (T_h about 0.25 year), a rather good simulation of measured data with the LAVSOE model should be expected. However, as earlier mentioned the ecosystem of the lake is out of balance with an extremely high grazing effect from zooplankton, and the model simulations indeed fail to reproduce the observed measurements of primary production, chlorophyll concentration and transparency.

Total N is simulated very well. Inorganic N is underestimated during late summer, which is a consequence of overestimating the phytoplankton biomass. Both total and inorganic P are underestimated, which reflects the fact that the external loading is underestimated or perhaps the measured values are too high.

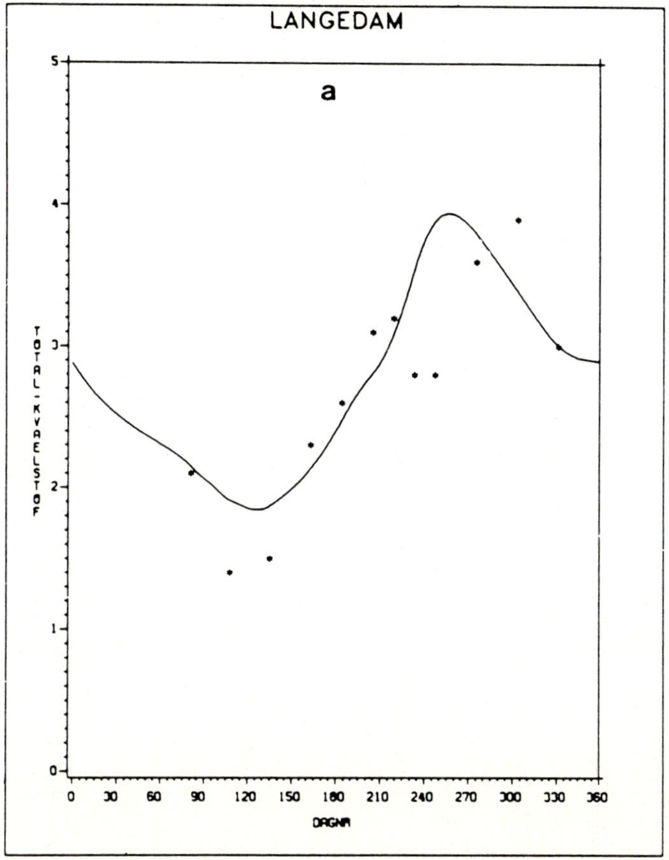

a) Total-P (mg/l)

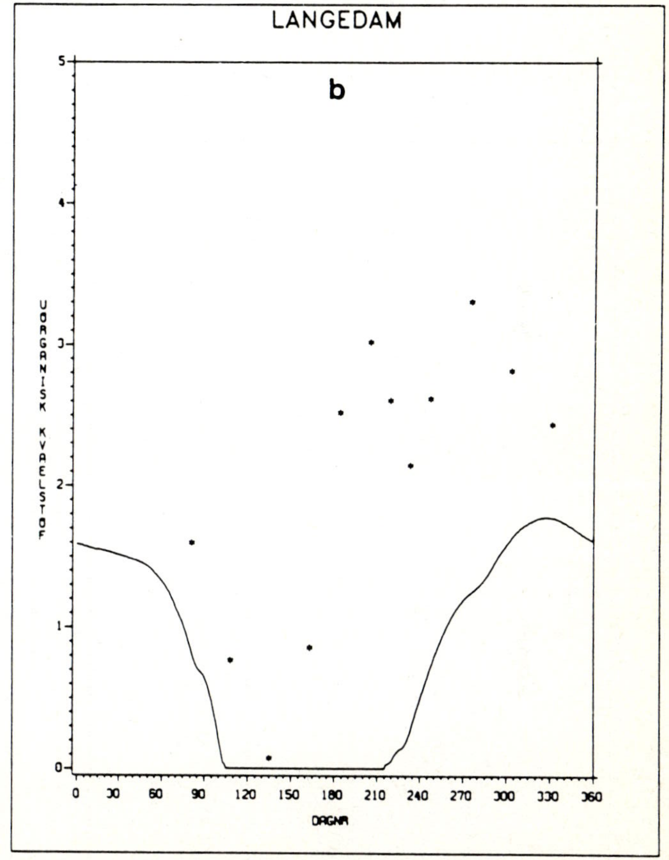

b) Inorganic, dissolved P (mg/l)

Figure 11 LAVSOE-simulation, Lake Langedam, 1979 (4)

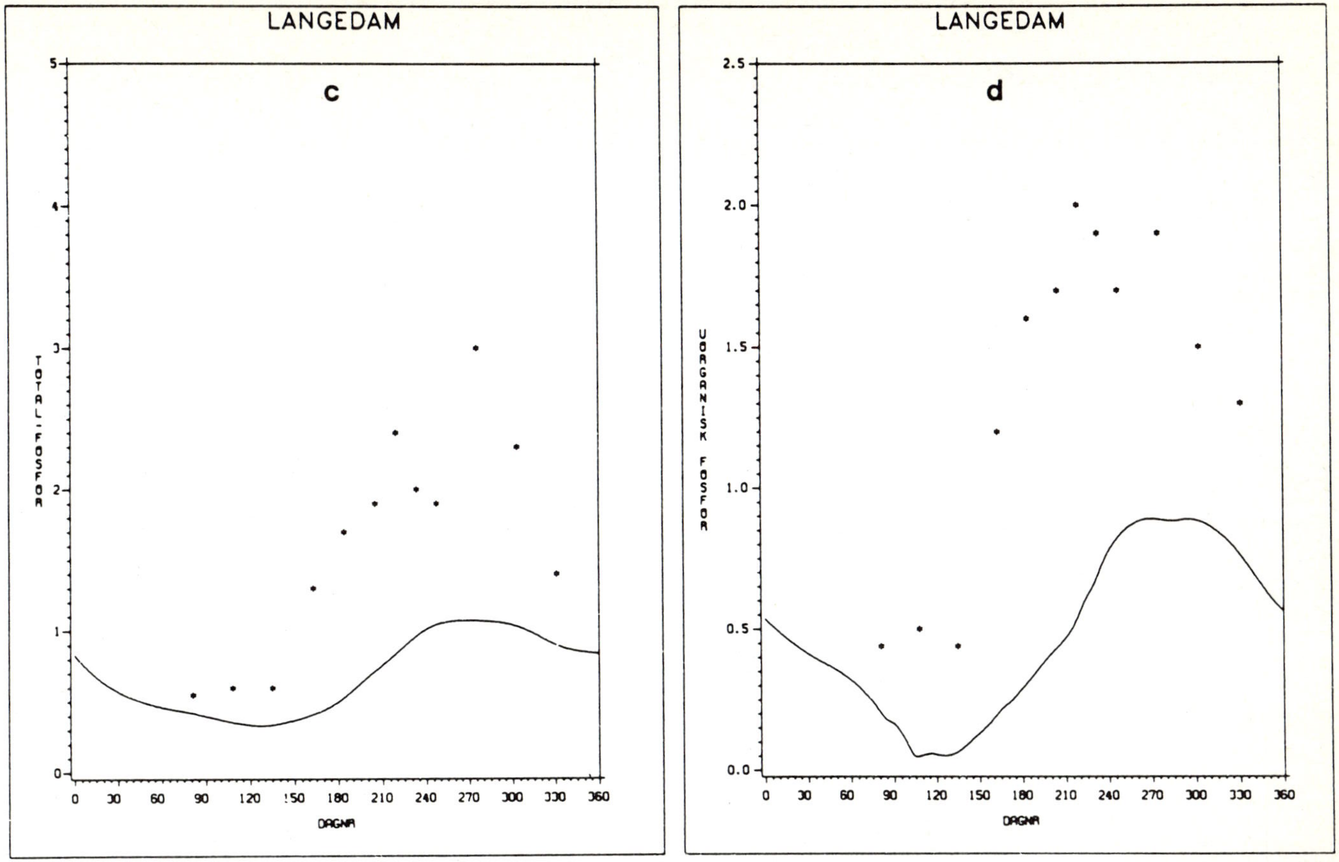

c) Total-N (mg/l)

d) Inorganic, dissolved N (mg/l)

Figure 11 LAVSOE-simulation, Lake Langedam, 1979 (4) (cont'd)

The results from altering the external loadings are shown in Table 6. The hypothetical zero-discharge situation (Alt. 3) shows some improvement compared to the average year, however, the realistic alternatives (Alt. 1 and 2) show practically no improvement in primary production and secchi depth. It could be predicted, however, that reestablishing a fish population and a more balanced ecosystem in Lake Langedam will cause the lake to change into a green, eutrophic appearance. This is, however, preferable to the present condition.

Lake Vibeengen

Finally, for demonstrating a better simulation of phytoplankton growth, examples from Lake Vibeengen are shown in Figure 12. During the simulated situation both N and P limitation occurs.

Discussion

Primary oxygen depletion in a lake as a function of CSO or SWR is in general not regarded as a major problem, but may in severe cases (shock loadings) be the reason for disrupting the fish fauna and, as a consequence, disturbing the ecological balance. The derived empirical relationship (Figure 7) is supposedly specific in terms of water temperature and rainfall characteristics.

A more general planning tool for describing these cause-effect relationships may be developed as a mathematical model. The basic principles may be similar to oxygen depletion in rivers from CSO as described by (7), and may be further simplified since aeration will be of limited importance. However, considerable amount of field observations may be required for calibration of decay rate constants.

Eutrophication modeling may be performed as demonstrated using the mathematical model LAVSOE, which is based on dynamic mass balances. This offers a compromise between simple input-output

Table 6 - Predictive simulations. Comparison of altered situations for external loadings to Langedam(4).

ALT. 1: Reduction of CSO to about 28% of average situation.

ALT. 2: Average CSO with 50% removal of suspended fractions of N and P.

ALT. 3: 0-discharge from CSO.

		1979	AVERAGE YEAR	AVERAGE + CSO STD.DIV.	ALT.1	ALT.2	ALT.3
External nutrient load	(kg P/year)	49	59	71	30	47	20
	(kg N/year)	234	259	292	183	229	156
Hydraulic retention time	(year)	0.25	0.23	0.20	0.29	0.23	0.33
Primary production	(g $C/m^2/year$)	200	549	559	500	510	478
Average secchi depth. Day 120-260	(m)	1.12	0.64	0.63	0.69	0.67	0.71
Minimum secchi depth	(m)	0.77	0.17	0.17	0.18	0.20	0.18
Nutrient limitation	Spring	P,N	P,N	P,N	P,N	P,N	P,N
	Summer	N	N	N	N	N	N
	Fall	N	N	N	N	N	N

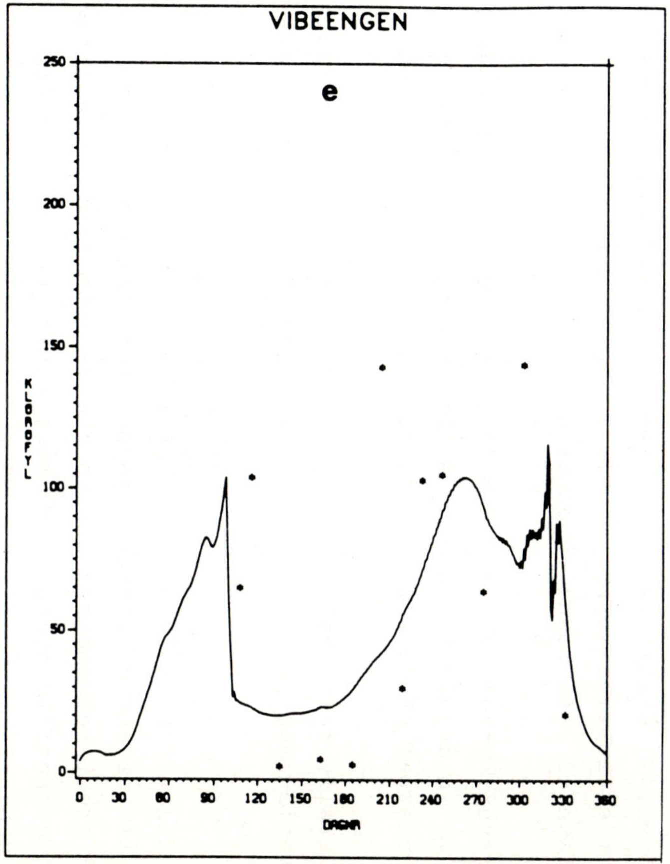

e) Chlorophyll (mg/l)

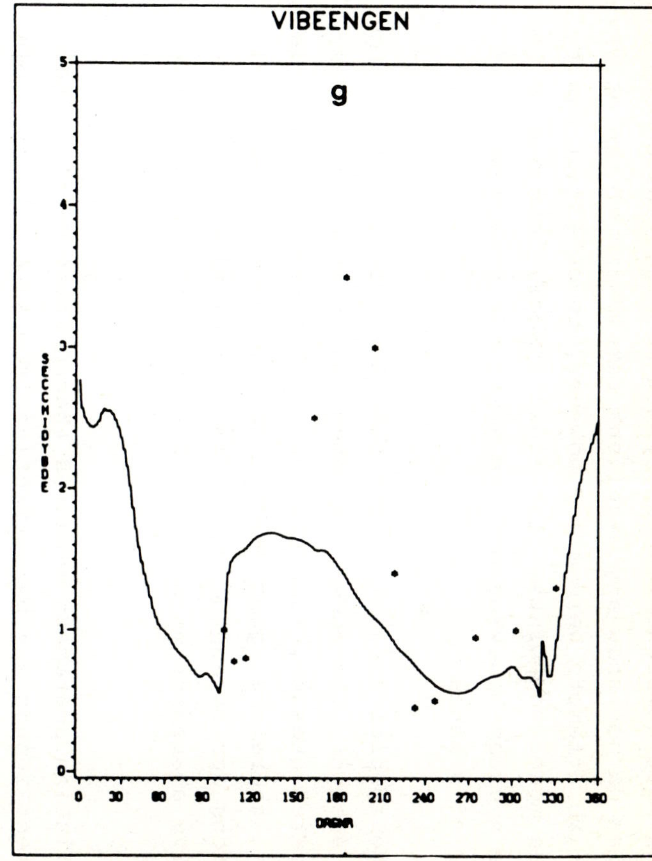

g) Secchi depth (m)

Figure 12 LAVSOE-simulation. Lake Vibeengen, 1979 (4). (Abcissa: Day no. from January 1).

models and more comprehensive ecosystem models. The demonstrated examples do not show any excellent simulation of the observed data. No attempts were made, however, to obtain an exact fit to the measured points. The model is only intended to reflect the average behaviour of the system on a time scale of the order of a month.

Taking the special conditions for Lake Birkerod and Lake Langedam into account, the model responses appear to be consistent with educated guesses, i.e. they are qualitatively reasonable. There are, therefore, good reasons to believe that the model calculations can be useful in estimating relative effects of alternative abatement measures such as reduction of nutrient loadings.

Other pollution abatement measures than decreasing the external nutrient inputs may be attractive from an economical point of view. Such restoration measures could be 1) aeration of lake water column, 2) stabilization or removal of sediment, 3) treatment of or effluating hypolimnotic water, 4) biological manipulation of fish fauna composition, etc..

At the present stage of knowledge it is not possible in general to quantify in detail the benefits obtained by a specific action on internal processes in a lake.

It is generally felt that planning tools for predicting effects of restoration measures for urban lakes need considerable improvement.

The investigations of 19 small urban lakes with different characteristics and different external loadings have led to quite different recommendations on the various lakes. In other words, site-specific measures to meet water quality criteria in small urban lakes receiving CSO and SWR will be more cost-effective than uniform technical means.

Acknowledgment

The entire investigation program described has been administrated and sponsored by Birkerød Municipality. The authors wish to thank Mr. Kurt Ojgaard and Mr. Bent Ohlsen for their cooperation, and the Birkerød Municipality for permission to publish the results.

References

1. Johansen, N.B., J.J. Linde-Jensen and P. Harremoës. Computing combined system overflow based on historical rain series. Proceedings of the third international conference on urban storm drainage, Vol. 3, pp 909-918, Goteborg, Sweden, June 4-8, 1984.

2. Birkerød kommune. Udledninger fra kloaknettet under regn i Birkerød kommune (Discharge to receiving waters from the sewer system during rain for the Municipality of Birkerød, In Danish). Birkerød Municipality, Technical Department, Birkerød, Denmark, Oct. 1982.

3. Water quality institute. Tilstandsundersøgelse af 20 so'r i Birkerød kommune 1979-80 (Monitoring survey of 20 lakes in Birkerød Municipality, in Danish). Report to Birkerød Municipality, July 1981.

4. Birkerød Municipality and Water Quality Institute. Regnvandsafledning i Birkerød kommune - konsekvenser og anbefalinger (Stormwater runoff in Birkerød Municipality - consequences and recommendations, in Danish). Published by Birkerød Municipality, 1984.

5. Nyholm, N. A mathematical model for microbial growth under limitation by conservative substrates. Biotechnol. Bioeng. 18: 147-165, 1976.

6. Nyholm, N. Kinetics of phosphate-limited algal growth. Biotechnol. Bioeng. 19: 467-492, 1977.

7. Hvitved-Jacobsen, T. The impact of combined sewer overflow on the dissolved oxygen concentration of a river. Water Research, Vol. 16, pp 1099-1105, 1982.

8. Larsen, D.P. and H.T. Mercier. Phosphorus rentention capacity of lakes. J.Fish.Board.Can. 33: 1742-1750, 1976.

9. Nyholm, N. A simulation model for phytoplankton growth and nutrient cycling in eutrophic, shallow lakes. Ecol. Modelling, 4: 279-310, 1978.

REALISTIC WATER QUALITY MODELING

Larry A. Roesner, Raymond Walton, John P. Hartigan
Camp Dresser & McKee Inc., Annandale, Virginia USA

Introduction

It is well known that the pollutant loads in urban runoff are highly variable in both time and space. The effect of these loads on receiving waters is to cause significant increases in pollutant concentrations over relatively short periods of time. Moreover, the magnitude of the concentration and the period of time that the concentration exceeds any specified level is variable depending on the characteristics of the storm, antecedent dry period, and the hydraulic and water quality condition of the receiving water itself when the storm occurs. The assessment of receiving water impacts is complicated further by the fact that the water quality-ecologic relationship for pollutants (besides nutrients) found in storm runoff are complex and not well understood.

The mathematical models that have been developed to simulate these receiving water responses vary greatly from the very simple statistical model proposed by Ditoro(1) to the extremely complex EXAMS model(2) which defines in great detail the generic biochemical processes that occur in a receiving water. Models in between these two extremes include: RWQM(3), a continuous receiving water model for dissolved oxygen, plus six other pollutants, that is driven by a time history of runoff quantity and quality generated by the model STORM(4,5); the HSPF model(6), which provides a continuous simulation of the pollutant buildup and runoff processes and attendant receiving water impacts; and DEM(7), a single event model that includes dynamic hydraulics and eleven interrelated water quality parameters.

NATO ASI Series, Vol. G10
Urban Runoff Pollution.
Edited by H.C. Torno, J. Marsalek, and M. Desbordes

Which of these models to use in a given application is always a subject for debate, and the one ultimately selected is based more often than not on the modeler's intuition and experience. This, of course, presents a problem for the inexperienced modeler whose intuition will be poor at best. In an effort to rectify this dilemma, a framework for model selection has been developed to identify and evaluate the importance of the physical/chemical/biological characteristics of the receiving water and to identify candidate models that will adequately represent those characteristics. The framework consists of eleven steps identified in Figure 1. Although this particular framework was developed for selection of estuary receiving water models, the methodology can be adapted to other water bodies. Each of the eleven steps is described below. For additional details, the reader is referred to Walton et al.(8).

Step 1 - Develop a Conceptual Model

The first step is critical and, unfortunately, is often overlooked. The starting point for this step is to define the objectives for the water quality modeling effort. Is this a scientific inquiry into how well we can simulate observed data set? Or is the wasteload allocation study aimed at identifying the allowable pollution load that can be contributed from a watershed? If it is a wasteload allocation, what pollutants are to be regulated. What will be the regulatory controls: maximum concentration, duration, total load, or a combination? What is the concern over these pollutants: their concentration in the receiving water, their effects on biota (long term, short term), and/or their accumulation in the bottom sediments? Once the objectives of the modeling exercise have been defined, water quality constituents and/or pollutants of interest have been stated, and the biota of interest identified, a conceptual model or diagram of the receiving water system can be developed.

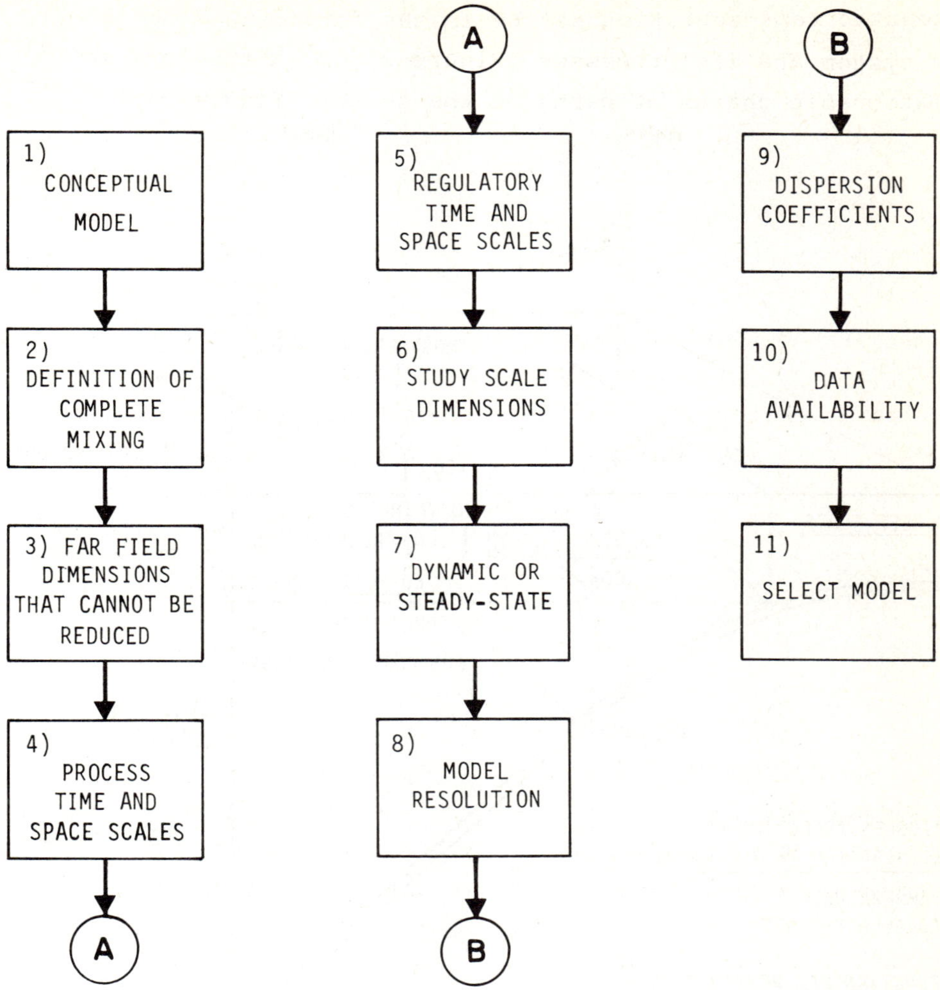

Figure 1. Framework for Model Selection

This can be done either as a set of equations describing the various physical/chemical/biological processes, a written description of them, or perhaps more usefully, a schematic representation of them. The figure is the most useful form, as one can readily visualize the system's processes. The schematic representation can be in the form of a diagram of the system and its processes (Figure 2) or in the form of relationship charts of parts of the system (Figure 3).

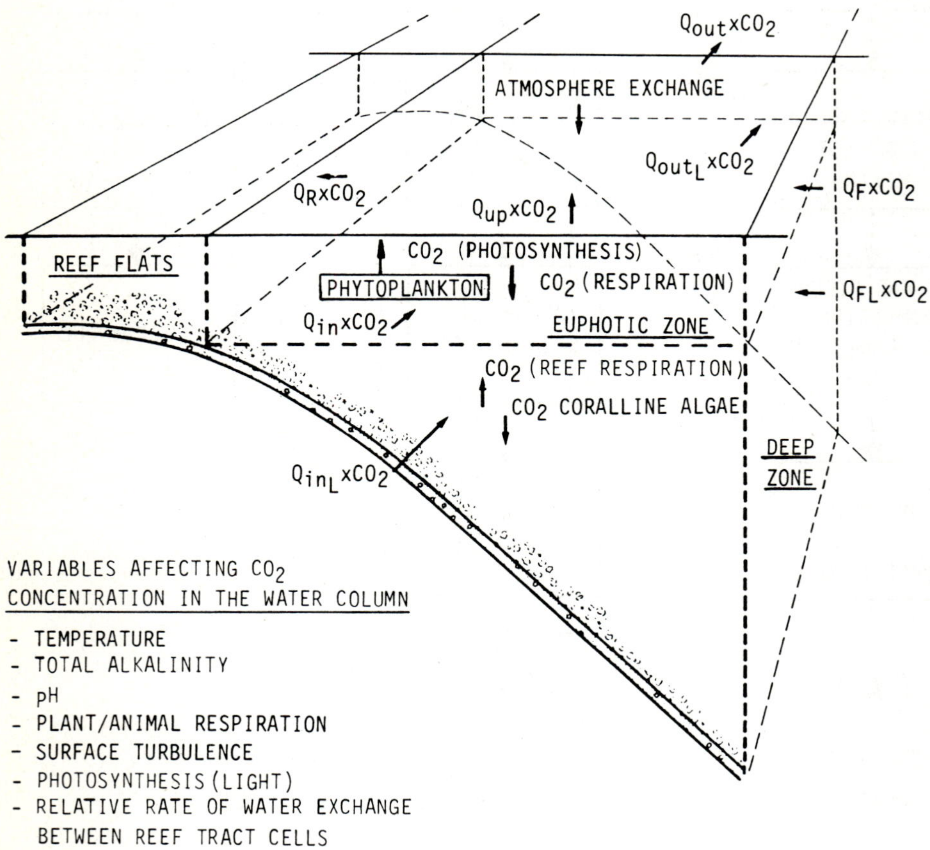

Figure 2.
Schematic Representation of CO_2 Mass Balance in Reef Tract

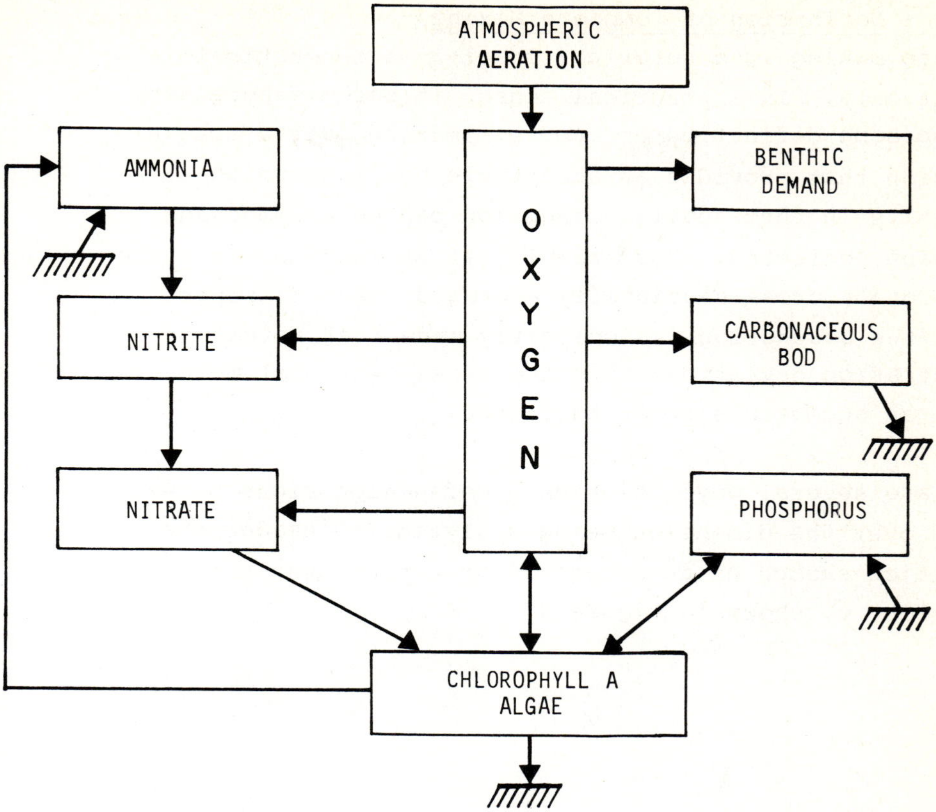

Figure 3. Major Constituent Interactions

The purpose of this step is to assimilate and present all the available knowledge of a system in a way that major processes and ecologic relationships can be evaluated for inclusion in the numerical model description. The conceptual model is the starting point from which systematic reductions in complexity can be made which will provide an adequate representation of the system, while meeting the objectives of the study.

Step 2 - Definition of Complete Mixing

Complete mixing in a numerical model is a theoretical concept only. In a practical sense, it becomes necessary to develop some definition of complete mixing over a spatial dimension that provides an acceptable point at which uniformity in that spatial dimension can be assumed and that dimension neglected. For example, if we wish to use a one-dimensional, cross-sectionally averaged, mass transport model, the assumption is implicitly made that actual concentration deviations from the cross-sectional mean are within an acceptable error tolerance.

There are several ways in which a definition might be established over the dimension being analyzed. Consider the definition sketch of an actual distribution over the lateral dimension, y, shown in Figure 4.

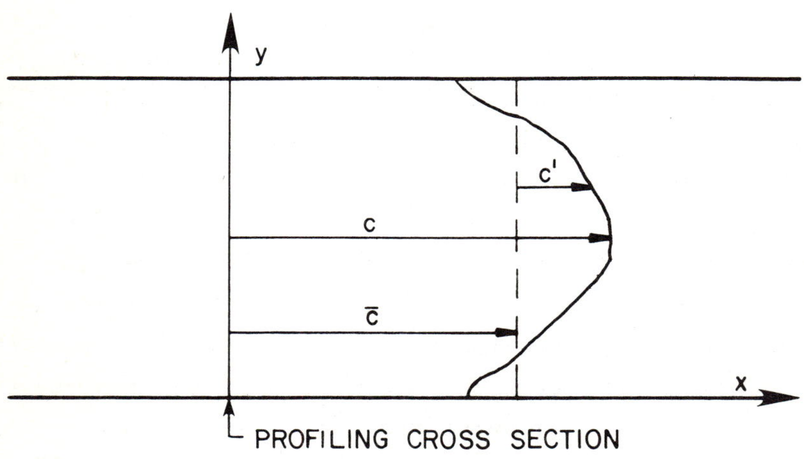

Figure 4. Lateral Concentration Distribution

Two definitions might be:

$$c'_{max} / \bar{c} < T_1 \quad \text{or} \quad c_{max}/ \bar{c} < T_2 \qquad (1)$$

where c'_{max} = maximum concentration deviation

c_{max} = maximum concentration value, and

$\bar{c}$ = average concentration over dimension, y,

and T_1, T_1 = acceptable error tolerance.

The problem is to assign suitable values to T_1 and T_2. Factors involved in selecting these values are errors in field measurements, errors in simulated values, and acceptable deviation from the mean value, $\bar{c}$.

Step 3 - Far Field Dimension Reduction

It is usually practical to choose a numerical model with as simple a description of the prototype physical/chemical/ biological processes as will yield sufficiently accurate results. A common approach to simplifying the analysis is to neglect one or more spatial dimensions (usually the vertical and/or lateral) over which the constituent being modeled can be assumed to be completely well-mixed from the definition developed in Step 2. Reductions in dimensionality, when justifiable, can realize considerable savings in model development/modification and simulation costs. As a first step to reducing dimensionality, one can consider the inverse condition: which dimensions cannot be neglected in the far field (at distances a long way away from the point(s) or discharge(s))? Considerations here include whether the system is stratified, in which case the vertical dimension must be included, whether flow reversals are observed, etc. This step is often straightforward and simplifies later analysis in which finer-scale fields are analyzed because it will limit such analyses to those dimensions over which the system might be considered to be well-mixed.

For estuarine modeling, we will assume that the longitudinal (x) dimension cannot be neglected (as is sometimes the case for reservoir modeling). The objective of this step, then, is to evaluate whether the lateral (y) or vertical (z) dimensions must be retained.

Step 3.1 - The Vertical Dimension

The most frequent cause of variations in the vertical direction is density stratification. The existence of vertical stratification can be observed in one of several ways: 1) salinity and/or temperature gradients; 2) tidal or residual velocity reversals; 3) dye cloud splitting and differential advection, and/or; 4) geomorphological classification.

The degree of stratification can be determined in a variety of ways. For example, the "Estuarine Richardson Number" R described by Fischer(9) is an estuary characterization parameter which is indicative of the vertical mixing potential of the estuary. If R is large (above 0.8), the estuary is typically considered to be strongly stratified and the flow dominated by density currents. If R is very small, the estuary is typically considered to be well-mixed and the density effects to be negligible.

Another method is to use the stratification-circulation diagram developed by Hansen et al.(10). The diagram requires the calculation of two parameters, the Stratification parameter and the Circulation parameter, which are based on measurements from a number of estuaries with known degrees of stratification. A third method for characterizing the degree of stratification is the calculation of the estuary number proposed by Thatcher et al.(11). Once the degree of stratification is determined by one of the above methods, we recommend the following criteria:

strongly stratified	- include the vertical dimension in at least a 2-layer model;
moderately stratified	- include the vertical dimension in a multi-layered model, or reserve judgment to the calculations in Step 6, and;
vertically well-mixed	- could neglect vertical dimension after calculation in Step 6.

Beyond the use of stratification indices, the analysis of vertical dimension reduction becomes more difficult and intuitive. However, the following criteria seem reasonable (See Figure 5):

tidal velocity reversals	- include vertical dimension in at least a 2-layer model;
residual velocity reversals	- include the vertical dimension in a multi-layered model or reserve judgment to the calculation in Step 6, and;
no observable reversals	- could neglect vertical dimension after calculation in Step 6.

If velocity profiles are not available, dye studies can be used. Dye studies simply replace the Eulerian observations of current meters with the Lagrangian movement of a dye cloud study. Again, quantitative analyses are difficult, but the following qualitative criteria seem reasonable (Figure 6):

Dye cloud separates and moves	- cloud is responding to a vertical flow reversal and moves as 2 or more distinct units, indicating the vertical dimension should be included in at least a 2-layer model;

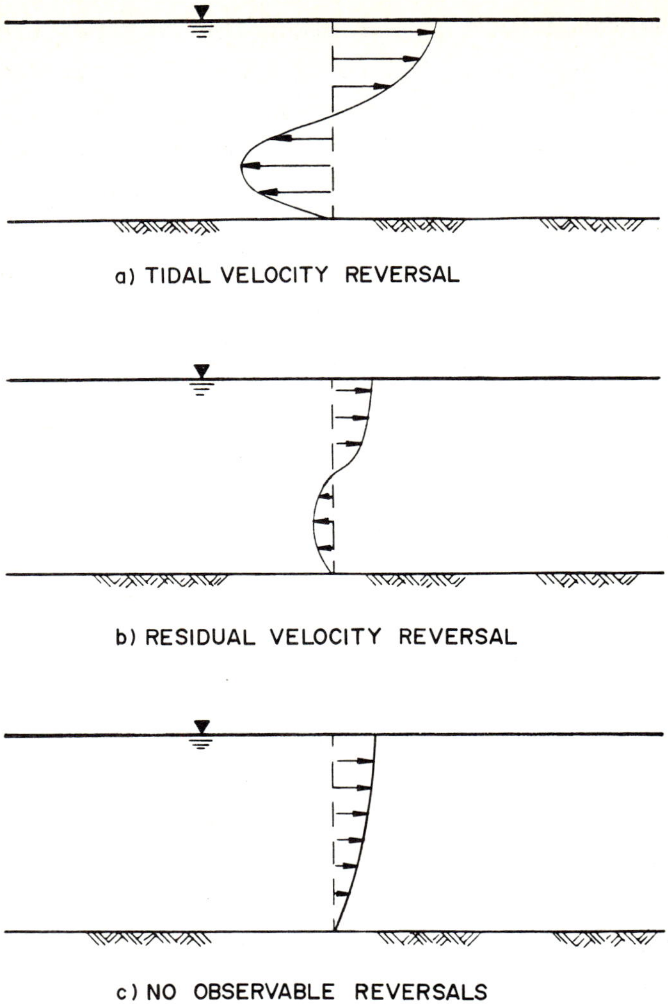

Figure 5. Vertical Velocity Profiles

Dye cloud spreads in non-Gaussian manner - some differential shearing is present and system should be studied using a multi-layer model, or reserve judgment to the calculations in Step 6, and;

Dye cloud moves downstream and diffuses in a Gaussian manner - little differential shearing present and system could be modeled neglecting vertical dimension after calculations in Step 6.

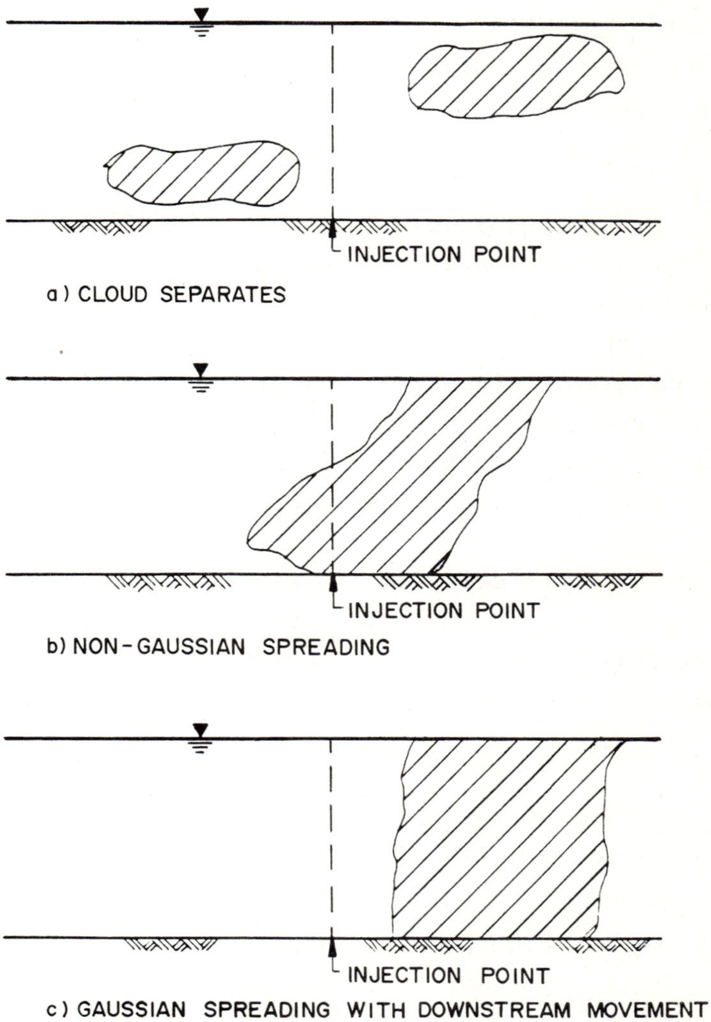

Figure 6. Vertical Dye Concentration Profiles

If little or no data is present, one can try to categorize the estuary within the basic definitions of Dyer(12). Over the years, a systematic structure of geomorphological classification has evolved. Dyer(12) and Fischer et al.(13) identify four groups:

- drowned river valleys (coastal plain estuaries);
- fjords;
- bar-built estuaries; and
- other estuaries that do not fit the first three classifications.

Using the geomorphological classification, the approach is to estimate the degree of stratification from known conditions in a geomorphologically similar estuary.

Step 3.2 - Lateral Dimension

When and under what conditions the lateral dimension in the far field can be neglected is more difficult to estimate than for the vertical dimension, although the same features can be considered. These are: 1) salinity and/or temperature gradients; 2) tidal or residual velocity reversals; 3) dye cloud splitting and differential advection; and/or 4) geomorphological classification.

The analyses in this step are the same as for the vertical dimension analyses (no equivalent to the stratification diagram exists) except that we are looking for lateral gradients and reversals.

Step 4 - Time and Space Scales of Important Processes

At this point we begin to narrow our attention to time and space scales compatible with physical/chemical/biological processes and at which resolution of predicted results is desired. It is important at this step to refer to the conceptual model of the important processes which should be included in the numerical model. This is necessary for two reasons. First, it provides a checklist of processes or variables included in the final numerical model selected. If the checklist is not completely satisfied, the shortfall can be used to assess the need for further model modification/development, or a compromise achieved and a revised conceptual model developed. Second, it can be used as a base to estimate rate coefficients used in defining process linkages in the numerical model.

An important consideration in selecting an appropriate numerical model is the question of dynamic versus steady-state modeling. It is intuitively evident that some measure of the ratio of physical-to-biochemical rates (perhaps defined by flushing time divided by a measure of kinetic rates, such as the shortest T_{90} rate) can be used to make this assessment.

In order to more closely investigate the possibility of reducing model dimensionality, we must be able to understand the time and space scales of the physical/chemical/biological processes included in the conceptual model of the prototype system, and their interrelationships. To do this, we will define only physical and chemical time and space scales as:

- physical: flushing time
- biochemical: growth, decay, and dieoff rate.

Step 4.1 Determine Flushing Time

The time that is required to remove pollutant mass from a particular point in an estuary (usually some upstream location) is called the flushing time. Long flushing times are often indicative of poor water quality conditions due to long residence times for pollutants. Flushing time, particularly in a segmented estuary, can also be used in an initial screening of alternate locations for facilities which discharge constituents detrimental to estuarine health if they persist in the water column for lengthy periods.

Factors influencing flushing times are tidal ranges, freshwater inflows, wind, and dispersion. All of these forcing functions vary over time, and may be somewhat unpredictable (e.g., wind). Thus, flushing time calculations are usually based on average conditions of tidal range and freshwater inflows, with wind effects neglected. A number of methods exist for computing flushing time (see Reference 8).

Step 4.2 - Compute Growth, Decay, or Dieoff Rates
A simple measure of biochemical process time scales is given by the rates of growth, decay, or dieoff, K. T_{90} is defined as the time to reduce or increase a constituent concentration by one order of magnitude (a 90 percent reduction or 100 percent increase). The relationship between the two is:

$$T_{90} = 2.3/K \qquad (2)$$

Of critical concern in selecting a water quality model is the ability to resolve the most rapid growth or dieoff for any constituent of concern. Thus, from a knowledge of the system's chemistry, one might select the minimum T_{90} or 1/2 the minimum T_{90} as the biochemical process time scale.

Step 5 - Regulatory Scales

It is important to make a clear statement of the purpose and objectives of a study. Frequently, such a statement is determined by local/state/federal regulations which, in turn may imply time and spatial scales that must be resolved by the model description. An example of this is the concept of an "allowable mixing zone" frequently defined in the vicinity of an outfall. The regulation may be so written that the numerical model must have sufficient spatial resolution to determine that: (a) the mixing zone length is not exceeded (thus requiring distance steps at least an order-of-magnitude less than this length); and (b) no violations occur at any time, which may necessitate a dynamic rather than steady state approach to modeling.

More often than not, the purpose of modeling is to evaluate compliance with federal or state regulations. Many of these regulations define time and space scales beyond which violations will not be permitted. For example, the regulation for a thermal discharge to a river may state that waters must return to within 2°C of the ambient temperature within 1,000 m downstream from this discharge. This clearly defines a length scale of 100 m (1,000/10), and a time scale of 100 m divided by the ambient river velocity.

Thus, if regulatory standards are a consideration in the model selection process, we would define a regulatory length scale, L_R as one-tenth of the minimum of the specified distance if known from regulations, length of the system to be modeled, or total length of system, and a time scale, T_R, as:

$$T_R = L_R/\bar{u} \qquad (3)$$

where $\bar{u}$ = velocity averaged over time and space scales.

Step 6 - Study Scale Dimension Reduction

Based on the information of Steps 1-5, we can now develop a more refined conceptual model that incorporates all the information of these steps. The purpose of step is to determine whether the receiving water model can omit spatial dimensions, other than those found to be non-negligible in Step 3, and still resolve all necessary physical/chemical/biological processes at the regulatory spatial scale.

As an example of this, consider wastewater discharged into a stratified estuary. Step 3 may show that stratification is important and thus the vertical variation cannot be neglected (or integrated). However, the question may still remain as to whether it is reasonable to neglect lateral variations and assume a laterally well-mixed system. If the regulations reviewed in Step 5 require a mixing zone of not more than 1000 m downstream of the outfall, we would wish to determine from this step whether the laterally well-mixed condition is reached before or after this distance. If it is reached before 1000 m, for example, it may be reasonable to neglect lateral variations, However, if it is reached afterwards, the lateral dimension is still required to be able to adequately compare simulated results with the regulatory standard.

In Step 3, we looked at the irreducability of dimensions at the far field scale. This included analyses and subjective indications that considered: 1) density stratification; 2) velocity data; 3) dye studies; and 4) topographic characteristics. These were included at the far field level because such data are usually either unavailable in the potential study area or are inadequate to make a reliable determination of dimension reductions at this scale. Normally any data in the study area will be part of a larger data set covering a much wider area.

Having identified the dimensions we should not neglect in Step 3, in Step 6 we determine whether any of the other dimensions can be neglected. Having probably exhausted our supply of data to make such determinations, we can rely on simple analytic solutions.

At this stage, we should have a fairly complete conceptual model of the system to be studied. This should include a definition of the minimum spatial resolution, L_S, defined as the minima of the process space scale, L_P, and the regulatory space scale, L_R:

$$L_S = \min (L_P, L_R) \tag{4}$$

There are three dimension reductions to be considered here: 1) cross-section area; 2) vertical dimension; and 4) lateral dimension.

There are two techniques described by Fischer et al.(13), that can be used to make these assessments: 1) Fischer's mixing length; and 2) analytic transport models.

Step 7 - Dynamic or Steady-State

Again based on the information of Steps 1-6, we can make a decision whether a steady-state model can be used or whether a dynamic model is required to achieve the required temporal resolution. An example of this is that the dissolved oxygen (DO) standard in many states requires no oxygen depletion below a given level at any time. In a tidal river, for example, in which processes can occur within or at the same period as a tidal cycle (many constituents will exhibit diurnal variations which are not in phase with the tidal cycle) a dynamic model may be required to provide adequate temporal and spatial resolution to make such a determination.

Steady-state modeling is inherently more economical than dynamic modeling. Another advantage is that it can perhaps be adapted to run on a microcomputer for relatively simple kinetic descriptions, thus making the approach more readily available to groups with computer hardware limitations. Finally, steady-state models are usually less complex, because the cumbersome hydrodynamic equations have been greatly simplified to include only the continuity equation.

There is a great danger, however, of choosing such an approach without considering the implications of the selection. It is intuitively argued that a steady-state approach is of little use in a closed-end canal system, for example, because there is no (or at least, very little) net advection towards the mouth. By contrast, one would frequently choose such an approach for a stream with no tidal effect. But what about the systems in between, for example, those that have both net downstream advection and tidal oscillations?

Unfortunately, there is no hard and fast rule that can be applied. Rather we must rely on our experience, plus some obvious physical/chemical process limits. It seems reasonable that a decision criteria would depend on the time scales of both physical and biochemical rates, most likely as a ratio. Physical time scales, T_p, will be defined as the minimum of the tidal period, T, and the flushing time, F:

$$T_p = \min (T,F) \tag{5}$$

The biochemical process time scale, T_c, is defined as the minimum of the kinetic rates, using the concept of an order-of-magnitude or 90% reduction, T_{90}, and any regulatory time scale, T_R, defined in Step 5:

$$T_c = \min (T_{90}, T_R) \tag{6}$$

Defining the ratio:

$$R = T_p/T_c \tag{7}$$

we would recommend the following criteria:

- If $R > 0.5$ - use a dynamic model
- If $R < 0.1$ - use a steady-state model
- If $0.1 < R < 0.5$ - selection should be based on other factors such as sensitivity of the study, reliability of available data, cost, etc.

Step 8 - Spatial and Temporal Resolution

The selection of space and time steps for the numerical model is intended to provide sufficient resolution to adequately describe the physical/chemical/biological processes within the prototype. However, in many cases, this selection must also include a consideration of model accuracy and stability.

Step 8.1 - Resolution

To ensure adequate temporal resolution of processes included within the model description, the time step, Δt, must be less than the time scale of the model, T_S, defined as the length scale, L_S, divided by a characteristic system velocity. Such a selection can be semi-intuitive. For example, consider the number of piece-wise continuous straight lines that can approximate a sine curve over a tidal cycle. In general, we would recommend:

$$\Delta t < 0.1\ T_S \qquad (8)$$

In a similar manner, the space step, characterized perhaps by its longitudinal value, Δx, must be adequate to resolve spatial distributions. In general, Δx should be smaller for larger concentration gradients, and larger for smaller gradients (if the numerical scheme can offer that type of flexibility). Higher concentration gradients are likely to be found near waste sources, and decrease away from them. To estimate spatial resolution, one can use the calculations of Step 6 and approximate the resulting distribution with a series of piece-wise lines until an adequate representation results. The minimum spatial resolution required, Δx_{min}, can be determined from the resulting construction (Figure 7).

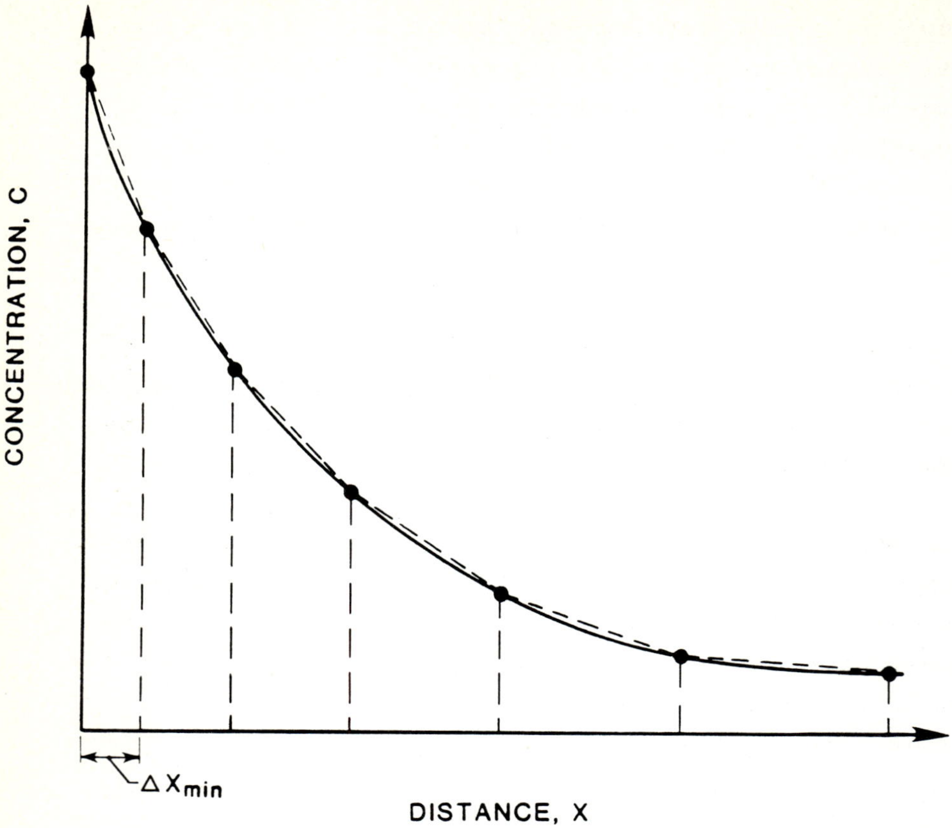

Figure 7. Approximation to Concentration Gradient

Step 8.2 - Stability

Most receiving water models use explicit schemes, i.e., variables at the new time level are calculated using known values at previous time levels. This leads to several common conditions that must be satisfied to ensure model stability (i.e., solutions remain within bounds and do not "blow up"). Furthermore, satisfying these conditions will often result in smaller time steps than would generally be needed from the standpoint of resolution alone.

These conditions, or criteria for an explicit one-dimensional model for example, are usually the minimum of a hydrodynamic criterion (Courant condition), a mass transport condition, and a dispersion condition, i.e,.:

$$\Delta t \leq \min \left[\left(\frac{\Delta x}{\sqrt{gh} \pm u} \right) , \left(\frac{\Delta x}{u} \right) , \left(\frac{\Delta x^2}{4E_L} \right) \right]$$

Similar conditions exist for 2 and 3-dimensional models, and other conditions, such as a friction term criterion, may also be required.

Some models use an implicit technique to approximate the governing equations. In these cases, parts or all of the model may be unconditionally stable, which means that the choice of the time step is not limited by stability considerations. Here, the time step should be chosen to provide adequate resolution of temporal processes.

Step 9 - Diffusion Coefficients

Diffusion coefficient selection for dynamic models is straightforward, in that the form used in any particular model selected will probably be adequate. Yet, it is very difficult to establish rigorous guidelines as to the most appropriate form. Our advice here is to ensure that the form used in a particular model has a reasonably sound basis. In most cases, as a dynamic model is complex, and usually well reviewed (choosing well-known models or agency-supported models can lead to increased confidence), one may generally assume that it contains an adequate description of the diffusion processes. Furthermore, diffusion in a dynamic model is not as critical as dispersion in a steady-state model, in which neglected processes are inherently lumped into the dispersion mechanism.

More care must be exercised when selecting the appropriate dispersion coefficient description to be used in a steady-state model. Such a selection, as has frequently been stated in this report, must be based on a sound knowledge of the physical/chemical processes at work and the inherent assumptions made when the hydrodynamic equations are simplified or time integrated.

There are three types of tidally-averaged dispersion coefficients, E_L^T, commonly used in steady-state models:

1. E_L^T = constant in time and space (9)
2. E_L^T = constant in time, variable in space (10)
3. $E_L^T = KRQ_R/A$ (11)

where K = constant
R = hydraulic radius
Q_R = river flow, and
A = cross-sectional area

Selection of the most appropriate form will be based on a consideration of the respective influences of river and tidal flows. In general, we would recommended that Eq. (10) be used for tidally dominated flows, Eq. (11) be used for river dominated flows, and Eq. (9) be used as little as possible or when no data is available for a more rational selection. Reference 8 contains methods for estimated E_L^T under various river flow and tidal conditions.

It should be noted that one equation for E_L^T may not apply everywhere in a given system. For example, in upstream areas the river flow tends to dominate and Eq. 11 can be applied. Downstream, however, tide may dominate and Eq. 10 can be applied. In the middle reaches, there is a balance between river and tidal flows and thus some value between the two extremes might be selected.

Step 10 - Data Availability

Data are required for two purposes--calibration and verification. Calibration is a process of adjusting model coefficients, until good agreement is achieved between simulated and observed values. The measures of "goodness-of-fit" may be either judgmental or statistical, using such approaches as relative errors, root-mean-square errors, and tests of statistical significance of variations, etc. Model verification then requires that a second data set be simulated, without adjusting model coefficients, to demonstrate that the model can accurately reproduce a (preferably) different set of conditions.

If sufficient data do not exist, a decision must be made to: (a) collect additional data to supplement existing data, or (b) to use the available data to perform model calibration only. This latter condition is not desirable because it reduces confidence in the model's ability to accurately simulate conditions other than those found during calibration. Perhaps the most desirable situation is to have three or four data sets covering a wide range of conditions to be used for calibration and verification.

Step 11 - Model Selection

Having completed Steps 1-10, the system to be studied should be comprehensively conceptualized. At this point, we are able to select the most appropriate model to do the job. As a basis for model selection, a "check list" should be prepared of desired model features, with room to evaluate the features of several candidate models. A sample list is shown in Table 1 as a guide.

The ability to perform this model selection depends not only on one's capability for evaluating Steps 1-10, but also on the knowledge of available numerical models that are

TABLE 1

EXAMPLE MODEL SELECTION CHECKLIST

Category	Item	Conceptual Model	Model 1	Model 2	Model 3
Dimensions	Longitudinal (x) lateral (y) vertical (z)				
Time Integration	Dynamic steady-state				
Dispersion Cofficient	Constant (x, t) tidal dominated river dominated				
Physical features	Coriolis nonlinear ac-celeration bottom friction wind shear variable water surface density				
Chemical constituent	Constituent a Constituent b				
Chemical kinetics	Kinetic 1 Kinetic 2				
Solution scheme	Finite difference Finite element link node				

possible candidates. Most "experienced" modelers will possess this knowledge, and one commonly held view is that only experieced modelers should use models. Certainly there is a great danger in an inexperienced modeler or team of modelers performing a study. However, the current trend is to consider the model as a "black box" and to use it with as much guidance as can be provided in reports such as this, authored by experienced modelers.

There are several ways one might approach the task of actual model selection:

1. Use models that are readily available to the user or lie within their realm of experience;
2. Perform a library search, perhaps using a computerized database, such as DIALOG;
3. Seek the advice of a knowledgeable modeler; and
4. Reference bibliographic and/or model specific reports and papers.

Most model users usually have both a basis of experience and access to knowledgeable modelers. Once Steps 1-10 have been performed, and several candidate models identified, final model selected is based on previous experience with various models. Factors such as ease of use, reliability, accuracy, availability, and economy can be weighed in choosing the model.

Without such a basis of knowledge, or when a particular study is beyond one's previous experience, model selection follows the procedures above. In particular, there are a number of bibliographic and model selection summary reports that one might turn to for information (e.g., EPA(14); Versar(15); Ambrose et al.(16)).

Summary

A rational procedure for selecting a receiving water quality model for urban nonpoint pollution studies has been presented. The procedure guides the model user through assessments of mixing conditions and time and space scales to make decisions about the need for a multidimensional representation and a dynamic simulation framework. Decisions about model complexity rely upon quantitative measures of physical/chemical/biological processes in the prototype receiving water. Although the model selection procedure was developed for estuary models, the methodology can be easily adapted to rivers, streams, and reservoirs.

References

1. Ditoro, D.M., Probability Model of Stream Quality Due to Runoff, ASCE Journal of Environmental Engineering Division, Vol. 110, No. 3, pp. 607-628, 1984.

2. Burns, L.A., Cline, D.M., Lassiter, R.R., Exposure Analysis Model System (EXAMS): User Manual and System Documentation, EPA-600/3-82-023, U.S. Environmental Protection Agency, Environmental Research Lab, Athens, Georgia, 1982.

3. Receiving Water Quality Model User's Manual, Hydrologic Engineering Center, Water Resources Support Center, U.S. Army Corps of Engineers, 1979.

4. Urban Storm Water Runoff "STORM", Hydrologic Engineering Center, U.S. Army Corps of Engineers, 1974.

5. Roesner, L.A., A Model for Evaluating Runoff-Quantity And Quality in Metropolitan Master Planning, Technical Memorandum No. 23, ASCE Urban Water Resources Research Program, ASCE, New York, 1974.

6. Johanson, R.C., Imhoff, J.C., and Davis, H.H., Users Manual for Hydrological Simulation Program - FORTRAN (HSPF), EPA-600/9-80-015, U.S. Environmental Protection Agency, Environmental Research Laboratory, Athens, Georgia, 1980.

7. Genet, L.A., Smith, D.J., and Sonnen, M.B., Computer Program Documentation for the Dynamic Estuary Model, Prepared for Systems Development Branch, EPA, Washington, D.C., by Water Resources Engineers, Inc., Walnut Creek, California, 1974.

8. Walton, R., George, T.S., and Roesner, L.A., Selecting Estuarine Models, Prepared for the Environmental Protection Agency, Washington, D.C., 1984.

9. Fischer, H.B., Mass Transport Mechanisms in Partially Stratified Estuaries, Journal of Fluid Mechanics, 53:672-67, 1972.

10. Hansen, D.V. and Rattray, M., New Dimension in Estuary Classification, Limnology and Oceanography, 11:319-326, 1966.

11. Thatcher, M.L. and Harleman, D.R.F., Predication of Unsteady Salinity Intrusion in Estuaries: Mathematical Model and User's Manual, Report No. 159, Parsons Lab., MIT, Cambridge, Massachusetts.

12. Dyer, K.R., Estuaries: A Physical Introduction, John Wiley and Sons, New York, 1973.

13. Fischer, H.B., List, E.J., Koh, R.C.Y., Imberger, J. and Brooks, N,H., Mixing in Inlands and Coastal Waters, Academic Press, New York, 1979.

14. Environmental Modeling Catalogue Abstracts of Environmental Models, U.S. Environmental Protection Agency, Prepared for Management Information and Data Systems Division, 1979.

15. Catalog of Waste Load Allocation Models for Toxic Compounds, Prepared for U.S. Environmental Protection Agency, Monitoring and Data Support Division, Washington, D.C., by Versar, Inc., Springfield, Virginia, 1983.

16. Ambrose, R.B., Najarian, T.O., Bourne, G. and Thatcher, M.L., Models for Analyzing Eutrophication in Chesapeake Bay Watersheds: A Selection Methodology, U.S. Environmental Protection Agency, Chesapeake Bay Program, Annapolis, Maryland, 1981.

CONTINUOUS SIMULATION MODELS TO EVALUATE URBAN DRAINAGE IMPACTS IN ONTARIO

D.G. Weatherbe
Water Resources Branch
Ontario Ministry of the Environment
135 St. Clair Ave., West
Toronto, Ontario M4V 1P5

Introduction

Water quality simulation model have been used in Ontario since the 1960's to establish the waste assimilation capacity of rivers to accept discharges of municipal and industrial point sources. A simple dissolved oxygen model was used, which accounted for biochemical oxygen demand and reaeration, as formulated by Streeter and Phelps (1). Benthic oxygen demand was added to the model, to account for the effect of sludge deposits below pulp and paper mill effluents. The model was deterministic and steady-state and was used to establish the effect of levels of effluent treatment and streamflow.

In the early 1970's the effects of photosynthesis and respiration of aquatic plants were added to the dissolved oxygen models (2,3) to produce diurnal variations in oxygen. This model was further extended for use in the Thames River, Ontario, as a continuous simulation model (4). However, in this example urban runoff loads were not modelled. This model was further extended and used in the Grand River, as described below.

Dynamic, continuous water quality modelling is seen to offer many benefits over steady state - event modelling. These are:

- explicit accounting of meteorological variations in flow, sunlight, temperature and rainfall; and man-made variations in effluent discharge rates and reservoir releases;

NATO ASI Series, Vol. G10
Urban Runoff Pollution.
Edited by H.C. Torno, J. Marsalek, and M. Desbordes

- o allowance for inputs to be deterministic, stochastic, or a combination of deterministic and stochastic;

- o ability to determine probability distribution of output variables representing pollution levels. This requires a statement of water quality standards in probabilistic terms.

- o Avoidance of the requirement to establish a design case or design event (storm).

Continuous modelling has matching costs including an increased complexity of algorithims dealing with varying flow conditions, extensive data requirements, and increased computational cost. Both benefits and costs are described in the examples discussed below.

Grand River Basin Water Management Study

The Grand River watershed in southwestern Ontario experienced problems of flooding, water supply limitations and water quality degradation. Seventy-two percent of the total population of 545,000 is primarily concentrated in five large urban centres. The estimated increase in urban population of from 200,000 to 300,000 by 2001 necessitates increased servicing for water supply and sewage disposal. Increased urban runoff and treated sewage loads would further degrade water quality.

The watershed study was carried out to develop plans to meet the objectives of reducing flood damages, providing adequate water supply and to maintain adequate water quality (5). Some projects under consideration would address all three objectives. For example, a multiple use reservoir can provide flood protection, serve as a water supply source, and increase low flows to offset pollution. Evaluation tools which would account for all aspects were developed.

The effect of urban runoff was evaluated by a combination of monitoring and simulation. Catchment monitoring provided data on pollutant concentrations and loadings of suspended solids, biochemical oxygen demand (BOD), nutrients, chloride, metals, polychlorinated biphenyls, and pesticides.

The data on BOD, suspended solids, nitrogen and total phosphorus were used to calibrate the STORM model (6). The model was used to provide continuous simulations of urban runoff at 1 hour time steps for present and future populations at the urban centres on the Grand River as shown in Figure 1.

The STORM model results were input to the continuous water quality model developed by the Ministry of the Environment for the study. The the Grand River Simulation Model - GRSM - was further developed from the Thames River case, referred to above, based on the dissolved oxygen model as formulated by O'Connor and Di Toro (2). The model accounts for dissolved oxygen (DO) deficit caused by carbonaceous and nitrogenous oxygen demand, and benthic oxygen demand, as well as the replenishment of oxygen due to reaeration. Photosynthetic production of plants and algae, and respiration are included. Day to day and seasonal growth, death and washout of three types of attached aquatic plants are included in an ecological subroutine - ECOL. In this subroutine growth is dependent on species, substrate, depth, sunlight intensity, sunlight penetration, and nutrient availability. Death is based on temperature and life cycle for each species. Washoff is a function of temperature and velocity.

The model was set up as a series of reaches and nodes as shown in Figure 2 (30). A steady-state version of the model - DOMOD7 - was also set up on this system.

Calibration and verification of the model were carried out using historical time series data on meteorological conditions, river hydrology, sewage treatment plant inputs, urban runoff inputs and boundary conditions of water quality. Dissolved oxygen predictions were compared to continuous data for several

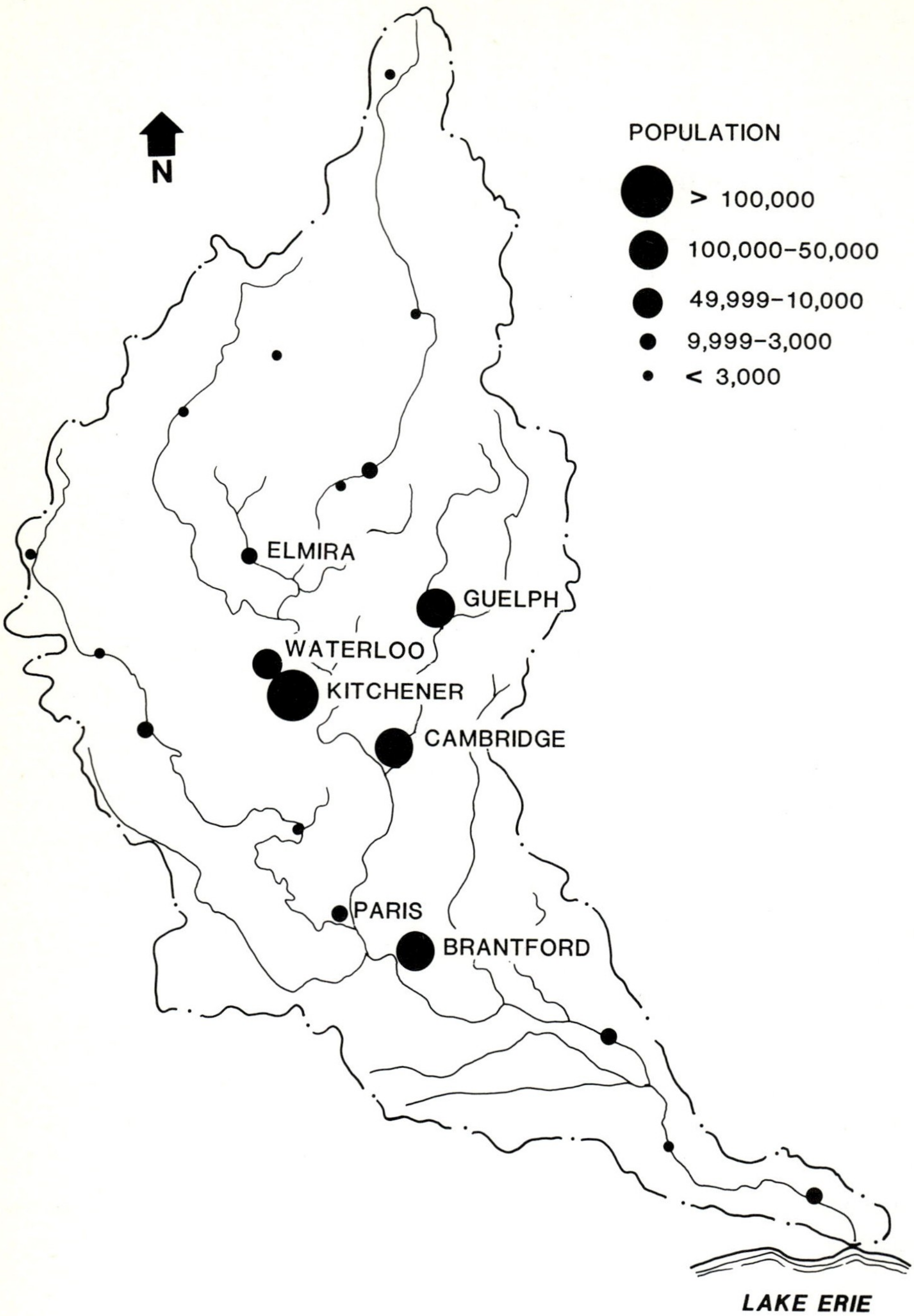

Figure 1 Major Urban Centres in the Grand R. Basin

locations in the study area as shown in Figure 3.

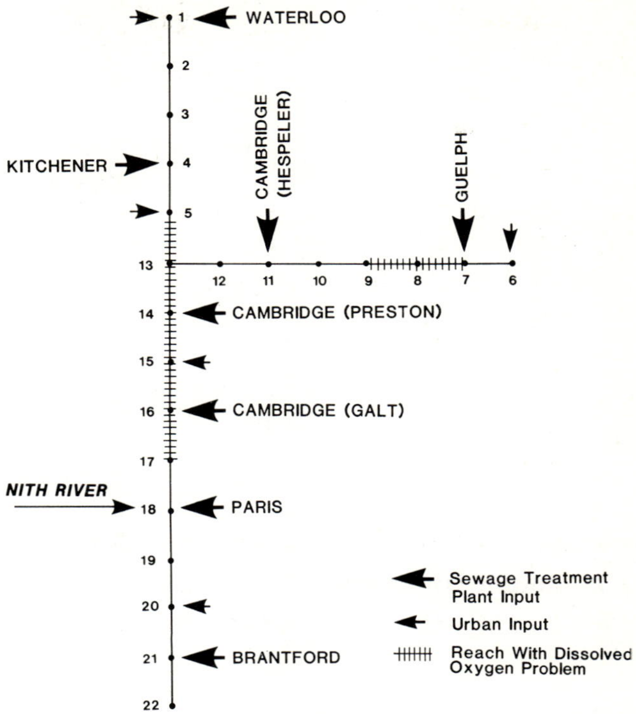

Figure 2 Grand River Simulation Model - Geometry of River System

The DOMOD7 was used to calibrate rate parameters measured during intensive surveys. DOMOD7 was also used to test the sensitivity of the model to some control options, such as sewage treatment levels. This minimized the use of the more expensive continuous simulation model.

The effect of urban runoff was estimated by running the model twice, once with urban storm inputs and once with the pollutants in runoff (but not flows) removed. The effect varied in different reaches. In one case, shown in Figure 4, dissolved oxygen levels showed only minor improvement when urban drainage pollutant loads were removed. In another

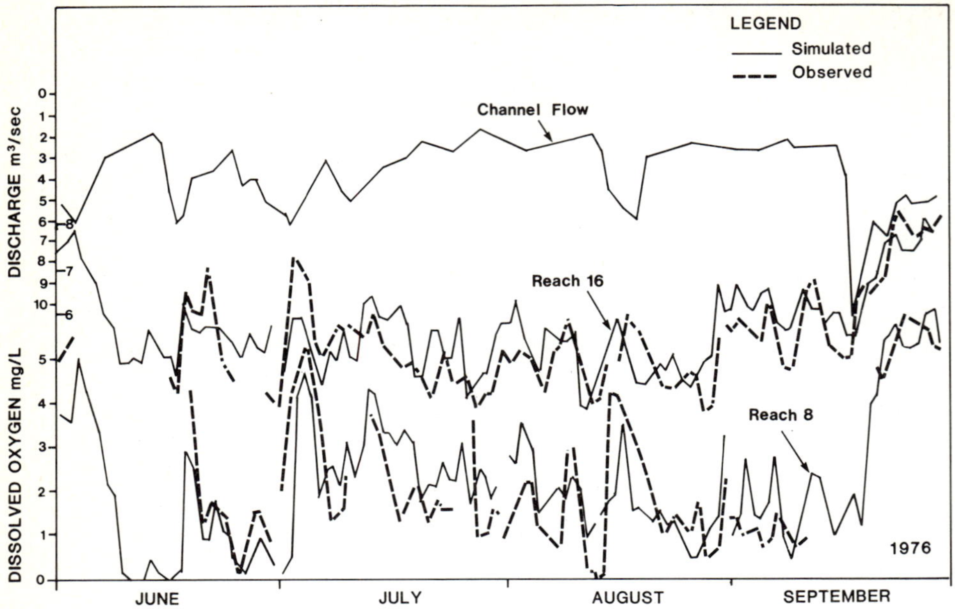

Figure 3 Comparison Between Observed and Simulated Daily Minimum Dissolved Oxygen Concentration for Two Locations on the Grand R.

location also shown in Figure 4, the effect was opposite - dissolved oxygen showed minor reductions without the urban runoff load. These inputs are explained by the relatively low loading of oxygen demanding material (BOD approximately 5 mg/L) and phosphorus (approximately 0.4 mg/L) compared to sewage discharges and agriculture. The dissolved oxygen regime in the Grand River is dominated by aquatic plants. Consequently the suspended solids in urban runoff inhibits attached algae growth in some reaches and offsets the effect of BOD and nutrient discharges.

It was concluded that the urban runoff impact on the dissolved oxygen was negligible. However, monitoring data did show a large impact in the smaller tributaries especially for bacteria, heavy metals and some pesticides (28).

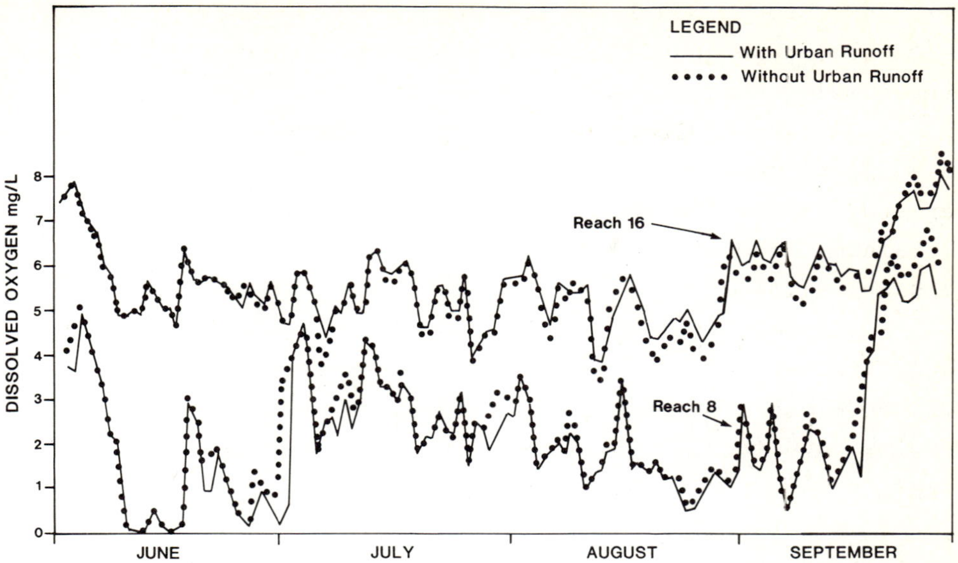

Figure 4 The Impact of Urban Runoff on Simulated Daily Minimum Dissolved Oxygen Concentration for Two Locations on the Grand R.

<u>Rideau River Stormwater Management Study</u>

A study of the lower Rideau River in the Regional Municipality of Ottawa-Carleton (including Ottawa - the nation's capital) was conducted in 1979 to 1983 (9). The purpose was to establish the causes of bacteriological water quality degradation in the urbanized reach, and to protect the river's use for swimming from impacts of future urbanization.

Several beaches in the lower urbanized portion had been closed for a number of years due to effects of combined storm and sanitary sewer discharges. A beach affected by newer suburban developments had been intermittently closed and, in order to protect this beach, new developments were required to include storm water treatment ponds. The need to size treatment pond requirements and to evaluate combined sewer control led to the study.

It was decided to model the upstream and downstream portions of the River independently because of the different hydraulic and water quality character of the two sections. The upstream section (above Hogs Back - see Figure 5) is operated as the Rideau Canal and thus is controlled. This section is characterized by low velocities, with a constant depth maintained. The lower section is free flowing in a series of pools and riffles.

The system was modelled using the US Army Corps of Engineers STORM model (6) for bacterial build-up and washoff from land surfaces, and the QUAL II model (10) for river transport and decay.

Modifications were made to the QUAL II model, by the engineering consultants working on the study, in order to allow varying base flow and loads from runoff (11). Since the hydraulic routing algorithm in the model was retained, the rapidly varying flow conditions occurring during runoff events was modelled in an approximate manner.

The model was calibrated and verified using intensive data collected in a single test reach. Hourly fecal coliform levels were collected over 800 hours at upstream and downstream ends of the reach and at a stream point source. Direct storm sewer loads were simulated by the STORM model calibrated to data collected from storm sewers at two locations.

The model was calibrated on the initial 200 hour sequence. A continuous lateral load of bacteria was added to the model in order to match observed and simulated data. A verification run was then made on the entire 800 hour sequence. Figure 6 illustrates a 10 day portion of the verification sequence. The degree of scatter which exists in the data, as expected in bacterial populations, was not simulated in the model. Also, diurnal variation in the observed data, possibly due to sunlight induced variations in die-off rates, was not simulated in the model. The simulation did match average daily values, and simulated peak levels and timing due to runoff events.

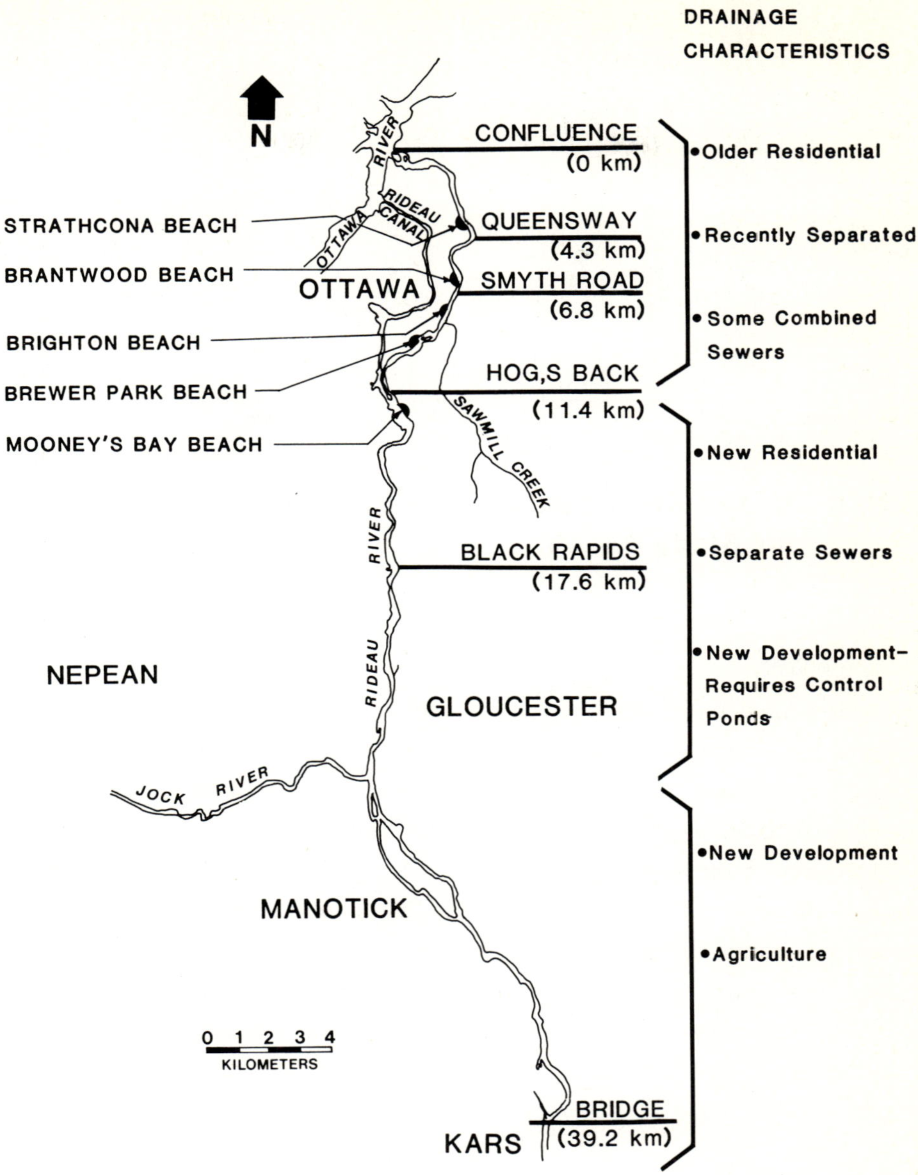

Figure 5 Rideau River Stormwater Management Study Area

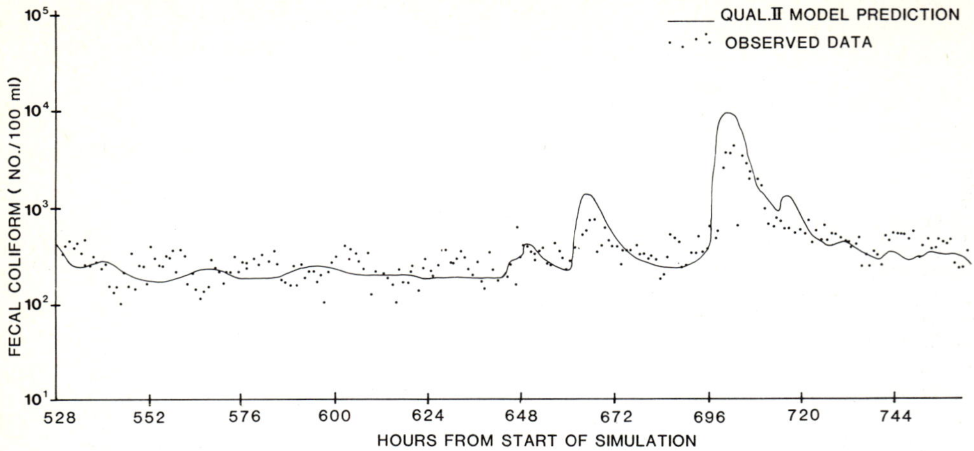

Figure 6 Rideau River Calibration and Verification Results for Study Reach

The model was used to evaluate the effect of treatment and control of combined sewers. Composite results for the lower Rideau are shown in Figure 7, which indicate the need for control of dry weather and wet weather sources. The study results are complicated by the confounding observation that an unknown source of bacteria exists in the river. Either unlocated continuous sources of bacteria are present from illegal sanitary sewage connections, or bacterial inputs from sediment sources are present. Future investigation is required to determine the source.

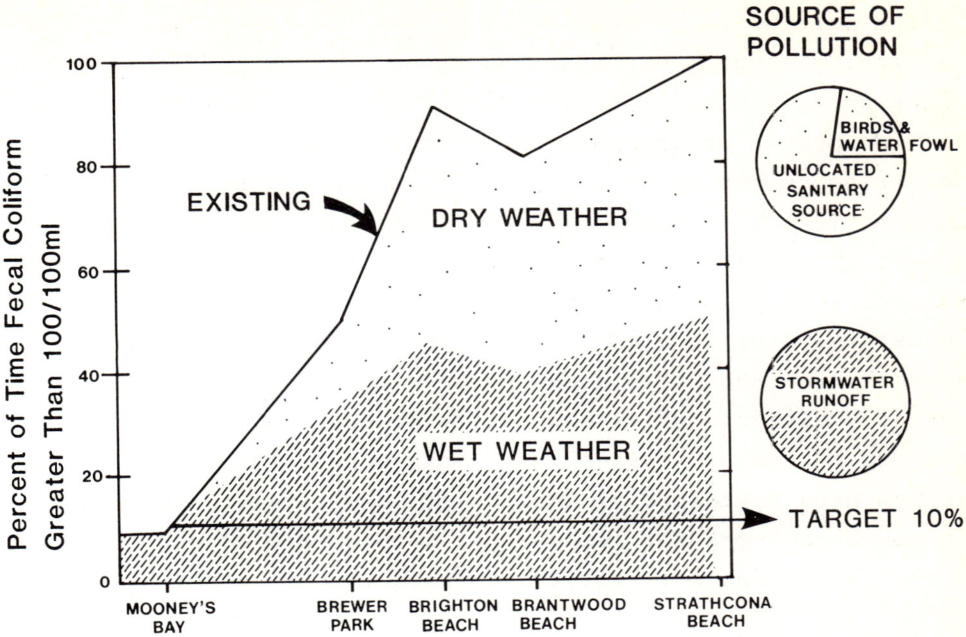

Figure 7 Average Duration of Exceedence of Bacteriological Swimming Standard Over Period June to August in the Rideau R.

Humber River Watershed

In 1981, studies of watersheds in the Toronto area were initiated as the Toronto Area Watershed Management Strategy Study (TAWMS). The objectives were:

- o to better define water quality conditions in the watersheds
- o to analyze causes of pollution
- o to develop cost effective measures for control of pollutants.

In the initial stage of the study, a review of water quality of the entire study area indicated water quality problems in the lower urbanized area of each watershed. Approximately 2.5 million people occupy the urban portion with heavy development occurring on the suburban fringe. Emphasis was placed on the Humber River watershed of 897 km^2, of which 15% is in the urbanized portion. Water quality impairment problems included bacteriological parameters, heavy metals, phosphorus, and some industrial organic chemicals and pesticides. Wet weather sources from the urban area appeared to account for the most severe violations of water quality objectives. Dry weather periods in the urban area, and background levels from the Upper Humber area contained violations of some parameters. The range of concentrations for lead, zinc, cadmium and copper are presented in Table 1, for three areas in the Humber.

The major activities of the Humber River Study include: characterization of pollutant sources in storm and combined sewers, evaluation of sources in the agricultural area in the Upper Humber, development and application of models to describe the system of pollutant sources and transport in the river; feasibility and cost analysis of control options; and evaluation of effectiveness of control options. The goal is to achieve acceptable water quality in the Humber River in a cost-effective manner, while reducing basement flooding.

A system model procedure was set up for the lower urbanized Humber River as described in Figure 8. The procedure includes models for urban runoff sources and for the river. The procedure is conceptually outlined in Figure 9 and described in more detail below (12).

Table 1: Humber River Metal Concentrations (ug/l)

Metal	Rural Humber R.			Urban Humber R.			Black Creek		
	Dry[1]	Wet[2]	Spring[3]	Dry	Wet	Spring	Dry	Wet	Spring
Lead									
Max	5	22	84	8	33	230	14	310	340
Mean	4	10	12	4	13	16	9	78	79
Min	3*	3*	3*	3*	3*	3*	5	6	3*
(25)[4]									
Zinc									
Max	17	46	76	42	70	250	48	430	290
Mean	9	20	20	9	29	29	33	115	62
Min	1*	2	1*	3	1*	1*	14	34	14
(30)[4]									
Copper									
Max	11	18	40	8	130	13000	20	72	70
Mean	8	12	17	7	20	132	16	26	25
Min	8	6	9	5	5	9	11	12	15
(5)[4]									
Cadmium									
Max	0.4	0.6	2.0	0.3	0.7	2.0	2.0	1.9	3.0
Mean	0.3	0.25	0.6	0.2	0.3	0.6	0.8	0.7	0.75
Min	0.2	0.2	0.2	0.2	0.2	0.2	0.2	0.2	0.2
(5)[4]									

1. Dry Weather period
2. Wet weather during runoff event
3. Spring snowmelt period
4. Number in bracket is Provincial Water Quality Objective in ug/l

* Detection limit

The model selected for the watershed is the Hydrologic Simulation Program - Fortran (HSP-F) model distributed by the US Environmental Protection Agency (13). HSP-F is a comprehensive hydrological water quality simulation model with the capability to simulate watershed physical processes such as runoff, infiltration, snow accumulation and melt, and pollutant washoff. The river processes include capability for sediment interactions, transport of conservative and non-conservative, substances and algal effects. The model includes a number of data analysis procedures which provides for handling and statistical

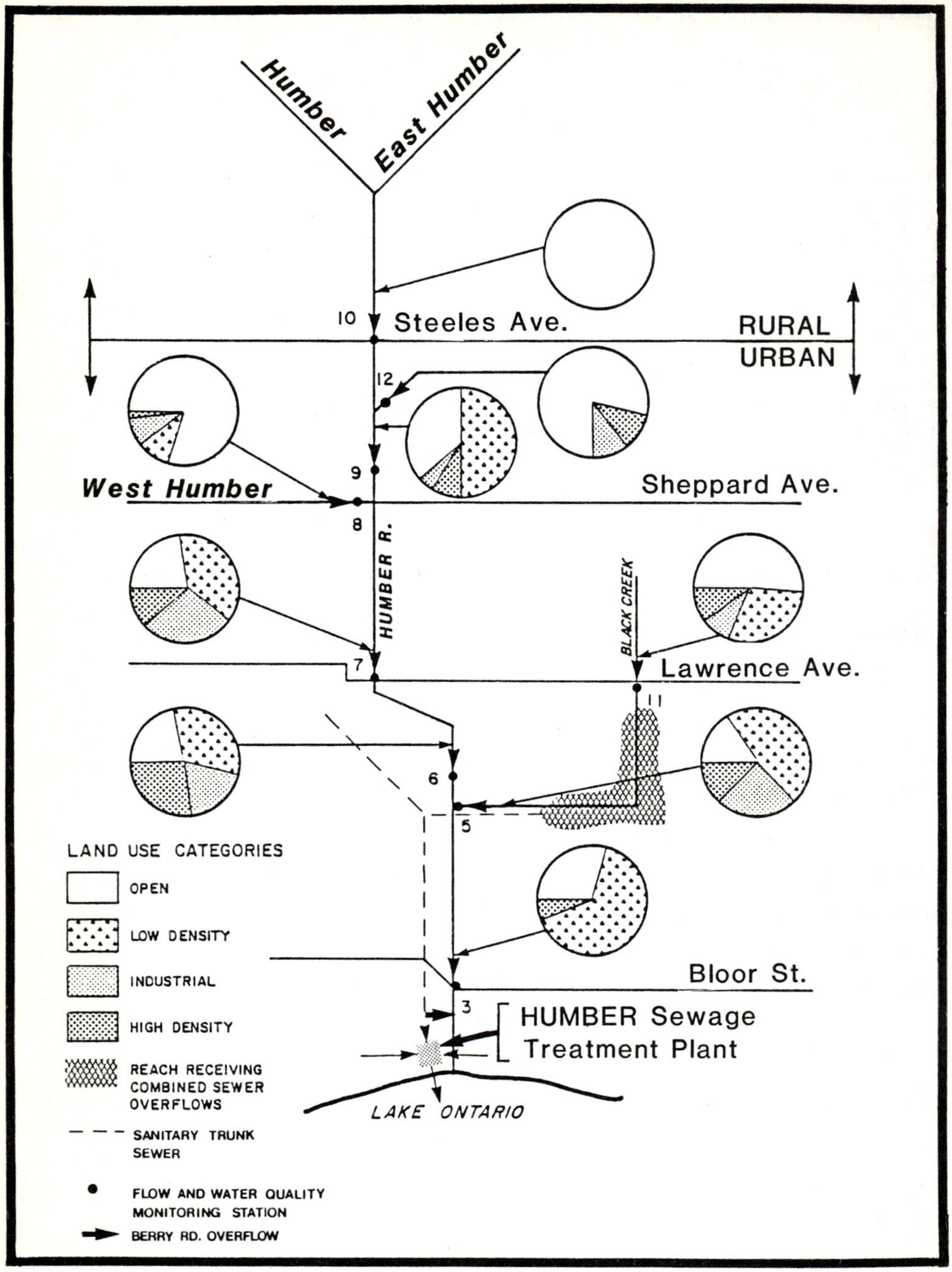

Figure 8 Humber River Study Area

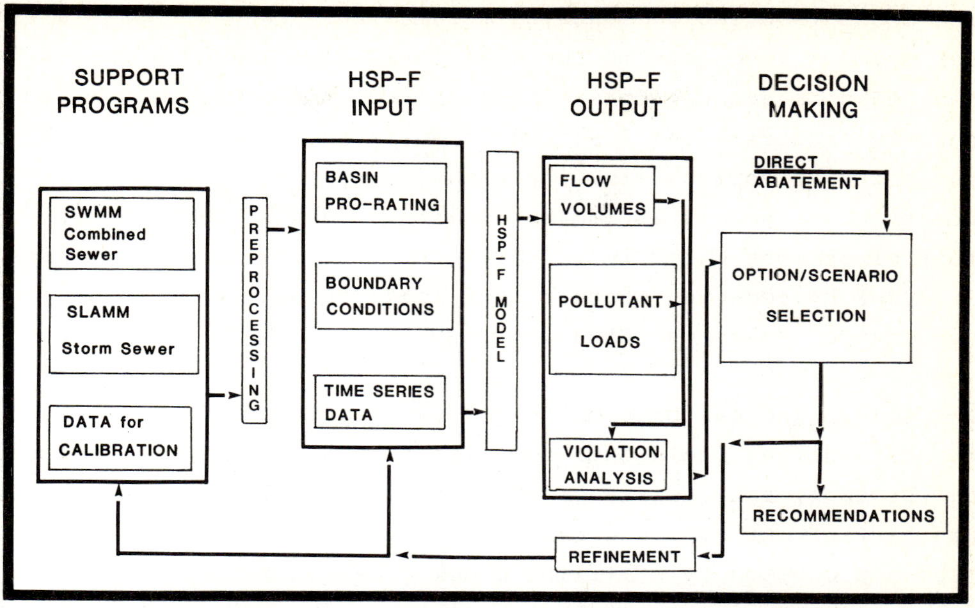

Figure 9 System Model Procedure for Humber River

presentation of the continuous time series required for input and produced as a result of the simulation.

The model accommodates both impervious and pervious land segments, which approximates processes occurring in urban drainage. Two other procedures were utilized to augment urban runoff processes modelling.

The EPA Storm Water Management Model (SWMM) (14) was utilized to model the complex hydraulics of combined sewer overflows. The model accounts for overflows from approximately 1330 hectares of combined sewer area servicing 81,000 people, as an external input to the HSP-F model.

The Source Load and Accumulation Management Model (SLAMM) was developed as part of this study, to provide an evaluation tool for stormwater controls at source - such as street sweeping.

The model is based on observed accummulation and washoff rates for various land uses in the Humber Watershed. HSP-F uses this same data to account for washoff and accumulation, as well as potency factors - the fraction of pollutants associated with sediments of various size fractions, compared to the soluble fraction. Since it is impractical to simulate storm runoff control options directly in HSP-F, control levels are simulated in HSP-F by reducing storm runoff loads by a treatment efficiency percentage, based on SLAMM results (15).

The set-up and calibration of HSP-F was carried out in several steps. Initial field data collection, model set-up, hydrological calibration, and quality calibration for total dissolved solids was carried out by consultants (16). Further calibration on metals proceeded by utilizing Black Creek as a simple system to debug the data input and the model.

The quality parameters selected for modelling include the metals copper, zinc and lead, and suspended solids. Cadmium was considered as well, since it also violated water quality objectives; however, fewer data were available, and conclusions about its occurrence can be inferred from results of the other metals. Analysis for bacteria is being carried out utilizing an independent procedure not described here. Organic contaminants and pesticides occurred so infrequently, that modelling was not considered feasible.

Calibration was carried out first on the hydrology of the system for the two full years, 1979 and 1980. These years were chosen as being representative of long term statistics of the system. Quality calibration was further validated for storm events in 1982. Figure 10 which gives results for the flow calibration, shows a good ability of the model to follow storm events and low flow periods.

Water quality calibration was carried out for the 1982 fall period. Results were considered adequate for each of the metals considered, although the first storm event was poorly simulated (Figure 11). It was concluded that precipitation input to the model did not represent actual conditions on the watershed for this event. In addition more data would have improved confidence in the model.

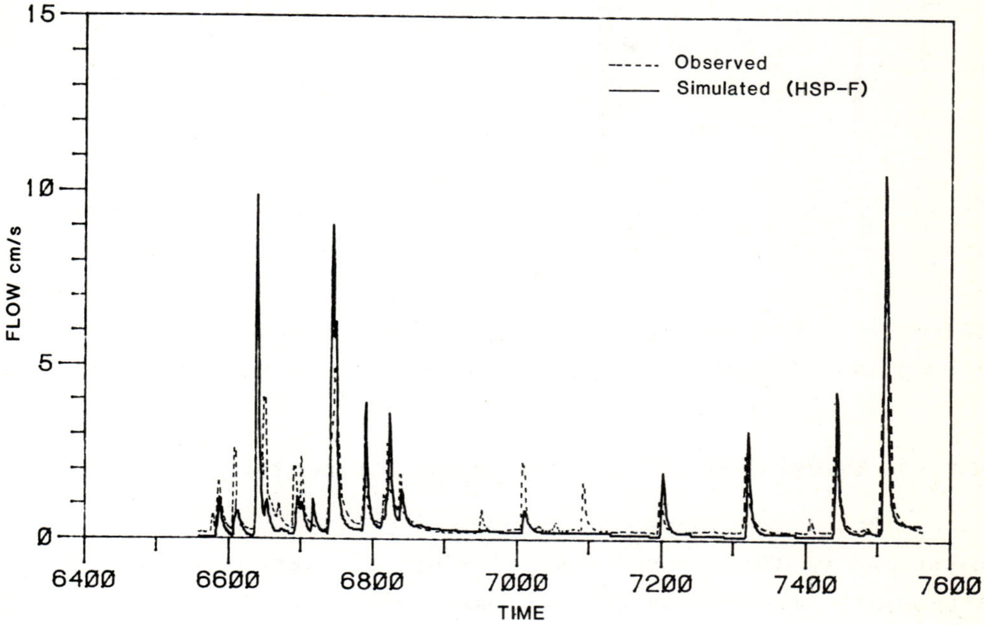

Figure 10 Model (HSP-F) Calibration on Flow for Black Creek Fall 1979

The calibrated model for Black Creek was next used to assess the sensitivity of the watershed to proposed urban pollution control scenarios. A one year simulation spanning April 1979 through March 1980 was used. The control scenarios considered included:

1. Existing condition - no controls
2. Combined sewer storage and diversion to treatment plant
3. Stormwater retention and treatment in industrial areas
4. Street sweeping of residential catchments (3 times per week)
5. Catchbasin cleaning (twice per year)
6. Disconnection of roof leaders in residential areas
7. Stormwater retention in residential areas
8. Abatement of dry weather discharges

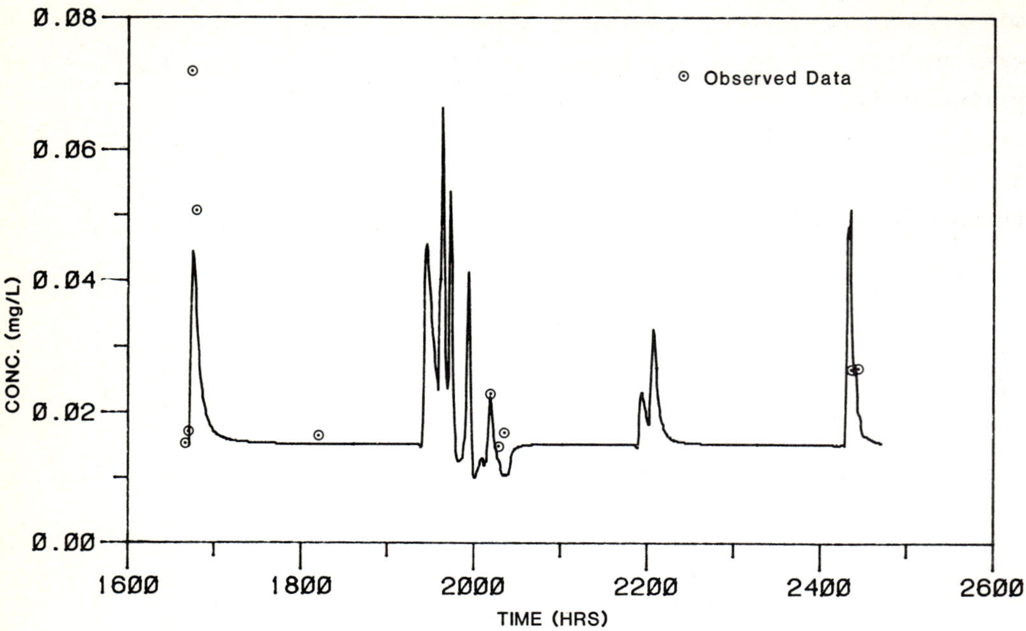

Figure 11 Model (HSP-F) Calibration on Copper for Black Creek

Evaluation of the control scenarios was conducted in a cumulative manner with the effects of each being combined with those previously considered. This limited the number of runs and allowed evaluation of the total effect of a maximum control effort on in-stream water quality. Since this was primarily a sensitivity analysis, only limited consideration was given to the feasibility of implementing the various scenarios. The selection of controls was based however on an *a priori* evaluation of the likelihood of control implementation, with the least likely controls being added last. However, the abatement of dry weather discharge would likely be implemented first (and not last as shown). In cases where the feasibility of implementation was questionable, such as retention ponds, relatively conservative estimates of the extent of control implementation were used.

Pollutant reductions for the various controls were determined through percent reduction of the "end of pipe" contribution to the stream in most cases. Two exceptions were: the CSO control which was simulated by complete elimination of CSO inputs; and roof leader disconnection, which was simulated by altering the pervious and impervious areas, thereby diverting the discharge of roof runoff and pollutants to lawns. The pollutant reductions attributable to the remaining control scenarious were based on work done by consultants for this study (15). The appropriate reductions are summarized in Table 2.

Table 2: Pollutant Reduction Efficiencies

Control	Assumption	Pollutant Reduction Assumed SS	Pb	Zn	Cu
Industrial Ponds	1%of contributing area dedicated to ponds	95%	95%	59%	20%
Street Sweeping	Residential Area 3x/week	3%	9%	3%	3%
Catchbasin Cleaning	2x/year	15%	15%	8%	8%
Residential Ponds	0.5% of contributing area dedicated to ponds	85%	87%	53%	20%

Reference 15

These pollutant reduction factors were adjusted to represent the extent of scenario implementation possible in a particular basin. For instance, if only 10% of a catchment consisted of industrial landuse only 0.95 x 0.10 = 0.095 of the suspended solids would be removed by industrial ponds.

The sensitivity analysis runs were examined to assess the impacts of the control scenarios on total annual load (Figure 12 for zinc, lead and copper), changes in the minimum, maximum and mean concentrations (Figure 13 for zinc) and

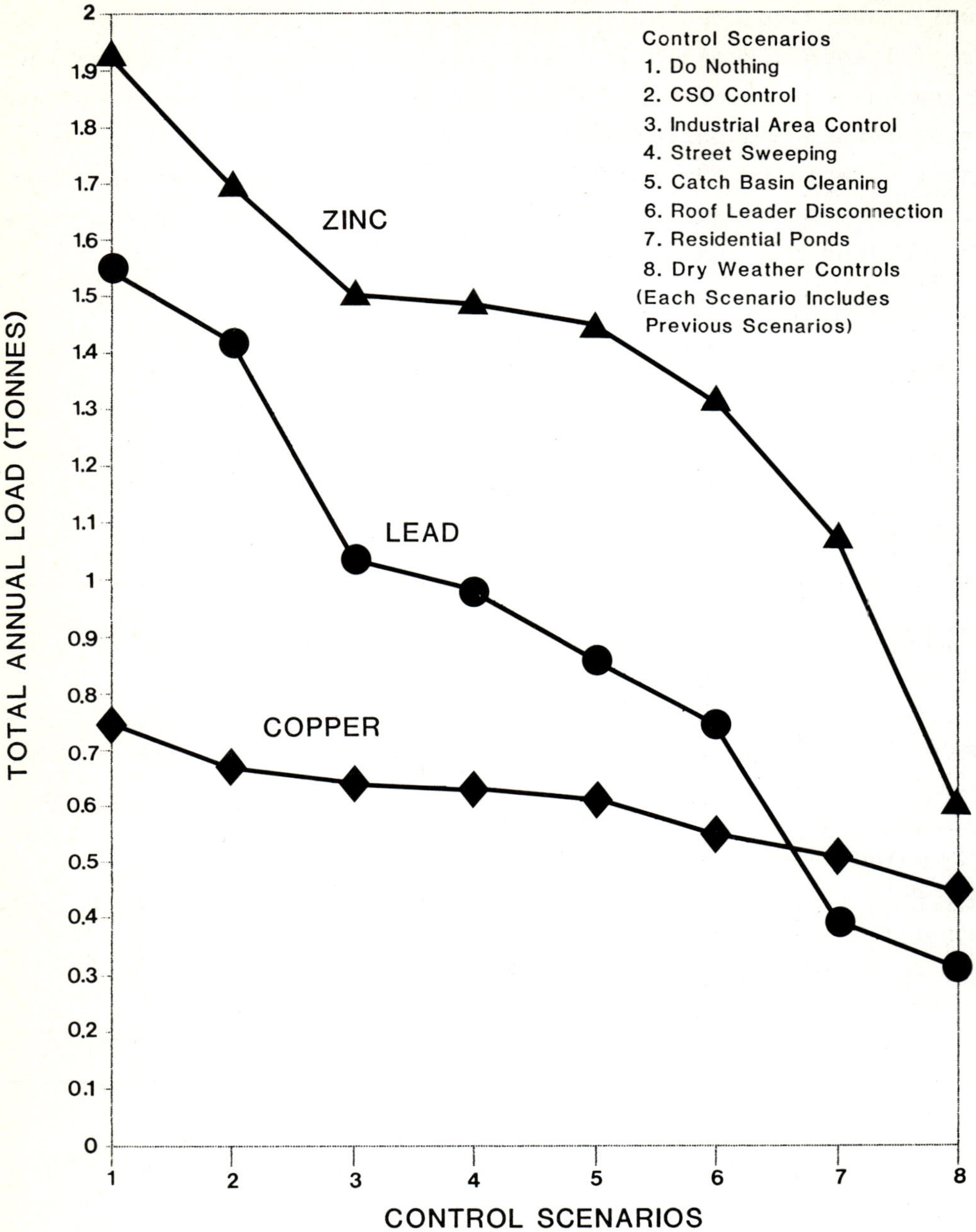

Figure 12 Cumulative Effect of Control Scenarios on Load of Metals to Black Creek

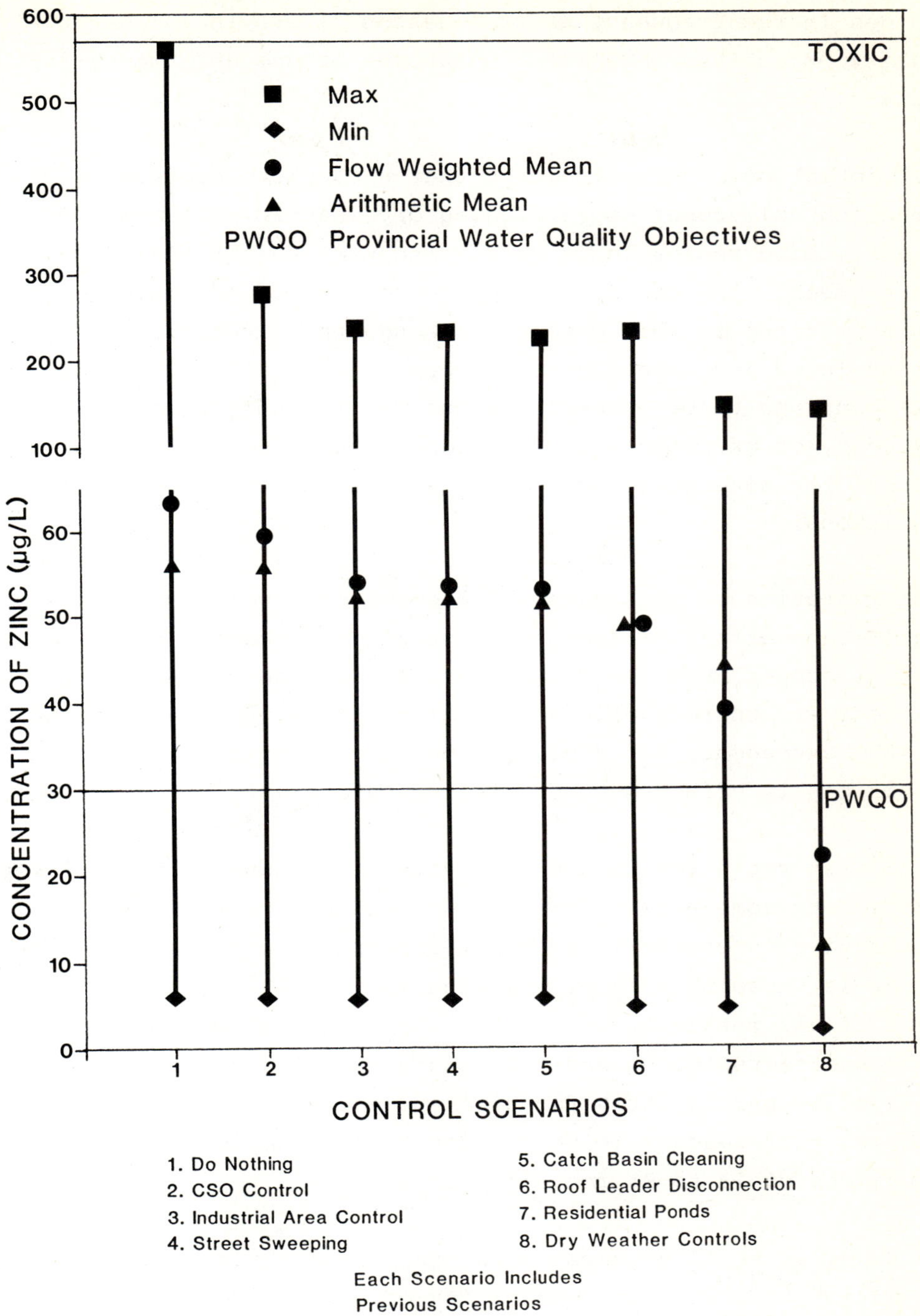

Figure 13 Effect of Control Scenarios for Black Creek

changes in the frequency of exceedences of Provincial Water Quality Objectives (PWQO) for durations of one and twenty-four hours.

The simulation results indicate that a complete program of urban control scenarios (excluding dry weather abatement of sewers) could reduce loads by 32% (copper) to 75% (Lead) on Black Creek. The controls will reduce maximum and mean concentrations to some extent, although these reductions will not be large for either zinc or copper. For all cases PWQO will continue to be exceeded during major runoff events. The frequency of exceedence may be significantly reduced for lead, but not for zinc or copper, which are dominated by dry weather conditions.

Implementation of dry weather abatement of sewers would have a significant effect, bringing the total load reductions up to 41% for copper, 61% for zinc, and 80% for lead. Furthermore, dry weather controls will have a major impact on the frequency of PWQO exceedence for zinc. Copper exceedences are not significantly effected by dry weather abatement.

Stormwater retention is the most effective means of control for lead and to some extend for zinc. Ponds are relatively ineffective in controlling copper inputs because most of the copper is in soluble form. Elimination of CSO will reduce loads of all metals by 8% to 12%. Conventional source controls such as streetsweeping and catchbasin cleaning are relatively ineffective particularly for zinc and copper. Disconnection of existing roof leaders is only moderately effective with reductions of 7% to 9% for all metals.

The study is still underway with the main activity being evaluation of costs and feasibility of control for several of the more favourable options. Simulation results will obviously aid in completion of the cost-effectiveness analysis which will guide decision makers.

Discussion and Conclusions

The benefits and costs of continuous simulation were previously described. The results of each application provide some insight into the utility of continuous modelling. Some additional cautionary notes are in order.

Continuous simulation utilizing large comprehensive models is expensive, time consuming and difficult to manage. Data requirements can be extensive, with difficulties experienced in co-ordinating efforts of various project groups. Discrepancies in rainfall and streamflow measurements, normally expected in a large network, become difficult to resolve in deterministic models which do not account for errors. Computational costs can be high. One run of the Humber River System on HSP-F, for two years and four parameters takes 12 minutes CPU time and costs approximately $500 on an IBM 370.

As described in this paper, different parameters present unique problems in simulation. All studies require similar hydrologic and hydraulic time series data which are relatively straight forward to collect. Water quality time series are more difficult to collect. Dissolved oxygen, conductivity, pH, and temperature are easily measured continuously and recorded in analog or digital form. This makes these parameters more amenable to continuous simulation since boundary conditions can be defined explicitly, and calibration and verification data can be compared directly with model predictions. In addition, confidence in results can be based on rigorous statistical tests. With other parameters which must be sampled discretely, extensive efforts must be made to collect quasi-continuous data as in the Rideau River Study. More extensive metals data than collected in the Humber River System are needed for adequate calibration and verification.

Simpler models make the task easier. The use of steady state models prior to continuous simulation models would take advantage of features of both kinds of models. This hierarchial approach was used in the Grand River study with good results.

Probabilistic models might be more suitable for use on bacterial parameters, since there is an apparent large component of random variation in observed results.

Models like the continous version of HYMO which incorporate rural and urban hydrology, water quality, pond performance and simple receiving water processes, may be more suitable for planning studies (17). The capability of personal computers to accommodate simpler models will reduce costs considerably.

Acknowledgements

The assistance of Zdenek Novak, Doug Andrews and Keith Willson of the Ontario Ministry of the Environment, is gratefully acknowledged in preparation of the section on the Humber River Watershed.

References

1. Streeter, H.W., and Phelps, E.B. A study of the pollution and natural purification of the Ohio River. Public Health Bulletin 146, U.S. Public Health Service. 1925.

2. O'Connor, D.J., and Di Toro, D.M., Analysis of dissolved oxygen variation in a flowing stream. in Advances in water quality improvement. University of Texas Press. 1968.

3. Weatherbe, D.G. Dissolved oxygen modelling of the Speed River, Ontario. M.A.Sc thesis submission. University of Waterloo, Waterloo, Ontario. 1977.

4. Weatherbe, D.G. A dynamic water quality simulation model for the Thames River. in Proceedings of conference on environmental modelling and simulation. Edited by Ott, Wayne R. EPA 600/9-76-016. July 1976.

5. Grand River basin water management study. Grand River implementation committee. Province of Ontario. Toronto, Ontario. 1982.

6. Storage, Treatment, Overflow, Runoff Model. U.S. Army Corps of Engineers. Hydrologic Engineering Centre. Davis, California. 1977.

7. Singer, S.N. and So, S.K. Urban non-point source pollution and control. Grand River basin water management study technical report number 28. Ontario Ministry of the Environment, Toronto, Ontario 1982.

8. Willson, K. et al. Water quality simulation models and modelling strategy. Grand River basin water management study technical report number 30. Ontario Ministry of the Environment. 1981.

9. Rideau River Stormwater Management Study. Ontario Ministry of the Environment. Toronto, Ontario. 1983.

10. Roesner, L.A., Monser, J.R., and Evenson, D.E., Computer program documentation for the stream water quality model, QUAL-II. Water Resources Engineers, Inc. Waltnut Creek, California, May 1973.

11. Rideau River stormwater management study. Technical report, phase II, stage 1. Gore and Storrie Ltd/Proctor and Redfern Ltd. February, 1982.

12. Weatherbe, D.G., and Novak, Z. Development of water management strategy for the Humber River. in Proceedings of conference on stormwater and water quality modelling. Edited by E.M. James and W. James, McMaster University, Hamilton, Ontario. September 1984.

13. Johanson, R.C., et al. Hydrological simulation program - fortran, users manual for release 8.0. EPA-600/3-84-0064. June, 1984.

14. Huber, W.C., et al. Storm water management model, users manual version II. EPA-670/2-75-017, March, 1975.

15. Pitt, R., and McLean, J. Humber River pilot watershed project. Gartner Lee Associates Ltd, for Ontario Ministry of Environment. Toronto, Ontario, 1985.

16. Interim report on Humber River and Toronto area water quality. Toronto area watershed management strategy study, technical report number 2. Acres Consulting Services Ltd. for Ontario Ministry of the Environment, Toronto, Ontario. December, 1923.

17. Rowney, A.C., and Wisner, P.E. An approach to analysis of stormwater detention pond systems for water quality control. University of Ottawa. Ottawa, Ontario. 1984.

DISCUSSION - RECEIVING WATER MODELING

In the discussion which followed the presentation of papers, several points were raised. These included:

1. Dr. Ferrera asked Dr. Roesner how he had arrived at the "R" rates described in his paper. Dr. Roesner replied that they were based on experience.

2. Dr. Geiger commented that when calibrating runoff models, it was possible to obtain plus or minus 10 percent results with water quantity, but much worse (plus or minus 50 percent or more) with water quality. Dr. Schilling notes that the 10% calibration accuracy mentioned by Dr. Geiger is probably an estimate for a well-surveyed research catchment with a well-maintained monitoring system such as the Munich-Harlaching catchment. Applications of models other than in research are often done without runoff data by using default parameters. In these cases, results are much less accurate.

3. Dr. Geiger also asked why, in Table 6 of Dr. Medina's paper which compared continuous simulations with data, the coefficient of skewness was very different even though other measures such as mean, variance or standard deviation showed good agreement. Dr. Medina replied that he was not sure, but that it was probably due to the fact that much of the data record was filled in with simulated data.

URBAN STORMWATER RUNOFF QUALITY MANAGEMENT: LOW-STRUCTURALLY INTENSIVE MEASURES AND TREATMENT

Richard Field
U. S. Environmental Protection Agency
Storm and Combined Sewer Program
Edison, NJ 08837
U.S.A.

Introduction

The Storm and Combined Sewer Program (SCSP) of the U.S. Environmental Protection Agency (EPA) has been involved tor 20 years in the development of a diverse technology including combined sewer overflow (CSO) and stormwater control technology, instrumentation, problem assessments, best management practices (BMP) development and evaluation, stormwater management models, sludge hansling and disposal methods, sewerline infiltration/ inflow (I/I) control, erosion control, and many others. The more important outputs of the Program will be highlighted, with emphasis on some recent developments and on separate runoff quality control.

Receiving Water Impact

Knowledge of the receiving water impacts resulting from urban wet-weather discharges is a basis for determining the severity of problems and for justifying control. Problem studies of receiving water impacts are described in a proceedings from a national conference (1) and in a journal paper (2).

Under certain conditions, storm runoff can govern the quality of receiving waters regardless of the level of dry-weather flow

NATO ASI Series, Vol. G10
Urban Runoff Pollution.
Edited by H. C. Torno, J. Marsalek, and M. Desbordes

(DWF) treatment provided. Based on national annual mass balance determinations, urban wet-weather oxygen demand loads are greater than the dry-weather loads from the same areas (3,4). Hence, control of runoff pollution can be a viable alternative for maintaining receiving water quality standards. However, documented case studies of urban runoff impacts are scarce.

The SCSP has had only partial success in finding urban storm flow generated receiving water impacts employing the conventional dissolved oxygen (DO) concentration criteria (2,5). The problem appears to be in the application of conventional dry-weather monitoring techniques to the unsteady-state flow regimes caused by storms. Further studies should be undertaken to resolve the anomalies between actual runoff loadings and observed receiving water impacts. The SCSP has been more successful in sediment analysis than in water column analysis for DO depletions. Direct evidence has been obtained (from the Milwaukee River Project) (6) of how a disturbed benthos depletes DO from the overlying waters. Studies have also shown that storm and CSO adversely affect sediment by toxics enrichment and resultant biological upsets (2, 7-9). Since particulate matter in untreated storm and CSO is larger, heavier, and in significant quantities when compared to treated sanitary effluent, more needs to be known about the fate and transport of settleable and separable materials. Hydrodynamic solids separation and sediment transport routines must be added to receiving water models to take care of the neglected or presently omitted significant particulate and bed flow fields.

A simplified continuous receiving water quality model has been developed for preliminary planning and screening of area-wide wastewater treatment alternatives in terms of frequency of water quality violations (10,11). Also, a general methodology has been developed for evaluating the impact of urban stormwater discharges on receiving waters and for determining the abatement costs for achieving various water quality goals (12). In addition, a methodology for defining criteria for wetweather quality standards has been developed (13-15). In recognition of an important gap in the developed methodologies, the duration of water quality standards vs. species survival was taken into consideration.

User References

Major emphasis has been placed on the development of state-of-the-art (SOTA) reports, manuals-of-practice (MOP's), and user's manuals.

The SOTA texts, user's guide and the assessment on urban stormwater technology are recommended documents (16,17). Separate engineering manuals are available for storm flowrate determination (18,19), porous pavement design (20-23), cost estimating (4,24,25), storm sewer design emphasizing quality control (26, 27), planning and design and guidance (28), and for conducting stormwater studies (29). Seminar proceedings with themes of "modeling, design, operation, and costs" have been published.

The SOTA document on particle size and settling velocity (30) offers significant information for solids treatability and their settlement in receiving waters, important areas overlooked in planning and design. An excellent film is being distributed by the General Services Administration's (GSA) National Audio-Visual Center which covers the EPA CSO Research, Development, and Demonstration Program, and in particular full-scale control technologies (31).

A report entitled, "Urban Stormwater Management and Technology: Case Histories" (32) presents 12 case histories which represent the most promising approaches to CSO and stormwater control. The case histories were developed by evaluating operational facilities that have significant information for future guidance.

Three illustrative methodologies for conducting stormwater facility planning have been published (12,33,34).

Management Alternatives

The first consideration in selecting management alternatives is the choice of where to attack the problem; at the source by land management, in the collection system, or offline by storage. Pollutants can be removed by treatment and by employing integrated systems combining control and treatment.

Land Management

Land management includes structural, semi-structural, and nonstructural measures for reducing urban and construction-site stormwater runoff and pollutants before they enter the downstream drainage system. Various concepts have been fostered, including:

- o Land use planning.
- o Natural drainage which will reduce drainage costs and pollution, and enhance aesthetics, ground water supplies, and flood protection,
- o Retention and drainage facilities, and other management techniques required for flood and erosion control which can be simultaneously designed for pollution control.

Retention on-site or upstream can provide for the multiple benefits of aesthetics, recreation, recharge, irrigation, or other uses.

Porous Pavement. Porous pavements provide storage, enhancing soil infiltration that can be used to reduce runoff and CSO. Porous asphalt-concrete pavements can be underlain by a gravel base course with whatever storage capacity is desired (Figure 1). Results from a study in Rochester, New York, indicated that peak runoff rates were reduced as much as 83 percent where porous pavement was used (35). The structural integrity of the porous pavement was not impaired by heavy load vehicles. Clogging did result from sediment from adjacent land areas during

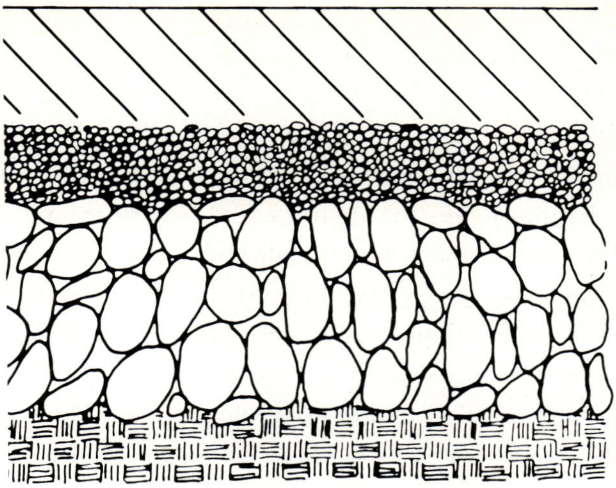

Figure 1. Porous Asphalt Paving Typical Section.

construction. It was relieved, however, by cleaning (flushing). The construction cost of a porous pavement parking lot is about equal to that of a conventional lot with stormwater inlets and subsurface piping.

A project in Austin, Texas, developed design criteria for porous pavements (21) and compared porous asphalt pavement to six other conventional and experimental pavements (22,23).

Surface Sanitation. Maintaining and cleaning urban areas can have a significant impact on the quantity of pollutants washed off by stormwater. Tests under real-world conditions in San Jose, California, showed that street cleaning can remove up to 50 percent of the total solids and heavy metals in urban stormwater with once or twice a day cleaning (36). Typical street cleaning programs of once or twice a month proved ineffective. Organics and nutrients could not be effectively controlled even with intensive cleaning.

In Bellevue, Washington, street cleaning proved ineffective; however, a modified regenerative-air Tymco street cleaner showed promise (37). Street cleaning is no panacea for stormwater pollution control (and is site-specific dependent upon climatic conditions), but if integrated with other methods, could reduce city-wide costs.

Chemical Use Control. Until assessment in 1971 (38), there had been limited research on highway de-icing effects. A search to define alternatives (39) confirmed the need for an economic impact analysis and for identifying a hydrophobic substance to reduce ice adhesion. The economic analysis has been conducted (40) and hydrophobic substances have been identified and investigated (41,42). Even though their costs appear greater than salt, when considering an estimated $3 billion annual damage to the environment, the costs are acceptable.

The 1973 assessment (43) identified the problems of sloppy salt storage practices and overapplication, resulting in manuals of practice (44,45) for improvement in these areas. These manuals which were recognized as highly significant, were reprinted by the Federal Highway Administration, and over 8,000 copies have been distributed. This work has prompted several states to enact legislation controlling salt application and storage.

Collection System Controls

Collection system controls refer to management alternatives for stormwater interception and transport. These include: improved maintenance and design of catchbasins, sewers, in-sewer (in-pipe) and in-channel storage, elimination of sanitary and industrial cross-connections, and remote flow monitoring and control.

Catchbasins. In a project conducted in Boston, Massachusetts, catchbasins were shown to be potentially quite effective for solids reduction (60-97 percent) (46-48). Removals of associated pollutants, such as chemical oxygen demand (COD) and BOD, were also significant (10-56 percent and 54-88 percent, respectively). To maintain the effectiveness of catchbasins for pollutant removal requires cleaning probably twice a year, depending upon conditions.

Regulators and Hydrobrakes. At present, there is a strong need to develop and have a reserve of control hardware for urban runoff control and to effectively reduce the associated high cost implications for conventional storage tanks, etc. The dual-functioning swirl flow regulator/solids concentrator has shown outstanding potential for simultaneous quality and quantity control (49). A helical type regulator/separator has also been developed based on principles similar to the swirl. It is felt that the swirl/helical type regulators, previously applied only to CSO, can also be installed on separate storm drains before discharge and the resultant concentrate flow can be stored in relatively small tanks, since concentrate flow is only a few percent of the total flow. Stored concentrate can later be directed to the sanitary sewer for subsequent treatment during low-flow or dry-weather periods, or if capacity is available in the sanitary interceptor/treatment system, the concentrate may be diverted to it without storage (50).

These methods of stormwater control may be more economical than building huge holding reservoirs for untreated runoff, and offer a feasible approach to the treatment of separately sewered urban stormwater.

A completed demonstration project in Boston, Massachusetts, evaluated the swirl and helical bend for separate stormwater treatment. A complete swirl/helical design textbook has been published (51).

Hydrobrakes, which have no moving parts, require no external energy source to operate and are not susceptible to blockages, have been demonstrated in Rochester, New York, (35), and Cleveland, Ohio (52 & 53). They can be used as upstream offline attenuators, and inline flow back-up devices. Conceptually, they allow flow release without orifice movement or closure at a constant predetermined rate regardless of upstream head.

New Concepts for Stormwater Control. Research efforts have shown that sanitary and industrial contamination of separate storm sewers is a nationwide problem. One response to this problem includes simple methods of checking for cross-connections. If found, they can be eliminated or the storm sewer can be dealt with as though it were a combined sewer, whichever is more feasible. Other new ideas include using storm sewers and storm drainage systems for in-channel storage applying CSO in-sewer storage and routing technology (54-56) and other storage facilities for ties into the existing sewage treatment system, thus making better use of facilities for overall water pollution control.

Storage

Because of high volume and variability associated with stormwater, storage is considered a necessary control alternative, and is the best-documented abatement measure. Project results and theory indicate storage must be considered at all times in system planning, because it allows for maximum use of existing dry-weather and other treatment plant facilities and results in the lowest cost in terms of pollutant removal.

Storage facilities may have auxiliary functions, such as sedimentation treatment, flood protection, sewer relief, flow transmission, and dry-weather flow equalization.

Storage concepts investigated include the conventional concrete holding tanks and earthen basins, and the minimum land requirement concepts of tunnels, underground and underwater containers, underground "silos", natural and mined under- and above-ground formations, and the use of abandoned facilities and existing sewer lines (57 & 58).

The in-receiving water flow balance method (Figure 2), is a recently developed storage alternative (59). In-receiving water storage facilities contain stormwater between plastic curtains supported by floating wooden pontoons and anchored to the receiving water bottom by concrete weights. After cessation of the overflow, pumps start automatically and the surrounding waterbody will enter the compartments and push the stormwater back toward the first compartment where it is pumped to the plant. Thus, the waterbody is used as a flow balance medium.

The storage method is low-cost due to the employment of low-cost materials (plastic and wood), the time required to install the unit (several days to months vs. months to a year for more structurally intensive alternatives), and the absence of land requirements. Studies show that costs could be about 5 to i5 percent of conventional concrete tank costs.

The facility, which was tested at three locations in Sweden, performed very satisfactorily, and was able to take ice and wind loads without adverse impact. A test in a harsh urban estuarine or marine site, such as the Fresh Creek Basin in New York City, New York, will be done soon.

A storage/sedimentation planning/design manual (60) is being finalized which includes the following:

- o Major Issues
  - applicability,
  - integration into existing treatment system,
  - costs, and
  - effectiveness as treatment/control device.

- o Secondary issues
  - siting impacts,
  - O&M requirements, and
  - sludge production and handling.

Treatment

Due to the adverse and intense flow conditions and unpredictable shock loading effects, it has been difficult to adapt existing treatment methods to storm-generated overflows, especially the microorganism-dependent biological processes. Physical/chemical treatment techniques have shown more promise than biological processes in overcoming storm shock loading effects. To reduce capital investments, projects have been directed toward high-rate operations approaching maximum loading.

Storm-flow treatment methods which have been demonstrated include physical-chemical, wetlands, biological, and disinfection (28). These processes, or combinations of these processes, can be adjuncts to the existing sanitary plant or serve as remote satellite facilities at the outfall.

Physical/Chemical Treatment. Physical processes/systems with or without chemicals, such as fine mesh screening, swirl degritting, fine mesh screening/high-rate infiltration (HRF), sedimentation, and fine-mesh screening/dissolved air flotation (DAF), have been successfully demonstrated. Physical processes have been shown to be important for storm-flow treatment because they are adaptable to automated operation, rapid startup and shutdown, high-rate operation, and resistance to shock loads. The high-rate processes/system (DAF, fine mesh screening, and HRF) are believed to be ready for municipal installation.

The swirl has also been developed for grit removal. The small size, high efficiency and absence of moving parts offer economical and operational advantages over conventional degritting facilities. A full-scale demonstration of a (16 ft diameter/ 11 mgd design flowrate) swirl degritter has been completed in Tamworth, Australia (61). Removal efficiencies confirmed

Figure 2. In-Receiving Water Flow-Balance Method

laboratory results. Compared with a conventional grit chamber, construction costs are halved, and operation and maintenance costs are considerably lower.

Disinfection. Because disinfectant and contact demands are great for storm flows (62), research has centered on high-rate applications by static and mechanical mixing, higher disinfectant concentrations (63-67), and more rapid oxidants, i.e., chlorine dioxide (64-67), ozone (63), and ultraviolet (UV) light; and onsite generation (63,68,69). Demonstrations in Rochester, New York (66); Syracuse, New York (67); East Chicago, Indiana (70); and Philadelphia, Pennsylvania (64), indicate that adequate reductions of fecal coliform can be obtained with contact times of two minutes or less by induced mixing and dosing with chlorine and/or chlorine dioxide. A pilot scale

UV demonstration with a contact time of less than ten seconds was conducted in New York City, New York.

The hypochlorite batching facility is still being used in New Orleans to protect swimming beaches in Lake Pontchartrain (69), and a brine hypochlorite generator is now being used in industry (68).

Treatment/Control Design Guidebook. A compilation of the SCSP's best research efforts in CSO treatment/control over the past 18 years has been published (28). Because of flow similarities it is felt that this can also be used as an important reference for urban stormwater treatment.

Treatment Process Performance. Treatment process costs and performance in terms of design influx rate (gpm/ft^2), BOD5 and SS removal efficiency is provided in Table 2. The high-rate performance of the swirl, and the microstrainer, screening/HRF and screening/DAF systems, is apparent when compared to sedimentation.

Sludge Solids

Sludge handling and disposal must be considered an integral part of treatment because they significantly affect the efficiency and cost of the total waste treatment system. A study quantified solids residue and sludge that would result from the treatment of separate urban stormwater on a nationwide basis and the associated treatability of these sludges (71,72).

Integrated Systems

The most promising and common approach to urban storm flow management involves the integration of control and treatment. Integrated systems are divided into storage/treatment, dual-use wet-weather flow/dry-weather flow facilities, and control/treatment/reuse.

TABLE 2. Wet-Weather Treatment Plant Performance Data

Device	Control Alternatives	Design Loading Rate (gpm/ft^2)	Removal Efficiency (%)	
			BOD_5	SS
Primary	Swirl Concentrator	60	25-60	50
	Microstrainer	20	40-60	70
	High-Rate Infiltration	24	60-80	90
	Dissolved Air Flotation	2.5	50-60	80
	Sedimentation	0.5	25-40	55
	Representative Performance		40	60
Secondary	Contact Stabilization		75-88	90
	Physical-Chemical		85-95	95
	Representative Performance		85	95

Storage/Treatment

When there is storage, there is treatment by settling, pump-back/bleed-back to municipal works, and sometimes disinfection. In any case, the break-even economics of supplying storage must be evaluated when treatment is considered. All of these storage/treatment concepts have been demonstrated at full scale.

Dual-Use Wet-Weather Flow/Dry-Weather Flow Facilities

The concept of dual-use involves maximum utilization of wet-weather facilities during non-storm periods and maximum utilization of dry-weather facilities during storm flows. Full-scale dual-use or high-rate trickling filters (73), contact stabilization (74-76), HRF (77), and equalization basins (78) have been demonstrated. Various municipalities are employing dual-use microscreening.

Control/Treatment/Reuse

Control/Treatment/Reuse is a "catch-all" for all integrated systems. A prime consideration should be various nonstructural and land-management techniques. In Mt. Clemens, Michigan, a series of three "lakelets" have been incorporated into a CSO treatment/park development (79). Treatment is being provided so that these lakes are aesthetically pleasing and allow for recreation and reuse for irrigation. A study of the reuse of urban stormwater for subpotable industrial and irrigational water supply purposes indicated that the stormwater would be less expensive than municipal potable water (80).

Recommendations for The Future

Receiving Water Impacts

Ties between receiving water quality and storm flow discharges must be clearly established and delineated. Quantification of the impairment of beneficial uses and water quality by such discharges is a major goal. Project results indicate the potential for significant impact of wet-weather flows on receiving waters. Control of stormwater runoff pollution can be a viable alternative for maintaining receiving water quality standards. However, the problems found seem to be site-specific in nature. Therefore, site-specific surveys are required which must consider the effects of larger materials and floatables near the outfalls. Based on results from these surveys, control may be warranted.

Toxics Characterization and Control/Treatment

Results from a limited in-house effort and the Nationwide Urban Runotf Program (NURP) indicate that urban stormwater runoff and CSO contain signiticant quantities of priority (toxic/hazardous) pollutants. Additional investigation of the significance of the concentrations and quantities of toxic pollutants with regard to their health effects or potential health effects and ecosystem effects is required. A need exists to evaluate the

the removal capability of alternative technologies for these toxics and to compare their effectiveness with estimated removal needs to meet water quality goals.

Sewer System Cross-Connections

Investigations have shown that sanitary and industrial contamination of separate storm sewers is an extensive nationwide problem. In practice, a significant number of separate stormmwater drainage systems actually function as combined sewer systems. Therefore, a nationwide effort on both the federal and local levels to alleviate the pollution impacts from discharges of these systems is required. It is better to classify such cross-connected drainage systems as combined systems for pollution control priorities.

Integrated Stormwater Management

The most effective solutions for wet-weather pollution problems must consider: (1) control of wet-weather pollution impacts in lieu of continuallY and blindly upgrading existing municipal plants; (2) structural vs. nonstructural techniques; (3) integrating dry- and wet-weather flow systems to make maximum use of existing drainage during wet conditions and maximum use of wet-weather control/treatment facilities during dry weather; and (4) the segment or bend on the percent pollutant control vs. cost curve, in which incremental costs exceed incremental control efficiencies, subject to load discharge or receiving water requirements.

Flood and erosion control technology must be integrated with pollution control, so that the retention and drainage facilities required for flood and erosion control can be simultaneously designed or retrofitted for pollution control. If land management and nonstructural techniques are maximized and integrated, costs for the extraction of pollutants from storm flows in the potentially more costly downstream plants will be reduced.

Institutional and Socio/Economic Conflicts

Some of the most promising opportunities for cost-effective environmental control are multipurpose in nature. However, there are institutional problems that hinder their implementation. First, the autonomous federal and local agencies and professions involved in flood and erosion control, pollution control, and land management and environmental planning must be integrated at both the planning and operational levels. Multiagency grant coverage must be adequate to stimulate such an approach. For example, the EPA would have to join with the Corps of Engineers, Soil Conservation Service, Department of Transportation, and perhaps other federal agencies, as well as departments of pollution control, sanitation, planning, and flood control at the local level.

Another problem is that construction grant (and other) incentives are geared towards structurally intensive projects which may counter research findings in the area of optimal solutions. Optimized wet-weather pollution abatement involves a city-wide approach including the integration of structural as well as low-structural controls. The low-structural measures are more labor-intensive. Construction grant funding does not presently address this expense and, accordingly, municipalities are discouraged from using them.

Conclusions

In general, on a mass basis, toxic compounds, bacteria, oxygen-demanding, and suspended and visual matter in urban stormwater are significant. Ignoring the problem because it appears to be too costly or too difficult to solve will not make the problem go away. The integrated approach to wetweather pollution control is the only way which is going to be feasible, economical and, therefore, acceptable. Potentially tremendous savings can be derived from wet-weather pollution control research fostering integrated solutions. Better advantage needs to be taken of proven technology.

References

1. Urban Stormwater and Combined Sewer Overflow Impact on Receiving Water Bodies. Proceedings of National Conference. Report No. EPA-600/9-80-056. United States Environmental Protection Agency. Florida. 1979.

2. Field, R., and Turkeltaub, R. Urban Runoff Receiving Water Impacts: Program Overview. ASCE 107(EE1):83-100 Journal of the Environmental Engineering Division. 1981.

3. Sullivan, R.H., et al. Nationwide Evaluation of Combined Sewer Overflow and Urban Stormwater Discharges. Volume I., Executive Summary. Report No. EPA-600/2-77-064a. United States Environmental Protection Agency. 1982.

4. Heaney, J.F., et al. Nationwide Evaluation of Combined Sewer Overflows and Urban Stormwater Discharges, Volume II, Cost Assessment and Impacts. Report No. EPA-600/2-77-064b. United States Environmental Protection Agency. 1977.

5. Keefer, N., et al. Dissolved Oxygen Impact from Urban Storm Runoff. Report No. EPA-600/2-79-156. United States Environmental Protection Agency. 1979.

6. Meinholz, M.A., et al. Verification of the Water Quality Impacts of Combined Sewer Overflows. Report No. EPA-600/2-79-155. United States Environmental Protection Agency. 1979.

7. Pitt, R.E. Sources of Urban Runoff Pollution and its Effects on an Urban Creek. Report No. EPA-600/2-82-090. United States Environmental Protection Agency. 1982.

8. Pitt, R.E. Sources of Urban Runoff Pollution and its Effects on an Urban Creek. Report No. LPA-600/S2-82-090. United States Environmental Protection Agency. 1982.

9. Tomlinson, R.D., et al. Fate and Effects of Particulates Discharges by Combined Sewers and Storm Drains. Report No. EPA-600/2-76-145. Unitec States Environmental Protection Agency. 1980.

10. Medina, M.A. Level III: Receiving Water Quality Modeling for Urban Stormwater Management. Report No. EPA-600/2-79-100. United States Environmental Protection Agency. 1979.

11. Medina, M.A. River Quality Model for Urban Stormwater Impacts. Report No. EPA-J-81-234. United States Environmental Protection Agency. 1981.

12. Moffa, P.E., and Karanik, J. Methodology for Evaluating the Impact and Abatement of Combined Sewer Overflows: A Case Study of Onondaga Lake New York. Report No. EPA-600/8-80-048. United States Environmental Protection Agency. 1980.

13. Mancini, J.L. Development of Methods to Define Water Quality Effects of Urban Runoff, Report No. EPA-600/S2-83-125. United States Environmental Protection Agency. 1983.

14. Mancini, J.L. Development of Methods to Define Water Quality Effects of Urban Runoff. Report No. EPA-600/2-83-125. United States Environmental Protection Agency. 1983.

15. Mancini, J.L. A Method for Calculating Effects on Aquatic Organisms of Time-Varying Concentrations. Water Research 17(10):1355-1362. 1983.

16. Lager, J.A., et al. Urban Stormwater Management and Technology: Update and User's Guide. Report No. EPA-600/8-77-014. United States Environmental Protection Agency. 1977.

17. Lager, J.A., and Smith, W.G. Urban Stormwater Management and Technology: An Assessment. Report No. EPA-670/2-74-070. United States Environmental Protection Agency. 1974.

18. Brater, E.F., anc Sherrill, J.D. Rainfall-Runoff Relations on Urban and Rural Areas, Report No. EPA-670/2-75-046. United States Environmental Protection Agency. 1975.

19. Yen, B.C., and Chow, V.T. Urban Stormwater and Runoff Determination of Volumes and Flowrates. Report No. EPA-600/2-76-116. United States Environmental Protection Agency. 1976.

20. Field, R., et al. Porous Pavement Research: Development and Demonstration. Transportation Engineering Journal of ASCE 108:244-258. 1982.

21. Diniz, E. Porous Pavement: Phase I. Design and Operational Criteria. Report No. EPA-600/2-80-135. United States Environmental Protection Agency. 1980.

22. Goforth, G.F. Stormwater Hydrological Characteristics of Porous Conventional Paving System. Report No. EPA-600/2-83-106. United States Environmental Protection Agency. 1983.

23. Goforth, G.F. Stormwater Hycrological Characteristics of Porous Conventional Paving System. Project Summary No. EPA-600/S2-83-106. United States Environmental Protection Agency. 1983.

24. Benjes, H.H. Cost Estimating Manual-Combined Sewer Overflow Storage Treatment", Report No. EPA-600/2-76-286. United States Environmental Protection Agency. 1976.

25. Benjes, H.H., and Field, R. Estimating Sewer Overflow Facility Costs. Water and Engineering, Sept. 1978. pp.56-62.

26. Sonnen, M. Abatement of Deposition and Scour in Sewers, Report No. EPA-600/2-77-212. United States Environmental Protection Agency. 1977.

27. Kaufman, H.L., and Lai, F.H. Conventional and Advanced Sewer Design Concepts for Dual Purpose Flood and Pollution Control: A Preliminary Case Study. Elizabeth, New Jersey. Report No. EPA-600/2-78-090. United States Environmental Protection Agency. 1978.

28. Field, R., and Weisman, D. A Planning and Design Guidebook for Combined Sewer Overflow Control. Report No. EPA-600/2-82-084. United States Environmental Protection Agency. 1982.

29. Wuelschleger, R.E., et al. Methodology for the Study of Urban Storm Generated Pollution and Control. Report No. EPA-600/2-76-145. United States Environmental Protection Agency. 1976.

30. Dalrymple, R.J., et al. Physical and Settling Characteristics of Particulates in Storm and Sanitary Wastewater. Report No. EPA-670/2-75-011. United States Environmental Protection Agency. 1975.

31. Field, R. Stormwater Pollution Control: A New Technology. 28-Minute Film. Distributed by General Services Administration (GSA). National Audio Visual Center. Washington, D.C. 20409.

32. Lynard, W., et al. Urban Stormwater Management and Technology: Case Histories. Report No. EPA-600/8-80-035. United States Environmental Protection Agency. 1980.

33. Heaney, J.F., et al. Stormwater Management Model: Level I. Preliminary Screening Procedures. Report No. EPA-600/2-76-275. United States Environmental Protection Agency. 1976.

34. Heaney, J.F., et al. Stormwater Management Model: Level I. Comparative Evaluation of Storage/Ireatment and Other Management Practices. Report No. EPA-600/2-77-083. United States Environmental Protection Agency. 1977.

35. Murphy, C.B., et al. Best Management Practices: Implementation. Report No. EPA-905/9-81-002. United States Environmental Protection Agency. 1981.

36. Pitt, R.E. Best Management Practices: Implementation. Report No. EPA-905/9-81-002. United States Environmental Protection Agency. 1979.

37. Pitt, R.E. Characterization and Controlling Urban Runoff Through Street and Sewerage Cleaning EPA-600/2-85/038 (NTIS No. PB 186500). March 1985.

38. Struzeski, E. Environmental Impact of Highway Deicing. Report No. 11040GKK06/71. United States Environmental Protection Agency. 1971.

39. Murray, D.M., and Figerman, M.R. A Search: New Technology for Pavement Snow and Ice Control. Report No. EPA-R2-72-125. United States Environmental Protection Agency. 1972.

40. Murray, D.M., et al. An Economic Analysis of the Environmental Impact of Highway Deicing. Report No. EPA-600/2-76-105. United States Environmental Protection Agency. 1976.

41. Ahlborn, G.H., ano Poehlmann, H.C. Jr. Development of a Hydrophobic Substance to Mitigate Pavement Ice Adhesion. Report No. EPA-600/2-76-242. United States Environmental Protection Agency. 1976.

42. Krukar, M., and Cook, J.C. Optimization and Testing of Highway Materials to Mitigate Ice Adhesion. Report No. EPA-600/2-78-035. United States Environmental Protection Agency. 1978.

43. Field, R., et al. Water Pollution and Associated Effects from Street Salting. Report No. EPA-R2-73-257. United States Environmental Protection Agency. 1973.

44. Richardson, D.L., et al. Manual for Deicing Chemicals: Storage and Handling. Report No. EPA-670/2-74-033. United States Environmental Protection Agency. 1974.

45. Richardson, D.L., et al. Manual for Deicing Chemicals: Application Practices. Report No. EPA-670/2-74-045. United States Environmental Protection Agency. 1974.

46. Aronson, Gerald L., et al. Evaluation of Catchbasin Performance for Urban Stormwater Pollution Control. Report No. EPA-600/2-83-043. United States Environmental Protection Agency. 1983.

47. Aronson, Gerald L., et al. Evaluation of Catchbasin Performance for Urban Stormwater Pollution Control. Report No. EPA-600/S2-83-043. United States Environmental Protection Agency. 1983.

48. Field, R. An Overview of the U.S. Environmental Protection Agency's Storm anc Combineo Sewer Collection System Research. Water Resources 16:859-870; June 1982.

49. Field, R., et al. Treatability Determinations for a Prototype Swirl Combined Sewer Overflow Regulator/Solids-Separator. Progress Water Technology 8:81-91; 1977.

50. Field, R., and Masters, H.E. Swirl Device for Regulating and Treating Combined Sewer Overflow, Report No. EPA-625/2-77-012. United States Environmental Protection Agency. 1983.

51. Sullivan, R.H., et al. Swirl and Helical Bend Pollution Control Devices: Design Manual. Report No. EPA-600/8-82-013. United States Environmental Protection Agency. 1982.

52. Matthews, T.M., et al. Hydrobrake-Regulated Storage Systems for Stormwater Management. Report No. EPA-600/2-83-097. United States Environmental Protection Agency. 1983.

53. Matthews, T.M., et al. Hydrobrake-Regulated Storage Systems for Stormwater Management. Report No. EPA-600/S2-83-097. United States Environmental Protection Agency. 1983.

54. Watt, T.R., et al. Sewerage System Monitoring and Remote Control. Report No. EPA-670/275-020. United States Environmental Protection Agency. 1975.

55. Dispatching Systems for Control of Combined Sewer Control. Metropolitan Sewer Board. St. Paul, Minnesota. Report No. 11020FAQ03/71. United States Environmental Protection Agency. 1971.

56. Leiser, C.P. Computer Management of a Combined Sewer System. Report No. EPA-670/2-74-022. United States Environmental Protection Agency. 1974.

57. Field, R., and Struzeski, E.J. Management and Control of Combined Sewer Overflows. Journal Water Pollution Control Federation 44(7):1393-1415. 1972.

58. Field, R. Urban Runoff Pollution Control: State-of-the-Art. Journal of the Environmental Engineering Division, ASCE 101:107-125. 1975.

59. Soderland, H., et al. Flow Balancing Method for Storm-Water and Combined Sewer Overflow. ISBN 91-540-3765-4: D17:1-27. 1982.

60. Smith, W.G., et al. Storage/Sedimentation Facilities for Control of Storm and Combined Sewer Overflow Design Manual. Report No. Pending (USEPA Contract No. 68-03-2877).

61. Shelley, G.J., et al. Field Evaluation of a Swirl De-gritter at Tanworth, N.S.W., Australia. Report No. EPA-600/2-81-063. United States Environmental Protection Agency. 1981.

62. Field, R., et al. Proceedings of a Workshop on Microorganisms in Urban Stormwater. Report No. EPA-600/2-76-244. United States Environmental Protection Agency. 1976.

63. Glover, G.E., and Herbert, G.R. Microstraining and Disinfection of Combined Sewer Overflows-Phase II. Report No. EPA-R2-73-124. United States Environmental Protect:on Agency. 1973.

64. Maher, M.B. Microstaining and Disinfection of Combined Sewer Overflows-Phase III. Report No. EPA-670/2-74-049. United States Environmental Protection Agency. 1974.

65. Moffa, P.E., et al. Bench-Scale High-Rate Disinfection of Combined Sewer Overflows with Chlorine and Chlorine Dioxide. Report No. EPA-670/2-75-021. United States Environmental Protection Agency. 1975.

66. Drehwing, F.J., et al. Combined Sewer Overflow Abatement Program, Rochester, New York. Volume II. Pilot Plant Evaluations. Report No. EPA-600/2-79-031b. United States Environmental Protection Agency. 1979.

67. Drehwing, F.J., et al. Disinfection/Treatment of Combined Sewer Overflows, Syracuse, New York. Report No. EPA-600/2-79-134. United States Environmental Protection Agency, 1979.

68. Leitz, F.B., et al. Hypochlorite Generator for Treat-of Combined Sewer Overflows. Report No. 11023DAA03/72. United States Environmental Protection Agency. 1972.

69. Pontius, U.R., et al. Hypochlorination of Polluted Stormwater Pumpage at New Orleans. Report No. EPA-670/2-73-067. United States Environmental Protection Agency. 1973.

70. Connick, D.J., et al. Evaluation of a Treatment Lagoon for Combined Sewer Overflow. Report No. EPA-600/S2-81-196. United States Environmental Protection Agency. 1981.

71. Carr, D.J., et al. Characteristics and Treatability of Urban Runoff Residues. Report No. EPA-600/2-82-094. United States Environmental Protection Agency. 1982.

72. Carr, D.J., et al. Characteristics and Treatability of Urban Runoff Residues. USEPA Project Summary No. EPA-600/S2-82-094. United States Environmental Protection Agency. 1982.

73. Homack, P., et al. Utilization of Trickling Filters for Dual Treatment of Dry and Wet Weather Flows. Report No. 670/2-73-071. United States Environmental Protection Agency. 1973.

73. Benedict, A.H., and Roelfs, V.L. Dry-Wet Weather Treatment of Municipal Wastewater of Clatskanie, Oregon. Report No. EPA-600/2-81-061. United States Environmental Protection Agency. 1981.

74. Benedict, A.H., and Roelts, V.L. Joint Dry-Wet Weather Treatment of Municipal Wastewater of Clatskanie, Oregon. USEPA Project Summary No. EPA-600/S2-81-061. United States Environmental Protection Agency. 1981.

75. Agnew, R.W., et al. Biological Treatment of Combined Sewer Overflow at Kenosha, Wisconsin. Report No. EPA-670/2-75-019. United States Environmental Protection Agency. 1975.

76. Innerfield, H., and Forndran, A. Dual Process High-Rate Filtration of Raw Sanitary Sewage and Combined Sewer Over-Flows. Report No. EPA-600/2-79-015. United States Environmental Protection Agency. 1979.

77. Welborn, H.L. Surge Facility for Wet- and Dry-Weather Flow Control. Report No. EPA-670/2-74-075. United States Environmental Protection Agency. 1974.

78. Mahida, V.U., and DeDecker, B.J. Multi-Purpose Combined Sewer Overflow Treatment Facility, Mt. Clemens, Michigan. Report No. EPA-670/2-75-010. United States Environmental Protection Agency. 1975.

79. Field, R., and Fan, C.Y. Industrial Reuse of Urban Stormwater. Journal of the Environmental Engineering Division, ASCE 107(EE1):171-189; 1981.

STRUCTURAL MEASURES FOR RUNOFF QUALITY MANAGEMENT

Peter Stahre, First Engineer
Malmo Gatukontor, Box 2500
S-200 12 Malmo
Sweden

Introduction

As municipal wastewater treatment is upgraded, discharges of urban stormwater and combined sewer overflows are emerging as significant sources of surface water pollution. If the pollution aspect of urban runoff is considered in the planning of stormwater systems, quality management can often be incorporated in the flood control strategies. However, in many cases we must deal with pollution discharges from existing sewer systems. The possibilities of quality management in such cases will be heavily restricted, especially in densely populated areas where it sometimes can be very hard to find cost-effective solutions to the pollution problem.

Decisions on the necessary degree of quality management must always be based on impacts on the receiving waters from urban wet-weather discharges. Knowledge of the amount of discharge alone is not sufficient for decisions of which measures shall be taken.

Quality Management Alternatives

Quality management of storm-generated urban runoff can be accomplished in a number of different ways. Measures can be taken upstream of the collection system (source control), in the collection system (in-line and off-line control) or at the downstream end of the collection system (downstream control).

NATO ASI Series, Vol. G10
Urban Runoff Pollution.
Edited by H. C. Torno, J. Marsalek, and M. Desbordes

The control measures can basically be of three different kinds.

- o Structural measures
- o Semi-structural measures
- o Non-structural measures

As indicated by the name, structural measures include those measures where a new structure is built into the drainage system. The construction of storage basins and treatment facilities are examples.

When semi-structural measures are involved, it is in most cases a matter of installing some form of regulator device in the system which requires only a limited amount of construction work.

Non-structural measures do not involve any building activity at all. Instead they are preventive in nature, such as street cleaning to reduce the pollution load in the stormwater or sewer flushing to reduce the first flush effect.

Table 1 gives some examples of what structural, semi-structural and non-structural measures can be used at different locations in a sewer system. In this presentation only structural measures will be dealt with. Three main groups of structural measures can be distinguished:

- o Storage
- o Treatment
- o Integrated systems

In the following discussion, these main groups will be described in more detail.

Table 1 Examples of structural, semi-structural and non-structural measures for urban runoff quality management

	Structural measures	Semi-structural measures	Non-structural measures
Source control	o Porous pavement o Percolation facilities o Open ponds	o Inlet control devices o Local infiltration of stormwater	o Street cleaning o Land-use planning o Improved catchbasin maintenance
In-line/Off line control	o Sewer separation o Storage tanks o Open ponds	o Flow regulator devices o Movable gates	o Sewer flushing o System maintenance
Downstream control	o Storage tanks o Open ponds o Treatment facilities	o Movable gates o Remote control systems	o Improved outlet structure maintenance

Quality Control by Storage

General

Temporary storage is most frequently used for urban runoff quantity control. A storage facility can, however, also provide quality control. In this connection one can distinguish between direct and indirect quality control.

By retaining, buffering and attenuating stormwater runoff in a storage basin, reduction of both downstream flooding problems and combined sewer overflows can be achieved. The reduction of the discharge of pollution to the receiving waters is in this case the result of a changed flow pattern and not of any reduction of the amount of pollution in the water. This case is referred to here as indirect quality control.

A storage basin also provides some level of direct quality control. This is accomplished by detaining the flow long enough for physical or biochemical processes to occur.

In this presentation only direct quality control in storage facilities will be dealt with. Activities in this field started about ten years ago when the multi-purpose concept for detention basins was first introduced. Today a lot of research work is being carried out on mathematical modelling of the sedimentation process in storage facilities. Theoretical understanding of the sedimentation process is of greatest importance, as we are dealing with intermittent and unsteady flows as well as spatial and temporal variations of the pollution load in the water. As was the case when we first started to model the hydraulic function of detention basins, field data is lacking to verify the sedimentation models on a general basis. In the future we can foresee much more activity concerning collection of field data.

Design considerations

The major pollution reduction mechanism associated with storage is sedimentation. Other mechanisms incorporated with storage are infiltration/percolation and different types of biological activities.

If a storage basin is intended for direct quality control it is necessary to consider this in the planning and design of the facility. The most important factors in this connection are:

- o Technical configuration and detailed design
- o Pollution load and settling characteristics

As a consequence of the increasing interest in quality control by storage much more attention must be paid to the technical configuration in the future. The technical configuration of a storage basin is to a great extent dependent on local conditions. With respect to the location in the drainage system, storage facilities can be categorized as shown in Figure 1.

Where direct quality control is concerned, open ponds are the most common type of storage facility. The main reason for this is that the water normally is kept in the pond for quite a long time. Other examples of direct quality control storage facilities are rock tunnels and concrete basins.

In the planning of a storage basin for quality control special attention must be paid to the detailed design of the facility. Examples of design criteria that will accomplish this are:

- o Length/width ratio greater than 2:1.
- o Inlet and outlet structures at extreme ends of the storage basin.
- o Baffles or flow retarders at the inlet.
- o Sand and grit chamber at the inlet.

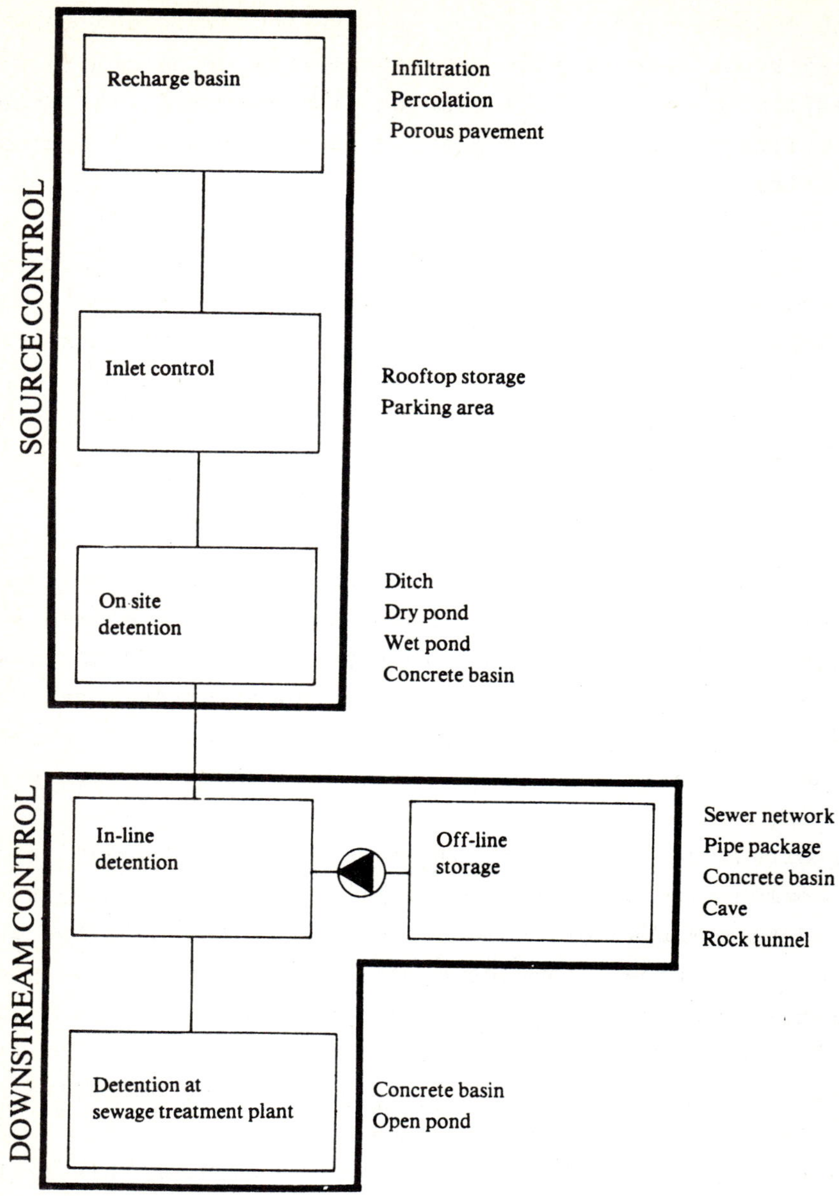

Figure 1 Scheme of how storage facilities can be catagorized according to their technical configuration and their location in the sewer system

The field investigations which have been carried out indicate that the settling characteristics can vary considerably from one site to another. The settling characteristics also change in time during a storm. To be able to simulate the sedimentation effect it is necessary to perform settling analysis for each location where quality control by storage is to be implemented.

Many pollutants of concern in quality control, such as phosphorus, heavy metals and petroleum-based organics, seem to have a high affinity for the suspended solids in the stormwater runoff.

Reported Removal Efficiencies

The pollution removal efficiencies in storage basins can vary very much. As a rule it reaches a maximum under steady state, quiescent conditions. Unsteady storm-generated flows generally produce velocity, and in some cases also temperature, gradients in the storage basin which influence the trap efficiency in a negative way.

The best removal efficiencies have been reported from large ponds with long detention times. In such cases 70-80 per cent reductions of suspended solids are not unusual.

The removal efficiencies in smaller storage basins are very sensitive to the technical configuration and the design of the facility. In some cases even negative removal efficiencies have been reported. The explanation for this is that sediments are resuspended from the bottom of the basin due to unfavourable currents.

Discussion

Storm sewer discharges tend to be quite heterogeneous, with a make-up of particulates changing from one moment to the next. Traditional sedimentation theory reflects idealized situations to which a large number of modifiers must be applied to give a true picture of the real conditions.

The inlet structure should be designed in such a way that the incoming water to the basin is spread over the whole cross section. In many cases it can be advantageous to arrange a sand and grit chamber connected to the inlet structure. By this means, the removal of deposits from the rest of the basin can be facilitated.

To control the outflow from the basin some form of outlet structure is needed. The outlet structure can be designed in a number of different ways, such as orifices, overflow weirs, special regulator devices, etc. The outlet structure is a very important part of the storage facility. By adjusting the outlet structure it is possible to completely change the flow pattern through the basin. It is therefore important that the outlet be designed in an appropriate way. Some recommendations in this respect have been presented by the ASCE Task Committee on Detention Basin Outlet Structures.

Besides the technical configuration and the design, the removal efficiency of a storage basin is dependent on the pollution load in the incoming water and the settling characteristics of the associated pollutants.

In general the removal efficiency increases with increased content of suspended solids in the water. It is important to consider this when comparing results from different storage facilities.

The settling characteristics of the pollutants in the water also play an important role in removal efficiency. Factors of special interest in this connection are:

- o Particle size distribution.
- o Particle density.
- o Pollutants associated with the particles.

A great deal of research work is now being carried out to build up mathematical models for simulating the trap efficiency in sedimentation basins. To be able to do this properly it is necessary to verify the models with extensive field investigations. A theoretical approach is of only limited value for practical applications.

Quality Control by Treatment

General

Quality control of storm-generated runoff can also be accomplished by treatment of the water in local treatment plants. Due to irregular flow conditions it is normally not possible to adapt traditional dry-weather treatment methods, especially when microbiological processes are involved. A treatment plant for quality control of urban runoff must fulfil the following requirements:

- Automatic operation.
- Intermittent operation.
- Rapid start-up and shut-down.
- High rate operation.
- Resistence to shock loading.

This is possible with physical and chemical treatment processes. Very few wet-weather treatment plants have been installed so far and most of the existing facilities are of pilot-plant character. Consequently, the practical experience with the technique is still very limited.

In the following shall be given two examples of how quality control by treatment can be accomplished for storm-generated urban runoff.

Fine Mesh Screening

Microscreens have traditionally been used in water treatment technology, and for polishing secondary sewage treatment plant effluents. In the U. S., full-scale microscreening units have been tested with good results for the treatment of combined sewer overflows. Suspended solids removals of about 50 per-cent have been reported.

By using microscreeners in conjunction with high-rate filters or dissolved air flotation, and by adding chemical coagulants, the removal efficiency can be further improved. In the U. S., full-scale screening/flotation facilities have been tested, in which suspended solids removals of 70 per cent and BOD removals of 55 per cent have been reported.

Disadvantages with fine mesh screening are that the mechanical equipment contains moving parts and that the process is sensitive to clogging of the screens if the treated water contains oil and grit.

Hydrodynamic Separators

The swirl solids concentrater is a simple and robust construction with no moving parts. The treatment of the water is accomplished by a "swirl" action in the facility which imparts liquid-solid separation. Tests in the U. S. indicate suspended solids and BOD removals of at least 50 per cent.

Full-scale tests with swirl concentrators show very promising results regarding removal efficiencies as well as good operation and maintenance experience. Before designing a swirl, it is important to check that the settling characteristics of the water that shall be treated are appropriate for this type of installation.

A helical type of solid separator has also been developed. This is based on principles similar to the swirl.

Discussion

Very few full-scale treatment plants for quality control of storm-generated urban runoff have been installed so far. Long-term experience with operation and maintenance have not been documented for any facility. One reason for this, of course, lies in the fact that the technique is rather new.

It would be of great interest to make an international survey of how many of the installed treatment facilities that have survived the first critical years of testing, are still in operation. An evaluation of the operation and maintenance of these facilities would give a good view of the feasibility of the new technique.

Integrated Systems for Quality Control

The integration of storage and treatment is here referred to as integrated systems. This is the most flexible approach to quality control of storm-generated urban runoff. The following two types of facilities can be distinguished:

1. Wet-weather treatment plant where the flows are equalized in a storage basin before entering the plant.

2. Dry-weather treatment plant, especially designed so that it can handle also a certain amount of wet-weather flow.

By equalizing the flow in a storage basin before entering the treatment facility, one is not limited to high rate processes only. Chemical treatment in small package plants is an example.

The use of dry-weather treatment plants for handling of wet-weather flows can be accomplished in a number of different ways, depending on the local conditions. The advantage to this solution is that the plant is continuously fed with sewage water.

Integrated systems are the most promising approach to urban wet-weather quality control. One reason for this is that they are normally based on well-documented treatment processes.

References

1. De Groot William (ed.) Stormwater Detention Facilities. Proceedings of an ASCE conference at New England College, New Hampshire, August 2-6, 1982.

2. Balmer, Peter, Malmqvist, Per-Arne, Sjoberg, Anders (ed.) Urban Storm Drainage. Proceedings of the Third International Conference on Urban Storm Drainage, Chalmers University, Sweden, June 4-8, 1984.

3. Stahre, Peter. Flow Equalization in Sewer Systems. Design Manual. Swedish Council for Building Research. Report T13:1981 (In Swedish).

4. American Public Works Association. Urban Stormwater Management. Special Report No. 49, 1981.

5. Field, Richard. The USEPA Office of Research and Development°s View of Combined Sewer Overflow Control. Third International Urban Storm Drainage Conference, Sweden, 1984.

6. Urbonas, Ben. Summary of Findings by ASCE Task Committee on Detention Outlets. Proceedings of the Third International Conference on Urban Storm Drainage, Sweden, 1984.

7. Knott, G.E. and Taylor, A.J. The Use of Detention Tanks for Flow Attenuation: A Review of Current Practice. Water Research Centre. External Report No. 171E, 1985.

THE SWIRL CONCENTRATOR AS AN URBAN RUNOFF TREATMENT DEVICE

Eivind Lygren and Torbjorn Damhaug
Norwegian Institute for Water Research (NIVA)
P.O.Box 333, Blindern
0314 Oslo 3
Norway

Introduction

The SWIRL concentrator is a small, highly effective device for flow control and primary treatment of combined sewer overflows. Swirl concentrators similar to the version developed by EPA (1) have been adopted by several sanitary districts in Norway.

The purpose of the following paper is to assess the swirl concentrator as a device for urban runoff treatment, with special emphasis on highway runoff under Nordic climatic conditions. This assessment is based on two projects recently carried out by NIVA. One of the projects covered the aspects of highway runoff pollution (2) while the other dealt with experiments with a swirl concentrator applied to combined sewer overflows (3,4).

Pictures of swirl concentrators installed in combined sewers in Norway are shown in Figure 1.

NATO ASI Series, Vol. G10
Urban Runoff Pollution.
Edited by H. C. Torno, J. Marsalek, and M. Desbordes

Figure 1. Swirl concentrator in combined sewer systems.

Characterization of Suspended Solids

The efficiency of the swirl separation process is sensitive to such suspended solids characteristics as particle size distribution, particle shape, density, etc. Thus, a method of wastewater characterization related to swirl concentrator performance was required. Among the different approaches considered, a method based on particle sedimentation velocity was selected as the main reference test (Figure 2).

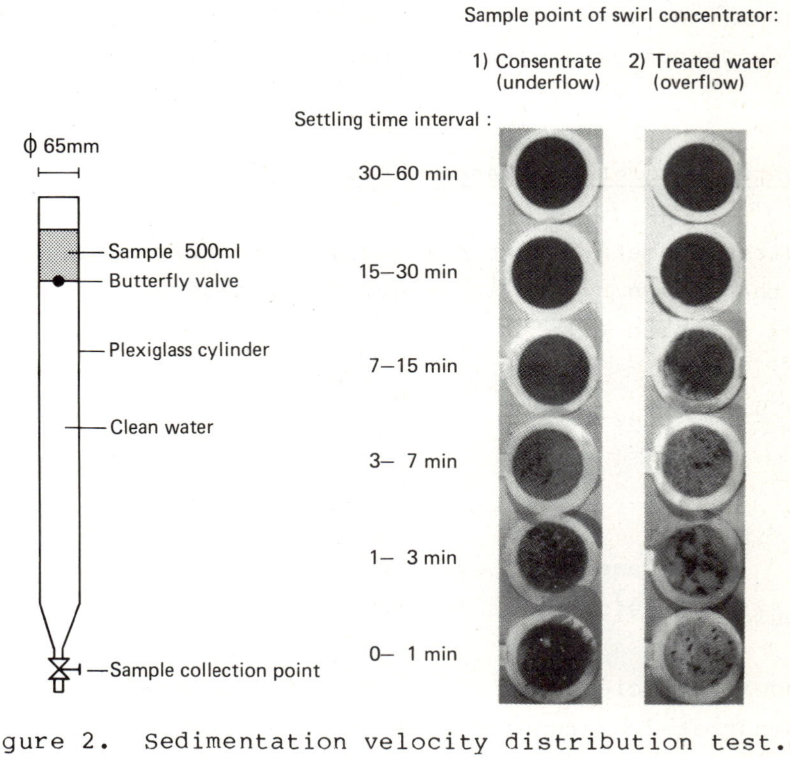

Figure 2. Sedimentation velocity distribution test.

A transparent cylinder of 60 cm length and 46 mm inner diameter is filled with clean temperated water up to a butterfly valve at the upper end. A 500 ml sample is then added at the top of the cylinder. After the sample has been properly stirred, the butterfly valve is opened and particles in the sample settle through the body of clean water. Water samples containing settled suspended solids are collected from the bottom cone at various lapsed times ranging from 1 to 60 minutes. The samples are analyzed for total and volatile suspended solids (Figure 2). This test yields the distribution of particle sedimentation velocity in the range from 1 to 60 cm/min.

This method is meant to serve as an indirect measure of the wastewater treatability in a swirl concentrator. Thus, it will be necessary to document the correlation between the sedimentation velocity distribution and the swirl concentrator performance. In order to study the mass balance of suspended solids in swirl system, tests on samples from the inlet, outlet and concentrated underflow should be carried out simultaneously.

Swirl Concentrator Performance Tests

The required diameter of an EPA swirl concentrator, as a function of the design hydraulic capacity, is described by the following equation (3):

$$D = 0.91\ (Q_{design}/9.1 \times 10^{-3})^{0.4}$$

where:

D = diameter of swirl unit (m)
Q_{design} = design hydraulic capacity (m^3/s)

The removal efficiency of different particle fractions has been demonstrated by running a series of experiments on a 1.17 m diameter swirl concentrator. The results of these experiments, supplemented with data from two actual operating concentrators, are shown in Figure 3 as typical removals of suspended solids versus sedimentation velocity distribution for different flow rates. As can be seen, swirl treatment efficiency is noticeably reduced at flowrates exceeding Q_{design}.

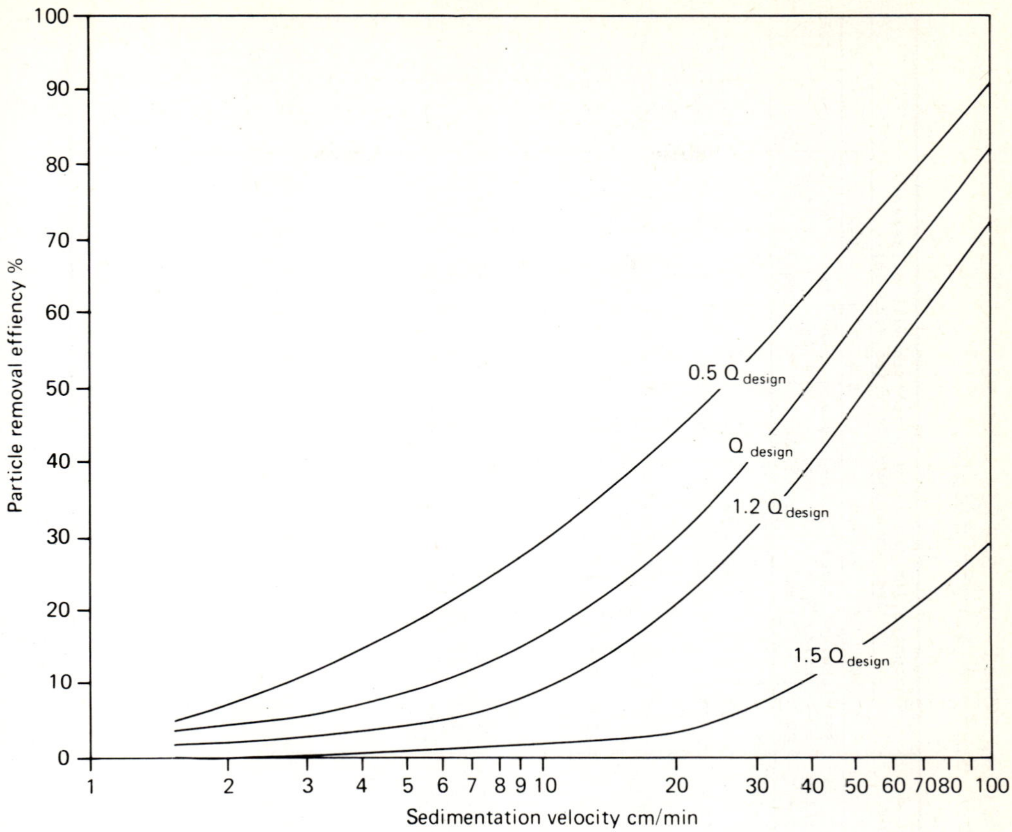

Figure 3. Swirl concentrator performance versus particle sedimentation velocity and hydraulic load for combined sewer overflows.

Treatment of Highway Runoff

The Norwegian Public Road Administration financed a 4-year programme aimed at characterizing the nature and fate of pollutants in highway runoff (2). Runoff quality from a four-lane highway with a daily traffic density of 8000 cars and a precipitation of 800 mm/year was investigated. The results indicated that a great portion of the pollutants were produced in the winter period, and transported as dust to the

Table 1. Total Suspended Solids, Total Dry Solids and Sedimentation Velocity Distribution of Road Surface Runoff Samples.

Number	Collection date	Rainfall = r Snowmelt = s	Total suspended solids		Total dry solids		Concentration (mg/l) of particles with different sedimentation velocities (cm/min)									
							>60		8.5-60		2-8.5		1-2		<1	
		r/s	total mg/l	volatile %	total mg/l	volatile %	total mg/l	volatile %	total mg/l	volatile %	total mg/l	volatile %	total mg/l	volatile %	total mg/l	volatile %
1.2	17.09.80	r	175	26	289	29	-	-	-	-	-	-	-	-	-	-
2.2	30.09.80	r	388	11	502	17	33	21	43	26	26	20	33	23	250	-
3.2	07.10.80	r	251	-	392	-	37	-	44	-	26	-	29	-	115	-
4.2	10.10.80	r	174	-	337	-	50	-	17	-	15	-	22	-	70	-
5.2	21.10.80	r+s	270	-	370	-	165	-	35	-	12	-	10	-	48	-
6.2	28.10.80	s	230	-	1100	-	67	-	25	-	25	-	28	-	85	-
7.2	24.11.80	s	600	-	2100	-	104	-	69	-	81	-	102	-	244	-
8.2	04.02.81	s	1428	-	4069	-	181	-	104	-	79	-	99	-	965	-
9.2	20.03.81	s	1669	13	5430	5	6	-	123	-	79	-	85	-	1376	-
10.2	02.04.81	s	948	17	2032	10	86	11	114	16	96	15	97	14	555	-
11.2	24.04.81	r	1600	11	2100	-	123	11	723	12	291	13	144	14	319	-
12.2	20.05.81	r	532	12	700	-	113	13	112	15	70	22	42	21	195	-
13.2	29.05.81	r	793	-	700	-	457	6	121	10	75	9	140	10	-	
14.2	30.06.81	r	485													
16.2	11.02.81	s	670													
17.2	02.03.82	s	285													
18.2	28.03.82	s	2110													
19.2	13.05.82	r+s	594													
1.1	17.09.80	r	179	13	383	15	-	-	-	-	-	-	-	-	-	-
2.1	30.09.80	r	355	12	475	16	99	7	47	21	30	21	35	18	144	-
3.1	07.10.80	r	162	-	228		23	-	11	-	17	-	14	-	97	-
4.1	10.10.80	r	178	-	350		43	-	17	-	16	-	17	-	85	-
11.1	24.04.81	r	2420	7	3334	-	1500	5	477	8	123	12	125	13	300	15
12.1	20.05.81	r	584	11	718	-	220	7	90	16	44	20	50	16	130	-
13.1	29.05.81	r	650	-	820	-	73	27	50	16	46	16	104	13	377	-

near surroundings, where they accumulated in the snow or were deposited on the ground. One of the reasons for high production of solids and associated pollutants is the use of studded winter tires in Norway.

Only the fine fractions are tranported during the snowmelt periods, while the coarse particulate matter remains deposited just below the snow banks. This material will be flushed away during the first severe rainfall runoff event after the snowmelt period.

Table 1 shows the total suspended solids, total dry solids and sedimentation characteristics of flow-proportional composite samples collected during a two-year period. Each sample represented from 5 to 15 mm of precipitation, either as rainfall or as snowmelt equivalent.

Table 2 shows the results of a long-term sedimentation test on one snowmelt and one rainfall runoff sample.

Table 2. Long term settling tests with snowmelt and rainfall runoff samples.

Settling time	Raw water sample	Supernatant suspended solids mg/l								
		1 h	2 h	7 h	1 day	2 d	5 d	6 d	8 d	15 d
Snowmelt runoff (2.04.81)	930	360	270		150	110	70			2
Rainfall runoff (13.05.82)	594	46		16	12	6	5	3	2	

Snowmelt runoff contains many fine particles that require days rather than hours to settle. In contrast to this, the rainfall runoff contains mostly coarse and rapid settling material.

The performance of a swirl separator for highway runoff was demonstrated in experiments at NIVA's research station near Oslo. The experimental unit was a 1.17 m diameter separator with a design hydraulic capacity of 17 l/s. The duration of each experiment was about 10 minutes.

Highway runoff for experimental purposes was prepared by adding different amounts of snowbank melt water to a diluted combined sewer overflow with a low content of particulate matter. Results from the highway runoff experiments are shown in Figure 4.

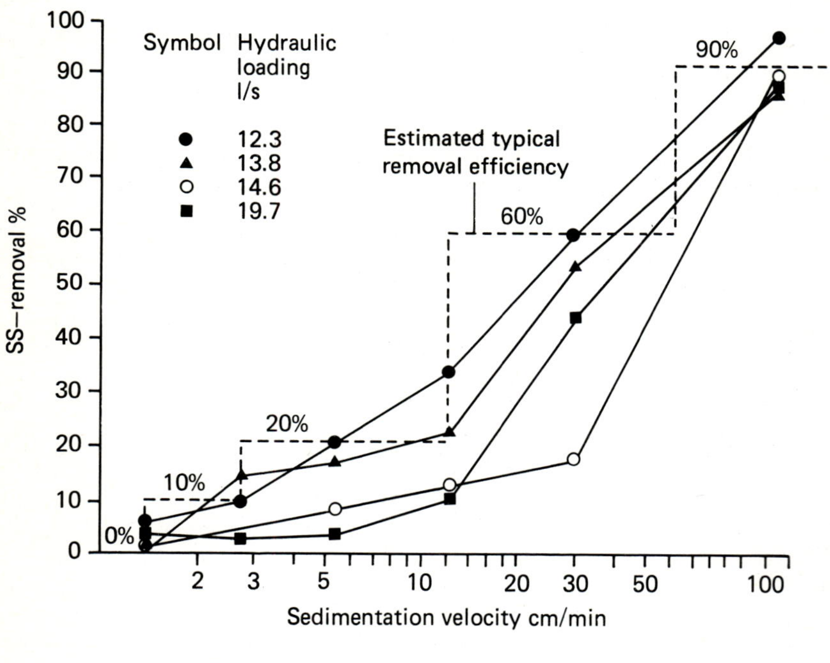

Figure 4. Results of swirl concentrator experiments for highway runoff.

The estimated removal efficiency within the different ranges of settling velocities is indicated in Figure 4 by the dotted line.

Figure 4 shows that results for runoff agree with the combined sewer experiments (Figure 3).

Prediction of Swirl Concentrator Performance on Highway Runoff

Based on the removal efficiency distribution in Figure 4, it is possible to estimate the overall suspended solids removal for a known inflow settling velocity distribution.

Calculations of expected treatment efficiency for all runoff samples listed in Table 1 are shown in Table 3. The calculted removal efficiency of total suspended solids shown in Table 3 varies from 6% to 69%.

Two examples of calculated efficiencies of the swirl concentrator in treatment of highway runoff are shown in Figure 5. Figure 5 (a) (24.04.81) indicates a typical rainfall runoff situation with coarse material predominant, while the snowmelt runoff, Figure 5 (b) (02.04.81), is dominated by the finer fractions. Under equal operating conditions this change in wastewater characteristics leads to a drop in the estimated overall suspended solids removal from 68% to 18%.

Discussion

This paper highlights the influence of particle characterics on swirl separator efficiency. The particle characterization test described in the paper is meant to facilitate the comparison of results from various swirl separator experiments and to assist in forecasts of the probable efficiency for specific wastewaters.

It is questionable whether a batch settling test may be used to characterize processes in a swirl separator, where hydrodynamic

Table 3. Predicted Treatment Efficiency of the Swirl Concentrator for Highway Runoff

Collection date	Rainfall = r Snowmelt = s r/s	Total suspended solids Total			Recidual suspended solids (mg/l) within different sedimentation velocity intervals (cm/min)									
					>60		8.5-60		2-8.5		1-2		<1	
		Total	Rem.		Total	Rem.* 90%	Total	Rem.* 60%	Total	Rem.* 20%	Total	Rem.* 10%	Total	Rem.* 0%
		mg/l	mg/l	%	mg/l	mg/l	mg/l	mg/l	mg/l	mg/l	mg/l	mg/l	mg/l	mg/l
30.09.80	r	388	64	16	33	30	43	26	26	5	33	3.3	250	-
07.19.80	r	251	67	27	37	33	44	26	26	5	29	2.9	115	-
10.10.80	r	174	60	34	50	45	17	10	15	3	22	2.2	70	-
21.10.80	r+s	270	174	64	165	150	35	21	12	2	10	1.0	47	-
28.10.80	s	230	83	36	67	60	25	15	25	5	28	2.8	85	-
24.11.80	s	600	161	27	104	94	69	41	81	16	102	10.2	244	-
04.02.81	s	1428	251	18	181	163	104	62	79	16	99	9.9	965	-
20.03.81	s	1669	104	6	6	5	123	74	79	16	85	8.5	1376	-
02.04.81	s	948	174	18	86	77	114	68	96	19	97	9.7	555	-
24.04.81	r	1600	617	39	123	111	723	434	291	58	144	14.4	319	-
20.05.81	r	532	187	35	113	102	112	67	70	14	42	4.2	195	-
29.05.81	r	793	513	65	457	411	121	73	75	15	140	14.0	-	-
30.09.80	r	355	126	36	99	89	47	28	30	6	35	3.5	144	-
07.10.80	r	162	32	20	23	21	11	7	17	3	14	1.4	97	-
10.10.80	r	178	54	30	43	39	17	10	16	3	17	1.7	85	-
24.04.81	r	2420	1674	69	1500	1350	477	286	123	25	125	12.5	300	-
20.05.81	r	584	266	46	220	198	90	54	44	9	50	5.0	130	-
29.05.81	r	650	115	18	73	66	50	30	46	9	104	10.4	377	-

* Removal estimated from figure 4.

Figure 5. Examples of predicted suspended solids removal from highway runoff by the swirl concentrator.

a) Example 1 - A rainfall runoff case (24.04.81)

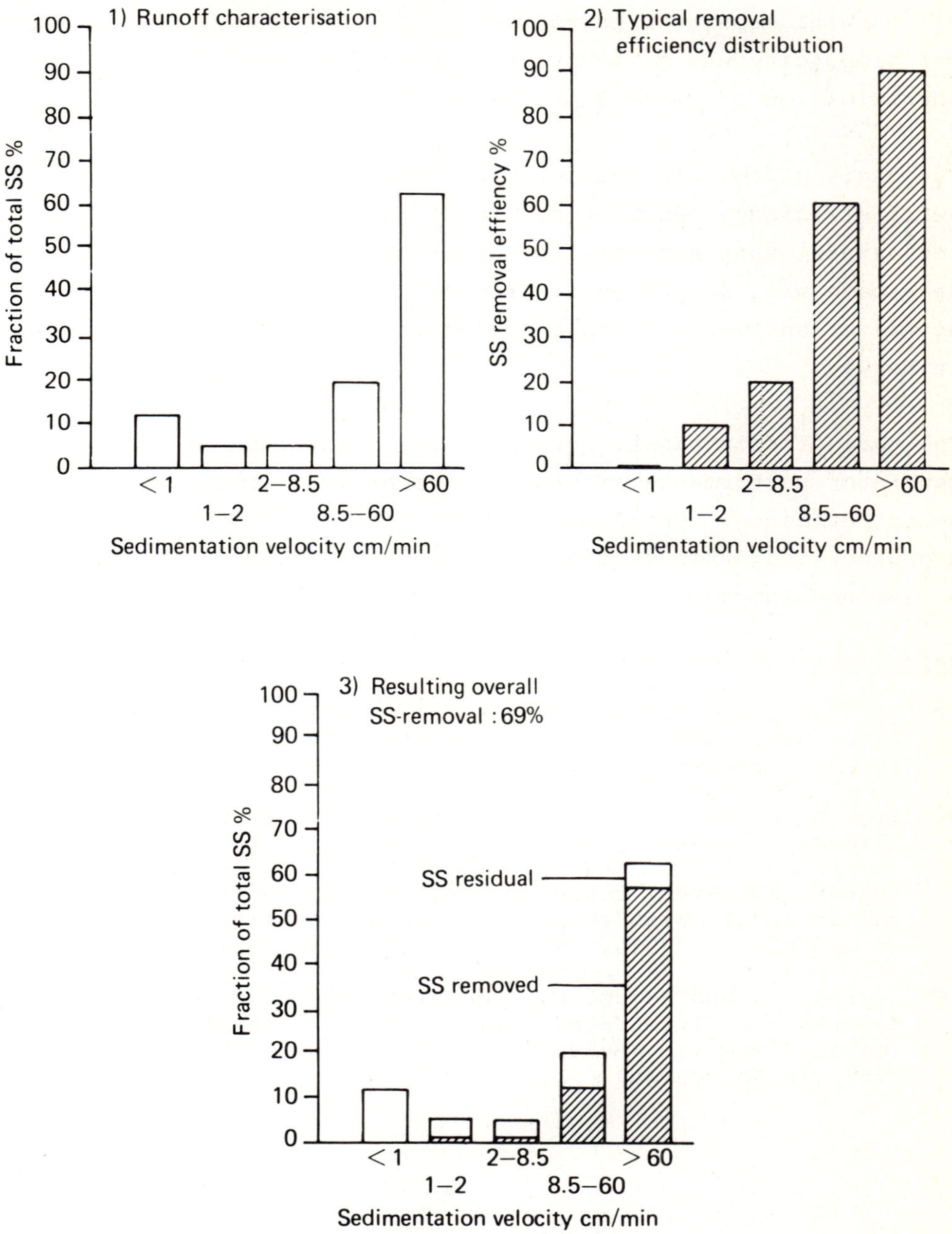

conditions and driving forces are obviously different. In that sense, a modified centrifuge test would probably be a better approach. However, the authors feel that any characterization test would require an empirical conversion of results to establish swirl separator performance. In this study, the demand for simplicity was given priority over the criteria of close approximation of removal processes.

The basis of correlation between the settling velocity and the swirl efficiency has been developed through series of laboratory experiments and in-situ measurements. The statistical data base will be gradually improved, as the settling characterization test will be included in future swirl projects in Norway.

The part of this paper concerning the prediction of swirl concentrator performance is based on relatively limited experimental evidence. Until more broad-scale experimental results are generated, this section must be regarded as an example of an engineering approach.

References

1. Sullivan, R.H. et al. The Swirl Concentrator as a combined sewer overflow regulator facility. USEPA Report No. EPA-RZ-72-008. 1972.

2. Lygren, E. and Gjessing, E. Highway Pollution in a Nordic Climate. VA-Report No. 3/84. 0-79024, NIVA.

3. Lygren, E. Swirl concentrator. Separation of suspended solids and floatable matter in urban runoff. VA-Report No. 13/80. 0-79024, NIVA Des. 1980 (In Norwegian).

4. Lygren, E. and Wedum, K. Vortex chamber and swirl concentrator. Integrated flow control and treatment of combined sewer. VA-Report No. 3/82. 0-79090, NIVA Apr. 1982 (In Norwegian).

A CONCEPTUAL MODEL OPTIMIZING THE MAINTENANCE OF GRIT CHAMBERS IN COMBINED SEWERS

D. Dartus and Alquier M.
Institut de Mecanique des Fluides de Toulouse
Equipe Ecoulements Diphasiques Solide-Liquide
2, rue Charles Camichel, 31071 Toulouse Cedex
France

Introduction

Combined sewer flow consists of a liquid and a solid phase which are interacting constituents, even as they form a mixture, and it is the liquid which carries the solid along with it. Urban pollution control is largely directed at separating these two components by retaining the polluting solid.

As far as combined sewers are concerned, this separation can, to some degree, be achieved by using grit chambers. Their behavior was studied in a reduced-scale laboratory model, and then compared with results obtained from chambers actually in use. These grit chambers fall under the responsibility of the Direction Departementale de l'Equipement de la Seine-Saint-Denis (near Paris), which provided the field measurements (5,6).

The approach presented in this paper is different from the one adopted by most other researchers, in that a conceptual (probability) model was used (1,2).

Scale Model (3)

The model is basically a rectangular tank, with dimensions of length, width and depth of 2000 x 400 x 300 mm respectively. As shown in Figure 1, four different configurations were studied, and the response after instantaneous injection of solid particles was investigated over a range of discharges.

NATO ASI Series, Vol. G10
Urban Runoff Pollution.
Edited by H.C. Torno, J. Marsalek, and M. Desbordes

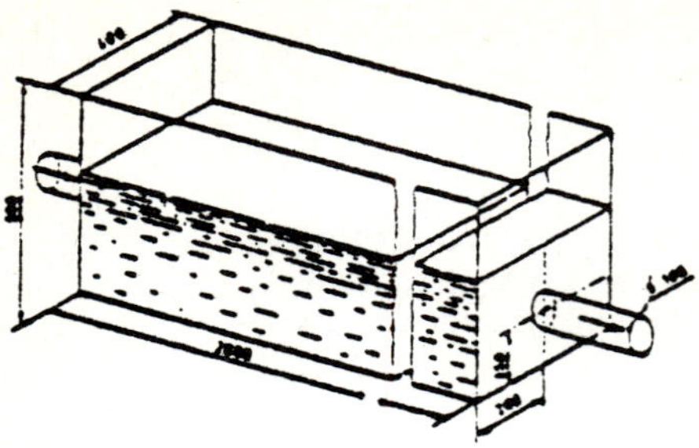

Case 1 - Without Obstacle

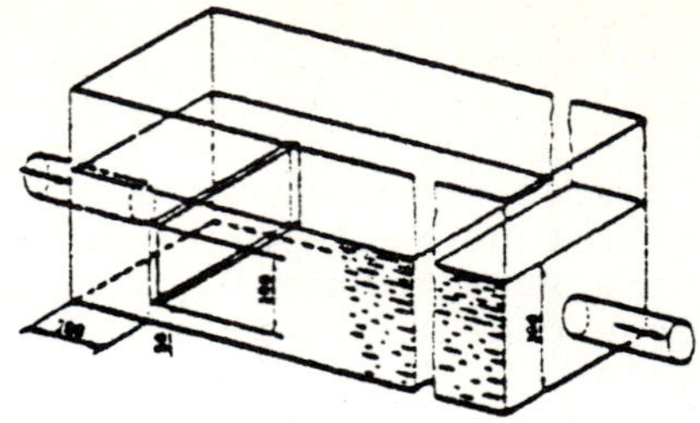

Case 2 - Baffle

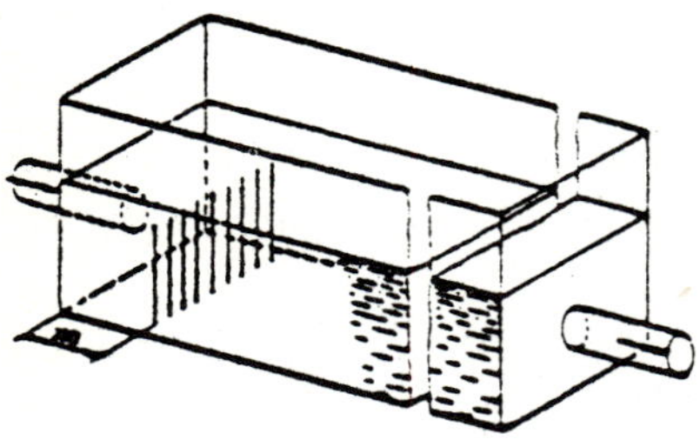

Case 3 - Bars

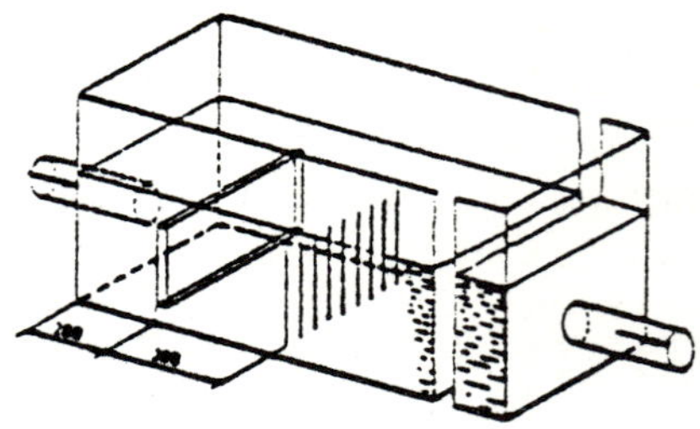

Case 4 - Baffle+Bars

Figure 1 - Different configurations

Conceptual Model

For all experiments carried out, the response function after instantaneous particle injection consisted of two distinct parts (see Figure 2):

a. A quick response corresponding to the process of deposition and related to the mean velocity as well as to the turbulence level inside the chamber.
b. A slow response corresponding to the resuspension of previously deposited particles and related to the macroturbulence.

Subsequent modeling of this behaviour pattern assumed that any one particle can only be formed in one of the following three states:

a. In suspension.
b. Leaving the tank.
c. Settled.

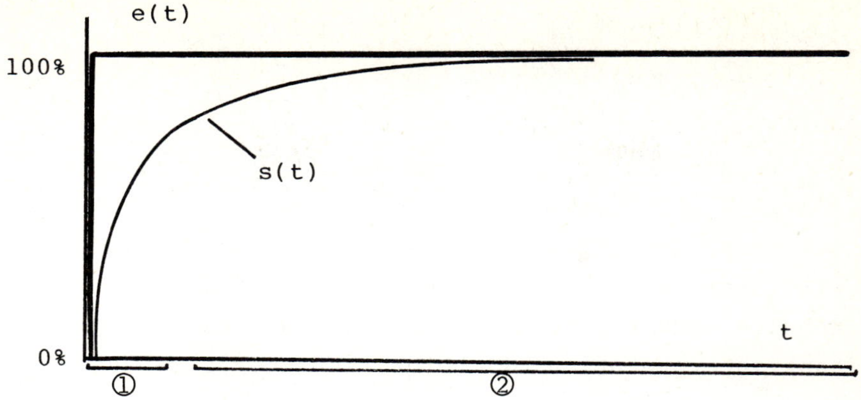

Figure 2 - General response after instantaneous injection

The transition probabilities for passing from one state to the other are , and , and are shown in Figure 3.

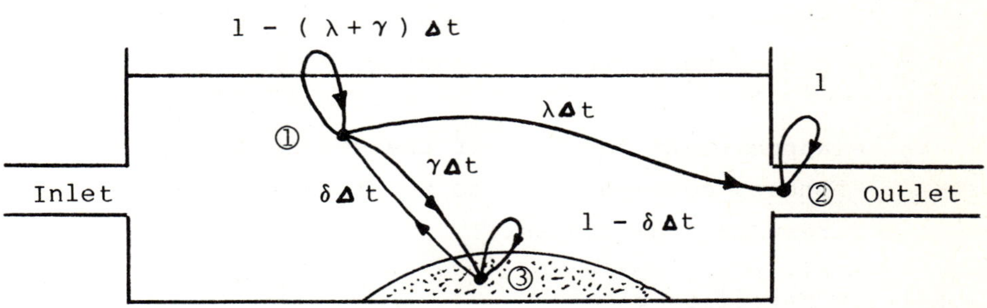

Figure 3 - Modeling particle behaviour

For such a system of instantaneously injected particles, it can be shown that the probablity, $P_i(t)$, of a particle being in state i at time t is :

$$P_1(t) = \frac{\alpha \lambda_1 e^{-\lambda_1 t} + (1-\alpha) \lambda_2 e^{-\lambda_2 t}}{\alpha \lambda_1 + (1-\alpha) \lambda_2}$$

$$P_2(t) = 1 - \alpha e^{-\lambda_1 t} - (1-\alpha) e^{-\lambda_2 t}$$

$$P_3(t) = \frac{\alpha(1-\alpha)(\lambda_1 - \lambda_2)}{\alpha\lambda_1 + (1-\alpha)\lambda_2}\left(e^{-\lambda_2 t} - e^{-\lambda_1 t}\right)$$

where α, λ_1 and λ_2 are functions depending on λ, γ and δ.

It can easily be seen that $P_2(t)$ is the transit time distribution, TTD, and that it allows best-fit smoothing of experimental results (see Figure 4). It also provides a physically significant interpretation of the grit-chamber's behaviour:

- α = rate of particles leaving the tank in the rapid phase. Therefore $(1-\alpha)$ is the rate of particles that remain deposited during this initial phase.
- $1/\lambda_1$ = approximate duration of the first quick-response phase, that is to say the velocity with which suspended particles are leaving.
- $1/\lambda_2$ = approximate duration of the second slow-response phase, corresponding to the flow's ability to resuspend particles.

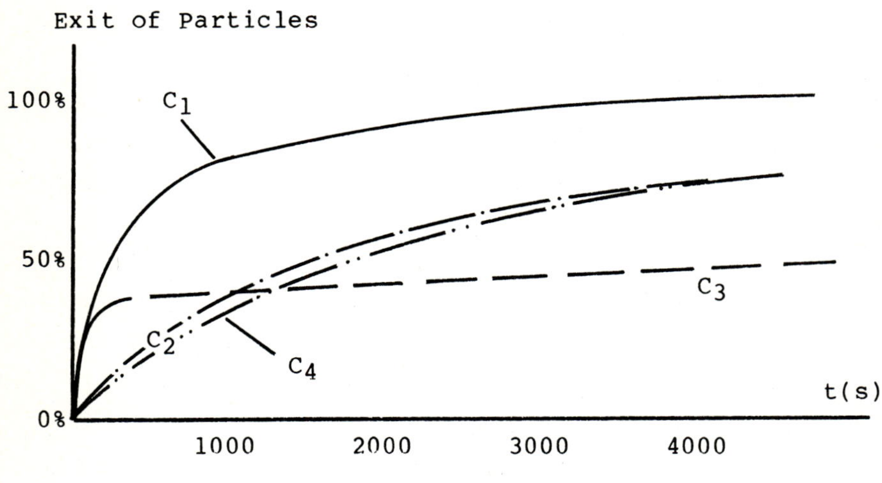

Figure 4 - Instantaneous injection - Flow = 7 l/s

As indicated in Figure 5, it is now possible to derive, for a certain time, t, after the instantaneous injection of particles, the percentage of particles being suspended or deposited.

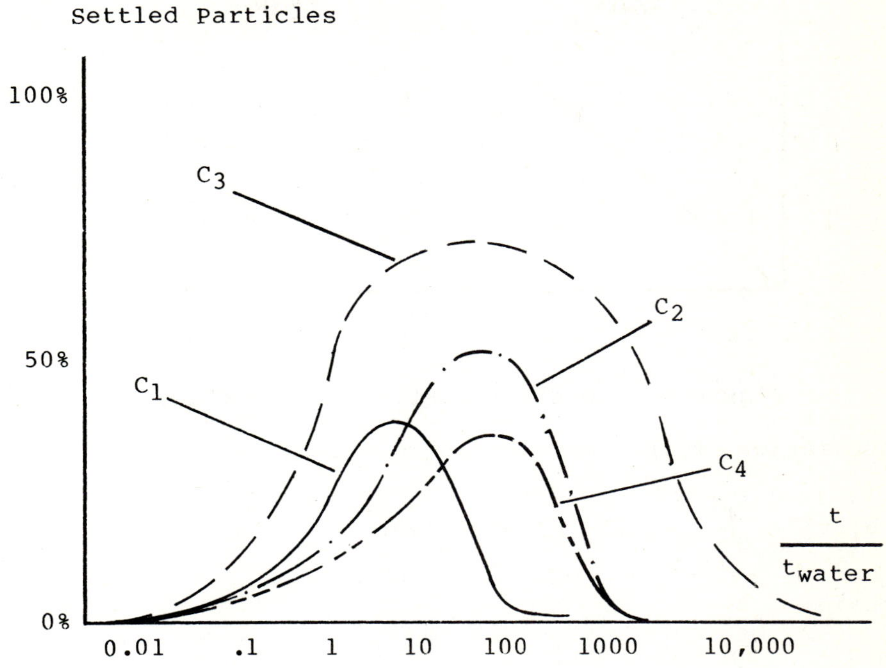

Figure 5 - Instantaneous Injection - Flow = 7 l/s

<u>Constant Continuous Injection and the Model</u> (4)

Assuming that dynamic characteristics do not vary with the tank's rate of filling, the parameters α, λ_1 and λ_2 must be constant. Therefore, only mean values valid for the whole filling cycle and constant sediment discharge can be obtained. This situation is shown in Figure 6, where e(t) represents the accumulation curve of particles entering the grit chamber, and $e_i(t)$ is the curve for any state i. The calculations are then carried out without taking into consideration the solid discharge. After multiplying $e_i(t)$ by the solid discharge (in l/s or Kg/s) at the entrance, the actual mass or volume of deposited material can be calculated.

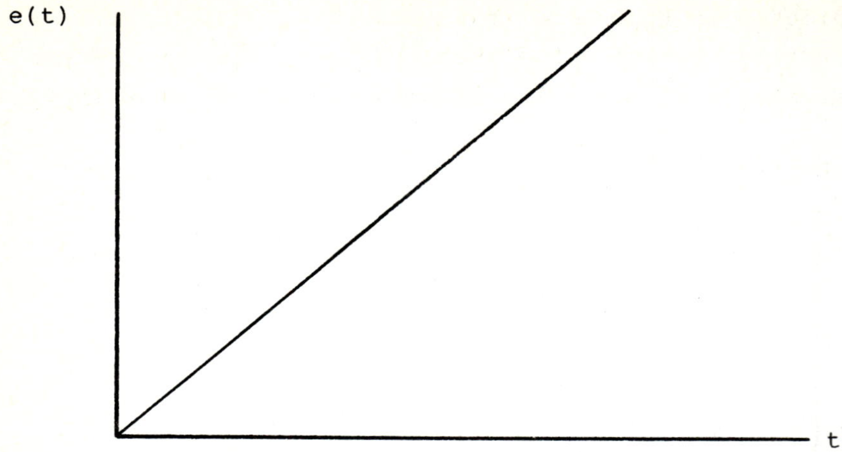

Figure 6 - Constant continuous injection

By integration of the convolution relation:

$$e_i(t) = \int_0^t P_i\,(t-\tau).e'\,(\tau).d\tau$$

We obtain:

$$e_1(t) = \frac{\alpha(1-e^{-\lambda_1 t}) + (1-\alpha)(1-e^{-\lambda_2 t})}{\alpha\lambda_1 + (1-\alpha)\lambda_2}$$

$$e_2(t) = t - \frac{\alpha}{\lambda_1}(1-e^{-\lambda_1 t}) - \frac{1-\alpha}{\lambda_2}(1-e^{-\lambda_2 t})$$

$$e_3(t) = \frac{\alpha(1-\alpha)(\lambda_1-\lambda_2)}{\alpha\lambda_1 + (1-\alpha)\lambda_2}\left(\frac{1-e^{-\lambda_2 t}}{\lambda_2} - \frac{1-e^{-\lambda_1 t}}{\lambda_1}\right)$$

From this, the chamber's yield can be derived as a function of time, as shown for our experiments in Figure 7.

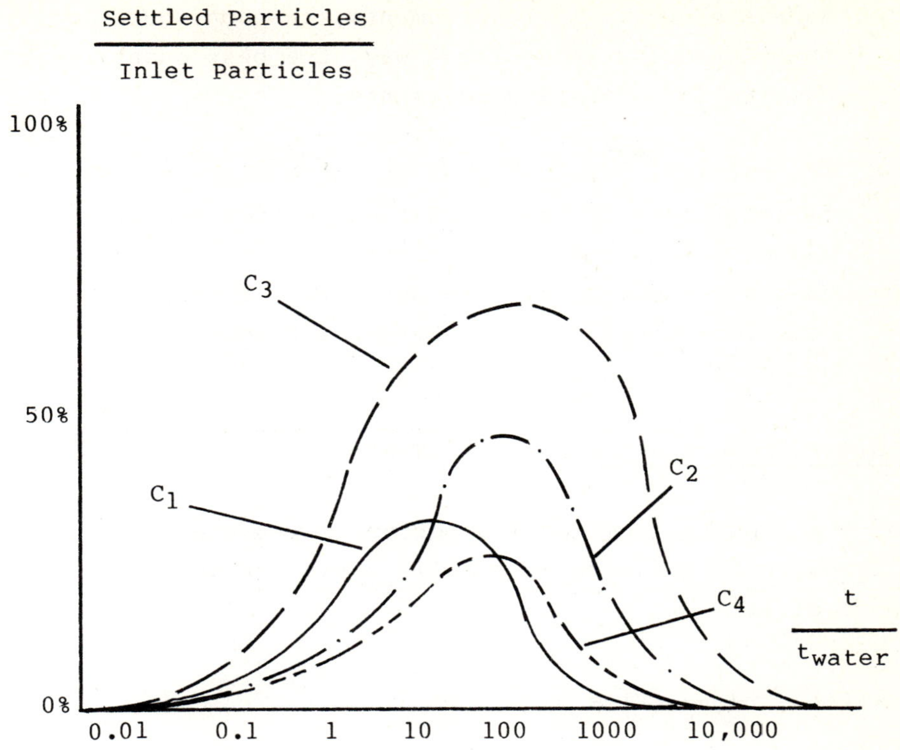

Figure 7 - Continuous Injection - Flow = 7 l/s

Analysis of Scale-Model Results

Without going into detail, Figures 5 and 7 show clearly that:

a. For instantaneous injection, regardless of the configuration used, all particles leave the tank sooner or later.

b. For continuous injection, there exists a maximum volume of deposition. Once this is obtained, sediment inflow is balanced by outflow.

Regarding the usefulness of inserting flow obstacles inside the grit chamber:

a. As was shown in our experiments, the configuration "bars" showed, in a global way, the best quantitative results in depositing sediments.

b. The vertical baffle board, by increasing the sedimentatation rate in the initial phase (see Figure 4), causes a subsequent detrimental effect, in that it favors resuspension. Its use should therefore be restricted, but only for temporal flood-attenuation purposes.

Real Grit Chambers and the Model (5,6)

One of the results from model experiments was that $\lambda_1 \gg \lambda_2$. This leads to a considerable simplification of the term $e_3(t)$, which becomes:

$$e_3(t) = \frac{1-\alpha}{\lambda_2}(1-e^{-\lambda_2 t})$$

and which represents a very good approximation of the "outflow" curves.

The water authority DDE-SSD provided the following field measurements for three grit chambers:

- Daily rates of suspended solids.
- Density of the deposition.
- Water content of the deposition.
- Tank-filling graphs.

After smoothing the measured filling graphs of different grit-chambers by means of a function of the $\alpha[1-e^{-bxt})$ type, the maximum rate of filling, the value $(1-\alpha)$ and λ_2 were determined (see Figure 8 and Table 1).

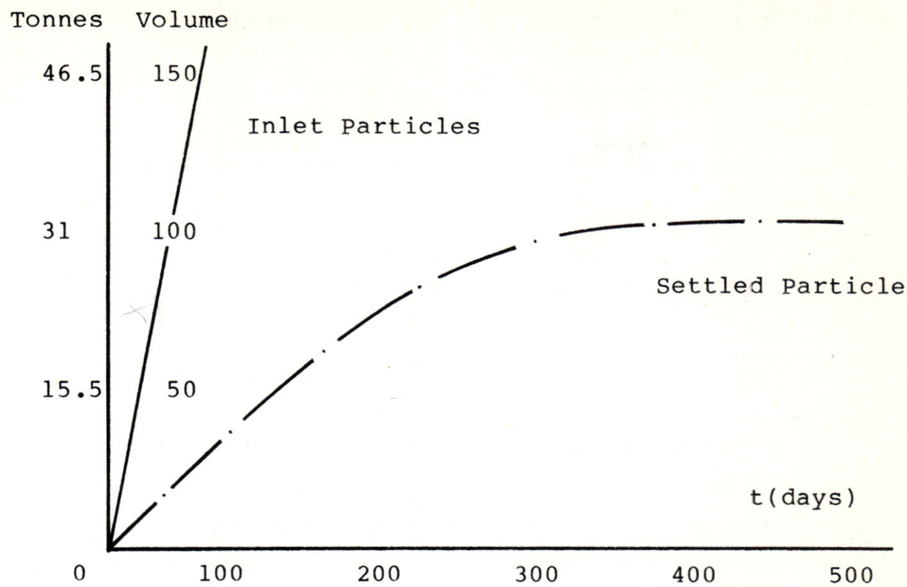

Figure 8 - Results for Chamber Number 2

Interpretation of Results

Among those grit chambers listed in Table 1, a perceptible difference exists between chamber 1 and chambers 2 and 3.

- In chamber 1 there are some 30% of particles that short-circuit the tank, whereas in chambers 2 and 3 there are 80% to 90%.

- The maximum filling rate of chamber 1 is 60%. This is because the chamber is overdimensioned with respect to the inflow of suspended solids. The trapping rate, however, is excellent.

- A 123% maximum-filling rate for chamber 2 is much greater than that of chamber 1, and resulted in sediment deposition as high as the inspection platform.

Table 1 - Measures and Results for Three Real Grit Chambers

<table>
<tr><td colspan="2">CHAMBER</td><td>1</td><td>2</td><td>3</td></tr>
<tr><td colspan="2">DIMENSIONS
L x l x h (m)</td><td>17 x 2.6 x 1.5</td><td>22.5 x 2 x 1</td><td>20 x 2.6 x 1.5</td></tr>
<tr><td colspan="2">Suspended Solids
(Kg/day)</td><td>441</td><td>1823</td><td>5254</td></tr>
<tr><td colspan="2">Density of Deposition
(Kg/m^3)</td><td>1820</td><td>1320</td><td>1610</td></tr>
<tr><td colspan="2">Water Contents</td><td>18 %</td><td>48 %</td><td>28 %</td></tr>
<tr><td rowspan="8">TANK FILLING</td><td rowspan="4">Number of days</td><td>80</td><td>35</td><td>35</td></tr>
<tr><td>24 %</td><td>28 %</td><td>26 %</td></tr>
<tr><td>190</td><td>100</td><td>65</td></tr>
<tr><td>35 %</td><td>66 %</td><td>30 %</td></tr>
<tr><td rowspan="4">Percentage of filling</td><td>270</td><td>140</td><td>105</td></tr>
<tr><td>44 %</td><td>80 %</td><td>46 %</td></tr>
<tr><td>480</td><td rowspan="2"></td><td>185</td></tr>
<tr><td>56 %</td><td>72 %</td></tr>
<tr><td colspan="2">RATE OF FILLING</td><td>61 %</td><td>123 %</td><td>107 %</td></tr>
<tr><td colspan="2">$1 - \alpha$</td><td>70 %</td><td>16 %</td><td>11 %</td></tr>
<tr><td colspan="2">λ_2
(1/day)</td><td>0,0051</td><td>0,0075</td><td>0,0059</td></tr>
</table>

- The average residence time of a particle in the chamber is on the order of four to seven months $(=1/\lambda_2)$.

Finally, to improve the chambers' maintenance, the curve

$$r(t) = \frac{\text{particles trapped in the tank}}{\text{mass of particles in the influent}}$$

is used which represents yield as function of time (see Figure 9):

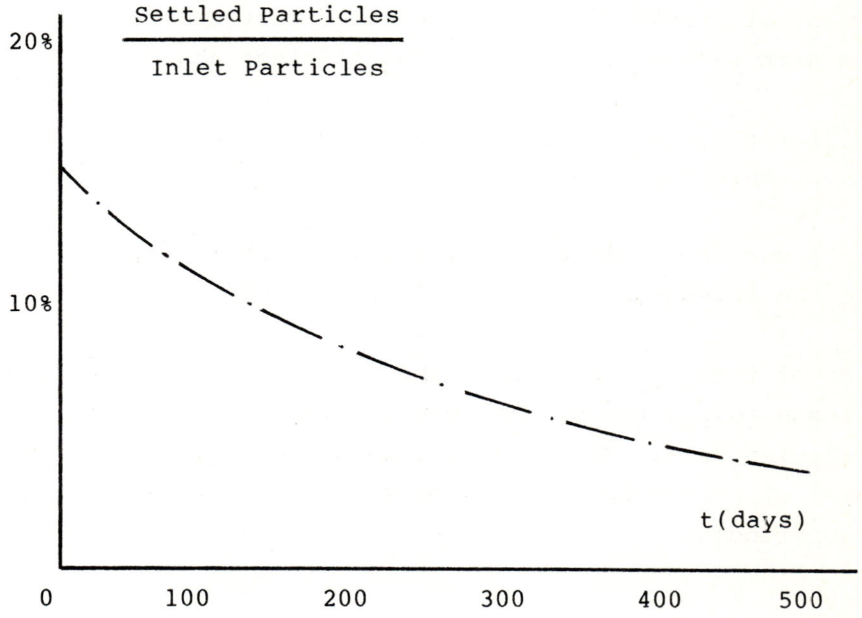

Figure 9 - Rate for Chamber Number 2

We can see from this plot that the trapping rate is maximum just after having been emptied. But this rate remains quite high for about the first year, at the end of which the trapping rate will still be about 50% of its initial value.

As a guideline one could say that a grit chamber having a maximum obtainable filling rate of less than 100% (of the theoretical design value) should be emptied at 80% to 90% of the maximum obtainable value. On the other hand, if the maximum obtainable value is larger than the design value, the chamber should be emptied at 80% to 90% of the design value. For the three chambers under study this would mean:

Chamber	1	2	3
Emptying every	1 year	5 to 6 months	8 to 10 months

Conclusions

The conceptual probability model revealed that the functioning of a grit chamber consists of two distinct phases:

a. A first, quick response, phase, very much of a short circuiting.

b. A second, slow-response, phase, which is related to the resuspension of deposited particles.

Insertion of flow obstacles into the chamber was studied, and it was found that a row of bars would always increase the chambers' trapping rate. The model was then used to predict trapping rates of prototype grit chambers, and to define optimal emptying periods.

References

1. Clements and Price. Stochastic model of motion of solid particles. Journal of the Hydraulics Division. ASCE Sept. 1967.

2. Hazen. On sedimentation. Trans. ASCE 52-45, 1904 (reprinted in J. Proc. lnst. Sew. Pur., 1962, p.521.

3. Dartus, D. Etude et modelisation de l'influence des grosses structures tourbillonnaires sur les performances d'un appareil de decantation. These de Docteur-Ingenieur, INPT, ENSEEIHT, 1982.

4. Dartus, D. Utilisation d'un modele probabiliste global pour l'etude du fonctionnement d'une chambre a sable reelle. Rapport de contrat RC-83-001, 1983.

5. Direction Departementale de l'Equipement de la Seine Saint-Denis, arrondissement operationnel N° 3. Etude du fonctionnement de chambres de dessablement sur le reseau unitaire. Rapport.

6. Dircction Departementale de l'Equipement de la Seine Saint-Denis, arrondissement operationnel N° 3. Etude du fonctionnement de chambres de dessablement sur le reseau unitaire. Annexes.

STANDARDIZATION OF DETENTION POND DESIGN FOR PHOSPHORUS REMOVAL

Ben Urbonas, Chief, Master Planning Program
Urban Drainage and Flood Control District
2480 W. 26th Avenue, Suite 156B
Denver, CO 80211 USA

William P. Ruzzo, Vice President
WRC Engineering, Inc.
1660 S. Albion Street, Suite 500
Denver, CO 80222 USA

Introduction

The Denver area is located in the semi-arid western region of the United States. It receives an average of 380 millimeters of precipitation a year, with approximately 80 percent occurring as intense, short duration, thunderstorms that produce flash floods. Denver is a rapidly urbanizing area for which records of rainfall with simultaneous runoff have been gathered since 1969. As a result of the Environmental Protection Agency's National Urban Runoff Program, water quality of urban runoff was studied in the Denver region between 1979 and 1982. The availability of local data and investigations formed a sound basis for estimating quality and the the impacts of urban runoff on receiving waters.

The foremost impact by urban runoff on the receiving waters in the Denver region was identified as the acceleration of the eutrophication process in some of the lakes and reservoirs. Eutrophication is a natural process that can be accelerated by the total organic carbon (TOC), nitrates (N03) and phosphorus (P) found in runoff from urbanized areas. Studies (14) have shown that the total load of phosphorous and nitrogen compounds can increase by factors of 10 to 40 as the land

NATO ASI Series, Vol. G10
Urban Runoff Pollution.
Edited by H.C. Torno, J. Marsalek, and M. Desbordes

changes from undisturbed meadows to an urban community. Figure 1 clearly illustrates such a trend. Studies in other cities of the United States (12) also show similar trends. In addition, runoff from agricultural land has been shown to have higher loads of these three constituents than is found in runoff from undisturbed lands, which may be due to soil erosion and the use of fertilizers.

The rate of eutrophication of lakes and reservoirs can be sometimes controlled by controlling the amount of phosphorous, nitrogen or total organic carbon in the standing body of water. Of these three constituents, control of phosphorous is probably most practical and may be most cost-effective. The need to control eutrophication was suggested by evidence (16) showing that excessive levels of algae in lakes adversely impact their recreational uses such as swimming, boating, and aesthetics and often determine the type of fish that inhabit a lake.

To protect the recreational uses of the Cherry Creek reservoir in Denver, the State of Colorado mandated that developing lands located in the watershed tributary to the reservoir provide phosphorous controls. To accomplish the stated goal of maintaining 0.035 mg/l growing season average of phosphorus in the lake, all wastewater treatment plants are required to provide advanced treatment. In addition, all urbanized areas are required to remove at least 50 percent of the phosphorous found in stormwater. To achieve the specified removals, various options are being investigated. One option, developed by the authors, provides a standardized design for detention ponds followed by sand filtration. The remainder of the paper discusses this technique and how it was developed.

Literature Review

As a result of the Environmental Protection Agency's National Urban Runoff Program (1,12) detention was found to be effective for removing phosphorous from urban runoff and "wet" ponds were

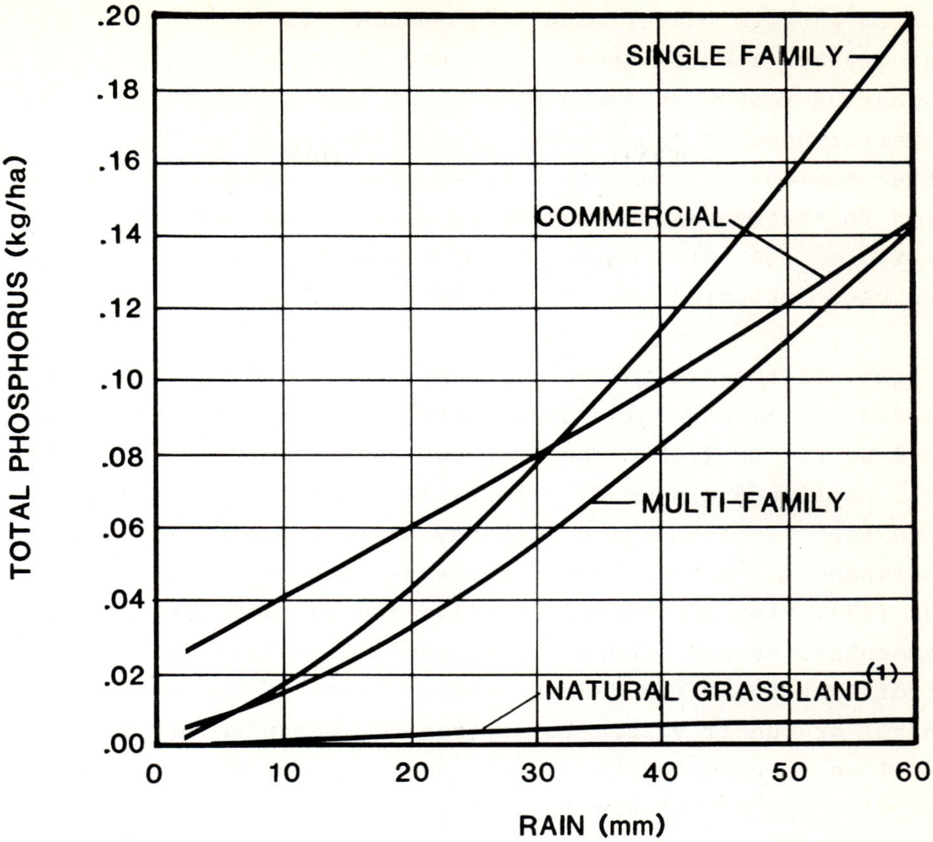

FIGURE 1

EVENT MEAN UNIT-AREA LOADING OF TOTAL PHOSPHOROUS VS. RAINFALL IN DENVER.

(1) OBSERVATIONS OF RUNOFF AND LOADS FOR LESS THAN 25mm OF RAINFALL WERE EXTREMELY VARIABLE.

Reference (14)

found to perform better than "dry" ponds. Long term detention (i.e. greater than 10 hours) was required before significant phosphorous reduction was achieved. A lack of local data on the performance of the ponds in Denver was also identified and further monitoring and data analysis were recommended. In addition to the Denver program, much data was developed and analyzed on the national level and reported in several references used for this paper (1,2,9,18).

The state-of-the-art in non-point source pollution control has advanced to where we have text books and design manuals published on the subject. The American Public Works Association (APWA) published a special report, (15) setting forth specific design recommendations for outlets, basin geometry and other appurtenances. A text book, Stormwater Management in Urbanizing Areas (17), also identified specific design requirements. These recommendations and others are summarized in Table 1. A concept of "settleability design storms", namely, the amount of rainfall or runoff volume to be treated, was also introduced. The values recommended by various agencies for the "settleability design storm" are presented in Table 2.

The advantages of a "wet" pond, namely, a pond which has a relatively constant body of water and a "dry" pond, a pond that drains totally, were discussed in several references. All investigators (1,5,8,10,11,12,14) preferred the "wet" pond instead of the "dry" pond for treatment of runoff. The advantages cited were longer detention times, minimal short circuiting of flows from the inlet to the outlet, greater settleability under quiescent condition, and the biological action promoted by wet ponds.

Because only a limited number of urban runoff treatment facilities were studied in the United States, the operations and maintenance aspects of pond design are not discussed in the literature. The maintenance experiences in Fairfax County Virginia (8) resulted in several design changes. The outlet

TABLE 1
SUMMARY OF WATER QUALITY POND DESIGN RECOMMENDATIONS

Item	Recommendation	Reference
Pond Shape	Wedge Shaped	15
Length to width ratio	2:1 to 3:1	8,10,15
Pond depth	greater than 3 feet	15
	greater than 3.5 feet	2
Outlet location	surface withdrawal	10,15
	riser pipe with holes	8
Short circuiting of flo	provide baffles and locate outlet furthest from inlet	15
Wet or dry pond	wet pond is better	1,5,8,10, 11,14,18
Detention times	greater than 10 hours	14
	greater than 16 hours	9
	greater than 32 hours	18
	greater than 40 hours	8
Additional treatment	promote filtration and/or promote infiltration	8,11,14

TABLE 2
SUMMARY OF RECOMMENDATIONS FOR
"SETTLEABILITY DESIGN STORMS"

Source	Design Criteria	
	Runoff Amount	Rainfall Amount
Fairfax County, Virginia (8)	22.9 mm	--
Delaware Raritan Canal Comm. (17)	--	31.8 mm in 2 hrs.
South Florida Flood Management District (17)	25.4 mm	3 yr. storm
Austin, Texas (6)	12.7 ?m	--
Central Florida(7)	12.7 mm	--

originally used in Fairfax consisted of an underdrain pipe on the pond bottom to promote filtration, but considerable clogging problems were encountered. The outlet was modified to have a riser pipe with many small perforations surrounded by a gravel pack. Other suggestions in the literature included provisions in the outlet to permit complete drainage of the pond (15) and a concrete base for the riser to facilitate sediment removal and to minimize vandalism (8).

The detention time to adequately treat runoff was found to vary widely as reported by various investigators. As can be seen in Table 1, values ranged from 10 hours to greater than 40 hours. This range was due to the pollutant being treated and the variations in data. Whereas suspended solids could be reduced 60% to 75% after 32 hours, reductions in Biological Oxygen Demand, phosphorous, copper, zinc and nickel varied from 20% to 70% (18) for the same period of detention.

Whipple and Hunter (18) specifically investigated the settleability of phosphorous in samples of urban runoff. Using settling tubes they found that 25% to 70% of phosphorous was removed after 32 hours. Randall (9,10) reported that even though phosphorous has a high affinity for adsorption to particulate matter, sedimentation alone may not be sufficient to protect receiving waters. Grizzard (5) reported that settleable phosphorous attaches itself to very small particles which may account for the lack of treatment by sedimentation alone.

Estimating Performance

Information on the performance of detention ponds for the removal of phosphorous from stormwater was provided by several investigators. Much of the information regarding the performance of detention in removing phosphorous for individual runoff events is summarized in Table 3. It is clear that detention alone cannot guarantee a 50 percent removal rate

TABLE 3
EFFECTIVENESS OF DETENTION
FOR REMOVAL OF PHOSPHOROUS

Source	Percent Reduction In Total P	Comments
Driscoll (2)	3 to 79	Observed removals were less than estimated
Whipple and Hunter (18)	25 to 68	Laboratory settleablility tests
Randall, et. al. (9,10)	42 to 85	Laboratory settleablility tests, after 48 hours
	11 to 79	after 12 hours

TABLE 4
ANALYSIS OF THE DENVER RAINFALL GAGE (3)

Storm Size	Accum. Percent of Events	Accum. Percent of Rainfall
0-13 mm	88.9	45
13-25 mm	95.0	64
25-38 mm	97.9	79
38-51 mm	99.0	86
51- mm	100.0	100

for phosphorous. Driscoll (2), after analysis of data from various projects in the United States (12) developed performance charts to estimate the average annual reductions of pollutants. These charts indicate that the detention element by itself, as proposed by the authors, is capable of removing at least 50 percent of the annual phosphorous load. Nevertheless, the authors are not convinced that 50 percent removal of the annual phosphorous load can be assured in the Denver climate using only detention. The policy in the Cherry Creek basin requires at least 50 percent removal of the non-point phosphorous load. One has to be cautious in how data and its interpretations in the technical literature are utilized in design. Before a design configuration for phosphorous removal is selected, the authors recommend that the following items be considered:

General

1) Is the pH of the runoff somewhat alkaline (i.e. greater than 7.0) which keeps the total phosphorous from becoming soluble phosphorous? Fortunately, this is the typical case for urban runoff in Denver, however, if the rainfall becomes more acid as is feared, this may not be the case in the future.

2) Inflow into the pond should not short circuit the pond on its way to the outlet.

3) Phosphorous removal needs to occur from runoff during a variety of rainfall distributions and intensities, snow melt, temperatures, pond site and watershed conditions.

4) Removal of phosphorous depends, to a large extent, on the removal of small suspended solids to which it is adsorbed.

Dry Ponds

1) Phosphorous will remobilize if vegetation is permitted to grow and decay on pond bottom between runoff events.

2) Normal low flows from urban watersheds resulting from lawn watering, car washing, water recreation, sub-drainage of buildings, etc. will not be detained by a "dry pond".

3) Runoff from small rainstorms, low intensity rainstorms of extended duration and from snowmelt will not be detained long enough to permit significant settlement of small solids containing phosphorous. Possibility of short circuiting the pond is greatly increased as this low rate of runoff virtually passes through the pond undetained.

4) Runoff events exceeding the pond volume will cause a portion of the runoff to bypass the pond.

Wet Ponds

1) Normal low flows and runoff from small rainstorms, low intensity protracted rainstorms and snowmelt will receive treatment not normally provided by "dry ponds".

2) A portion of runoff from larger storms will "bypass" the pond volume.

3) Algae growth in the permanent pool between storms can remobilize phosphorous for the next runoff event.

4) Depletion of dissolved oxygen in the lower layers of a wet pond can cause remobilization of phosphorous.

5) Pond short circuiting by inflows due to density or temperature stratification is possible.

The above considerations indicate a need for some other treatment in addition to detention alone if 50 percent removal is to be assured. For these and other reasons, the authors suggested that the detention pond be followed by a sand filter. Filtration was considered an effective second step to remove some of the suspended particulate not removed by the pond or remobilized within the pond. The removal rates for the standardized design were estimated as follows:

1) Dry Ponds:	Removed by Settlement	30%
	Removed by Filtration	25%
	Total	55%
2) Wet Ponds:	Removed by Settlement	40%
	Removed by Filtration	25%
	Total	65%

These estimated efficiencies are not expected to be constant for all storms over a full annual cycle of meteorological events. It is expected, however, that the proposed design will meet the 50 percent average annual removal goal. A testing program is being established and, hopefully, field performance data will be available within the next two years to show how the standardized design will actually perform in the field.

Establishment of Design Parameters

To meet the stated goal for the Cherry Creek watershed, namely, removing at least 50 percent of the phosphorous in urban runoff, the authors decided to develop a standardized approach for the design of detention facilities. On-site detention for the control of runoff peaks is a requirement in Colorado for the unincorporated portions of all counties in the State.

Adding on stormwater quality enhancement capabilities to such ponds is practical and can be achieved with little or no additional demand for land area. The standardized designs assume that stormwater detention will need to be provided to control the runoff quantity, but the basic design can also be used without such additional detention.

To select the design event, the authors drew upon the work by late Jon Colt of the Adams County Planning Department. Mr. Colt analyzed a 20 year record of recorded rainfall at the Denver National Weather Service rain gage. Table 4 summarizes his findings (3) and shows that almost 90 percent of all storms had 13 millimeters or less rainfall. Although the larger storms account for over 50 percent of the recorded rainfall, their frequency of occurrence is only ten percent. In addition, the National Urban Runoff Program (12) characterized the Denver region as having event mean rainfall depth of 5 millimeters, which agreed reasonably well with Mr. Colt's estimate of 6 millimeters.

Based on Mr. Colt's findings, it was reasoned that a phosphorous removal pond needs to control approximately 12.7 millimeters of runoff from the impervious surfaces in the tributary basin. Studies in Denver (14) also show that practically no runoff occurs from the grassed surfaces during storms of 26 millimeters or less; therefore, sizing of detention ponds to intercept runoff from only the impervious surfaces appears to be appropriate. It is expected and hoped that this criteria will, in fact, provide reasonably good phosphorous removal rates for storms producing up to 38 millimeters of rainfall.

After evaluation of the literature, the rainfall characteristics in Denver and the various concerns described earlier, the authors chose the following basic design parameters for the standardized design.

1. Provide a detention volume equal to 12.7 millimeters of runoff from the impervious surfaces in the watershed.

2. Provide an outlet to drain the detention volume in approximately 40 hours and use a perforated riser pipe for an outlet.

3. Provide a follow-up sand filter (masonry sand gradation) with underdrains. The surface area of the filter bed is to be based on a maximum hydraulic loading of 0.61 m3/sec /ha.

4. Allow both wet and dry ponds, but encourage wet ponds.

5. Require the pond length to be at least twice the pond width to reduce short circuiting.

6. Provide a baffle structure near the inflow point to diffuse the inflow currents.

7. Permit the water quality enhancement facilities to be a part of on-site or regional stormwater quantity management ponds.

In addition to the above parameters, the standardized design has to be relatively simple to construct and has to provide for relatively low level of operation and maintenance. The clogging of underdrains experienced in the original outlet design in Fairfax County (8) suggested a need for a separate filtration system downstream of the pond. This arrangement is expected to result in an extended life for the filter bed since much of the suspended solid load will be removed by sedimentation in the pond. Another technique also used in Fairfax County, that of wrapping the perforated pond outlet riser in filter fabric, was rejected because it was discovered the fabric had to be replaced frequently because of clogging. Other maintenance factors that were incorporated included accessibility for cleaning and repairs, flat pond side slopes (i.e. 4-horizontal to 1-vertical), and access by manholes or cleanouts to all underground pipes.

Consolidation of individual on-site detention facilities is strongly recommended to promote more reliable maintenance and institutional commitment. Although on-site detention is often promoted as a solution to stormwater management problems, proliferation of such facilities in large numbers practically guarantees that many of them will not receive adequate maintenance. On the other hand, consolidated regional facilities have a much greater visibility and, as a result, can expect better upkeep.

Standardized Design

Simple guidelines for the design of the pond, outlet and filter were developed that leave little room for misunderstanding. Such simplification and standardization reduces the possibility of individual designers misunderstanding which basic elements are needed and how they should be designed. There remains, however, considerable latitude for creativity in how each site is graded, shaped and incorporated into the overall land development site plan.

The 12.7 millimeters of runoff from impervious surfaces can be calculated using,

$$V_Q = 1.27AI \tag{1}$$

in which,

V_Q = required pond volume in cubic meters

A = area of the tributary watershed in hectares

I = impervious portion of tributary watershed in percent

It is recommended that this water quality volume be located, wherever possible, within the 100-year flood routing facilities normally required for stormwater management in the Denver area. Figure 2 depicts a typical layout of a water quality pond which includes such features as a baffle near the inflow, a minimum length to width ratio of two, a separate water quality outlet and a sand filter downstream of the pond.

Figure 3 illustrates an outlet riser for a "dry" pond. A gravel pack around the outlet is intended to screen out larger debris that can clog the riser perforations. Although there is no data to verify it, the gravel may also provide some biological treatment from growth on the rocks. When the pores in the gravel pack become clogged, it can easily be replaced.

Of note is the standardized sizing and spacing of the perforations. The standardized design contains tables relating the numbers and sizes of perforations and riser diameters to assist the designers in selecting the most efficient outlet configuration. Since the discharge rate through the perforated outlet riser depends on the net area of the perforations in each row and the depth of water, the variable discharge rate was integrated and reduced to a design chart. Figure 4 permits the designer to determine the standardized outlet configuration and insures that the detention volume will drain, for the most part, in approximately 40 hours. Performing the necessary calculations without the assistance of Figure 4 can be very tedious and time consuming.

The water quality detention pond outlet releases its flow onto a sand filter consisting of a 30 centimeter thick layer of mortar sand over a gravel/pipe underdrain. Sizing of the filter bed depends on the allowable unit loading rate for the filter material which is built into equation 2. Thus, the surface area for the filter can be calculated using,

$$AF = 1.65\ Q_Q \tag{2}$$

in which AF = surface area of sand filter in hectares

Q_Q = peak discharge rate from the water quality pond in m3/sec.

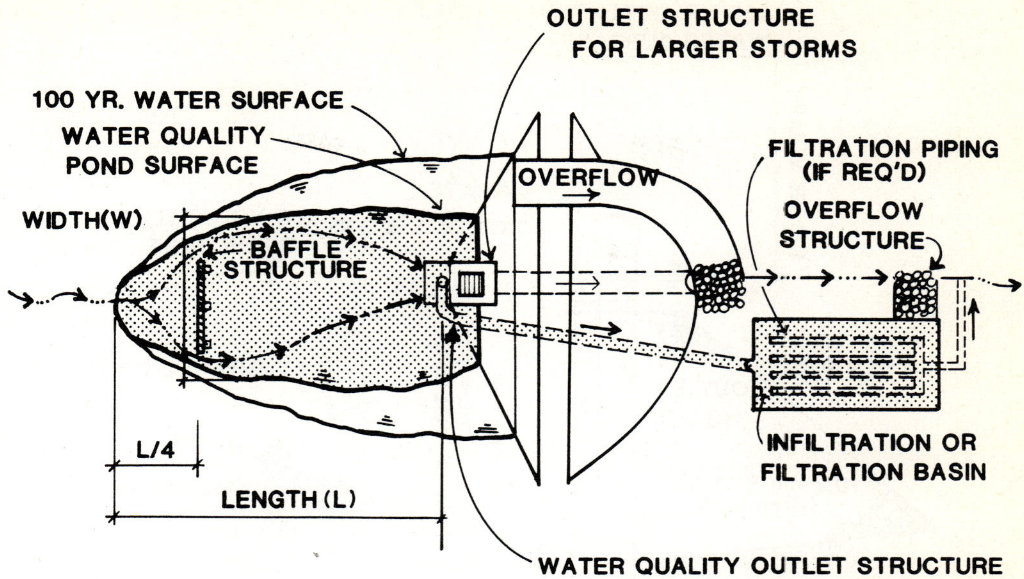

FIGURE 2
STANDARIZED POND LAYOUT FOR
TREATING FOR PHOSPHOROUS

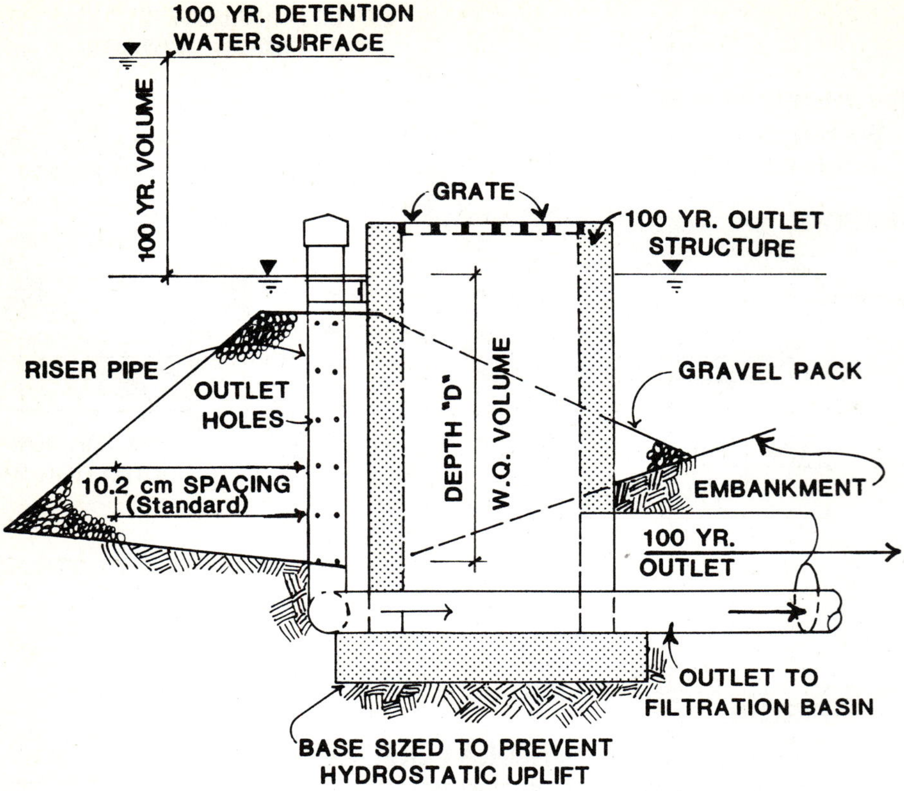

FIGURE 3
STANDARIZED OUTLET FOR A "DRY" POND

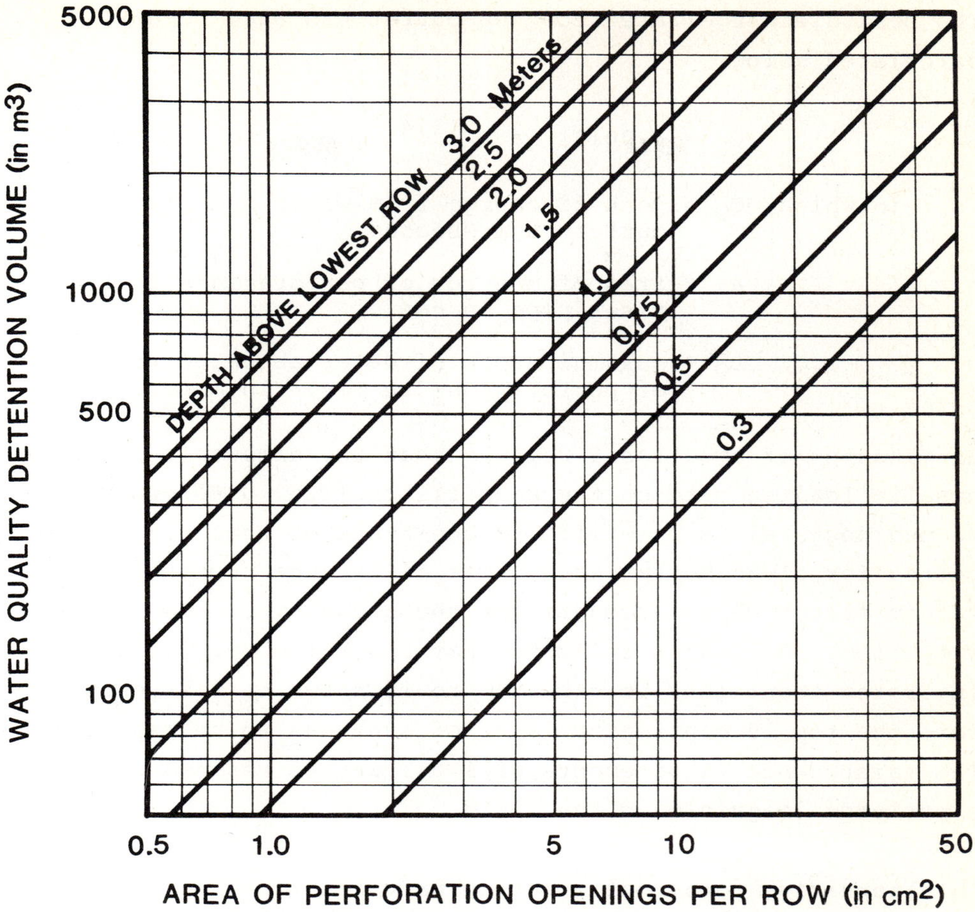

FIGURE 4

RELATIONSHIP BETWEEN POND VOLUME, PERFORATION AREA AND DEPTH OF POND.

The peak discharge rate from the outlet can also be easily calculated using,

$$Q_Q = 0.00191 \ a \ D_Q^{\ 1.4} \tag{3}$$

in which, Q_Q = peak discharge from water quality outlet in m^3/second

a = area of one row of perforations in the riser pipe in cm^2

D_Q = maximum depth of water above bottom row of perforations in riser in meters.

Examination of equation 2 reveals that the allowable unit hydraulic loading rate on the sand filter (i.e. 0.61 cubic meters/second/hectare) is one-fifth to one-tenth of what can be expected for clean mortar sand. The difference represents a safety factor which accounts for the clogging of the filler with time. Estimates indicate that under the permitted loading rates the filter is expected to last two to six years before the top 15 centimeter layer of sand has to be replaced. The maintenance requirements will be further refined after experience is gained in the field.

An alternative pond configuration provides for a permanent pool of water having a volume equal to 13 millimeters of runoff from all tributary impervious areas in the watershed. This is a preferred configuration since the existing field data (NURP) indicate "wet" ponds outperform "dry" ponds. Figure 5 illustrates the outlet for a "wet" pond which basically consists of all the same components except that the gravel pack around the outlet is not required. It is also expected that the "wet" ponds will provide greater aesthetic appeal.

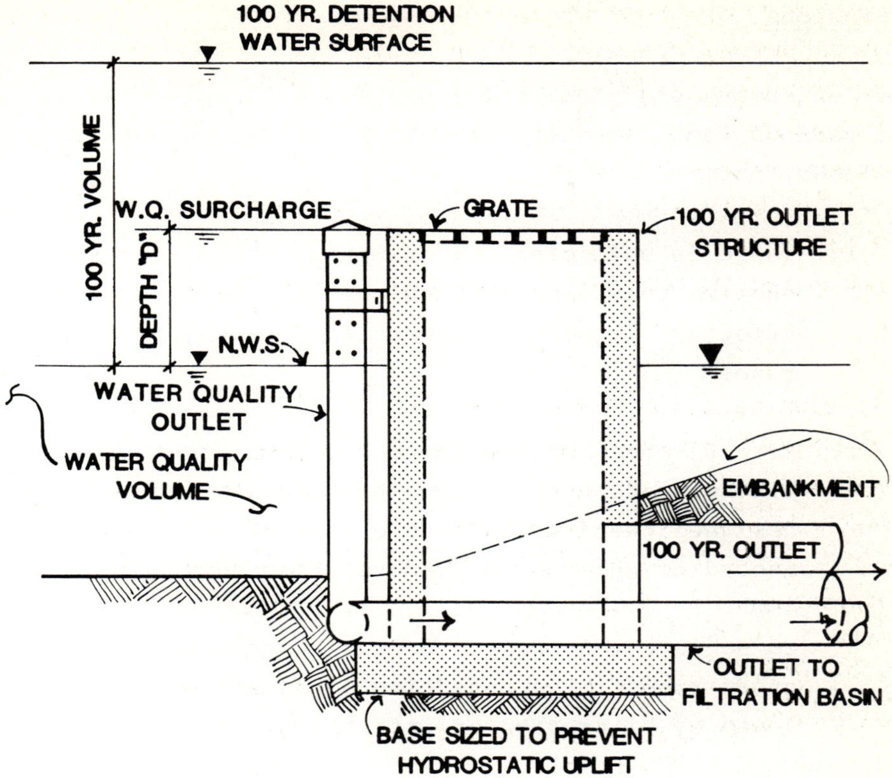

FIGURE 5
STANDARDIZED OUTLET FOR A "WET" POND

Conclusions

After extensive study of literature and information received from various local agencies in United States the authors have concluded that,

1. There is sufficient experimental and practical data showing that properly designed detention can be an effective method for improving the quality of urban runoff.

2. Effectiveness of phosphorous removal from urban runoff using detention alone has been varied. Properly designed "wet" ponds have a consistently better track record of phosphorous removal than "dry" ponds.

3. To achieve a 50 percent removal of phosphorous from urban runoff, the detention pond will need to be followed by filtration or infiltration of the stormwater.

4. A standardized design is suggested for the Denver area to provide designers and administrators with an easy to follow design approach that can fit into the on-site detention requirements for the control of runoff quantity.

5. Field testing of the suggested design and research of the basic physical, biological and chemical processes in Urban runoff water quality treatment ponds is needed before the existing state-of-the-art can significantly move forward.

References

1. Athayde, D., Nationwide Urban Runoff Program (NURP), APWA Reporter, American Public Works Assoc., (April 1984) p 10-11

2. Driscoll, E.D., Performance of Detention Basins for Control of Urban Runoff Quality, International Symposium on Urban Hydrology, Hydraulics and Sediment Control, University of Kentucky, Lexington, Kentucky, July 1983

3. Erosion and Sedimentation Control Planning Manual, Adams County Planning Dept., Adams County, Colorado, 1982

4. Griffin, D. M., et. al., Efficient Design of Stormwater Holding Basins used for Water Quality Protection, Water Research, Great Britain, Vol. 14, 1980, p 1549-1554

5. Grizzard, T. J., et. al., An Evaluation of Stormwater Management Ponds for Control of Urban Runoff Pollution, Virginia Polytechnic Institute and State University, Blacksburg, Virginia, August 1982

6. Personal Communication, George Oswald, Director of Watershed Management Division, Dept. of Public Works, Austin, Texas, September 12, 1984

7. Personal Communication, Marty Wanelista, University of Central Florida, April 1985

8. Preliminary Design Manual for BMP Facilities, Dept. of Environmental Management, Fairfax County, Virginia, August 1980

9. Randall, C.W., et. al., Urban Runoff Pollutant Removal by Sedimentation, Proceedings of EF/ASCE Conference Stormwater Detention Facilities Planning, Design Operation & Maintenance, Henniker, New Hampshire, August 1-6, 1982

10. Randall, C.W., Basic Consideration for the Utilization of Stormwater Detention Ponds for Water Quality Control, Virginia Polytechnic Institute and State University, Blacksburg, Virginia, August 1982

11. Randall, C.W., Detention Pond Outlet Design for Pollution Control, Virginia Polytechnic Institute and State University, Blacksburg, Virginia, August 1984

12. Results of the Nationwide Urban Runoff Program, Water Planning Division, United States Environmental Protection Agency, Washington, D.C., December, 1983

13. Ruzzo, W.P., Arapahoe/Douglas County Draft Storm Drainage Criteria Manual - Technical Memorandum No. 3, WRC Engineering, Inc., Denver, Colorado, September 28, 1984

14. Urban Runoff Quality in the Denver Region, Denver Regional Council of Governments, Denver, Colorado, September 1983

15. Urban Stormwater Management, Special Report No. 49, American Public Works Assoc., Chicago, Illinois, 1981

16. Wallen, W.W. Jr., Eutrophication Factors and Objectives for Cherry Creek Reservoir, Concord, Massachusetts, April 1984

17. Whipple, W., et. al., Stormwater Management in Urbanizing Areas

18. Whipple, W., Hunter, J.V., Settleability of Urban Runoff Pollution, Journal Water Pollution Control Federation, Vol. 53, Number 12, (December 1981) p 1726-1731

DISCUSSION - RUNOFF QUALITY MANAGEMENT

In the discussions which followed the presentation of papers, the following points were raised by the participants:

1. Mr. Jacquet observed that on the Seine-St. Denis project they had a swirl concentrator which did not function properly because it was oversized, and as a result did not overflow frequently, causing sedimentation in the system which was then flushed out during overflow events. It was his opinion that such operational considerations were much more critical to actual operational performance than detailed refinements in the hydraulic design of swirl concentrators.

2. Dr. da Cunha asked about several factors which could influence runoff quality management. These included:

    a. The effects of the increased use of unleaded gasoline - Mr. Urbonas replied that, in Denver, there had been dramatic reductions in lead concentrations in runoff. Other participants indicated similar experience.

    b. The effects of such practices as the use of porous pavement and rooftop storage - Mr. Urbonas and Mr. Field both noted that while they are effective, institutional problems frequently render them ineffective after a very short time. Mr. Urbonas believes that we do not have enough police power to insure continuous control, and that control at the source is better.

c. Control of corrosion - Mr. Field noted that salt control (controls on storage facilities) seems to be the only practical possibility.

d. The use of electro-magnetic separators - Mr. Field stated that they are a very effective control technology, able to effect very high removal rates. They are expensive and energy-intensive, however, though they are competetive with other advanced waste treatment processes.

3. Dr. da Cunha also asked about the effectiveness of dissolved-air floatation in treating runoff - Mr. Field replied that energy costs are higher, but that one can achieve 3-5 times higher loading rates. This can be very useful in situations where available space restricts the size of settling facilities, or with existing tanks to effect a capacity increase.

4. Mr. B. N. Jacobsen asked Dr. Damhaug if there had been any improvements in swirl concentrator design, particularly the elimination of the center cone (a problem because it is the most expensive part of the device) and substituting a side weir. Dr. Damhaug replied that there had not. Mr. Jacobsen also asked if the side weir had been considered in the original designs, and, if so, why it was rejected. Dr. Damhaug replied that it had not been considered. Mr. Ellis noted that work done in the United Kingdom

indicates that the low-side weir does not perform too well, although high-side weirs are better. Low-side weirs allow longitudinal roller action in the main channel which mixes polluting solids throughout the flow and thus leads to high pollutant concentrations in the overflow. High-sided weirs (with throttles) ensure that the first spill occurs at the prescribed flow setting, and also ensures that flow into the chamber is subcritical. This requires long weir lengths, a stilling zone of 4D in length at the upstream end of the chamber, a scumboard behind the weir, etc.. Mr. Field noted that minor design changes in the swirl concentrator can radically effect device performance, and cautioned against such changes. He also noted that side weir overflow is used in some cases where is it necessary to relieve flows greater than about 2 times design flow - as a skimming device to cut off excess flow and allow the device to operate on at least a portion of the flow.

5. Dr. Stahre asked Mr. Dartus if they would be working with the Seine-St. Denis project to collect data to demonstrate the sedimentation tank designs described in his paper. Mr. Dartus replied that they would.

6. Dr. Schilling asked Mr. Urbonas whether phosphorus was retained in the soil under infiltration basins. The reply was that in Denver and elsewhere the soils seem to trap much of the phosphorus. This generated considerable additional discussion, with several individuals noting that trap efficiency was very dependent on soil type, and that breakthrough could occur in time. Dr. Lijklema pointed out that some clays, for instance, form aluminum phosphates which are quite stable.

7. Dr. Roesner noted that wet ponds should be designed with sufficient depth (2-3 meters) to prevent light penetration and resultant weed growth.

8. Mr. Jacquet asked whether the dry and wet pond designs in Denver included some provisions for periodic sludge removal. Mr. Urbonas replied that filter cleaning and sludge removal is included, and that the cleaning frequency will be a function primarily of upstream sediment controls. Dr. Huber reinforced this, noting that dredging is usually necessary if ponds are to continue to improve water quality. Several individuals pointed out that some data shows negative pollutant removal efficiency in ponds, due probably to what happens in the ponds between storm events.

URBAN RUNOFF QUALITY MANAGEMENT BY REAL-TIME CONTROL

Wolfgang Schilling
Institut fuer Wasserwirtschaft
Universitaet Hannover
Callinstrasse 32
3000 Hannover 1
Federal Republic of Germany

Introduction

Consider an automobile factory representing an investment of $100 million. Designed to produce a peak of 1000 cars per day, it produces at most times only about 50 cars per day. Sometimes bottlenecks and accidents occur due to the non-synchronized assembly lines, and 20% of the production is lost to the scrap heap. Sometimes parts of the production line are inundated with supplies at the same time that other parts cannot produce due to shortages. However, no one knows exactly what is happening in the plant since it runs without supervision.

Consider now a combined sewer system (CSS) representing an equal investment of $100 million. Designed to carry the 5-year storm flow, it mostly carries dry weather flows (DWF) of no more than 5% of the design flow. Since the treatment plant can only handle about twice the dry weather flow, the plant is sometimes overloaded with dry weather peaks and after even small rains the overflow regulators are actuated, resulting in combined sewer overflows (CSO). Occasionally, even basement flooding results, although upstream detention tanks are not filled.

The conclusion? A ridiculous factory, but a common sewer system. Why are they operated so differently? Why is "performance" almost a religion in private industry? Is it

NATO ASI Series, Vol. G10
Urban Runoff Pollution.
Edited by H. C. Torno, J. Marsalek, and M. Desbordes

really questioned in the public sector? Did we overlook a cost effective methodology to solve our urban runoff pollution (URP) problems, and do we now face a high-tech revolution in URP management, or is the idea of real-time control (RTC) of wastewater transport systems damned to fail in any case? Which are and have been the driving forces behind the development of real-time control systems (RTCS) in urban runoff management? Is this tremendous market only waiting to be served? RTC is applied in industries such as oil refinery, paper and pulp, manufacturing, power production, traffic control, and last, but unfortunately not least, the military complex. What about urban drainage?

Urban Storm Drainage and Water Pollution Management

Scope of the Problem

The early drainage systems in modern Europe were storm sewers. In the 19th century these sewers were more and more used to remove wastes, and thus gradually turned into combined sewers. For almost a century people seemed to live happily with their drainage systems and their later implemented interceptors and treatment plants. In the industrialized countries this peaceful situation changed within a relatively short period during the last two decades. What makes this time so different from "the good olde days"?

1. Apart from undesired exceptions economic growth is exponential with the exponent greater than the one of population growth (documenting the increasing standard of living). It is every society's expressed policy to maintain this growth pattern. The economy of a country can be viewed as a reservoir, with the production of goods being inflow, their use the storage, and their consumption (i.e. wastes) the outflow. The limited lifetimes of all goods explain why - in the long run - the exponentially increasing inflow implies exponentially increasing outflow.

2. The outflow capacity of this reservoir does not increase. On the contrary, the potential of receiving waters to remove pollution has even deteriorated by such means as waterway realignment and improved land drainage. Especially in Europe, the density of population and industrialization made it clear that simply "getting rid of the wastewater" is no longer acceptable for URP management. This policy mainly shifts problems downstream.

3. Pollution used to be biodegradable leftovers from humans, animals, and old style production. Most of the pollutants were chemicals that existed in the environment. This led to the implicit assumption that nature would ultimately be able to handle them. Modern production, however, creates a diversity of artificial products of which the effects on the environment are hardly known. Biodegradation does not necessarily apply to many of the modern pollutants. Their long term behavior in ecosystems is not well understood.

4. There is no parameter such as "pollution". One at least has to consider batch indicators such as floatables, suspended solids (SS), biological oxygen demand (BOD), chemical oxygen demand (COD), nutrients (i.e. phosphorus, nitrogen), fecal bacteria, and toxics (i.e. lead, cadmium, organic hydrocarbons) to roughly describe the quality of water. Unfortunately, sources as well as impacts of pollutants show a tremendous diversity which hardly can be explained, let alone predicted, even for old style pollutants. With respect to new pollutants the situation is worse. Sometimes major disasters such as the Minamata disease in Japan (mercury) and the Seveso accident in Italy (dioxin) are needed to make the public aware of how dangerous a waterborne substance can be. Through these tragic instances mankind learns about possible effects of hazardous substances, one out of thousands each time. Public concern then focuses on these substances, with the danger of neglecting all others.

This is the framework in which the URP problem is imbedded. Urban runoff should be defined generally enough to include a number of important sources and routes of pollutants. It is certainly not sufficient to define it exclusively as rainfall-induced outflows from storm sewers. A better definition might be: all runoff out of urban areas passing from discrete outlets into surface waters. This definition allows the inclusion of important pollution sources such as direct surface runoff, storm sewer discharges, combined sewer overflows, sanitary sewer diversions, and treatment plant effluents and by-passes.

Combined Versus Separate Sewer Systems

CSS collect sanitary sewage and stormwater runoff within a single conduit. Separate sewer systems employ one conduit for stormwater runoff and one for sanitary sewage. In existing systems, however, these simple definitions almost never apply. Combined sewers often carry additional sewage from upstream sanitary sewers increasing the strength of occasional CSO. Sometimes combined sewers are even used to transport sludge from upstream to a downstream treatment plant. Sanitary sewage is sometimes discharged into storm sewers through false connections causing continuous "stormwater" runoff and pollution. Conversely, sanitary wastewater flow patterns sometimes show a striking similarity with combined flow patterns, i.e. peak flows during rains. In the worst case a separate system might actually consist of two combined sewers. The smaller one transports sewage to the treatment plant and the larger one discharges directly into the receiving water. While false connections cause storm dependent inflows in sanitary sewers, jammed tide gates can cause large amounts of backflowing river water into CSS. Sanitary as well as combined sewers usually suffer from the problem of infiltration (through cracks, worn out joints).

Intended deviations from the genuine designs also exist. To reduce stormwater loading roof leaders might be disconnected from a combined sewer. To reduce the cost of separation, sometimes only road runoff is connected to new storm sewers leaving roof drainage hooked up to the old combined sewers. Rather than allowing sanitary sewers to discharge into a CSS, it might be worthwhile to use parallel combined and sanitary sewer lines or even to install a sanitary sewer within the combined sewer.

Economics and performance of treatment processes require that only sewage with high enough pollutant concentrations be treated. Secondary treatment capacity is usually only 1.5 to 4 times DWF. Excess flow, be it sanitary/infiltrated or combined, may be treated by primary processes alone or diverted directly to the receiving water. During a storm up to 95% of the organic load may bypass the treatment plant.

Although this happens only at approximately 6% of the time and about 10 to 100 times per year, the annual load discharges from CSO in the United States are 22% of the total urban BOD and 44% of the total SS discharges [25]. If the "best practicable dry weather treatment" technology (i.e. secondary treatment) is applied, CSO's even cause 37% of annual BOD and 64% of annual SS discharges, Tab.1. It is estimated [62] that between 40% and 80% of the total organic load from urban catchments originates from sources other than treatment plant effluents. These numbers show that secondary treatment of DWF can only be part of a solution to improve the performance of CSS.

CSS "combine" the problems of sanitary and storm sewers including:

- Very low dry weather flows during the evening hours, causing sedimentation, septic conditions, and corrosion of pipes.

- Very high flows during storms (up to 100 times DWF), causing CSO and flooding with wastewater.
- High concentrations of organic pollution (50% of raw sanitary sewage).
- High concentrations of solids, heavy metals, pesticides, and other toxicants from surface washoff.
- Particularly high first flush concentrations, due to both surface washoff and sewer sediment resuspension.
- Good settleability of CSO-SS in receiving waters.

The variety of flow and pollutant concentration patterns in CSS complicates treatment procedures. However, if storage can be activated and if the number of CSO locations can be reduced, most of the pollutants can be captured in CSS.

Treatment activities can be concentrated at a few locations and thereby ease monitoring and operation of the entire drainage system. CSO have to be allowed, in order to protect the collection system, the interceptors, and the treatment plant from surcharging. However, CSO often occur although there is no actual need for this protection, since there is still idle capacity. This is a major motivation for the implementation of RTC.

Table 1. Pollution Loads Into Receiving Waters in the United States from Combined Sewer Overflows (CSO) and Treatment Plant Effluent (TPE)

	Data 1973	Best Possible Dry Weather Treatment	BPDWT incl. Primary CSO Treatment	BPDWT incl. Secondary CSO Treatment
CSO-BOD	0.8 (22%)	0.7 (37%)	0.3 (20%)	0.1 (8%)
TPE-BOD	2.8 (78%)	1.2 (63%)	1.2 (80%)	1.2 (92%)
CSO-SS	2.1 (44%)	2.1 (64%)	1.0 (45%)	0.4 (25%)
TPE-SS	2.7 (56%)	1.2 (36%)	1.2 (55%)	1.2 (75%)

[in million tons]

Since CSO only occur during short periods, conclusions about their impact on receiving waters must not only be drawn by looking at annual loads, but also at their intermittent pattern of occurrence. Features where CSO may produce major disruptions to aquatic ecosystems include

- Stagnant or low throughflow in the receiving water.
- Upstream and downstream CSO into small streams or CSO into dead end canals, ("fish traps").
- Low dilution ratio compared to upstream flow, or large catchment area compared to upstream catchment.
- Low background pollution allowing for a large variety of aquatic life.
- High water temperatures causing low dissolved oxygen (DO) concentratations in receiving water.
- Extended dry periods implying high URP concentrations from surface washoff and combined sewer sediments in cases of CSO.
- High intensity storms producing high flow rates and causing resuspension of receiving water sediments by scour.

Therefore, in the worst, but by no means unrealistic, case present pollution (e.g. sanitary sewage) is strengthened with short term sediments (from CSS) and long term sediments (from receiving water). In [43] it is documented that observed pollutant concentrations in the Milwaukee River could only be explained by a combination of effects described above. Sediment scour in receiving waters affected by CSO or storm runoff is also reported in [30] and [37].

In the 15 years between 1961 and 1975, 19% of the reported fish kill incidents in the United States and 42% of the amount of fish killed were due to municipal sources (mostly sewerage systems). 72% of the incidents happened during the summer months, with low DO as the main cause [20]. Thirteen percent of the beaches in the United States have to be closed temporarily because of pollution caused by urban runoff [28].

Receiving waters within urban regions mostly have a water quality that does not allow direct body contact uses. Nevertheless, these waters are often important parts of in-city recreation facilities. Although aquatic life within these waters might not be highly developed, the public is usually sensitive to any further deterioration of water quality. Esthetic disruptions caused by CSO might be floatables or resuspension of septic sediments by hydraulic scour. Incidents of this kind often result in public outcries, although no direct economic benefit can be attached to esthetic criteria. Since public reaction seems to be more sensitive to rapid changes (e.g. a fish kill incident) than to generally bad conditions (e.g. no fish at all), water quality improvement should not only allow for enhanced acquatic life but also be stable enough to maintain it over time.

Drawbacks of Static Solutions and Passive Operation

Any approach to solve a URP problem has to start with upgrading the system to its initially intended purpose. Upgrading techniques include:

- Better maintenance (flushing, fixing defective regulators, correct diversion settings).
- Infiltration/inflow analysis with subsequent repair of cracks, worn out joints, breakdowns, blocked tide gates).
- Sedimentation analysis to identify conduit reaches with cumulative sedimentation caused by low flows, inverse slopes, etc. [10], [65].
- Hydraulic analysis to identify low capacity reaches which are prone to frequent flooding, and high capacity reaches which could be activated for storage.
- Reduction of the number of CSO locations.
- Expansion of treatment plant capacity for secondary treatment of at least peak DWF.

Although this upgrading is necessary to avoid poor system performance, it will not solve the URP problem. The most rigorous "upgrading" technique of CSS is sewer separation where a new storm sewer is usually constructed. In the United States and Canada this approach has often been adopted. Meanwhile, a number of important disadvantages resulted in a more cautious application.

- Separation is generally an extremely expensive solution.
- It is very disruptive, especially within downtown areas (which are the common CSS catchments).
- Separation cannot be achieved within a short period, because of its cost and technical complexity.
- There is increasing evidence that stormwater runoff from urban catchments contributes large amounts of especially undesired pollutants; [35], [36]. It is the largest source of SS and lead; [26], and usually contains significant concentrations of other heavy metals, nutrients, hydrocarbons, phenoles, and pesticides [23], [58].
- Stormwater and CSO SS settle better than effluent SS from treatment plants. Therefore, the former accumulate close to outfalls and cause long term hazards to the receiving water [58].

Separation is now regarded cost-effective only in a few cases. For new developments, however, in Europe as well as in North America, mostly separate systems are applied.

Opposite from high-cost separation is the other extreme, "best management practices" (BMP), which include a number of low-cost, high commitment techniques for URP abatement [24], [35], [36] such as

- improved street sweeping and catchbasin cleaning,
- erosion control and porous pavement,
- ponding of streets, parking lots, and parks,
- wastewater source control,

- disconnection of roof leaders,
- rainwater barrels, etc.

Although most BMP are source controls and therefore conceptually superior to end-of-pipe controls, their implementation requires much personal effort, enthusiasm for details, and close collaboration between different departments of municipalities.

A widely-used approach to abate URP is construction of additional storage. Storage functions reliably and serves the two-fold purpose of flood protection and pollution reduction. Whereas off-line storage is only activated if a present flow level is exceeded, in-line storage is part of the conduit and thereby permanently used. Although expensive, detention is conceptually simple and usually works. But to what extent? The simple concept of a gravitationally governed in-line tank, as shown in Fig.1, demonstrates the difference between passive (e.g. orifice) and active control (e.g. motorized gate) of the outflow. Passive control not only requires a larger volume of storage, but the the tank is also available later for the next storm. In real world

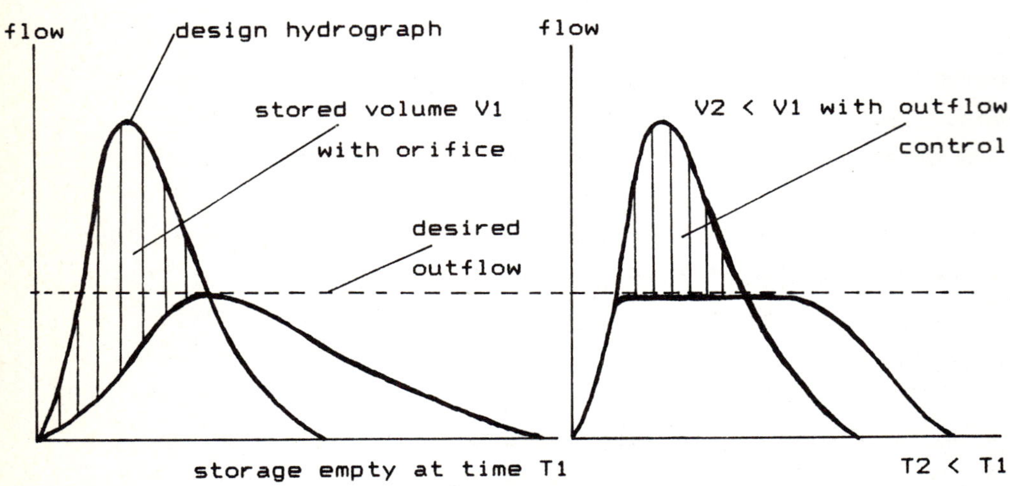

Fig. 1 In-Line Storage With and Without Outflow Control

systems with distributed storage, situations can arise where unfavorably distributed loading results in even higher peak flows and CSO than without detention tanks.

All static measures for pollution abatement have one disadvantage: They only perform optimally for one loading pattern, namely the design hydrograph. For all other loadings, static systems with passive operation necessarily have to operate suboptimally. Therefore, the concept of permanently achieving maximum effectiveness of a wastewater transport system is a driving force for the development of RTCS.

Real Time Control of Combined Sewer Systems

Definitions

A CSS is operated in "real-time control" mode if, and only if, process data currently monitored in the system is used to operate regulators to achieve better systems performance during the actual process. Synonyms for RTC are "on-line control", "operational control", and "real time operation".

For the operation of complex systems (e.g. with multiple detention tanks) the notion of "systems control" becomes important. Real time systems control is applied if the measurements allow determination of the state of the entire system. The data is transferred to some central intelligence, where decisions are made how to operate regulators with respect to the current and future states of the entire system. Synonyms for systems control are "integrated control", "strategic control", and "global control".

Ultimate performance might be reached if "optimal control" is applied where control decisions are made to achieve the maximum benefit of a pre-specified system performance criterion. "Automatic control" means that operational decisions are made by machines rather than humans. In a "computer control" system ("softwired" RTCS) a pre-programmed computer decides how to activate regulators. In a "supervisory" RTCS operators make strategic decisions on set points (i.e. desired flow rates), whereas the local control actions (i.e. operating a gate) are automatically actuated by a controller. A "manual" RTCS is a system in which all actions are initiated by operators. In a "management information system" (MIS) data is transfered to, and processed at, a central location, but there is no hardware to immediately put control decisions into effect.

In a RTCS at lease one set of the following elements apply:

- A sensor (e.g. level gage) that monitors the ongoing process.
- A regulator (e.g. pump or gate) that manipulates the process.
- A controller that decides how to exert the regulator to bring the process back to its desired (set point) value.
- A telemetry device that carries the field data from the sensor to the controller, and the instructions of the controller back to the regulator.

These four elements form a "control loop" which is common to every RTCS, Fig.2. Conceptually a RTCS can be conveniently separated into three layers [45], including:

- The "optimization" layer.
- The "direct control" layer.
- The adaption" layer.

In the optimization (strategic control) layer current and forecasted process data is used to find and adjust set points, in order to achieve optimum systems performance at all times. Either formal algorithms or heuristic decisions based on operator experience can be applied. In the direct (local, process) control layer process forecasts are compared to their actual values. As soon as the deviations exceed a limit, set

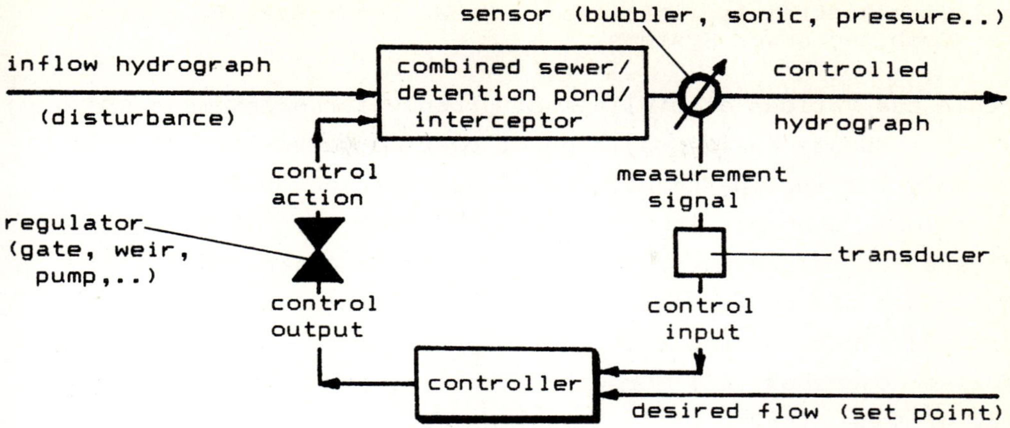

Fig.2. Typical Control Loop in a Combined Sewer System

points have to be recalculated (adapted, updated) according to the new information. In a "deterministic" control algorithm the state of the system is assumed to be known with certainty. A "stochastic" control algorithm takes quantitatively into account that the underlying process model does not exactly represent reality.

The following terms are primarily related to the direct control layer. In a feedback or "reactive" loop, controls are actuated depending on the measured deviation of the controlled process from the set point. Unless there is a deviation, a feedback controller is not actuated. A feedforward or "predictive" controller anticipates the immediate future values of these deviations using a process model. It then activates controls ahead of time to avoid the deviations. Its performance is limited by accuracy of the underlying process model. A "feedback/feedforward" controller avoids this disadvantage of a pure feedforward controller.

Real-Time Control Applications in Urban Water Systems Other Than Combined Sewer Systems

Within the various disciplines of urban water management RTC of water supply has a tradition of over 50 years [41], and is probably leading in the widespread application of modern operational concepts in urban water systems. In 1982, 259 process control computers were installed in Japan [46]. In some cases the entire water system, including pumping, transfer, purification, and distribution, is controlled. An RTCS is also described [46] for water transfer and supply for the city of Rome, Italy, with as many as 10,200 data points permanently scanned at 163 outstations. Applications of RTC in water supply systems are also reported to exist in other countries such as Denmark [46], United States [63], United Kingdom [46], [64], Switzerland [46], and Germany [44]. Attempts to develop concepts of on-line optimization for water supply control date back as far as 1963 [8], [14], [18], [22]. Control decisions are primarily made heuristically by operators, however [9]. With the apparent exception of some Japanese systems [19], RTC optimization is not applied yet.

Much later than in the water supply field, wastewater treatment plants were equipped with RTCS. In 1984, 17 plants in the UK had RTC equipment [57]. The 150 process computers installed in German treatment plants are mostly used for data acquisition but only few for RTC [40]. A nine-loop automatic control system in a Milwaukee treatment plant is described in [27]. An excellent overview on the state-of-the-art of RTC in wastewater treatment can be found in the proceedings of three international workshops on instrumentation and control [19], [31], [32]. Although models for treatment processes have existed for quite a while [2], [34], optimized on-line control of wastewater treatment has not been applied up to this time.

Applications of RTC in other urban water systems than water supply, wastewater treatment, and CSS are not so frequent.

Still important are stormwater drainage and urban flood control [15], [16], [42], and water quality monitoring, [6]. Since the flexibility to actually control these systems is quite restricted (no storage and/or no regulators) they serve mostly only as MIS.

Political and Administrative Environment

During the 1960's the public in both Western Europe and North America rapidly developed concern about the integrity of the natural environment. Subsequently, different countries passed legislation to ensure water quality protection. The Congress of the United States, for example, passed the "Clean Water Act" in 1972, requiring that:

- BOD and SS concentrations of treatment plant effluents be limited to 30 mg/l and/or 85% removal by 1983, and
- No more pollutant discharges into receiving waters be allowed by 1985.

At the same time, various countries launched research programs on URP, for which outstanding results have been published, for example, in three IAHR/IAWPRC conferences [5], [29], [67].

The first projects to demonstrate the feasibility of RTC in CSS were also initiated at that time [13], [38], [65]. Although URP research provided a wide spectrum of pollution abatement techniques, their actual application beyond the demonstration scale has fallen behind. A major reason is the fact that URP abatement programs are usually financed with grants up to 95% coming from sources other than the polluting city. For example, the United States Environmental Protection Agency (EPA) used to support eligible projects with 75% (now 55%) of the investment cost, but did not support any operation and maintenance (O&M) cost. Naturally, all projects were designed to reach a minimum O&M cost level, almost regardless of the investment cost, (an extreme example being separation). However, RTC of CSS is a typical example for a low investment, high O&M cost URP abatement technique.

Recently the thrust into URP abatement has slowed down, more, however, in the United States than in Europe. The reasons are, among others that:

- Groundwater pollution and "acid rain" as competing hazards have drawn increasing attention (and funds).
- Supply-oriented economic policies led to the reduction of available federal funds.
- Supply-oriented economic policies tend to shift responsibility to lower levels of government.
- Deregulation policies tend to shift responsibility to lower levels of government.
- Mis-perception of the true cost for URP abatement led to public upset about the effectiveness of pollution abatement programs in general [60].

This new political environment will certainly not encourage new massive URP abatement projects such as separation and deep tunnels. However, depending on continuing public pressure and/or enforcement of water quality standards, the tighter financial framework might even be a stimulus to routinely apply low investment URP abatement techniques such as RTC.

The implementation of a RTCS is also strongly influenced by features of local administration. Typically, various cities in a metropolitan area form a "Sanitary District" which is governed by a "Municipality of Metropolitan..." This configuration is perfect to manage a "clean" separate sewer system (storm and sanitary sewers by the cities, interceptors and treatment plants by the sanitary district). However, difficulties arise in the operation of "dirty" separate systems (i.e. with false connections, infiltration/inflow, high stormwater pollution) and CSS. Typically, the sanitary district is mainly interested in reducing the treatment plant throughput to reduce O&M cost, whereas the city's concern is basement flooding. Even if a sanitary district is willing to accept additional combined sewer stormflow, the district's usable storage might be very small. The cities operating large trunk sewers, on the contrary, are not eager to activate this storage since this might increase the flooding hazard. Different drainage authorities operating in a metropolitan area usually assure each other that, whatever measures are applied, backwater will not occur. Obviously, these conditions are not in favor of the implementation of RTC, where the prefered in-line storage is activitated by backwater.

The management of urban drainage systems is imbedded in public service legislation and regulations. Their main difference from the free market economy is that productivity increases have no benefit to those who achieve them. "Economic" solutions therefore are not identical to those which would originate in the framework of a competitive market.

On the operations level, an interesting discrepancy between the planning/design stages of a project and its operation/maintenance can be observed. Whereas the former is executed by engineers and technicians, operations is usually carried out by personnel who frequently know little about "what it's all about". This is often used as an argument that RTC cannot be successful because the appropriate personnel are lacking. Indeed, performance limits of existing RTCS are sometimes due to the fact that operators do not understand the purpose of the system. Operation of a RTCS requires intensive communication between all divisions and levels of a drainage agency, a management task that is extremely difficult to accomplish, especially in large organizations.

Hardware in Real Time Control Systems

Design Concepts

The basic elements of a RTCS have been introduced above as sensor, regulator, telemetry, and controller. They are also discussed in the Workshop paper by Delattre et al. The arrangement and interconnection of multiple control loops determines the design concept of the RTCS. Generally, RTCS can be divided into two groups; "decentralized" and "centralized" systems. The groups are distinguished as to where the intelligence that provides the (strategic) control decisions on appropriate set points is located. A supervisory or manual control system is necessarily a centralized system.

The first step into RTC is usually a data logging system (decentralized) or a data acquisition system (centralized), where process data is sampled and stored either on-site or at a central location, respectively.

If the sensors are each connected to a regulator and a controller, thus forming isolated control loops, the RTCS is a local control system. Here, the set points are fixed and cannot be changed during the process. Sometimes these are still "analog" RTCS where all data is transmitted in analog form. Any kind of computer control, however, requires a "digital" RTCS.

Long time delays (i.e. flow time between the regulator and the sensor) can require that local control loops be augmented or interconnected. An example is a feedforward/feedback loop where the controller uses the measured signals from both the controlled variable (e.g. storage outflow) and the disturbance (e.g. storage inflow) to determine the control action (e.g. gate opening). Another way to interconnect control loops is cascaded control, where the output of a downstream controller is taken as the set point for an upstream controller. These control concepts obviously require connecting the various loops by wire.

In the centralized layout, control is through a direct digital control (DDC), where a central computer replaces all controllers of the system. A DDC system requires wiring between all sensors to the computer and back to all regulators. It relies heavily on the availability of the central intelligence. Therefore, a DDC system is usually backed up by "hardwired" local control loops (relay controls) that maintain simplified operation when communications or the computer fail.

One RTC simplification is "bang-bang" or "I/O" control, where regulators can only work in two positions (i.e. open/closed, on/off), as compared to "continuous" control where all regulator positions between a maximum and a minimum setting can be reached.

The disadvantage common to all RTC concepts described above is that set points are fixed. Since their optimum values depend on the actual loading of the system, extra intelligence has to be added to provide their "adaption". If a centralized system already exists (e.g. a DDC system) additional hardware requirements are relatively minor. Since forecasts are needed to optimally specify the control strategy (i.e. the time sequence of setpoints), additional flow level monitors and rainfall monitors are required upstream. If the operation has to be automatic, extra computer capacity is needed for the strategic software (i.e optimization procedures, expert system).

The state-of-the-art of RTCS design is a combination of decentralized and centralized systems. Microprocessors function as local controllers, with set points updated by a central intelligence (i.e. computer and/or operator). This "distributed" control concept combines the flexibility of a DDC system with the reliability of independently-working multiple control loops.

With the recent development of powerful microcomputers it is foreseeble that these will be installed as local controllers at the regulator sites. At the same time they will communicate with each other to determine the optimal control policy. In such a "decentralized integral control system" the central computer would eventually be obsolete.

Sensors

Although a large variety of sensors do exist, only very few fulfill the requirements for CSS. These include [55]:

- A wide measurement range (level: few cm to few m, flow: Qmax = 100 Qmin).
- Resonable accuracy for routine operation on-site.
- Suitable for on-site calibration.
- Rugged, foolproof, maintenance-friendly.
- Sensing free-surface, pressurized, as well as reverse flow.
- Unaffected by humidity, solids, foam, aggressive chemicals.
- Suitable for manhole installation.
- Inexpensive.

Sensors in a RTCS also have to indicate malfunctions automatically and be suitable for remote data transmission. Although manufacturers usually guarantee good accuracy, repeatability, sensistivity, linearity, resolution, availability, etc. under laboratory conditions, these performance indices are definitely less favorable in routine operation. A monitoring system is a data processing chain where each link (i.e. sensor, transducer, data transmission system, data processing system) is subject to random noise. Measurement errors have to be defined as the integral errors originating along this chain, and not just the error of the sensor output. Although, for example, a gate opening indicator and a bubbler would have an accuracy of 5% each, the overall flow measurement error with the gaging device might easily be ten times as high.

Water level sensors are presently the only sensors which are routinely applied in CSS. The following principles are used:

- With a bubbler a constant rate of air, or other gas, is injected into the wastewater close to the bottom of the conduit. The measured backpressure is proportional to the water depth above the air outlet.

- An air column pressure sensor measures the water depth through the compression of a static air column in a pipe. The pipe is hanging in the wastewater and is slightly inclined into the flow direction to prevent clogging.
- In a water pressure sensor a diaphragm/spring system is used to sense the water pressure acting on the sensor, thereby indicating the water depth.
- With a sonic level sensor the travel time of ultrasound from the probe to the water level and back is converted into the water depth.

Other water level sensors such as conductivity probes, capacitance probes, and floats are not suitable for CSS.

Only a few flow sensors are available to measure combined sewer flow directly. Their high cost, inaccuracy, or limited range have prevented their routine application so far. The most commonly applied "flow" measurement technique is the conversion of a level measurement into a flow rate using flow formulas such as the Manning equation. Here, errors result from the fact that:

- The flow is proportional to the 8/3 power of the level, hence any level measurement error is amplified.
- The actual water level gradient is unknown.
- Roughness and gradients of old conduits are difficult to assess, unless these constants are calibrated [33].
- Unknown sediments considerably reduce the cross sectional area [10].
- Pressure flow cannot be handled at all.

More accurate techniques combine depth and velocity measurements. Methods applied for velocity measurement use:

- The Doppler effect, where the frequency shift of ultrasound passing through the wastewater is proportional to the flow velocity [1].
- Electromagnetic induction where the conductor "wastewater" induces a voltage by passing a magnetic field that is generated by coils above the conduit [47], [61]. This technique is very accurate for pressurized flow (e.g. through an inverted siphon).

In RTCS, regulators can be used for flow measurements. Structures such as perpendicular sharp-crested weirs (bascule gates), sluice gates, constrictions, and vortex valves can be calibrated, and in conjunction with level sensors (vortex: pressure sensor) used for flow monitoring.

Because of the difficulties and high costs of in-sewer measurements, a second type of urban drainage system analysis is traditionally applied - rainfall-runoff modelling. Based on the physical and conceptual understanding of the runoff generating processes, rainfall data is computationally transformed into runoff and conduit flows. Raingages are therefore an important part in a URP monitoring network. Furthermore, the time delay between rainfall and runoff can be conveniently used to provide short-term runoff forecasts. Point measurements of rainfall intensities are straightforward, with errors less than 10% for moderate to high intensity storms [54]. A major source of uncertainty, however, is the spatial variablity of rainfall, which is impossible to detect with the usual raingage densities [49]. A large number of raingages is required to keep the errors of computed runoff resulting from undetected rainfall below 10% [52]. Most types of raingages feature a collector funnel leading precipitating water into a calibrated container. Either the rising velocity of a float, the increasing weight of a bucket, or the time interval between the "tips" of a tipping bucket indicate the actual rain intensity. A different principle involves the formation of "norm" drops through a small pipe below the collector funnel. A highly accurate, though expensive, raingage with extreme temporal resolution is a drop spectrometer that measures the impact of each rain drop falling on a diaphragm. Knowing the terminal falling velocity of raindrops and their size, the rainfall intensity can be determined. The most commonly applied type seems to be the tipping-bucket raingage.

Especially suited for supervisory or manual RTCS are qualitative rainfall measurements by radar [17], [42]. Every major airport operates a radar that can scan an area of approximately 30 km radius for approaching storms. This data can be telemetered to a RTC center. In the USA and UK, quantitative radar data is available by user subscription. A number of cities either experiment with, or already apply, radar to aid operations, i.e. Detroit, Chicago, Milwaukee, Seine St-Denis/France, Ruhr district/Germany.

Water quality sensors play a very minor role in RTC of CSS. Except for some explosives monitors and quality sensors in receiving waters (i.e. for SS, pH, DO, temperature, redox potential, conductivity), quality sensors are not routinely used.

Also important are status indicators such as the mercury switch, diaphragm switch, thermal dispersion sensor, or proximity sensor, although they do not have a continuous measurement range. They are widely used in hardwired control loops as backup systems.

Regulators

The best known regulators for combined sewer flows are pumps. Wastewater pumps are the most flexible, but also most expensive, flow regulators. Pumping stations in CSS are usually equipped with a number of constant-/or variable-speed radial pumps to handle the wide range of flow rates. If deep interceptors are used for in-line storage, the sump levels can vary considerably. In these cases the proper selection of the optimal operating points of the various pumps is important. Otherwise, cavitation and excessive wear and tear can occur. Especially if large flows have to be pumped, measures for transient suppression have to be carefully designed to avoid upstream damage due to emergency shut off and subsequent surges and water hammer.

Other remotely controllable regulators in CSS include [3], [4]:

- Gates (i.e. sluice, radial, sliding).
- Weirs (i.e. perpendicular, side-spill, leaping).
- Air-regulated siphon weirs.
- Air-regulated inverted siphons.
- Valves (i.e. plug, knife, butterfly, vortex).
- Inflatable fabric dams.

Regulators in CSS are often very large and custom designed. However, some basic principles are common to all successful designs.

1. Regulators are fail-safe designed, in that malfunction of vital parts results in an acceptable functional decline of the system ("graceful degradation"). For example, sluice gates would have by-passes, weirs would move into the lowest position in case of a power failure, etc.

2. All parts exposed to sewage and the sewer atmosphere are drastically simplified and corrosion resistant. The preferable material is stainless steel. Acceptable materials are cast iron, aluminum, and plastic.

3. Sensitive parts are placed in an appropriate environment, i.e. extra-dehumidified vault for hydraulic and electric machinery, extra-dehumidified and heated vault for transducers, telemetry equipment, etc.

4. All parts of a regulator station (including gates, sensors, motors, etc.) are accessible, maintainable, and exchangeable.

5. Vital functions of regulators can be remotely supervised from the control center.

Data Transmission

Any regulator featuring external power needs a special data transmission system. If very short distances apply between the sensor/transducer, the controller and the regulator/motor, data transmission can be analog (pneumatic or electric). If the signals are frequency modulated, transmission distances can be increased. Here, the voltage or current transducer output is transformed into a frequency. For catchment-wide RTCS, digital data transmission is increasingly applied. Advantages compared to analog transmission are that:

- Some transducers have already digital output (e.g. tipping bucket raingage).
- Digital data is suitable input for digitl computers,
- Transmission reliability (against noise) can be as high as desired.
- High information transmission rates (bits per second, bps) can be obtained.

If analog data is to be transmitted in digital form, it first has to be decoded with an analog/digital (A/D) converter. There, the analog signal is sampled with a frequency higher than twice the maximum process frequency to avoid loss of information. The sampled signal is then approximated by the closest binary number given a selected word length. For example, 1% conversion accuracy requires a 7 bit word length. The words are arranged and transmitted in blocks. Each block consists of words indicating the beginning of a transmission, source, number, type (request/transmit), block length, the information, the termination of the transmission, and control bytes for data processing.

Transmission can be done by wire or wireless. However, in most of the European countries wireless transmission is restricted to mobile transceiving stations. Transmission by wire uses either privately owned, leased, or dialed public telephone lines. Leased lines are generally used, to take advantage of the services offered by the telephone companies. Dialed lines are preferred if data transmission over long distances is not required at all times, e.g. for security checks, transfer of stored data, or rain data transmission. If public telephone lines are to be used, data has to be fed into an approved modem which connects the RTCS with the telephone network. The modem provides the appropriate frequency (tone) modulation of phone transmission.

Several logical channels can be transmitted in one physical channel if multiplexing is applied. With frequency multiplexing a freqency band, and with time multiplexing a time slot, is assigned to each logical channel.

Required capacities of transmission channels depend on the number of data points, the scanning frequency, and the amount of information per scan and data point. A standard for data transmission in public telephone lines is 1200 bits per second. Typically, a sluice gate regulator station has on the order of 10, a pumping station on the order of 100 data points.

Controllers and Process Computers

Mechanical or pneumatic controllers are widely used in industry. However, these types of analog controllers are difficult to modify remotely. Time-varying set point adjustment is practically unfeasible in topographically large RTCS. With the development of digital computers many analog controllers could be replaced by one central computer. This also would allow a more flexible controller calibration, interconnection of control loops, and set point adjustment. With the recent advent of inexpensive microprocessors the vulnerability of such a system could be overcome by implementing a central minicomputer and several local microcomputers. Although they have separated tasks they can back each other up. Typically, the on-site microprocessor ("programmable field panel", "remote terminal unit") controls and coordinates all functions of an outstation, including:

- Acquisition of measurement data.
- Processing (smoothing, filtering, etc.) of data.
- Status, functional, and limit checks.
- Temporary data storage.
- Control computations and output.
- Receive and report data to and from the central station.

To perform these tasks the microcomputers are equipped with input/output interfaces for local/manual operation, A/D converter, modem, energy supply and backup, real time clock, and storage for software and temporary data recording.

Differences between microcomputers and central process computers are gradual. In distributed RTCS their tasks are somewhat interchangeable and gradually shift towards the on-site microcomputers. However, a number of major tasks usually remain for the central computer.

1. System wide data acquisition, long term storage, and data management is performed through the central computer. This data base provides the performance record of the RTCS, and is used for all kinds of further planning.

2. Usually the operator interfaces are controlled by the central computer. It might also be used for interactive simulation/optimization software ("decision support software").

3. In the case of automatic systems control (optimized control, expert system), set points are adjusted by the central computer.

4. Since the central computer acquires system wide rainfall and flow data, it can be used to generate automatic forecasts of system inflows using rainfall forecasting and rainfall-runoff models.

Central process computer hardware is similar to microcomputers. Generally, they feature more user-friendly operator interfaces, higher computing capability, disk and magnetic tape storage, and more input/output devices such as teleprint, hardcopy, plotter, etc. For reliability reasons and/or background computations, two central computers are often employed ("front end"/ "host").

Real Time Control Software

Operational Concepts

This section discusses the concepts of how to operate a given system of RTC hardware. Operational options depend on the existing hardware. A data logging system provides data that can only be used to simulate, for example, a planned RTCS. Without regulators these simulations cannot be verified. A central data acquisition system, furthermore, allows monitoring of runoff processes in real time (without the capability of manipulation, however). Central monitoring of drainage systems is usually the starting point of RTC, because it reveals immediately how unfavorable storage and transport capacities are used unless active control is applied.

If local control loops are not interconnected control, options are still very limited. If operational data is collected with on-site data loggers it is tedious to improve the calibration of controllers, e.g. avoid instability such as "gate hunting", avoid transient generation, or increase control speed. Set points might also be corrected to provide better average systems performance. However, individual set point selection for each event is unfeasible since, in a local control system, this requires manually changing the set points at the regulator site. If operational data is centrally acquired, it is easier to suprvise the system. If problems arise, maintenance crews can be dispatched without time delays.

Fixed set points, of course, are a disadvantage, as is demonstrated in Fig.3 and Fig.4. If a storage tank can be controlled to a specified outflow, there is only one inflow hydrograph (namely the second) for which the available storage volume is mobilized with maximum success. For a larger storm the peak flow remains the same (compared with "no storage"). For a smaller storm only little water is stored and the tank is not fully utilized.

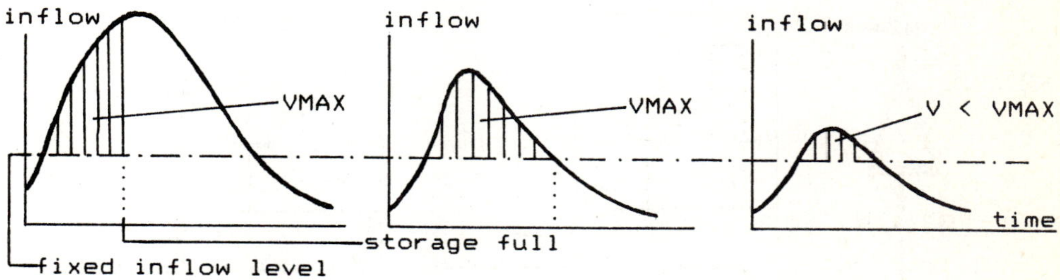

Fig. 3. Effectiveness of Off-Line Storage with Fixed Inflow Level.

Consider also the cascade of two tanks, each of them actively controlled to release not more than maximum downstream conduit capacity. Fig. 4 shows that this control policy, with fixed release, results in downstream overloading (CSO). A better strategy with time-varying release avoids this. Note that for the better solution no additional capacities are needed. These examples show that the operational flexibility is substantially enhanced if local control loops can be manipulated from a central location. This can be achieved in a number of ways.

1. Control loops are hardwired into the control center. Measurements are displayed, and regulators are either actuated by operators or electric circuits. It is difficult to change the parameters of the controllers. Since only limited information is available, operators need full understanding (or rather "feeling") of the dynamics of both the control and the drainage system. In the automatic mode on/off control with fixed set points is applied.

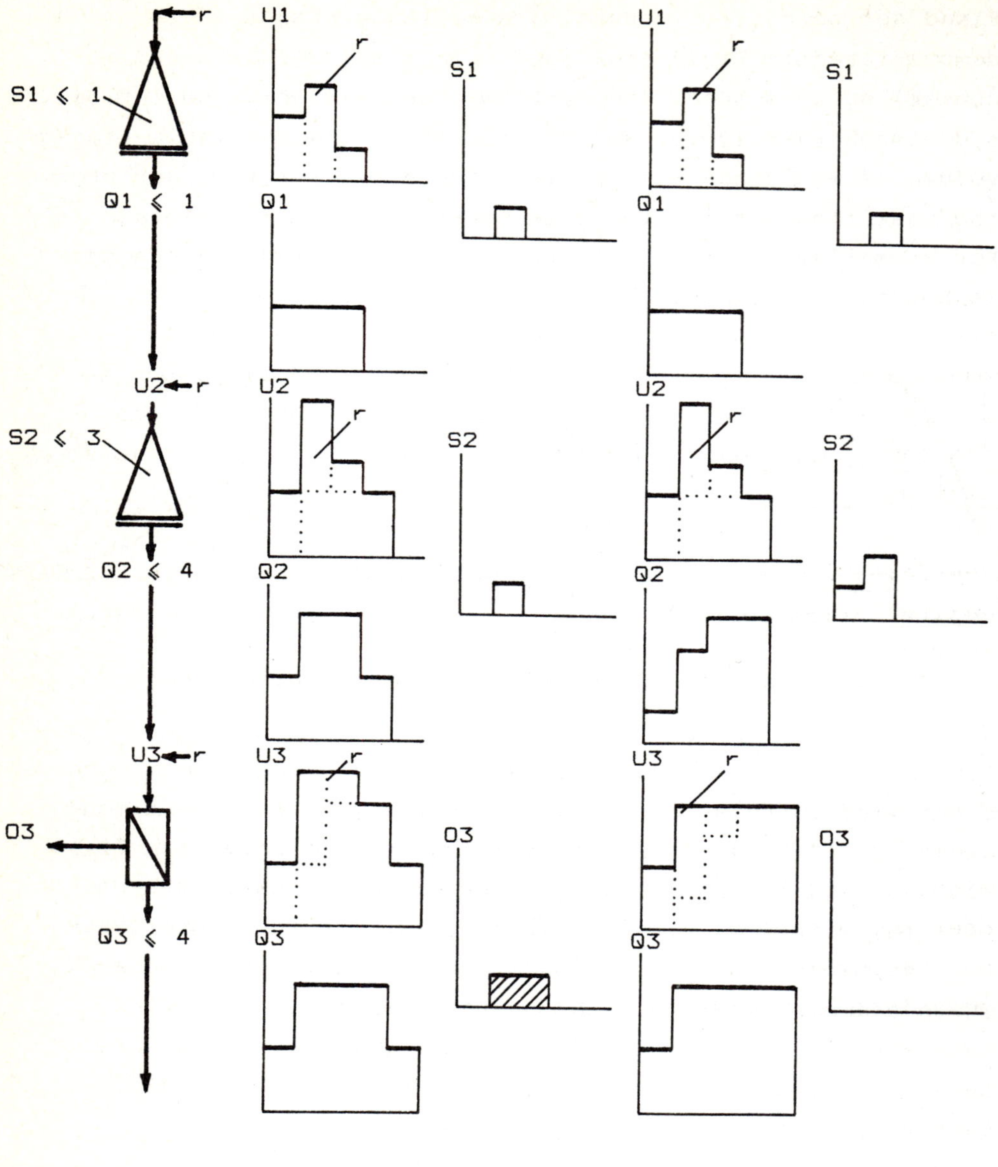

Fig. 4. Advantages of Systems Control (right) Versus Local Control (left) for a Storage Cascade.

2. Control loops are softwired into a computer. In such a direct digital control (DDC) system any control parameter can be changed rapidly by keying in aporopriate commands. Loops can easily be interconnected.

Operators do not actually operate the regulators. They merely supervise and adjust set points. If the computer is large enough, it can be used for on-line computation of flow profile, control scenarios, or even flow forecast and set point optimization.

3. Local control is softwired using on-site microprocessors (distributed control). Operators and/or a central minicomputer only supervise the performance of the system. Whenever the center is not operational, the local loops work independently with pre-selected set points. The options are the same as with DDC, however the reliability of the system is improved.

The following sections will discuss the fundamentals of how controllers actually, work and how set points can be selected in real time.

The Local Level: Controllers at Work

Consider the regulation of the gate opening of a detention outlet to maintain constant outflow. A flow level meter might be located downstream of the gate. Whenever the gate opening is changed the flow shows a typical reaction. Based on the physical behavior of the system (i.e. its differential equation) these reactions can be of a proportional (P), integral (I), or derivative (D) type (Fig.5). They can also be a combination thereof, or display time delay between regulator action and sensor reaction.

The simplest controller behavior is that of a P-controller. Here the signal to actuate the regulator u is proportional to the difference e between the measured variable and the set point. At time t:

$$u(t) = KP\ e(t)$$

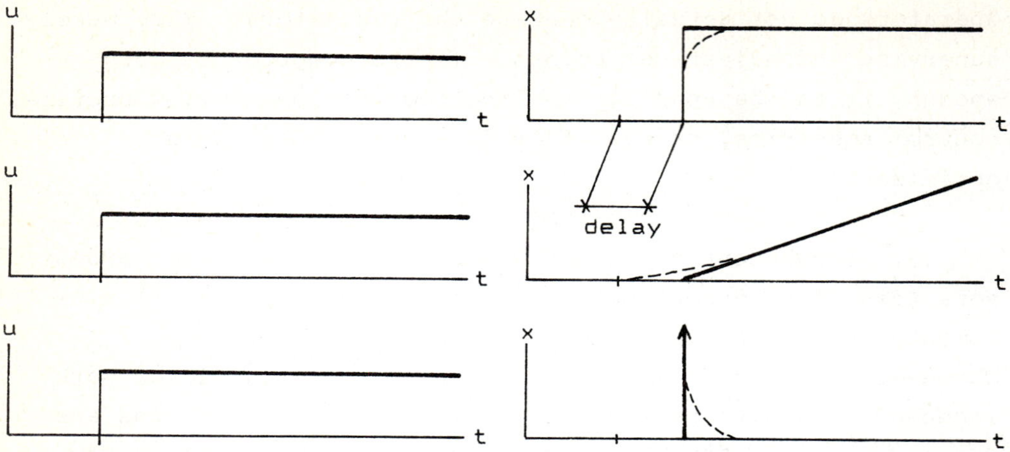

Fig.5. Proportional (P), Integral (I), and Derivatives (D) Responses of Controlled Variables x to Instantaneous Regulator Action u.

The intensity of the controller reaction is given by the proportional gain KP. Too low a KP makes the controller slow, a value too high causes "over reaction" of the regulator known as "instability". A simple example for a P-controller is a float actuated gate, where the gate displacement is proportional to the float motion, Fig.6. The disadvantage of a P-controller in a loop with pure P-behavior is that constant deviations from the set point ("offset") cannot be avoided. In Fig.6, for example, levels and flow might represent an equilibrium state with the set point level of 2.0 m. If a disturbance occurs (say, the outflow is increased to 3.0 m^3/s) the P-controller will open the inflow gate to 0.75 m and a new pond level of 1.25 m will be obtained. Thus, the offset would be 0.75 m. Other disturbances or proportional gain will yield different offsets.

The output of an I-controller depends on the past history of set point deviations, namely:

$$u(t) = KI \sum_{i=0}^{t} e(i)$$

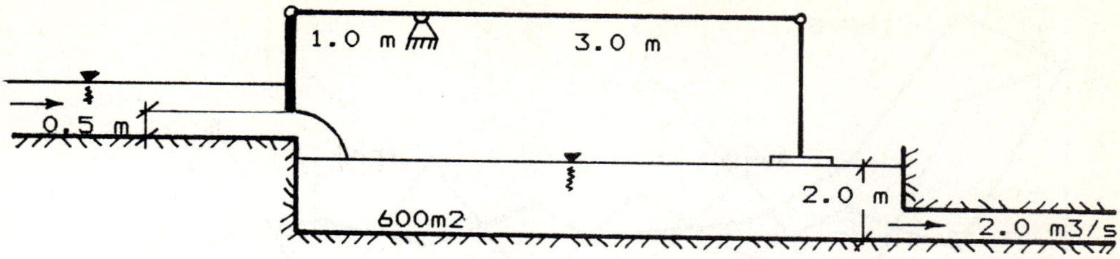

Fig. 6. Proportional Water Level Controller.

An I-controller reacts slower than a P-controller. However, it always forces the process back to the setpoint. Fig.7 displays the behavior of the controlled variable "water level h" in an outflow chanel of a pond, regulated by a sharp-crested weir. A higher KI accelerates the controller (KI=1.0). Too high a KI, however, causes instability (KI=1.5)

The disadvantages of P- and I- controllers can be avoided by combining them. The output of a PI-controller is:

$$u(t) = KP\ e(t) + KP\ T/TN \sum_{i=0}^{t} e(i)$$

with TN called the "reset time" and T being the time step between regulator actions. Fig.8 displays level reactions of the inflow/level-loop for the pond shown in Fig.6. Only the PI-controllers allow fast and precise control to the set point level of 2.0 m. In this case the combination of KP=2.0 and KI=1.0 seems to be the optimum parameter set.

If a loop involves long time delays, it might be worthwhile to make the controller more "nervous". If derivative action is added to the PI-controller the output of the resulting PID-controller is:

$$u(t) = KP\ e(t) + KP\ T/TN \sum_{i=0}^{t} e(i) + KP\ TV/T\ (e(t)-e(t-1))$$

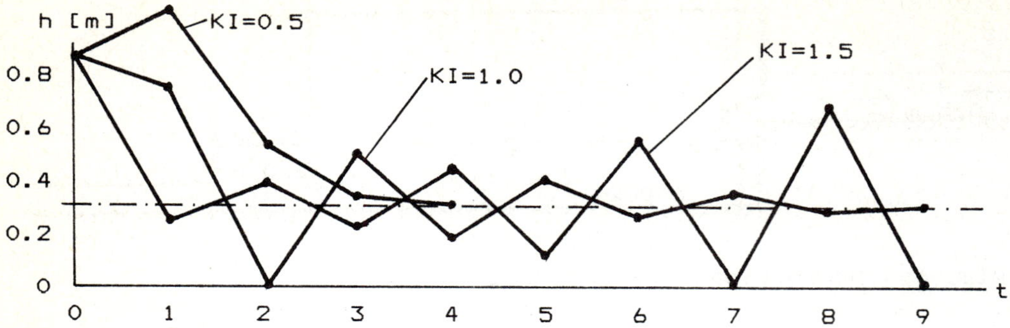

Fig. 7. Integral Control of a Conduit Level in a Pond Outlet

The "rate time" TV > 0 results in a short over-reaction of the controller. Controller calibration, i.e. selection of type and parameters, can be done through real or simulated experiments.

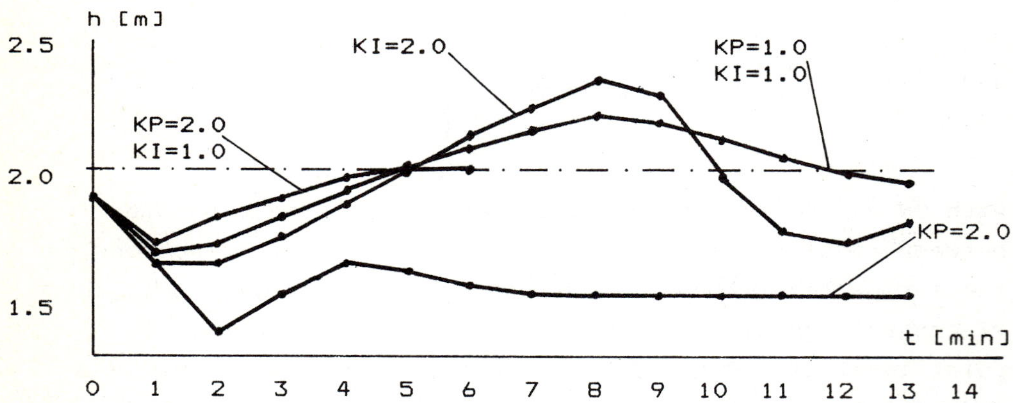

Fig. 8. Proportional, Integral, and Proportional/Integer Control of the Pond Level in Fig. 6.

1. Given a regulator displacement the reaction of the controlled variable is measured or computed, Fig.9.

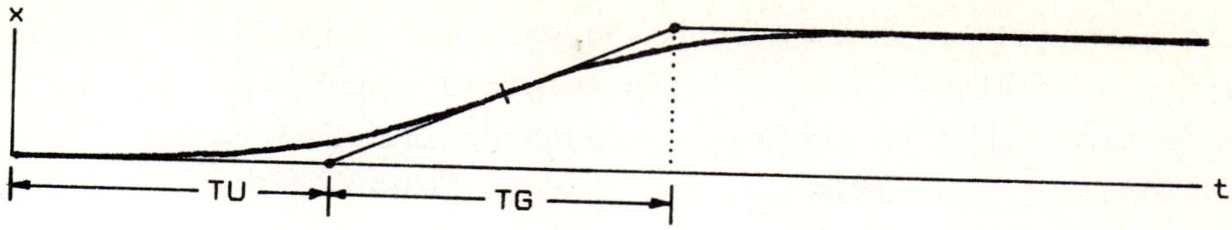

Fig. 9. General Response of Control Variable After Instantaneous Regulator Action.

2. If the reaction is partly integral, a P-controller will yield no offset. A preliminary estimate of KP can be:

KP = 0.3 TG/TU

3. In most cases (TG/TU > 7) PI-controllers are applied. The following parameters can be used.

KP = 0.3 TG/TU
TN = 1.2 TG

4. For loops with 3< TG/TU <7, PID-controllers are superior to PI-controllers. Here, initial parameters can be chosen as:

KP = 0.6 TG/TU
TN = 1.0 TG
TV = 0.5 TU

5. If the controlled process is nonlinear (e.g. gate flow as a function of gate opening, channel flow as a function of water level), the controller parameters are only valid in the vicinity of a reference point (e.g. a particular flow rate or water level). For other reference points other sets of parameters have to be found.

6. Once implemented, the behavior of a controller has to be tested. If possible, full scale experiments over the whole range of control variables have to be carried out to ensure that instability cannot occur. Otherwise, regulators

would permanently be activated and wear out very fast. During routine operation, the initially selected control parameters can be fine-tuned to eventually reach optimum controller behavior.

The Strategic Level: How to Set Set Points

In the preceeding section it has been shown how a controller adjusts a regulator to achieve minimum set point deviation. This section discusses options for determining these set points. Fig.3 and Fig.4 demonstrate that this is a crucial part of operating a RTCS.

Obviously, the simplest possibility is to keep set points constant. For example, this option might be advisable at the inflow of a treatment plant. Generally, however, optimum set points vary with each loading pattern. If these patterns display periodic variabiity, set point selection can simply follow a time schedule. A typical example is a water supply system where demand patterns show a well known regularity. RTC of CSS, however, is control of transient stormflows which show no regular pattern. Here, the optimum strategy differs for every storm.

Set points can be determined either heuristically or automatically. The latter approach, although widely discussed in research, is hardly ever applied in RTC of CSS. Heuristic control simply means that an operator decides upon set points based on his experience. The rationale behind these decisions can be very complex and is not necessarily consistent. Intuition and "feeling" plays a major role. For example, the author talked to an older operator who feels his back aching when rainfall can be expected. He uses these "forecasts" to pump down a large collection system and thereby create storage for the expected stormflow. In case of an actual storm an operator might look out of the window and judge how the storm would continue. In case of emergency he can immediately seek assistance, advice maintenance crews, etc. An experienced operator will probably carry out near-optimum control. He disregards (like a good chess player) all control options that are possible but not advisable.

Heuristic control also has serious drawbacks. Operators are not motivated, per se, to maximize storage, minimize CSO, avoid flooding, etc., all at the same time. Like in planning and design they do not have the automatic benefit of good operation but possible trouble from bad operation. Therefore, they will simplify their multiobjective task to the simple purpose of "minimize my trouble". Once an operator leaves his job, his experience will also be gone. His successor will make mistakes all over again. Experience with stormwater RTC can only be gained slowly. It rains approximately 5% of the time. An operator spends approximately 20% of the time on his job. Hence, he is really exposed to a stormwater RTC problem only 1% of the time. Unless his job is integrated into treatment plant operation, 2 to 4 operators working in shifts have to be educated, trained, and paid just for stormwater operation. In conclusion, there are arguments in favor of automatic systems control.

Two major possibilites exist to determine set points automatically, e.g. "expert systems" and "optimized control".

Expert systems are basically a matrix of "if-then" decisions, in which all possible control situations are included. This procedure might also be referred to as "pattern control" or "scenario control". Here, the optimal control strategy for each pattern has already been computed and is stored in a computer.

Expert systems have great flexibility in formulating different scenarios in that quantitative (e.g. m^3/s flow), qualitative (e.g. heavy rain), deterministic (e.g. storage full) and probabilistic information (rain likely) can be combined. If subjective statements are to be included, the expert system obviously can only be used interactively with an operator. Its disadvantage is the sheer number of possible scenarios that have to be examined. No matter how many are included, one can never be sure that an important one is not missing. Expert systems are extremely application-oriented and seldom transferable to other RTCS. In fact, their name implies the idea that the control decisions are not found through a model of the physical system, but through a model of an expert the operator). Expert systems can be formulated either by documenting and copying operator reactions, or by running multiple optimization runs off-line and storing their results for later on-line use.

The second option of automatic set point selection is optimized control, where the problem is formulated as a mathematical optimization problem. This is the only way to compare the actual RTC performance with the best possible. It requires that one formulate and quantify:

- The operational objectives of the drainage agency in a single equation ("objective function").
- The physical dynamics of the drainage system (i.e. maximum and minimum flows and levels, conduit routing) in a set of equations ("constraints").
- The present and future inflows to the system ("right hand side" of the constraints).

Different optimization techniques are discussed in more detail in the workshop paper by Patry. Applicable techniques include dynamic programming (DP) and its derivatives, linear programming (LP) and its derivatives, and calculus of variation [7], [45], [53], [59], [66].

A perfect method for RTC optimization does not exist. Either only small problems can be solved (DP), the true optimum cannot be guaranteed (derivatives of DP, calculus of variation), or the problem has to be linearized (LP). Common to any optimization technique are a number of disadvantages:

1. Operational objectives of an agency might be monetary, intangible, and/or conflicting. It is very difficult to lump them all into one equation. Multiple objective optimization techniques, however, cannot be fully automatic. They require interaction with a decision maker to find best compromise solutions ("Pareto optimum").

2. If a single objective function has been specified it is usually of a mixed integer/continuous, nonlinear, and nonmonotone type. Since powerful optimization techniques are not available for this kind of objective the function has to be further simplified.

3. Optimum control requires perfect forecasts over the entire control horizon, which is the time required to bring the system back to its initial state. Automatic storm forecasts, however, are very uncertain; [48], [50]. If subjective forecasts are more reliable (which has not been shown yet) heuristic control would have a major advantage.

4. Flow routing has to be simplified to allow application of standard optimization techniques. This involves spatial and temporal aggregation, and linearization. The effects of these simplifications on final control performance, although surely negative, have not been quantified yet.

Despite the substantial amount of research that has been devoted to optimized operation of water systems, these techniques are rarely applied in real world systems, whereas heuristic control and expert systems are occasionally applied.

Applications of Real-Time Control in Wastewater Transport Systems

A Survey in the United States and Canada

This section is based on a 1984/85 survey using a questionnaire and personal communications [51]. Table 2 shows how differently agencies recognize their past and present URP problems, mostly in ranked sequence. Infiltration/inflow and dry weather CSO are mentioned quite often, indicating basic upgrading and treatment capacity needs.

Table 2. Problems with Combined Sewer Systems.

	wet weather overflow	dry weather overflow	over-designed	corrosion	detention basin not effective	infiltration/inflow	septic conditions	structural breakdown	sediments	flooding
Akron, OH	3					2			1	
Albany, NY	1					2			3	1
Boston, MA		1				2		4	3	6
Chicago, IL	x									x
Cleveland, OH	1	2								3
Detroit, MI	1	2	9	8		4	6	7	3	5
Edmonton, Alta.	2	4				3				1
Evansville, IN	7	2				6	5	3	4	1
Grand Rapids, MI	1				4	3				2
Hamilton, Ont.	3	2	1	1		1	1	1	1	3
Indianapolis, IN		1				4		3		2
Lima, OH	2					3			4	1
Milwaukee, WI	1	2								
Minneapolis-St.Paul,MN	1	2				4				3
Montreal, Que.	2	1				4				3
Omaha, NE	2	1						3	4	5
Ottawa, Ont.	x									
Peoria, IL	x									
Philadelphia, PA		x				x		x	x	x
Providence, RI	1	5		9	7	4	8	2	3	6
Rochester, NY	2							3	4	1
San Francisco, CA	x									
Seattle, WA	2	1								3
St. Louis, MO	2	1				2		2	2	
Toledo, OH		3				2			4	1
Toronto, Ont.	1									2
Vancouver, B.C.	1					x		x		x
Washington, DC	1	2								3

In Table 3 the approaches currently being pursued to solve these problems are listed. According to this rather non-representative survey, separation, additional storage, and RTC of the existing systems are the most frequently applied approaches.

Table 3. Application of Combined Sewer Overflow Abatement Techniques

	sewer grouting	better maintenance	real time control	surface retention	source controls	combined sewer overflow treatment	regular flushing programs	new detention basins	new collector mains with fixed regulators	new collector mains	separation
Akron, OH		x									
Albany, NY								x	x	x	x
Boston, MA		x			x	x		x		x	x
Chicago, IL			x							x	
Cleveland, OH			x								
Detroit, MI			x								
Edmonton, Alta.								x	x	x	
Evansville, IN			x								
Grand Rapids, MI			x					x	x		
Hamilton, Ont.						(x)				(x)	
Indianapolis, IN	x				x		x				
Lima, OH			x								
Milwaukee, WI			x							x	x
Minneapolis-St.Paul,MN			x								x
Montreal, Que.			x							x	x
Omaha, NE			x						x	x	x
Ottawa, Ont.						x			x		x
Philadelphia, PA		x	x				x			x	
Rochester, NY			x			x			x	x	
San Francisco, CA			x			x				x	
Seattle, WA			x								
St. Louis, MO			(x)								
Toledo, OH				x	x	x		x			x
Toronto, Ont.			x						x		x
Vancouver, B.C.			(x)					(x)			
Washington, DC			x			x				x	

Table 4 gives a comprehensive review on some characteristic data for existing RTCS, some of which are described below. Surprisingly, these systems have almost nothing in common. Successful RTCS exist both in large and small catchments.

Available storage, as well as population served, differs by orders of magnitude. Apparently RTC is feasible and effective for a wide variety of CSS.

Some Implemented Systems in the United States

The Detroit Metro Water Department (DMWD) serves 11 communities with 35% of the Michigan population, and large industrial inflows [65]. The rather flat sewers divert CSO into the Detroit, Rouge and Huron rivers via 110 outfalls. The controllable in-line storage, spread over the entire catchment area is only 8.3 m^3/ha. DMWD operates an extensive flow level monitoring system at 247 sites using very simple and apparently reliable air pressure cells. Operation routinely relies on National Weather Service radar data which allows them to pump down the water level in the system up to three hours before an approaching storm. The conditions of 70 tide gates are monitored to detect blockage and other operational problems. The RTCS is operated in 24-hour supervisory control, together with the water supply system. The RTCS has achieved a reduction in CSO duration from 5% to 1.3%. Central automatic control is not yet implemented.

The Municipality of Metropolitan Seattle has operated a RTCS for more than 10 years now to reduce the CSO into Puget Sound (Pacific Ocean), the Duwamish River and Lake Washington [38]. The system can be run in local automatic, central supervisory, or central automatic control modes. Original plans to adjust "rule curves" (i.e. the expert system) as a function of local rainfall have never been implemented. Average CSO volumes for different modes of control are as follows; 7.0 million m^3 for "no RTC", 4.4 for local control, 1.9 for central automatic

Table 4. Characteristic Data of Some Real Time Control Systems.

	Lima	Seattle	Detroit	San Francisco	Cleveland	Chicago	Minneapolis-St.Paul	Philadelphia	Montreal
Total area [km2]		873	1670	97.2	430	2260	4120	258	365
Combined sewer area	15	52.8	636	96.2	184	972	77.3	170	273
Area slope	flat	mod	flat	steep	mod	flat	mod	mod	mod
Population (Mil.)	0.047		3.2		1.3	5.25	1.7	2.25	1.9
Overflow structures	5	81	110	39	>500	645	120	184	125
Storage volume (100m3)		34	120.4	530	235	11230			
Storage volume [m3/ha]	22.7	22.7	8.3	69.5		115.5			7
Rainfall monitors	0	11	41	27	25		9	13	10
Water level monitors	13	70	247	<120	100	77	44	7	50
Monitored overflow locations	5	37	110	0		211	18	4	14
Out-stations	9	36			69		38		
Data points	300	800	475		1500	1900	139		>444
RAM storage [kByte]	2x52	45	24	16	2x>128		24	16	2x?
Controlled regulators (w/o p.s.)	8	16	10	0	34	64	6	3	14
Pumping stations (p.s.)	1	21	17	2		3	0		1

control, and 1.5 for central supervisory control. Planning is currently underway to upgrade the system, and to replace the old computer hardware by state-of-the art-technology. Because of the success of the RTCS, plans for separation have been deleted. Additional benefits of the system, according to operating personnel, have been reduced flooding and monitoring, and source detection of gasoline spills.

The Northeast Ohio Regional Sewer District operates major trunk facilities, interceptors, and treatment plants for the city of Cleveland and 33 suburban communities [12]. Before remedial measures, almost every rain created CSO at some 600 overflow points into the Cuyahoga River and Lake Erie. The RTCS was initiated in 1975 with 3 regulators, and a greatly expanded system became operational in 1983. The system now controls some 50% of the original CSO volume. Operation to date has been local automatic with central supervision. Centrally adjustable controls are actuated by on-site microprocessors. The control center has a back-up computer which is intended for later simulation and optimization of controls. Operator interfaces include color CRT's with graphic displays, "trending" software, and alphanumeric and dedicated keyboards. In terms of hardware, this system is one of the most advanced, but the period of performance is too short for a comprehensive evaluation of success.

The City of Lima, Ohio operates a RTCS for its CSS that covers some 15 km^2; [11]. It and Seattle are the only two with central automatic operation. However, both systems do not employ real-time modeling of the rainfall/runoff process. There are 8 sluice gate regulators that are operated from the central control facility. The system has been operating for some 5 years without major problems. The treatment plant inflows are regulated during dry weather by the control gates. After storms, the system is flushed by automatically opening interceptor gates, beginning downstream. After-hours, the system is run automatically without supervision.

Applications of Real Time Control in Europe

It is more difficult to obtain an overview of RTC applications in Europe because of the many different countries and administrative regulations involved. If a sump level controlled pump is already defined as a RTCS this approach has quite a long history. For example, large parts of The Netherlands are below sea level. There, virtually every drop of water has to be pumped, mostly from CSS. Integrated systems, of course, are not so frequent. Apparently three large RTCS exist in The Netherlands, in Amsterdam, Utrecht, and Eindhoven.

After the re-organization of water resources management and administration in England, the number of integrated RTCS is steadily increasing. Solely dedicated to wastewater control, for example, are the systems in Newcastle-upon-Tyne, Birmingham/Wolverhampton, and Grimsby/Cleethorpe.

The county of Seine St-Denis, close to Paris, has operated a large RTCS for more than a decade. This system is mainly referred to in the workshop paper by Delattre et al. Other existing or planned RTCS in France include the counties of Hauts de Seine, Val de Marne, and the cities of Nancy and Bordeaux.

Some German states enforces specific CSO regulations. These have resulted in the construction of literally thousands of small detention tanks to catch the first flush of pollutants in CSS. Some 20% are equipped with controllable regulators. Quite a few employ some form of integrated control, mainly to avoid downstream CSO caused by upstream tank releases. Large supervisory RTCS examples can be found in the cities of Bremen, Hamburg, Munich, Mannheim, Nuremerg, and in the Ruhr district (Emschergenossenschaft).

Discussion and Conclusions

There is growing evidence that a major portion of pollutants found in receiving water is URP. Due to the diversity of sources and routes, abatement techniques generally have moderate efficiency at relatively high cost. RTC of CSS offers a potential way of reducing the CSO problem substantially with little investment. Apart from pollution abatement, RTC can create side-benefits such as less flooding, improved wastewater treatment, in-sewer sediment control, and supervision, understanding, and record-keeping of systems performance.

Virtually any CSS would seem to be worthy of real-time control. The market offers few but proven hardware elements to equip a sufficiently reliable RTCS. Current designs are remarkably similar to each other. They typically include:

- Point raingages and water level sensors.
- Pumps, inflatable dams, and other stainless steel regulators.
- Leased telephone line telemetry.
- Distributed digital control system with on-site microprocessors and dual central minicomputers.

Despite the availability of hardware, there probably do not exist more than 50 large integrated systems with possibility for operator interference and automatic systems control (out of thousands of CSS in North America and Western Europe).

The vast majority of operational concepts are simple local control schemes. Some of the RTCS with integrated hardware are still on (or have reverted to) this operational level. Only two RTCS exist in North America with automatic systems control, with not many more in Europe. Despite decades of research and publications, advanced operational methodologies apparently have not found their way to application. There is no need for futher research into even more involved mathematical techniques, but there is a great need for continuing education, expert meetings, joint research/application projects and enhanced communications between scientists, city engineers, and operators. Otherwise "cannot engineering" on one side and "hyper-enthusiasm" on the other will continue to rule progress in this field.

These more technical aspects (i.e. hardware and software) again have minor significance if the administrative framework is not tailored to provide equal opportunity for both massive construction and "hi-tech" approaches. Without appropriate project funding arrangements (investment versus operation costs), by-laws to clarify inter-agency responsibilities (district versus city), and crystal clear operational objectives, RTC can never succeed over such brute force approaches as separation and deep storage tunnels.

Finally, without the proper political environment URP might not even be regarded as a problem. Either public pressure (very efficient) or regulatory action is needed to get started with URP abatement. Only few regions in Western Europe and North America already have specific and comprehensive URP regulations. Compared to treatment plant effluent standards the matter is very complicated and an area of continuing discussion.

No law demands RTC, no regulatory agency encourages it, consultants do not necessarily propose it, it is not part of civil engineering tradition, and direct economic pressure is usually not visible in the public service realm. So, why does RTC in CSS exist at all? Whenever there is a successful RTCS a small group of engineers and technicians can be found, eager to communicate (e.g. with the operators), willing look over the borders of the own field, backed by executives with a firm goal/flexible means policy, and persuasive towards the operational staff. Thus, it is the human enthusiasm and curiosity in a positive management environemnt that keeps it all going.

References

1. Anderson, R.J., Bell, S.S., van der Heyden, W.H., Genthe, W.K. Wastewater flow measurement in sewers using ultrasound. United States Environmental Protection Agency. Washington DC. Report No. 600/2-76-243. Nov.1976.

2. Andrews, R.J., Stenstrohm, M.K., Buhr, H.O. Control systems for the reduction of effluent variability from the activated sludge process. Porg.Wat. Techn. 1(1). 1976. pp. 41-68.

3. APWA. Combined sewer regulator overflow facilities. United States Federal Water Quality Administration. Washington DC. Report No.11022 DMU 07/70. July 1970a.

4. APWA. Combined sewer regulation and management. United States Federal Water Quality Administration. Washington DC. Report No.11022 DMU 08/70. July 1970b.

5. Balmer, P., Malmqvist, P.-A, Sjoeberg, A. (eds.). Urban storm drainage. Gothenburg, Sweden. June 1984.

6. Bech, M.B. Operational water quality management: beyond planning and design, Executive Report No.7. International Institute for Applied Systems Analysis. Laxenburg. Austria. 1981.

7. Bradford, B.H. Optimal storage in a combined sewer system. J.Wat.Res.Plann.Man.Div. ASCE. 103(WR1). 1977. pp. 1-15.

8. Brock, D.A. Closed-loop automatic control of water system operation. J.AWWA, 55(4). Apr. 1963. pp. 467-480.

9. Brock, D.A. Optimization: cost saving, not pretty pictures. InTech. May 1984. pp.31-32.

10. Broeker, H.W. Impact of depositions on sewer operation. in: Balmer, P. et al.(eds.). Urban storm drainage. Gothenburg, Sweden. June 1984. pp.879-886.

11. Brueck, T.M., Knudsen. D.I., Peterson, D.F. Computer-based implementation of combined sewer control. J.W.PCF. 54(2). 1982 pp. 198-205.

12. Buczek, T.S., Chantrill, C.S. A computer-based system for reduction of combined sewer overflow in a metropolitan wastewater collection system. 57th Ann. WPCF Conf. New Orleans. Oct. 1-4. 1984.

13. Callery, R.L. Dispatching system for Control of combined sewer flows, United States Environmental Protection Agency. Washington DC. Report No. WQO-11020-FAQ-03/71. 1971.

14. Codner, G.P. A dynamic programming approach to the optimization of a complex urban water supply scheme. Australian Water Resources Council. Canberra. Techn. Paper No.47. 1979.

15. Cosgriff, G.O. Computer controlled telemetry network for a water supply, sewerage and drainage system. Melbourne and Metropolitan Board of Works. Mar. 1978.

16. Curtis, D.C. Automated community flood warning system. ASCE Hydraul.Div.Conf. MIT. Cambridge, MA. Aug. 9-12. 1983.

17. Damant, C., Austin, G.L., Bellon, A., Osseyrane, M., Nguyen, N. Radar rain forecasting for wastewater control. J. Hydraul.Div. ASCE. 109(2). Feb. 1983. pp.293-297.

18. DeMoyer, R., Horwitz, L.B. A system approach to water system modeling and control. Lexington Books. Lexington, MA. 1975.

19. Drake, R.A.R. (ed.). Instrumentation and control for water and wastewater treatment and transport systems. Pergamon Press. Oxford. England. 1985.

20. EPA. Fish kills caused by pollution. United States Environmental Protection Agency. Washington DC. Report No. 440/4-7-011. Apr. 1978.

21. EPA. Needs survey, 1982. United States Environmental Protection Agency. Washington DC. Report No. 430/9-83-002. Mar. 1982.

22. Fallside, F., Perry, P.E. Hierarchical Optimization of a water-supply network. Proc. IEEE. 122(2). Feb. 1975. pp.202-208.

23. Field, R. Combined sewer overflow techniques in the USA. in: Balmer, P. et al(eds.). Urban storm drainage. Gothenburg, Sweden. June 1984. pp.1333-1356.

24. GAO. Large construction projects to correct combined sewer overflows are too costly. Comptroller General of the United States. US General Accounting Office. Washington, DC. Report No. CED-80-40. Dec. 1979.

25. Giggey, M.D., Smith, W.G. National needs for combined sewer overflow control. J.Env.Eng.Div. ASCE. 104(EE4): 351-366. 1978.

26. Graham, P.H. 1978 needs survey. United States Environmental Protection Agency. Washington DC Report No. 430/9-79-003. Feb. 1979.

27. Grinker, J.R., Meagher, R.F. Computer controlled operation of an activated sludge plant. J.WPCF. 56(7). July 1984. pp. 823-829.

28. Heaney, J.P., Huber, W.C., Lehman, M.E. Nationwide assessment of receiving water impacts from urban stormwater pollution. United States Environmental Protection Agency. Washington DC. Report No. 600/2-81-025. Feb. 1981.

29. Helliwell, P.R. (ed.) Urban storm drainage. Pentech Press. London/Plymouth. England. Apr. 1978.

30. Hvidved-Jacobsen, T., Harremoes, P. Impact of combined sewer overflows on dissolved oxygen in receiving waters. in: Yen, B.C. (ed.). Urban stormwater hydraulics and hydrology. Water Resources Publications. Littleton, CO. 1982. pp. 1592-1600.

31. IAWPR. Instrumentation and control for water and wastewater treatment and transport systems. Prog.Wat.Techn., 9(5/6). 1977.

32. IAWPR. Practical experiences of control and automation on wastewater treatment and transport systems. Prog.Wat. Techn. 13 (8...11). 1981.

33. Jacobsen, P., Harremoes, P. The significance of head loss paremeters in surcharged sewer systems. in: Balmer, P. et al. (eds.). Urban storm drainage. Gothenburg, Sweden. June 1984. pp.167-176.

34. Keinath, T.M., Waniliesta, M. (eds.). Mathematical modeling for water pollution control processes. Ann Arbor Science. Ann Arbor, MI. 1975.

35. Lager, J.A., Smith, W. Urban stormwater management and technology: an assessment. United States Environmental Protection Agency. Washington DC Report No. 670/2-74-040. Dec. 1974.

36. Lager, J.A., Smith, W.G., Lynard, W.G., Finn, R.M., Finnemore, E.J. Urban stormwater management and technology: update and user's guide. United States Environmental Protection Agency. Washington DC. Report No. 600/8-77-014. Sep. 1977.

37. Lavallee, P., Lessard, P., Villeneuve, J.-P. Water quality variations in running waters due to combined sewer overflowing. in: Balmer, P. et al. (eds.). Urban storm drainage. Gothenburg, Sweden. June 1984. pp.761 -769.

38. Leiser, C.P. Computer management of a combined sewer system. United States Environmental Protection Agency. Washington DC. Report No. 670/2-74-022. July 1974.

39. Lindholm, O.G. Dry weather deposition in combined sewers. in: Balmer, P. et al. (eds.). Urban storm drainage. Gothenburg, Sweden. June 1984. pp.861-868.

40. Lohmann, J. State of automation at wastewater treatment plants in the F.R.G. in: Drake, R.A.R. (ed.). Instrumentation and control for water and wastewater treatment and transport systems. Pergamon Press. Oxford. England. 1985. pp.9-17.

41. McPherson, M.B. Feasibility of the metropolitan water intelligence system concept. ASCE Urban Water Resources Research Program. Techn.Memo.No.15. Dec. 1971.

42. McPherson, M.B. Integrated control of combined sewer regulators using weather radar. United States Environmental Protection Agency. Washington DC. Reprot No. 600/2 -81-041. Mar. 1981.

43. Meinholz, T.L., Kreutzberger, W.A., Harper, M.E., Fay, K.J. Verification of the water quality impacts of combined sewer overflows. United States Environmental Protection Agency. Washington DC. Report No. 600/2-79-155. Dec. 1979.

44. Naumann, J. The Kladow municipal waterworks in Berlin. Aqua. No.1. 1981. pp.41-45.

45. Papageorgiou, M. Automatic control strategies for a combined sewer system. J.Env.Eng. Div. ASCE. 109. 1983. pp.1385-1402.

46. Proceedings of the International Water Supply Conference. Specialized Subject 24. Zurich. Switzerland. Sep. 1982.

47. Rowse, A.A. Measurement of flows in part filled sewer pipes using the electromagnetic technique. Proc.Int.Conf.Plann.Constr.Maint.Oper.Sewerage Systems. Reading. England. Sep.12-14. 1984. pp.419-426.

48. Schilling, W. Short term flow forecasting for real time stormwater control. in: Balmer, P. et al.(eds.). Urban storm drainage. Gothenburg, Sweden. June 1984a. pp.1111-1120.

49. Schilling, W. Effect of spatial rainfall distribution on sewer flows. Water Science and Technology, 16(8/9). 1984b. pp. 177-188.

50. Schilling, W. Real-time estimation and forecasting of spatially distributed areal rainfall. Water Science and Technology. 16(8/9). 1984c. pp.327-348.

51. Schilling, W. Application of real time control in combined sewer systems. Interim Report. Department of Civil Engineering. Colorado State University. Fort Collins, CO. Dec. 1984d.

52. Schilling, W., Fuchs, L. A quantitative assessment of uncertainties in stormwater modeling. in: Balmer, P. et al. (eds.). Urban storm drainage. Gothenburg, Sweden. June 1984. pp.625-634.

53. Schilling, W., Semke, M. Combined sewer control in real time - Bremen, West Germany. in: Drake, R.A.R. (ed.). Instrumentation and control for water and wastewater treatment and transport systems. Pergamon Press. Oxford. England. 1985. pp.183-189.

54. Sevruk, B. Methods of correction for systematic error in point precipitation measurement for operational use. World Meteorological Organization. Geneva. Switzerland. Operational Hydrology Report No. 21. WMO publication 589. 1982.

55. Shelley, P.E., Kirkpatrick, G.A. Sewer flow measurements - a state-of-the-art assessment. United States Environmental Protection Agency. Washington DC. Reprot No. 600/2-75-027. Nov. 1975.

56. Sullivan, R.H., et al. Nationwide Evaluation of Combined Sewer Overflows and Stormwater Discharges. United States Environmental Protection Agency. Washington DC Report No. 600/2-77-064. Sept. 1977.

57. Tomlinson, E.J., Hamilton, I.M., Williamson, K. Process management systems at sewage treatment works. Water Pollution Control. 83(2). 1984. pp.172-183.

58. Tomlinson, R.D., Bebee, B.N., Heyward, A.A., Munger, S.G., Swartz, R.G. Fate and effect of particulates discharged by combined sewers and storm drains. United States Environemtnal Protection Agency. Washington, DC. Report No. 600/2-80-111. Aug. 1980.

59. Trotta, P.D., Labadie, J.W., Grigg, N.S. Automatic control strategies for urban stormwater. J. Hydraul.Div. ASCE. 13(HY12). Dec. 1977. pp.1443-1459.

60. Tyler. P.E. Costly monuments to idealism now lie rusting in the mud. Washington Post. May 5, 11, 12, 13, 14. 1981.

61. Valentin, F. Continous discharge measurement for the transition between partly filled and pressurized conduit flow in sewerage systems. in: IAWPR. Practical experiences of control and automation in wastewater treatment and water resources management. Prog.Wat.Techn. 13(8). 1981. pp.81-87.

62. Vitale, A.M., Sprey, P.M. Total urban water pollutants: the impact of stormwater. US Council of Environmental Quality. Oct.1974.

63. Water and Wastes Engineering. Special issue on instrumentation, monitoring and process control. Feb. 1979.

64. Water and Waste Treatment Journal. Special Issue. 27(7). Quality. Oct.1974.

65. Watt, T.R., Skrentner, R.G., Davanzo, A.C. Sewerage system monitoring and remote control. United States Environmental Protection Agency. Washington DC. Report No. 670/2-75-020. May 1975.

66. Wenzel, H.G., Labadie, J.W., Grigg, N.S. Control strategy development study - San Francisco wastewater master plan. J.Wat.Res.Plann.Man.Div. ASCE: 102(WR1). Apr. 1976. pp.117-135.

67. Yen, B.C.(ed.). Urban stormwater hydraulics and hydrology. Water Resources Publications. Littleton, CO. 1982.

PERFORMANCE OF HARDWARE COMPONENTS FOR REAL TIME MANAGEMENT OF SEWER SYSTEMS

J.M. Delattre and A. Bachoc
DDE Seine-Saint-Denis
99, Avenue du General de Gaulle
93110, Rosny-sous-Bois
France

Guy Jacquet
Cergrene
La Courtine, B.P. 105
93194, Noisy-le-Grand Cedex
France

Introduction

Real-time management of sewer systems requires a comprehensive approach to wastewater and runoff pollution processes, urban hydrology and unsteady sewer hydraulic phenomena. This ought to be a complementary new response to engineering works, when a sewer network has reached an advanced stage of complexity over a long period of intensive urbanization. The ultimate goal is an optimal control of wastewater and storm flows through a sewer network.

This paper deals with a case study, in Seine St Denis County, where the objective is a gradual development of a monitoring and control logic system through a 5-year plan. For almost 10 years, the Seine St Denis County program has been considered in some way as a full scale experiment. In 1980, a decision was made to go further on a long term project that should include more than 100 telemetry and/or control stations by 1987. A new department, whose function was focused on real-time management, was created in January 1983. It was a technological challenge, that is now changing the human resources policy within the County Public Agency. After 2 years of experimentation, we are in the operational startup stage of sewer man-

NATO ASI Series, Vol. G10
Urban Runoff Pollution.
Edited by H. C. Torno, J. Marsalek, and M. Desbordes

agement. So far, our evaluation of the performance of hardware components may appear biased, nevertheless we could define our operational criteria satisfactorily at the most important stations.

Providing statistics on reliability and accuracy in sewer management is a need for which no satisfactory response is available at this time, but talking about on-going adjustments or improvements might be a realistic and workable approach.

This paper will first review hardware components currently used in a real-time system. Then we will describe typical breakdowns and failures as well as the strategies selected to avoid most of them during the design and implementation stages of the overall system. Finally we will define our needs for more research and our expectations from new technology.

The Seine St Denis Remote Supervisory Control System

Figure 1 - The Seine St Denis Sewer Network

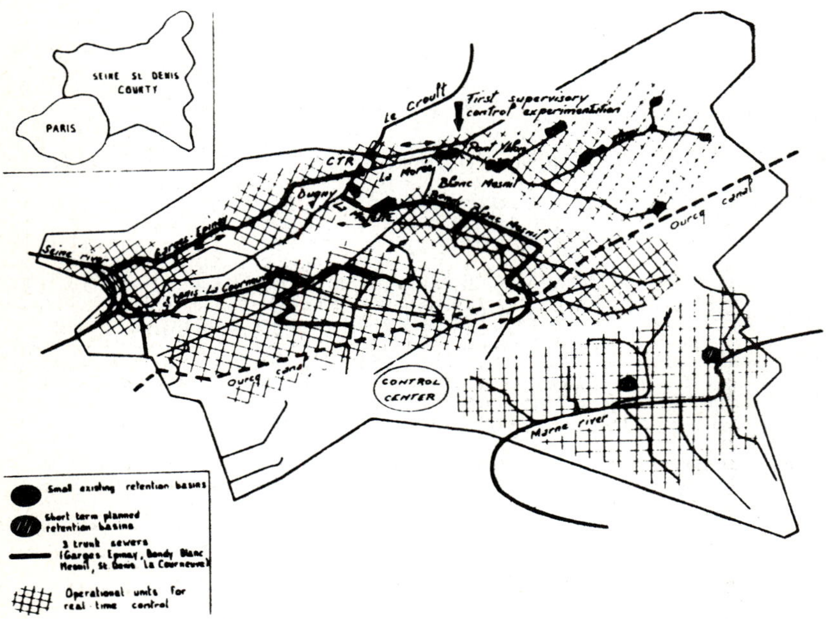

The sewer network of Seine St Denis County (Figure 1), managed by the County Public Agency, includes a great number of heterogeneous sewers, up to a total length of 400 miles, most of them old and combined. During the past 10 years, 3 trunk storm sewers, 2.5 to 3.5 m diameter, and 10 retention basins, have been constructed in the central area of the sewer network and in new urbanized districts.

Since 1975, a dynamic wave model (Caredas), developed by Sogreah, has been used to improve the feasibility study of any project related to the main sewers. More than 100 rain and water level recording gauges have been installed and survey studies have been carried out to gain a more comprehensive knowledge of the main sewers hydraulic response and quality behaviour. The idea of real-time monitoring came out gradually, along with a necessary program for increasing discharge or storage capacities, and for controlling downstream flows.

A 20-year development plan was prepared in 1976, which included the fundamentals of a monitoring system. A first experimental system, initiated in 1974 and focused on a few selected sites, provided useful training for the final project.

Recently, a new gradual approach to a control logic system, based on rainfall predictive capabilities, was introduced. Four reasons called for such a development:

- Increasing frequency of damaging floods.
- A flat topography, commonly 1m/Km, and extensively urbanized watersheds.
- A high-density population, i.e. 1,500,000 persons on 25,000 ha, largely predominant impervious areas, and very limited facilities for enlarging capacities or bypassing existing sewers.
- An uneven spatial distribution of rainfall during summer storms, allowing an optimal use of the available capacities for each particular event.

At the first stage of system operation in 1986, real-time management will be applied to 3 retention basins in the central area of the sewer network, since protection against flooding is the first priority (the overall system already inludes water quality control capabilities).

So far, long-term financial backing since 1975 and the creation of a new department in 1983 (10% of the actual sewer division of the county) have given steady support to develop real-time control (Fig 4). Within 2 years, we expect that much of the overall project will be achieved. From now on, we will introduce new management techniques and improve human resources policies to insure longterm reliability of the overall project.

The Conceptual Structure of the Monitoring and Control Logic Systems

In 1980, a broader comprehensive approach led us to start a new sewer monitoring project, with the aim of achieving 4 goals:

1. Decrease of frequency and volume of floods, through optimal decisions on downstream controlled flows.
2. Better control and measurement of wastewater flowrates during dry weather periods, particularly at the most important pumping stations and discharge sites.
3. Reduction of overflow pollutant discharges (average and peak loads) into receiving rivers and recreational retention basins, during wet weather and storm events, particularly in summer.
4. Better protection for workers in their daily inspection and maintenance activities.

The remote supervisory control system has been organized as a hierarchical scheme (Figure 2). Its execution started in 1983

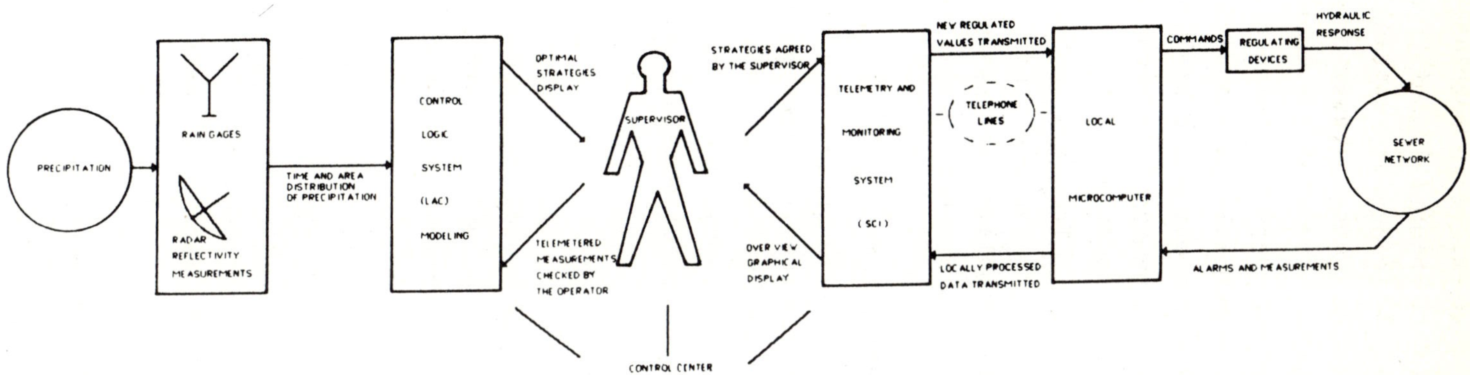

Fig. 2: THE CONCEPTUAL STRUCTURE OF THE REMOTE SUPERVISORY CONTROL

under a 5-year development program. When completed in 1988, the system should control 130 stations, 70% of them just transmit alarms and measurements, while 30% are considered as regulating stations that can perform certain tasks or computations automatically. Figure 3 gives an idea of how local stations are configured. While microcomputers manage on-site measurement and supervision, and generate suitable gate positions, a computer at the control center works out an optimal decision strategy for regulated flows. The supervisor checks the input and output data and then transmits selected flow controls to local stations for execution.

The control center includes 2 complementary systems: a telemetry and monitoring system, and a control logic system.

- Telemetry and monitoring (SCI)

The SCI controls scanning operations, manages coded messages to and from local stations, operates a data bank, handles alarms and warnings, generates graphical displays, and executes continuous and daily printings of messages and stored data.

- Control logic system (SAC)

The development of such a system required an adequate implementation of advanced computer technology (hardware and software facilities). The fundamentals of adaptive strategies are:

- A rainfall prediction capability, based on radar technology.
- Recursive simulations of rainfall-runoff relations and unsteady sewer hydraulics, including time-varying regulated flows.
- Dynamic programming algorithms, providing optimal strategies at each 20 min time interval.

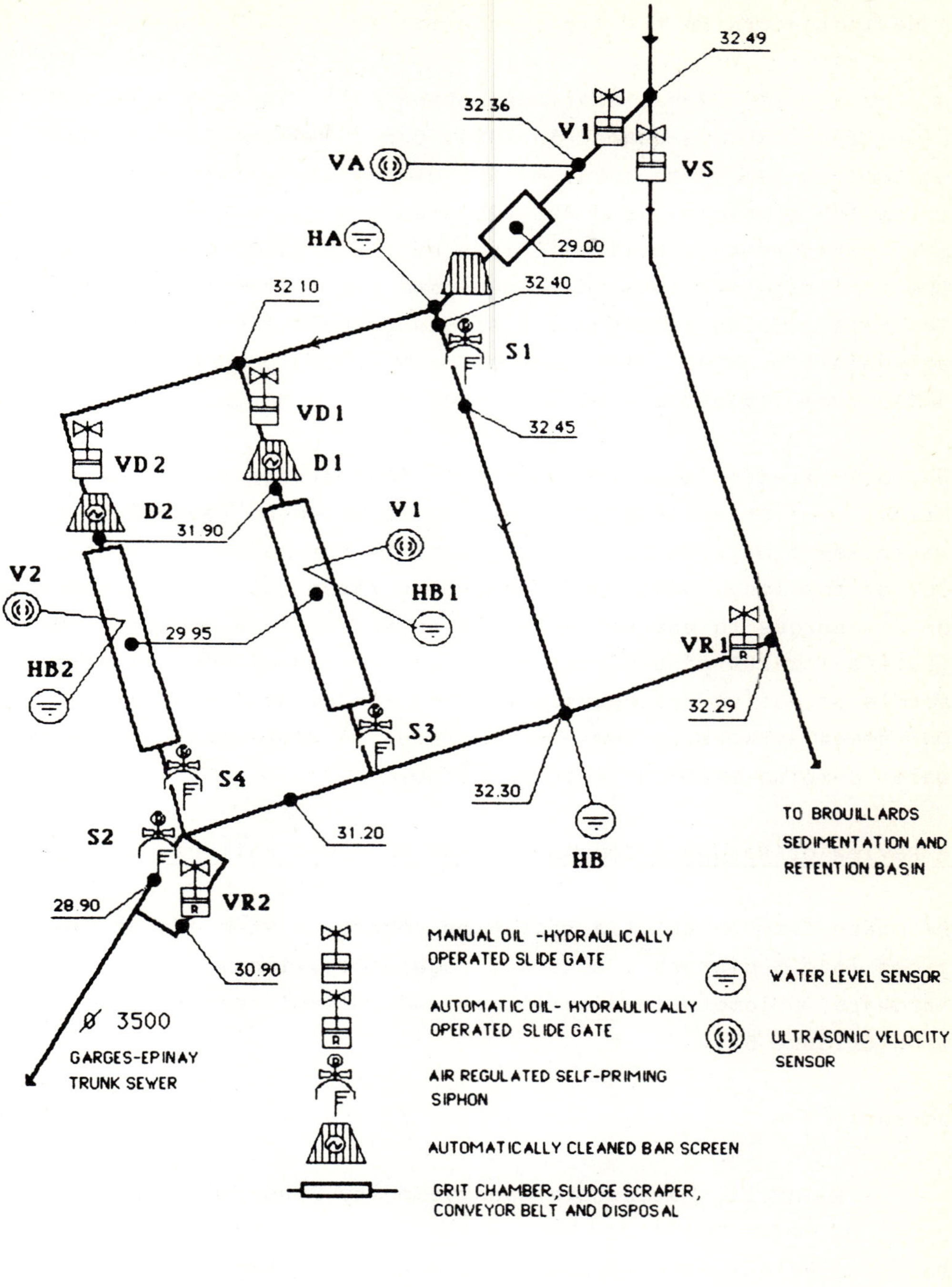

Fig. 3: **PRIMARY TREATMENT AND REGULATION STATION OF DUGNY**

Some Indicators on the Implementation Program of the Project

Starting from 6 local stations connected to an experimental simplified control center in 1983, our objective is for local control in December, 1986, of 50% of the projected local stations (65 stations), and 80% (25 stations) of the more complex stations that will be driven by the control logic system. The remaining stations to be equipped after 1986 mainly refer to small pumping stations and single purpose electrically-driven slide gates. The control center development was scheduled over 1 year, and will be operating by March, 1986.

We have selected some indicators to demonstrate the progress of our project by 1987 (Figure 4). Measurement (automatic rain, water level or flow recording gages) is the major task for 30% of the local stations. Real-time water quality monitoring of discharges to the Seine and Marne rivers is a specific objective for 20% of the stations. Protection of workers as a single objective has been introduced in 20% of the stations. Our sewer network system also includes 28 wastewater and stormwater pumping stations with 80 pumps.

A Review of Hardware Components and Observed Failures

Hardware components of a real-time control system can be divided into 5 classes : sensors, regulators, local process hardware, telemetering facilities and control center equipment (Figure 5).

Sensors

- Rainfall, Water Level and Velocity Sensors

The discussion in this paper will not extend beyond our own experience. Reliability of ultrasonic velocity measurement seems to decrease when water contains air bubbles (turbulent flows) or high suspended solids concentrations (rising branch of hydrographs). For example, any velocity greater than 1.3 m/s

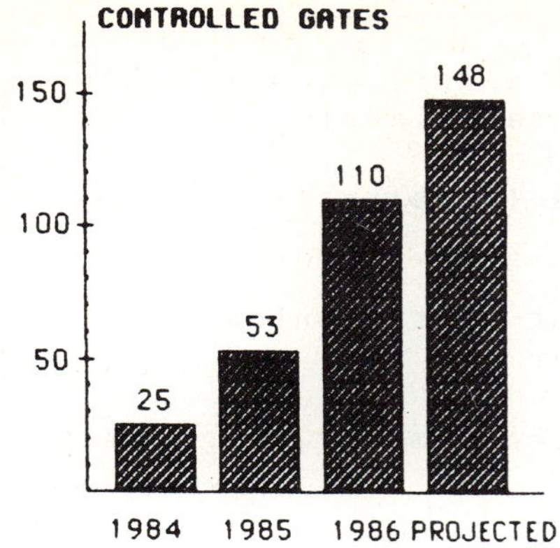
CONTROLLED GATES
150
100
50
25
53
110
148
1984
1985
1986
PROJECTED

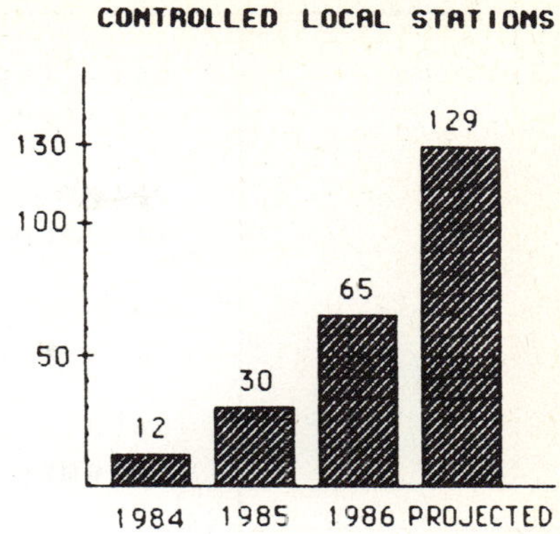
CONTROLLED LOCAL STATIONS
130
100
50
12
30
65
129
1984
1985
1986
PROJECTED

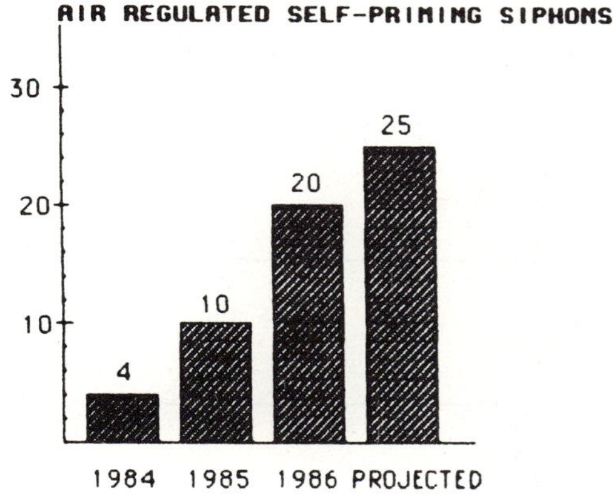
AIR REGULATED SELF-PRIMING SIPHONS
30
20
10
4
10
20
25
1984
1985
1986
PROJECTED

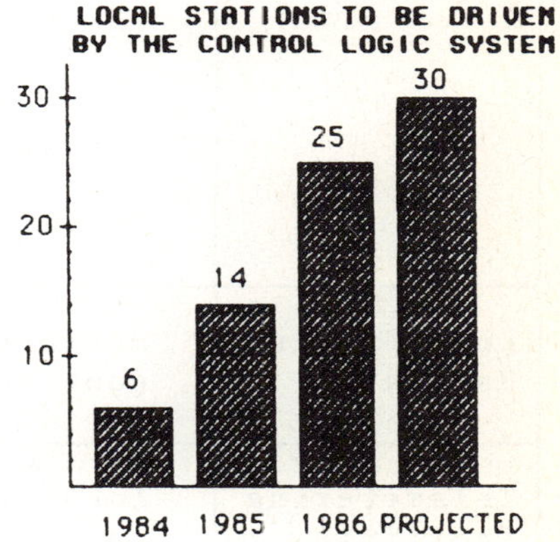
LOCAL STATIONS TO BE DRIVEN
BY THE CONTROL LOGIC SYSTEM
30
20
10
6
14
25
30
1984
1985
1986
PROJECTED

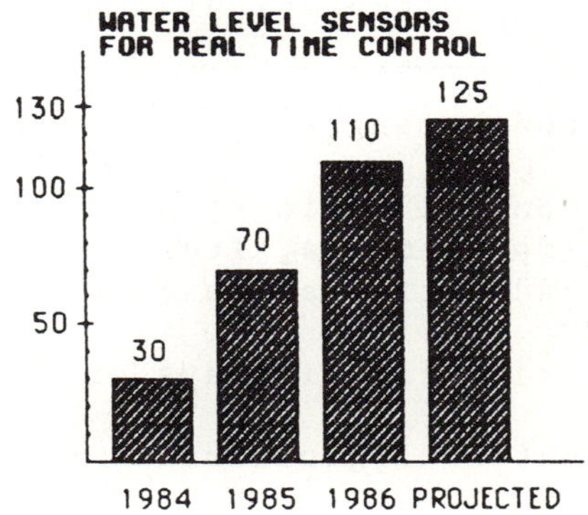
WATER LEVEL SENSORS
FOR REAL TIME CONTROL
130
100
50
30
70
110
125
1984
1985
1986
PROJECTED

Sensors	Rain tipping-bucket gages, coupled with on site memory and calculator water level sensors (piezoresistive and ultrasonics) Velocity measurement (ultrasonics), coupled with calculator gate positions (potentiometer, angle measurement)
Regulators	electric-drive spindle operated gates oil hydraulically operated gates air-regulated self-priming siphons inflatable fabric dams
Local process Hardware	microcomputer controller circuit
Telemetering facilities	local telemetry box telephone support (leased or dialed public telephone lines) direct calls from local stations and scanning frequency of the control center
Control Center	monitoring system (interfaces) Control logic system (CPU performance improvements, iterative 20 mn simulation and optimization runs) graphical displays of local measurements meteorological radar display

Fig. 5: CLASSIFICATION OF HARDWARE COMPONENTS

could not be measured upstream of siphon S1 at Dugny (Figure 3). However, ultrasonics water level sensors perform very well when they are located above retention basin water surfaces. Most of our water level sensors are piezoresistive. We only deal with buried closed sewers flowing under pressure for extreme rainfall events, so that our choice went to submerged sensors.

It should be said that an observed 5-10 cm accuracy range on 4 m water height is quite satisfactory for water level measurements in a real time control system.

- Gate Position Sensors

At this time, more problems come from gate position sensors, particularly potentiometers coupled with chains. Their error is usually 4% or more. This old device, though reliable, often leads to random measurement hystereris that cannot be allowed for short travel slide regulating gates. However, when it is replaced by a waterproof integrated circuit electric motor, a 2% precision can be reached. A direct angle measurement on regulating radial gates was installed last year and is giving satisfactory results (1% precision).

Sensors are often located in sewer appurtenances, which is a corrosive and humid environment with constrained access. Therefore telemetry is important to check sensor performances. Since 1983, we have had roughly an 8% annual failure rate, mainly caused by lightning and floating material impacts. This cannot be considered as a mean reliability index during a development stage of a project.

Regulators

- Gates

Our approach is to use waterproof electric drives for controlled gates, and oil-hydraulically operating devices for regulating gates. For flow regulation purposes, we commonly use radial gates, some of them acting as adjustable sharp-crested weirs. Electric drives include a declutching hand-wheel operation for security, in case of electrical failure. Hydraulic commands for regulating gates come from high pressure tanks, so that a predefined safe position can be obtained in case of electrical failure during regulation. Most of the problems of gate regulators come from errors in construction of concrete chamber side walls and resulting leakage associated with improper gate seating.

- Air-Regulated Self-Priming Siphons

In 1981, the Seine St Denis Public Agency experimented with the self-priming siphon technique as a sewer regulator. Until then it was used on spillways or in water treatment plants to keep upstream water levels within a narrow narrow fixed range. Air-regulating devices were mechanical (small air inlet or float-operated air entrance valve) or mechanical-pneumatic (a circular air-entrance check valve including a vertical spiral spring is regulated through a dynanamic equilibrium between siphon back pressure and a selected bubble pressure).

The use of siphons instead of gates in sewer networks resulted from construction and maintenance considerations, but at the time there was not much information available on regulation performance. The Seine St Denis siphons are concrete and inverse U shaped, with steel lining in some cases. Negative pressure inside the siphon is controlled through a top air entrance pipe connected to a small chamber above the pavement that contains air regulating devices. The inverse U-shape, though hydraulically poor, is easily implemented by contractors, particularly in deep wells, since much of this hydraulic structure is concrete, without moving parts.

Regulation (water level or negative air pressure inside the siphon) is obtained by controlling air entrance instead of acting directly on a sewer gate. Therefore, corrosion is a smaller problem, energy consumption is lowered (pneumatic small valve regulation versus high pressure hydraulic jack), maintenance access is easier and less costly, the regulator is really waterproof, and regulation components are standard industrial products.

Most of the Seine St Denis siphons have been built in retention basins and at the inlet structures of deep stormwater trunk sewers. We have had satisfactory performance for 4 siphons within a 10-20 cm level accuracy, but the small number of significant storm events during 1984 and 1985 does not allow any firm final conclusion. Moreover, one should be careful in evaluating air-regulated self-priming siphons, because more research is required on air-water mixing and entrainment phenomena inside the siphon, and on transfer function formulations through this hydraulic structure.

In 1984, we chose an electric-pneumatic controller, e.g. electric for signal processing and pneumatic for valve drive (short response time). Vacuum pumps, controlled by the control center, allow an anticipatory priming of siphons. Safety devices have been added to force closing of the air regulating valve. In addition, positive air pressure inside the siphon caused by simultaneous upstream and downstream water level rises is avoided by additional air relief valves.

To conclude, let us say that the use of air-regulated self-priming siphons for real time control of sewer systems is an attractive positive new experience that looks promising. It should be improved through more research and on site operational evaluation, and the Seine St Denis County is working towards these objectives.

- Inflatable fabric dams

In 1980 and 1982, 3 air filled inflatable dams were installed to control wastewater discharges of deep sewers to receiving rivers. They were broken or pulled away from the sidewalls because they could not withstand the oscillating forces caused by upstream and downstream waterwaves. This failure led us to try a water inflatable dam in 1982. We had more success with this technology, after leakage problems at the pipe-dam connection were solved in 1984. Its width is 2.5 m and its height is 1.2 m. It controls overflow discharges from a combined sewer to a 2.5 m diameter trunk sewer. Control is based on filling and emptying volumes of fresh water within a low pressure recirculating system. All hydraulic and electric equipment is installed in a separate room disconnected from the sewer to ease maintenance.

However, the use of water inflatable dams as regulators (equivalent to an adjustable weir) is still limited, because the mean height of inflatable dams cannot be measured accurately. It has a bumped shape and a position hysteresis due to lateral and bottom rubber-rubber friction and shear stress. A rough indirect estimate of height might be achieved using water volume conservation equations within the recirculating system.

Before extending the use of this regulator, we want to make sure that it is definitely reliable. So far, the long term observed data over 15 months look promising. Nevertheless, more technological analysis is required to define it as an industrial standard product. This should include a cost-effectiveness evaluation.

Local Process Hardware

We used 2 different microcomputers in 1984, because we were suffering very frequent failures on the 4 units we had in 1983. Then, at the beginning of 1985, we decided to select an available industrial type. We finally considered that quality of electronic components was a more important requirement than mathematical processor performance.

Our department develops its own software algorithms in assembler language, works out the emulation process and checks them on a in-house test system before the first on-site runs. A close cooperation between the hydraulic and the electrical engineers is resulting in some standardization of algorithms, and preserves the possibility of changing part of the existing programs at any time, which is important at the implementation stage.

Controller circuit design requires an accurate prior definition of the local station operation, which means a sufficient knowledge of local hydraulics and of the remote control objectives. Standard technical requirements have been defined by the sewer management department. Main failures are detected at first operation checks, and most of them come from improper electrical connections. Other problems come from inadequate checking operations, until maintenance crews acquire a working knowledge of the local station and establish a routine operation.

Telemetering Facilities

We selected a standard telemetry box. At this time, 20 have been installed at local stations. Integrated circuits generally consist of 1 or 2 instantaneous water level measurements, 16 or 24 warning contactors, and up to 8 gate opening/closing contactors; 4 of them have a double capacity. Major failures are caused by lightning. We now plan to improve power supply protection against lightning and stray currents by using alternately one of two batteries: one providing power supply, the other being recharged.

Through a connection to a local microcomputer, the SCI will receive mean 5-minute data as a result of local calculations. At this time, we have no experience with telemetered 5-minute data. In case of computer failure, or when the local station has no computer, only instantaneous water level measurements are telemetered to the control center. Any change at local stations produces a telemetered warning message to the control center, and the complete electrical state of the station is displayed at the monitoring system.

Control Center

The control center equipment is very dependent on the selected structure of the control logic system. At this time, our experience comes from the existing monitoring system, connected to 4 retention basins, 10 pumping stations, 2 siphons and 1 inflatable dam.

This monitoring system is currently used to control 20 local stations. It has also provided excellent support for training, and it helped in new monitoring system design. A first training experience with rainfall starting time and duration predictions from a radar image started in June 84. At the present time we largely use this technique for daily inspection tasks. An experiment on an electrically driven slide gate was also conducted. Two daily telemetered opening/closing operations were executed during 100 days. We had a 3% failure rate, half of them from the computer, the other half from the telephone line. We receive messages when any failure or alarm appears at local stations. Scanning operations on 10 stations are also conducted as follows: 2 scans/day when dry weather conditions prevail, 15 min scanning for a rainfall intensity greater than 10 mm/2h from radar predictions on a TV graphical display.

Simultaneously combining routine maintenance on site with telemetry checking at the control center has been an efficient way for detecting and rapidly solving failures at local stations during the first operation tests.

The Seine St Denis is finalizing the assembly phase of the

control center system with contractors. Coupling a computer (Bull SPS 9) with a real-time process-oriented hardware system led to a costly and time-consuming interface, but we finally solved it.

Much work is being carried out to achieve a logical and efficient use of graphical displays (water levels and flows versus time). Failures and alarms will automatically generate real-time modifications of input data in the control logic system. A programmed SCI process will start telephone calls to maintenance crews. In absence of central supervision, the local stations will execute a predefined regulating operation, inferred from previous off-line simulations.

The control center should operate at all times, even in case of electrical failure. A decision was therefore made to design an auxiliary power supply using a gasoline-engine generator.

Probably no standard can be achieved for any future control logic system, because it should reflect specific performance criteria. Monitoring systems have been developed for industrial applications, so it is easier to get standard reliable products.

Design Criteria and Reliability Approach for Sewer System Operation

Reliability approach

Statistically-based evaluation of reliability appears biased when applied to sewer systems, because environmental conditions, maintenance performance, selection criteria of hardware components have probably as much impact on reliability as the technological quality of the component itself. It takes a long time to get a satisfactory on-site operation routine, particularly for sewer systems in which regulation performance is highly dependent upon rainfall intensities or hazardous discharges.

Reliability criteria have been included through all development stages of the overall Seine St Denis project:

- Define the conceptual structure of the real time system.
- Create simultaneously a technical staff and electric/ electronic maintenance crews.
- Select standard hardware components.
- Progressively change maintenance and operation practices.

Maintenance Trends

Maintenance is viewed as a major response to hardware failure. It should be reinforced either through maintenance contracts, or through scheduled on site checking tasks. The control center is located at the main office, in optimal environmental conditions. Electronic equipment is standard and more attention is focused on it. Maintenance contracts provide steady support for the reliability of the monitoring and control logic systems.

Communications between the control center and local stations depend completely upon telephone lines. We try to obtain close cooperation with the local public telephone company.

One main weakness of real-time control of sewer systems is the low-frequency of regulator use, mainly for stormwater sewers. Therefore, periodically exercising manual operations at local stations and checking controllers, regulators and sensors, are complementary requirements of the scanning operation from the control center. Also, there seems to be a simultaneous occurrence of high return-period rainfall events and control system failures, particularly electronic protection shutdowns and power supply breakdowns. That makes static hydraulic works, or safety devices, a necessary ultimate alternative for extreme floods.

Components of local stations are located in sewer appurtenances (underground rooms or small chambers above the pavement).

Access constraints (location at congested cross-roads, small size manholes), environmental limitations (corrosion, leakage from pavement runoff, loosely waterproofed stuffing boxes, etc), and difficult maintenance conditions led us to prefer chambers above the pavement. An adequate design prevents vandalism as well.

Design of Local Stations and Performance of Real-Time Control

Hydraulic design of local stations is carried out by the sewer management department. This department provides all mechanical, electric and electronic equipment, and also checks on site performance. It does not enter into engineering design, construction or contract administration, in order to save more time for sewer management tasks.

This strategy is giving good results. Initial operational failures almost disappeared, more standard products could be introduced, it was easier to define clear local station operation schemes, and department personnel gained on-site technical training experience.

Let us examine the Molette retention basin as a case study (Figure 6). Hydraulic design agreed with 5 objectives:

- Control of downstream flows.
- No water level above elevation 35.00 in the retention basin.
- Sedimentation of combined sewer overflows.
- Detention of wastewater flows diverted from combined sewers.
- Permanent pond for any expected future recreational use.

Allocating overflows to 3 different basins according to rainfall intensity and combining different types of hydraulic structures at the downstream left corner of the impervious sedimentation and retention basin brought a flexible response

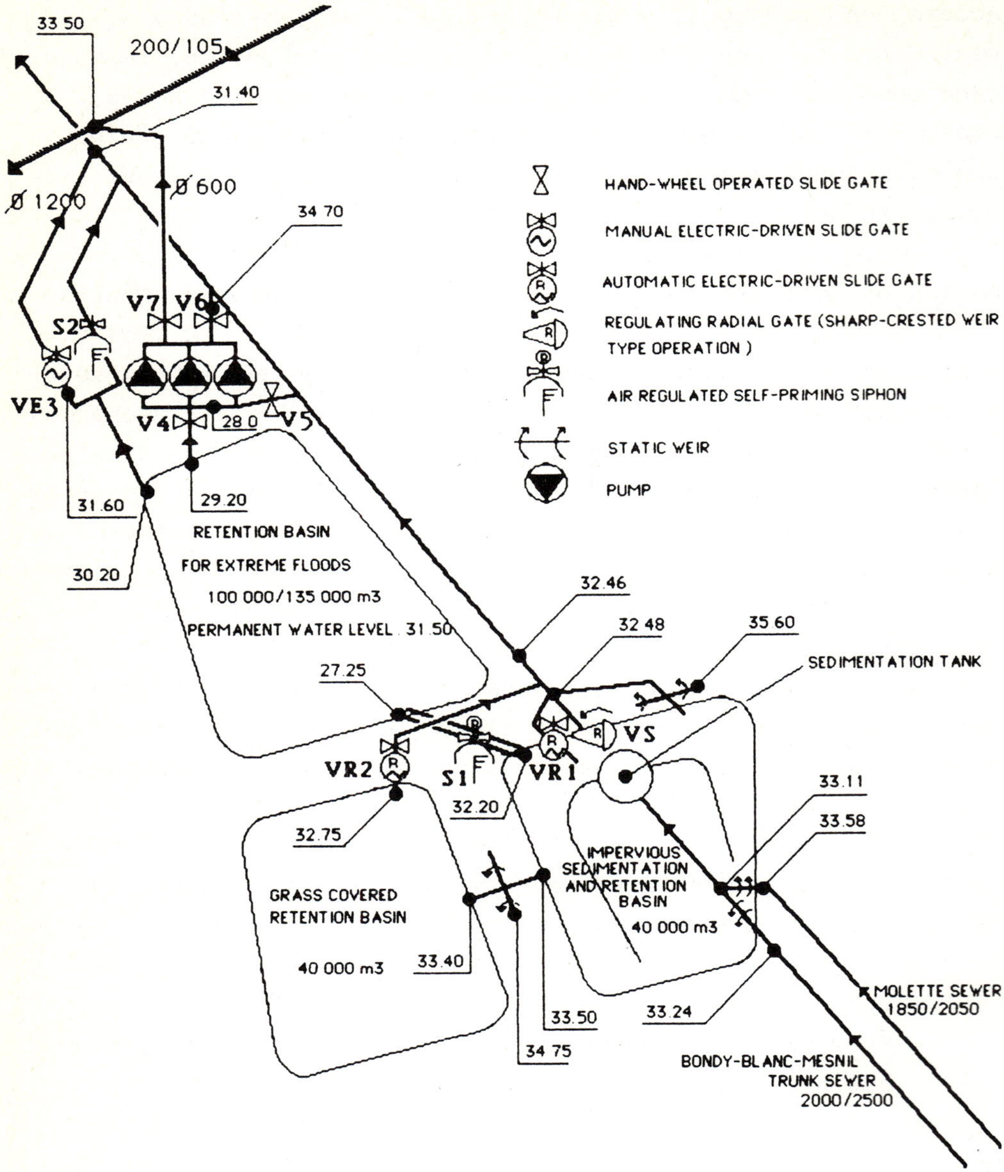

Fig. 6: MOLETTE RETENTION BASIN PROJECT

to multipurpose real-time control.

Only an iterative design approach, including operation criteria checked by simulation models, could improve the hydraulic design. Hydraulic structures include:

- An oil-hydraulically operated regulating gate VS to control downstream flows according to supervisor strategies.
- 2 electric-drive slide gates VRI and VR2, for emptying the 2 upstream basins when allowed by the local microcomputer.
- 3 static weirs
  - Upstream, to generate a low velocity flow on both sides of the retention basin.
  - On the central embankment, to discharge overflows from the sedimentation basin to the grass covered retention basin.
  - Downstream, as spillway during extreme floods.
- A siphon Sl divided in 2 siphons, one of them regulated, the other one acting as a spillway to the downstream retention basin.
- A siphon S2, associated with a locally operated slide gate VE3, to discharge possible overflows stored in the permanent pond.
- A dual-purpose pumping station:
  - To lower the permanent water level and increase the storage capacity of the permanent pond by 35 000 m^3.
  - To discharge diverted dry weather wastewaters to a combined sewer after detention in the sedimentation basin.

Hardware components for Real-Time Control accounted for 5% of the overall investment, but significant benefits are expected for maintenance (labour cost reduction due to less routine visits and manual tasks).

Safety Criteria for Real-Time Control

The simultaneous implementation of an intensive local station equipment program and control center during 1985-1986, with flood control as a short term objective, led the Seine St Denis county to a careful strategy of local operation and supervisory control.

Because of the low frequency use of regulators, we have almost no back-up power supply, nor have we duplicated hardware components. Therefore, we decided to deal with 4 levels of flow control strategies at local stations, depending upon possible failures:

1. Execution of flow control strategies sent by a supervisor from the control center to local stations.

   - Full satisfactory performance of the control logic system.
   - Suggested strategies come from on-line modeling, but should be validated by a supervisor.

2. Execution of the local flow control strategy.

   - No order coming from the control center: no supervisor, or no new flow control sent by the supervisor, or telephone line failure over a predefined duration.

3. Return of regulators to a predefined hydraulically safe position.

   - Used in case of microcomputer, sensor or oil-hydraulic regulator failures.
   - Partially or totally static hydraulic behaviour.
   - In some cases, this return to a predefined hydraulic safe position can be constrained by local hydraulic conditions, generally selected warning water levels measured by sensors, or electric contactors.
   - Limited risks, depending upon hydraulic design and rainfall event.

- If these failures follow a telephone line failure, the local station works out a return strategy but the control center will not know it.

4. Uncontrolled flows

- Sudden long duration power failure at local station.
- Electric-drive gates stay at their last position, vacuum pumps at siphons do not start, but telemetry goes on through battery power supply, oil-hydraulic gates and air-regulating valves on siphons go to hydraulically safe positions.
- SCI alarms are transmitted to maintenance crews, for immediate predefined on-site manual operation

The control logic system should know and take into account the real state of all sensors and regulators at any time, even in case of failures. Input data for simulation and optimization models (SAC) should agree with telemetered data processed by the monitoring system (SCI), and regulator behaviour should be the same at local stations and within the models.

At a first stage of experimentation, we want to demonstrate that the overall supervisory control system can perform satisfactorily. Within an operational unit of interacting local stations, supervisory control is logically assumed to be more efficient than any independent local control.

However, we are aware of all existing limitations in forecasts and hardware reliability during extreme events. Therefore, in 1986 we will be using a very constrained optimization scheme with limited variations from predefined local flow controls. We expect to enlarge the range of flow control strategies in the future, depending on improvements in forecasts, hardware reliability, and training experience of supervisors.

Conclusions

Real-time control is a new approach to sewer management that looks promising and feasible whenever it refers to telemetry and monitoring systems for independent local station operation or simplified reactive algorithms. Coupling a recursive-adaptive logic system to a monitoring process for stormwater management requires a more complex design to preserve reliability at any step, from rainfall prediction to on-site regulation.

A first experiment will be conducted in the Seine St Denis county in 1986, on 3 retention basins, using radar rainfall starting time and duration predictions coupled with 8 telemetering rainfall gages. We are now engaged in an intensive implementation of the overall project, both control center and local station equipment. A large scale evaluation of hardware performance will be carried out in 1987.

We expect conclusions from on-going research related to radar calibration and rainfall-runoff modeling. More research is required to solve uncertainty in control logic, and to improve the reliability of telemetry support, regulating siphons, velocity and water quality sensors.

OPERATIONAL ALGORITHMS FOR APPLICATION IN REAL-TIME CONTROL OF COMBINED SEWER SYSTEMS

Gilles G. Patry
McMaster University
Department of Civil Engineering &
Engineering Mechanics
Hamilton, Ontario L8S 4L7

Introduction

Over the last few decades, water pollution abatement programs have been responsible for significant and sometimes dramatic improvements in the water quality of many receiving water bodies. However, as point sources of pollution are being eliminated through attainment of secondary and sometimes tertiary treatment of municipal and industrial wastewaters, other less apparent sources of pollution are becoming increasingly threatening to many beneficial water uses.

Storm-related pollution in general, and combined sewer overflows (CSO) in particular, have been observed to contribute significant amounts of pollutants to receiving water bodies, including large quantities of oxygen consuming materials, heavy metals, and petroleum hydrocarbons (1). In fact, in its report to Congress, the United States Environmental Protection Agency (U.S. EPA) established that, on a runoff event basis, CSO were the major sources of biochemical oxygen demand (BOD), suspended solids (SS), nutrients and lead to receiving water bodies. On an annual basis, the investigation further indicated that CSO and urban stormwater runoff remained the major sources of SS and lead (2).

A further indication of the magnitude of urban nonpoint sources of pollution is provided by the U.S. EPA's 1978 Needs Survey (3) in which it is estimated that $26 billion would be required to control pollution from CSO, while $62 billion would need to be allocated for the control of stormwater runoff from existing urbanized areas. In light of present economic times and based on

NATO ASI Series, Vol. G10
Urban Runoff Pollution.
Edited by H. C. Torno, J. Marsalek, and M. Desbordes

the increasing awareness of society towards environmental issues, particularly with respect to beneficial water uses, solutions to the problem of storm-related sources of pollution will need to be technically innovative as well as economically attractive. Recently, Torno (4) indicated that

> "... an area that promises to expand rapidly in the future is the application of models in the automatic control of urban drainage systems. ... (These) models will form an integral part of any such systems, where they are used simply for providing an understanding of system responses, or for developing actual control algorithms under which direction the system will actually be operated".

Accordingly, the purpose of this paper is to review some of the theoretical and practical aspects related to the implementation of real-time control strategies in urban water resources systems. After establishing the need for real-time control of CSO, current control strategies will be reviewed from a theoretical as well as from a practical point of view. Recent developments in water quality and quantity forecasting will be discussed briefly. Finally, the paper will review some of the 'real-world' constraints in operational control of CSO and suggestions as to areas of further research will be discussed.

Need for Real-Time Control of Combined Sewer Systems

The purpose of this section is to examine the potential benefits of real-time control of combined sewer systems, particularly in large metropolitain areas. Only recently has it been realized that the treatment of point sources of pollution, originating from municipal and industrial discharges, may not bring about the expected improvements in receiving water quality. In fact, water pollution abatement programs centered solely on point sources of pollution usually fail to address a significant fraction of the urban water quality problem. In cities that provide secondary treatment of municipal sewage, storm-related sources of pollution have been reported to account for between 40 to 80% of the annual total oxygen consuming materials. Moreover, it has been observed that during intense storms, 95% of the BOD load to receiving water bodies is contributed directly by runoff (5). The interested reader is referred elsewhere (2,3,6) for a thorough analysis of the pollution loads from storm-related sources.

Impacts from Storm-Related Sources of Pollution

Considering that storm-related sources of pollution can be important sources of organic matter, sediments, nutrients, microorganism contamination, trace metals, and some trace organics, it is not unusual for beneficial water uses to be threatened by surface runoff in general, and CSO overflows in particular (6). Depending on the characteristics and sensitivity of the receiving water body and on the time-history of the pollutant loads to the receiving waters, CSO can have a wide range of impacts. Heaney et al. (7) identified the impacts from storm-related sources of pollution at three levels: a) policy or master plan violations; b) criteria or standards violations; and c) actual environmental degradation or loss of beneficial water uses such as fish kills, beach closing, etc. They also noted that such impacts were very difficult to isolate owing to the complex structure and inter-relationships within the urban hydrological cycle, because of the 'masking' of the impacts by other sources.

Few studies have ever been conducted that specifically document the loss of beneficial water uses on CSO. Usually, impact assessment, when studied is limited to examining the dissolved oxygen (DO) response following CSO, as in the case of Hvited-Jacobsen (8) who reports observing an immediate and a delayed impact on the DO concentration of a small stream following CSO. Meinholz et al. (9) found that a significant portion of the DO deficit following CSO could be attributed to the scouring of sediment oxygen consuming materials near the combined sewer outfalls. Klemetson et al. (10) examined the movement and effects of CSO sediments on the beneficial water uses of the receiving waters using the Colorado State University sediment transport model. Their experiments on the movement of hypothetical CSO sediment loads emphasized the importance of low flow deposition around the outfall and subsequent scouring under flood flow conditions (1 in 5-year flood flows). Finally, Pitt and Bozeman (11) examined the sources and effects of urban runoff pollution on the Coyote Creek near San Jose, California. Their results indicated that various urban consituents, especially organics and heavy metals, may be responsible for many of the adverse biological conditions observed in Coyote Creek. The non-urban section of the creek was found to support a comparatively diverse assemblage of aquatic organism, while the urban portion generally lacked diversity and was dominated by pollution tolerant species.

Combined Sewer Overflow Abatement Technologies

Turkeltaub and Field (12) found that following upgrading of dry-weather flow treatment, CSO pollutant loads were the next most cost-beneficial pollutant loads to control. A variety of strategies have been suggested to reduce the pollutant load from CSO on receiving waters, thereby minimizing the impacts on the intended beneficial water uses. Traditionally, abatement of pollution from CSO has been grouped in three main categories: a) source controls; b) collection system controls; c) off-line storage treatment and control.

Source controls are designed to reduce the amount of pollutant entering the sewer system. They are usually non-structural labor-intensive in nature and include: a) improved street cleaning programs; b) combined sewer flushing; and c) catch basin cleaning. Collection system controls are designed to improve the system's effectiveness in storing and conveying the flows in order to minimize the opportunities for CSO. Flow reduction, sewer separation, and in-line storage are typical collection system control measures. Ward (13) reported the results of a case study in which the volumes of CSO were significantly reduced through in-system storage control. Finally, off-line storage and treatment facilities ranging from sedimentation basins to carbon adsorption processes can be used to remove pollutants from CSO (6). However, because of the multiplicity of overflow points and the usually high capital investments required, storage-treatment facilities represent more the exception than the rule in CSO abatement programs. Selection of the most appropriate CSO control measures is usually based on a cost-benefit analysis, bearing in mind that the ultimate objective is to prevent, within limits, the loss of beneficial water uses along the receiving waters. Accordingly, general guidelines and removal efficiency type of requirements are particularly inappropriate for the management of non-point sources of pollution.

Real-Time Control Strategies

In general, abatement of pollution from CSO is regarded by many localities as requiring the addition of extensive storage, transport and treatment facilities to their systems with correspondingly high annual operation and maintenance expenditures. In accordance with the 1979 suggestion of the Genral Accounting Office, indicating that the less costly solutions should be explored to handle the CSO problem (14), more innovative alternative need to be examined.

Real-time integrated operation of interceptor regulators, of the type shown in Figure 1 (14), has been suggested as a means to provide a potentially cost-effective CSO control alternative (4,15,16,17). In fact, automation of combined sewer systems is part of collection systems controls discussed earlier, and can also be integrated into storage-treatment control alternatives if it is needed, to protect the intended beneficial water uses of the receiving water body. To achieve this ultimate objective, several indirect but more explicit criteria have been suggested (18,19), including: a) minimization of the number of overflows; b) minimization of the volume of overflows; c) reduction of the 'first-flush' effect; d) minimization of the weighted mass of a specified pollutant or group of pollutants; e) minimization of the mass of a specified pollutant or group of pollutants above a certain threshold value. However, none of these criteria can guarantee that loss of beneficial water use will not occur. At best, these criteria have only an indirect relationship with the utlimate objective, and are more the result of mathematical convenience than anything else.

As far as real-time control strategies are concerned, Trotta et al. (20) identified two basic types of automatic control modes: reactive and predictive (or anticipatory). In reactive mode, the system is operated in response to its state as the storm progresses over the cathment. In predictive mode, the system is

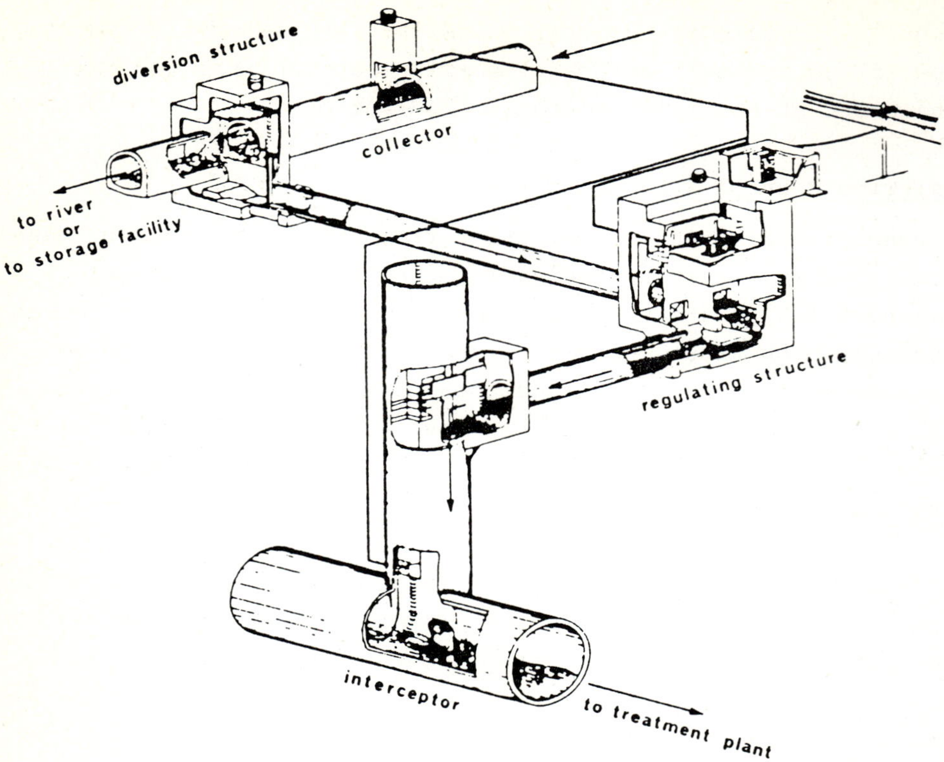

Figure 1. Typical CSO Regulating Structure

operated in response to the anticipated state of the system prior to the actual occurence of the rainfall event, and in response to both the actual and predicted states of the system once the event has started. In addition to these two modes of operation, many municipalities operate CSO regulators under simple local control, in which flows from the combined sewer collector are diverted to the interceptor through a fixed capacity regulator. Each of these three modes of operation will be discussed in more detail below.

Before proceeding to the description of the various control strategies, allocation of flows around a regulating structure will be discussed briefly. Figure 2 shows the flow elements around a typical regulating structure with off-line storage facilities. Flows from the combined sewer collector j, Q_{Cj}, are diverted in whole or in part to the sanitary sewer interceptor, Q_{Rj}, or to the receiving water body as overflow, Q_{Oj}. If off-line storage facilities are available, flows that would normally be diverted to the receiving waters would be directed to the storage reservoir, Q_{Sj}. Depending on the state of the system, flows from the storage facility could either be returned to the interceptor, Q_{sj}, or overflow, Q_{Oj}. The allocation of flows at a regulating structure can be quite complex and is a function of the selected level of control. Factors to be considered include: a) regulator capacity, Q_{rj}; b) available interceptor capacity, i.e., interceptor capacity, Q_{Ij}, minus actual interceptor flow at time t; c) use of possible off-line storage facilities; d) state of the combined sewer flows, i.e., quality and quantity; e) interactions with other catchments downstream of the outfall; and f) intended beneficial water uses along the receiving waters.

Simple Local Control

This particular control strategy uses constant capacity regulators. Flows from the combined sewer collector are diverted to the interceptor up to the capacity of the regulator, any excess being diverted to the receiving water body as overflow or to an off-line storage facility (Figure 3). Flow to the interceptor is thus controlled exclusively by the geometry of the regulator orifice, which is not allowed to vary during a storm event. Flow

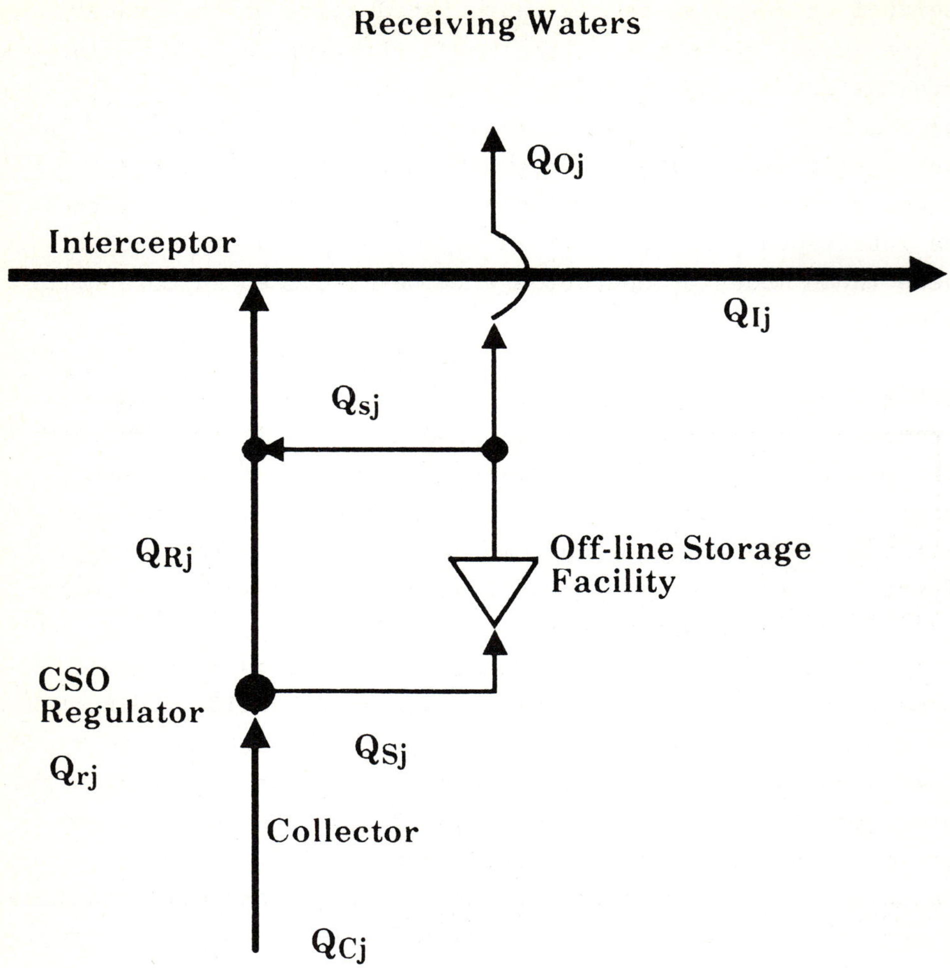

Figure 2. Flow Components at a Regulating Structure

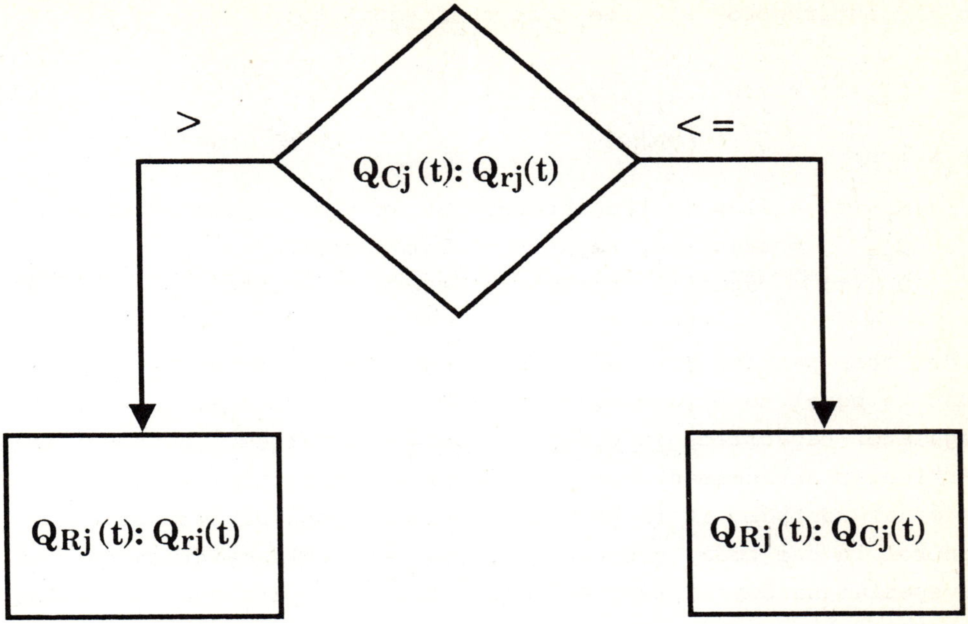

Figure 3. Decision Algorithm for Simple Local Control

to the interceptor at time t is thus equal to:

$$Q_{Rj}(t) = \min [\ Q_{rj},\ Q_{Cj}(t)\ ] \qquad [1]$$

in which:

$Q_{Rj}(t)$ = flow to the interceptor at time t from catchment j;
Q_{rj} = regulator capacity for catchment j;
$Q_{Cj}(t)$ = capacity of combined sewer collector j.

Under this particular control strategy, the interceptor capacity must be equal to or greater than the sum of the upstream regulator capacities in order to avoid surcharges. Although this particular arrangement does not offer any opportunity for external control, it is probably the most popular form of CSO control in use today because it requires little maintenance and no special hardware. Simple local control is particularly suited for those areas where the impacts from CSO are not likely to be significant.

Reactive Control (Local Automatic Control)

Under this type of control startegy, flows diverted to the interceptor are a function of the available downstream interceptor capacity. When no off-line storage facilities are available, flows are assigned in accordance with the decision algorithm presented in Figure 4. Flow to the interceptor at time t is given by:

$$Q_{Rj}(t) = Q_{Ij}^{max} - Q_{Ij-1}(t) \qquad [2]$$

in which:

Q_{Ij}^{max} = interceptor capacity downstream of regulator j;
$Q_{Ij-1}(t)$ = interceptor flow in reach j-1 at time t.

Such that:

$$Q_{Rj}(t) \quad \min [\ Q_{rj},\ Q_{cj}(t)\ ] \qquad [3]$$

where $Q_{cj}(t)$ is the flow in collector j at time t. A result of this mode of operation is to minimize, for example, the volume of overflows by maximizing interceptor use. A typical objective

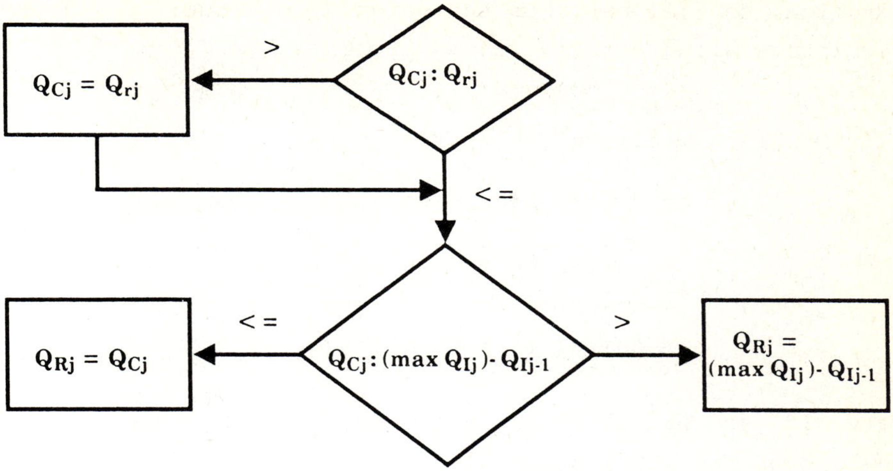

Figure 4. Decision Algorithm for Local Automatic Control (without off-line storage)

function, $OF_R(t)$, based on the weighted volume of overflows for N regulators along a reach can be expressed as:

$$OF_R(t) = \text{minimize} \left\{ \sum_{j=1}^{N} w_j(t)\, Q_{Oj}(t) \right\} \qquad [4]$$

in which:

$w_j(t)$ = weighting factor for reach j at time t;
$Q_{Oj}(t)$ = overflow to reach j at time t;
N = number of regulators along thereach.

Ward (13) and Brueck et al. (21) have presented case studies describing the application of one form of reactive control to minimize the volume of CSO through in-system storage.

Predictive Control

Implementation of reactive control strategies alone does not guarantee that the impacts from CSO will be minimized throughout the course of a rainfall event, because operation of the system is based solely on the state of the system up to time t. Under anticipatory control, a typical objective, OF_P, formulated to minimize the weighted volume of overflows, can be expressed as:

$$OF_P = \text{minimize} \left\{ \sum_{t=1}^{T} \sum_{j=1}^{N} w_j(t)\, Q_{Oj}(t) \right\} \qquad [5]$$

in which T is the duration of the rainfall event. It should be noted that:

$$\sum_{t=1}^{T} [OF_R(t)] \geq OF_P \qquad [6]$$

Accordingly, efforts to minimize the impacts from CSO require that predictive control strategies be implemented in a way that the cause and effect relationship from all regulators be integrated in a control algorithm prior to the allocation of overflows. Figure 5 describes the elements of an idealized predictive control strategy. Three levels of action are identified: forecast, optimization, and impact assessment, all of which would need to be carried out in real-time. Selected elements of the strategy will be discussed below.

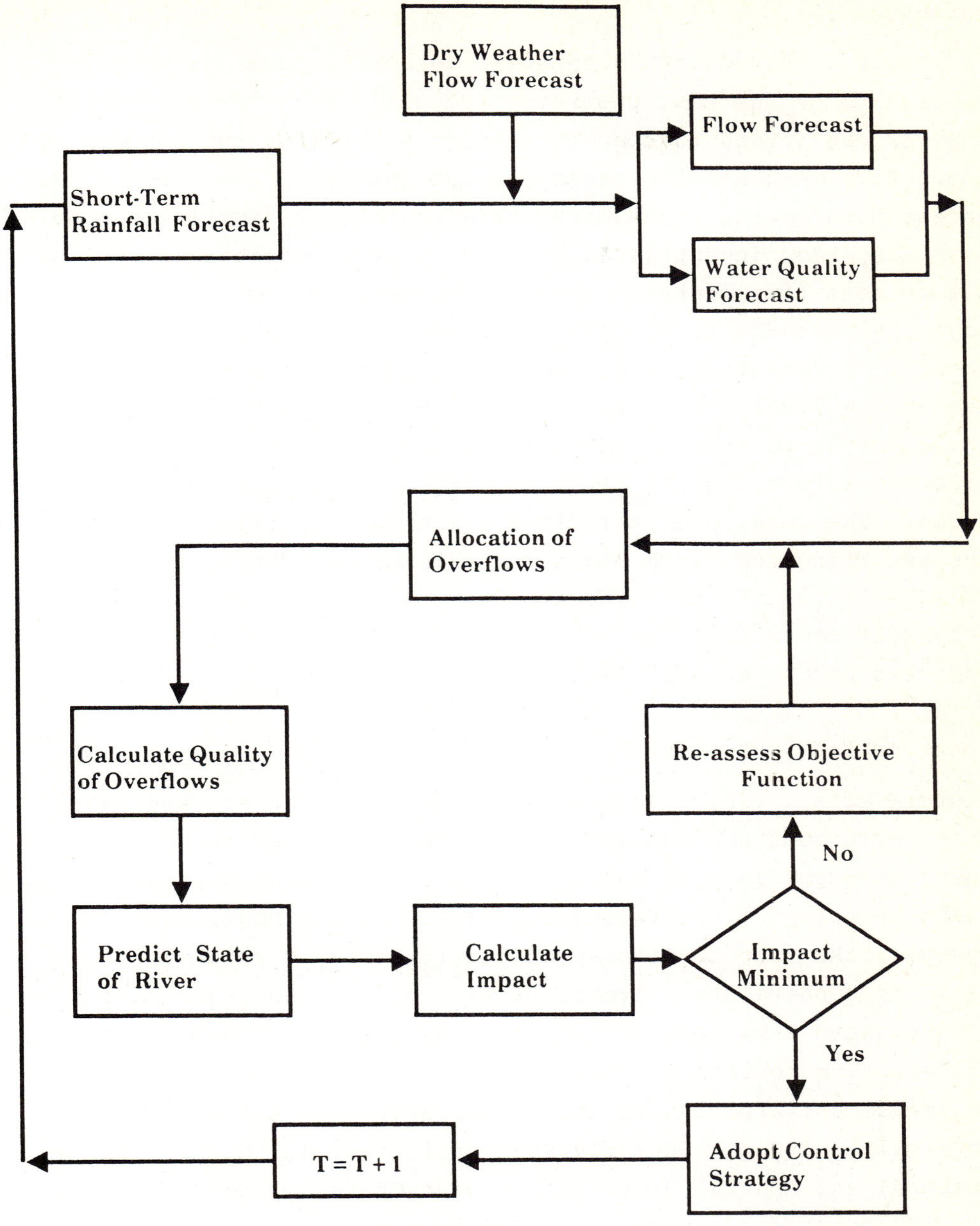

Figure 5. Decision Algorithm for Predictive Automatic Control

Forecast

In the area of short-term rainfall forecast, radar forecasting appears to be the most promising (22,23). Very few cities however can afford to operate a radar solely for the purpose of short-term rainfall forecasts for CSO control. Consequently, more modest forecasting procedures, based on simple regression models have also been investigated and have proven potentially applicable to real-time control of combined sewer systems (24). More recently, Labadie et al. (25) examined the worth of short-term rainfall forecasting in CSO control. The authors attempted to define the levels of forecast error that can be tolerated before it is better to abandon adaptive control policies utilizing forecast information in favor of simple reactive control methods. The results of a limited number of experiments using the San Francisco North Shore Outfalls Consolidation Project (SFNSOCP) indicate that the expected forecast model errors are generally lower than the error threshold obove which reactive policies become more attractive.

Optimization

Labadie et al. (26) have discussed decomposition and aggregation techniques and applied them to the San Francisco project. The system consists of multiple reservoirs operated so as to minimize street flooding and overflows. Their model did not however take into consideration interceptor and regulator capacity constraints. Trotta et al. (20) on the other hand investigated a variety of optimization techniques under reactive and adaptive control strategies. The authors concluded that some degree of anticipation or storm forecasting is important in controlling CSO particularly when it is desired to control the temporal and spatial distribution of overflows. Nguyen Van et al. (27) tested a dynamic programming algorithm in an exploratory study of CSO control for the Communauté Urbaine de Montréal. Labadie et al. (16) proposed a dynamic programming approach linked to an unsteady flow routing model. Finally, Patry (19) presented a linear programming model for the control of CSO with off-line storage facilities.

Impact Assessment

As indicated in Figure 5 a fully integrated CSO control strategy requires some measure of real-time impact assessment. This element of the overall control strategy is certainly the most challenging. To the author's knowledge no model has ever been tested or developed for real-time impact assessment for CSO control. Béron et al. (28) developed a water quality index that could be used for real-time control of CSO in order to assess whether beneficial water uses had been violated.

It is evident that the strategy proposed in Figure 5 relies heavily on the availability of dependable real-time forecasts of the state variables of the system, namely rainfall, runoff, and water quality parameters. The state variables then serve as independent variables to the CSO optimization algorithm. As reported by Coffee et al. (16), San Francisco is attempting to develop a similar form of CSO control. In their final report, the authors concluded that:

> "Integrated reactive-predictive control strategies appear to hold promise but their cost effectiveness is highly sensitive (among other things) to ... the accuracy of the prediction of rainfall and runoff."

The authors go on to recommend that simple predictive models be developed for use in reactive-predictive control. Finally, in a study on the feasibility of integrated control of combined sewer regulators using weather radar, McPherson (15) concluded that:

> "... no advantage over local automatic (reactive) dynamic control can be obtained by integrated operation of regulators unless expected flow rates to the interceptor can be estimated in advance ..."

If losses of beneficial water uses are to be minimized, then water quality forecasts must also form an integral part of any predictive strategy as indicated in Figure 5.

Real-Time Water Quality and Quantity Forecasts

It is not difficult to appreciate the complexity of the urban hydrological cycle, even when only a few of the major processes are considered, as shown schematically in Figure 6. The flow of matter from one subsystem to another is represented by the vector

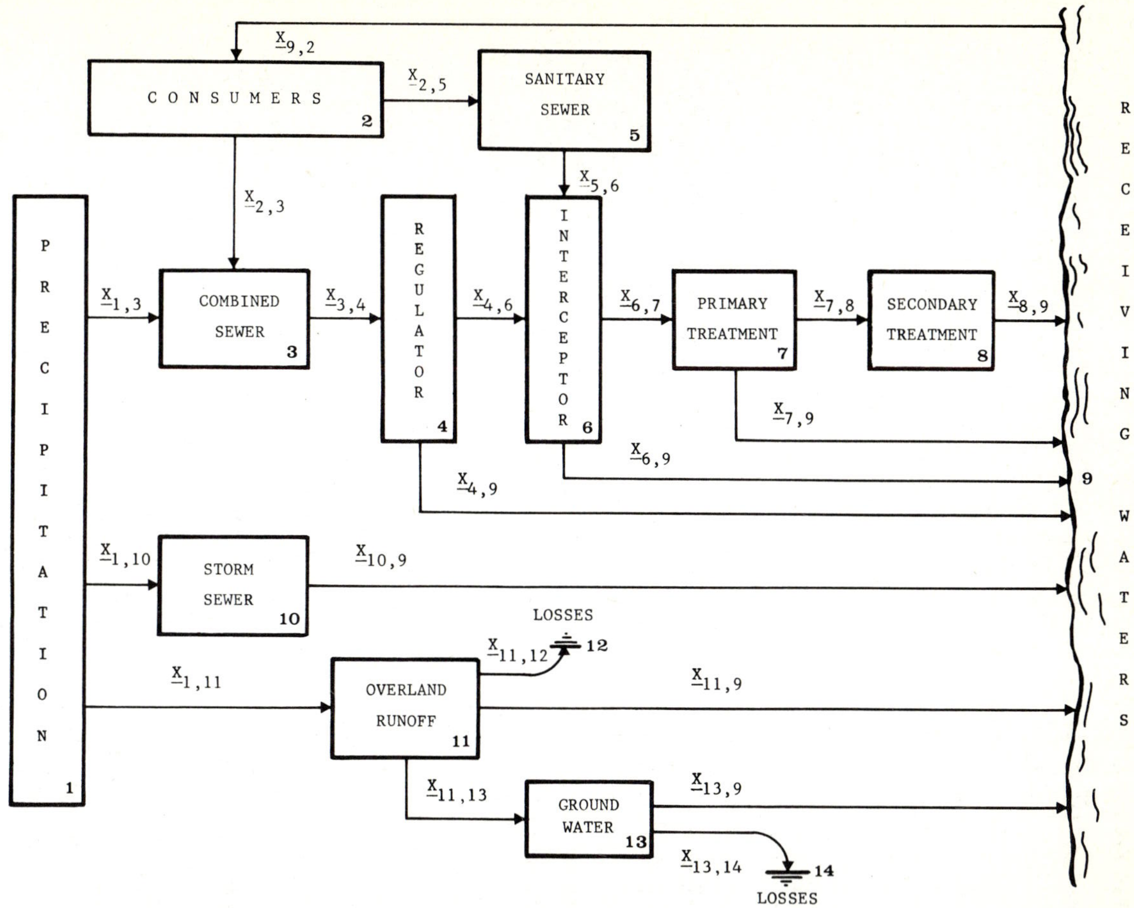

Figure 6. Simplified Representation of the Urban Hydrological Cycle

flux $\underline{X}_i j$, where i and j refer to the originating and terminal processes, respectively. Ideally, one would like the flux vector to consist of the mass flow rates of those water quality parameters needed to assess and predict the state of the receiving water body (node 9). Under the most desirable conditions, operational urban water quality management would require a description of the dynamics and inter-relationships among the various processes, to assess the time-history of the flux vectors $\underline{X}_{i9}$ to the receiving waters. The previous section emphasized the need for reliable, robust forecasts in a predictive CSO control strategy. The question that this section tries to address is 'how should the dynamics of the system and subsystem be defined?'. In addition to the basic causal relationships, the proposed forecast models should ideally consider the importance of nonlinearity, time-variance in the system dynamics, and the presence of measurement noise and errors in both input and output variables. Moreover, real-time hydrological forecast models must satisfy certain minimal requirements to be successful:

a) the models should be adaptive, viz. model parameters should be able to adjust to reflect the recently observed state of the system as additional data are being made available;
b) parameter estimation of the deterministic and stochastic model components should be performed recursively in real-time so as to provide immediate response to the time-varying characteristics of the catchments; and
c) real-time forecasting models should be computationally efficient and adaptable to microcomputer operation.

Characterization of a system's dynamic response usually falls in one of two broad classes: a) statistically-based models, also referred to as input-output models; and b) physically-based models also referred to as conceptual, mechanistic or internally descriptive models.

Statistically-Based Approach

Statistical or input-output models are designed to describe the dynamics of a process based on the history of the system's inputs and outputs, and on the stochastic disturbance experienced by the system (29). The key feature of this approach is that it assumes no *a priori* knowledge of the physical or internal relationships involved in the prototype, except for the requirement to find a proper functional relationship between the system's inputs and its output. Single-input single-output (SISO) linear processes can generally be represented by the following difference equation:

$$y(t) = \frac{A(z^{-1})\, z^{-d}}{B(z^{-1})} u(t) + \frac{D(z^{-1})}{C(z^{-1})} e(t) \qquad 7$$

in which u(t) and y(t) are sequence of measured input and output, respectively; z^{-1} is the backshift operator defined by $z^{-1}u(t)=u(t-1)$; d is the time delay between the input and the response variable; e(t) is a sequence of independent normal random variables; and $A(z^{-1})$, $B(z^{-1})$, $C(z^{-1})$, and $D(z^{-1})$ are polynomials in the backshift operator of degrees na, nb, nc, and nd, respectively. Equation 7 is represented in block diagram in Figure 7. Typically, in a single-input single-output rainfall-runoff model u(t) would represent rainfall, y(t) would represent the flow and d would be the lag between the excess rainfall and the runoff (29). The structural identification of rainfall-runoff models of the type described by Equation 7 are described elsewhere (30,31,33).

Input-output models are generally considered when a) the biophysico-chemical processes involved are either unknown, ill-defined, or not measurable; and/or b) the cause and effect relationships for the system are so complex, the number of parameters so great, and the data requirements so large that a conceptual formulation would be impractical. Patry and Mariño (30,31,32,33) have investigated the application of recursively estimated models of the type described above to water quality and quantity forecasts in urban for the purpose of real-time CSO control. In effect, the models were found to yield flow forecasts with a mean absolute percent error (MAPE) ranging from 5 to 17% for lead times of 5 to 60 min. Similar models were applied to suspended solids (SS) and chemical oxygen demand (COD) forecasts in combined sewer systems. The MAPE in SS forecasts ranged from 20% to 58%, for lead times ranging from 5 to 60 min, while the MAPE for COD ranged from 11% to 59%.

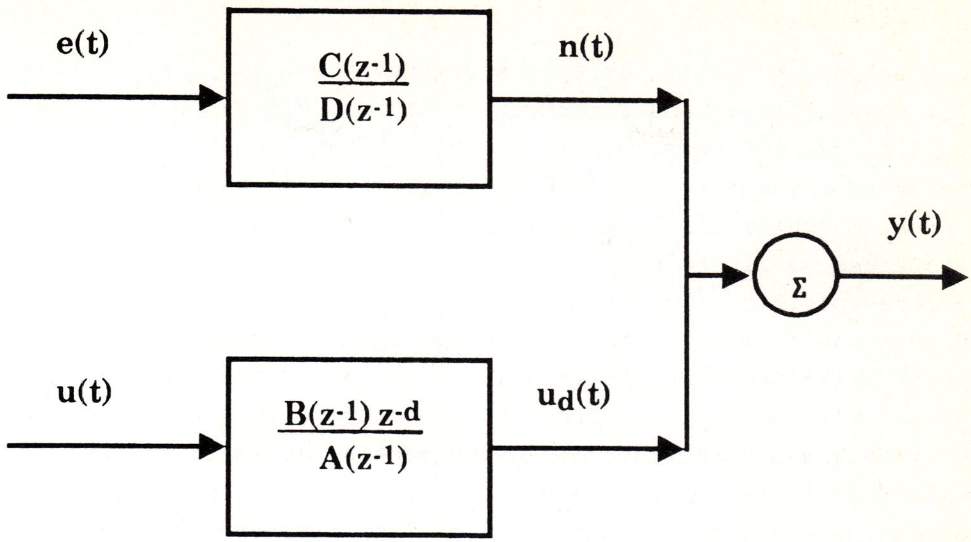

Figure 7. Statistically-Based Models of the ARMAX Type

Physically-Based Models

Input-output models of the type described previously suffer from a few drawbacks: a) model parameters have little or no physical significance, and consequently are of limited use to make inferences about the nature of the physical processes involved; b) input-output models are site-specific and thus cannot be transferred easily to other catchments; b) input-output models are data-specific, i.e., they are valid only within the range of observations from which they were developed, in other words, they are not suited for extrapolation.

The drawbacks of statistically-based models have motivated the development of conceptual models. Under this particular approach, modeling of the water quantity and quality response of a catchment is based on some _a priori_ knowledge of the biophysico-chemical interactions amont the different processes of the system. These models are in effect _micro-abstractions_ of the real-world that attempt to reproduce a selected number of processes that are believed to govern the flux of matter among the various elemental storage components, conditional on the principles of continuity and conservation of mass (34).

Linear conceptual models formulated for real-time forecasting purposes can generally be represented by the following general discrete state-space formulation:

$$\underline{x}(t) = A\,\underline{x}(t) + B\,\underline{u}(t) + D\,\underline{w}(t) \qquad [8]$$

in which $\underline{x}$ is a state vector, $\underline{u}$ an input vector, $\underline{w}$ a noise component, and A, B, and D are known conformable matrices. The dot in Equation 8 refers to the time derivative. The observation equation is given by:

$$\underline{y}(t) = C\,\underline{x}(t) + \underline{n}(t) \qquad [9]$$

in which $\underline{y}$ is the observed measurement of the output, C is a known observation matrix, and $\underline{n}$ a sequence of independent measurement errors. In addition to being more universally applicable than statistically-based models, the major advantage of physically-based models is that they can provide considerably more insight into the mechanisms of the system's performance (29). On the other hand, physically-based models of the type described above are usually more difficult to identify, and typically require the specification of a large number of parameters subject to considerable error. Moreover, such model also require a more extensive data base. In the interest of providing the _most comprehensive and universal model_ this particular approach often produces what can be referred to as _over-parameterized models_, for which parameter identification is more a _knob-turning_ exercise than a true assessment of the physical characteristics of the parameters.

Patry and Mariño (32,33) have described the application of recursively estimated conceptual models to water quality and quantity forecasts in urban catchments. In general, the results measured in terms of the MAPE in water quality and quantity forecasts compared favourably with the simpler statistically-based models.

Constraints and Features of Real-Time Forecasting

Data Availability

Modeling for simulation or design purposes differs significantly from real-time forecasting, particularly in the context of CSO control. Adaptive operational urban water quality management implies, among other things, the availability of an appropriate data base to be collected in real-time, so that estimates of the various dynamic model parameters be updated at every time step.

While data collection and analysis for water quality control purposes have gained considerable momentum over the last decade, it is generally agreed that only limited data are available for real-time model building purposes. The reasons for this lack of adequate urban water quality data are many:

a) the technology for real-time water quality monitoring is limited to a handful of parameters, many of which have only an indirect relationship to most beneficial water uses. To alleviate the problem, investigators have examined the possibility of correlating readily measurable water quality parameters to more significant parameters not measurable in real-time [e.g., turbidity versus BOD; turbidity or conductivity versus SS (34,35)].

b) field conditions, whether they be in combined sewer systems or on a river channel, can hardly be described as ideal data collecting environments. Moreover, sample representativity, while rarely assessed, usually contributes an important source of error in the few data bases that are available.

c) finally, the costs involved in water quality data collection and analysis of combined sewer flows are usually quite high.

Computation Efficiency

The concern to this point has been to define the structural adequacy of the mathematical models needed to describe the dynamics of a system for real-time CSO control. A further requirement for the implementation of real-time control of CSO is concerned with computation efficiency, as it relates to both hardware requirements and computation time. In fact, it is feasible to assume that implementation of operation control strategies of the type shown in Figure 5 could require that the decision process (i.e., optimal allocation of overflows based on the predicted state of the system) be repeated every 5 to 15 minutes. Experience in modeling of the urban rainfall-runoff process using the full equations of flows can easily create a computational burden. Consequently, the tools of analysis for use in real-time CSO control need to be reliable and computationally efficient in terms of both hardware requirements and time of computation.

Measurement and Model Errors

As indicated previously, the development of a real-time CSO control algorithm can be very challenging. In addition to the many processes involved, the problem is complicated by the fact that many of the parameters are subject to large measurement errors (e.g., rainfall assessment). The propagation of measurement and model errors is not well understood in real-time rainfall-runoff modeling. In fact, there appears to be a definite need to provide a quantitative analysis of the effects of errors, including model errors, measurement errors, and calibration errors (37). Moreover, there is considerable interest in providing a probabilistic characterization of hydrological output variables for use in real-time (36).

Conclusions

This paper has provided a review of operational algorithms for real-time control of combined sewer overflows. While it may be possible to define, at least conceptually, the desired characteristics of an operational control algorithm very few municipalities have, so far, demonstrated the benefits of real-time CSO control over other more conventional CSO abatement technologies. The mathematical formulation of a predictive CSO control strategy is complicated among other things by the need for reliable water quality/quantity forecasts. Finally, operational CSO control is affected by the limited availability of real-time water quality data. Moreover, the propagation of measurement and model errors is not very well understood.

Acknowledgements

The research reported in this paper was funded by the Natural Science and Engineering Research Council of Canada.

References

1. Ammon, D. and R. Field. Potential of urban stormwater impacts based on comparative analysis of wet and dry weather pollutant loads. In: Proceedings Urban Stormwater and Combined Sewer Overflow Impact on Receiving Water Bodies. U.S. EPA, Municipal Environmental Research Laboratory, EPA-600/9-80056, pp. 502-522, 1980.

2. U.S. EPA. Report to Congress on Control of Combined Sewer Overflow in the United States. Office of Water Program Operations, EPA-430/9-78-006, MCD Report No. 50., 226 pp., 1978.

3. U.S. EPA. 1978 Needs Survey. Cost methodology for control of combined sewer overflow and stormwater discharge. Officie of Water Program Operations, EPA-430/9-79-003, FRD Report No. 3, 474 pp., 1979.

4. Torno, H.C. Urban runoff management: future modeling. Journal of Technical Councils, ASCE, 106(TC1):165-169, 1980.

5. Pisano, M.A. Nonpoint sources of pollution: a federal perspective. Journal of the Environmental Engineering Division, ASCE, 102(EE3):575-586, 1976.

6. U.S. EPA. Benefit analysis for combined sewer overflow control. Environmental Research Information Center, EPA-625/4-79-013, 55 pp., 1979.

7. Heaney, J.P., W.C. Huber and M.E. Lehman. Nationwide assessment of urban stormwater impacts on receiving water bodies. In: Proceedings Urban Stormwater and Combined Sewer Overflow Impact on Receiving Water Bodies. U.S. EPA, Municipal Environmental Research Laboratory, EPA-600/9-80-056, pp. 407-436, 1980.

8. Hvited-Jacobsen, T. The impact of combined sewer overflows on the dissolved oxygen concentration of a small stream. In: Proceedings Urban Stormwater and Combined Sewer Overflow Impact on Receiving Water Bodies. U.S. EPA, Municipal Environmental Research Laboratory, EPA-600/9-80-056, pp. 245-259, 1980.

9. Meinholz, T.L., W.A. Kreutzberger, M.E. Harper and K.J. Fay. Verification of the water quality impacts of combined sewer overflows. U.S. EPA, Municipal Environmental Research Laboratory, EPA-600/2-79-155, 202 pp., 1979.

10. Klemetson, S.L., T.N. Keeefer and R.K. Simons. Movement and effect of combined sewer overflow sediments in receiving waters. U.S. EPA, Municipal Environmental Research Laboratory, EPA-600/2-80-126, 148 pp., 1980.

11. Pitt, R. and M. Bozeman. Sources of urban runoff pollution and its effects on an urban creek. U.S. EPA, Municipal Environmental Research Laboratory, EPA-600/S2-82-090, 7 pp., 1982.

12. Turkeltaub, R. and R. Field. Overview of urban runoff receiving water impacts. Paper presented at the ASCE National Conference, May 11-15, 11 pp., 1981.

13. Ward, R.C. Combined sewer overflow control by in-system storage. Paper presented at the 1982 WPCF Conference October, 28 pp., 1982.

14. Paquin, G. Wastewater interception on the Communauté urbaine de Montréal. In: Proceedings Stormwater Management Model Users Group Meeting, H.C. Torno (Editor), U.S. EPA, Office of Air, Land, and Water Use, EPA-600/9-79-026, pp. 135-153, 1979.

15. McPherson, M.B. Integrated control of combined sewer regulators using weather radar. U.S. EPA, Grant No. R806702, 89 pp., 1980.

16. Coffee, H.C., D.E. Evenson, P.R. Giguere, G.T. Handa, C.A. Phanartzis and L.A. Roesner. Demonstrate real-time automatic control of combined sewer systems. Phase 1 Report, U.S. EPA, 168 pp., 1979.

17. Labadie, J.W., D.M.Morrow, and Y.H. Chen. Optimal control of unsteady combined sewer flow. Journal of the Water Resources Planning and Management Division, ASCE, 106(WR1):205-223, 1980.

18. Desbordes, M. Modèles de gestion dans le domaine de l'assainissement urbain. Université des Sciences et Techniques du Languedoc. Laboratoire d'hydrologie mathématique, 17 pp., 1978.

19. Patry, G.G. A linear programming model for the control of combined sewer system with off-line storage facilities. Canadian Water Resources Journal, 8(1):83-105, 1983.

20. Trotta, P.D., J.W. Labadie, and N.S. Grigg. Automatic control strategies for urban stormwater. Journal of the Hydraulics Division, ASCE, 103(HY2):1443-1459, 1977.

21. Brueck, T.M., D.F. Peterson, and J.A. Nye. Computer-based implementation of combined sewer control. Journal Water Poll. Control Federation, 54(2):198-205, 1982.

22. Bellon, A. and G.L. Austin. The evaluation of two years of real-time operation of a short-term precipitation forecasting procedure (SHARP). Journal of Applied Meteorology, 17(12):1778-1787, 1978.

23. Huff, F.A., J.L. Vogel, and S.A. Changnon. Real-time rainfall monitoring-prediction system and urban hydrologic operations. Journal of the Water Resources Planning and Managment Division, ASCE, 107(WR2):419-435, 1981.

24. Nguyen, V.T.V., M.B. McPherson, and J. Rousselle. Feasibility of storm tracking for automatic control of combined sewer system. ASCE Urban Water Resources Research Program. Technical Memorandum No. 35, 29 pp., 1978.

25. Labadie, J.W., R.C. Lazaro and D.M. Morrow. Worth of short-term rainfall forecasting for combined sewer overflow control. Water Resources Research, 17(5):1489-1497, 1981.

26. Labadie, J.W., N.S. Grigg and B.H. Bradford. Automatic control of large scale combined sewer systems. Journal of the Environmental Engineerig Division, ASCE, 101(EE1):27-39, 1975.

27. Nguyen Van, H.N., M. Osseyrane, and M.B. McPherson. Integrated control of combined sewer regulators. Journal of the Environmental Engineering Division, ASCE, 108(EE6):1342-1360, 1982.

28. Béron, P., L. Valiquette and G.G. Patry. Indices de qualité des eaux du G.R.E.M.U. Ecole Polytechnqiue de Montréal, GREMU-79/03, EP-80-R-11, 80 pp., 1979.

29. Beck, M.B. Dynamic modelling and control applications in water quality maintenance. Water Research, 10(7):575-595, 1976.

30. Patry, G.G. and M.A. Mariño. Parameter identification of time-varying noisy difference equations for real-time urban runoff forecasting. Journal of Hydrology, 72:25-55, 1984.

31. Patry, G.G. and M.A. Mariño. Sensitivity and application of difference equation models to real-time urban runoff forecasting. Journal of Hydrology, 72:57-66, 1984.

32. Patry, G.G. and M.A. Mariño. Prévision de la qualité de l'eau dans les bassins de drainage unitaires. Science et Technique de l'Eau, 17(2):159-162, 1984.

33. Patry, G.G. and M.A. Mariño. Real-time forecasting of water quantity and quality in urban catchments. Water Science and Engineering Paper No. 3011, University of California, Davis, California, 1983.

34. Matalas, N.C, J.M. Landwehr, and M.G. Wolman. Prediction in water management. In: Scientific Basis of Water Resource Management, National Academy Press, Washington, D.C., pp. 118-127, 1982.

35. Drozier, C. Contrôle de la pollution diffuse: corrélation entre différents paramètres de qualité des eaux d'égouts mixtes en temps de pluie. Mémoire de maitrise en ingénierie, Section Environnement, Ecole Polytechnique de Montréal, 124 pp., 1977.

36. Lessard, P. Etude de la qualité des eaux issues des réseaux unitaires en temps de pluie. Mémoire de maitrise ès sciences appliquées, Ecole Polytechnique de Montréal, 135 pp., 1981.

37. Potter, S. and G.G. Patry. Runoff modelling under noise-corrupted rainfall conditions. Proceedings, Annual Conference Canadian Scociety for Civil Engineering, May 27-31, Saskatoon, pp. 233-251, 1985.

DISCUSSION - REAL-TIME CONTROL

In the discussion which followed the presentation of the papers, several points were raised. They included:

1. Dr. Medina asked Dr. Patry about the definition of a water-quality index for real-time control, pointing out that water quality is a phenomena with a very long time constant, and fundamentally imcompatible with the time scales for real-time control. Dr. Patry replied that this was difficult to determine, but that he would probably approach it in conventional terms.

2. There was extensive discussion about where real-time control is needed. Dr. Schilling pointed out that if one believes that storage has an overall positive effect on receiving water quality (positive = less combined sewer overflow; negative = lower treatment efficiency), then real-time control is useful for pollution abatement because it creates storage. Dr. Lijklema also noted that in certain cases excess storage is available in wastewater treatment facilities, and that real-time control would allow its effective use. Dr. Schilling voiced strong agreement with this, stating in addition that the only logical place for a real-time control center is at the treatment facility, where you can take advantage of trained personnel already on site.

3. Mr. Jacquet pointed out that real-time control was most useful in the Seine-St. Denis project in the reduction of the use of relatively small dual-purpose detention ponds. For longer term in-stream pollution control, they will use monitoring and a micro-computer system to determine whether such treatment measures as in-stream aeration are appropriate.

4. Dr. James observed that a real-time control system can provide enormous amounts of system performance data which can be extremely valuable in situations where the sewer systems "fail" (which is inevitable since runoff cannot be controlled 100 percent of the time) and litigation results. Data acquired centrally in a real-time control system is also very useful for off-line studies because it is often highly discretized in time and, by definition, truly synchrnized (i.e. has no time errors).

5. Dr. James asked Mr. Delattre if their radar was accurate enough for real-time control. The reply was that some radar is used, but that it is correlated with observed rainfall measurements. Mr. Jacquet also noted that accurate rainfall forecasts are not necessary in real-time control situations, and that such things as the time of onset of the rainfall front are much more important. Dr. Patry stated that this was consistent with his experience in Montreal, which has been researching the use of radar for some time.

6. Dr. Geiger stated that he felt that establishing the effects of real-time control on wastewater treatment plant performance must be included. For the primary stage for the sedimentation process the

quantitative fluctuations seem to be most important and most likely to upset the facility, whereas for secondary treatment processes, influent quality variations are the most important.

Therefore, a properly working real-time control system would be a system that alleviates quantitative fluctuations into treatment plants to prevent primary stage upsets, and thereby also prevents "first flushes" from entering the secondary stage.

7. Mr. Weatherbe noted that there is nothing in the Canadian regulations to provide incentives for real-time control, and asked if other countries had the same sort of problem. Dr. Schilling replied that in Germany, due to a proliferation of detention basins which are not being properly utilized, real-time control has increasing support.

8. Dr. Huber asked if there were any real-time simulation models actually being used in control applications. Dr. Patry and Dr. Schilling both stated that there were not, although some are in the developmental stages.

9. Dr. Geiger emphasized that real-time control can only be carried out successfully when the quality objectives are incorporated and the question of responsibility in case of failures is clarified.

FUTURE RESEARCH NEEDS

Harry C. Torno
U. S. Environmental Protection Agency
Science Advisory Board (A-101-F)
Washington, D.C. 20460

One of the major objectives of the Workshop was to identify those areas where future research is still needed, or where work needs to be done to move technology which has been demonstrated in the laboratory to practical applications in the field. This was accomplished in the Workshop in two ways. First, many of the principal review papers highlighted those areas where our knowledge is weak. More important, a session devoted to future research was held on the last day of the Workshop. Each participant had been asked earlier in the week to list the four or five most important research needs from their individual points of view. 37 lists were received from the 46 participants. These lists were then analyzed, and the needs aggregated. The following list is that aggregation, and indicates, in priority order, the needs which the participants felt were most important (the number at the right indicates the number of participants who listed the need among their most important).

1. The understanding of the processes and mechanisms, including the related mathematical models, of pollutant accumulation and transport in urban runoff. -13

2. The effects of metals and other toxic pollutants on aquatic biota, including the means to monitor these effects. -8

NATO ASI Series, Vol. G10
Urban Runoff Pollution.
Edited by H. C. Torno, J. Marsalek, and M. Desbordes

3. The impacts of urban runoff pollution on receiving waters, including the impairment of beneficial uses. This includes the modeling of these impacts. -8

4. Better understanding of the processes of sedimentation and scour of solid particles in sewers. -8

5. The development of standards for sampling, analysis and data reporting for urban runoff pollutants. -8

6. The performance of stormwater and combined sewer overflow treatment devices. -7

7. The economic and institutional impacts of pollution control programs. -6

8. The development of stormwater and combined sewer overflow discharge criteria based on receiving water impacts. -6

9. The operation and maintenance of sewer and control technology systems, including the effects of O&M on system performance. -5

10. The impacts of infiltrated urban runoff on ground water. -4

11. Improved techniques for rainfall forecasting, including the use of radar. -4

12. Identification of sediment/water interactions in both sewers and receiving waters. -4

13. Identification and quantification of sources of error and uncertainty in model results. -3

14. Development of criteria for model calibration and verification. -3

15. The effects of real-time control systems on system performance and improved water quality -3

16. Case studies where complete planning, design, construction, operation and maintenance of storm and combined sewer systems can demonstrate effects on receiving water quality. -3

17. Impacts of stormwater and combined sewage on the performance of wastewater treatment facilities. -3

There were seventeen other needs expressed, but none of them was mentioned more than twice.

The aggregated list was presented to the participants at the session on research needs, and in the discussion which followed, there were several additional points made.

1. Dr. Geiger noted that data collection standards should look to the future, with the goal of achieving some common standards. Dr. Ellis (who is Chairman of the European Water Pollution Control Association group looking at data standardization) expressed the opinion that the European community is not far enough advanced to develop and implement such standards. He believes that it is enough for now to try evaluate data characterization in the various countries and see what, if any, commonality exists.

2. Dr. Roesner observed that we still do not have the technical capability to control runoff quality, even though there is a great push, politically, to pass laws and enact regulations on nonpoint source pollution control. In response, Mr. Urbonas expressed the need to standardize the sample collection, handling and analysis process, because he felt that much of the sample variability that is observed could be attributed to these sources, rather than to any inherent variability in the samples themselves.

3. Dr. Hamilton and Dr. Lijklema urged that the control of pollutants at their source should be looked at further. Dr. Lijklema further urged that pollutants be looked at across media, and from the point of generation to the point of ultimate disposal. Mr. Barnwell echoed the latter point, stating that it was particularly true with respect to toxic chemicals.

4. Dr. Geiger pointed out that much more work is needed in the identification and use of surrogate measures to predict impacts on water quality, particularly methods relating conservative pollutants to toxic pollutants.

5. Dr. James pointed out that while we may not directly do the work ourselves, we should be increasingly aware of work done by health researchers on the effects of pollutants, and ought to coordinate our activities with them. Mr. Urbonas and several others made the same point.

6. Dr. Gonenc observed that, in developing countries, they still have a great lack of data on urban runoff quality, and are therefore unable to make any useful comparisons with values reported in the literature.

7. Mr. B. N. Jacobsen observed that studies have shown that concentrations of toxic pollutants are, in general, far below acute effects levels for organisms in the biota. However the real problem seems to be chronic effects and bio-accumulation. He suggested beginning long-term programs of monitoring the effects on receiving water biota close to relevant urban runoff discharges.

8. Dr. Ferrara observed that by applying some common sense we can save a good deal of time and money collecting data on toxic pollutants (by, for instance, evaluating the volatility and solubility of these compounds, so as to estimate whether or not we would expect to find them in the liquid phase of urban runoff).

9. Mr. Metaxatos presented a theoretical comparison between combined sewer overflows and separate urban runoff in a typical Greek city of 10,000 population. His study indicated that the separate storm discharges had a greater effect on receiving waters, and demonstrated the need for treatment of separate urban storm runoff prior to discharge.

9. There was a suggestion made by Mr. Urbonas that the Workshop proceedings reflect a consensus by the participants that there was an urgent societal need for the control, at the source, of toxic compounds and heavy metals. After extensive discussion, it was the group's conclusion that this should not be included in the Proceedings.

The aggregated list (items 1-17 at the beginning of this section) is very interesting in a number of respects. When compared, for instance, to a similar list developed at an International Symposium in Urban Hydrology held in Baltimore, Maryland (U.S.A.) in June, 1983 (1), one notices several significant differences. There now seems to be much more interest in the effects of urban runoff on receiving waters, whereas two years ago this was hardly mentioned. There is also now much more emphasis on determining the economic and social impacts of urban runoff control programs, although Dr. Ellis commented that in Europe, as well as in Japan and Australia, more work needed to be done on the evaluation of source control options (particularly infiltration systems). There are indications in the literature (2) that work has begun in this area, irrespective and independently of the work that has been completed in the United States. In the United Kingdom and Europe, such approaches are very much involved with social, political and professional attitudes as much as by economic costs or system durability.

There is much less emphasis now, on the other hand, in the validation and verification of models. All of this seems to reflect a growing sophistication among researchers and practitioners in the field, and also reflects our increased knowledge about many aspects of the problem. We seem to feel comfortable, for instance, with the fact that models will never give us the "answers", but are rather a tool to aid in analysis and design. We recognize that it is not enough to say that urban runoff is a source of pollution, but we must also be able to describe, in quantitative terms, the impacts of urban runoff on receiving waters. We are getting more comfortable with the application of statistical techniques to urban runoff problems. We are increasingly emphasizing the use of real-time control as a means of maximizing the capability of existing systems to store and treat urban runoff.

We are still, however, far from a complete understanding of the processes by which pollutants accumulate on the surface, are washed from the surface (and in some cases redeposited) and are transformed as they travel through conveyance systems to receiving waters. Studies of particulates, for instance, under the scanning electron microscope suggest that aggregation mechanisms exist about which we have little or no knowledge. We are not able to relate pollutant accumulations, as observed in the outfalls, in urban areas to varying land uses, and even less able to quantify these accumulations, even though the work of Sartor and others has provided us with some general listings of pollutant loadings from categorized urban land uses.

It is clear then that much more research needs to be done, and hopefully this list will be a useful guide to researchers in deciding where to concentrate their future efforts.

All of the individual needs comments submitted by participants have been collected, and are available to anyone who wishes to do a more detailed evaluation of the research needs expressed. In addition, the entire research needs session was taped, and the tapes are also available from the author for review.

References

1. Delleur, J. W. and Torno, H. C. (Eds.) Urban hydrology. Proceedings of an international symposium on urban hydrology, Baltimore, Maryland, May 30-June 2, 1983. ASCE, 1983.

2. Balmer, P., Malmquist, P.-A. and Sjoberg, A. (Eds.) Proceedings of the third international conference on urban storm drainage, Goteborg, Sweden, June 4-8, 1984. Chalmers University of Technology, 1984.

LIST OF PARTICIPANTS

CANADA

Mr. Jiri Marsalek (Co-Chairman)
National Water Research Institute
867 Lakeshore Road
Burlington, Ontario L7R 4A6

Dr. Donald H. Waller
Technical University of Nova Scotia
P. O. Box 1000
Halifax, Nova Scotia B3J 2X4

Mr. Don Weatherbe
Ontario Ministry of the Environment
135 St. Clair Avenue West
Toronto, Ontario M4V 1P5

Dr. Jean Rousselle
Departement de Genie Civil
Ecole Polytechnique
Case Postale 6079, Succursale "A"
Montreal, Quebec H3C 3A7

Prof. Jean-Pierre Villeneuve
Universite du Quebec
Institut National de la Recherche Scientifique
2700, rue Einstein
Case Postale 7500
Sainte Foy, Quebec G1V 4C7

Dr. Gilles Patry
Department of Civil Engineering
McMaster University
1280 Main Street West
Hamilton, Ontario L8S 4L7

Dr. William James
Department of Civil Engineering
McMaster University
1280 Main Street West
Hamilton, Ontario L8S 4L7

M. Pierre Lavallee
Universite du Quebec
Institut National de la Recherche Scientifique
2700, rue Einstein
Case Postale 7500
Sainte Foy, Quebec   G1V 4C7

DENMARK

Prof. Thorkild Hvitved-Jacobsen
Department of Environmental Engineering
Aalborg University Centre
Sohngaardsholmsvej 57
DK-9000 Aalborg

Mr. Bo Neergaard Jacobsen
Water Quality Institute
11, Agern Alle
DK-2970 Horsholm

FRANCE

Dr. Michel Desbordes   (Co-Chairman)
Laboratoire d'Hydrologie Mathematique
Universite des Sciences et Techniques du Languedoc
34060 Montpellier CEDEX

Mr. Jean-Claude Deutsch
Technical Service of Urbanisation
Ministry of Urbanisation and Housing
64, rue de la Federation
75015 Paris

Mr. Denis Dartus
Institut de Mecanique des Fluides de Toulouse
2, rue Charles Camichel
31071 Toulouse CEDEX

Dr. Jean-Claude Hemain
L.C.P.C.
Division Eau
B. P. 19
44340 Bouguenais

Mr. Jean-Marie Delattre
DDE Seine St. Denis, Gestion des Eaux
99 Avenue du General de Gaulle
93110 Rosny sous Bois

Mr. Guy Jacquet
CERGRENE-ENPC
La Courtine
BP 105
93194 Noisy le Grand

Dr. Eric Servat
Laboratoire d'Hydrologie Mathematique
Universite des Sciences et Techniques du Languedoc
34060 Montpellier CEDEX

M. F. Bruant
Institut de Recherches Hydrologiques
10, rue Ernest Bichat
54000 Nancy

FEDERAL REPUBLIC OF GERMANY

Dr. Wolfgang Geiger
Lehrstuhl fur Wassergutwirtschaft
und Gesundheitsingenieurwesen
Technical University of Munich
Am Coulombwall
8046 Garching/Munich

Dr. Wolfgang Schilling
Institut fur Wasserwirtschaft
University of Hannover
Callinstrasse 32
D-3000 Hannover 1

GREECE

Mr. Takis Metaxatos
Analipseos 29
VRILISSIA 152 35
Athens

ITALY

Prof. Alessandro Paoletti
Istituto di Idraulica
Politecnico di Milano
20133 Milano

Ing. Mario Giulianelli
Istituto Ricerca sulle Acque
Consiglio Nazionale delle Ricerche
Via Reno 1
00198 Roma

THE NETHERLANDS

Dr. Lambertus Lijklema
Department of Water Pollution Control
Wageningen Agricultural University
De Dreijen 12
6703 BC Wageningen

Ir. Frans H. M. van de Ven
IJsselmeerpolders Development Authority
P. O. Box 600
8200 AP Lelystad

NORWAY

Dr. Torbjoern Damhaug
Norwegian Institute for Water Research
Postbox 333
Blindern
0314 Oslo 3

PORTUGAL

Mrs. Maria Rafaela Pinto de Abreu
Hydraulics Department
National Laboratory of Civil Engineering
Ministry of Housing and Public Works
Avenida do Brasil, 101
1799 Lisboa CODEX

TURKEY

Dr. I. Ethem Gonenc
Technical University of Istanbul
Insaat Fakultesi
Cevre Muh. Bolumu
Ayazaga / Istanbul

UNITED KINGDOM

Mr. J. B. Ellis
Urban Pollution Research Centre
Middlesex Polytechnic
Queensway, Enfield
Middlesex EN3 4SF

Dr. Christopher Pratt
Department of Civil and Structural Engineering
Trent Polytechnic
Burton Street
Nottingham NG1 4BU

Dr. R. S. Hamilton
School of Applied Science
Middlesex Polytechnic
Queensway, Enfield
Middlesex EN3 4SF

UNITED STATES

Mr. Harry C. Torno (Co-Chairman)
U. S. Environmental Protection Agency
Science Advisory Board (A-101F)
Washington, D.C. 20460

Dr. Wayne C. Huber
Department of Environmental Engineering Sciences
A. P. Black Hall
University of Florida
Gainesville, Florida 32611

Dr. Raymond Ferrara
Department of Civil Engineering
Lafayette College
Easton, Pennsylvania 18042

Mr. Richard Field
U. S. Environmental Protection Agency
Storm and Combined Sewer Research Program
Edison, New Jersey 08837

Mr. Ben Urbonas
Urban Drainage and Flood Control District
2480 West 26th, Suite 156-B
Denver, Colorado 80211

Dr. Miguel Medina
Department of Civil Engineering
Duke University
Durham, North Carolina 27706

Mr. Michael Terstriep
Illinois State Water Survey
2204 Griffith Drive
Champaign, Illinois 61820

Mr. Thomas Barnwell
Environmental Research Laboratory
U. S. Environmental Protection Agency
Athens, Georgia 30613

Dr. Larry Roesner
Camp Dresser and McKee
2701 Maitland Center Parkway
Maitland, Florida 32751

Mr. Harry N. Tuvel
American Society of Civil Engineers
345 East 47th Street
New York, New York 10011

INTERNATIONAL ORGANIZATIONS

Dr. Luis Veiga da Cunha, Science Administrator
Scientific Affairs Division
North Atlantic Treaty Organization (NATO)
B-1110 Brussels
BELGIUM

OTHER COUNTRIES

SWEDEN

Mr. Jan Falk
Swedish Environment Protection Board
Box 1302
S-171 25 Solna
Sweden

Dr. Per-Arne Malmquist
VIAK
Molndalsvagen 85
S-412 85 Gothenburg
Sweden

Dr. Peter Stahre
Malmo Gatukontor
Box 2500
S-200 12 Malmo
Sweden

FINLAND

Dr. Risto Laukkanen
Soil and Water, Ltd.
Italahdenkatu 2
SF-00210 Helsinki 21

AUTHOR INDEX

AUTHOR INDEX (continued)

SUBJECT INDEX

NATO ASI Series G

Vol. 1: **Numerical Taxonomy.** Edited by J. Felsenstein. 644 pages. 1983.

Vol. 2: **Immunotoxicology.** Edited by P.W. Mullen. 161 pages. 1984.

Vol. 3: **In Vitro Effects of Mineral Dusts.** Edited by E.G. Beck and J. Bignon. 548 pages. 1985.

Vol. 4: **Environmental Impact Assessment, Technology Assessment, and Risk Analysis.** Edited by V.T. Covello, J.L. Mumpower, P.J.M. Stallen, and V.R.R. Uppuluri. 1068 pages. 1985.

Vol. 5: **Genetic Differentiation and Dispersal in Plants.** Edited by P. Jacquard, G. Heim, and J. Antonovics. 452 pages. 1985.

Vol. 6: **Chemistry of Multiphase Atmospheric Systems.** Edited by W. Jaeschke. 773 pages. 1986.

Vol. 7: **The Role of Freshwater Outflow in Coastal Marine Ecosystems.** Edited by S. Skreslet. 453 pages. 1986.

Vol. 8: **Stratospheric Ozone Reduction, Solar Ultraviolet Radiation and Plant Life.** Edited by R.C. Worrest and M.M. Caldwell. 374 pages. 1986.

Vol. 9: **Strategies and Advanced Techniques for Marine Pollution Studies: Mediterranean Sea.** Edited by C.S. Giam and H.J.-M. Dou. 475 pages. 1986.

Vol. 10: **Urban Runoff Pollution.** Edited by H.C. Torno, J. Marsalek, and M. Desbordes. 893 pages. 1986.